ADVANCES IN CRYSTALLIZATION PROCESSES

Edited by **Yitzhak Mastai**

Advances in Crystallization Processes
Edited by Yitzhak Mastai

Published by InTech
Janeza Trdine 9, 51000 Rijeka, Croatia

Publishing Process Manager Vana Persen

Technical Editor Teodora Smiljanic

Cover Designer InTech Design Team

Additional hard copies can be obtained from orders@intechopen.com

Advances in Crystallization Processes, Edited by Yitzhak Mastai
p. cm.
ISBN 978-953-51-0581-7

For EU product safety concerns:
IN TECH d.o.o., Prolaz Marije Krucifikse Kozulić 3, 51000 Rijeka, Croatia,
info@intechopen.com or visit our website at intechopen.com.

Meet the editor

Professor Yitzhak Mastai was born in 1966 in Tel Aviv, Israel. He obtained his B.Sc in physical chemistry from Bar-Ilan University in 1989 and received his PhD from the Weizmann Institute of Science with Prof. Gary Hodes on nanomaterials synthesis (1999). He then went to the Max Planck institute of colloids and interfaces for 3 years, as postdoctoral fellow to work with Prof. M. Antonietti and Prof. H Cölfen on biomimetic chemistry and chiral polymers. In 2003 joined the staff of the chemistry department at Bar-Ilan University, where he is currently a Professor at the institute of nanotechnology at Bar-Ilan University leading the nano chirality laboratory. Prof. Mastai's earlier interests included nanomaterials synthesis and characterization. His current research is focused on the synthesis and analysis of chiral nanosurfaces, chiral self-assembled monolayers and polymeric chiral nanoparticles. Mastai has published more than 100 scientific articles and book chapters on various aspects of nanomaterials and chirality at the nanoscale.

Contents

Preface

Crystallization is certainly among the most studied processes in science and also of great practical importance. This is because crystals are the pillars of modern technology. Without crystals, there would be no electronic industry, no photonic industry, no fiber optic communications, solid state lasers, non-linear optics, piezo-electric, electro-optic or crystalline films for microelectronics and computer industries. In addition, crystallization is an elemental separation technique one of the most simple self-assembly processes to create order from the atomic to the macroscopic scale. Finally, crystallization creates beautiful crystals of esthetical value, which fascinate humankind already for centuries. Crystallization is an interdisciplinary subject covering physics, chemistry, material science, chemical engineering, metallurgy, crystallography, mineralogy, etc.

It is not astonishing that crystallization processes are already studied for a long time, beginning with alchemy and in a systematic, scientific fashion since the end of the 18th century. In the past few decades, there has been a growing interest on crystal growth and crystallization processes, particularly in view of the increasing demand of materials for technological applications. One might think that a process of such scientific and technological importance is well known down to the finest details after such intense studies for more than a century, but this is not true. Understanding crystallization from the atom level is still rather restricted, and it is the same for the interface of a crystal with solvent and other dissolved compounds.

The overall purpose of this book is to provide timely and in-depth coverage of selected advanced topics in crystallization. This text book goes into considerable detail concerning the many elements of knowledge needed to understand both quantitatively and qualitatively advanced subjects in crystallization process. The articles for this book have been contributed by the most respected researchers in this area and cover the frontier areas of research and developments in crystallization processes incorporating most recent developments and applications of crystallization technology. Divided into five sections this book provides the latest research developments in many aspects of crystallization including: chiral crystallization, crystallization of nanomaterials and the crystallization of amorphous and glassy materials. It is our hope that you, as readers, will find this book useful for your work. If so, this will be the nicest reward for us.

I wish to thank all the authors for their contributions for this book and it is my pleasure to acknowledge the assistance of Ms Vana Persen for her assistance during the write-up of this book and is preparation in final format.

Finally I would like to acknowledge my family who has supported me through all the years of doing science. To my wife Dina von Schwarze and to our children Michael, Reut and Noa for their patience and support.

Prof. Yitzhak Mastai
Department of Chemistry and the Institute of Nanotechnology
Bar-Ilan University, Ramat-Gan,
Israel

Section 1

Chiral Crystallization

1

Separation of the Mixtures of Chiral Compounds by Crystallization

Emese Pálovics[2], Ferenc Faigl[1,2] and Elemér Fogassy[1*]
[1]Department of Organic Chemistry and Technology, Budapest University of Technology and Economics,
[2]Research Group for Organic Chemical Technology, Hungarian Academy of Sciences, Budapest, Hungary

1. Introduction

Reaction of a racemic acid or base with an optically active base or acid gives a pair of diastereomeric salts. Members of this pair exhibit different physicochemical properties (e.g., solubility, melting point, boiling point, adsorbtion, phase distribution) and can be separated owing to these differences. The most important method for the separation of enantiomers is the crystallization. This is the subject of this chapter.

Preparation of enantiopure (ee~100%) compounds is one of the most important aims both for industrial practice and research. Actually, the resolution of racemic compounds (1:1 mixture of molecules having mirror-imagine relationship) still remains the most common method for producing pure enantiomers on a large scale. In these cases the enantiomeric mixtures or a sort of their derivatives are separated directly. This separation is based on the fact that the enantiomeric ratio in the crystallized phase differs from the initial composition. In this way, obtaining pure enantiomers requires one or more recrystallizations. (Figure 1).

The results of these crystallizations (recrystallizations) of mixtures of chiral compounds differ from those observed at the achiral compounds. Expectedly, not only the stereoisomer in excess can be crystallized, because the mixture of enantiomers (with mirror image relationship) follows the regularities established from binary melting point phase diagrams, and ternary composition solubility diagrams, respectively, as a function of the starting enantiomer proportion. According to this fact, we talk about conglomerate behaviour when the enantiomeric excess is crystallized, and racemate behaviour when it remains in the mother liquor.[1]

At the same time, there are some enantiomeric mixtures having racemate properties (based on binary phase diagram) which show conglomerate behaviour during its purification by fractionated precipitation. Always the enantiomeric excess is crystallized independently from the starting isomeric composition. This is explained by the kinetic crystallization of the enantiomeric excess.[2] Consequenlty, if the enantiomeric purity obtained after recrystallization or by other partial crystallization (as the result of splitting between the two phases) is

* Corresponding Author

plotted against the initial enantiomeric composition, either racemate or conglomerate behaviour is expected, or a conglomerate like curve is obtained (hereinafter: kinetic-conglomerate) (Figure 2).[3]

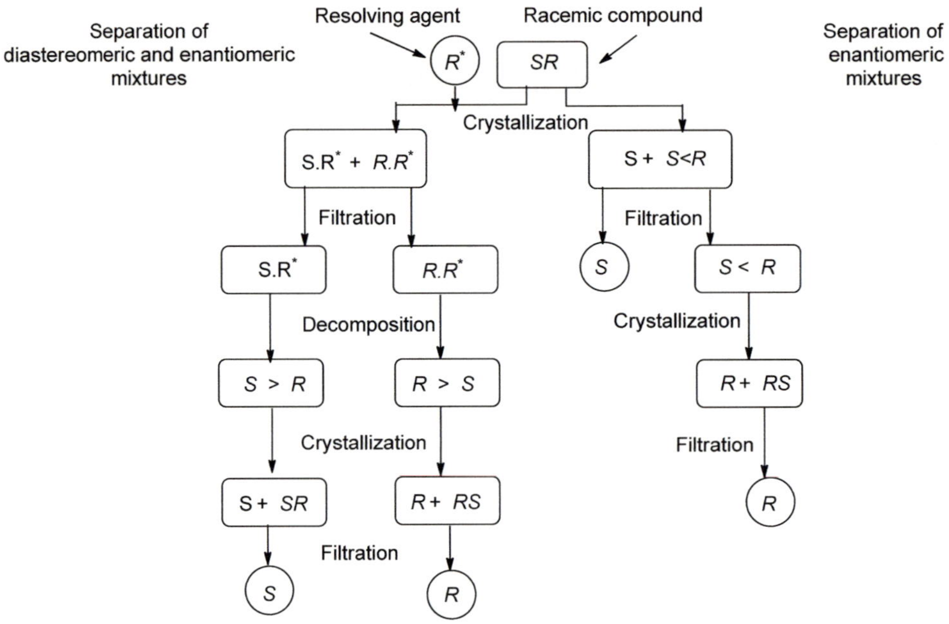

Fig. 1. The essential processes of enantiomeric separation starting from the racemic compound

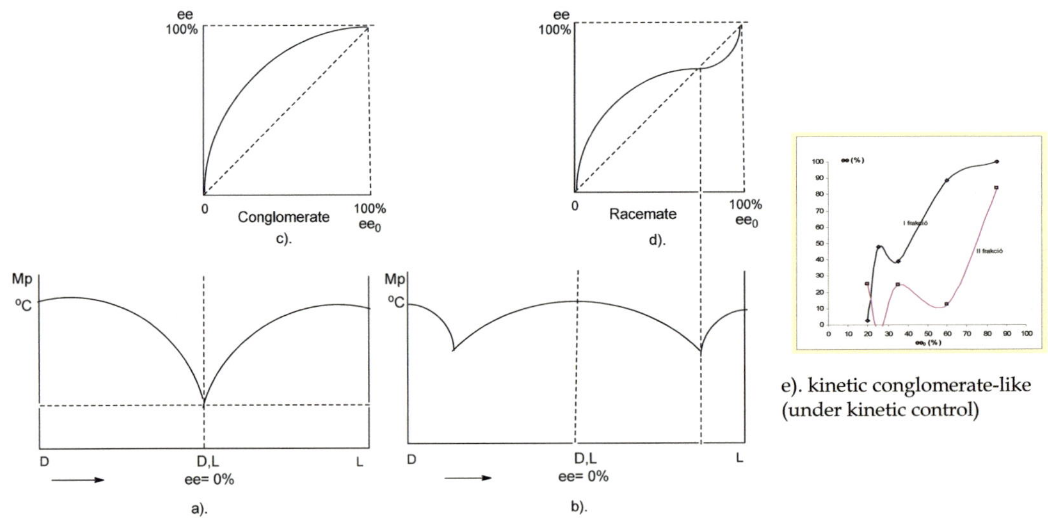

Fig. 2. The composition of enantiomeric mixtures obtained by crystallization (ee) in function of starting enantiomeric purity (ee_0). a). conglomerate melting point diagram; b). racemate melting point diagram; The diagrams of expected results of c). conglomerate crystallization, d). racemate crystallization, e) kinetic-conglomerate like crystallization

The above mentioned behaviour depends on the substituents of the base skeleton of the racemic compound. It means that different substituents can determine the change in behaviour. For example, the formyl -, acetyl - and propionyl-derivatives of phenyl alanine (**FoPA, AcPA, PPA**, see below) show different behaviour.

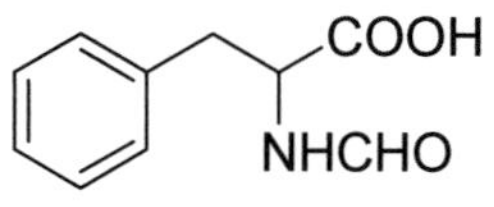

FoPA

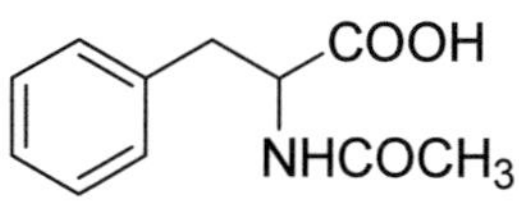

AcPA

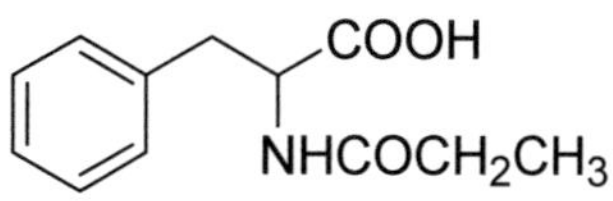

PPA

racemate conglomerate kinetic conglomerate

At separations based on the distribution between solid and another (liquid or melt) phase formation of a crystalline material is needed. The chiral discriminating effect can be explained by the presence and reactions of homo- and heterochiral aggregates (the simplest members of them are the dimers), and by their different physical and chemical properties. Thus in the solution or in the melt of racemic mixtures of enantiomers (S and R) two homodimers (SS and RR) and one heterodimer (SR) should be present, while in a hypothetic enantiomeric mixture (S>R, ee = 50%) presence of two dimmers, namely SS and SR, is expected. Considering that homo- and heterodimers are in diastereoisomeric relation with each other, their solubility and reactivity are different.[4]

$$3S,3R \rightleftharpoons S.S + R.R + S.R$$

$$3S, R \rightleftharpoons S.S + S.R$$

The behaviour of a mixture of chiral compounds, in which the resolving agent is a substituted derivative of one of the enantiomers (which are also present in the mixture in racemic form) is similar to the behaviour of the above mentioned enantiomeric mixtures. Under such conditions real diastereoisomeric pairs are formed containing an enantiomer and the structurally similar resolving agent, respectively.

If the racemic compound (mixture of enantiomers in 1:1 ratio) is reacted with equimolar amount of a derivative of one of the enantiomers (e.g. R*, having opposite chemical character) a quasi-enantiomeric mixture is formed.

$$\underbrace{S.R + R^*R^*}_{ee_0 \sim \text{ quasi } 50\%} \rightleftharpoons \underbrace{SR^* + RR^*}_{\text{diastereomers}}$$

In these cases non-symmetric conglomerate or racemate biner phase diagrams are expected and „quasi-conglomerate" (*RR**) or „quasi-racemate" (*SR**) precipitate may be obtained. In the former case the enantiomeric purity of R or S in the crystalline diastereoisomeric salt (*RR** or *SS**) can be anything, but in the latter („quasi-racemate") case an enantiomeric purity (ee) usually reflects the eutectic composition for the enantiomeric mixture of original racemic compound.

The resolving agent has a decisive role at the separation of diastereoisomers. Previously, we referred to the fact that the resolving agent (having related structure with one of the enantiomers of the racemic compound) may form well-crystallizing diastereoisomer. But usually separations based on crystallization of diastereoisomers obtained from chiral reagents with significantly different structures are performed. One of these frequently used reagents is the dibenzoyltartaric acid (**DBTA**), which is suitable for the resolution of racemic bases. It was observed, that it has also an ability to form relatively stable molecular complexes with racemic alcohols, and its Ca and Mg salts form coordinative complexes both with alcohols and phospholenes. In these cases pure enantiomers can also be obtained by fractionated precipitation of the diastereoisomeric complexes.[5,6]

It is assumed based on the results of the resolutions (both via salt-formation, and complex-formation) that the composition of the crystalline diastereoisomer is determined by the properties of the enantiomeric mixture of the racemic compound even if the resolving agent has not a similar structure to the racemic compound.

The enantiomeric mixture obtained by recrystallization (fractionated precipitation) and decomposition of diastereoisomers usually needs further purification to obtain the pure (single) enantiomer. The above mentioned behaviour of the stereoisomeric mixtures is expected at these enrichment processes, too.

However, during the purification of enantiomeric mixtures (having diastereoisomeric relationship and –behaviour of the homo- and heterochiral aggregates) both the kinetic and thermodynamic control can be observed, and certainly these phenomena can be observed at the (re)crystallization of diastereoisomers as well.[7] At the same time in case of reciprocal resolution (one of the enantiomers of the original racemic compound is applied as resolving agent for resolution of the racemic mixture of the initial chiral agent) the former behaviour is expected inevitably (namely the previously experienced kinetic- or thermodynamic control can be observed).

So, the crystallization rules (based on expected chiral-chiral interaction) of enantiomeric mixtures can be applied even if two or more chiral compounds are present.

However, the mode of separation of the crystalline fraction is very important. For example, the separation of enantiomers can be effectuated even if a mixture from the melt of enantiomeric mixtures is crystallized. The crystalline material can be separated from the residue by filtration, distillation, sublimation or extraction.[8]

A similar separation can be achieved even if the amount of the resolving agent applied is less than an equivalent to the racemate. Then the enantiomeric mixture can be separated from the crystalline diastereisoomer by sublimation or distillation, in accordance with the above mentioned methods.

The dielectrical constant of the solvent (if solvent is used at the resolutions) changes the formation, composition and enantiomer recognition of the crystalls.[9] The composition of crystalline diastereoisomers is also influenced by the pH of the reaction mixture.[10]

The purity (*de*) of the diastereoisomer can be improved using a mixture of structurally related resolving agents. It is often referred as "Dutch resolution" in the literature.[11]

If the diastereoisomeric salt can not be separated by fractionated precipitation, it is feasible to get its crystalline solvate by fractionated precipitation from a solvate forming solution.[12] When the solvent, unsuitable for separation of the diastereoisomers, contains structurally (partly) similar compound(s) to the solvate forming solution,[13] the separation of enantiomers became feasiable by fractionated precipitation of the diastereoisomeric salt. It is particularly striking, when the wanted crystal composition is obtained by using a structurally similar (either with racemic compound or resolving agent) achiral compound.[14]

It is possible that the recrystallization from an adequate solvent does not give pure enantiomers, but a good enrichment can be often achieved by fractionated precipitation.[15]

The racemate and conglomerate like behaviour of enantiomers can also be observed both at the recrystallization and at the liberation of the enantiomers from their salts formed with achiral compounds.[16]

Based on own and other's experimental results some conclusions can be drawn according to the fractionated crystallization of chiral compounds.

The regularities of crystallization of the systems containing at least two chiral compounds can not be described by linear correlations.[17] At the same time, the behaviour of the enantiomeric mixtures of the racemic compound will be one of the determining factor of the behaviour of the systems containing chiral compounds.

To characterize the obtained fractions not only their yield (Y), but the enantiomeric purity of the products (*ee*) is also relevant. Therefore it is advisable to use the parameter F characterizing the efficiency of the procedure (F=ee*Y).[18] This factor is actually the yield of enantiomeric excess separated. (Earlier this factor was marked with the letter S, but it could refer to the configuration even if it was marked by a different font.)

Hereinafter, the possibilities of enantiomer and diastereoisomer separations based on crystallization are shown by known examples.

At first examples will be given for separation of mixtures, containing two or more chiral compounds in solventless conditions or using solvent, respectively.

Among the next examples we refer to the results derived from earlier considerations, too. We demonstrate that behaviour of these enantiomeric and diastereoisomeric mixtures raised from interactions between chiral compounds are not allways the expected ones. The shown examples relate mainly to calculable regularity, but often a minor modification in the molecular structure can induce significant change of results. For justification, the two essential perception of Pasteur[19], are shown at first, then the consequences of the „slight changing" of the molecular structures will be presented in the light of our present knowledge.

It was recognized by Pasteur, that the enantiomers of racemic tartaric acid can be separted by induced crystallization, if the supersaturated aqueous solution of its sodium-ammonium salt was seeded with the crystals of a pure enantiomer. In this case significant amount of the pure enantiomer could be find in the solid phase and a mixture of enantiomers remaind in the saturated solution in which the other enatiomer was in excess. That other enantiomer could be crystallized in pure form from the suprasaturated solution of this latter enatiomeric mixture. This experiment demonstrated the fact, that the excess of an anantiomer can crystallize from an enantiomeric mixture of the above mentioned tartaric acid salt. According to our present knowledge, it means that the enantiomers of sodium-ammonium salt of tartaric acid form a conglomerate in their mixtures.

When the same experiment, namely crystallization of the non-racemic enantiomeric mixture of sodium-ammonium tartarate, was effectuated at a temperature above 27 °C, the racemic fraction (the racemate) crystallized, because the sodium-ammonium salts of racemic tartaric acid have a racemate like behaviour at around 30 °C. In this case the derivative of tartaric acid (the mixed salt) was suitable for fractioned enantiomeric separation, but only at a lower temperature than 27 °C (influence of temperature).

Pasteur also recognized, that more efficient separation of the enantiomers of racemic tartaric acid could be achieved by application of another chiral base (Quinotoxine (**Q**)) as resolving agent to the enantiomers of tartaric acid was obtained a better enantiomeric separation. In this case a diastereoisomeric salt ((*R*,*R*)-**TA.Q.6H$_2$O**) crystallized while the better soluble diastereoisomeric salt((*S*,*S*)-**TA.Q**) remained in solution.

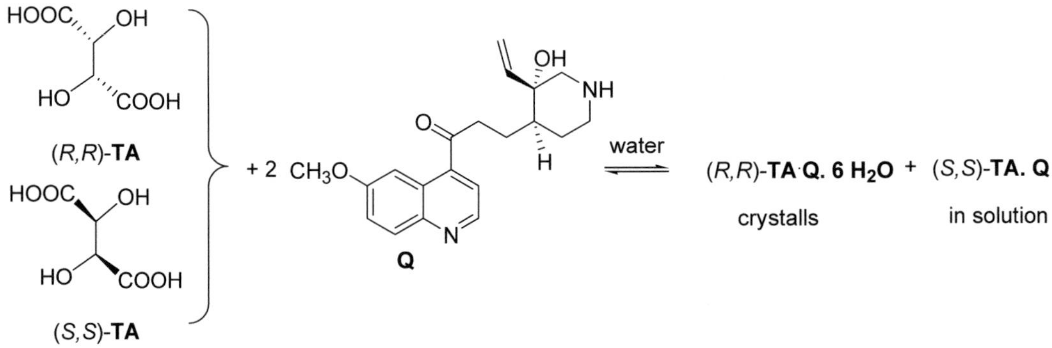

When cinkotoxine (a chiral base, similar to quinotoxine, without CH$_3$O- substituent) was used, the other tartaric acid enantiomer ((*S*,*S*)-**TA**) crystallized in the diastereoisomeric salt.

It is supposed that the crystal solvate was a decisive role during crystallizationof the first salt or the lack of methoxy substitution (H-bridge acceptor) on common molecular configuration of the resolving agent changed its enantiomeric recognition ability in the second case. Keeping these experimental results in mind, the known methods of enantiomer separation *via* crystallization will be discussed in the next sections

2. Crystallization without any solvent

Both the enantiomeric mixtures and the mixtures of true diastereoisomeric pairs and free enantiomers can form crystalline and melt fractions, which are separable by an adequate method.

2.1 The crystallization of the enantiomeric mixtures

When the non-racemic mixture of enantiomers may be melted without decomposition, depending on whether the homochiral or the heterochiral associations are more stable, the enantiomeric excess (conglomerate type) or the racemic fraction (racemate type) crystallizes from the melt. Sometimes these phases may be separated by filtration.

2.1.1 Separation with filtration

2.1.1.1 Conglomerates

In such cases the enantiomeric excess is crystallized with a higher purity than that of in the initial composition, in a certain range of temperature. For example, the common intermediate of the synthesis of several prostaglandines is a lactone (**PGL**). Its enantiomeric mixtures (ee_0) can be enriched by crystallization (ee_{solid}) from melt, while the racemic ratio (liquid residue, ee_{liquid}) can be recovered for repeated resolution.[20]

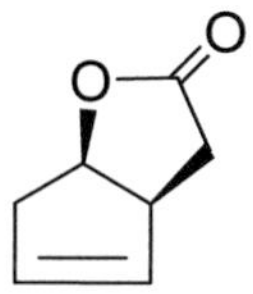

(1*S*,5*R*>1*R*,5*S*)-**PGL**

crystallization from melt → at 0-5 ^{0}C

ee_0 %	ee_{solid} %	ee_{liquid} %
21.5	69.2	6.2
29.2	72.5	2.6
52.3	76.9	29.2
79.5	93.0	54.2

2.1.1.2 Racemates

In racemate forming enantiomeric mixtures the enantiomeric excess with a higher enantiomeric purity than that the initial composition remains in the melt and the crystalline phase formed has a lower *ee*. For instance, at the crystallization of the enantiomeric mixture of *trans*-chrysanthemic acid (**CHRA**) from melt the liquid phase contains a higher purity fraction than the ee_0 value of the initial mixture.[21]

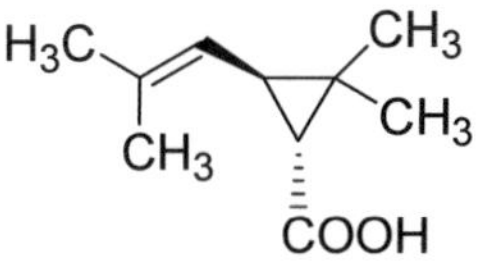

(1*R*>1*S*)-**CHRA**

crystallization from melt → at 0-10 ^{0}C

ee_0 %	ee_{solid} %	ee_{liquid} %
42.6	15.5	83.3

At the crystallization of melts of racemate forming enantiomeric mixtures the eutectic composition usually determinates the composition of the crystallized mixture and the oily residue. That eutectic composition can be known from the binary melting point phase diagram. When the initial isomeric composition (ee_0) is higher than the eutectic composition, the pure optical isomer cam be crystallized. By way of illustration the eutectic composition (it is approximately at *ee* 40%) of flumequine intermediate (**FTHQ**) cannot be enriched, but by starting from a mixture of ee_0 > 40%, significant enrichment can be achieved, especially because it has a „conglomerate like" behaviour.[22]

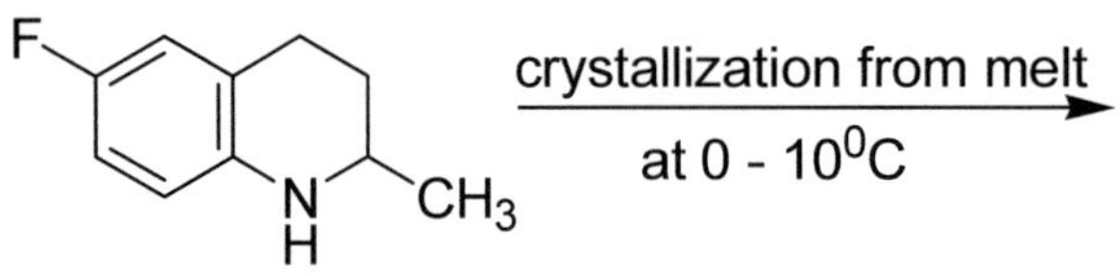

(*S*>*R*)-**FTHQ**

ee_0%	ee_{solid}%	ee_{liquid}%
40	40	40
50	92	47

2.1.2 Partial crystallization, distillation

When a part of an enantiomeric mixture is transformed into a crystalline salt and the residue can be distilled, the solid phase may contain either the racemic composition or the excess of the enantiomer. In the former case the distilled fraction should be the enriched mixture relative to ee_0, but the less pure mixture can be found in the distilled fraction when the $ee_0 > ee_{Eu}$. This is applied even if the salt of the enantiomeric excess proportion is the more stable, as well. An example for this phenomenon is the purification of the enantiomeric mixture of phenylethylamine (**PEA**). The mixture was treated with less than an equivalent amount of dicarboxylic acid (oxalic acid, succinic acid, fumaric acid or phthalic acid = **DCA**) (see Table 1, the molar ratio: **PEA/DCA**~ 3,3/1,25), and the excess of the amine was distilled. The **PEA-DCA** salt behaved as a racemate because $ee_{deszt} > ee_0$ In case of the salts of oxalic- and malonic acids this is changed above of eutectic composition.[23]

	Salt forming acid	ee_0 (%)	$ee_{distillate}$ (%)	$ee_{residue}$ (%)
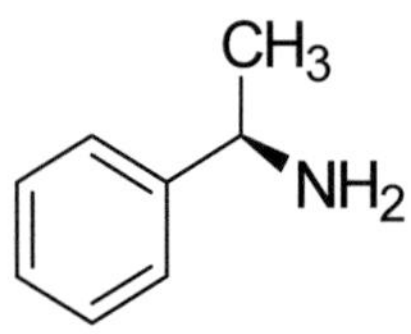 (R>S)-**PEA**	COOH | COOH Oxalic acid	12.5	26.6	7.8
		25.0	44.6	18.5
		50.0	63.6	45.5
		75.0	65.7	78.1
		87.5	71.5	92.8
	CH_2COOH | CH_2COOH Succinic acid	12.5	51.2	0.0
		25.0	63.1	12.3
		50.0	62.7	45.8
		75.0	57.4	82.9
		87.5	49.9	100
	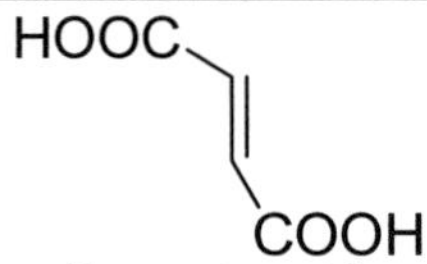 Fumaric acid	12.5	41.8	2.7
		25.0	68.5	10.5
		50.0	95.4	34.9
		75.0	96.3	67.9
		87.5	97.6	84.1
	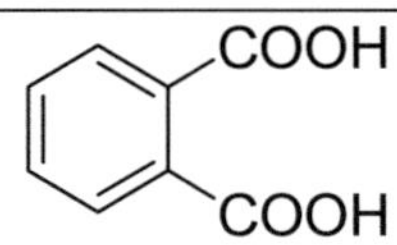 Phthalic acid	12.5	42.5	2.5
		25.0	84.2	5.3
		50.0	94.6	35.1
		75.0	94.4	68.5
		87.5	95.0	85.0

Table 1. The enrichment of enantiomeric mixtures of (R>S) – **PEA** via partial salt formation followed by distillation.

2.2 Crystallization of diastereoisomers

In this chapter we discuss on such separations in which the racemic compound is reacted with a resolving agent, so the mixture contains three chiral compounds: the two enantiomers and another chiral compound. At this time the more stable diastereoisomer crystallizes and it can be separated from the non racemic enantiomeric mixture by several methods.

2.2.1 Separation by filtration

In this case the resolving agent should be added to the racemic compound and the mixture should be warmed until it melted then is should be cooled while one of the diastereoisomers crystallizes from the melt. The crystalline material can be separated by filtration from the residue of the melt (as in case of enantiomeric mixtures). An example for this method is the resolution of racemic menthol (**MEN**) with O,O′-dibenzoyl-(*R*,*R*)-tartaric acid ((*R*,*R*)-**DBTA**). The crystalline molecular complex (diastereoisomer) contains the (1*R*,2*S*,5*R*)-menthol (the L-menthol (L-**MEN**)), while the remained melt is enriched in the other enantiomer[24]

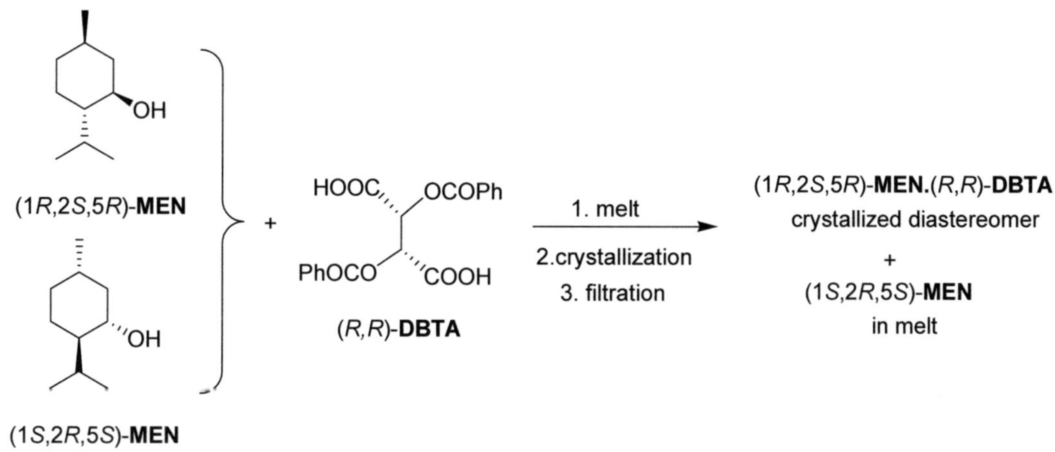

2.2.2 Separation by distillation

If the racemic compound is reacted with half an equivalent amount of resolving agent, the enantiomeric mixture remained after crystallization can be separated by distillation. Such a method was accomplished at the resolution of *N*-methyl-phenylisopropylamine (**MA**) by (*R*,*R*)-dibenzoyl-tartaric acid ((*R*,*R*)-**DBTA**). In this case the (*S*)-**MA** was distilled off beside of crystalline (*R*)-**MA.**(*R*,*R*)-**DBTA** salt. [25]

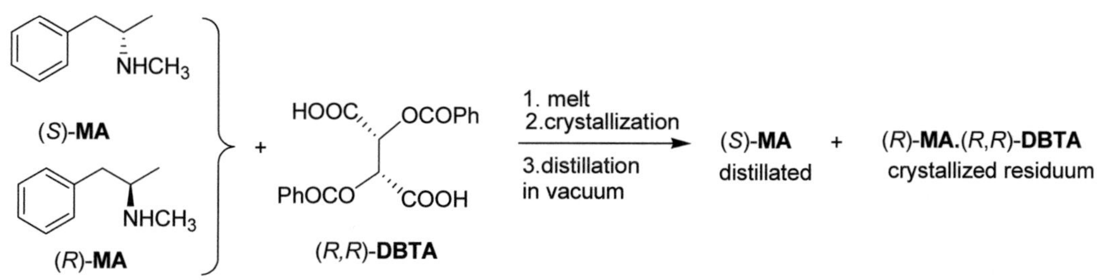

When the racemic phenylisopropylamine (**A**) is resolved with (*S*)-*N*-phthaloyl-phenylethylamine (**PPEA**) according to the abover mentioned method, further separation can be accomplished starting from the residue of the first distillation. When the temperature is increased, the residual crystalline diastereoisomeric salt decomposes and the liberated enantiomer ((*R*)-**A**), constituting the diastereoisomeris salt) can also be distilled off while the resolving agent is transformed into a phthalimide derivative.[26]

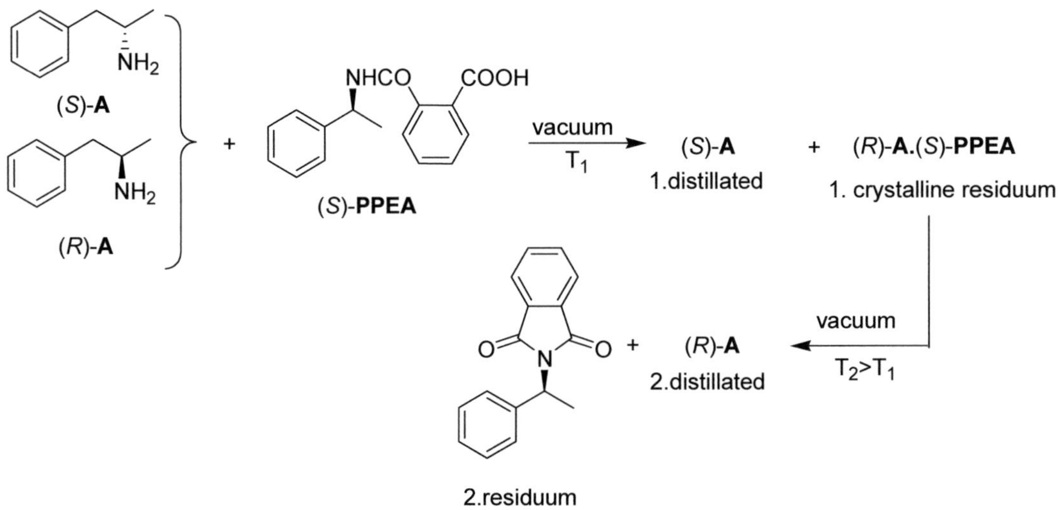

The situation is essentially the same when the racemic **A** is resolved by (*R,R*)-TADDOL (**TAD**). The first crystalline residuum is the molecular complex of (*R*)-**A**.(*R,R*)-**TAD**. At higher temperature this complex is decomposed and the regenerated **TAD** remais in the flask as the final residuum. [28]

2.2.3 Separation by sublimation

Sublimation method for indirect separation of enantiomers is very similar, in principle to the above mentioned distillation-based procedure. In the next example the diastereoisomeric molecular complexes could be separated using the significant difference between their thermal stability.

The solid phase reaction between the racemic compound (*transz*-2-iodo-cyclohexanol = **IC**)) and the resolving agent ((*R,R*)-**DBTA)** occurs during a long-term staying.[27] Then the enantiomers could be separated from the mixture by two sublimation steps.

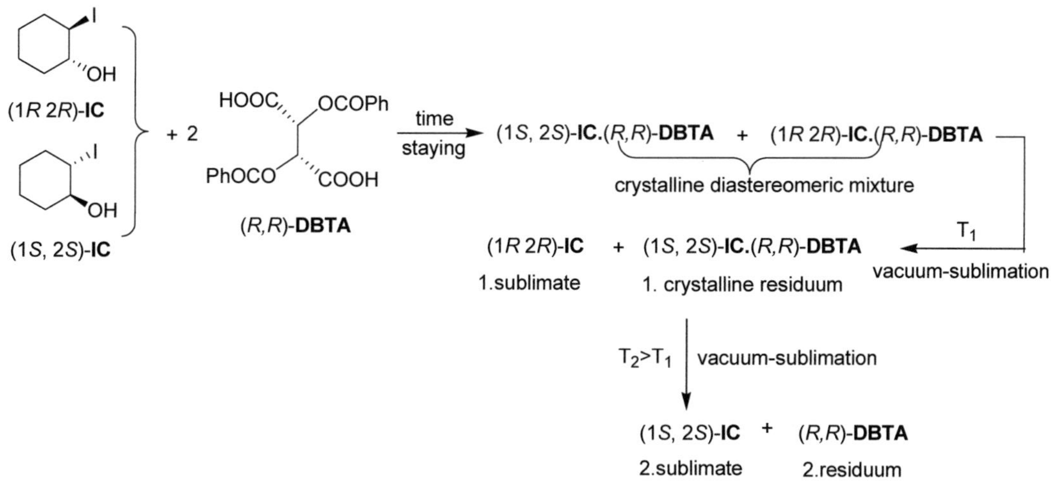

2.2.4 Separation by a supercritical fluid (carbon dioxide)

We think that this method for separation of the crystalline diastereoisomeric salt and the enantiomeric mixture can also be placed to the methods where the crystallization is accomplished without a solvent, and based on the instantiated examples we can declare that both the diastereoisomeric salt and molecular complexes may be separated by supercritical carbon dioxide.

When the racemic acid or base is treated with less than an equivalent amount of chiral compound (resolving agent) one can obtain a good enantiomer separation if crystallization occures in the mixture. For example, at the resolution of racemic ibuprofen (**IBU**) with (*R*)-**PEA,** the free enantiomer (*S*)-**IBU** can be separated from the salt ((*R*)-**IBU.**(*R*)-**PEA)** by extraction with a supercritical fluid, most often with carbon dioxide[28]

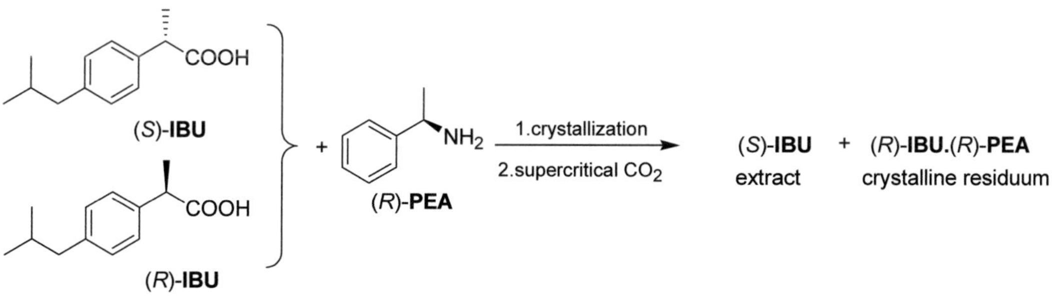

The pure enantiomers of racemic 1,2-cyclohexane-diol (**CHD**) could be obtained with the same method using (*R,R*)-Tartaric acid (**TA**) as resolving agent. In this case the (*S,S*)-**CHD** isomer could be extracted and the crystalline molecular complex of (*R,R*)-**CHD** and (*R,R*)-**TA** remained in the solid phase.[29]

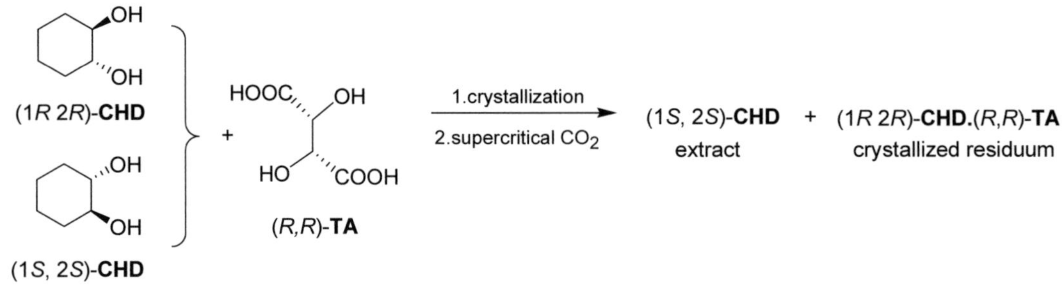

3. Crystallization from a solvent

The majority of the enantiomer separation methods based on crystallization involves solvent or solvent mixtures. Crystallizations from supersaturated solutions can be used both at the separation of enantiomeric mixtures, and at the separation of diastereoisomeric mixtures.

Comparison of the variable parameters of that method with the conditions of the above outlined solventless processes show that one have to take into consideration the characteristics of the applied solvent or solvent mixtures during optimization of such a resolution process. Furthermore, chiral or achiral additives can also be applied which improve without exception the efficiency of the resolution (F=ee*t).

3.1 Crystallization of enantiomeric mixtures

The most frequently applied versions of the classical enantiomeric enrichment method (effectuated by crystallization from a solution) are presented below.

3.1.1 Recrystallization

The optical isomers exist in the supersaturated solutions as a mixture of homochiral and heterochiral associations. Therefore both the conglomerate or the racemate like behaviour can be observed during the crystallization, similar to the before discussed solventless crystallization methods

3.1.1.1 Conglomerates

When a nonracemic enantiomeric mixture is recrystallized from a solvent, and the enantiomeric excess (*ee*) of crystalline phase is always higher than the initial composition (ee_0), then the compound (or its crystallized derivative) is conglomerate forming material. For example, when the enantiomeric mixture of dilthiazem hydrochloride (**DIL.HCl**) is recrystallized from ethyl acetate, the enantiomeric excess crystallizes in a very high purity.[30]

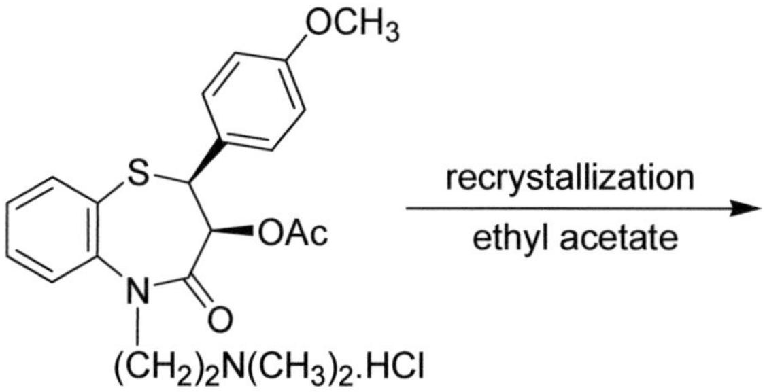

ee_0 %	ee_{solid} %	ee_{liquid} %
66	100	0

(*S,S* > *R,R*)-**DIL** . HCl

3.1.1.2 Racemates

At the recrystallization of racemate forming (the most common) enantiomeric mixtures the enriched enantiomeric mixture usually can be recovered from the mother liquor after filtration of the crystalline fraction having near to racemic composition. However, the composition of the crystalline material can be changed when the initial enantiomeric purity is higher then the eutectic composition. A certain example is the recrystallisation of the enantiomeric mixtures of tofizopam (**TOF**) from ethyl acetate. Almost racemic composition crystallized from small or medium pure enantiomeric mixtures. However the sole enantiomer crystallized from the solution when the initial composition was higher than the eutectic one (ee_0>85%).[31]

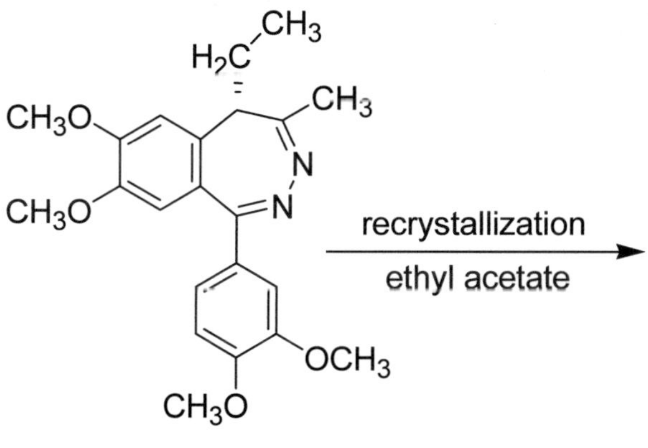

ee_0%	ee_{solid}%	ee_{liquid}%
53	9	95
61	10	92
85	98	79
89	100	64

(*S*>*R*)-**TOF**

The above example demonstrates that one can prepare the enantiomerically pure product from almost any samples having medium ee_0, with two recrystallizations if the enantiomeric mixture recovered from the filtrate of the frst crystallization is applied as starting material of the second recrystallization. This technique was adapted at the purification of the hydrochloride salt of a flumequine intermediate (**FTHQ.HCl**), too.[23]

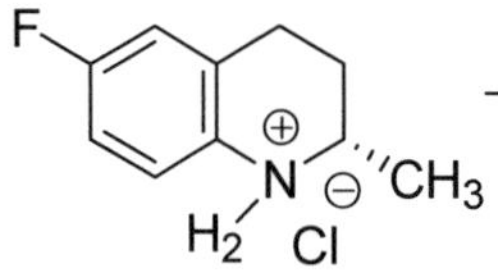

(*S*>*R*)-**FTHQ.HCl**

recrystallization → water

ee_0%	ee_{solid}%	ee_{liquid}%
42.1	10.3	81.4
95.7	97.0	93.8

3.1.2 Fractionated precipitation.

In several cases the recrystalllization fails to result in enantiomeric enrichment. The solution for the purification of such enantiomeric mixtures may be the fractionated precipitation.[32] In order to carry out such a purification step the enantiomeric mixture or its derivative (e.g. its salt formed with an achiral reagent) is dissolved in a solvent. Then a part of the dissolved material is liberated from its salt (or a salt is prepared from a part of it) with a reagent in a way that the liberated fraction (or salt) can precipitate from the solution, while the other proportion of the initial mixture remains in the solution. For example, the salts (e.g. water-soluble) of acids or bases are dissolved and a part of the free acid or base is precipitated by addtion of less than equimolar achiral acid or base.

3.1.2.1 Conglomerates

Similar to the experiences shown at the recrystallization of enantiomeric mixtures, the conglomerate behaviour can also be observed at fractioned precipitation. An application of this method was effectuated at the resolution of racemic tisercine (**TIS**) with half an equivalent of (*R,R*)-tartaric acid. The (*S*)-**TIS** enantiomer, the active pharmaceutical ingredient. remained in the filtrate of the diastereoisomeric salt formation process. Consequently, it contaminated with its mirror imge isomer. The enantiomeric enrichment was accomplished by selective precipitation. The (*S*>*R*)-**TIS** mixture was dissolved in water as a hydrochloric acid salt, then less than an equivalent amount of potassium hydroxide was added to the solutionin order to liberate the excess of (*S*)-**TIS**. The pure (*S*)-**TIS** base precipitated from the solution and an almost racemic hydrochloride salt remained in the solution..[33]

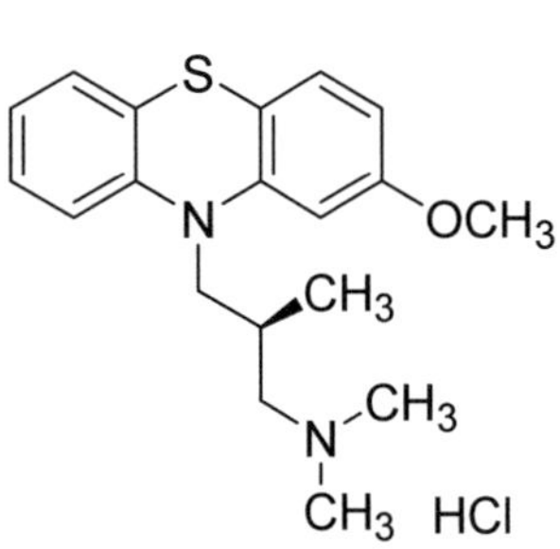

(*S*>*R*)-**TIS**.HCl

aquaeous sol. → non eq. KOH aqueous sol.

ee_0%	ee_{solid}%	ee_{liquid}%
25	100	5
67	100	7

The conglomerate-like behaviour during the fractionated precipitation was observed in case of several chiral acids, too. A good example is the enantiomeric enrichment of *N*-acetyl-phenylalanine (**AcPA**). The pure (*R*)-**AcPA** isomer was obtained in two steps from its nonracemic enantiomeric mixtures.[2]

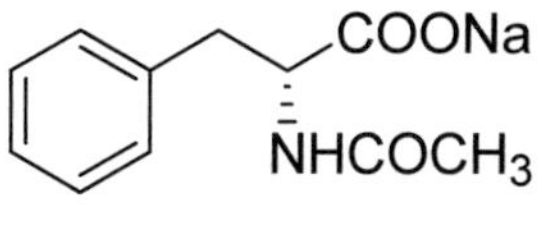

(*R*>*S*)-**AcPA.Na**

aqueous sol. → addition of non eq. aqueous HCl sol.	ee_0%	ee_{solid}%	ee_{liquid}%
	45.0	59.6	6.0
	78.2	100	6.7

3.1.2.2 Racemates

As it was already discussed in the above cases, enantiomeric enrichment of racemate forming enantiomeric mixtures result in an almost racemic crystalline phase when the starting enantiomer purity is smaller than the eutectic composition ($ee_0 < ee_{eu}$), but highly enriched mixture or the sole enantiomer can be crystallized when the initial composition is bigger than the eutectic one ($ee_0 > ee_{eu}$)

This type of selective precipitation was applied in the cases of *N*-acyl-aminoacids when their recrystallization from water failed. However, additon of less then an equimolar amount of hydrochlororic acid to the solution of their sodium salt get good enrichment. It is therefore not surprising that considerable enrichment could be attained with a mixture of ee_0 = 49.6 %, while two stage precipitation starting from an aqueous solution of **FoPA.Na** with ee_0 = 73.4% (composition near the eutectic point), did not resulted in significant enrichment.[11]

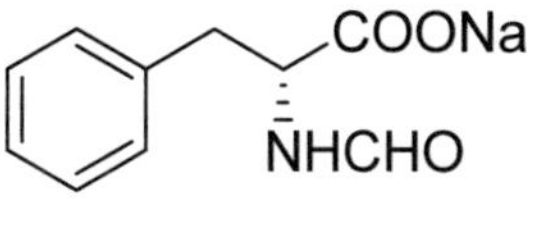

(*R*>*S*)-**FoPA.Na**

aquaeous sol. → addition of non eq. aqueous HCl sol.	ee_0%	ee_{solid}%	ee_{liquid}%
	49.6	19.3	75.0
	73.4	70.8	75.2

The situation was almost the same at the separation of the enantiomeric mixtures of *N*-acetyl-phenylglycine (**AcPG**) where ee_{eu} could not be exceeded ($ee_{eu} \approx 86$ % in both cases, according to the binary phase diagrams).[34]

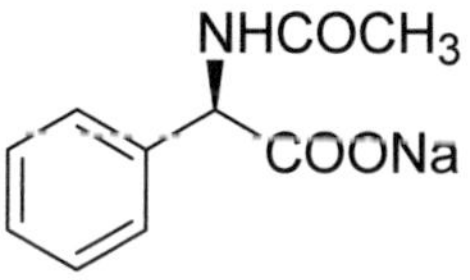

(*R*>*S*)-**AcPG**

aquaeous sol → addition of non eq. aqueous HCl sol.	ee_0%	ee_{solid}%	ee_{liquid}%
	63.6	52.6	86.2
	79.5	78.8	86.6

3.1.2.3 Kinetic control

The above mentioned examples (**FoPA** and **AcPG**) demonstrated both the conglomerate and racemate behaviour during fractionated precipitation of enantiomeric mixtures under thermodynamic control. However a new type of crystallization order was observed when the propionyl derivatives of phenylalanine and phenylgylcine (**PPA** and **PPG,** respectively) were examined. This type of compounds presented different behaviour during the enantiomeric enrichment processes than expected on the basis of their binary (melting point/composition) phase diagrams. Binary phase diagram of *N*-propionyl-phenylglycine (**PPG**) indicated conglomerate type behavior, while that of *N*-propionyl-phenylalanine (**PPA**) was a racemate type with ee_{eu} 59%. The results of selective precipitations contradicted the anticipations based on these findings. In these case the results of selective precipitations were rather similar to a conglomerate like behaviour. This can be explained with a faster crystallization of enantiomeric excess than the expected racemic proportion.[35]

Selective precipitation of (*R*>*S*)-*N*-propionyl-phenylalanine (**PPA**) was effectuated starting from the aqueous solution of its sodium salt. Addition of less than equimolar amount of hydrochloric acid resulted in the crystallization of the excess of (*R*)-**PPA** to such an extent, that in certain cases the usually remaining racemic composition become unbalanced, too (therefore the (S) enantiomer is enriched in the filtrate)

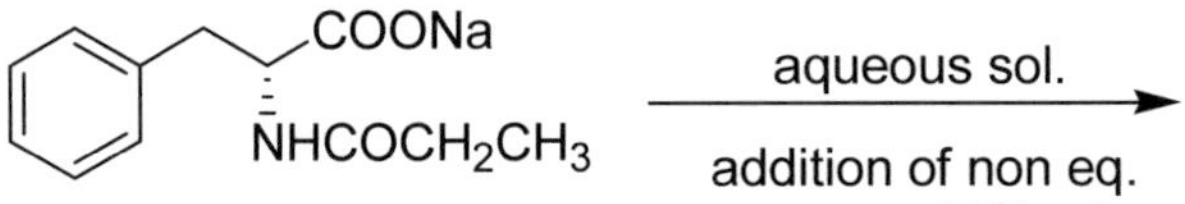

(*R*>*S*)-**PPA**

ee_0%	ee_{solid}%	ee_{liquid}%
20	2.5	24.8
25	47.7	-4.3
35	39.0	24.3
60	88.7	12.4
85	99.9	83.3

Essentially the same phenomenon was observed at the fractionated precipitation of the enantiomeric mixture of (*R*>*S*)-*N*-propionyl-phenylglycine (**PPG**).

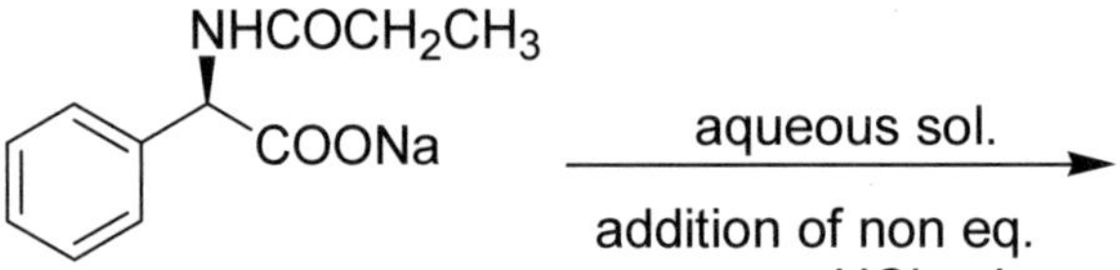

(*R*>*S*)-**PPG**

ee_0%	ee_{solid}%	ee_{liquid}%
10	4	17
12.5	27	-3.6
15	38.2	-4.8
25	27	1.2
35	59	0.9
50	84	-0.4
75	90	16
90	92	76

In light of these experimental data it should be mentioned, that during the purification of enantiomeric mixtures of *N*-acyl-aminoacids not only the suitable methods but also the subtituent on the molecule skeleton (acyl group) may determine the productivity of enrichment.

3.2 Crystallization of diastereoisomers

The similarities between the enriching processes starting from enantiomeric mixtures or diastereoisomeric mixtures were already discussed above at the presentation of the solvent free methods. It is an interesting question: whether the widely used fractionated precipitation from a solvent will change this trend at the diastereoisomeric mixtures?

The diastereoisomeric systems discussed bellow contain salts, molecular complexes and coordinative complexes as well.

3.2.1 The salt-forming resolving agents

3.2.1.1 Structurally similar resolving agents

Comparison of the results of the enantiomeric enrichments mentioned above let us to conclude that the difference between the enantiomeric excesses of the crystalline and the liquid phases is the greatest when the starting enantiomeric purity is near 50% ($ee_0 \approx 50\%$). In the cases of ten separtions made by crystallization this enrichment ($\Delta ee = ee_{cryst}-ee_{liq}$) was between 34-95%, (on the average 60%), while the average of the six greatest difference was 70%. This means that the purification of enantiomers by crystallization is a wery fruitful route (independently from the method applied (from melt to using a solvent) to enantiopure materials.

Another approach is the resolution of an enantiomeric mixture with a structurally similar resolving agent. It is the situation when one of the enantiomers of the racemic compound is transformed (with a minimal chemical transformation) into a reagent able to form diastereoisomeric salt with the initial racemic compound. It can be done if the aminoacid is *N*-acylated and one of the pure enantiomers is esterified, or if its carboxylic group is transformed into an amide, or changed with methyl group, respectively.

For example, the racemic *N*-acetyl-phenylglycine (**AcPG**) can be reacted with methyl (*S*)-phenylglycinate (**MePG**). The resolution can be treated as a recrystallization of a quasi enantiomeric mixture with ee_0=50% from water when the less soluble diastereoisomeric salt (namely, the heterochiral quasi-racemic mixture) crystallized containing (*R*)-**AcPG** in good enantiomeric excess (*ee*: 79%).

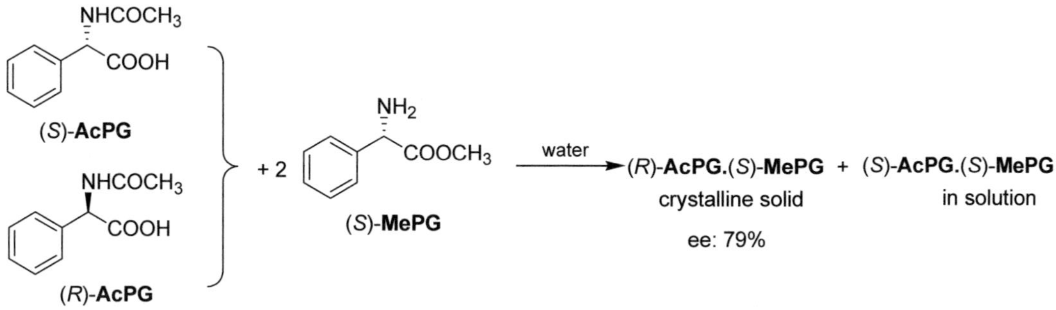

If the substituents of the compounds would be removed (Ac from **AcPG** → R and S and Me from **MePG** → S), the composition of the mixture would be

$$1R + 3S \equiv RS + SS$$

Namely, the starting mixture would be an "enantiomeric mixture with ee= 50%". Therefore the above shown crystalline diastereoisomeric salt is a quasi-racemate.

The situation was almost the same when the *N*-acetyl-phenylalanine (**AcPA**) was reacted with methyl (*S*)-phenylalaninate (**MePA**) according to the preceding resolution. The obtained crystalline diastereoisomeric salt also will be a quasi-racemate, but its enantiomeric excess (ee: 93%) was even higher than that of the former case.

It was also observed, that the average *ee* and F values of diastereoisomeric salts obtained at the resolutions (in number 28 resolutions) of racemic *N*-acetyl-phenylalanine and -phenylglycine (**AcPA** and **AcPG**) with structurally similar bases (resolving agents) correspond to the eutectic composition (ee_E value) of the adequate racemate (in these cases the enantiomers of the racemic compounds form racemates). Consequently, if the resolution of a racemic compound is accomplished with structurally similar resolving agents (in water) the purity of the enantiomer obtained from the crystalline diastereoisomeric salt corresponds to the biner phase diagram of the enatiomeric mixture or converge to the (experimental) eutectic composition. In other words, the characteristics of the resolved compounds strongly influence or even determinate the composition of the diastereoisomeric salts crystallized during the resolutions.[36]

In five cases among the resolutions of six racemic compounds with structurally similar resolving agents the crystallized diastereoisomeric salts have shown quasi-racemate behaviour (thus these were quasi-resolutions). At the same time in one case, during the resolution of *N*-formyl-phenylalanine (**FoPA**), crystallization of a quasi-homochiral diastereoisomer was observed. The observed quasi-racemate/quasi-conglomerate = 5/1 ratio has also been in acccordance with the earlier described data on the similar ratio of the racemates and conglomerates among the true enantiomeric mixtures,[37]

On the basis of the above discussed experimental data we concluded, that the behaviour of diastereoisomers follows the behaviour of their constituent enantiomeric mixtures. We have also observed that the similarity between the molecular structures of a racemic compound and a resolving agent has a positive effect on the enantiomeric separation. It also may be established, that a „derivative resolving agent" (use of optically active derivatives of the racemate as resolving agent) would be the optimal agent for separation of any racemic compound, but the best derivative should be found on experimental way.

3.2.1.2 Resolving agents with diverse molecular structures

According to the previously mentioned, the appropriate resolving agents for a racemic compound should be found among the structurally similar compounds because of the higher probability of the formation of well-fitting associates (salts or complexes) from that type of resolving agents and one of the isomers of racemic compound.

Of course, such a well-fitting chiral compound may be found among several other compounds, as well. They are the well-known and traditional resolving agents. Such a universal resolving agent is the O,O′-dibenzoyl-(*R,R*)-tartaric acid (**DBTA**) (mentioned several times previously). Presumably the above discussed findings, that the efficiencies of resolutions using structurally similar resolving agents may be determined by the behaviour of the enantiomeric mixtures of the racemic compound, may also be expanded on that diastereoisomeric salts forming resolutions where the resolving agent has „non-similar" molecule structure (particularly in cases of crystallizations from water).

In the simplest cases the generally used methods follow the first ever resolution by salt formation accomplished by Pasteur (fractionated precipitation method). In this case the racemic compound and the resolving agent are dissolved in a solvent, in a molar equivalent amount. The less soluble diastereoisomeric salt crystallizes (during cooling) and it can be separated from the other diastereoisomeric salt (remained in solution) by filtration. The enantiomeric mixtures are isolated from the diastereoisomers and they are enriched to the desired *ee* values by applying one of the previously discussed methods.

It has to be mentioned, that the enantiomeric mixture found in the crystalline diastereoisomer is usually much purer than that is isolated from the filtrate.

Pope and Peachey[38] have been recognized that if the half of the resolving agent necessary for the better soluble diasteroisomer is replaced by an achiral reagent (having the same chemical character as it is for the resolving agent) the less soluble diastereoisomer will be crystallized and the other isomer remains in the solution as a salt of the applied achiral additive. This method not only economizes a half part of the resolving agent, but the efficiency of resolution can also be improved.

Hereafter some examples of the above mentiuoned procedures, used in the pharmaceutical industry, will be presented.

The racemic intermediate of chloramphenicol (**AD.**HCl) is obtained as a hydrochloride salt during the synthesis, and it is dissolved in water (the free base is low-soluble in water). In this case the resolving agent (O,O'-dibenzoyl-(*R,R*)-tartaric acid *N,N*-dimethylamide (**DBTADA**)) is practically insoluble in water but its ammonium salt is well soluble and it can be prepared easily. The two aqueus solutions are mixed so that the molar ratio between the racemic compound and the resolving agent sould become 1/0.5. The diastereoisomeric salt containing the desired enantiomer (*R,R*)-**AD** crystallizes while the hydrochloride salt of (*S,S*)-**AD** remains in solution.

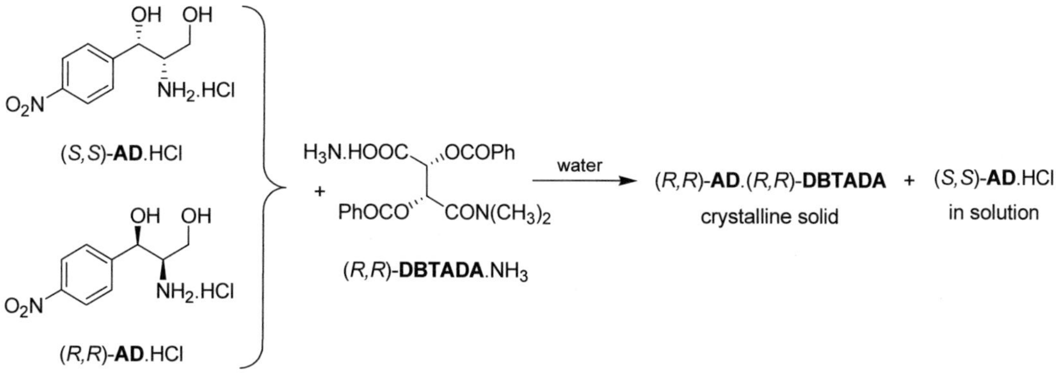

The next example one is of the first resolutions of a racemic intermediate of prostaglandines. During the synthesis it is isolated as the water-soluble sodium salt ((*R,S*)-***cis*-Na**). Half an equivalent amount of (*R*)-1-phenylethylamine (**PEA**) neutralized with hydrochlorid acid is added to the aqueous solution of (*R,S*)-***cis*-Na**. The (*S*)-**cis.**(*R*)-**PEA** crystallizes while (*R*)-***cis*-Na** remains in the solution.

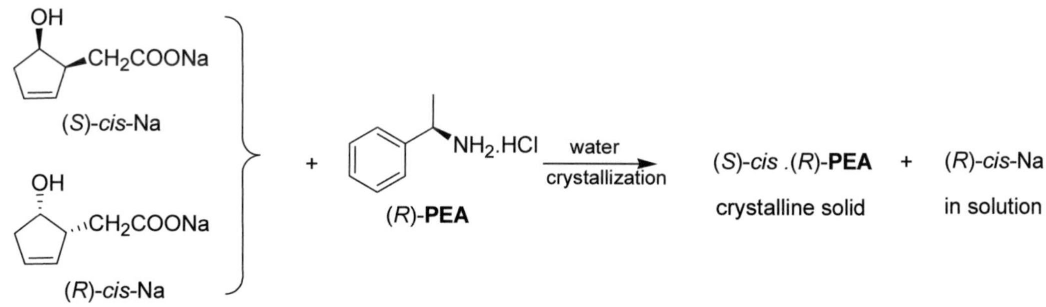

It may happen frequently, that only one enantiomer of the resolving agent is available, or one of the two enantiomers is significantly cheaper. For this reason mainly (*R,R*)-tartaric acid is used industrially among of tartaric acid enantiomers and their derivatives. It is the case, for example, at the resolution of a racemic intermediate (**AN**) of α-methyl-**DOPA**. The resolution of racemic **AN** is performed in the aqueous solution of its hydrochloric acid salt

(**AN**.HCl) by half an equivalent amount of mono sodium salt of (*R*,*R*)-tartaric acid (**TA**.Na). The crystalline salt of undesired (*R*)-**AN.TA** precipitates. After filtration of the diastereoisomeric salt the pure (*S*)-**AN**.HCl can be crystallized from the mother liquor by salting out (addition of sodium chloride into the filtrate) while the racemic **AN.**HCl remains in solution. Since the enantiomeric excess crystallized from the filtrate we can speak about conglomerate behaviour. However, the compound may be a racemate forming material, if the purity of the (*S*>*R*)-**AN**.HCl enantiomeric mixture is higher than the eutectic composition (ee_{eu} for **AN**.HCl).

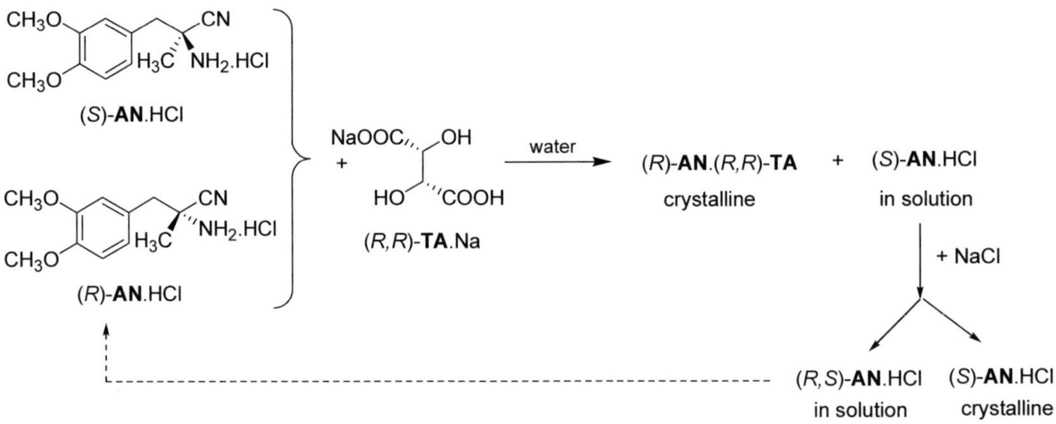

3.2.1.3 Mixture resolving agents

The so called "Dutch resolution" [39] utilize the recognition whereas a part of the original resolving agent can be replaced by another chiral compound with similar chemical character and under such circumstances (when the resolving agent really is a mixture of structurally related molecules) a diastereoisomer (containing usually an enantiomer of the racemate in very high purity) crystallizes from the supersaturated solution. The advantage of this resolution method is that the efficiency may often be superior to the use of any of the resolving agent components alone, and sometimes it works even if one of the reagents alone is unsuitable to form crystallizing diastereoisomeric salt.

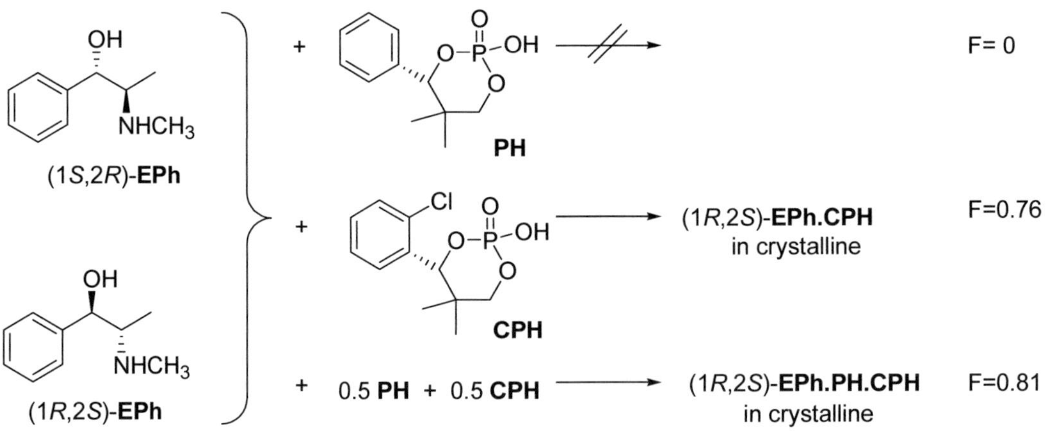

In the following example three kind of resolution of racemic ephedrine (**EPh**) via diastereoisomeric crystallization will be presented, as before.[40] The favourable resolving agent is a chiral diester of phosphoric acid containing 2-chloro-phenyl substituent (**CPH**), while the structurally similar phosphoric acid containing unsubstituted phenyl group (**PH**) is an unsuitable resolving agent.

3.2.1.4 Mixture of a resolving agent and a structurally related achiral molecule

The above mentioned observation, namely a part of the resolving agent can be replaced by a structurally related compound which is unsuitable for separation of the enantiomers means in general, that the presence of a structurally similar but achiral reagent may also improve the results of a resolution.

For example, when *N*-acetyl-phenylalanine (**AcPA**) is resolved with half an equivalent amount of (*R*)-phenylglycine amide (**PGA**) in presence of sodium hydroxide both the enantiomeric purity and the productivity were lower than when instead of sodium hydroxide half an equivalent amount of benzylamine (**BA**) was applied (**BA** has a more related structure with the resolving agent than sodium hydroxide). Its presence improved significantly the crystallization of (*S*)-**AcPA.**(*R*)-**PGA** diastereoisomeric salt, as well as the efficiency of the resolution

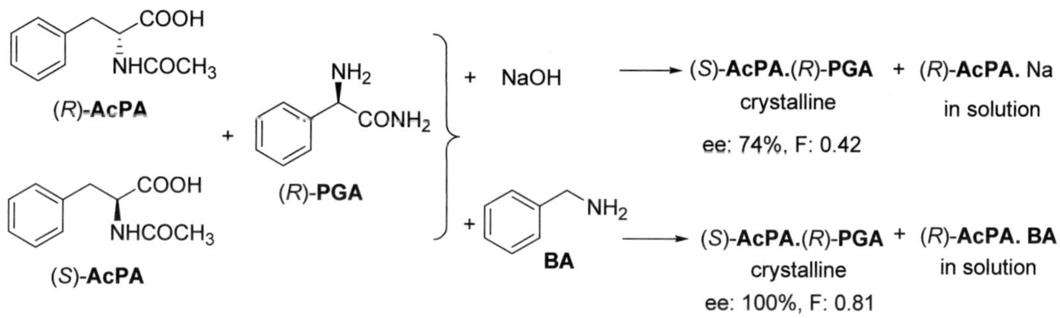

The use of an achiral compound with related molecular skeleton may also be advantageous, if its structure is related with structure of the racemic compound. For example, if the resolution of racemic **AcPA** is carried out using (*R*)-**PEA,** the crystallized diastereoisomeric salt contains almost racemic **AcPA**, but addition of the achiral analogous phenoxy-acetic acid (**POAA**) to the racemic compound, result in a very good enantiodiscrimination. The achiral additive crystallizes quickly together with resolving agent and, during the long standing, the **POAA** is exchanged for (*S*)-**AcPA** enantiomer in the crystal.

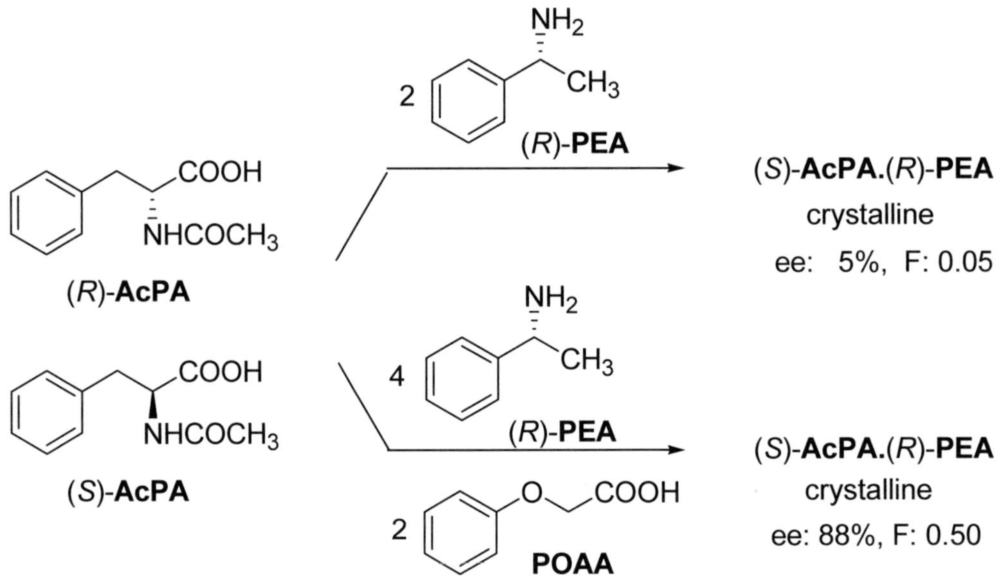

3.2.2 Influence of solvents

In the majority of the above discussed fractionated precipitations or crystallizations of diastereoisomeric salts water was used as a solvent. The experimental observation of decades is that the dissociation in solvent of diastereoisomeric salts and of salts formed with achiral auxiliaries in water or in other protic solvents, is considerably influenced by the proton concentartion (pH value). Naturally this value influences both the quantity and the isomeric purity of the crystallized salt, namely the efficiency of the resolution.[41]

There are also examples in the literature, when the configuration of the enantiomer in the crystalline diastereoisomeric salt and/or the efficiency of the resolution are determined by the polarity of solvent (the dielectric constant characteristic for this), respectively.

In other cases, the use of solvent mixtures could influence which diastereoisomer can be found in the crystalline phase. In such cases, either the crystallizing diastereoisomer or the one remained in the mother liquor form stable solvate with one of the solvents. After a short summary of the theoretical background, it will be shown via examples, how can we use the influence of pH, solvent polarity, and solvate formation, respectively, to improve the efficiency of enantiomeric separation.

3.2.2.1 The influence of pH

The pH dependence of resolutions based on fractionated crystallization of diastereoisomeric salts was described by a thermodynamical equilibrium model.[33]

The equilibrium model of a resolution of DL racemic base with an acidic resolving agent (RH) can be outlined bellow.

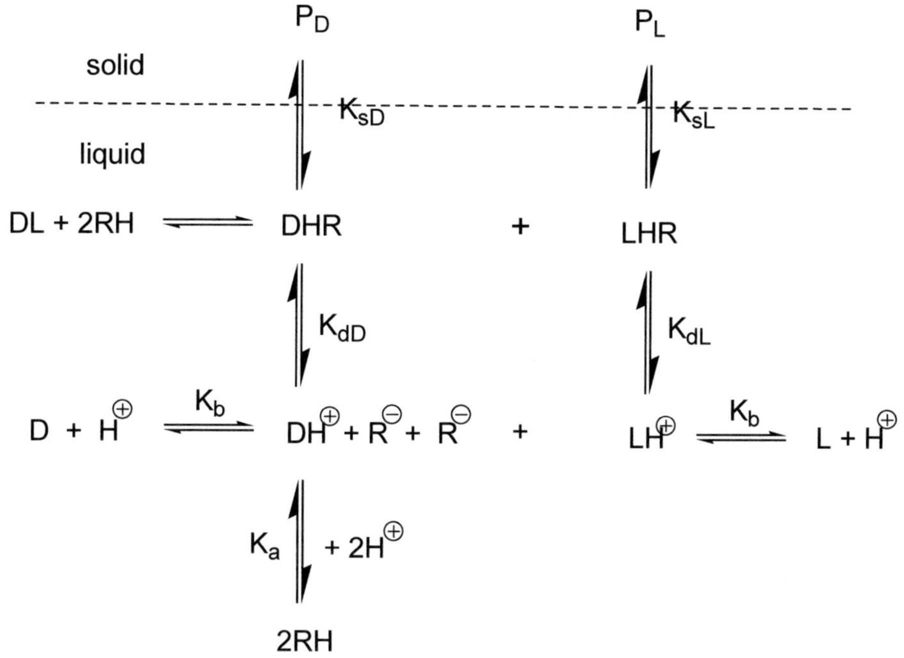

Where K_s = salt solubility constant, K_d = salt dissociation constant, K_b = base dissociation constant, K_a = acid dissociation constant, P_D and P_L are the precipitated diastereoisomeric salts, respectively.

The efficiency of the resolution (S = F, see above in point 1)[14] can be calculated by means of the next formula, using the known thermodynamic constants data, and the concentrations of the starting compounds (c_o = concentration of the racemic compound, $[RH]_o$ =concentration of resolving agents) and of $[H^+]$ = proton concentration, as well.

$$0.5c_oS = K_{sL} - K_{sD} + \left(1 + \frac{K_b}{[H^+]} + \frac{[H^+]}{K_{RH}} + \frac{K_b}{K_{RH}}\right)\left(\frac{K_{sL}K_{dL} - K_{sD}K_{dD}}{[RH]_0 - 0.5c_0y - (K_{sL} + K_{sD})}\right)$$

From the relationship deduced on basis of the model the regularity of several „rule of thumb" based on experimental observations were justified. A fundamental conclusion is that the efficiency of resolution is the function of both the hydrogen ion concentration and the initial concentration of the resolving agent.

a. With derivation of the equation, describes the thermodynamic equilibrium model, can be justified that in so far as $K_{dL} > K_{dD}$, the minimum value (S_{min}) of the $S = f(H^+)$ function at $[H^+] = \sqrt{K_b K_{RH}}$ that corresponds to the proton concentration provided by hydrolysis of the neutral diastereoisomeric salt; namely that concentrations occur at the equivalent resolution discovered by Pasteur: It menans that this type of resolution results in the smallest separation efficiency. It is worth to mention, that the essential condition, whereas the dissociation constant of the rather soluble salt is bigger, is realized at the majority of resolutions. (if $K_{dL}>K_{dD}$ is not valid, application of the equivalent amount of resolving agent gives the best result.)

In the light of the above deduction the experimental observation, that the tartaric acid (**TA**, as a dibasic acid) is an efficient resolving agent is not surprising because in the most cases the hemitartarates crystallize and one carboxylic group remains „free". This additional acidic group increases the proton concentration compared to the proton concentration of the solution of a neutral salt (salt of a monobasic acid).

In practice, an excess (10-20%) of the resolving agent used to be added to the racemic compound and, of course, the excess shifts the pH value of the solution as well.

b. Based on the equation of the thermodynamic equilibrium model, the optimum relative concentration of the resolving agent can be determined. It means that the efficiency of S is maximal when the denominator of the second member of multiplication in the equation tends to zero, namely $S \rightarrow S_{max}$ if $[RH]_o = 0.5c_oy + (K_{sL} + K_{sD})$. If the solubility constants of the diastereoisomers are low, and we tend to obtain the highest yield ($y \approx 1.0$), then $[RH]_o \sim 0.5c_o$, consequently using half an equivalent amount of resolving agent will result in the highest efficiency. At the same time it is the theoretical argument of the successfulness of the method disclosed by Pope-Peachey.[33]

The dramatic influence of the *p*H of the medium was first encountered at the separation of optical isomers of *cis*-permethrinic acid [42] carried out with half an equivalent amount of (*S*)-2-benzylaminobutanol ((*S*)-**BAB**). The resolution started in the presence of a 25 mol % excess of sodium hydroxide. In such a medium the diastereoisomeric salt containing the almost pure (*S*)-**CPA** crystallized (*ee*: 96%) with a yield of 27% counted to the amount of racemic acid. After removal of the precipitate by filtration, the excess alkali was neutralized with a counted amount of hydrochloric acid. That time the resolving agent remained in the filtrate crystallized with (*R*)-**CPA** enantiomer as diastereoisomeric salt and in the second mother liquor an almost racemic **CPA-Na** salt was found (it can be recycled into the resolution).

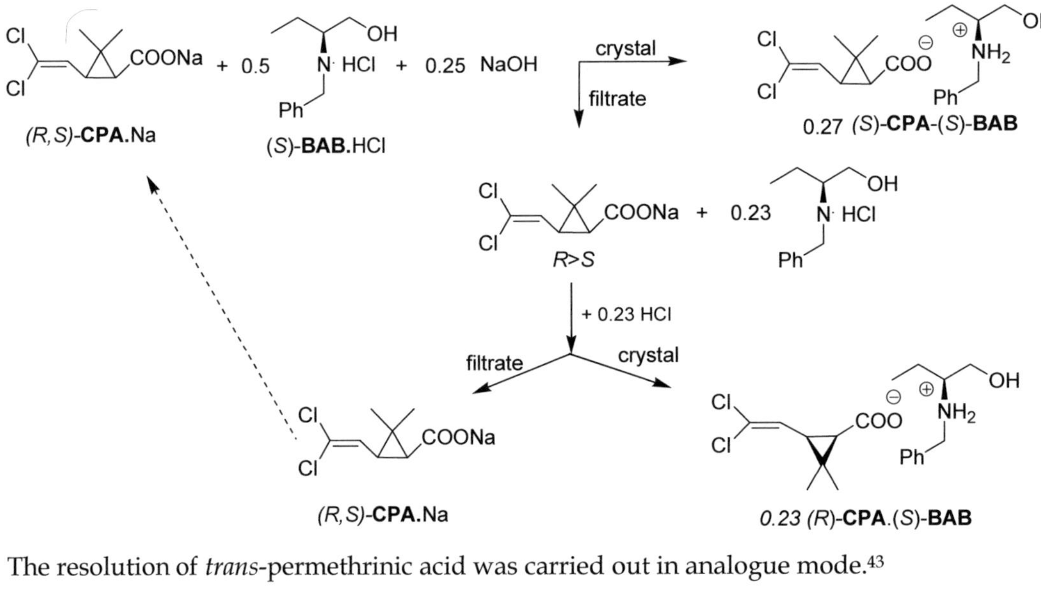

The resolution of *trans*-permethrinic acid was carried out in analogue mode.[43]

The behaviour of racemic compounds may also be influenced with the choice of pH value of the aqueous solution. The isoelectronic point of 2-phenylglycine (**PG**) is at neutral pH value (Ip = 7.0). The free racemic aminoacid with (*S*)-camphorsulfonic acid (**CSA**) forms a well crystallizing salt in water, but the amino acid in the precipitated diastereoisomer have almost racemic composition. When the crystallization is started from an aqueous solution containing half an equivalent amount of **CSA** and equivalent amount of hydrochloric acid (total 50% excess of acid to the PG as base!), the crystalline salt contains (*S*)-2-phenylglycine in high enantiomeric purity (and in good yield), while the almost pure (*R*)-2-phenylglycine was obtained from filtrate by crystallization of its hydrochloric acid salt.[44]

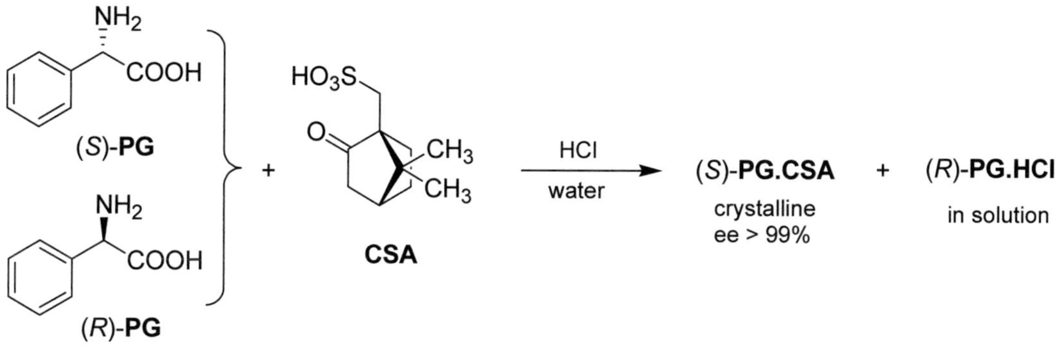

The resolution of racemic free phenylalanine (**PA**) was also carried out with high efficiency, if near the half an equivalent amount of resolving agent (O,O′-dibenzoyl-*(R,R)*-tartaric acid (**DBTA**) or the optically active *N*-benzoyl-d-phenylalanine (**BPA**)) was applied in aqueous methanolic solution which contained more than 0.5 equivalen amount of hydrochloric acid, too.[45]

3.2.2.2 Influence of dielectric constant

In the previous chapter the crucial role of the pH value has been shown via examples.

It was also recognized that tha polarity of the solvents or solvent mixtures (quantified by the dielectric constant of solvents) can also determinate the composition of the crystallized diastereoisomeric salt.[4] For example, in the resolution of α-aminocaprolactam (**AC**) with (*S*)-*N*-tosylphenylalanine ((*S*)-**TsPA**) in various solvents the (*S*)-**AC** was predominant when the dielectric constant of the solvent was ε<27 or ε>62. However, the salt of the antipode (*R*)-**AC** crystallized when the solvent was in meium polarity (29<ε<58).[46,47,48]

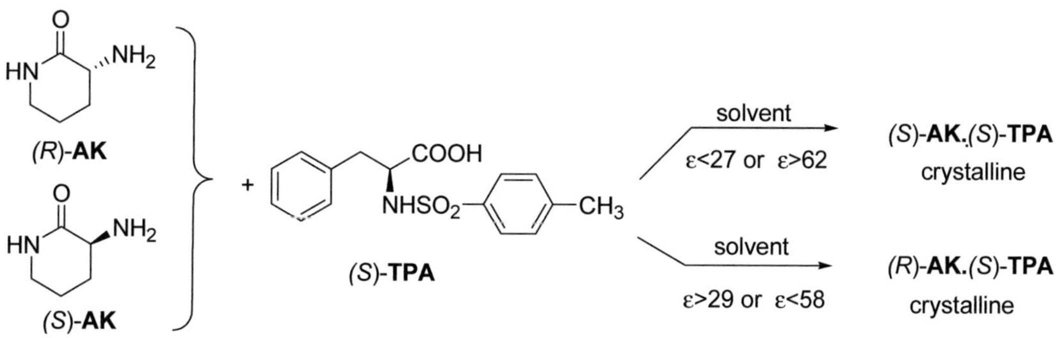

That observation can be generalized: systematic variation of the solvent, the diastereoisomer containing the desired enantiomer can be brought to crystallization. (Using the adequate solvent one of the diastereoisomers is crystallized from the mixture of diastereoisomers, and the other one can be crystallized from another solvent.)

For example, if the resolution of flumequine intermediate **(FTHQ)** is resolved with O,O′-di-para-toluoyl-(*R*,*R*)-tartaric acid ((*R*,*R*)-**DPTTA)** in presence of acetic acid the (*R*)-**FTHQ** enantiomer crystallizes in the salt, while the (*R*)-**FTHQ** enantiomer forms a better crystallizing salt when isopropyl alcohol is used as a solvent.[49]

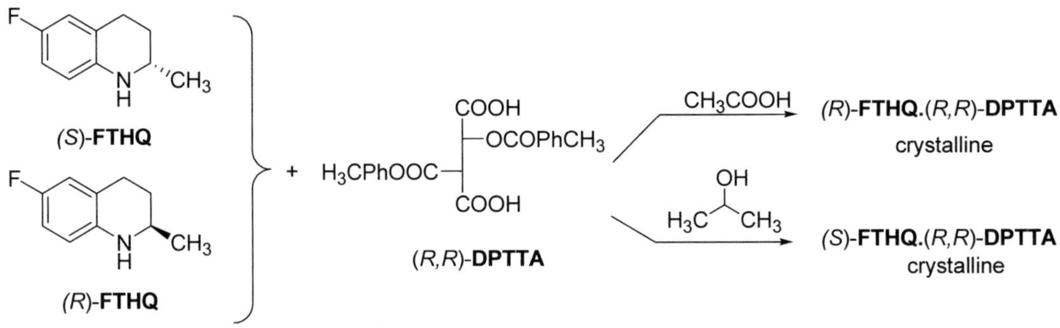

3.2.2.3 Resolution in two immiscible solvents

One variant of the resolution methods is that salt formation is carried out in a system of two immiscible solvents (e.g. water and dichloromethane). In this case the more stable diastereomeric salt separates, while the free enantiomer dissolves in one of the phases (usually the organic one).[50,51]

This manner was used at the resolution of the anxiolytic tofisopam (**TOF**) with half equivalent of (*R*,*R*)-**DBTA.** It was accomplished in a water–chloroform system. The diastereomeric salt containing the (*R*)-**TOF** enantiomer was crystallized from the biphase solvent, while free (*S*)-**TOF** was recoverable from organic phase.[52,53]

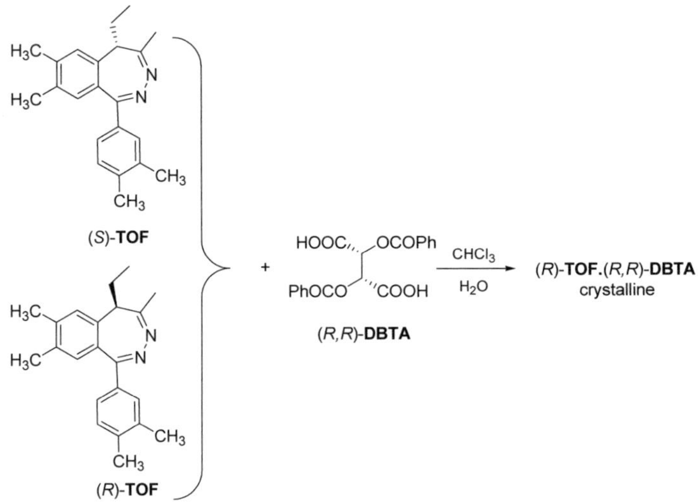

3.2.2.4 Presence of solvates and compounds with similar effect

It can be frequent observed at the fractionated crystallization of diastereomeric salts that the crystalline diastereomer forms solvate with the solvent. As an example can be mentioned the first ever resolution by salt formation accomplished by Pasteur wher the "d-quinotoxine-d-tartrate" obtained was a hexahydrate.[54] Of course it can be found numerous similar examples in the literature.[55]

It is not rare the phenomenon that the selective solvate formation of diastereomers gives a much better separation or the solvate formation is essential for successful resolution. It is true because only in this case crystallizes the diastereomer. E.g. in methanol *trans*-(*R*)-chrysanthemic acid (*trans*-(*R*)-**CHRA**) forms with (*R,R*)-2-*N,N*-dimethlyamino-1,3-propanediol [(*R,R*)-**DMAD**] crystals containing the solvent. If the resolution was carried out in another solvent (e.g. in methyl isobutyl ether) best results were obtained if some methanol was added, because it permitted the precipitation of the methanol solvate (is crystallized the (*R*)-***CHRA***.(*R,R*)-**DMAD**.CH_3OH).[17]

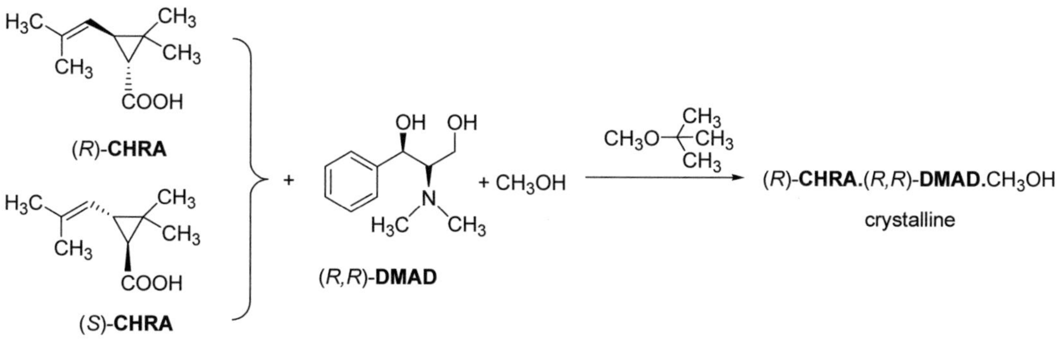

3.2.2.4.1 When the separation of diastereomers is possible just from solvate forming solvent

The separation of enantiomers of racemic amlodipine (**AML**) with (*R,R*)-tartaric acid by fractionated crystallization of diastereomers from the common solvents was without success, but in **DMSO**, however, the salt of (*S*)-**AML** crystallizes as a **DMSO** solvate.[56] Even more surprisingly, in *N,N*-dimethyl-acetamide it is the solvate of the (*R*)-**AML** salt which crystallizes in good yield and high purity.[57]

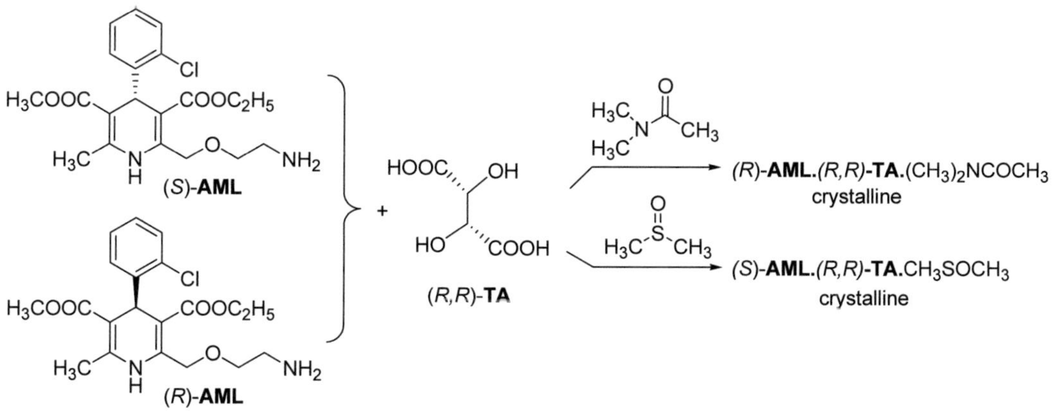

3.2.2.4.2 Presence of achiral compound

It is well known separation of diastereomers when the enantiomeric purity reachable using a resolving agent may be significantly improved if the crystallization of diastereomeric salt is promoted by crystallization of an achiral compound having analogue molecular structure with a part of the resolving agent molecules.

This phenomenon was observed at the resolution of racemic α-phenylethylamine (**PEA**) using (*R*)-*N*-(1-phenyl-ethyl)-glutaric acid ((*R*)-**PEGA**) as resolvong agent. From acetone was crystallized diastereomeric salt containing (*S*)-**PEA** with ee: 58%. If is added an equivalent of carbamide to the mixture the enantiomeric excess of the isolated (*S*)-**PEA** was 90%.[58]

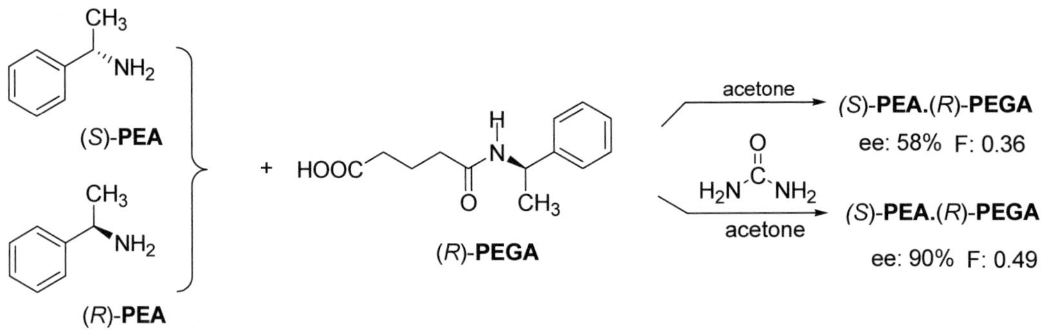

3.2.3 Resolving agents forming diastereomer complexes

The difference betwee, solubility of diastereomeric salts, their solvent deppendence justify that although the ions of salt estabilish strong interactions, neverthless the formation of secondary interactions between the compounds containing these ions (formation of diastereomeric molecular complexes) is the reason of differential solubility of diastereomeric salts in the respective solvent. Naturally, it is not by chance that the diastereomeric salt of dibenzoyl-tartaric acid (**DBTA**) are well separable and it is a very suitable resolving agent for a wide range of racemic basis. It means that the enantiomers of racemic compounds which do not contain either acidic or basic functional groups form diastereomerically releted molecular complexes with **DBTA**. These complexes can be separeted by fractionated crystallization. Such several alcohols with various molecular structures could be resolved, e.g. the resolution of racemic menthol (**MEN**) was carried out with (*R,R*)-**DBTA** in solution, too. In hexane insoluble reagent contacting the dissolved racemic menthol forms crystalline diastereomeric molecular complex.[59] (as was shown earlier at the crystallization of diastereomers from melt).

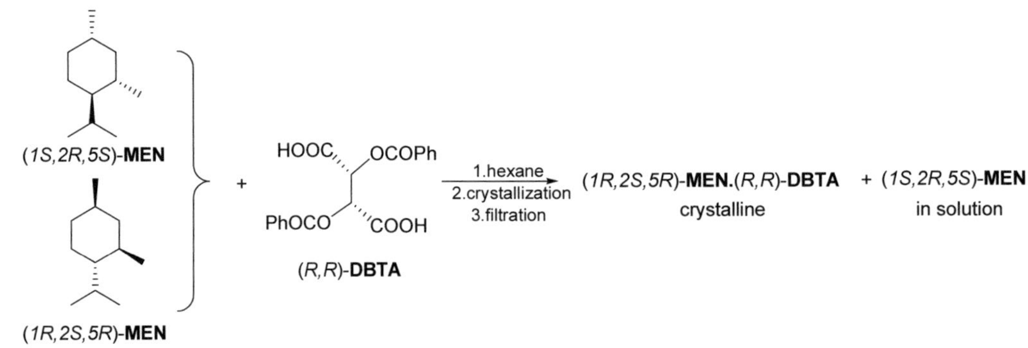

From the crystalline diastereomer could be isolated the L-**MEN** with a good enantiomeric purity (ee: 83%).

The widely applied **TADDOL** (**TAD**) was also a suitable reagent of several diastereomeric molecular complex, for example enantiomers of phospholene oxides (**PHO**) could be obtained with high optical purity after the fractionated crystallization and decomposition of the precipiated diastereomers.[8]

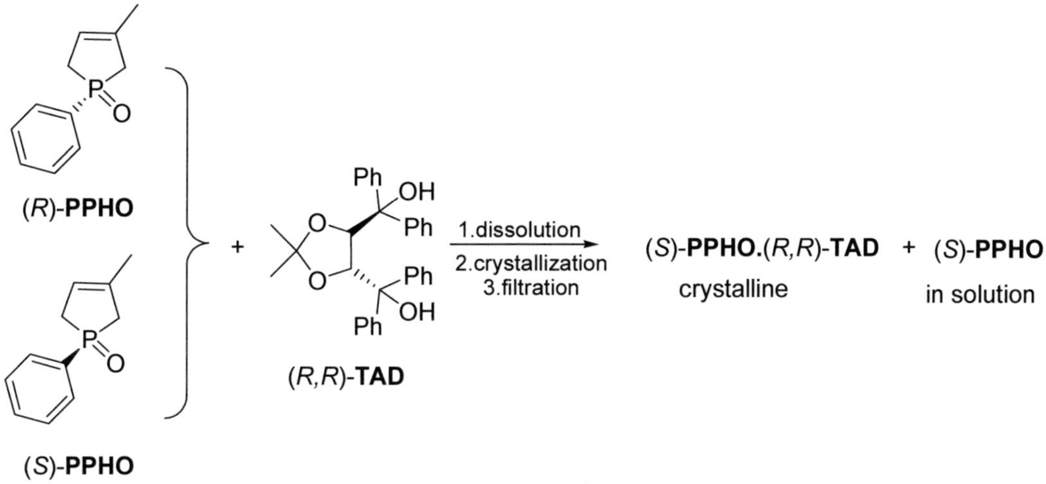

3.2.4 Resolving agents forming coordinative complexes

It is well known that the coordanative complex of several metal is well crystallize. The calcium salt of dibenzoyltartaric acid (**DBTAC**), was used be quite a few for separation of diastereomeric coordinative complex by fractionated crystallization. For example the separation of derivatives of phospholene oxides (previously mentioned) was carried out using (**DBTAC**).[9]

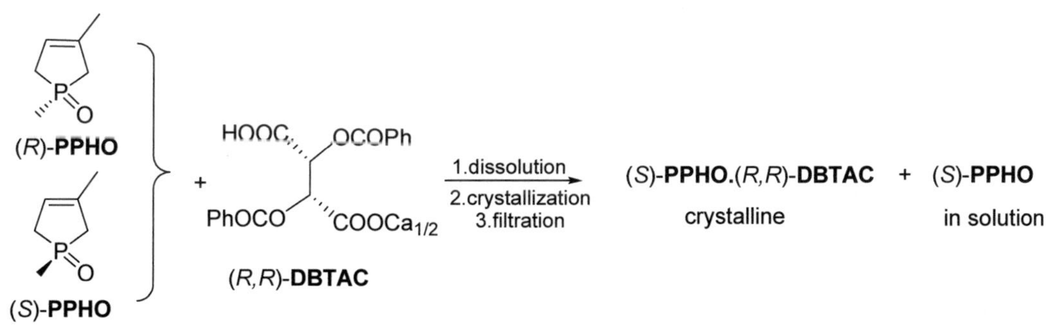

3.2.5 Time of crystallization

Although the methods based on fractionated crystallization suggest that during crystallization the less soluble diastereomer is crystallized, but at the enantiomeric separations realized by fractionated crystallization, or precipitation, respectively, was observed in some cases that instead of expected racemate behaviour was crystallized the enantiomeric excess (conglomerate like behaviour), namely its faster crystallization determines the separation (kinetic conglomerate).

3.2.5.1 Thermodinamic control

In general, at the fractionated precipitation if is expected a conglomerate like behaviour on base of binary phase diagram of diasteroemeric mixture, the filtration is efffectuated when the quantity of crystalline precipitate is constant. However if the difference between the crystállization rate of diastereomers is not large enough, can be occured that both diastereomers are crystallized in near similar quantity. In this case we cold not talk about enantiomeric enrichment. If instead of quick filtration is given time for stabilition of thermodynamic equilibrium, in turn is obtain a separation with good result. This can be demonstrate via resolution of tamsulosin intermediate **(TAI)** using as resolving agent **DBTA** in the mixture of water and ethanol, when at the end of crystallization was not carried out enantiomweric enrichment, but if the crystalline mixture was allowe to stand for 2 days, was reached a very high enantiomeric excess.[60]

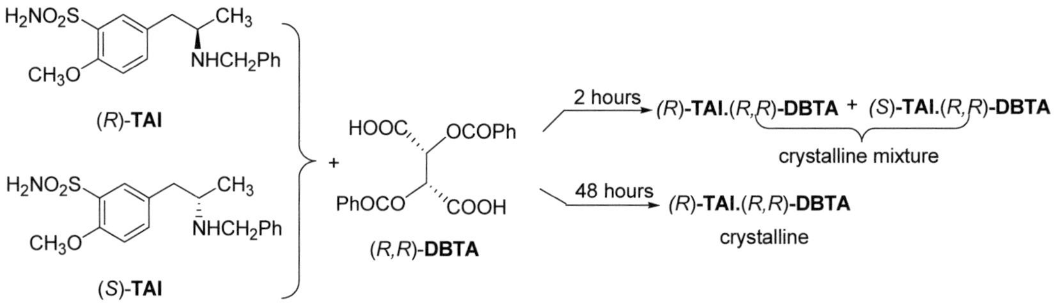

Similar phenomenon was also observed at the resolution of racemic oxirane derivatives, containing two terciary aminogroups.[61]

3.2.5.2 Kinetic controlled crystallization

Accordingly to the previously examples would appear to be advantageous that the time of crystallization to be more longer but if one of the diastereomers crystallizes quickly and the other is wrong solvable, then neverthless its opposite is also favourable.

Thus in the resolution of the intermediate of flumequine **(FTHQ)** in ethyl acetate with di-paratoluyl-tartaric acid (*R*,*R*)-**DPTTA,** after a crystallizing of 5 minutes the (*R*)-**FTHQ**.(*R*,*R*)-**DPTTA** salt predominates, but on standing 3 weeks its antipode (*S*)-**FTHQ**.(*R*,*R*)-**DPTTA** will be in excess in the crystals.[48]

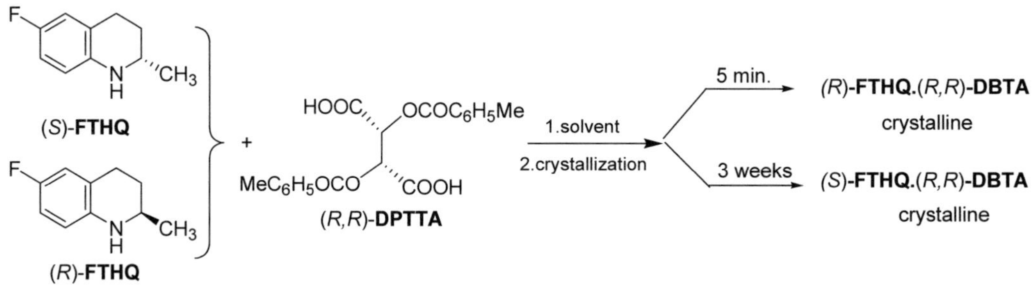

In the most cases it is not necessary to effectuate so quickly the filtration of crystalls, because the stabilition of thermodinamic equilibrium need a longer period. This is so even if the racemic compound of resolving agent is resolved using one of enantiomers of racemic compound.[62]

The resolution of *N*-formyl-phenylalanine (**FoPA**) with (*S*)-phenylethylamine (**PEA**) in water after 2 hours gave diastereomeric salt with high purity, but this enantiomeric excess diminished when the filtration of the precipitated salt was carried out after a longer standing (1 week).

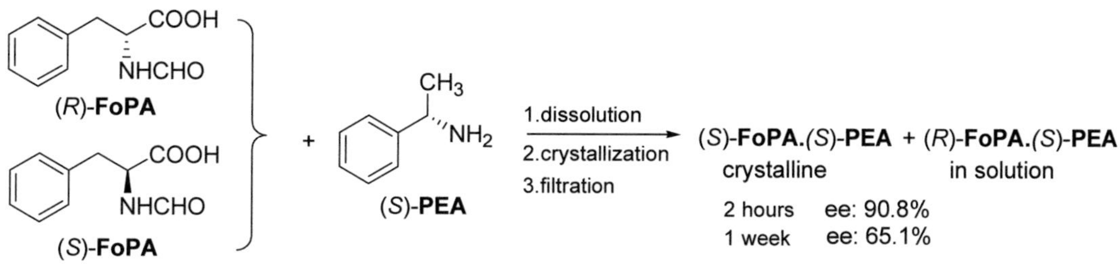

If the reolution of racemic **PEA** is accomplished with (*S*)-**FoPA** (reciprocal resolution), in function of crystallization time practically the same thing was happened as expected on basis of the foregoing, but the results were not identical because at the normal and the reciprocal resolutions the systems are in diastereoisomeric relationship.

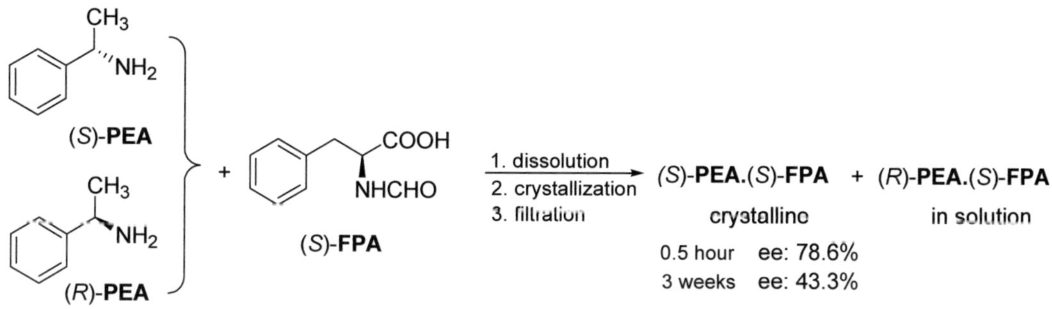

4. Conclusions

The behaviour of the enantiomeric mixtures can be interpretated by formation of their homo- and heterochiaral aggregates having diastereomeric property. Due to spontaneous formation of these aggregates the non-racemic mixtures of enantiomers can be enriched by processes required some selective crystallization. The running of biner meltingpoint/composition phase diagrams (conglomerate or racemate), and the kinetic controlled crystallizations (kinetic conglomerates), respectively, determine the fact, that the crystalline phase contains the enantiomer or the racemic proportion. These regularities are also reflected by the running of curves where the enantiomeric purity of the product obtained (ee) during the selective crystallizations and precipitations were plotted against the initial composition (ee_0).

At the separation of racemic compounds using another chiral material (namely resolution) the behaviour of enantiomeric mixture forming the racemate determines the efficiency of resolution. If the racemic compound and resolving agent have similar molecule structure, the diastereomers form quasi-racemate, and quasi-conglomerate respectively.

If the structures of racemic compound and resolving agent are not considered structurally related compounds, at separation of the diasteremers also exist the complementarity between compounds form the diastereomer, which is in advanced insured in case of structurally similar compounds. In all probability the results of separation of diastereomers are also determined by the behaviour of enantiomeric mixtures of racemic compound and resolving agent.

The probability of this statement is underlined near the experimental data, the fact that the conditions, separation methods, typical behaviours observed at the separation of diastereomers based on the fractionated crystallization were similar to the experienced one at the separation of mixtures containing only enantiomers.

The rulings of chiral-chiral recognitions are valid even if the aim is to isolate by crystallization a chiral compound, having a higher purity than the initial one ($ee>ee_0$), from a mixture containing more than two chiral component.

5. Acknowledgements

The authors thank the financial support of the Hungarian OTKA Fundation (T 75236) for E. Fogassy)

This work is connected to the scientific program of the "Development of quality-oriented and harmonized R+D+I strategy and functional model at BME" project. This project is supported by the New Hungary Development Plan (Project ID: TÁMOP-4.2.1/B-09/1/KMR-2010-0002).

6. Abbreviations

A	phenylisopropylamine
AC	aminocaprolactam
AcPA	acetyl-phenylalanine
AcPG	acetyl-phenylglycine
AD	aminodiol
AML	amlodipine
AN	aminonitrile
BA	benzylamine
BAB	benzylaminobutanol
BPA	benzoyl-d-phenylalanine
CHD	cyclohexane-diol
CHRA	chrysanthemic acid
CPA	permetric acid
CPH	chloro-phenyl substituent
CSA	camphorsulfonic acid
DBTA	dibenzoyltartaric acid
DBTAC	calcium salt of dibenzoyltartaric acid

DCA	dicarboxylic acid
DIL	diltiazem
DMAD	dimethlyamino-1,3-propanediol
DPTTA	di-para-toluoyl-tartaric acid
EPh	ephedrine
FoPA	formyl-phenylalanine
FTHQ	flumequine
IBU	Ibuprofen
IC	Iodo-cyclohexanol
MA	methyl-phenylizopropylamine
MEN	menthol
MePA	phenylalanine methyl ester
MePG	phenylglycine methyl ester
PA	phenylalanine
PEA	Phenylethylamine
PEGA	phenyl-ethyl-glutaric acid
PG	phenylglycine
PGA	phenylglycine amide
PGL	Prostaglandine
PHO	phospholene oxides
POAA	phenooxy-acetic acid
PPA	propionyl-phenylalanine
PPEA	phthaloyl-phenylethylamine
PPG	propionyl-phenylglycine
Q	Quinotoxine
TA	Tartaric acid
TAD	Taddol.
TAI	tamsulosin intermediate
TIS	Tisercine
TOF	Tofizopam
TPA	tosylphenylalanine

7. References

[1] a) Eliel, E. L.; Wilen, S. H.,: *Stereochemistry of Organic Compounds* Eds.; Wiley New York, 1994 b).Roozeboom, H. W. B.: *Z. Physik. Chem.* 1899, *28*, 494.

[2] Bereczki, L.; Pálovics, E.; Bombicz, P.; Pokol, G.; Fogassy, E.; Marthi, K.:.*Tetrahedron: Asym.* 2007, *18*, 260

[3] Faigl, F., Fogassy, E.; Nógrádi, M.; Pálovics, E.; Schindler, J.; *Tetrahedron: Asymm.*, 2008, *4*, 519

[4] Scheldon, R. A. *Chirotechnology*, Marcel Dekker Inc. N.Y 1993

[5] Novák, T; Schindler, J; Ujj, V; Czugler, M; Fogassy, E; Keglevich, Gy: *Tetrahedron: Asymmy.* 2006, 2599.

[6] Ujj V; Schindler J; Novák T; Czugler M; Fogassy E; Keglevich Gy: *Tetrahedron: Asymm.* 2008, 1973-1977

[7] a).Blackmond, D.G., Klussmann, M.: *Chem. Commun.*, 2007, 3990-3996. b).Perry, R.H., Wu, C., Nefliu, M., Cooks, R.G.: *Chem. Commun.*, 2007, 1071-1073. c).S. I. Goldberg, *Orig Life Evol Biosph.*, 2007, *37*, 55-60.

[8] Faigl, F.; Fogassy, E.; Nógrádi, M.; Pálovics, E.; Schindler, J.: *Organic & Biomolecular Chemistry* 2010, *8*, 347-359.

[9] Sakai, K.; Sakurai, R.; Yuzawa, A.; Hirayama, N. *Tetrahedron: Asymmetry*, 2003, 14, 3716

[10] Fogassy, E.; Faigl, F.; Ács, M.; Grofcsik, A. *J. Chem. Res.*,1981, S 11, 346./ Fogassy, E.; Faigl, F.; Ács, M.; Simon, K.; Kozsda, É.; Podányi, B.; Czugler, M.; Reck, G. *J. Chem. Soc. Perkin Trans. 2.* 1988

[11] Kellogg, R.M.; Nieuwenhuijzen, J.W.; Pouwer, K.; Vries, T.R.; Broxterman, Q.B.; Grimbergen, R.F.P.; Kaptein, B.; La Crois, R.M.; de Wever, E.; Zwaagstra, K.; van der Laan, A.C. *Synthesis*, 2003, 1626.

[12] Schindler, J.; Egressy, M.; Bereczki, L.; Pokol, G.; Fogassy, E.; Marthi, K. *Chirality*, 2007, 19, 239.

[13] a). Hung. Pat. no. 214720, *Chem. Abs.* 1995, 124, 117097; b).US Pat. 02133894 2001, *Chem. Abs.* 2001, 139, 90595

[14] a). Weissbuch, I., Lahav, M., Leiserowitz L. *Advances in Crystal Growth Research*, Page: 381-400, 2001. b).Fogassy, E.; Faigl, F.; Pálovics, E.; Schindler, J. *11th International Conference of Chemistry Cluj*, ed. Hung. Techn. Sci. Soc. of Transylvania 2005, 357.

[15] Pálovics, E.; Schindler, J.; Borsodi, J.; Bereczki, L.; Marthi, K.; Faigl, F.; Fogassy, E.: *Műszaki Szemle*, 2007, 39-40, 48

[16] a). Kozma, D.; Böcskei, Zs.; Simon, K.; Fogassy, E.: *J. Chem. Soc. Perkin Trans 2*, 1994, 1883 b). Nemák, K.; Ács, M.; Kozma, D.; Fogassy, E.: *J. Therm. Anal.* 1997, 48, 691.

[17] Pálovics, E.; Fogassy, E.; Schindler, J.; Nógrádi, M.: *Chirality*, 2007, *19*, 1

[18] Fogassy, E.; Lopata, A.; Faigl, F.; Darvas, F.; Ács, M.; Tőke, L.: *Tetrahedron Lett.* 1980, *21*, 647.

[19] Pasteur, L. *Acad. Sci.*, 1848, 26, 535.

[20] Ács, M.; Pokol, G.; Faigl, F.; Fogassy, E.: *Termal. Anal.* 1988, *33*, 1241.

[21] Kozsda, K.R.; Keserű, Gy.; Böcskei, Zs.; Szilágyi, J.; Simon, K.; Bertók, B.; Fogassy, E.: *J. Chem. Soc. Perkin Trans 2*, 2000, 149.

[22] Bálint, J.; Egri, G.; Vass, G.; Schindler, J.; Gajáry, A.; Friesz, A.; Fogassy, E.: *Tetrahedron: Asymm.* 2000, *11*, 809.

[23] Kozma, D.; Simon, H;. Kassai, Cs.; Madarász, J; Fogassy, E.: *Chirality* 2001, *13*, 29.

[24] Simon, H.: PhD disz. 2003

[25] a). Kozma, D.; Madarász, Z.; Kassai, Cs.; Fogassy, E. *Chirality*, 1999, 11, 373.b). Hung. Pat. no. 212 667

[26] Ács, M.; Mravik, A.; Fogassy, E.; Böcskei, Zs. *Chirality*, 1994, *6*, 314

[27] Kassai, Cs: PhD diss. BME 2000

[28] Fogassy, E.; Ács, M.; Szili, T.; Simándi, B.; Sawinsky, J.: *Tetrahedron Letters*, 1994, *35*, 257-260 .

[29] Molnár, P.; Thorey, P.; Bánsághi, Gy.; Székely, E.; Poppe, L.; Tomin, A.; Kemény, S.; Fogassy, E.; Simándi, B.:. *Tetrahedron: Asymmetry*. 2008, *19*, 1587.

[30] Gizur, T.; Harsányi, K.; Fogassy, E.; *J. Pract. Chem.* 1986, *336*, 628.

[31] Fogassy, E.; Ács, M.; Tóth, G.; Simon, K.; Láng, T.; Ladányi, L.; Párkányi, L.: *J. Mol. Structure* 1986, *147*, 143.
[32] Fogassy, E., Faigl, F., Ács, M.: *Tetrahedron,* 1985 41, 2841.
[33] Jacob, R.M.; Reginier, G.L: DE 1050035 1958
[34] Pálovics, E.:PhD theses. http://www.doktori.hu/index.php?menuid=193&vid=2140&lang=EN
[35] Faigl, F.; Fogassy, E.; Nógrádi, M.; Pálovics, E.; Schindler, J.: *Organic & Biomolecular Chemistry* 2010 *8,* 347-359.
[36] Pálovics, E., Faigl, F., Fogassy, E.: *New Trends and Strategies in the Chemistry of Advanced Materials* (eds.: Tudose, R., Muntean, S.G.), Mirton, Timisoara, 2010, pp 20-26
[37] Jaques, J.; Wilen, S.H.; Collet, A.: *Enantiomers racemates and resolution,* John Wiley-and Sons, New York, 1981
[38] Pope, W.J.; Peachey, S.J. *J. Chem. Soc.,* 1899, 75, 1066.
[39] Kellogg, R.M.; Nieuwenhuijzen, J.W.; Pouwer, K.; Vries, T.R.; Broxterman, Q.B.; Grimbergen, R.F.P.; Kaptein, B.; La Crois, R.M.; de Wever, E.; Zwaagstra, K.; van der Laan, A.C. *Synthesis,* 2003, 1626.
[40] Faigl, F.; Fogassy, E.; Nógrádi, M.; Pálovics, E.; Schindler, J.: *Tetrahedron: Asymmetry* 2008,*19,* 519-536.
[41] Fogassy, E.; Faigl, F.; Ács, M.; Grofcsik, A.: *J. Chem. Res. (S)* 1981, *11,* 346 (1981); *(M)* 1981, *11,* 3981. pg. 9
[42] Fogassy, E.; Faigl, F.; Ács, M.; Simon, K.; Kozsda, É.; Podányi, B.; Czugler, M.; Reck, G. *J. Chem. Soc. Perkin Trans. 2.* 1988, 1385-1392. Hung. Pat. no.188.255.
[43] Simon, K.; Kozsda, É.; Faigl, F.; Fogassy, E.; Reck, G. *J. Chem. Soc. Perkin Trans. 2.,* 1990, 1395. Hung. Pat. no. 197.866, 1985.
[44] Tőke, L.; Lonyai, P.; Fogassy, E.; Acs, M.; Csermely, Gy.; Szenttornyai, A.; Faigl, F.:Hung. Pat HU 24599, 1983, CAN 99:140389
[45] Ács, M.; Faigl, F.; Fogassy, E.: Pat. No. WO 8503932, 1985, CAN 104:168835.
[46] Sakai, K.; Sakurai, R.; Yuzawa, A.; Hirayama, N. *Tetrahedron: Asymmetry,* 2003, 14, 3716.
[47] Sakurai, R.; Sakai, K. *Tetrahedron: Asymmetry,* 2003, 14, 411.
[48] Sakai, K.; Sakurai, R.; Hirayama, N. *Tetrahedron: Asymmetry,* 2004, 15, 1073.
[49] Bálint, J.; Egri, G.; Kiss, V.; Gajáry, A.; Juvancz, Z.; Fogassy, E. *Tetrahedron: Asymmetry,* 2001, 12, 3435.
[50] Ács, M.; Fogassy, E.; Faigl, F. *Tetrahedron,* 1985, 41, 2465.
[51] Kozma, D.; Fogassy, E. *Synth. Comm.,* 1999, 29, 4315.
[52] Hung. Pat. no. 179452, 1978, *Chem. Abs.* 1978, 97 6331
[53] Fogassy, E.; Ács, M.; Toth, G.; Simon, K.; Láng, T.; Ladányi, L.; Párkányi, L. *J. Mol. Struct.,* 1986, 147, 143.
[54] Woodward, R.B.; Doering, W.E.: *J. Chem Soc.,* 1945, 67, 860.
[55] Jaques, J.; Wilen, S.H.; Collet, A. *Enantiomers racemates and resolution* Wiley-Interference, N.Y. 1991
[56] Hung. Pat. no. 214720, *Chem. Abs.* 1995, 124, 117097
[57] US Pat. 02133894 2001, *Chem. Abs.* 2001, 139, 90595
[58] Schindler, J.; Egressy, M.; Bereczki, L.;Pokol, Gy.; Fogassy, E.; Marthi, K.: *Chirality* 2007, *19,* 239-244.
[59] Kassai, Cs.; Juvancz, Z.; Bálint, J.; Fogassy, E.; Kozma, D.: *Tetrahedron,* 2000, *56,* 8355.
[60] Gizur, T.: Hun Pat. 202963 2002, CAS 140 253341

[61] Faigl F.; Thurner A.; Farkas F.; Proszenyák Á.; Valacchi M.; Mordini A.: *Arkivoc,* 2004 (vii), 53-59.
[62] Pálovics, E.; Schindler, J.; Faigl, F.; Fogassy, E.: *Tetrahedron: Asymm.y.*2010. *21* ,2429-2434

2

Crystallization on Self Assembled Monolayers

Michal Ejgenberg and Yitzhak Mastai
Bar Ilan University
Israel

1. Introduction

Over the past two decades, self assembled monolayers (SAMs) (Love *et al.*, 2005; Smith *et al.*, 2004; Ulman, 1996) have been extensively studied due to their many applications in various fields such as electrochemistry (Eckerman *et al.*, 2010), biosensors (Nyquist *et al.*, 2000), protein separation (Chun and Stroeve, 2002) and enantiomer separation (Mastai, 2009). SAMs are organic assemblies formed by the adsorption of molecules from solution or from the gas phase onto the surfaces of solids. These molecules spontaneously organize into highly ordered, crystalline (or semicrystalline) two dimensional films. The molecules composing the SAMs adsorb to the surface through a "headgroup", a functional group with a high affinity for the solid surfaces. Two of the most widely studied systems of SAMs are alkanethiols adsorbed on metals, including gold, silver and platinum and alkylsilane chains formed on silicon dioxide surfaces, including glass and mica. In these cases, the "headgroups" are thiols, which have a high affinity for metal surfaces and silanes, which have a high affinity for silicon-dioxide surfaces. SAMs have become so popular since they offer a unique combination of physical properties that allow fundamental studies of interfacial chemistry, solvent molecule interactions and self-organization. Their well-ordered arrays and ease of functionalization make them ideal model systems in many fields.

One of the important advantages of SAMs is that they can be prepared in the laboratory by dipping the desired substrate in the required solution for a specified time followed by thorough washing with the same solvent and drying, often using a jet of dry nitrogen. Gas-phase evaporation of the adsorbate can also form good monolayers, although structural control is difficult. Several factors affect the formation and packing density of the self assembled monolayers including the solvent, temperature, nature of adsorbate, adsorbate concentration and the nature and roughness of the substrate. SAM substrates range from planar surfaces to highly curved surfaces (which will not be discussed here). The most common planar substrates for alkanethiol SAMs are thin films of metals supported on flat surfaces, such as glass, silica wafers and mica. These substrates can be easily prepared using methods like physical vapor deposition (PVD) and chemical vapor deposition (CVD). In some cases, an additional layer of titanium or chromium between the solid and the metal is needed in order to improve the adhesion between them. The composition of the thin films (grain size, etc.) is affected by the properties of the metal used (for example, melting point), the solid surface roughness and the experimental conditions. This, in turn, affects the organization and density of the SAM. The most common alkylsilane substrates are silicon dioxide surfaces such as glass, which are usually pre-treated with sulfuric acid.

The self assembled adsorbate has a great influence on the SAM outcome and can be tailored according to the desired SAM properties. The molecule used can possess a number of functional groups in addition to the molecule's headgroup. These functional groups can be distributed within the monolayer interior and located at the terminus of the molecule. Manipulation of the monolayer interior affects its degree of order and how easily electrons are conducted through it. For example, the molecular chain length and the steric crowding of the organic groups affect the density of the organic layer and the tilt angle of the molecule away from the surface normal. In general, longer chains and less robust organic groups yield denser, more organized SAMs, allowing high degrees of van der Vaals interactions (and in some cases, hydrogen bonds) with the neighboring molecules. The molecular constituent exposed at the SAM surface is critical to the SAM's interfacial properties. It affects the surface's general hydrophobic/hydrophilic character, adhesive properties and reactivity. In addition, it determines the surface interaction with other molecular species that come in contact when placed on the surface of the SAM.

SAMs are of prime technological interest, as the presence of molecules chemically bound to the surface renders the properties of the modified interface (i.e., wetting, conductivity, adhesion, and chemistry) to be entirely different than those of the bare substrate. The incorporation of functional moieties such as chromophores, electroactive groups, or molecules that can bond within the SAM (i.e., covalent cross-linking between adjacent molecules or non-covalent hydrogen bonding) enable capabilities in sensing, electron transfer, molecular recognition, and other areas (Smith *et al.*, 2004).

This review will focus on the use of SAMs in crystallization processes. We will begin with a short introduction on crystallization on SAMs. Then, we will review the latest advances in crystallization on patterned SAM's and effects of SAMs on crystal morphology and crystal polymorphism. This chapter will also include a description of chiral SAMs and their role in enantioselective crystallization.

2. Crystallization on SAMs

Crystallization refers to the formation of solid crystals from a solution, melt or more rarely, directly from gas. Crystallization consists of two processes - nucleation and crystal growth. Nucleation is the formation of small clusters of molecules. These clusters may re-dissolve in the crystallizing solution or go on to become a crystal, depending on their size. Beyond a critical size, they are stable and form crystal nuclei. Crystal growth refers to the subsequent addition of molecules to the formed nuclei. The nucleation can be homogenous, occurring spontaneously or heterogenous, induced by foreign particles such as dust particles or vessel walls. Primary nucleation occurs in the absence of crystals in solution whereas secondary nucleation occurs on seeds or existing crystals present in the crystallizing solution.

The ability to control crystallization is a critical requirement in many technologies such as in the food, pharmaceutical and chemical industries. Crystallization parameters such as particle size, particle shape, particle morphology and polymorph selectivity determine the crystal properties and uses. The solution concentration, crystallization time, crystallization temperature, solvent and crystallization vessel all have an effect on the crystal parameters. In the past few decades, researchers have been busy searching for new ways to control crystallization. Self assembled monolayers are showing great promise in this field. The

SAMs interact with the crystallizing molecules and thus, affect their organization, ultimately affecting the crystallization outcome. SAMs have been reported to affect crystal orientation, morphology, polymorphism and crystal size. The interaction between SAMs and molecules in solution has usually been rationalized on the basis of hydrogen bonding and/or strong ionic interactions.

2.1 Effects on crystal orientation

During crystallization, the SAM constituent exposed at the SAM surface interacts with molecules present in the crystallizing solution. These interactions are specific and in turn, can cause the stabilization of specific crystal faces. When this happens, the crystals grow in a specific orientation, causing changes in crystal morphology. Many researchers have reported specific crystal orientation on SAMs, brought about by interactions of the crystallizing molecules with SAMs.

For example, Lee *et al.* employed SAMs of rigid thiols on gold surfaces in order to investigate the effects of interfacial molecular recognition on nucleation and growth of L-Alanine and DL-Valine crystals (Lee *et al.*, 2002). L-alanine crystallizes from water in the orthorhombic space group $P2_12_12$ (a=6.025Å, b=12.324Å and c=5.783Å), with bipyramidal morphology, dominated by the {020}, {120}, {110} and {011} growth forms (Figure 1).

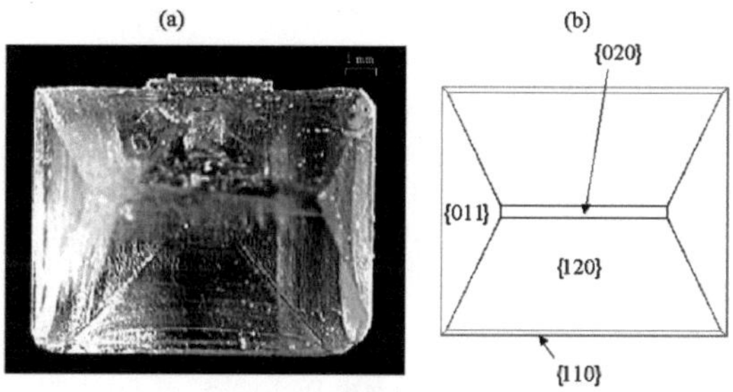

Fig. 1. L alanine crystal grown from aqueous solution. (Lee *et al.*, 2002)

In order to examine the effect of the SAMs on L-Alanine crystallization, L-Alanine was crystallized from solutions and also onto a variety of rigid SAMs of thiols: 4'-methyl-4-mercaptobiphenyl, 4'-hydroxy-4-mercaptobiphenyl and 4-(4-mercaptophenyl) pyridine. Powder X-ray diffraction patterns and interfacial angle measurements of the L-alanine crystals are shown in Figure 2. In methyl-terminated SAMs, L-alanine selectively nucleated on the (020) plane on the SAM surface while on the OH-terminated SAMs, L-alanine nucleated on an unobserved (200) side face. In both cases, the area of each crystal face was substantially larger than the other faces on the crystal. Crystallization of L-alanine on 4-(4-mercaptophenyl) pyridine resulted in the (011) face as the plane corresponding to nucleation. The preferential interaction was explained on the basis of hydrogen bonding between the pyridine surface and the amino and methyl groups protruding out of the (011) plane. Figure 2 reveals that L-alanine crystals nucleating on SAM surfaces crystallize in an orthorhombic space group with similar unit cell dimensions. However, the functionalized SAMs have an effect on the nucleating plane and ultimately, on L-alanine crystal growth.

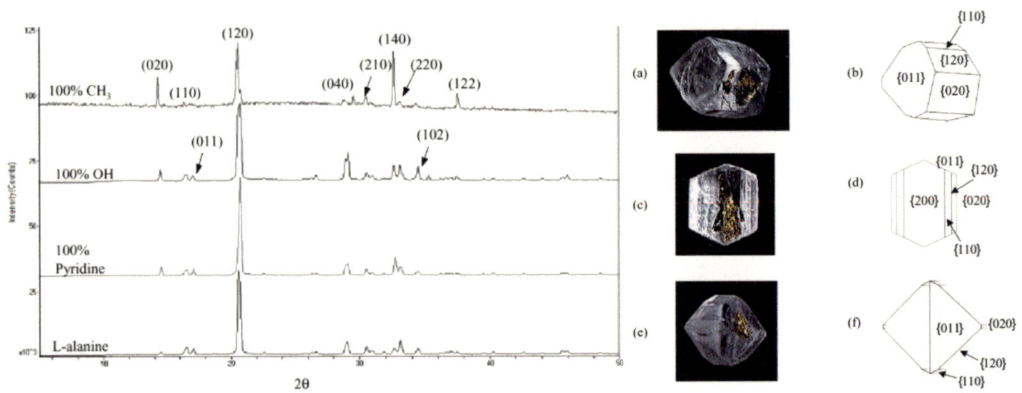

Fig. 2. X-ray diffraction spectra of L alanine nucleated on functional SAMs compared with L alanine crystallized from aqueous solution (left) and crystallographic images and morphology of L-alanine crystallized on (a,b) 4′-methyl-4-mercaptobiphenyl SAMs, (c,d) 4′-hydroxy-4-mercaptobiphenyl SAMs and (e,f) 4-(4-mercaptophenyl) pyridine SAMs - (right). (Lee *et al.*, 2002)

DL-Valine crystallizes in the monoclinic space group $P2_1/c$ with unit cell dimensions (a=5.21Å, b=22.10Å, c=5.41Å and β=109.2°). In aqueous solution, DL-valine crystallizes as hexagonal platelets dominated by a slow-growing (020) flat face with (100), (002) and (202) side faces. DL-valine was crystallized on 4′-methyl-4-mercaptobiphenyl, 4′-hydroxy-4-mercaptobiphenyl and 4-(4-mercaptophenyl) pyridine. On the pyridine and OH-terminated thiols, DL-valine crystals nucleated from the flat (020) plane, whereas in methyl terminated SAMs, the fast growing (100) face was the nucleating plane and the hexagonal platelet crystals were oriented perpendicular to the SAM surface (Figure 3). Again, the surface chemistry of the SAMs led to different interfacial interactions and thus oriented the crystallization of amino acids. In both cases, hydrogen bonding was responsible for the epitaxial crystallization of the amino acid on the monolayer substrates. Molecular modelling studies were also undertaken to examine the molecular recognition between the monolayer and the crystallographic planes and were in good agreement with the experimental results.

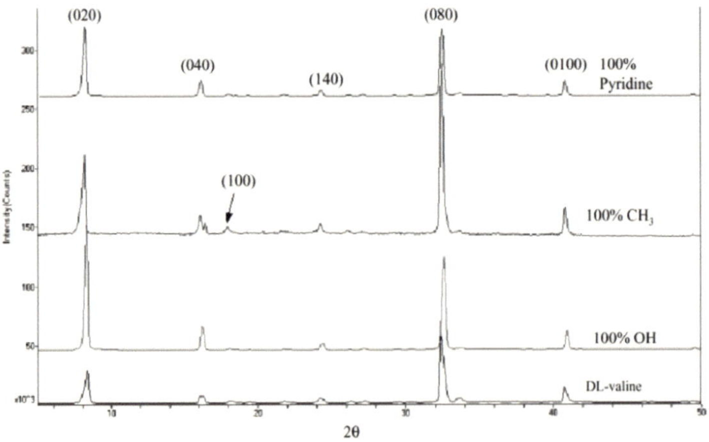

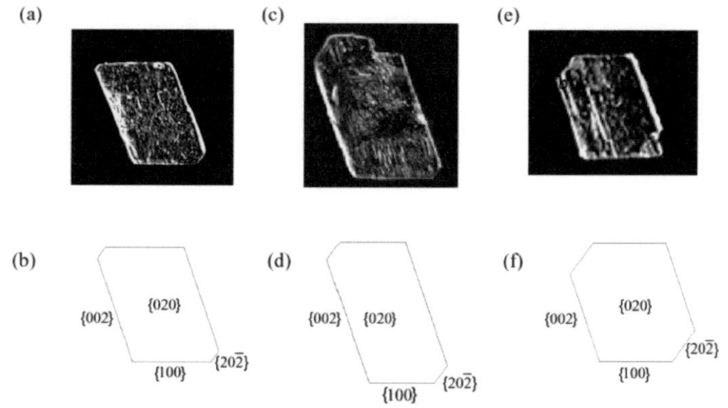

Fig. 3. X-ray diffraction spectra of DL valine nucleated on functional SAMs compared with DL valine crystallized from aqueous solution (top) and crystallographic images and morphology of DL valine crystallized on (a,b) 4-(4-mercaptophenyl) pyridine SAMs, (c,d) 4'-hydroxy-4-mercaptobiphenyl SAMs and (e,f) 4'-methyl-4-mercaptobiphenyl SAMs – (bottom). (Lee *et al.*, 2002)

An additional example for the effect of SAMs on crystallization can be demonstrated in the work of Aizenberg and co-workers, who studied the growth of oriented calcium carbonate crystals on X-terminated alkanethiol SAMs ($HS(CH_2)_nX$) (Aizenberg *et al.*, 1999). The SAMs were terminated with a variety of functional groups ($X=CO_2^-$, SO_3^-, PO_3^{2-}, OH, $N(CH_3)_3^+$,CH_3) and two different metal support films were used - Ag and Au. The calcium carbonate crystals were characterized by a variety of techniques - optical microscopy, SEM with image analysis and X-ray diffraction. Their results showed that SAMs terminated in acidic functionalities were found to induce nucleation more effectively than control surfaces of bare metal films, whereas SAMs terminated in CH_3 or slightly basic $N(CH_3)_3$ actually inhibited their nucleation. The orientation of the crystals was highly homogeneous for each surface and depended on the functional group exposed at the SAM surface. The authors achieved a very high level of control over the orientations of the crystals. They were able to selectively nucleate calcite from six crystallographic planes. This was the first time that calcite had been grown in such a variety of orientations in an artificially controlled system.

2.2 Effects on polymorphism

Polymorphism is the ability of a molecule to crystallize in more than one packing arrangement. Polymorphs (Chieng et al., 2011) may possess significantly different structural and physical properties and therefore polymorphism has broad practical implications in solid state chemistry, materials science and pharmacology. Many methods have been developed to control polymorphism including tailor made additives (Yokota et al., 2006), surfactants (Garti and Zour, 1997) and varying solvents and temperature (Weissbuch et al., 2005). Recently, self assembled monolayers have also been shown to control crystal polymorphism.

Hiremath *et al.* applied SAM templates in order to control the polymorphism of 1,3-bis(*m*-nitrophenyl) urea (MNPU – Figure 4). (Hiremath *et al.*, 2005) Previously studied, MNPU was found to have at least three different polymorphs - α, β and γ. The polymorphs are visually distinguishable, but often crystallize concomitantly from aqueous ethanol solutions. The crystal structures of the α and β polymorphs were first reported by Etter et al (Etter *et. al*, 1988, 1990, as cited in Hiremath *et al.*, 2005). Bernstein *et al.* (Bernstein *et al.*, 2005, as cited in Hiremath *et al.*, 2005) obtained the crystal structure of the γ phase and also discovered a previously unknown anhydrous δ phase.

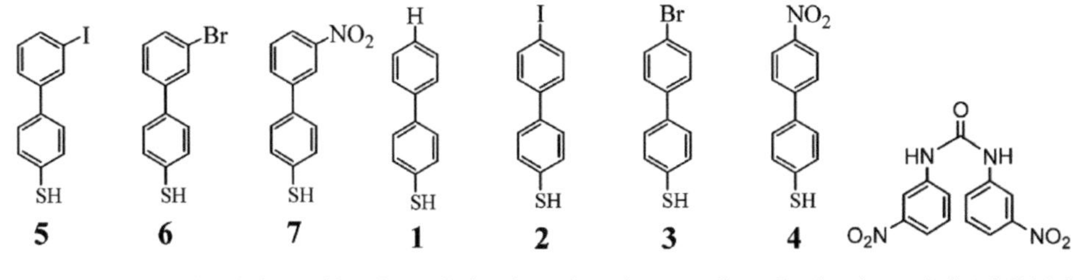

Fig. 4. MNPU (right) and biphenyl thiol molecules used in the biphenyl thiol SAMs (1-7) - (left). (Hiremath *et al.*, 2005)

Initially, Hiremath *et al.* obtained the pure polymorph phases by varying solvents and temperature. The α-MNPU was obtained from crystallization in ethanol at 60°C, β-MNPU was obtained from crystallization in ethyl acetate at room temperature. The γ form was obtained by re-crystallization of the α or β forms in 95% ethanol. Then, MNPU was crystallized on a variety of biphenyl thiol SAMs (Figure 4) and two types of control surfaces: bare gold substrates and 1-pentanethiol, 1-octanethiol and 1-dodecanethiol SAMs. The biphenyl thiols were chosen for two reasons - their close packing affords a more perpendicular molecular alignment at the surface (smaller tilt angles) and the slightly larger size of the 2D lattice dimensions makes them potentially more compatible with typical unit cell dimensions. The crystallizations were conducted in saturated ethanol or ethyl acetate solutions of MNPU at room temperature or 60°C. The results obtained from the crystallizations are summarized in Table 1. Crystals of MNPU only appeared on SAMs 1-3. No crystallization of MNPU was observed on SAMs 4-7 or on the control surfaces. SAMs of 1, 2 and 3 were each found to serve as selective nucleating templates for a single MNPU phase. Crystals of α-MNPU formed on monolayers of 1, β-MNPU grew selectively on 2 and γ-MNPU grew on 3. The crystallizations were repeated numerous times and the results were always consistent. The resultant crystal form obtained was independent of the solvent or temperature conditions explored. In addition, the crystals grown on SAMs 1-3 all adopted preferred orientations with respect to the SAM templates, illustrating the importance of the interface in the nucleation process. The observed orientations were rationalized on the basis of two-dimensional lattice matching, complementary functional group interactions and dipole moments across the SAM/crystal interface.

The work of Dressler and Mastai provides an additional example of polymorphism control by SAMs. Dressler and Mastai crystallized the metastable α-L-Glutamic acid on L-Phenylalanine SAMs (L-AAPP, Figure 5) (Dressler and Mastai, 2007). L-Glutamic acid is known to form two polymorphs - the metastable α-form and the stable β-form. It was previously shown that the stable β-form emerges from the metastable α-crystals through a kinetic process. It was also shown that the α-polymorph converts rapidly to the β-form at 45°C, but at ca. 15°C, the conversion process is very slow and almost nonexistent, so that the α-polymorph is obtained exclusively. Dressler and Mastai demonstrated that the α-form can be obtained at room temperature, without the need for crystallization at low temperatures.

SAM	initial solute phase	solvent	temp (°C)	growth
1	α, β, or γ or any combination of phases	ethanol or ethyl acetate	25 or 60	α
2	α, β, or γ or any combination of phases	ethanol or ethyl acetate	25 or 60	β
3	α, β, γ or any combination of phases	ethanol or ethyl acetate	25 or 60	γ
4, 5, 6, 7, bare gold, or alkanethiol controls	α, β, γ or any combination of phases	ethanol or ethyl acetate	25 or 60	none

Table 1. Crystal of MNPU grown on SAMs 1-7 including the initial solution phase, solvent and temperature crystallization parameters. (Hiremath *et al.*, 2005)

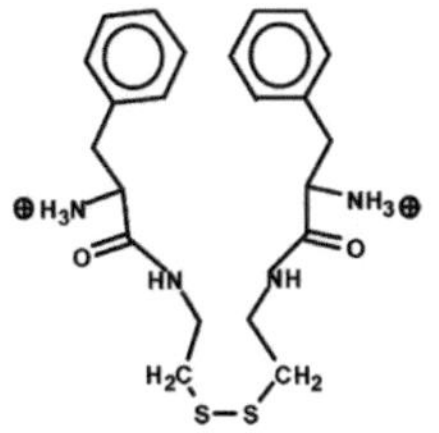

Fig. 5. L-AAPP (Dressler and Mastai, 2007)

Substrates for SAM preparation were prepared by vacuum deposition of thin gold films onto glass slides. The gold substrates were then immersed in 2mM L-AAPP ethanolic solutions overnight in order to receive the L-AAPP SAMs. L-AAPP was chosen as the molecule for self assembly since L-phenylalanine has already been shown to promote the stabilization of the α-form of L-glutamic acid in solution. In order to examine the effect of the SAMs on L-glutamic acid crystallization, a series of crystallization experiments were performed on the surfaces and in solutions. The crystals were then identified using various techniques (Figure 6 and Figure 7). The X-ray diffraction and Raman spectra reveal that the crystals grown on the SAMs are of the α-form and those from solution are β-L-glutamic acid. From X-ray diffraction, It is also evident that L-glutamic acid crystallizes on the surfaces with preferential crystal growth along the <111> crystal plane. It is also clear from the SEM images that the polymorphs possess very different crystal habits. Crystals grown on the SAM surfaces reveal well-ordered large prismatic structures typical of α-L-glutamic acid, whereas crystals from solution expose plate like structures characteristic of the β-form. These results verify the stabilization of the α-polymorph of L-glutamic acid on the L-phenylalanine terminated SAM. The stabilization is explained in the following manner: β-L-glutamic acid emerges from the (011) plane of α-L-glutamic acid. L-Phenylalanine attaches to the (011) plane of the α-crystals, thereby inhibiting and preventing the α to β transformation.

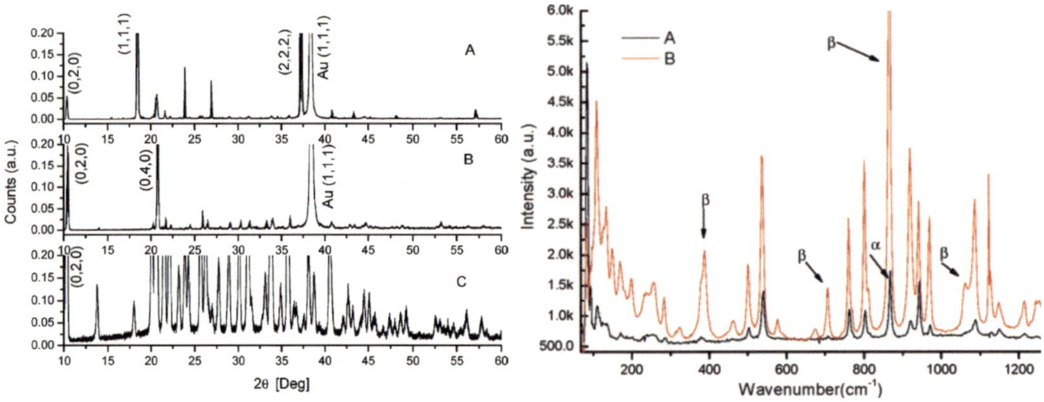

Fig. 6. X-ray diffraction of L glutamic acid crystals (A) grown on L-AAPP SAMs, (B) grown on Au and (C) grown from solution – (left) and Raman spectroscopy of L glutamic acid crystals (A) grown on L-AAPP SAMs (α-polymorph) and (B) grown from solution (β-polymorph). (Dressler and Mastai, 2007)

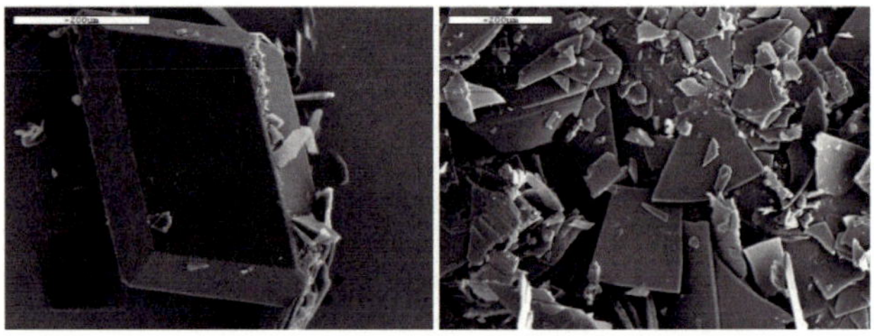

Fig. 7. Scanning electron microscopy images of L glutamic acid crystals (scale bar=200μm): (A) crystal morphology of crystals grown on the surface of L-AAPP SAMs and (B) crystal morphology of crystals grown from solution. (Dressler and Mastai, 2007)

Recently, Ejgenberg and Mastai demonstrated that self assembled monolayers based on S-leucine (Figure 8) can be used for polymorphism control of DL glutamic acid (Ejgenberg and Mastai, 2011). It is known that DL-glutamic acid can crystallize in two forms: the conglomerate form, known as anhydrous DL glutamic acid and the racemic compound, known as DL glutamic acid monohydrate. Ejgenberg and Mastai were able to afford anhydrous DL glutamic acid under conditions where the monohydrated DL glutamic acid is thermodynamically more stable.

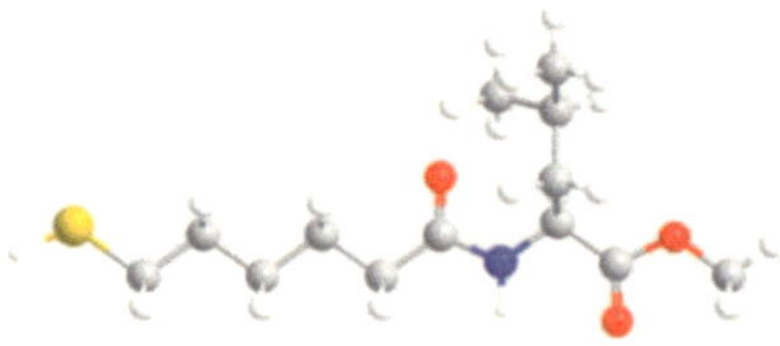

Fig. 8. S-leucine methyl ester covalently attached to 6-mercaptohexanoic acid (S-LMHA). Red atoms = oxygen, blue =nitrogen, gray = carbon, yellow = sulfur, white = hydrogen. (Ejgenberg and Mastai, 2011)

The chiral SAMs were prepared on ultra-flat gold surfaces. Gold films of 50 nm thickness were deposited on mica substrates using a high vacuum sputtering technique. The chiral

SAMs were then formed on the gold surfaces by immersing the gold substrates in 0.1M solutions of S-LMHA in ethanol overnight. In order to examine the effect of the chiral SAMs on the DL-glutamic acid crystallization, DL-glutamic acid was crystallized from aqueous solution onto the S-LMHA chiral SAMs. DL-glutamic acid was also crystallized in solution, under the same conditions. The crystals from solution and those grown on the chiral SAMs were examined using X-ray diffraction (XRD), scanning electron microscopy (SEM), micro-Raman and differential scanning calorimetry (DSC). It should be mentioned that crystallization or crystal growth did not occur on the bare gold surfaces.

The X-ray diffraction spectrum of crystals from solution is shown in Figure 9 and corresponds to monohydrated DL-glutamic acid, namely the racemic compound, as reported in the literature. The monohydrate of DL-glutamic acid crystallizes in an orthorhombic unit cell (space group Pbca) with the following parameters a=9.08, b = 15.40, c = 10.61 (A°) and $\alpha = \beta = \lambda = 90°$. On the other hand, the X-ray diffraction patterns of crystals grown on S-LMHA that were taken directly on the chiral surface show a typical X-ray diffraction spectrum of the conglomerate, anhydrous DL-glutamic acid. The X-ray diffraction spectrum of conglomerate DL-glutamic acid corresponds to that reported in the literature with unit cell parameters (in A°): a=5.16, b=17.30, c=6.95, $\alpha = \beta = \lambda = 90°$ (orthorhombic unit cell, space group P212121).

The anhydrous and monohydrated forms of DL-glutamic acid have different crystal habits and therefore their formation was also studied using SEM. Figure 9 displays SEM images of crystals collected from solution and crystals grown on the S-LMHA SAMs. The crystals

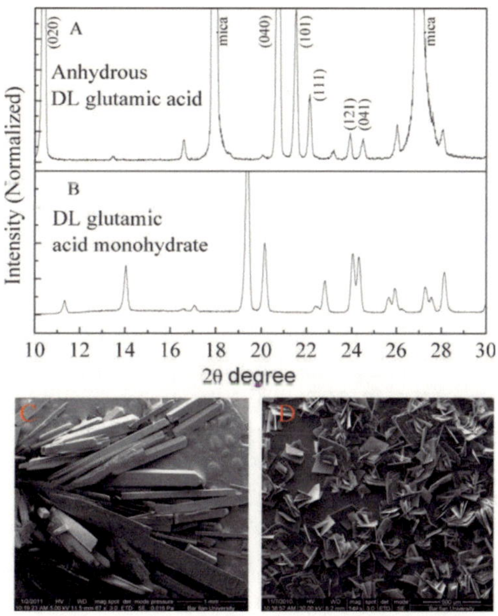

Fig. 9. X-Ray diffraction spectra of (A) DL-glutamic acid crystals grown on S-LMHA chiral SAMs (anhydrous DL-glutamic acid) and (B) crystals grown in solution (DL-glutamic acid monohydrate) and SEM images of (C) crystals grown in solution (scale bar = 1 mm) (DL-glutamic acid monohydrate) and (D) crystals grown on S-LMHA chiral SAMs (scale bar = 500 mm) (anhydrous DL-glutamic acid). (Ejgenberg and Mastai, 2011)

grown from solution reveal typical needle-like morphology, characteristic of monohydrated DL-glutamic acid. The crystals are fairly large and non-uniform in size (1–4 mm long and 0.2–0.5 mm wide). However, crystals grown on the S-LMHA SAMs expose plate like morphology characteristic of anhydrous DL-glutamic acid. They are uniform in size with approximately 250 X 250 μm² and are much smaller than the crystals from solution.

In order to further characterize the conglomerate and racemic compound DL-glutamic acid crystals, micro-Raman measurements were carried out. The Raman spectrum of DL-glutamic acid crystallized on the S-LMHA surface is shown in Figure 10 with main peaks at 867, 1295, 1339 and 1405 cm-1. The Raman spectrum of DL-glutamic acid crystallized from solution (Figure 10) displays main peaks at 855, 919 and 1314 cm-1. These spectra correspond to the Raman spectra of conglomerate and racemic glutamic acid as reported in the literature. The Raman measurements of DL-glutamic acid crystallized on the chiral surface match those of the conglomerate form of DL-glutamic acid.

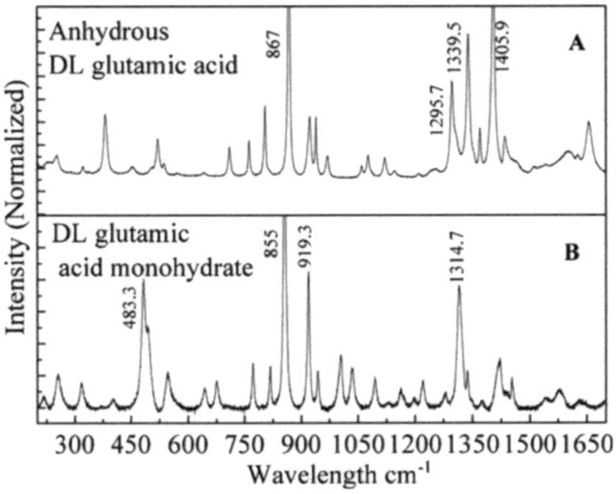

Fig. 10. Raman spectra of (A) DL-glutamic acid crystals grown on S-LMHA chiral SAMs (anhydrous DL-glutamic acid) and (B) crystals grown in solution (DL-glutamic acid monohydrate). (Ejgenberg and Mastai, 2011)

Further evidence for the formation of the conglomerate DL-glutamic acid crystals on the chiral SAM surfaces was obtained from differential scanning calorimetry (DSC) (Figure 11). The DSC thermogram of the DL-glutamic crystals grown on the S-LMHA SAMs exhibits a sharp melting point at 185.4 1C. The DSC thermogram of the crystals from solution shows two endothermic peaks at 112.7 1C and 184.7 1C. These results can be explained in the following manner. The crystals grown on the chiral SAMs are anhydrous DL-glutamic acid and therefore exhibit one peak, while those from solution are monohydrated DL-glutamic acid and therefore contain two peaks: the first endotherm is due to the loss of water molecules and the second sharp endotherm is due to the melting of anhydrous DL-glutamic acid.

In conclusion, S-LMHA SAMs were used to stabilize the conglomerate form of DL-glutamic acid crystals. The results offer a powerful tool in the development of processes for controlling chiral polymorphic systems and can further develop into a novel method for chiral resolution by crystallization.

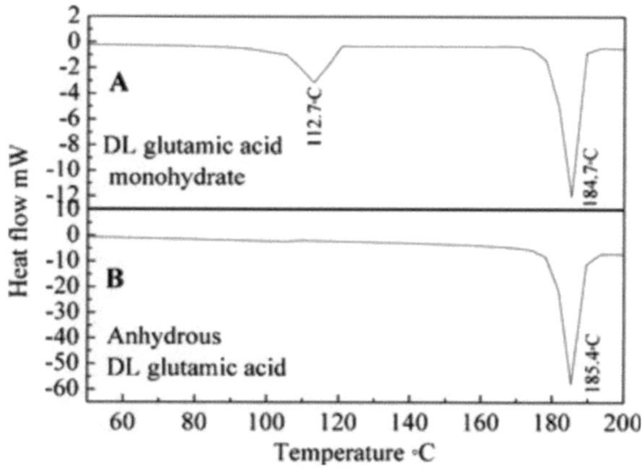

Fig. 11. DSC thermograms of (A) DL-glutamic acid crystals grown in solution (DL-glutamic acid monohydrate) and (B) crystals grown on S-LMHA chiral SAMs (anhydrous DL-glutamic acid). (Ejgenberg and Mastai, 2011)

2.3 Crystallization on patterned SAMs

Patterned SAMs are composed of two or more SAMs which are deliberately distributed on specific areas on the surface. They can be fabricated in different ways including micro-contact printing and dip-pen nanolithography. Patterned SAMs have been used to attract particular nanostructures like ribbons or wires to particular areas on the surface. They have also been used in the functionalization of biosensors in which the localized SAMs have an affinity for specific cells and proteins. Crystallization on patterned SAMs where the crystals grow in localized areas has also been demonstrated.

In a fascinating study, Aizenberg *et al.* demonstrated that crystallization could be restricted to well-defined regions on micro-patterned SAMs (Aizenberg *et al.*, 1999). The substrates were used in order to control the crystallization of calcite. Calcite is a stable polymorph of calcium carbonate, with space group R3c, α = 4.99Å and c = 17.06Å. Calcite has been extensively studied and its crystallization is relatively easy to perform.

Self assembled monolayers having areas of different nucleating activity were patterned on metal substrates. The self assembled monolayers were patterned by microcontact printing with an elastomeric stamp that had a relief structure consisting of a square array of raised circles: as inks, 10mM solutions of $HS(CH_2)nX$ ($X=CO_2H$) in ethanol were used. The surface was then washed with a 10mM solution of $HS(CH_2)_{15}CH_3$ in ethanol to passivate the areas that had not contacted the stamp. The patterned substrates were supported upside down in the crystallizing solutions to ensure that only particles grown on the SAM would be bound to the surface.

Figure 12 shows a low magnification image of calcite crystals formed on one of the patterned SAMs. It is clear that crystallization is restricted to well defined, CO_2 - terminated regions and does not occur on the CH_3 - terminated areas. Interestingly, by adjusting parameters such as the density and size features on the stamp, Aizenberg *et al.* could control important crystallization characteristics, including the location and density of nucleating regions on the surface, the number of crystals nucleated in each region and the crystallographic orientation of the crystals (Figure 13). The following mechanism was suggested: the rate of nucleation on SAMs terminated in polar groups is faster than on methyl-terminated SAMs. As soon as crystal growth begins in a polar region, mass transport to the growing crystals depletes calcium and carbonate ions over the local methyl-terminated region to the point of under saturation.

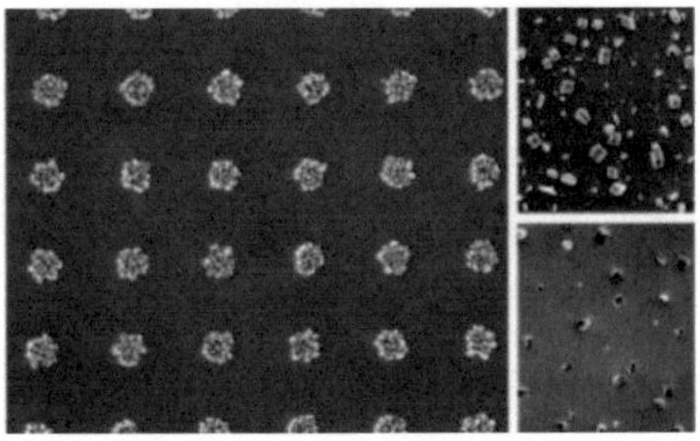

Fig. 12. Scanning electron micrograph of a patterned surface – calcite crystals appear on circles of $HS(CH_2)_{15}CO_2H$ in a background of $HS(CH_2)_{15}CH_3$. (Aizenberg *et al.*, 1999)

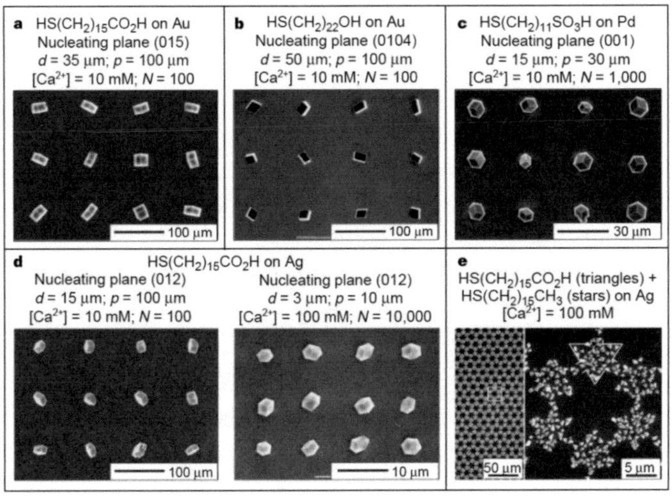

Fig. 13. Ordered two dimensional arrays of calcite crystals. (Aizenberg *et al.*, 1999)

Recently, Lee *et al.* investigated the concomitant crystallization of glycine polymorphs on patterned self assembled monolayers (Lee, I.S. *et al.*, 2008). Six distinct polymorphic forms of glycine are known in the literature; three of them – α, β and γ - are formed at ambient environment. At room temperature, γ-glycine is the thermodynamically most stable form. However, in neutral aqueous solution, the α-form is normally obtained. The γ-form is typically obtained from acidic and basic solutions. β-glycine is the least stable of the three forms. It can be obtained from ethanol-water mixtures, but readily converts to α-glycine in the presence of water or upon heating.

Patterned subtrates were designed in the following manner: metallic gold islands were formed by evaporating titanium followed by gold through brass meshes onto glass substrates. The dimensions and patterns of the islands depended on the size and shape of the holes on the mesh. Square shaped islands in a variety of sizes were prepared. 4-Mercaptobenzoic acid (4-MBA) was self assembled onto the gold surface while octadecyltrichlorosilane (OTS) was applied to the exposed glass surface.

The crystallization experiments were conducted by immersing and withdrawing the patterned substrates from aqueous glycine solutions of three different pHs (neutral, acidic and basic) and concentrations. Solution droplets were created on the patterned surface and each droplet served as an independent crystallization trial. More than 2000 islands per substrate were analyzed to achieve the statistical accuracy of the polymorph distribution of crystals formed on the substrate.

pH	concentration (M)	size of island (μm)	total samples	crystallization time (min)[a]	α-form (%)	β-form (%)	γ-form (%)
3.40	1.60	500	2000	8	36.3	0.4	63.3
	2.40	500	2000	not measured	54.1	10.4	35.5
	3.20	500	2000	25	66.0	7.8	26.2
5.90–6.20	1.60	500	2000	5	90.3	7.3	2.4
	2.40	500	2000	5	88.3	5.8	5.9
	3.20	500	3000	10	92.2	2.9	4.9
10.10	1.60	500	2000	100	58.2	6.4	35.4
	2.40	500	2000	220	68.2	10.3	21.5
	3.20	500	2000	450	83.8	4.9	11.3

[a] The crystallization time corresponds to the time when crystals are observed on 99% islands of the patterned substrate.

Table 2. Polymorph distribution of glycine for different pHs and solution concentrations. (Lee, I.S. *et al.*, 2008)

The polymorph distribution of glycine crystals formed on the patterned substrates with respect to the solution concentration and pH are summarized in Table 2. For the neutral aqueous glycine solution, the α-form was preferred at all concentrations, but its frequency significantly decreased at acidic and basic solutions. Consequentially, the percentage of γ-glycine increased in the acidic and basic solutions, without any large difference in the polymorph distribution of the β-polymorph. The increased percentage of the γ-polymorph in acidic and basic solutions was rationalized in the following way: the acidic and basic glycine solutions contain charged glycine species which reduce the possibility of the formation of α-glycine. Table 3 summarizes the polymorph outcome of glycine for different pHs and island sizes. As shown in previous reports (Lee, A.Y. *et al.*, 2005, 2006), metastable forms (β-glycine) were consistently more frequently observed with smaller islands. However, using acidic and basic glycine solutions, the frequency of the thermodynamically most stable γ-glycine increased with decreasing island size. This can be attributed to the increased proportion of the charged glycine species caused by the decrease/increase in the pH of solution for acidic/basic solutions during evaporation.

pH	concentration (M)	size of island (μm)	total samples	crystallization time (min)[a]	α-form (%)	β-form (%)	γ-form (%)
3.40	3.20	250	3000	10	11.6	14.1	73.3
		500	2000	25	66.0	7.8	26.2
5.90–6.20	3.20	250	3000	5	94.5	5.4	0.1
		500	3000	10	92.2	2.9	4.9
10.10	3.20	250	2000	60	52.4	17.1	30.5
		500	2000	450	83.8	4.9	11.3

[a] The crystallization time corresponds to the time when crystals are observed on 99% islands on the substrate.

Table 3. The polymorph outcome of glycine for different pHs and island sizes. (Lee, I.S. *et al.*, 2008)

In similar studies, Lee *et al.* reported that the size of the glycine crystals is controlled by the dimensions of the metallic gold islands or the concentration of the solution (Lee, A.Y. *et al.*, 2005). Moreover, it was observed that the high energy unstable β-form of glycine crystallizes on small metallic islands, in contrast to large islands, where the α-form is favoured. The increased frequency of the high energy form (β-glycine) with decreasing feature sizes is a result of the high supersaturation that is generated from fast solvent evaporation. (Lee, A.Y. et al., 2006)

3. Enantioselective crystallization

The crystallization of racemic molecules is very similar to the crystallization of achiral molecules. However, because of their chirality, racemic molecules can form different types of crystals with different compositions. If the crystal lattice contains equal left and right handed molecules arranged in an ordered manner, the crystal is heterochiral and referred to as a racemic compound. In the case where the crystal lattice is composed of only one enantiomer (left or right), the crystal is homochiral and referred to as a conglomerate. In nature, racemic compounds greatly outnumber conglomerates.

Chiral resolution is the process of the separation of racemates into their enantiomers. Many methods have been developed for this purpose. The most common is the use of chiral stationary phases in HPLC. However, large scale chiral separation is based on the classical crystallization method, which incorporates diasteriomeric transformation and chiral seeding. Even today, new methods of chiral crystallization are being designed. Lahav et al. developed a method, called "tailor-made" additives, which has been exploited for the kinetic resolution of conglomerates (Addadi, 1982). These additives have molecular structures which are very similar to one of the enantiomorphs of the substrate crystals. Consequentially, the inhibitors stereoselectively adsorb to the surfaces of one of the enantiomorphs, delaying or preventing its growth. As a result, enantiomeric excess is achieved. It is important to mention that in order to acheive chiral resolution by crystallization, the racemate system must spontaneously resolve upon crystallization to a conglomerate.

4. Chiral SAMs and enantioselective crystallization on chiral SAMs

As mentioned above, SAMs are organic assemblies formed by the adsorption of molecules from solution or the gas phase onto the surface of solids. If the adsorbed molecules are chiral, the self assembled monolayer is also rendered chiral. The chirality of the molecule can be distributed within the monolayer interior or located at the terminus of the molecule. However, the chirality is only expressed when the chiral constituent is exposed at the monolayer surface. Chiral SAMs are used in chiral systems. The SAMs can be used to specifically interact with chiral species, such as proteins or amino acids. Chiral SAMs have been used in enantioselective crystallization. In this case, a racemic solution of a chiral molecule is crystallized on a chiral SAM. The chiral SAM serves as a nucleating surface for one of the enantiomers, thereby increasing its crystallization on the SAM. Thus, enantioselective crystallization is achieved.

Dressler and Mastai studied the enantioselective crystallization of glutamic acid on self assembled monolayers of cysteine (Dressler and Mastai, 2007). The chiral SAMs were prepared by immersing gold covered glass slides in aqueous solutions of cysteine (10mM) for 2 hours. These SAMs were then characterized by X-ray diffraction, micro-Raman measurements, X-ray photoelectron spectroscopy, contact angle and ellipsometry.

In order to study the interactions between the chiral crystals and the chiral SAMs, pure enantiomers of glutamic acid were first crystallized on the L and D cysteine films. Interestingly, enantiomers of glutamic acid with identical chirality to that of the cysteine SAMs did not grow on the SAMs in a specific orientation and were unchanged from crystals grown in solution (Table 4). On the other hand, enantiomers with chirality opposite to that of the cysteine films grew in a preffered orientation. In either case, morphological differences between crystals grown on the SAMs and those grown in solution were not very significant.

	(+)-L cysteine film		-D cysteine films		Pure solution	
Crystal system	Crystal habit	Crystal orientation	Crystal habit	Crystal orientation	Crystal habit	Crystal orientation
(+)-L glutamic acid	Plate like	No specific orientation	Plate like	[020]	Plate like	No specific orientation
(−)-D glutamic acid	Plate like	[020]	Plate like	No specific orientation	Plate like	No specific orientation
rac-glutamic acid	Rectangular (well oriented)	[220]	Plate like	[220]	Needle like	No specific orientation

Table 4. Crystal habits and crystal orientation of DL, D and L glutamic acid crystallized on cysteine films and in pure solution. (Dressler and Mastai, 2007)

The chiral cysteine SAMs were then utilized for the enantioselective crystallization of glutamic acid. For this purpose, L and D cysteine SAMs were immersed in supersaturated solutions of DL-glutamic acid for ca. 2 hours. The crystals were then characterized using various techniques. It is important to note here that glutamic acid grows as a monohydrate crystal from water and therefore has a different structure from its enantiomers which were discussed previously. Figure 14 presents SEM images of glutamic acid monohydrate crystals grown on L and D cysteine surfaces and grown from solution. Crystals grown from solution have a needle-like morphology with an average size of 300 μm whereas crystals grown on the cysteine surfaces grow in well ordered arrays of rectangular/plate like shaped crystals with a typical size of 10-30 μm. X-ray diffraction measurements of the crystals showed preffered orientation of the glutamic acid crystals (Table 4).

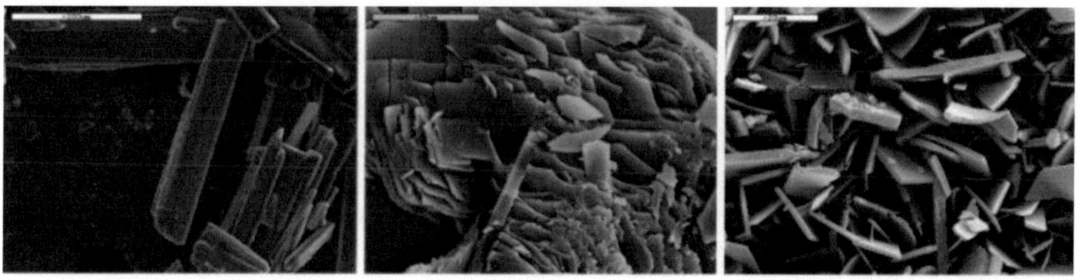

A (scale bar=200μm) **B** (scale bar=50μm) **C** (scale bar=20μm)

Fig. 14. SEM images featuring the crystal morphology of rac-glutamic acid crystallized (A) from solution, (B) on a chiral D-cysteine surface and (C) on a chiral L-cysteine surface. (Dressler and Mastai, 2007)

The enantiomeric excess of the crystals grown on the chiral SAMs was investigated using polarimetry. The crystals were gently removed from the chiral films, dissolved in 5N HCl solution and their optical activity was measured. The results show that the enantiomeric excess of glutamic acid crystals collected from the L-cysteine films is about 31% e.e. enriched with D-glutamic acid (Table 5). The crystallization on the D-cysteine films resulted in 27% e.e. enriched with L-glutamic acid. The optical activity of the crystallization solution was also measured (at different intervals). The optical activity results measured from the solution and from the crystals are in good agreement with each other.

	(+)-L cysteine film		(−)-D cysteine film	
Crystal system	e.e in solution (%)	e.e of crystals on chiral surface (%)	e.e in solution (%)	e.e of crystals on chiral surface (%)
rac-glutamic acid	15 -L after 45 min	31 -D	12 -D after 45 min	27- L

Table 5. Enantiomeric excess of crystals crystallized in solutions and on chiral cysteine surfaces. (Dressler and Mastai, 2007)

Dressler and Mastai also studied the crystallization of histidine on the chiral cysteine films. They demonstrated that the cysteine films induce a modification in histidine morphology and lead to specific orientation of the histidine crystals. They also found that the L and D cysteine surfaces were active in chiral discrimination. Again, the enantiomeric excess was ca. 30%.

Singh *et al.* used chiral self assembled monolayers (Figure 15) as resolving auxiliaries in the crystallization of valine (Singh *et al.*, 2010). When starting with racemic solutions, the crystals obtained on the chiral SAMs contained one of the valine enantiomers in excess. The enantiomer obtained in excess was always of opposite chirality to the monolayer used. The chiral resolution was enhanced as a result of decreasing supersaturation. In addition, Singh et. al also monitored the crystallization of valine on chiral SAMs beginning with initial solutions containing one of the enantiomers in excess. When starting with a solution containing 50% ee, crystals of one pure enantiomer were obtained on the SAMs whereas mixtures of the pure enantiomer and the racemic compound were obtained in control experiments, not containing the chiral SAMs. The enantiomer obtained on the chiral SAMs was the enantiomer initially present in excess, regardless of the chirality of the monolayer being used.

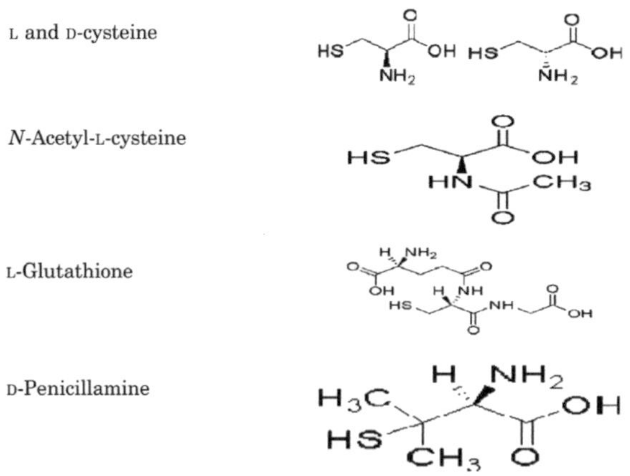

Fig. 15. Chemical structures of the chiral self assembled monolayers. (Singh *et al.*, 2010)

Banno *et al.* studied the enantioselective crystal growth of leucine on self assembled monolayers with covalently attached D or L-leucine molecules (Banno *et al.*, 2004). The monolayers were formed in the following way: gold substrates were prepared by vapour deposition of gold on quartz crystals. The substrates were flame annealed and subsequently immersed in 2 mM ethanolic solutions of 11-mercaptoundecanoic acid (MAU). Leucine was then covalently attached to the MAU monolayer. The crystals on the SAM were grown by immersing the substrates in saturated solutions of D, L or DL-leucine.

In order to identify the leucine crystals grown on the monolayers, X-ray diffraction and QCM (quartz crystal microbalance) measurements were utilized. Figure 16 shows XRD patterns of the leucine SAMs after immersion in a 175 mM D or L-leucine solution for 3 hrs. For the D-leucine SAM, the diffraction peak was observed only after it was immersed in the D-leucine solution, whereas no peak appeared when it was immersed in the L-leucine solution. With the L-leucine SAM, exactly the opposite results were obtained. The diffraction angles of the two observed peaks were identical to eachother and equal to 6.07°. The above mentioned cross inversion between D and L-leucine strongly suggests that the crystallization on the Leucine SAM is highly enantioselective. The enantioselectivity was also confirmed by the increase in surface mass resulting from the grown leucine crystals on the modified SAM, detected by QCM.

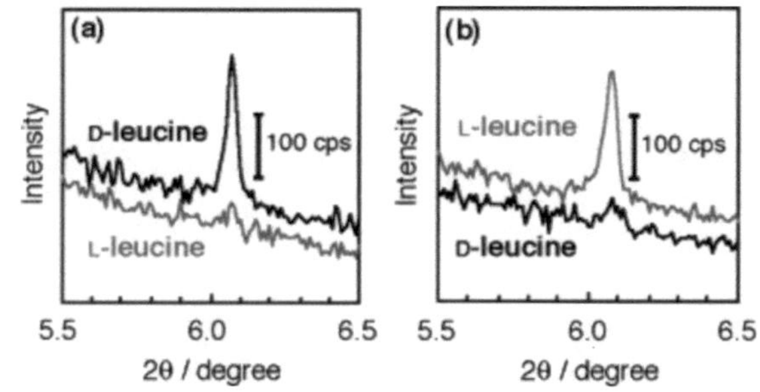

Fig. 16. XRD patterns of (a) D and (b) L-leucine SAMs after immersion in pure enantiomeric leucine solutions. (Banno *et al.*, 2004)

The leucine monolayers were also immersed in saturated DL-leucine solutions. As shown in Figure 17, a diffraction peak was observed at 6.34° for both D and L-leucine SAMs. This happens because in the presence of equimolar amounts of D and L enantiomers of leucine, the racemic crystal tends to form instead of either one of the two pure enantiomeric crystals.

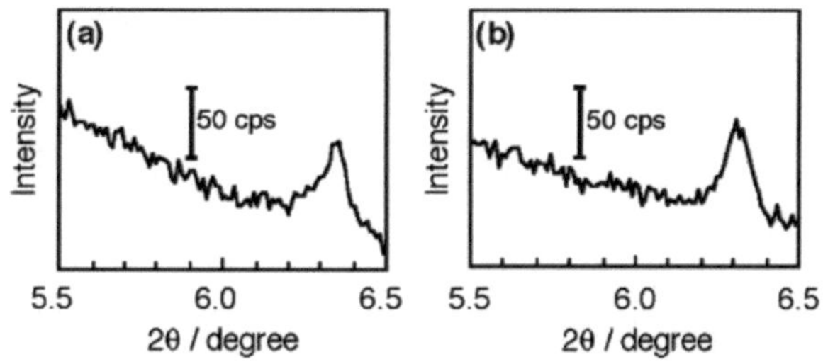

Fig. 17. XRD patterns of (a) D and (b) L-leucine SAMs after immersion in racemic leucine solution. (Banno *et al.*, 2004)

It should be noted that the primary interaction between the enantiomers attached on the SAMs and prenucleation aggregates formed in solution is due to hydrophobic bonding or Van der Waals interaction. It is assumed that crystal growth proceeds only when the chirality of the prenucleation aggregate of pure enantiomers formed in solution is the same as that of the attached enantiomer. In racemic solution, on the contrary, most of the prenucleation aggregates are considered to be in DL-form, resulting in the growth of the racemic crystalline phase on both D and L-leucine attached SAMs.

5. Conclusions

In this book chapter the preparation, structure, properties and applications of self-assembled monolayers (SAMs) in crystallization processes have been briefly reviewed. We have reviewed a variety of applications of SAMs in crystallization which can be catalogued into either pure crystallization methods or through the expression of chiral molecular interactions of SAMs.

Although two decades have passed since the discovery of SAMs, it is still a very active and broad research area that is almost impossible to review comprehensively. The fundamental questions of adsorption, structure, phases, and phase transitions have been thoroughly studied in the past, but still several issues remain unresolved. A large part of present and future work is aimed at the modification and fictionalization of surfaces by SAMs for different applications, with molecular and biological recognition being the most dynamic. Since SAMs are not so much a specific class of compounds, but rather a very flexible concept with virtually unlimited potential for applications, we expect that the area of SAMs will continue to prosper.

6. Acknowledgment

M. Ejgenberg would like to acknowledge the BIU President's scholarship and the ministry of science and technology program for the promotion of women in science and technology for funding.

7. References

Addadi, L., Weinstein, S., Gati, E., Weissbuch, I., & Lahav, M. (1982). Resolution of conglomerates with the assistance of tailor-made impurities. Generality and mechanistic aspects of the "rule of reversal". A new method for assignment of absolute configuration. *Journal of the American Chemical Society*, Vol. 104, 4610-4617

Aizenberg, J., Black, A.J., & Whitesides, G.M. (1999). Control of crystal nucleation by patterned self-assembled monolayers. *Letters to Nature*, Vol. 398, 495-498

Aizenberg, J., Black, A.J., & Whitesides, J.M. (1999). Oriented growth of calcite controlled by self assembled monolayers of functionalized alkane thiols supported on gold and silver. *Journal of the American Chemical Society*, Vol. 121, 4500-4509

Banno, N., Nakanishi, T., Matsunaga, M., Asahi, T., & Osaka, T. (2004). Enantioselective crystal growth of leucine on a self assembled monolayer with covalently attached leucine molecules. *Journal of the American Chemical Society*, Vol. 126, 428-429

Chieng, N., Rades, T., & Aaltonen, J. (2011). An overview of recent studies on the analysis of pharmaceutical polymorphs. *Journal of Pharmaceutical and Biomedical Analysis*, Vol. 55, 618–644

Chun, K.Y., & Stroeve, P. (2002). Protein Transport in Nanoporous Membranes Modified with Self-Assembled Monolayers of Functionalized Thiols. *Langmuir*, Vol. 18, 4653-4658

Dressler, D.H. & Mastai, Y. (2007). Chiral crystallization of glutamic acid on self assembled films of cysteine. *Chirality*, Vol. 19, 358-365

Dressler, D.H., & Mastai, Y. (2007). Controlling polymorphism by crystallization on self assembled multilayers. *Crystal Growth & Design*, Vol. 7, No. 5, 847-850

Eckermann, A.L., Feld, D.J., Shaw, J.A., & Meade, T.J. (2010). Electrochemistry of redox-active self-assembled monolayers. *Coordination Chemistry Reviews*, Vol. 254, 1769-1802

Ejgenberg, M. & Mastai, Y. (2011). *Chemical Communications*, DOI: 10.1039/c1cc14952k.

Garti, N. & Zour, H. (1997). The effect of surfactants on the crystallization and polymorphic transformation of glutamic acid. *Journal of Crystal Growth*, Vol. 172, 486-498

Hiremath, R., Basile, J.A., Varney, S.W., & Swift, J.A. (2005). Controlling molecular crystal polymorphism with self assembled monolayer templates. *Journal of the American Chemical Society*, Vol. 127, No. 51, 18321-18327

Lee, A.Y., Ulman, A., & Myerson, A.S. (2002). Crystallization of amino acids on self assembled monolayers of rigid thiols on gold. *Langmuir*, Vol. 18, 5886-5898

Lee, A.Y., Lee, I.S., Dette, S.S., Boerner, j., & Myerson, A.S. (2005). Crystallization on confined engineered surfaces: a method to control crystal size and generate different polymorphs. *Journal of the American Chemical Society*, Vol. 127, 14982-14983

Lee, A.Y., Lee, I.S., & Myerson, A.S. (2006). Factors affecting the polymorphic outcome of glycine crystals constrained on patterned substrates. *Chemical Engineering and Technology*, Vol. 29, No. 1, 281-285

Lee, I.S., Kim, K.T., Lee, A.Y., & Myerson, A.S. (2008). Concomitant crystallization of glycine on patterned substrates: the effect of pH on the polymorphic outcome. *Crystal Growth & Design*, Vol. 8, No. 1, 108-113

Love, J.C., Estroff, L.A., Kriebel, J.K., Nuzzo, R.G. & Whitesides, G.M. (2005). Self assembled monolayers of thiolates on metals as a form of nanotechnology. Chemical Reviews, Vol. 105, No. 4, 1103-1169

Mastai, Y. (2009). Enantioselective crystallization on nanochiral surfaces. *Chemical Society Reviews*, Vol. 38, 772-780

Nyquist, R.M., Eberhardt, A.S., Silks III, L.A., Li, Z., Yang, X., & Swanson, B.I. (2000). Characterization of self-assembled monolayers for biosensor applications. *Langmuir*, Vol. 16, 1793-1800

Singh, A. & Myerson, A.S. (2010). Chiral self assembled monolayers as resolving auxiliaries in the crystallization of valine. *Journal of Pharmaceutical Sciences*, Vol. 99, 3931-3940

Smith, R.K., Lewis, P.A., & Weiss, P.A. (2004). Patterning self assembled monolayers. *Progress in surfaces science*, Vol. 75, 1-68

Ulman, A. (1996). Formation and structure of self assembled monolayers. *Chemical Reviews*, Vol. 96, 1533-1554

Weissbuch, I., Torbeev, V.Y., Leiserowitz L., & Lahav, M. (2005). Solvent effect on crystal polymorphism: Why addition of methanol or ethanol to aqueous solutions induces the precipitation of the least stable β form of Glycine. *Angewandte Chemie International Edition*, Vol. 44, 3226-3229

Yokota, M., Doki, N., & Shimizu, K. (2006). Chiral separation of a racemic compound induced by transformation of racemic crystal structures: DL-Glutamic acid. *Crystal Growth & Design*, Vol. 6, No. 7, 1588-1590

3

Asymmetric Reaction Using Molecular Chirality Controlled by Spontaneous Crystallization

Masami Sakamoto and Takashi Mino
Chiba University
Japan

1. Introduction

Organic crystals have been utilized by organic chemists to identify materials by comparing melting point and mixed melting with known compounds, and to obtain pure materials by recrystallization. Furthermore, many attractive aspects of the crystals have been discovered since the latter half of the 1960's. For example, when organic molecules aggregate and form a crystal, a new property such as the electric conductivity or nonlinear optics characteristic develops by forming a peculiar arrangement style, and these crystals are used as electronic and optical materials. In addition, chemistry using the key information recorded in the molecule in the crystal has been developed. Molecular motion, which was cluttered with a variety of conformations in solution, is considerably controlled in the crystal, and molecules are arranged in a specific conformation depending on the closest-packing and molecular interactions in many cases. Therefore, a different molecular conformation will often be afforded in the crystal and in solution. In some cases, information for chirality recorded in the crystal is particularly interesting.[1] Conglomerate crystals (racemic mixture) afforded from racemic compounds are also chiral crystals, which are used for preferential crystallization for optical resolution of racemate.[2] On the other hand, compounds with unstable axial chirality, even in achiral in solution, may provide an axial chirality in a crystal where the molecular arrangement and molecular conformation are controlled. Using the chiral properties of the crystal, asymmetric synthesis without an external chiral source from achiral substrates has been studied by many research groups. This approach is also broadly recognized as an absolute asymmetric synthesis.

Since the 1970's, the combination of chiral crystallization and the solid-state photoreaction has provided many successful examples of absolute asymmetric synthesis.[3] In these reactions, achiral materials adopted chiral arrangement only by spontaneous crystallization, and optically active products are obtained from the topochemically controlled reaction with high enantiomeric excesses (Figure 1).[4] This method incurs a problem in the crystallization of achiral molecules in chiral space groups, albeit rare and unpredictable. However, in recent years, crystal engineering and the solid-state reaction to a variety of new systems has progressed to such an extent that it can now be regarded as an important branch of organic chemistry. Furthermore, some unique ideas involving solution chemistry utilizing the chiral crytals have also been developed. The achievement of an asymmetric synthesis starting from an achiral reagent and in the absence of any external chiral agent has long been an intriguing challenge to chemists and is also central to the problem of the origin of optical activity on Earth.

We have designed many molecules suitable for intramolecular photochemical processes, and have reported the asymmetric photochemical reaction in the solid-state. Please refer to the literature because many reviews are provided about this reaction example.[5)]

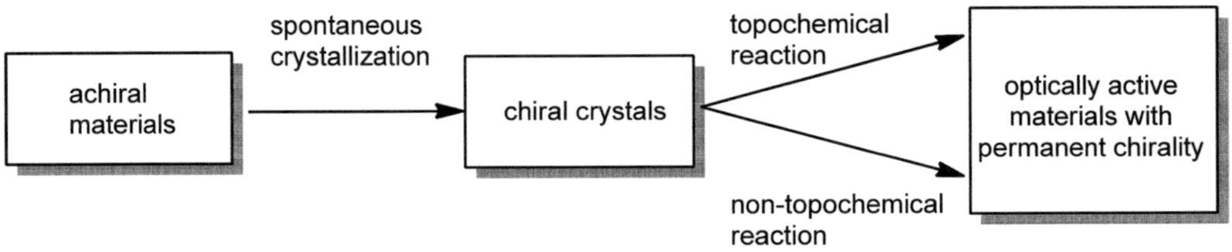

Fig. 1. Absolute asymmetric synthesis using chiral crystals

However, this asymmetric synthesis has yet to be recognized as the peculiar example, and there are many issues to be resolved that play a part in synthetic organic chemistry. The greatest problem is the small ratio of substrates forming a chiral crystal, and its inability to adapt to all crystalline organic compounds. Another problem is only applicable to this technique for topochemical reaction. If the crystalline state is broken according to the progress of the reaction, molecular chirality in the crystal would disappear in an instant. Exceptionally, an asymmetric amplification reaction using a chiral crystal surface dialkylzinc developed by Soai yielded product of high optical purity.

Soai *et al.* reported the catalytic asymmetric automultiplication in the addition of the dialkylzinc reagents to pyrimidine aldehydes without decrease in the optical purity of the product. The 1,2-addition in the presence of a small amount of optically active material resulted in enhancement of the *ee* value in the product formation.[6)] The asymmetric reaction was widely spread to the use of chiral crystals as a chiral trigger. The reaction promoted on the surface of the crystal of quartz gave optically active carbonyl addition product in 97% *ee*. Furthermore, the reaction with chiral crystals of $NaClO_3$, benzoylglycine, and cocrystals resulted in the formation of product with high *ee* value.

Håkansson *et al.* reported that two six-coordinate Grignard reagents crystallized as conglomerates and racemized rapidly in solution. Enantiopure Grignard reagents were reacted with butanal to give alcohol in up to 22% *ee*.[7)]

As the first example of asymmetric synthesis using chiral crystals involving solid-gas reaction,[3a)] some other interesting examples of solid-gas reaction using chiral crystals were reported. Reaction of chiral crystals of chalcone derivative with bromine in connection with rearrangement gave optically active dibromide in 8% *ee*.[8)]

Gerdil *et al.* reported two examples involving the solid-gas reactions of inclusion complexes of tri-*o*-thymonide with alkene or epoxycyclopentanone. The complex crystallized in a chiral fashion, and the reaction with singlet oxygen or hydrogen chloride gave products up to 22% *ee*.[9)]

The optical purity of the product was not satisfactory even in the reduced conversion rate of the reaction. To solve this problem, we explored a new methodology using molecular chirality in crystals as a source of chiral memory in solution (Figure 2).[10] The provisional molecular chirality derived from chiral crystals can be effectively transferred to optically active products with various asymmetric reactions in fluid media. Two requirements must be met for asymmetric synthesis to become possible: chiral crystallization of the starting materials and slow racemization at a controlled temperature.

If we can effectively use the chirality of the molecules expressed by spontaneous crystallization, asymmetric synthesis using chiral crystals, which has been restricted to the topochemical solid-state photoreaction, becomes a general synthetic method playing the part of organic synthesis.[11] In this review, a general method of chiral crystallization and a new advanced asymmetric synthesis using the chiral crystals in a non-topochemical process will be described.

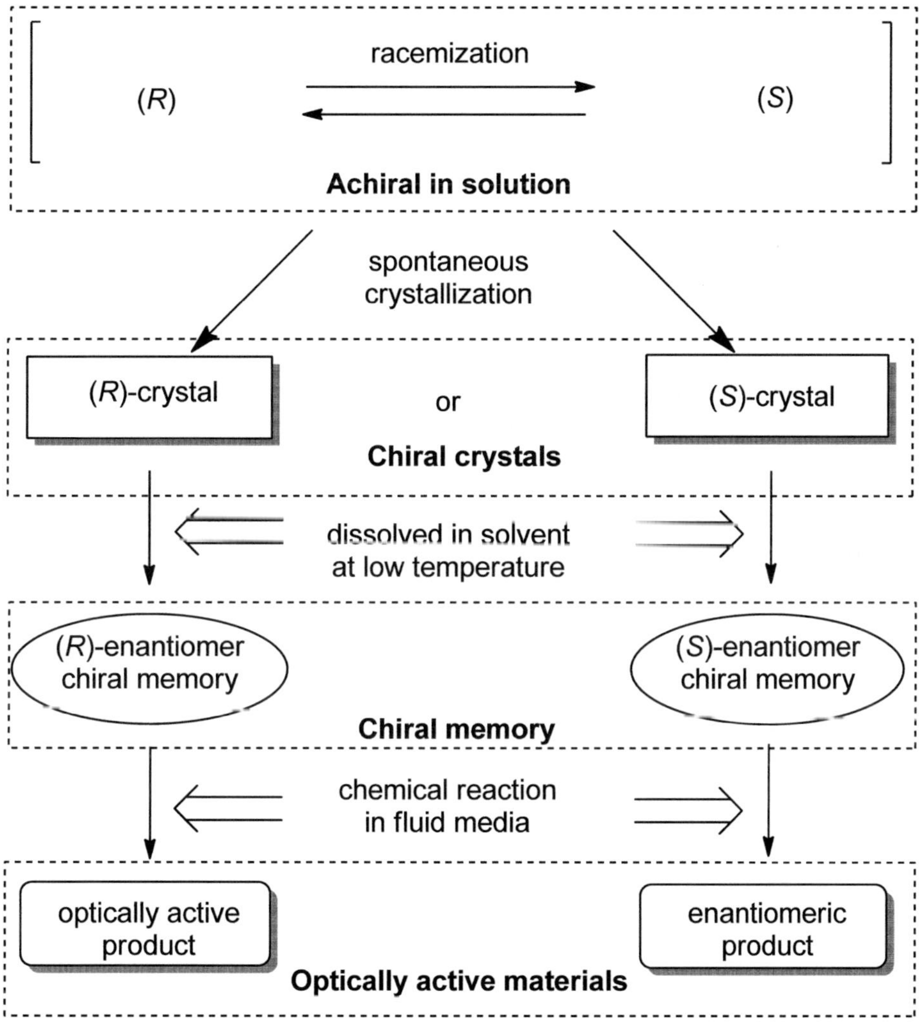

Fig. 2. Non-topochemical Asymmetric synthesis using chiral memory derived from the chiral conformation in the crystal.

2. Generation and amplification of chirality by crystallization

Optically active molecules must crystallize into chiral space groups, but a racemic mixture in solution may either aggregate to form a nonchiral racemic crystal or undergo a spontaneous resolution where the two enantiomers segregate into a conglomerate of enantiopure crystals. Achiral molecules may crystallize into either a nonchiral or a chiral space group. If they crystallize into a chiral space group, the achiral molecules reside in a chiral environment imposed by the lattice. Most achiral molecules are known to adopt interconverting chiral conformations in fluid media, which could lead to a unique conformation upon crystallization. Crystals that have chiral space groups are characterized by being enantiomorphous. They exist in right-handed and left-handed forms that may or may not be visually distinguishable. In spite of impressive work on crystal engineering, predictions on a correlation between crystal symmetry and molecular structures are still hard to make.[12)]

Chiral crystals, like any other asymmetric object, exist in two enantiomorphous equienergetic forms, but careful crystallization of the material can induce the entire ensemble of molecules to aggregate into one crystal, of one-handedness, presumably starting from a single nucleus (Figure 3). However, it is not uncommon to find both enantiomorphs present in a given batch of crystals from the same recrystallization.

For achieving asymmetric synthesis, we should begin with a compound crystallizing in any one of the chiral space groups. Of the 230 distinct space groups, the most commonly occurring are $P2_1/c$, P-1, $P2_12_12_1$, $P2_1$, $C2/c$, and $Pbca$, the chiral ones being $P2_12_12_1$, $P2_1$, P_1 and C_2.[13)]

The asymmetric crystallization of achiral compounds is stimulated by autoseeding with the first crystal formed. Although the chiral sense of the spontaneously formed chiral crystals cannot be predicted, seed crystals of the preferred chirality can be added in a more practical procedure to obtain one enantiomorph of a crystal.

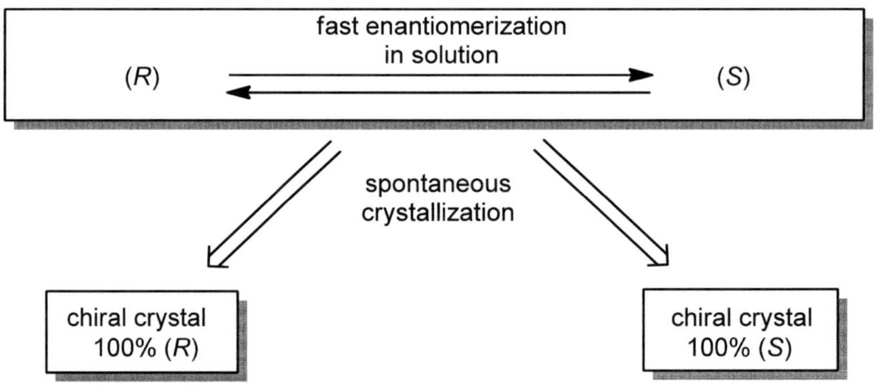

Fig. 3. Chiral crystallization with fast enantiomerization

3. Chiral memory effects of molecular conformation

When we start the study for chiral memory and select a molecule with the ability to hold for a certain amount of time the molecular conformation in the crystal lattice, we are focused on the molecules with the aromatic amide scaffold; this is because they can be synthesized many derivatives with variety substituents, and show considerable stability for the subsequent reaction. Molecules **1-4,** as shown in Figure 4, were designed to consider the development of subsequent asymmetric reactions. Imide **2** has an axial chirality based on the C-N bond; on the contrary, in compounds other than **2**, there is C-C bond axial chirality between the aromatic ring and amide groups. The racemization is inhibited to some extent by the steric repulsion between the aromatic ring and substituent on the amide nitrogen atoms of the planar structure. However, depending on the substituents on the aromatic ring and nitrogen atom, the rotational barrier of the N-C(=O) bond, except in special cases, is 20 ~ 24 kcal mol^{-1}, and the bond can rotate slowly even at room temperature.[14] In these substrates, when N-C(=O) bond rotation occurs, the racemization spontaneously proceeds depending on the Ar-(C=O) bond rotation; therefore, the maximum free energy of activation for racemization is also controllable. By changing the size and electronic property of the substituents, we can finely tune the energy of racemization of the aromatic amide.

Azumaya *et al.* also reported an example of retention of the molecular chirality when the chiral crystal of 1,2-bis(N-benzoyl-N-methylamino)benzene **5** was dissolved in cold $CDCl_3$.[15] The BIPHOS ligand **6** crystallizes as a conglomerate, and a single crystal in CH_2Cl_2 at -78°C reacts with $[PdCl_2(CH_3CN)_2]$ to give the enantiomerically pure complex $[PdCl_2(biphos)]$. Tissot *et al.* reported an example of catalytic asymmetric synthesis where the enantiomerically pure complex was used in the catalytic asymmetric allylic substitution of 1,3-diphenylprop-2-enyl acetate with the anion of dimethyl malonate to give product in 93% yield and with 80% *ee*.[16]

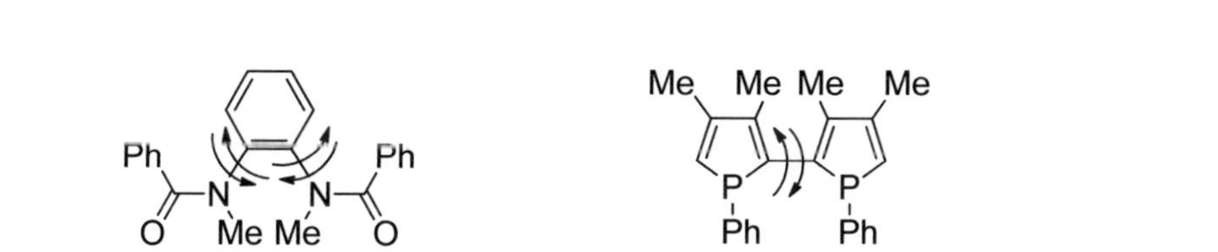

1 **2** **3** **4**

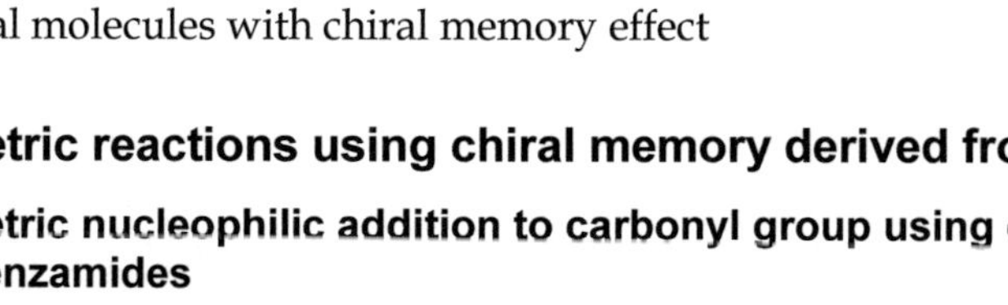

5 **6**

Fig. 4. Achiral molecules with chiral memory effect

4. Asymmetric reactions using chiral memory derived from chiral crystals

4.1 Asymmetric nucleophilic addition to carbonyl group using chiral crystals of 2-benzoylbenzamides

When 2-benzoyl benzamides **1a-d** with various sizes of substituents on the nitrogen atom were synthesized and the crystal structure was analyzed by X-ray crystallographic analysis, it was revealed that three substrates **1a-c** afforded chiral crystals (Figure 5).[17] As mentioned, achiral materials generally yield chiral crystals in about 10%; however, the designed molecules with the polarity and their round shapes crystallized in the chiral space group in very high proportion.[18] In these substrates, the two carbonyl groups are unable to form a planar conformation, but formed a twisted up and down structure, and chiral crystals were obtained by forming crystals of either enantiomer. If the substrate had a relatively fast rate of racemization, the same enantiomeric crystals were obtained in the flask by the usual recrystallization from a solution. It is easy to obtain large quantities of the desired enantiomer crystals by adding seed crystals of one enantiomer during crystallization.

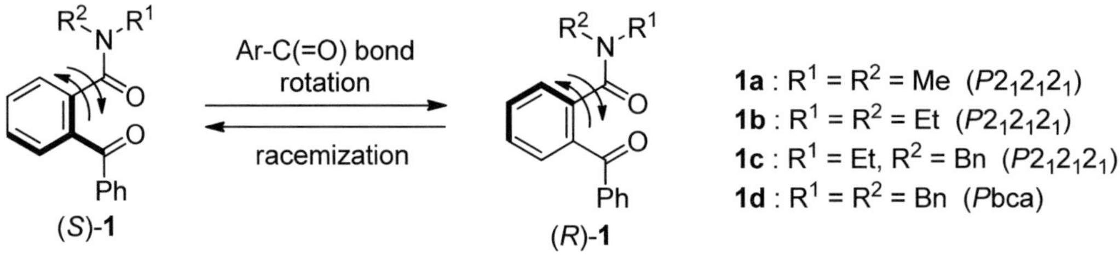

Fig. 5. Achiral 2-benzolybenzamides studied for chiral crystallization and asymmetric synthesis.

When the chiral crystals were dissolved in a solvent at room temperature, the molecular chirality disappeared in an instant. How long chiral molecular conformation is retained after the crystals were dissolved in a cold solvent? There are several techniques to measure the rate of racemization, and it can be obtained from the decay of optical rotation or CD spectra by longer life to some extent. However, these substrates have very fast rates of racemization, and the Cotton effect was not observed by the CD spectra immediately after dissolving the crystals in a solvent of -80°C by using a cryostat apparatus. Therefore, the rate of racemization for **1b** was determined using VT-NMR techniques. Methylene protons of the diethyl group on the nitrogen atom of **1b** are observed at the two different positions of anti and syn at room temperature; however, the protons in the diastereomeric relationship will be observed as four peaks at low temperatures. Activation free energy of racemization was calculated from the chemical shift, the coalescence temperature was 10.3 kcal mol^{-1} in toluene at -60°C, and the half-life could be reached in 2.2 x 10^{-3} sec. The half-life is very short; however, asymmetric synthesis of a sufficiently high stereoselectivity was expected to be achieved by selecting a stereospecific reaction faster than racemization.

Nucleophilic addition reactions to the carbonyl group were examined. When the chiral crystals of **1** were added to a cooled toluene solution containing butyllithium, the adduct **7** was obtained in high yield. Optical purity of the product was determined after induction to phthalide **8** by treatment with acetic acid (Table 1). The product from **1a** showed optical activity of 17% *ee* because of the significantly faster rate of racemization of **1a** having small steric hindrance of dimethyl groups on the nitrogen atom. On the other hand, when using the chiral crystals of **1b** and **1c**, the product could be obtained with optical purity of more than 80% *ee*. As a matter of course, when using the crystals of racemic **1d**, **7d** and **8d** were obtained as racemates. Whereas compound **1** showed short-lived chiral memory, it was enough to react with butyllithium. These results prepared unprecedented asymmetric synthesis involving nucleophilic reaction of carbonyl groups using chiral memory with C-C axial chirality.

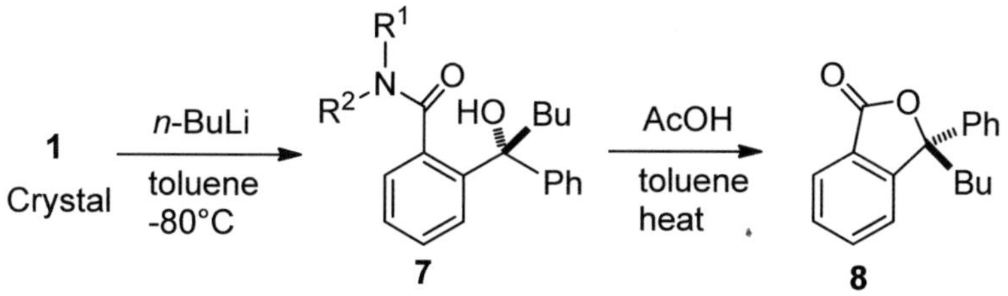

amide	R^1	R^2	space group	yield (%) of **8**	*ee* (%) of **8**
1a	Me	Me	$P2_12_12_1$	82	17
1b	Et	Et	$P2_12_12_1$	82	82
1c	Et	Bn	$P2_12_12_1$	83	80
1d	Bn	Bn	*Pbca*	87	0

Table 1. Reaction using the crystal of 2-benzoylbenzamides with n-BuLi

4.2 Asymmetric reactions using chiral memory derived from asymmetric imides

We reported the asymmetric reaction using chiral memory owing to the C-N axial chrality derived from the chiral crystal of acyclic imide (Figure 6).[19] The imide nitrogen atom has a planar structure, similar to the sp^2-hybridized amide nitrogen atom, and can form multiple structures of *EE*, *EZ*, *ZE*, and *ZZ*.[14] In imide **2**, the tetrahydronaphthyl group (TENAP) on the nitrogen atom exhibited almost perpendicular conformation to the plane of the imide. Depending on the direction of the TENAP group, each conformation enters an enantiomeric relationship. Racemization rate of the substrate could be determined by measuring the decay of the CD spectra using a cryostat apparatus by dissolved chiral crystals in a cold solvent. Activation free energy of racemization of this imide was 18.4 kcal mol^{-1} in THF at -20°C, which indicates that the chiral molecular conformation could be retained long enough after the chiral crystals were dissolved in a solvent. We tried the reaction of butyllithium with chiral molecules using this chiral memory. The reaction site is the benzoyl group of the side of the imide plane, and was reacted at -80°C; **9** and **10** were obtained as optically active forms of 83% and 81% *ee*, respectively. Furthermore, when provisional chiral molecule was irradiated in THF at -60°C, optically active oxetanes **11** and **12** were isolated.[19]

4.3 Development of chiral crystallization and asymmetric reactions using the chiral crystal of naphthamides

To perform this asymmetric synthesis, 2-alkoxy-1-naphthamides **3** were chosen because the bond rotation between the naphthalene ring and the amide carbonyl corresponds to enantiomerization of **3**, and the rate is considerably affected by the substituent of both the naphthalene ring and the amide group (Figure 7).[20] Naphthamides with a bulky group such as the *N,N*-diisopropyl amide group have stable axial chirality, which is utilized in many asymmetric synthesis.[21] Therefore, achiral naphthamides possessing a relatively compact-size amide group derived from pyrrolidine and piperidine, **3a** and **3b**, were prepared for our purpose. The X-ray single crystallographic analyses of the crystals revealed that both amides tend to have almost the same molecular conformation; remarkably, each carbonyl group twists almost orthogonally to the naphthalene plane.

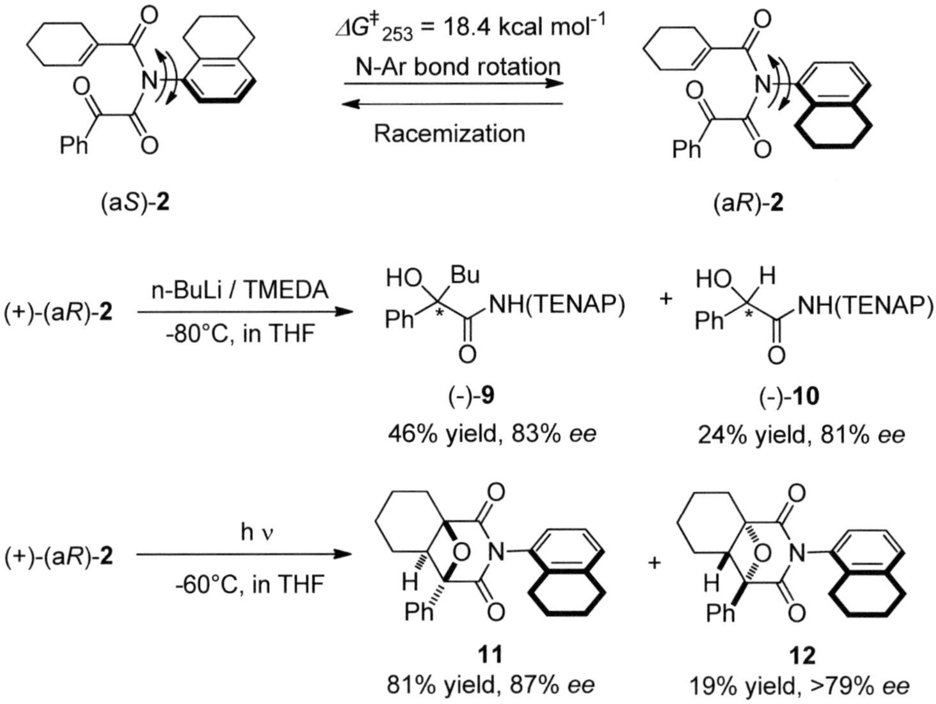

Fig. 6. Chiral crystallization of achiral acyclic imide and asymmetric synthesis using the chiral memory

Fortunately, both **3a** and 3**b** crystallized in a chiral space group, $P2_12_12_1$, and the constituent molecules adopted a chiral and helical conformation in the crystal lattice. The rate of racemization of **3** after dissolving the chiral crystals in a solvent was measured based on the changes in the CD spectra. The activation free energy ($\Delta G^{\neq}$) was calculated from the temperature dependence of the kinetic constant between 5 and 15°C as 21.2 ± 0.2 kcal mol^{-1} in THF. These facts indicate that the racemization of **3b** is too fast to be resolved in the usual manner. The lifetime can be lengthened by lowering the temperature so that the racemization is sufficiently slow and the reaction can be used to accomplish asymmetric synthesis.

The crystals of naphthamide **3** used for the asymmetric reaction were prepared by stirred crystallization from the melt.[22] The samples were completely melted at 120°C, which greatly exceeds their melting points (mp of **3b**: 110-112°C), cooled and solidified by lowering the temperature by stirring to 100°C. High reproducibility of both the chiral crystallization and asymmetric reaction was achieved by this method; however, the direction of the optical rotation of the photoproduct was inconsistent and appeared randomly. Of course, the desired crystals of **3** could be selectively prepared in large quantities by the addition of a corresponding seed crystal during the crystallization process.

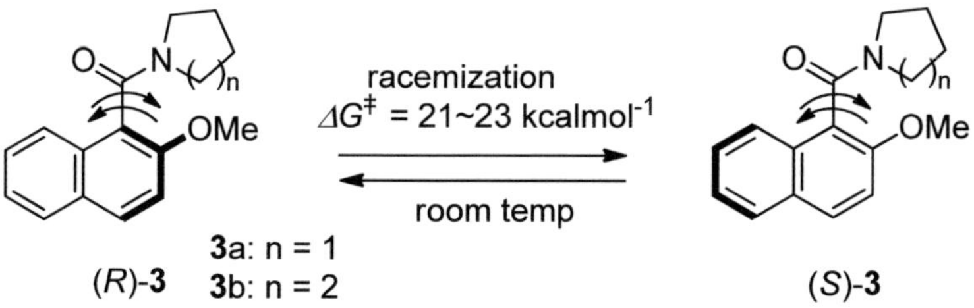

Fig. 7. Racemization of 2-alkoxy-1-naphthamides by Ar-(C=O) bond rotation

We examined the asymmetric reaction using the provisional molecular chirality of naphthamide **3b** (Figure 8).[23] After chiral crystals of **3b** were dissolved in a toluene solution containing diene (0.10 M) at -20°C, the solution was irradiated at the same temperature. Cyclobutane **13** was obtained in an optically active form of 76% *ee* at a 10% conversion of **3b**. Furthermore, with the increase of the conversion of **3b**, the *ee* value of **13** decreased rapidly. The rate of racemization of **3** at -20°C is very slow, and chirality was retained during irradiation. Why did the *ee* values decrease during the progress of the reaction? The amide directly racemized on irradiation via the twisted intramolecular charge transfer (TICT) mechanism from the singlet excited state.

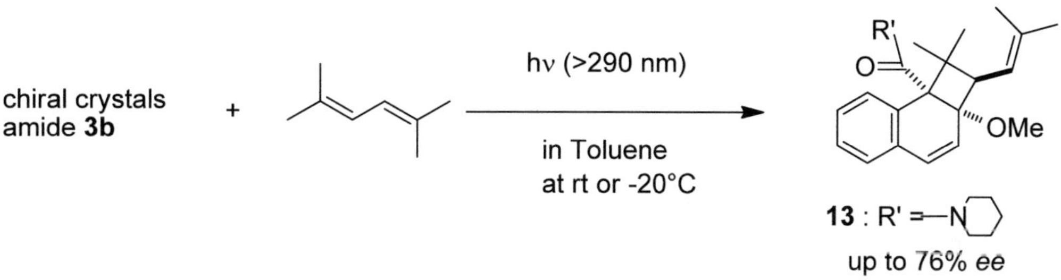

Fig. 8. Photochemical cycloaddition of naphthamide **3b** with 2,5-dimethylhexa-2,4-diene

We also reported the 4+4 cycloaddition of excited 9-cyanonaphthalene with the ground state of **3**, leading to an optically active 4+4 adduct with high enantioselectivity (Table 2).[20] Chiral crystals of **3b** was dissolved in a cooled THF solution of -20°C including 9-CNAN, and the solution was irradiated with an ultra-high pressure mercury lamp for 30 min. Only one cycloadduct **14** was obtained in 100% chemical yield, and the adduct showed optical activity of 95% *ee*. When we used a mixed solvent of MeOH and THF, almost the same *ee* value (94% *ee*) was obtained. Even at 20°C, we were able to obtain 29% *ee* of the adduct from the reaction in THF, in which the enantiomerization occurs competitively with the photocycloaddition. The rate of enantiomerization in alcoholic solvent is more suppressed than in THF, and the half-life is still 62.6 min. at 20°C. Surprisingly, we could obtain 88% *ee* of the product at 20°C.

For the stereochemistry of the reaction pathway, the (*S*)-proline derivatives were synthesized. The comparison of the absolute structure of the starting naphthamide and the adduct was studied, and it can be estimated that 9-CNNAP approaches from the vacant site of the carbonyl group to avoid the bulky substituents on the nitrogen atom (Figure 9).[24]

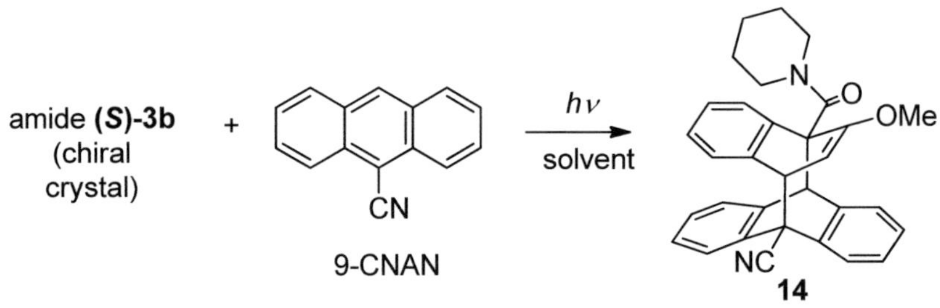

entry	solvent	temp (deg)	conc (mol L^{-1})	conv of **3b** (%)	ee of **14** (%)
1	THF	-20	0.05	52	95
2	THF	20	0.05	45	29
3	MeOH/THF	-20	0.025	54	94
4	MeOH/THF	20	0.025	40	88

Table 2. Asymmetric photochemical cycloaddition reaction of **3b** with 9-cyanoanthracene using the provisional axial chirality

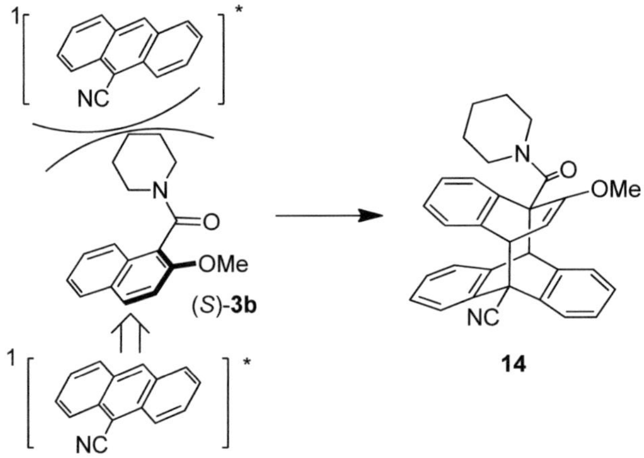

Fig. 9. Reaction course of the photocycloaddition of chiral conformation of (*S*)-**3b** with excited state of 9-CNAN

4.4 Chiral lock of the provisional chiral memory of naphthamides

Next, we examined how to lock the bond rotation affecting racemization of the amides **3**.[25] The racemization should be suppressed by substitution of the methoxy group at the 2-position of the naphthalene ring with the more bulky group.

A THF solution containing 3.0 eq. of *t*-BuLi was cooled to -80°C and was followed by the addition of chiral crystals of **3a**. After reaction for 1 h at the same temperature, the reaction mixture was treated in the usual manner. Analysis of the reaction product showed the formation of 2-*t*-butyl derivatives **15a** in a 97% yield and in an optically active form of 85% *ee*. When the chiral crystal of **3b** was used for the S_NAr reaction, nearly similar results were obtained. When the crystals of (+)-**3a** or **3b** were used for the S_NAr reaction, (+)-**15a,b** were obtained. Both amides **15a** and **15b** have stable axial chirality and did not racemize at room temperature for several months.

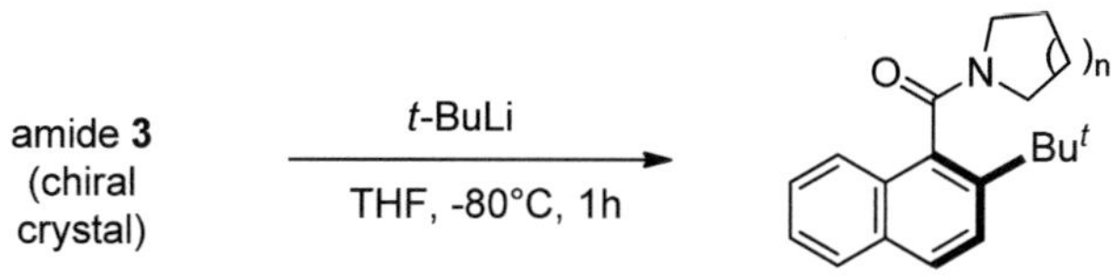

15

15a (n = 1) : 97% (85% *ee*)
15b (n = 2) : 72% (85% *ee*)

Fig. 10. Asymmetric S_NAr reaction of **3** with *t*-butyllithium using the provisional axial chirality

4.5 Kinetic resolution of racemic amines using provisional molecular chirality

We found that the kinetic resolution of racemic piperidines alkylated at the 2- or 3-position was performed using the provisional enantiomeric conformation of naphthamide **3a** derived from chiral crystals.[26] Chiral crystals of naphthamide **3a** were added to a THF solution of racemic lithium amides prepared by the reaction of substituted piperidines with *n*-BuLi at -80°C; the reaction mixture was stirred for 5 hours at -20°C because the substitution reaction did not proceed below -20°C.

When a 2.0 equimolar amount of 2-methylpiperidine was used, 30% of naphthamide **3a** was consumed, and 30% of **16a** was isolated. The reaction conversion was low; however, the reaction was very clean, and 70% of unreacted naphthamide was recovered. Fortunately, **16a** was obtained as the optically active form in 94% *ee*. Based on an increase of the amount of lithium amide to 5 eq., all naphthamide **3a** was consumed and converted to **16a** quantitatively with 81% *ee*. These results indicate that the 2.0 equimolar amounts of lithium amide form a stable intermediacy complex with naphthamide **3a**, and an extra amount of piperidine must be necessary to obtain product **16** in good yield. The intermediacy enantiomeric complex reacts preferentially with one enantiomer of the racemic peridine.

We diminished the amount of piperidines using such additives as diisopropylamine, TMEDA, and HMPA by displacing the complex ligand. When 3.0 eq. of LDA was added to the 2.0 eq. of the piperidine lithium amide, the substitution reaction occurred in a 100% conversion, and **3a** was isolated almost quantitatively in 73% *ee*. The addition of 1.0 eq. of TMEDA was also effective, and 80% of product was obtained with 80% *ee*. When 3.0 eq. of HMPA was used for the additives, all naphthamide was consumed, and **16a** was obtained in 90% with 69% *ee*.

When racemic 2-ethylpiperidine was reacted with the provisional chiral conformation of **3a**, similar results were obtained as the reaction with 2-methylpiperidine. However, when racemic 3-methylpiperidine was used as a nucleophile, the substitution reaction occurred quantitatively; however, the *ee* value resulted in 17%. The methyl group at the 3-position of the piperidine ring was insufficient to control the stereoselectivity because of the distance between the reacting nitrogen atom and the substituent. On the contrary, the substituent at the 2-positions worked sufficiently as strong chiral flags to control the stereoselectivity.

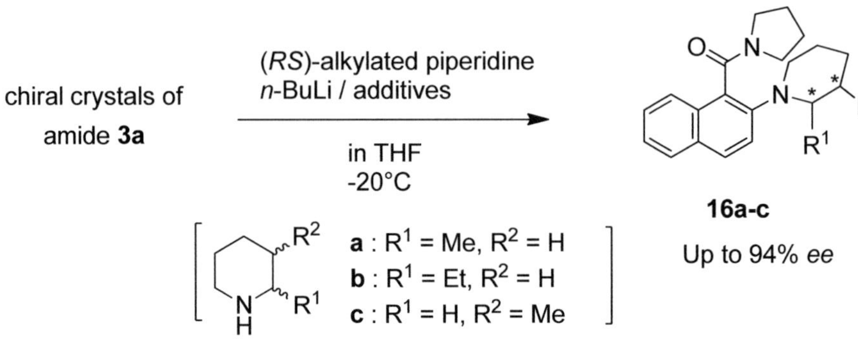

Fig. 11. Kinetic resolution of racemic piperidines by provisional molecular chirality of naphthamide **3a**.

4.6 Photosensitized 2+2 cycloaddition reaction using chiral memory

Achiral *N,N*,4-triethylcoumarin-3-carboxamide **4** crystallized in a chiral space group, $P2_12_12_1$ (Figure 12).[27] The X-ray single crystallographic analysis of the crystals revealed that the amides **12** tend to have a twisted conformation, of which the amide carbonyl group twists almost orthogonally to the coumarin chromophore. There are two enantiomeric conformations as shown in Figure 12 caused by the C-(C=O) bond rotation in fluid media; however, the chiral crystal is composed of a single enantiomer.

The rate of racemization was measured according to the changes in the CD spectra using a cryostat apparatus, and the activation free energy and the half-life were calculated. The racemization of **4** in THF was too fast at room temperature to determine the rate. On the other hand, when the crystals of **4** were dissolved in THF at 5°C, the half-life of racemization was 11.9 min. The half-life increased as the temperature was lowered, and $t_{1/2}$ was 30.5 and 82.0 min at the temperatures of 0°C and -5°C, respectively. The activation free energy ($\Delta G^{\neq}$) was calculated as the temperature dependence of the kinetic constant (5°C: 4.85 x 10^{-4}, 0°C: 1.89 x 10^{-4}, -5°C: 7.04 x 10^{-5}) to be 20.5-20.7 kcal mol^{-1}. In the case of the racemization in MeOH or DMF, the rate showed a considerably low activation free energy of 22.3-22.4 kcal mol^{-1}, and exhibited 20.2 and 23.6 min of half-life at 25°C, in MeOH and DMF, respectively. These facts indicate that the racemization of **4** is too fast to resolve in the usual manner; however, it can be controlled by lowering the temperature and with the selection of the solvent, and the lifetime becomes long enough for utilization in asymmetric synthesis.

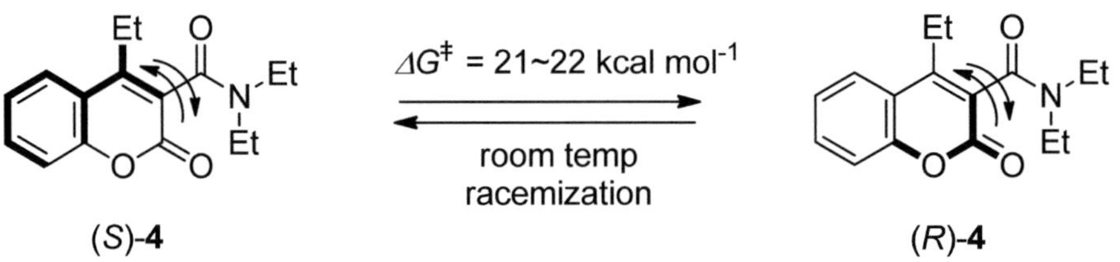

Fig. 12. Racemization of coumarinecarboxamide **4** owing to the rotation of the Ar-(C=O) bond

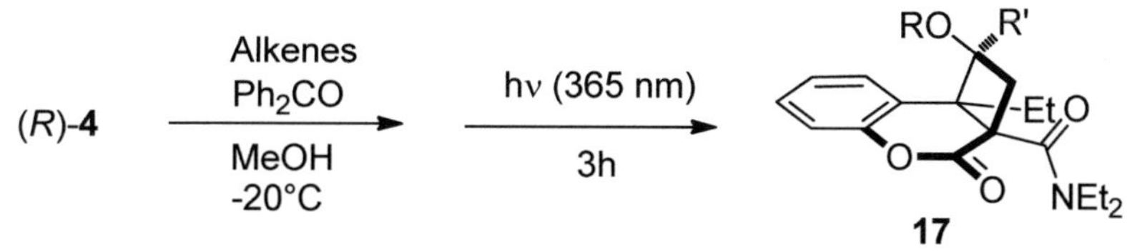

alkene	conv. of **4** (%)	yield of **17** (%)	*endo*-**17** : *exo*-**17**	*ee* of **17** (%) *endo*-**17**	*exo*-**17**
OEt	96	99	3 : 1	98	97
Me, OMe	74	98	100 : 0	99	-

Reaction conditions; **4** (0.02 mol/L), alkene (0.2 mol/L), BP (0.1 mol/L)

Table 3. Asymmetric photosensitized cycloaddition reaction of **4** with alkenes using the provisional axial chirality

When powdered crystals of **4** (0.02 mol L^{-1}) were dissolved in a cooled MeOH solution (-20°C) containing ethyl vinyl ether (0.1 mol L^{-1}) and irradiated at -20°C with a 365 nm line, optically active adducts **17** were obtained as would be expected. However, the *ee* values were affected by the conversion. High *ee* values of products (82% *ee* of *endo*-**17** and 86% *ee* of *exo*-**17**) were obtained by suppressing the conversion of 18%; however, the *ee* value decreased as the conversion increased. We measured the changes of the CD spectral of an MeOH solution of **4** using a cryostat apparatus by irradiating at -20°C, and it was confirmed that racemization of the starting amide **4** in the singlet-excited state had occurred. Therefore, we examined the triplet-sensitized photocycloaddition of **4** using benzophenone (BP) as a triplet sensitizer to avoid photoracemization from the singlet-excited state. When powdered crystals of **4** were dissolved to a cooled MeOH solution (-20°C) containing ethyl vinyl ether and benzophenone (0.01 mol L^{-1}) and were irradiated at -20°C for 3 h, high optical yields of adducts, 94% *ee* of both *endo*- and *exo*-**17**, were obtained. The use of 0.02 mol L^{-1} of benzophenone resulted in 100% conversion and a 98% chemical yield of adducts; furthermore, surprisingly high *ee* values of products, 96% *ee* of *endo*-**17** and *exo*-**17**, were obtained. Finally, 98% *ee* of *endo*-**17** and 97% *ee* of *exo*-**17** were isolated by increasing the concentration of benzophenone to 0.1 mol/L.

The asymmetric cycloaddition using the homochirality in the crystal was also performed by the use of 2-methoxypropene. In this case, only *endo*-**17** was obtained in an almost quantitative yield, and asymmetric synthesis with high *ee*, up to 99%, was performed.

This reaction provided the first example of photosensitized intermolecular cycloaddition with high enantiomeric excess using the homochirality in the crystal generated by spontaneous crystallization.

5. Diastereoselective reaction using memory effect of molecular chirality controlled by crystallization

Asymmetric chiral transfer techniques using chiral memory derived from chiral crystals can be applied to diastereoselective reaction (Figures. 13, 14). We used the aromatic amides **18-20** with an optically active proline goup as the chiral source to give the chiral memory capabilities. Four diastereomers are possible in these compounds owing to the rotational isomer of C(=O)-N and C-(C=O) bonds. When we used diastereomeric mixture in asymmetric reactions, we obtained the mixture of products reflecting the ratio of starting diastereomers, and high stereoselective reaction could not be achieved. However, when diastereomixture of amides were crystallized, epimerization owing to the bond rotation was promoted, and the mixture automatically converged to a single diastereomer by Crystallization-Induced Diastereomer Transformation (CIDT) (Figure 13).

5.1 Diastereoselective reaction using memory effect of naphthamides

In the NMR spectrum of naphthamide **18**, several peak pairs were observed derived from diastereomixtures, which suggested that this amide existed as a mixture of diastereomers in the solution.[24] After evaporating the solution to solidify the amide, NMR spectrum measured immediately after dissolving the solid to $CDCl_3$ showed only peaks derived from single diastereomers. This fact indicated that diastereomixture crystallized by converging to single crystalline diastereomer by simple solidification (Figure 15). Conformation in the crystal was determined as (*S,aR*)-**18-A** by single crystal X-ray crystallographic analysis. Moreover, this chiral molecular conformation in solution was also retained at low temperature for long enough time to be used as a chiral memory.

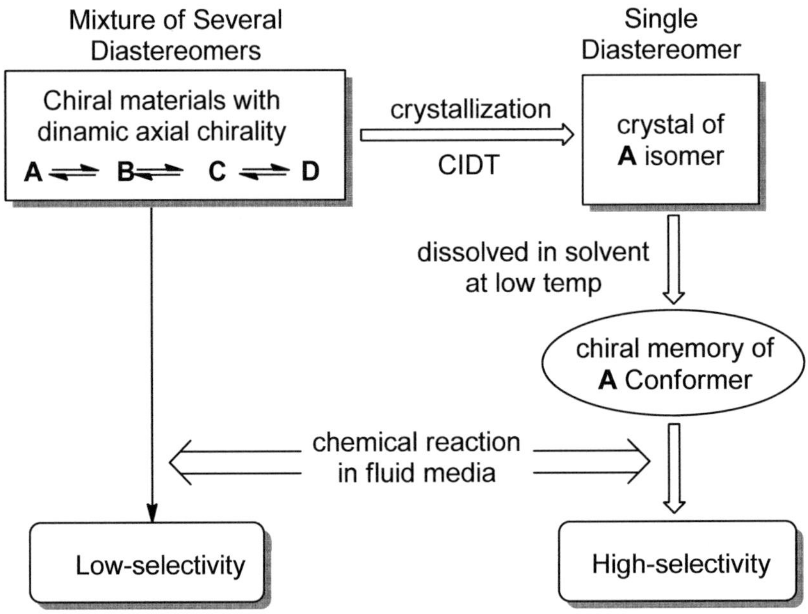

Fig. 13. Asymmetric synthesis using chiral memory derived from CIDT

Diastereoselective photocycloaddition reaction with 9-CNAN was examined. Irradiation of amides **18a-b** before solidification in the presence of 9-CNAN gave adducts **21a** and **21b** with the *de* values of 63% and 53%, respectively. On the other hand, when the crystals of amides obtained by CIDT were dissolved in THF at -20°C, 96% *de* of adduct **21a** and 100% *de* of **21b** were obtained. The molecular conformation of **18** was retained in a cold solution after dissoving the crystals, and could act as a chiral molecular memory; 9-CNNAP approached selectively from the less hindered site of carbonyl group of naphthamide.

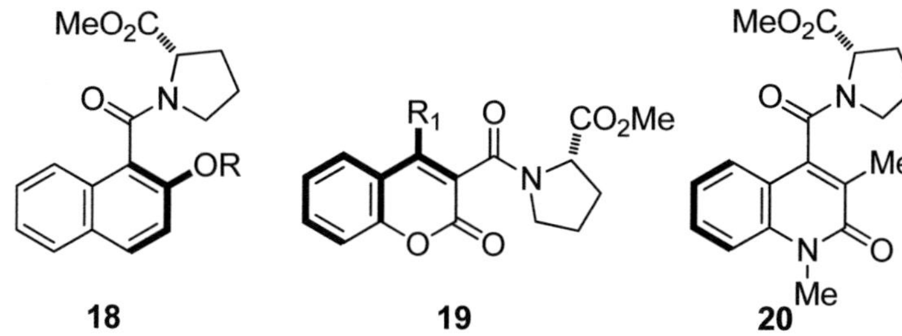

Fig. 14. Aromatic amides with chiral memory effect derived from CIDT

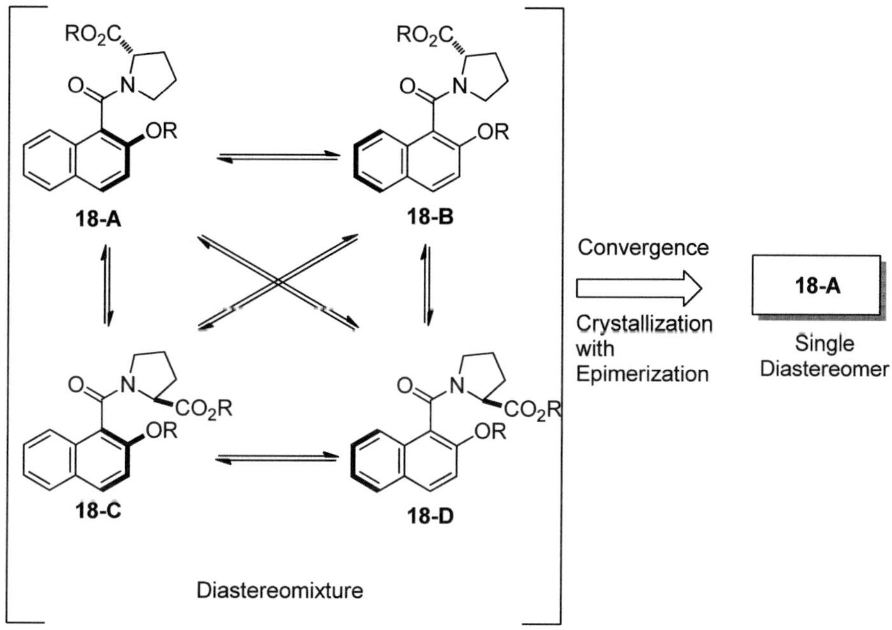

Fig. 15. CIDT of naphthamide with proline group

5.2 Diastereoselective reaction using memory effect of coumarinecarboxamides

We found that coumarin carboxamide **19** also existed as a mixture of diastereomers in solution, and could be converged to (*S,aR*)-conformer by solidification.[28] When the crystals of **19** were dissolved in MeOH at -20°C and irradiated in the presence of electron-rich alkenes, perfectly stereo-controlled 1:1 photoadduct **22** was obtained in 100% *de* (Figures 16, 17).

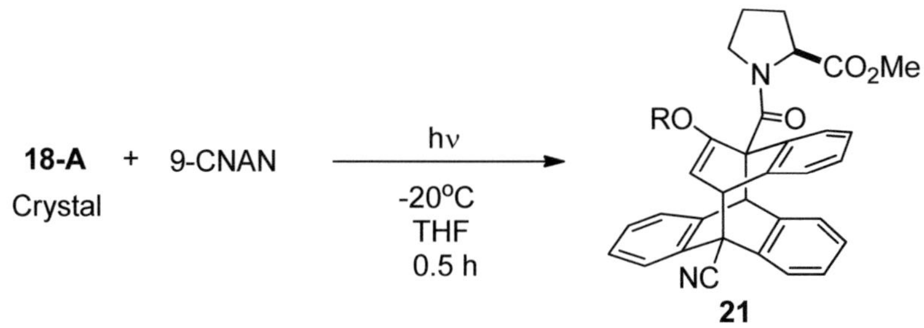

amide **18-A**	solidification	conv. of **18** (%)	yield of **21** (%)	*de* of **21** (%)
18a (R = Me)	before	77	>99	63
18b (R = Et)	before	70	>99	53
18a (R = Me)	after	62	>99	96
18b (R = Et)	after	70	>99	100

Table 4. Diastereoselective photocycloaddition using chiral memory derived from CIDT

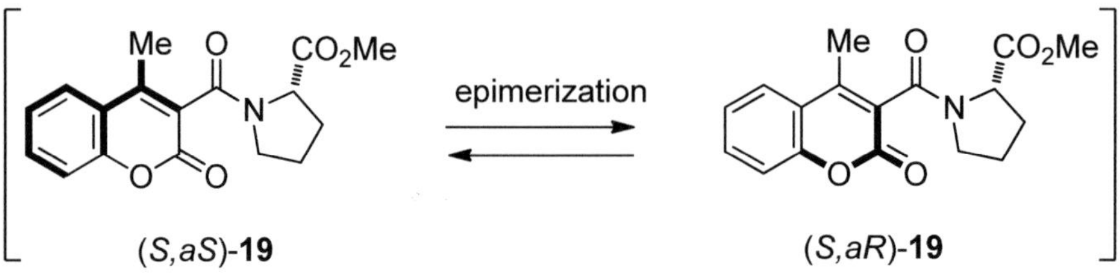

Fig. 16. Epimerization of coumarinecarboxamide with chiral handle

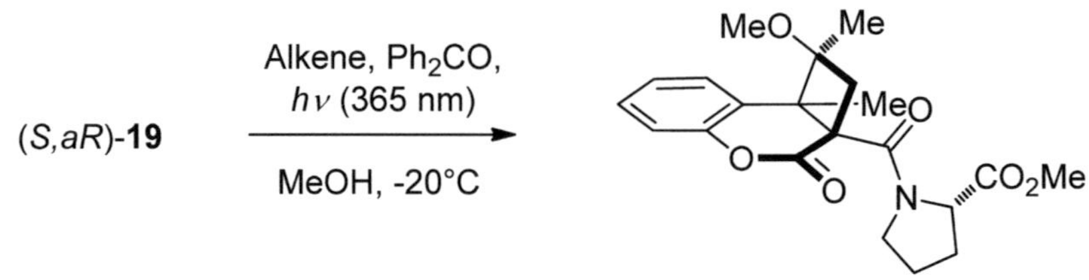

22 : 98% *de*

Fig. 17. Stereoselective photocycloaddition using chiral memory derived from CIDT

5.3 Diastereoselective reaction using memory effect of quinolonecarboxamides

2-Quinolone-4-carboxamide **20** possessing (*S*)-proline methyl ester as a chiral handle was chosen for the purpose.[29] Amide **20** exists as a mixture of two diastereomers in the ratio of a*R*:a*S* = 40:60 (diastereomeric excess, *de* = -20%) before crystallization (Figure 18). When the mixture was crystallized from a THF solution by evaporating solvent at 70°C, the minor isomer was easier to crystallize than the major isomer, and they converged to almost a single diastereomer in a ratio of a*R*:a*S* = 95:5 (*de* = 90%). Furthermore, recrystallization of the crystals once from ether easily gave 99% *de* of amide **20**. Axial chirality was stable for several days at room temperature; however, heating the chloroform solution of (*S*,a*R*)-**20** at 55°C for 12 hrs gave the exact same diastereomixture as before crystallization.

Next, the homochiral crystals were utilized for subsequent diastereoselective reactions. Diastereoselective 2+2 photocycloaddition reaction of amide **20** with methacrylonitrile was examined. The reaction with methacrylonitrile proceeded effectively, stereo-, and regiospecifically. The amide **20** before crystallization was irradiated in the presence of methacrylonitrile with a high-pressure mercury lamp at 20°C until most of the starting amide was consumed (2 hrs). The photochemical reaction occurred effectively, and 2 + 2 cycloadducts were obtained in 100% chemical yields; both diastereomers were *endo* isomers, minor (1*S*,2a*R*,8b*R*)-**23**, major (1*R*,2a*S*,8b*S*)-**23**, and the *de* value was -25%. Since epimerization was not observed at 20°C, it seems that the *de* value of the photoproducts should be attributed to the ratio of the diastereomers of the amide **20** before crystallization (-20% *de*).

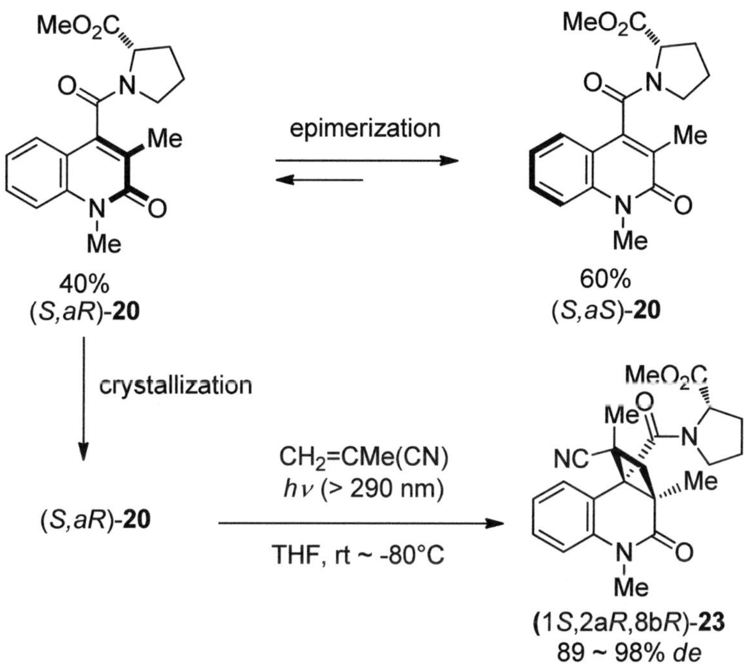

Fig. 18. Epimerization of quinolonecarboxamide with chiral handle and stereoselective photoaddition using chiral memory.

Furthermore, we examined a photocycloaddition reaction using the homochiral molecular conformation converged by CIDT. The crystals were expected to be composed of a single diastereomer of (*S*,a*R*)-conformation, and the epimerization in the solution caused by the bond rotation between the quinolone and the carbonyl group was restricted at room temperature. In other words, conformation in the crystals may be retained as frozen after dissolving them in the solvent at room temperature, and molecular homochirality can be effectively transferred to the products. The THF solution of (*S*,a*R*)-**20** containing methacrylonitrile was irradiated with a high-pressure mercury lamp for 2 hrs until the starting amide was consumed. When the reaction was performed at 20°C, two 2 + 2 adducts, major (1*S*,2a*R*,8b*R*)-**23** and minor (1*R*,2a*S*,8b*S*)-**23** were obtained in 99% yield. As expected, epimerization was strongly controlled at this temperature, and a high *de* of 89% was achieved. Even at low temperature, the reaction proceeded effectively, and after decreasing the temperature, the *de* values improved; the best *de* of 98% was obtained in the reaction at -80°C. The axial chirality evoked by crystallization directed the course of the approach of the reacting molecules, and a fully controlled diastereoselective intermolecular photocycloaddition reaction was performed.

6. Prospects for new asymmetric reaction using crystal

New aspects of chemistry are deployed by effectively utilizing the unique nature of crystals. In this chapter, a new approach to asymmetric chemistry using chiral crystals found in our research groups is introduced. We succeeded in transferring the chirality expressed by chiral crystallization of achiral substrate to products in a variety of reactions in homogeneous systems. Stereoselective reaction using chiral crystals was limited to the topochemical solid-state photoreaction, and the reactions expanded in a homogeneous system were able to achieve the highly stereoselective asymmetric reaction. Some readers may think that chiral molecules possessing memory effects introduced in this chapter are like acrobats using a very narrow energy band. Figure 19 shows an energy diagram for racemization of chiral molecules that express chirality by spontaneous crystallization or by Crystallization-Induced Enantiomer Transformation (CIET). We can effectively use molecules with

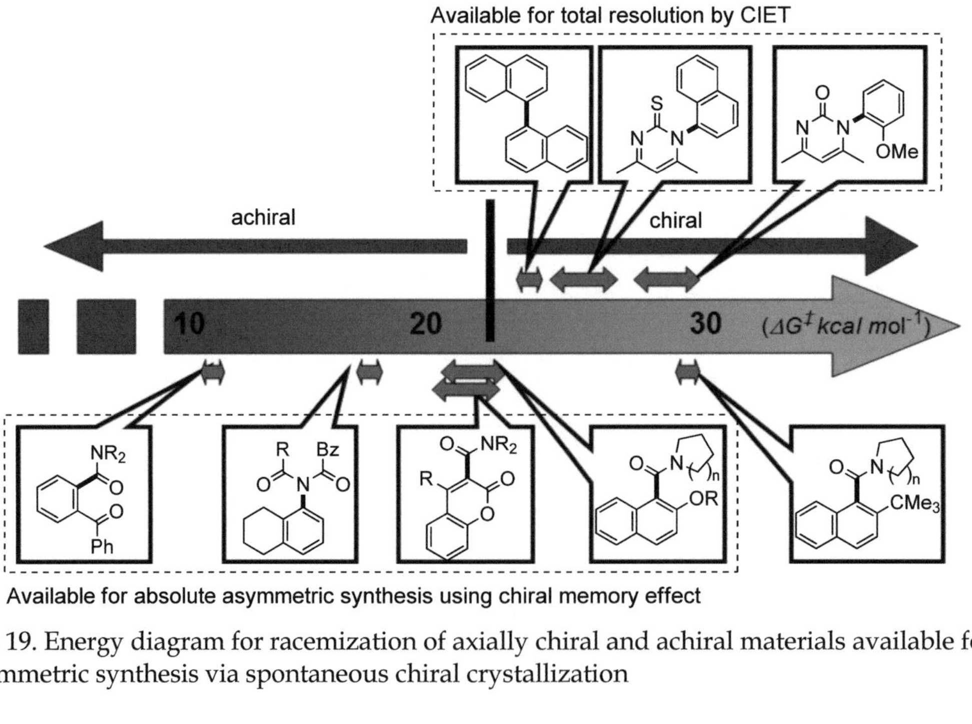

Fig. 19. Energy diagram for racemization of axially chiral and achiral materials available for asymmetric synthesis via spontaneous chiral crystallization

activation free energy of racemization from 10 to 30 kcal mol^{-1} for chiral memory. It is not an unusual technique, but a rather typical approach that can be widely utilized in asymmetric reactions. We have found many more applications to this method using chiral memory; we will present them in future work. In addition, this method could be applied to the diastereoselective reaction using chiral memory derived from CIDT. These findings suggest that the molecular information in the crystal can be widely applied to a variety of chemistry, in addition to organic asymmetric reaction.

7. References

[1] (a) Jacques, J. Collet, A. & Wilen, S. H. (1981). *Enantiomers, Racemates and Resolutions*, J. Jacques, (Ed.), Krieger Publishing, Malabar, FL; (b) Schmidt, G. M. J. (1971). Photodimerization in the solid state. *Pure Appl. Chem.*, Vol. 27, pp. 647-678; (c) Green, B. S.; Lahav, M. & Rabinovich, D. (1979). Asymmetric synthesis via reactions in chiral crystals. *Acc. Chem. Res.* Vol. 12, pp. 191-197; (d) Sakamoto, M. (1997). Absolute asymmetric synthesis from achiral molecules in the chiral crystalline environment. *Chem. Eur. J.*, Vol. 3, 684-689

[2] (a) Lu, Y.; Wang X. & Ching, C. B. (2009). Application of preferential crystallization for different types of racemic compounds. *Ind. Eng. Chem. Res.*, Vol. 48, pp. 7266–7275; (b) Barrett, P.; Smith, B.; Worlitschek, J.; Bracken, V.; O'Sullivan, B.; & O'Grady, D. (2005). A review of the use of process analytical technology for the understanding and optimization of production batch crystallization processes. *Org. Process Res. Dev.*, Vol. 9, pp. 348–355; (c) Collet, A.; Brienne, M.-J. & Jacques, J. Optical resolution by direct crystallization of enantiomer mixtures. *Chem. Rev.* Vol. 80, pp. 215-230

[3] (a) Penzien, K. & Schmidt, G. M. J. (1969). Reactions in chiral crystals: an absolute asymmetric synthesis. *Angew. Chem. Int. Ed.*, Vol. 8, pp. 608-609; (b) Green, B. S. & Heller, L. (1974). Mechanism for the Autocatalytic Formation of Optically Active Compounds under Abiotic Conditions. *Science*, Vol. 185, 525-527; (c) Addadi, L. & Lahav, M. (1978). Photopolymerization of chiral crystals. 1. The planning and execution of a topochemical solid-state asymmetric synthesis with quantitative asymmetric induction. *J. Am. Chem. Soc.*, Vol. 100, pp. 2838; (d) Evans, S. V. & Garcia-Garibay, M.; Omkaram, N.; Scheffer, J. R.; Trotter, J. & Wireko, F. (1986). Use of chiral single crystals to convert achiral reactants to chiral products in high optical yield: application to the di-p-methane and Norrish type II photorearrangements. *J. Am. Chem. Soc.*, Vol. 108, pp. 5648-5649; (e) Toda, F.; Yagi, M. & Soda, S. (1987). Formation of a chiral β-lactam by photocyclisation of an achiral oxoamide in its chiral crystalline state. *J. Chem.* Soc., *Chem. Commun.*, pp. 1413-1414; (f) Sekine, A.; Hori, K.; Ohashi, Y.; Yagi, M. & Toda, F. (1989). X-ray structural studies of chiral b-lactam formation from an achiral oxoamide using the chiral crystal environment. *J. Am. Chem. Soc.*, Vol. 111, pp. 697-699

[4] (a) Ramamurthy, V. & Venkatesan, K. (1987). Photochemical reactions of organic crystals. *Chem. Rev.*, Vol. 87, pp. 433-481; (b) Scheffer, J. R.; Garcia-Garibay, M. & Nalamasu, O. (1987). The influence of the molecular crystalline environment on organic photorearrangemens. In: *Organic Photochemistry*, Vol. 8, A. Padwa, (Ed.), pp. 249-347, Marcel Dekker, New York; (c) Vaida, M.; Popovitz-Biro, R.; Leiserowitz, L. & Lahav, M. (1991). Probing reaction pathways via asymmetric

transformations in chiral and centrosymmetric crystals. In: *Photochemistry in Organized and Constrained Media,* V. Ramamurthy, (Ed.) pp. 249-302, Wiley, New York; (d) Sakamoto, M. (2004). Absolute asymmetric photochemistry using spontaneous chiral crystallization In: *Chiral Photochemistry,* V.Ramamurthy, (Ed.), 415-446, Marcel Dekker, New York; (e) Sakamoto, M. (2004). Spontaneous chiral crystallization of achiral materials and absolute asymmetric transformation in the chiral crystalline environment In: *Enantiomer Separation,* F. Toda, (Ed.), 103-133, Kluwer Academic Publishers, Dordrecht, Boston and London

[5] (a) Sakamoto, M.; Hokari, N.; Takahashi, M.; Fujita, T.; Watanabe, S.; Iida, I. & Nishio, T. (1993). Chiral thietane-fused β-lactam from an achiral monothioimide using the chiral crystal environment. *J. Am. Chem. Soc.*, Vol. 115, p. 818; (b) Takahashi, M.; Sekine, N.; Fujita, T.; Watanabe, S.; Yamaguchi, K. & Sakamoto, M. (1998). Solid-state photochemistry of *o*-aroylbenzothioates: Absolute asymmetric phthalide formation involving 1,4-aryl migration. *J. Am. Chem. Soc.*, Vol. 49, pp. 12770-12776

[6] (a) Soai, K.; Shibata, T.; Morioka, H. & Choji, K. (1995). Asymmetric autocatalysis and amplification of enantiomeric excess of a chiral molecule. *Nature*, Vol. 378, pp. 767-768; (b) Shibata, T.; Yamamoto, J.; Matsumoto, N.; Yonekubo, S.; Osanai, S. & Soai, K. (1998). Amplification of a slight enantiomeric imbalance in molecules based on asymmetric autocatalysis: The first correlation between High enantiomeric enrichment in a chiral molecule and circularly polarized light. *J. Am. Chem. Soc.*, Vol. 120, pp. 12157-12158; (c) Kawasaki, T.; Jo, K.; Igarashi, H.; Sato, I.; Nagano, M.; Koshima, H. & Soai, K. (2005). Asymmetric amplification using chiral cocrystals formed from achiral organic molecules by asymmetric autocatalysis. *Angew. Chem. Int. Ed.*, Vol. 44, pp. 2774-2777

[7] (a) Johansson, A. & Håkansson, M. (2005). Absolute asymmetric synthesis of stereochemically labile aldehyde helicates and subsequent chirality transfer reactions. *Chem. Eur. J.*, Vol. 11 pp. 5238-5248; (b) Vestergren, M.; Eriksson, J. & Håkansson, M. (2003). Absolute asymmetric synthesis of "Chiral-at-Metal" Grignard reagents and transfer of the chirality to carbon. *Chem. Eur. J.* Vol. 9, pp. 4678-4686

[8] Garcia-Garibay, M.; Scheffer, J. R.; Trotter, J. & Wireko, F. (1988). Addition of bromine gas to crystalline dibenzobarrelene: An enantioselective carbocation rearrangement in the solid state. *Tetrahedron Lett.*, Vol. 29, pp. 1485-1488

[9] (a) Gerdil, R.; Huiyou, L. & Gerald, B. (1999). Organic reactions in the solid state: reactions of enclathrated 3,4-epoxycyclopentanone (=6-oxobicyclo[3.1.0]hexan-3-one) in tri-*o*-thymotide and absolute configuration of 4-hydroxy- and 4-chlorocyclopent-2-en-1-one. *Helv. Chim. Acta,* Vol. 82, pp. 418-434; (b) Gerdil, R.; Barchietto, G. B. & Jefford, C. W. (1984). Heterogeneous chirality transfer on photooxygenation. *J. Am. Chem. Soc.*, Vol. 106, pp. 8004-8005

[10] Sakamoto, M.; Mino, T. & Fujita, T. (2010). Generation and control of chirality by crystallization: Asymmetric synthesis using the crystal chirality in fluid media. *J. Synth. Org. Chem.*, Vol. 68, pp. 1047-1056.

[11] Sakamoto, M. (2006). Spontaneous chiral crystallization of achiral materials and absolute asymmetric photochemical transformation using the chiral crystalline environment. *J. Photochem. Photobiol. C: Photochem. Rev.* Vol. 7, pp. 183-196

[12] (a) Collet, A.; Brienne, M. & Jacques, J. (1972). Spontaneous resolution and enantiomeric conglomerates . *Bull. Soc. Chim. Fr.* pp. 127-142; (b) Matsuura, T. & Koshima, H. (2005). Introduction to chiral crystallization of achiral organic compounds: Spontaneous generation of chirality. *J. Photochem. Photobiol. C: Photochem. Rev.*, Vol. 6, pp. 7-24

[13] Belsky, V. K. & Zorkii, P. M. (1977). Distribution of organic homomolecular crystals by chiral types and structural classes. *Acta. Crystallogr. Sect. A*, Vol. A33, pp. 1004-1006

[14] (a) Noe, E. A. & Raban, M. (1975). Stereochemistry in trivalent nitrogen compounds. XXVIII. Conformational analysis and torsional barriers of imides and triamides. *J. Am. Chem. Soc.*, Vol. 97, pp. 5811-5820

[15] Azumaya, I.; Yamaguchi, K.; Okamoto, I.; Kagechika, H. & Shudo, K. (1995). Total asymmetric transformation of an N-methylbenzamide.. *J. Am. Chem. Soc.*, Vol. 117, pp. 9083-9084.

[16] Tissot, O.; Gouygou, M.; Dallemer, F. Daran, J.-C.B. & Balavoine, G.G.A. (2001). The combination of spontaneous resolution and asymmetric catalysis: A model for the generation of optical activity from a fully racemic system. *Angew. Chem. Int. Ed.*, Vol. 40, pp. 1076-1078

[17] Sakamoto, M.; Kobaru, S.; Mino, T. & Fujita, T. (2004). Absolute asymmetric synthesis by nucleophilic carbonyl addition using chiral crystals of achiral amides. *Chem. Commun.* pp. 1002-1003

[18] Sakamoto, M.; Sekine, N.; Miyoshi, H.; Mino, T. & Fujita, T. (2000). Absolute asymmetric phthalide synthesis via the solid-state photoreaction of N,N-disubstituted 2-benzoylbenzamides involving a radical pair intermediate. *J. Am. Chem. Soc.*, Vol. 122, pp. 10210-10211.

[19] Sakamoto, M.; Iwamoto, T.; Nono, N.; Ando, M.; Arai, W.; Mino, T. & Fujita, T. (2003). Memory of chirality generated by spontaneous crystallization and asymmetric synthesis using the frozen chirality. *J. Org. Chem.*, Vol. 68, pp. 942-946

[20] Sakamoto, M.; Unosawa, A.; Kobaru, S.; Saito, A.; Mino, T. & Fujita, T. (2005). Asymmetric photocycloaddition in solution of a chiral crystallized naphthamide. *Angew. Chem. Int. Ed., Vol.* 44, pp. 5523-5526

[21] (a) Clayden, J.; Lund, A.; Vallverdu, L. & Hellwell, M. (2004). Ultra-remote stereocontrol by conformational communication of information along a carbon chain. *Nature*, Vol. 431, p. 966-971; (b) Clayden, J. (1997). Non-biaryl atropisomers: New classes of chiral reagents, auxiliaries, and ligands? *Angew. Chem. Int. Ed. Engl.* Vol. 36, pp. 949-951

[22] (a) Pincock, R. E.; Perkins, R. R.; Ma, A. S. & Wilson, K. R. (1971). Probability distribution of enantiomorphous forms in spontaneous generation of optically active substances. *Science*, Vol. 174, pp. 1018–1020.; (b) Kondepudi, D. K.; Laudadio, J. & Asakura, K. (1999). Chiral symmetry breaking in stirred crystallization of 1,1'-binaphthyl melt. *J. Am. Chem. Soc.*, Vol. 121, pp. 1448-1451; (c) Sakamoto, M.; Utsumi, N.; Ando, M.; Saeki, M.; Mino, T.; Fujita, T.; Katoh, A.; Nishio, T. & Kashima, C. (2003). Breaking the symmetry of axially chiral *N*-aryl-2(*1H*)-pyrimidinones by spontaneous crystallization. *Angew. Chem. Int. Ed.* Vol. 42, pp. 4360-4363; (d) Sakamoto, M.; Yagishita, F.; Ando, M.; Sasahara, Y.; Kamataki, N.; Ohta, M.; Mino, T.; Kasashima, Y. & Fujita, T. (2010). Generation and amplification of optical activity of axially chiral *N*-(1-naphthyl)-2(*1H*)-pyrimidinethione by

crystallization. *Org. Biomol. Chem.*, Vol. 8, pp. 5418–5422; (e) Yoshioka, R. (2007). Racemization, optical resolution and crystallization-induced transformation of amino acids and pharmaceutical intermediates, *Top. Curr. Chem.* Vol. 269, pp. 83-132; (f) Brands, K. M. J. & Davies, A. J. (2006). Crystallization-induced diastereomer transformations. *Chem. Rev.* Vol. 106, pp. 2711-2733

[23] Sakamoto, M.; Yagishita, F.; Saito, A.; Kobaru, S.; Unosawa, A.; Mino, T. & Fujita, T. (2011). Asymmetric photocycloaddition of naphthamide with a diene using the provisional molecular chirality in a chiral crystal. *Photochem. and Photobiol. Sci.* Vol. 10, pp. 1387-1389

[24] Sakamoto, M.; Unosawa, A.; Kobaru, S.; Hasegawa, Y.; Mino, T.; Kasashima, Y. & Fujita, T. (2007). Diastereoselective photocycloaddition using memory effect of molecular chirality controlled by crystallization. *Chem.Commun.*, pp. 1632-1634

[25] Sakamoto, M.; Unosawa, A.; Kobaru, S.; Fujita, K.; Mino T. & Fujita, T. (2007). Asymmetric S_NAr reaction using the molecular chirality in crystal, *Chem. Commun*, pp. 3586-3588

[26] Sakamoto, M.; Fujita, K.; Yagishita, F.; Unosawa, A.; Mino T. & Fujita, T. (2010). Kinetic resolution of racemic amines using provisional molecular chirality generated by spontaneous crystallization. *Chem. Commun.*, Vol. 47, pp. 4267–4269

[27] Sakamoto, M.; Kato, M.; Aida, Y.; Fujita, K.; Mino, T. &Fujita, T. (2008). Photosensitized 2+2 cycloaddition reaction using homochirality generated by spontaneous crystallization. *J. Am. Chem. Soc., Vol.* 130, pp. 1132-1133

[28] Sakamoto, M.; Unosawa, A.; Kobaru, S.; Hasegawa, Y.; Mino, T.; Kasashima, Y.; & Fujita, T. (2007). Diastereoselective photocycloaddition using memory effect of molecular chirality controlled by crystallization. *Chem.Commun.*, pp. 1632-1634

[29] Sakamoto, M.; Sato, N.; Mino, T.; Kasashima, Y.; & Fujita, T. (2008). Crystallization-induced diastereomer transformation of 2-quinolone-4- carboxamide followed by stereoselective intermolecular photocycloaddition reaction. *Org. Biomol. Chem.*, Vol. 6, pp. 848-850

Section 2

Crystallization of Amorphous and Glassy Materials

4

Preparation of Na^+ Superionic Conductors by Crystallization of Glass

Toshinori Okura
Kogakuin University
Japan

1. Introduction

The use of glass-making processing is favorable for the fabrication of Na^+ conducting electrolyte tubes, which has been the key to the technological development of 1 MW Na/S secondary battery plants. However, the processing technique cannot be applied to well-known β- and β''-aluminas (e. g., $NaAl_{11}O_{17}$ and $NaAl_5O_8$) and Nasicons ($Na_{1+x}Zr_2P_{3-x}Si_xO_{12}$) because their high inclusion of Al_2O_3 or ZrO_2 brings about the inhomogeneous melting or crystallization from glasses. Alternatively, Nasicon-like glass-ceramics were synthesized using the composition with lower content of ZrO_2 ($m Na_2O \cdot x ZrO_2 \cdot y P_2O_5 \cdot (100-m-x-y) SiO_2$ [m=20, 30 mol%]), however, the conductivities (σ) attained were, at most, as high as $\sigma_{300}=2\times10^{-2}$ S/cm at 300°C with the activation energies (E_a) of ca. 30 kJ/mol. These low conductivities were attributed to the crystallization of the poorly conductive rhombohedral phase in these Nasicon-like materials. $Na_5YSi_4O_{12}$ (N5), which comprises 12-$(SiO_4)^{4-}$-tetrahedra membered skeleton structure (Fig. 1), is another Na^+-superionic conductor with $\sigma_{300}=1\times10^{-1}$ S/cm and E_a=25 kJ/mol. A pioneering work on N5-type glass-ceramics has been performed by Banks *et al.* on the family of N5-type materials by substituting Y with Er, Gd or Sm. However, their results were not completely satisfactory because of the relatively lower conductivities of $\sigma_{300}<2\times10^{-2}$ S/cm than the reported values of N5. This discrepancy may possibly have arisen from the occurrence of a less conductive metastable phase during crystallization, as is discussed below.

Contrary to the results of Banks *et al.*, the present authors have produced glass-ceramics with $\sigma_{300}=1\times10^{-1}$ S/cm and E_a=20 kJ/mol, which were based on the phosphorus-containing N5-type materials discovered in the Na_2O-Y_2O_3-P_2O_5-SiO_2 system. These N5-type materials have been obtained, as well as $Na_3YSi_3O_9$ (N3)-type materials, with the composition formula originally derived for N3-type solid solutions and expressed as follows,

$$Na_{3+3x-y}Y_{1-x}P_ySi_{3-y}O_9 \ (x<0.6,\ y<0.5) \tag{1}$$

With the aim of searching for more conductive glass-ceramic N5-type materials, the verification of the validity of the generalized composition formula

$$Na_{3+3x-y}R_{1-x}P_ySi_{3-y}O_9 \tag{2}$$

for the synthesis of other kinds of rare earth N5-type glass-ceramics was studied first. Formula 2 is rewritten with formula 3 according to the formula N5.

$$Na_{4(3+3x-y)/3}Y_{4(1-x)/3}P_{4y/3}Si_{4(3-y)/3}O_{12} \quad (3)$$

In relation to previous works, formula 2 was employed in this work, and formula 3 is referred to in the results. The trivalent ions employed here for R^{3+} were Sc^{3+}, In^{3+}, Er^{3+}, Gd^{3+}, Sm^{3+}, Eu^{3+}, Nd^{3+} and La^{3+} as well as Y^{3+}. These results are to be interpreted in terms of the effect of the rare earth ions on the crystallization of N5-type phase in glasses.

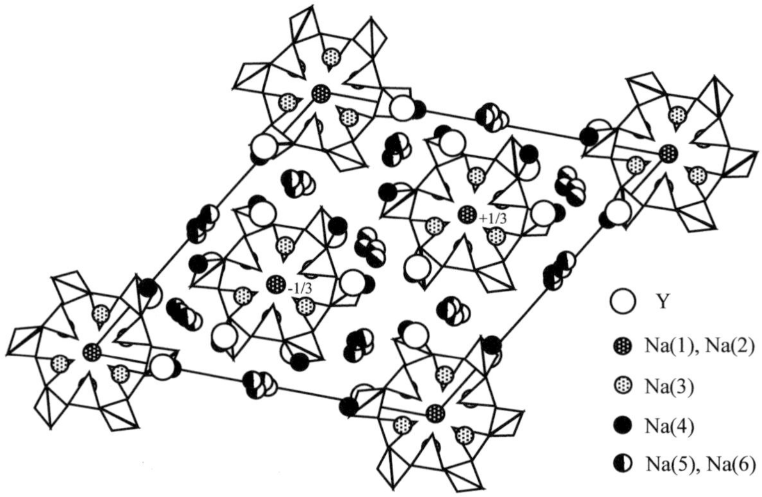

Fig. 1. Crystal Structure of $Na_5YSi_4O_{12}$. Projection of the $Na_5YSi_4O_{12}$ Structure on (100).

In the course of the fundamental studies on glass-ceramic $Na_{3+3x-y}R_{1-x}P_ySi_{3-y}O_9$, we have interestingly found the crystallization of those N3- and $Na_9YSi_6O_{18}$ (N9)-type phases as the precursors in the glasses. These are the analogues to the silicates N3 and N9 and therefore are the same members of the family of $Na_{24-3x}Y_xSi_{12}O_{36}$ as N5. Although we had also successfully synthesized those materials by the solid-state reactions of powders with the above composition of various sets of the parameters x and y, the metastability of those precursor phases had not been noticed in the synthesis. It has been observed that such precursor phases were transformed to the Na^+-superionic conducting phase on specimens with appropriate sets of x and y. The present paper will deal with the thermodynamic and kinetic study on the phase transformation of metastable phases to the stable phase with Na^+-superionic conductivity. The superiority of our present materials to the other silicate N5 will also be detailed based on the kinetic results.

The microstructure of a glass-ceramic, including neck growth among grains as well as grain size, is generally affected by the crystallization process. As the above mentioned devices utilize dc conduction properties of Na^+-superionic conductors, another aim was to study the microstructural effects on the conduction properties of a whole glass-ceramic. Special attention was paid to the analysis of grain boundary properties using the Na_2O-Y_2O_3-P_2O_5-SiO_2 system. For the analysis of grain boundary properties, as will be discussed below, composition dependences of the conductivity of sodium silico-phosphate glasses containing Y_2O_3 were also studied in the Na_2O-Y_2O_3-P_2O_5-SiO_2 system. For convenience, the present materials are abbreviated as NaRPSi taken from the initials of the Na_2O-R_2O_3-P_2O_5-SiO_2 system.

2. Materials

2.1 Preparation of glasses and glass-ceramics

Precursor glasses were prepared from reagent-grade oxides of anhydrous Na_2CO_3, R_2O_3 (R=Y, Sc, In, Er, Gd, Sm, Eu, Nd, La), $NH_4H_2PO_4$ and SiO_2; the mechanically mixed powders according to formula 2 or appropriate compositions shown below were melted at 1350°C for 1 h after calcinations at 900°C for 1 h. The melts were quickly poured into a cylindrical graphite, then annealed at 500°C for 3 h, giving NaRPSi glasses. The composition parameters studied were in the range of $0.2<x<0.6$ and $0<y<0.5$ of formula 2. As shown below, grain boundary conduction properties are discussed in relation to the properties of glasses. For the evaluation of the composition dependence of conductivity in Na^+ conducting glasses, various sodium yttrium silico-phosphate glass specimens with different atomic ratios of [Na]/[P+Si] and [Na]/[Y] were also prepared.

Crystallization was carried out according to the previous report; bulk glasses were heated with an increasing rate of 75°C/h to a temperature above ca. 50°C of the glass transition point, which had been determined in advance by differential thermal analysis (DTA). This pretreatment was done in order to obtain homogeneous nucleation. After the annealing for 1 h, specimens were heated at temperatures of 800 to 1100°C, depending on the composition, for 0.5 to 72 h, thereafter slowly cooled in a furnace with a decreasing rate of 150°C/h to room temperature. These quenched glasses or glass-ceramic specimens were polished down with 0.5 µm diamond paste, thereafter subjected to the conductivity measurements.

2.2 Measurements and characterization

Ionic conductivities were measured by the complex impedance method on cylindrical glasses or glass-ceramics of typically 15 mm in diameter and 2 mm in thickness. Electrodes were prepared by sputtering of gold on polished surfaces. The applied ac field ranged from 5 to 10 MHz in frequency. The temperature dependence of the conductivity was measured in a similar way at several temperatures ranging from room temperature to 350°C. The complex impedance or admittance loci of glass and glass-ceramics were analyzed by an equivalent circuit (Fig. 2), which was experimentally found to comprise one and two semicircles in NaRPSi glasses and glass-ceramics, respectively. The two intercepting points on the real axis are interpreted as the resistance of crystallized grains ($R_{G(c)}$) and the total resistance of grains and remaining glassy grain boundaries ($R_{GB(g)}$). Assume the complex admittance diagram shown in Fig. 3, where the parameters L_1 and L_2 are set here as the radii of the two arcs 1 and 2. Those parameters are related to one another as the following:

$$L_1 \propto 1/(R_{G(c)}+R_{GB(g)}) \tag{4}$$

and

$$L_2 \propto (1/R_{G(c)})-1/(R_{G(c)}+R_{GB(g)}) \tag{5}$$

Then,

$$L_2/L_1=R_{GB(g)}/R_{G(c)} \tag{6}$$

Therefore, in an ideal grass-ceramic where residual glass would have negligible influence on the total, arc 2 would be much smaller than arc 1, since $L_2/L_1 \rightarrow 0$.

Crystalline phases of glass-ceramic specimens were identified by X-ray diffraction (XRD) method. The lattice parameters of the N5-type hexagonal unit cell were calculated by a least-squares method using the XRD peaks of (054), (044), (134), (440) and (024). Glass-ceramics of Y^{3+}-contained NaRPSi were subjected to scanning (SEM) and transmission electron microscope (TEM) for microstructural analysis. Electron diffraction and compositional analyses were also performed to characterize the structure of the grain boundary.

For the description of a specific NaRPSi, R of the term will be replaced, respectively, with Y, Sc, In, Er, Gd, Sm, Eu, Nd and La as NaYPSi, NaScPSi, NaInPSi, NaErPSi, NaGdPSi, NaSmPSi, NaEuPSi, NaNdPSi and NaLaPSi for Y_2O_3, Sc_2O_3, In_2O_3, Er_2O_3, Gd_2O_3, Sm_2O_3., Eu_2O_3, Nd_2O_3 and La_2O_3.

3. Thermodynamic and kinetic study on the phase transformation

3.1 Composition dependence of precursor and high temperature stable phases

Fig. 4 shows the composition dependence of both the precursor phases and the high temperature stable phases of glass-ceramic NaYPSi on the maps of phosphorus-yttrium (P-Y, Fig. 4(a)), yttrium-sodium (Y-Na, Fig. 4(b)) and phosphorus-sodium (P-Na, Fig. 4(c)), where the variables on the abscissas and ordinals are expressed with the composition parameters $1\text{-}x$, y and $3+3x\text{-}y$ for yttrium, phosphorus and sodium, respectively. As reported before, N3- and N9-type NaYPSi glass-ceramics can be crystallized as the high-temperature stable phases at the regions of higher [Y] ($1\text{-}x>$ca. 0.8) and rather lower [Y] ($1\text{-}x<$ca. 0.55), respectively, in the [Y]-[P] relation.

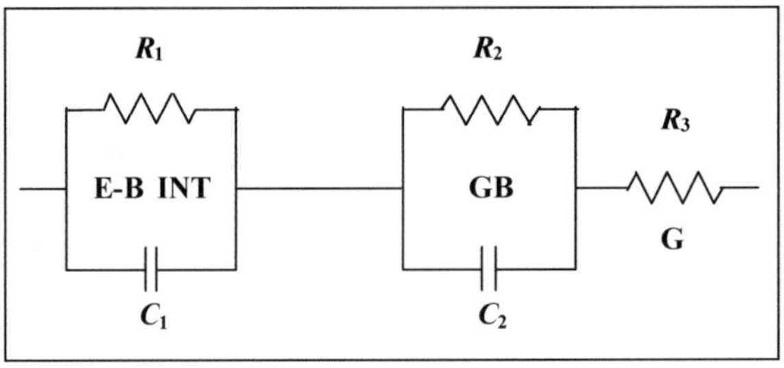

Fig. 2. Equivalent circuit employed for the admittance analysis.

E-B INT, GB, and G represent the electrode-bulk interface, grain-boundaries and grains, respectively, and (R_1, C_1), (R_2, C_2), and R_3 are their resistances and capacitances.

Concerning the precursor phases, only either N3- or N9-type NaYPSi was found in any composition, while N5-type NaYPSi was difficult to crystallize from glasses at low temperatures. It is also seen in the [P]-[Y] map (Fig. 4(a)) that, under a given phosphorus content ([P]<0.6) a composition with higher content of yttrium gives N3-type NaYPSi (○) as the precursor phase, while lower [Y] content results in N9-type phase (open square). The values of [Y] dividing the regions allowed for N3- and N9-type NaYPSi glass-ceramics decreased with increasing [P], and the boundary seems to locate slightly apart from the deduced line of [Y]=0.75-0.5[P] shown with the solid line. Around the boundary region N5-type NaYPSi can be obtained as the stable phase at high temperatures (solid marks of circle or square). In the [Y]-[Na] or [P]-[Na] relations (Figs. 4(b) and 4(c)), the region where

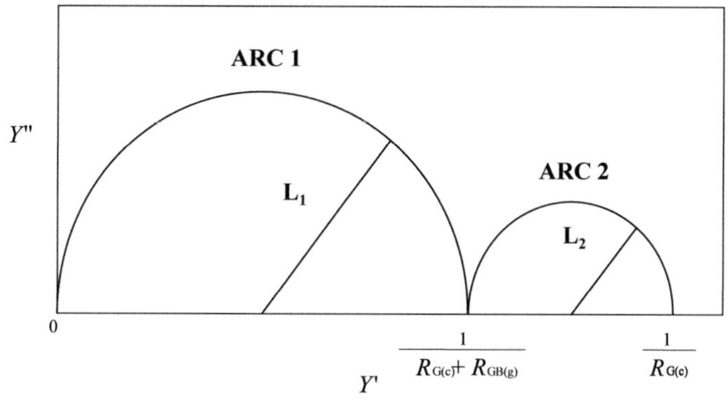

Fig. 3. An idealized diagram of complex admittance for glass-ceramics, in which arc 1 (ARC 1) and arc 2 (ARC 2) are related to the crystallized grains (G(c)) and remaining glasses (GB(g)). L_1, L_2, $R_{G(c)}$, and $R_{GB(g)}$ are, respectively, the radii of arcs 1 and 2, the resistances of G(c) and GB(g).

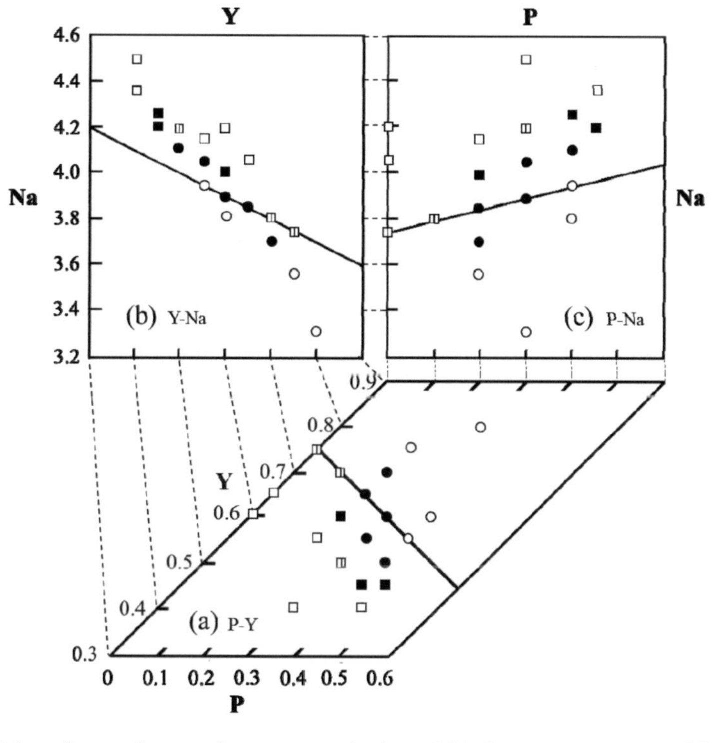

Fig. 4. Composition dependence of precursor (*pp*) and high temperature-stable phases (*sp*) of glass-ceramic NaRPSi on P-Y (a), Y-Na (b) and P-Na (c) maps, where precursor phases N3 and N9 are shown with *circles* and *squares*, respectively. High temperature-stable phases are shown in such a way that solid marks means that N5-NaRSi is the stable, while *open marks* indicate that the precursor phases are also stable even at high temperatures. Mixed phases are also shown: *open circle* pp = sp = N3; *filled circle* pp = N3, sp = N5; *open square* pp = sp = N9; *filled circle* pp = N9, sp = N5; *open split square* pp = N9, sp = N9 + N5

N5-type NaYPSi can be found as the high-temperature stable phase is found under ca. 3.6<[Na]<4.3. The effect of sodium content seems insignificant, because the value of [Na] is subordinately determined as [Na]=6-3[Y]-[P] (=3+3x-y) depending on the contents of both yttrium and phosphorus.

The above results may suggest that the [P]-[Y] relation dominates the region which is allowed for each NaYPSi at high temperatures. Considering this inference, we calculated the products of [P]×[Y] for all of the specimens. The values of [P]×[Y] were as follows (shown in Fig. 5); 0.16-0.25 for single phase N3-type NaYPSi, 0.14 for mixed phases of N3- and N5-type NaYPSi, 0.12-0.20 for single phase N5-type NaYPSi, 0-0.14 for the mixed phases of N5- and N9-type NaYPSi, and 0-0.17 for single phase N9-type NaYPSi, respectively. It was therefore deduced (Fig. 5) that the free energy of formation (ΔG_f) of N9-type NaYPSi would be the lowest in a lower region of [P]×[Y], N5-type NaYPSi may have the lowest ΔG_f in a medium [P]×[Y] region, and higher [P]×[Y] would lower ΔG_f of N3-type NaYPSi.

For a specimen in which N5-type NaYPSi is the stable phase at high temperatures, the aspect such as Fig. 6a would be illustrated in that ΔG of N3- or N9-type NaYPSi would be much smaller than that of N5-type NaYPSi near the crystallization temperature (T_c), and the value of N5-type NaYPSi would be lowered much less than of the two. Fig. 6b indicates the aspect that ΔG or N3- or N9-type NaYPSi stable.

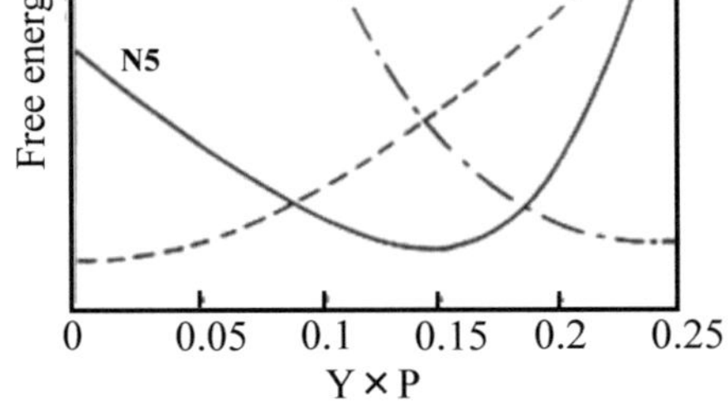

Fig. 5. Schematic figure of composition ([Y]×[P]) dependence of free energy of N5-, N3- and N9-type NaYPSi.

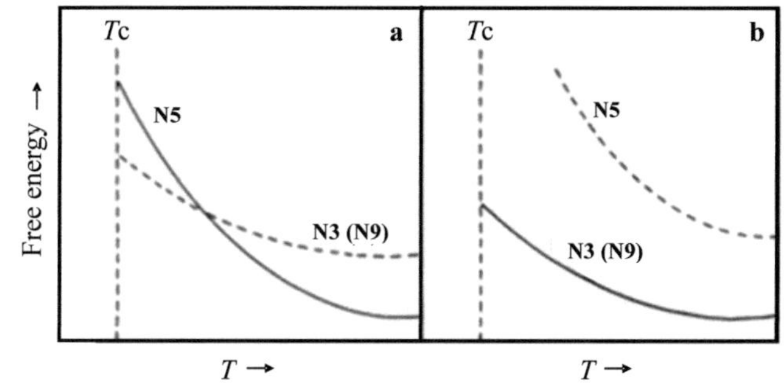

Fig. 6. Schematic figures of temperature dependence of free energy change of N5- and N3- or N9- type NaYPSi in the cases assuming N5- (a) and N3- (b) or N9-type (b) NaYPSi as the high temperature-stable phase, where *Tc* is the crystallization temperature.

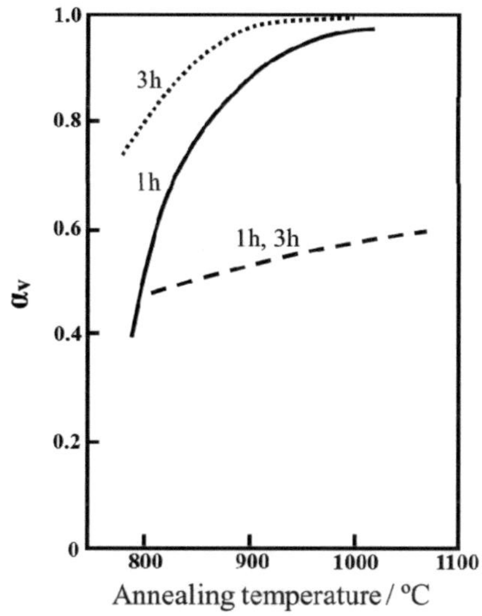

Fig. 7. Comparison of phase transformation rate (α_v) between specimens $Na_{3.9}Y_{0.6}P_{0.3}Si_{2.7}O_9$ (1h-annealing: (○); 3h-annealing: (●)) and $Na_{3.75}Y_{0.65}P_{0.3}Si_{2.7}O_9$ (1h-annealing: (□), 3h-annealing: (■)).

3.2 Kinetic effects of composition on the phase transformation

The kinetic effects of composition on the phase transformation are shown in Fig. 7, which compares the phase transformation rates of specimens $Na_{3.9}Y_{0.6}P_{0.3}Si_{2.7}O_9$ and $Na_{3.75}Y_{0.65}P_{0.3}Si_{2.7}O_9$. The transformation rate (α_v) of a precursor phase to the stable N5 phase was determined as the weight ratio of N5-type NaYPSi in a glass-ceramic specimen. The value of α_v was experimentally obtained from the relationship of weight ratio to XRD intensity ratio, which relationship had been made previously by XRD intensity measurement on specimens with given weight ratio of N5-type NaYPSi to metastable phases. It is seen that the composition $Na_{3.9}Y_{0.6}P_{0.3}Si_{2.7}O_9$ is superior to the other, for the N5 single phase NaYPSi was difficult to obtain in the latter specimen. In specimen

$Na_{3.9}Y_{0.6}P_{0.3}Si_{2.7}O_9$ a glass-ceramic of N5 single phase NaYPSi was easily obtained at a temperature higher than 900°C for only three hours. The composition $Na_{3.75}Y_{0.75}Si_3O_9$ (or $Na_5YSi_4O_{12}$) was inferior in the same meaning.

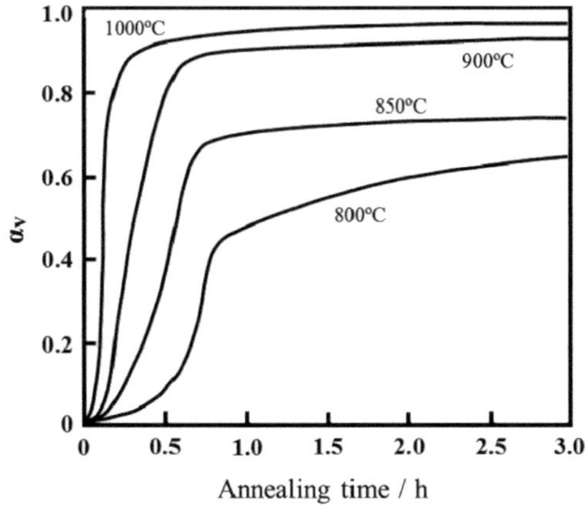

Fig. 8. Phase transformation rate (α_v) of N3- to N5-type NaYPSi on the specimen $Na_{3.9}Y_{0.6}P_{0.3}Si_{2.7}O_9$.

Fig. 8 shows the kinetic characteristics of phase transformation of the metastable phase of N3- to N5-type NaYPSi of specimen $Na_{3.9}Y_{0.6}P_{0.3}Si_{2.7}O_9$ at various temperatures. The transition rates, α_v, of the silicophosphate NaYPSi were much higher than those of the $Na_{3.75}Y_{0.75}Si_3O_9$ silicate material.

The results shown were analyzed with the Avrami empirical equation, $\alpha_v=1-\exp(-kt^n)$, where k is the rate constant, and n is a constant. The data on α_v obtained at the initial and intermediate stages gave a linear relationship between $\ln(\ln(1-\alpha_v)^{-1})$ and $\ln(t)$ with a correlation coefficient of more than 0.99. The Avrami parameter and rate constants obtained are summarized in Table 1. Based on the Arrhenius relationship (Fig. 9), $k=A\exp(-E_v/RT)$ with E_v as the activation energy and constants A and R, on those k values which increased with increasing temperature, we obtained an activation energy of 1.2×10^3 kJ/mol, suggesting that the phase transformation can be rather difficult to take place. An addition of phosphorus and the excess sodium seem effective to the promotion of the phase transformation.

4. Microstructural effects on conduction properties

4.1 Crystallization and phase diagram

As expected from the previously reported results on NaYPSi, the crystallization of the superionic conducting N5-type phase took place, depending both on the contents of [R] and [P], at temperatures of 800 to 1000°C in most NaRPSi glasses of Er to Sm except for scandium and lanthanum NaRPSi glasses. The N5 single phase region was wider for NaRPSi of smaller R, but was limited at the [P]≈0 region. The effect of phosphorus substitution for Si is important in the crystallization of N5-type phase. The composition 7

$$Na_{3.9}R_{0.6}P_{0.3}Si_{2.7}O_9 \quad (7)$$

was experimentally shown as the most appropriate composition for the crystallization of N5-type phase.

Anneling temp. (K)	Avrami modulus *n*	ln *k*
1073	2.61	-20.7
1123	1.94	-14.6
1173	1.39	-9.54
1223	0.75	-4.41

Table 1. Kinetic parameters of phase-transformation of N3- to N5-type NaYPSi of $Na_{3.9}R_{0.6}Si_{2.7}O_9$.

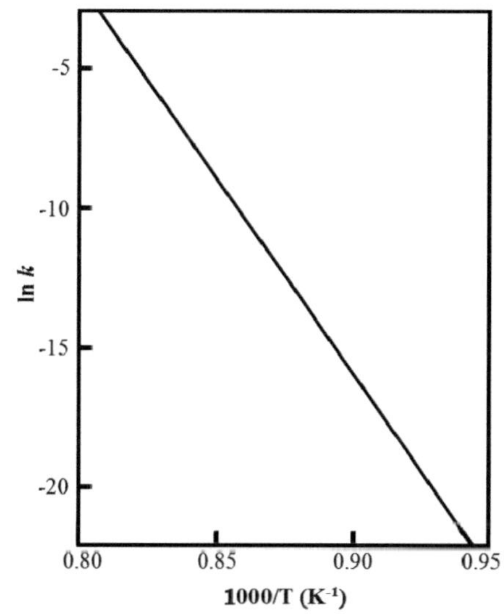

Fig. 9. Arrhenius-type plot of ln *k* with 1000/T of specimen $Na_{3.9}Y_{0.6}Si_{2.7}O_9$.

The relationship between the ionic radius of R^{3+} (r_R) and the hexagonal lattice parameters of N5-type single phase is consistent with the previous report on $Na_5RSi_4O_{12}$ (R=Sc-Sm) in the tendency that both lattice parameters increased with increasing r_R. The elongation of these lattice axes is attributed to the octahedral coordination of R^{3+} with the O^{2-} of SiO_4- or PO_4-tetrahedra of the 12-membered rings. The local structure around R^{3+} ions is to be further discussed below in relation to conduction properties. On the formation of N5-type single phase, the incorporation of excess sodium ions [4(3+3*x*-*y*)/3-5=(12*x*-4*y*-3)/3 in composition 3] and substitution of rare earth ions [1-4(1-x)/3=(4*x*-1)/3] must be accounted for in view of N5-type crystal structure.

Banks *et al.* have reported the values of σ_{300} as 5×10^{-3} to 1×10^{-2} S/cm for glass-ceramic $Na_5RSi_4O_{12}$ (R=Er, Y, Gd, Sm), which are as low as those of the mixed phase NaRPSi specimens. The single phase N5-type glass-ceramic was not obtained in the present work. Based on the above crystallization analysis, their glass-ceramic specimens are reasonably considered to suffer from phase inhomogeneity brought about by insufficient annealing. The formation of N5-type structure from the precursor glasses is a matter of crystallization kinetics, since single-phase N5 has been synthesized in single crystal or polycrystalline form based on the composition of N5. It is noted here that the precursor phases identified were N3- or N9-type. Both N3 and N9 are considered to form iso-structural with $Ca_3Al_2O_6$ to be comprised of the skeleton structure of 6-membered SiO_4-tetrahedra rings. It is generally known that phosphorus pentoxide acts as nucleating agent in the formation of glass-ceramics. It is therefore presumed at present that the substitution of an asymmetric PO_4-tetrahedron has the weakening effect on the bonding of the skeleton structure of 6-membered SiO_4-tetrahedra rings, resulting in the tendency to form the stable 12-membered structure.

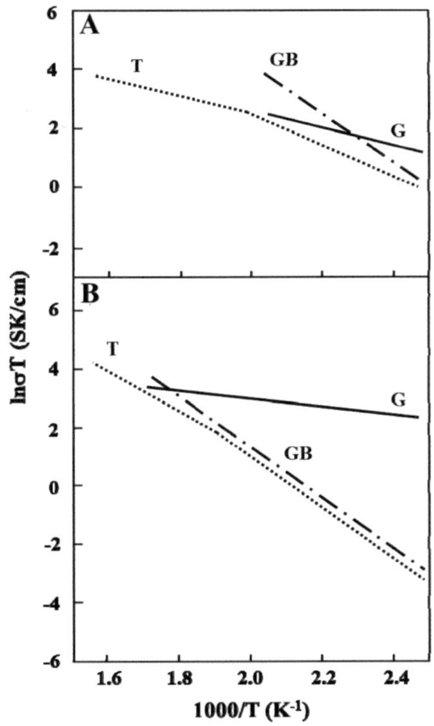

Fig. 10. The Arrhenius plots of the conductivities of grains (G), grain boundaries (GB) and the total bulk (T) of the glass-ceramic $Na_{3.9}Y_{0.6}P_{0.3}Si_{2.7}O_9$ (A) and $Na_{3.9}P_{0.3}Sm_{0.6}Si_{2.7}O_9$ (B).

R^{3+} (ions)	E_a kJ/mol	Conductivity (σ_{300}) S/cm	Crystalline phase
Sc	35.3	3.2×10^{-3}	N5-type + unknown
In	39.8	3.1×10^{-3}	N5-type + unknown
Er	26.9	3.6×10^{-2}	N5-type
Y	26.6	6.6×10^{-2}	N5-type
Gd	23.0	1.3×10^{-1}	N5-type
Eu	24.4	5.2×10^{-2}	N5-type
Sm	20.9	6.3×10^{-2}	N5-type
Nd	55.1	2.2×10^{-5}	Unknown
La	57.8	1.6×10^{-4}	Unknown

Table 2. Conduction properties of various NaRPSi glass-ceramics with composition $Na_{3.9}R_{0.6}Si_{2.7}O_9$.

4.2 Conduction properties of crystalline grains

The complex impedances and admittances of the measured NaRPSi glass-ceramics consisted of two semicircles below 300°C. The two intercepting points on the real axis are interpreted as the resistance of crystallized grains (R_G) and the total resistance of grains and remaining glassy grain boundaries (R_{GB}). Shown in Fig. 10 are examples of the temperature dependence Arrhenius plots made on the basis of the calculated conductivity values of grains and grain boundaries of the glass-ceramic NaYPSi ($Na_{3.9}Y_{0.6}P_{0.3}Si_{2.7}O_9$) and NaSmPSi ($Na_{3.9}Sm_{0.6}P_{0.3}Si_{2.7}O_9$), in which the geometrical ratios of thickness to surface area for grains were also used for convenience for those of grain boundaries because of their undefinable shapes. Table 2 summarizes the measured conductivities (σ_{300}) and the calculated activation energies (E_a) assigned for grains of the glass-ceramics with composition 7 of Sc to La, regardless of whether their crystalline phases are N5-type or not. The conductivities, σ_{300}, of single-phase NaRPSi specimens of Er to Sc range from 4×10^{-2} to 1×10^{-1} S/cm; in accordance the E_a falls in the range of 23 to 27 kJ/mol. In contrast, the mixed phase NaRPSi of Sc and In showed much smaller σ_{300} of 3×10^{-3} with an E_a of 35 to 40 kJ/mol, while non-NaRPSi glass-ceramics with unknown or mixed phases showed much lower conductivities of 1×10^{-5} to 1×10^{-4} S/cm with an E_a of 55 to 58 kJ/mol.

The tendency of the conduction properties in single-phase NaRPSi specimens is consistent with the reported result measured on the corresponding polycrystalline $Na_5RSi_4O_{12}$; σ increased with increasing r_R. The previous works have proposed a mechanism that rare earth ions, octahedrally coordinated with the non-bridging oxide ions of the 12-membered rings of silica tetrahedra, work to expand the conduction paths for Na^+ ions along the c-axis, which can explain the observed dependence of E_a on r_R in this work.

4.3 Structure and conduction properties of grain boundaries

As R_{GB} decreases rapidly with increasing temperature because of high $(E_a)_{GB}$ to a comparable value with R_G at 300°C (Fig. 10), the total conductivities (R_G+R_{GB}) are dominated by grain boundary conductivity. The grain size-dependence of σ_{300} is therefore explained by the decrease in the number of poorly conductive grain boundaries with increasing grain size.

The conduction properties of grain boundaries were strongly dependent on the annealing conditions, although those of the grains were little changed by annealing temperature and time. Glass-ceramics are generally composites consisting of crystallized grains and small amounts of residual glass (<1%). To compare the properties of grain boundaries with those of glasses, the conduction properties of sodium yttrium silicophosphate glasses with various compositions were measured. Unlike glass-ceramics the impedance loci of glasses were comprised of one arc, which indicates that there is no polarization arising from microstructural inhomogeneity. Based on the intercepting points on the horizontal axis, the composition dependence of conduction properties of σ_{300} and E_a were evaluated. The value of σ_{300} ranged from 1×10^{-4} to 5×10^{-3} S/cm and E_a increased from 53 to 67 kJ/mol with [Na] or [Na]/[Y]. These results are also in good agreement with those reported for the glasses in the Na_2O-Y_2O_3-SiO_2 system. The values of $(E_a)_{GB}$ of the specimens annealed below 950°C for shorter times correspond to those in the range of glasses, strongly suggesting that their grain boundaries are a glassy matrix. The above mentioned dependence of $(E_a)_{GB}$ on [Na_2O] is explained by the well-known tendency that the conduction properties of glasses are improved by increasing [Na_2O], which provides the increase of carrier Na^+ ions. The ratio of [Na]/[Y] is also an important parameter for the conduction properties, showing an effect on the conduction properties similar to [Na_2O].

In order to identify the structure of the grain boundaries of the specimen ($Na_{3.9}Y_{0.6}P_{0.3}Si_{2.7}O_9$) annealed at 800°C for 0.5 h, TEM analysis was performed both on grains and grain boundaries. The results show clear electron diffraction on grains, while not on grain boundaries. This fact confirms that the grain boundaries are amorphous. Compositional analyses were also performed, however, [Na] was difficult to determine because of the evaporation by electron ablation. It was also observed that the glassy phase was condensed at triple points enclosed by grains, and that neck growth among the grains was well developed. Thus, it is reasonable to consider that the grain boundaries annealed at lower temperatures are amorphous, while those annealed at higher temperatures for longer periods of time are poorly conductive crystalline compounds in the specimens.

5. Recent research on conductive improvement and structural control of glass-ceramic Na^+ superionic conductors

5.1 Preparation of crack free $Na_5YSi_4O_{12}$-type glass-ceramics containing the largest Sm^{3+} ions: Crystallization condition and ionic conductivity

Glass-ceramics of the phosphorus containing N5-type Na^+ superionic conductors have been developed by crystallization of glasses with the composition formula 2. The R elements have a significant effect on the crystallization of glasses, as well as on the conduction properties. To date, polycrystalline N5-type NaRPSi has been obtained with Sc, Y, Gd or Sm as the R element. The ionic radius of R (sixfold oxygen coordinated R) has been expected to have a significant effect on the crystallization of the phase. The reported results on the silicate ceramics show that the conductivity of the N5-type NaRPSi increases with increasing ionic radius of R, giving the order NaSmPSi>NaGdPSi>NaYPSi>NaScPSi. It can be expected that NaSmPSi is the most conductive. However, this order was not always true in glass-ceramics. Although most of the NaRPSi compounds were obtained as crack free bulky glass-ceramics (15 mm in diameter, 5 mm in thickness), NaSmPSi was difficult to prevent from cracking during crystallization. It was found that crack free NaGdPSi with larger Gd^{3+} ions was the most conductive; however, NaSmPSi with the largest Sm^{3+} ions was less conductive than NaYPSi with medium Y^{3+} ions. In the present study, the N5-type NaSmPSi ionic conductors were prepared by crystallization of glasses. The optimum conditions for crystallization were discussed with reference to the conduction properties and the preparation of crack free N5-type glass-ceramic NaSmPSi.

	a-b	b-c	c-d
(A)	1	7	17
(B)	3	8	10
(C)	6	8	5

a-b : Nucleation c-d : Crystal growth

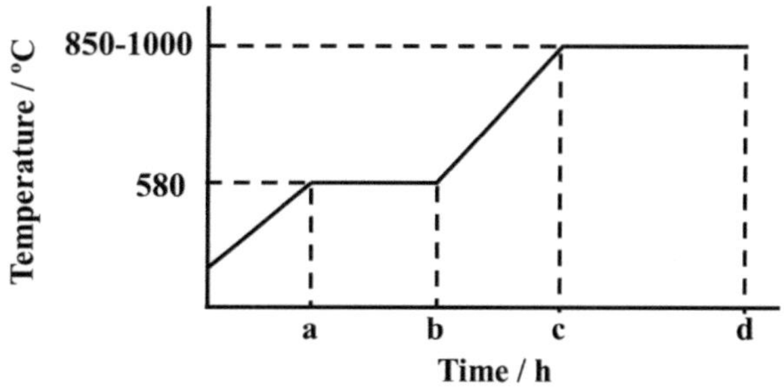

Fig. 11. Program of temperature and time for the production of NaSmPSi glass-ceramics.

Samples were prepared according to the chemical formula mentioned above of $Na_{3+3x-y}Sm_{1-x}P_ySi_{3-y}O_9$. The temperatures employed for nucleation and crystallization of glass specimens were also determined by the results of DTA analysis. Fig. 11 shows the program of temperature and time for the production of glass-ceramic NaSmPSi employed in the present work. The N5-type NaSmPSi ionic conductors were successfully produced by crystallization of glasses. Although the glass samples heated by the program pattern (A) shown in Fig. 11 broke during crystallization and the glass-ceramic NaSmPSi obtained by the pattern (B) was difficult to prevent from cracking during crystallization, most of the NaSmPSi compounds by the pattern (C) were obtained as crack free bulky glass-ceramics (the glass samples broke during crystallization when heating time for crystallization was over 5 h). Fig. 12 shows the phase-composition diagram of samples crystallized at 900°C by the pattern (C). The crystallization of the N5 single-phase glass-ceramic NaSmPSi was dependent strongly on the concentrations of both [R] and [P] (or x and y in the composition parameters) and the temperature for crystallization of glass specimens. Fig. 13 shows SEM photograph of microstructure of specimen with the $Na_{3.9}Sm_{0.6}P_{0.3}Si_{2.7}O_9$ composition heated at 900°C by the pattern (C). The grain size of the specimen was about 3-5 μm. The state of grain growth is promoted with increase of heating temperature and heating time for crystallization. Although grain growth may cause high conductivity, it was difficult to prevent the sample heated for a long time from cracking during crystallization. Studies are underway to produce a crack free sample. Conduction properties were measured by the ac two-probe method on cylindrical glass-ceramics of typically 15 mm in diameter and 2 mm in thickness with an LF impedance analyzer. Electrodes were prepared by sputtering of gold on polished surfaces. The applied ac field ranged from 5 to 10 MHz in frequency. The temperature dependence of the conductivity was measured in a similar way at several temperatures ranging from room temperature to 350°C. Table 3 summarizes the conduction properties of the N5-type glass-ceramic NaSmPSi specimens. It was found that NaSmPSi containing the largest Sm^{3+} ions was less conductive than NaYPSi with medium Y^{3+} ions as the grain sizes of the presented specimens were very small.

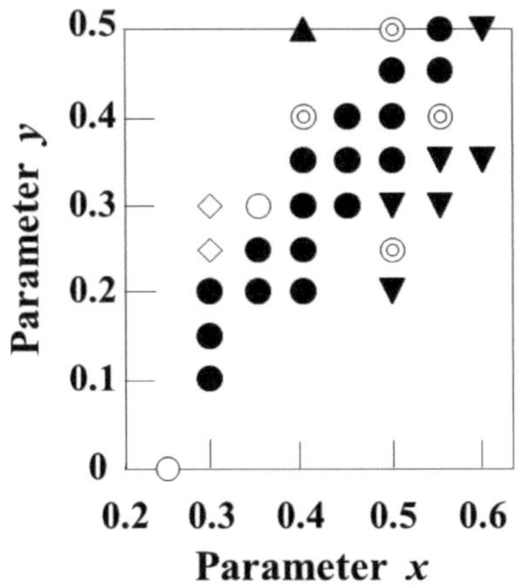

Fig. 12. The diagram of phase-composition of NaSmPSi glass-ceramics crystallized at 900°C.
● $Na_5YSi_4O_{12}$ (N5) type, ▲ $Na_3YSi_3O_9$ (N3) type, ▼ $Na_9RSi_6O_{18}$ (N9) type
○ N5+N3, ◎ N5+N9, ◇ N3+N9

Fig. 13. SEM photograph of the specimen with $Na_{3.9}Sm_{0.6}P_{0.3}Si_{2.7}O_9$ composition heated at 900°C by the pattern (C).

Mix proportion		σ_{300}	E_a / kJ•mol^{-1}		
x	y	/ 10^{-1} S•cm^{-1}	T	G	G. B.
0.40	0.30	0.238	27.6	17.9	51.4
0.45	0.40	0.408	30.4	18.8	95.7
0.50	0.35	0.352	19.5	15.8	
0.50	0.40	0.478	29.3	16.6	-------

σ_{300} : conductivity at 300 °C
E_a : activation energy (T : total G : grain G. B. : grain boundary)

Table 3. Conduction properties of N5-type NaSmPSi glass-ceramics.
Heat-treatment: 900°C, 5 h

5.2 Composition control of silico-phosphate glass-ceramics

5.2.1 Ionic conductivities of Nasicon-type glass-ceramic superionic conductors in the system Na_2O-Y_2O_3-XO_2-SiO_2 (X=Ti, Ge, Te)

Our phosphorus containing compositions have been confirmed superior to the mother composition of $Na_5RSi_4O_{12}$, especially in the production of the single-phase glass-ceramics. Considering the inference, our main work has recently been focused on the synthesis of various glass-ceramics with single-phase $Na_5RSi_4O_{12}$. In the present study, the glass-ceramics of the titanium-, germanium- or tellurium-containing $Na_5RSi_4O_{12}$-type (R=Y) Na^+-superionic conductors (N5YXS) from the glasses with the composition $Na_{3+3x}Y_{1-x}X_ySi_{3-y}O_9$ (X=Ti; NYTiS, Ge; NYGeS, X=Te; NYTeS) were prepared, and the effects of X elements on the separation of the phase and the microstructural effects on the conduction properties of glass-ceramics were discussed.

The glass-ceramics have been obtained under the appropriate sets of the parameters x and y of the composition formula $Na_{3+3x}Y_{1-x}X_ySi_{3-y}O_9$ ranging in x=0.1~0.55 and y=0.1~0.45. The precursor glasses were made by melting stoichiometric mixtures of reagent-grade powders of anhydrous Na_2CO_3, Y_2O_3, (TiO_2, GeO_2 or TeO_2) and SiO_2 at 1300~1400°C for 1 h, followed by annealing for several hours at an optimum temperature. The N5YXS ionic conductors were successfully produced by crystallization of glasses. Figs. 14, 15 and 16 show the diagrams of phase-composition-crystallization temperature of NYTiS, NYGeS and NYTeS glass-ceramics, respectively. The phase formed was dependent on composition and crystallization temperature. N5YTiS, N5YGeS and N5YTeS are obtained as a stable phase at high-temperatures. The crystallization of N5 single phase is strongly dependent both on the contents of yttrium and (titanium, germanium or tellurium) ions (or the values x and y correspond to the composition parameters in $Na_{3+3x}Y_{1-x}X_ySi_{3-y}O_9$). N3 and N9 phases can be crystallized as the high-temperature stable phases at the regions of higher [Y] and rather lower [Y], respectively. The combination of x and y was most varied in N5YGeS and more limited in the order of N5YTeS>N5YTiS. Table 4 summarizes the conduction properties of the N5 glass-ceramics with $Na_{3.6}Y_{0.8}Ti_{0.2}Si_{2.8}O_9$, $Na_{4.2}Y_{0.6}Ge_{0.3}Si_{2.7}O_9$ and $Na_{4.2}Y_{0.6}Te_{0.3}Si_{2.7}O_9$ compositions. Their conductivities and activation energies are of the order of 10^{-2} S/cm at 300°C and of 15 to 24 kJ/mol, respectively. The conductivity decreases giving the order N5YGeS>N5YTeS>N5YTiS. It is considered that this order corresponds to the N5 single phase region.

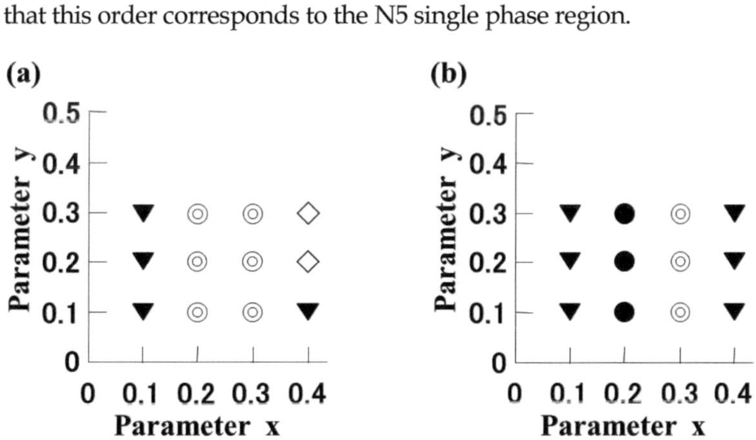

Fig. 14. The diagrams of phase-composition of NYTiS glass-ceramics heated at 900°C (a) and 1000°C (b) for 5 h. ● $Na_5RSi_4O_{12}$ (N5) ▼ $Na_9RSi_6O_{18}$ (N9) ◎ N5+N9 ◇ N3+N9

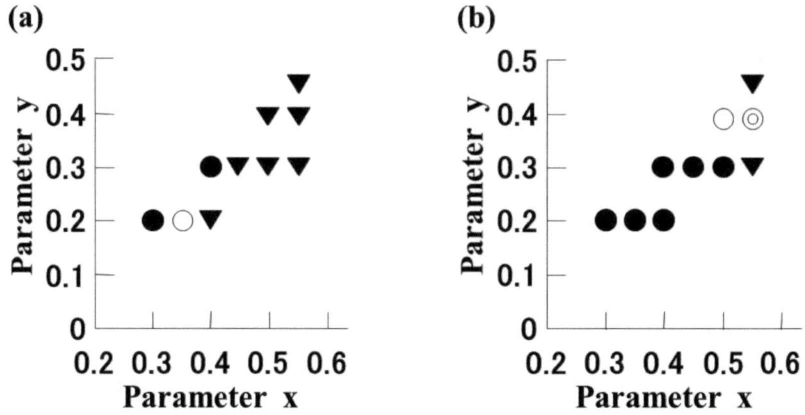

Fig. 15. The diagrams of phase-composition of NYGeS glass-ceramics heated at 900°C (a) and 1000°C (b) for 5 h. ● $Na_5RSi_4O_{12}$ (N5) ▼ $Na_9RSi_6O_{18}$ (N9) ○ N5+N3 ◎ N5+N9

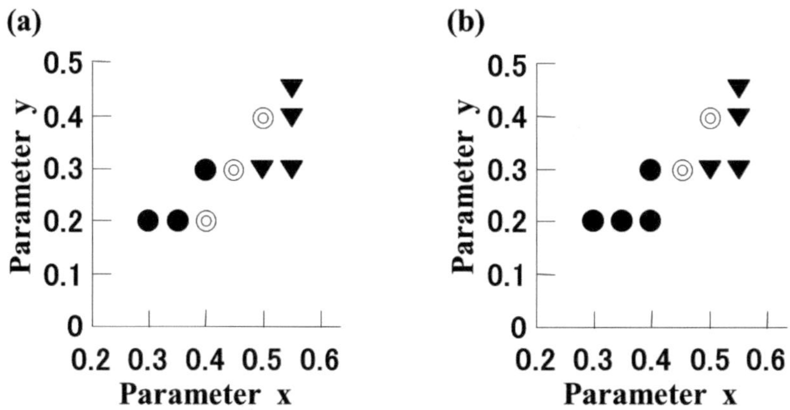

Fig. 16. The diagrams of phase-composition of NYTeS glass-ceramics heated at 900°C (a) and 1000°C (b) for 5 h. ● $Na_5RSi_4O_{12}$ (N5) ▼ $Na_9RSi_6O_{18}$ (N9) ◎ N5+N9

Specimen	Heat-treatment		σ_{300}	E_a / kJ•mol^{-1}		
	Temp.	Time	/10^{-2}•S•cm^{-1}	T	G	GB
NYTiS	1000	5	2.5	15.5	19.8	10.0
NYGeS	900	5	4.0	21.1	17.3	44.0
	1000	5	4.5	24.1	19.3	56.6
	1000	24	6.7	22.9	20.6	67.2
NYTeS	900	5	3.2	19.8	18.7	43.7
	1000	5	4.6	21.8	20.5	56.6

σ_{300} : conductivity at 300°C
E_a : activation energy (T:total G:grain GB:grain boundary)

Table 4. Conduction properties of the N5 glass-ceramics with $Na_{3.6}Y_{0.8}Ti_{0.2}Si_{2.8}O_9$, $Na_{4.2}Y_{0.6}Ge_{0.3}Si_{2.7}O_9$ and $Na_{4.2}Y_{0.6}Te_{0.3}Si_{2.7}O_9$ compositions.

5.2.2 Effect of substitution of Si with V and Mo on ionic conductivity of $Na_5YSi_4O_{12}$-type glass-ceramics

Glass-ceramics of the vanadium- or molybdenum-containing N5-type Na^+-superionic conductors were prepared by crystallization of glasses with the compositions $Na_{3+3x-y}Y_{1-x}V_ySi_{3-y}O_9$ (NYVS) or $Na_{3+3x-2y}Y_{1-x}Mo_ySi_{3-y}O_9$ (NYMS), and the effects of V or Mo elements on the separation of the phase and the microstructural effects on the conduction properties of glass-ceramics were discussed.

The glass-ceramics have been obtained under the appropriate sets of the parameters x and y of the composition formulas $Na_{3+3x-y}Y_{1-x}V_ySi_{3-y}O_9$ or $Na_{3+3x-2y}Y_{1-x}Mo_ySi_{3-y}O_9$ ranging in x=0.3~0.5 and y=0.1~0.4. The precursor glasses were made by melting stoichiometric mixtures of reagent-grade powders of anhydrous Na_2CO_3, Y_2O_3, V_2O_5, MoO_3 and SiO_2 at 1400°C for 1 h, followed by annealing for several hours at an optimum temperature. Shown in Fig. 17 are the diagrams of phase-composition-crystallization temperature of the glass-ceramic specimens with the $Na_{3.9}Y_{0.6}V_{0.3}Si_{2.7}O_9$ (A) and $Na_{3.7}Y_{0.7}Mo_{0.1}Si_{2.9}O_9$ (B) compositions. N5-type NYVS and NYMS are obtained as a stable phase at high-temperatures. The crystallization of N5 single phase is strongly dependent both on the contents of yttrium and (vanadium or molybdenum) ions (or the values x and y correspond to the composition parameters in $Na_{3+3x-y}Y_{1-x}V_ySi_{3-y}O_9$ or $Na_{3+3x-2y}Y_{1-x}Mo_ySi_{3-y}O_9$). N3 and N9 phases can be

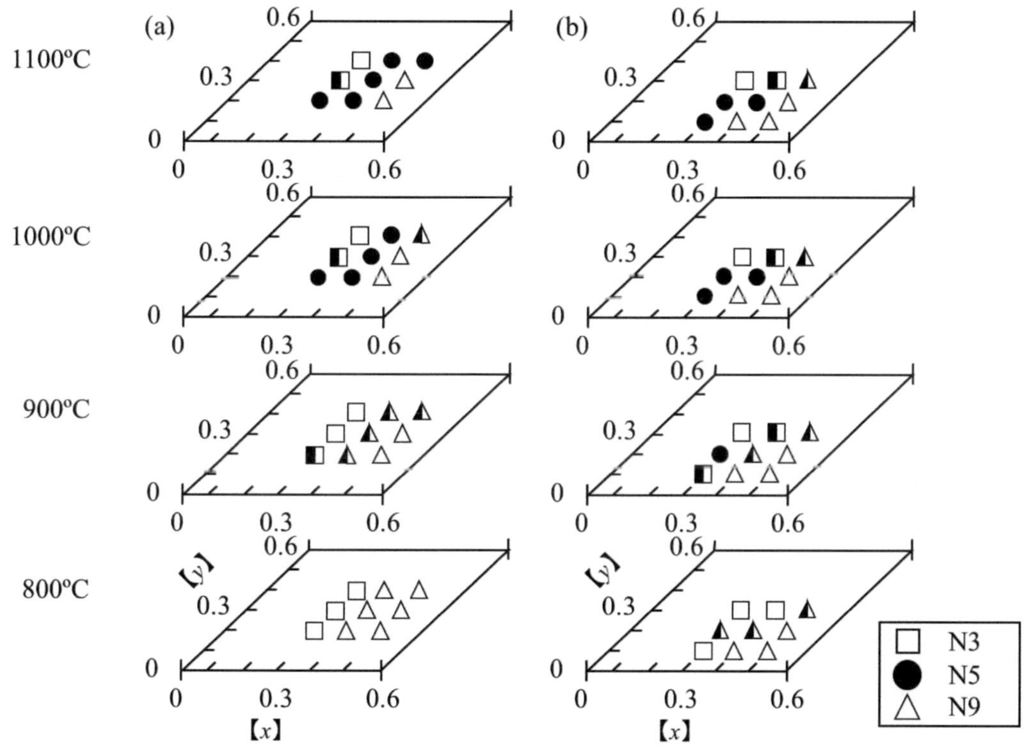

Fig. 17. The diagrams of phase-composition-crystallization temperature of NYVS (a) and NYMS (b) glass-ceramics crystallized at 800~1100°C.
● $Na_5RSi_4O_{12}$ (N5) □ $Na_3RSi_3O_9$ (N3) △ $Na_9RSi_6O_{18}$ (N9) ◧ N5+N3 ◭ N5+N9

crystallized as the high-temperature stable phases at the regions of rather lower [Y] and higher [Y], respectively. The total conductivities and the activation energies are summarizes in Table 1. The total conductivities of the specimens (A) and (B) were 0.87×10^{-2} and 3.58×10^{-2} S/cm at 300°C, respectively, and the activation energies of those specimens were 38.1 and 21.8 kJ/mol, respectively. The combination of x and y was most varied in N5-type NYPS and more limited in N5-type NYVS and NYMS. The conductivity decreases giving the order NYPS>NYMS>NYVS. It is considered that this order corresponds to the N5 single phase region. We assume that the effect of the substitution of Si with V or Mo should be to bring about the difference of homogeneity in the N5 ring structure. The total and electronic conductivities and the Na^+ ionic transport numbers of the specimen (A) determined by Wagner polarization method are summarizes in Table 5. The ionic transport numbers of the specimen (A) were nearly 0.9, while those of the specimen (B) were nearly 1. It is considered that about 10% of total conduction is electronic conduction (hopping conduction by transition metal vanadium) in the specimen (A). This result can explain following facts; the conductivity of the specimen (A) are lower than other N5 conductors.

(A)

Temp. / °C	σ / 10^{-1} S·cm^{-1} G	GB	T	E_a / kJ·mol^{-1} G	GB	T
150	0.43	0.19	0.13			
200	0.44	0.73	0.28	10.5	47.1	27.5
250	0.62	1.97	0.47			
300	1.24	2.93	0.87	40.5	31.7	38.1
350	2.33	5.38	1.63			

σ : Conductivity E_a : Activation Energy G: Grain GB: Grain Boundary T: Total

(B)

Temp. / °C	σ / 10^{-1} S·cm^{-1} G	GB	T	E_a / kJ·mol^{-1} G	GB	T
150	1.24	1.27	0.63			
200	2.11	4.68	1.45		41.1	29.2
250	3.33	9.45	2.46	22.5		
300	4.81	13.91	3.58		19.1	21.8
350	6.49	15.95	4.61			

σ : Conductivity E_a : Activation Energy G: Grain GB: Grain Boundary T: Total

Table 5. Total conductivities and activation energies of the glass-ceramic specimens $Na_{3.9}Y_{0.6}V_{0.3}Si_{2.7}O_9$ (A) and $Na_{3.7}Y_{0.7}Mo_{0.1}Si_{2.9}O_9$ (B).

5.3 Structure and conduction properties of $Na_5YSi_4O_{12}$-type glass-ceramics synthesized by bias crystallization of glass

Glass-ceramics of the phosphorus containing N5-type Na^+ superionic conductors were prepared by bias crystallization of glasses with the composition $Na_{4.05}Y_{0.55}P_{0.3}Si_{2.7}O_9$ in an electric field. The conditions for bias crystallization are discussed with respect to the microstructure and the conduction properties.

The precursor glasses were made by melting stoichiometric mixtures of reagent-grade powders of anhydrous Na_2CO_3, Y_2O_3, SiO_2 and $NH_4H_2PO_4$ at 1350°C for 1h, followed by annealing for several hours at an optimum temperature. The annealed specimens were heated up to 900°C in an electric field for bias crystallization. The thermostable heating holder was produced in order to do the crystallization in a direct current electric field. This holder is made of alumina and platinum. Glass samples (5 mm × 5 mm × 8 mm) were held between the platinum plates, and crystallized in the electrical field using 1 V/mm. The thermal treatment was the same as that used in conventional crystallization in nonelectric field.

The microstructure was investigated with SEM. The grain length of the cross section which is parallel with the electric field direction was 10~15 nm, and it was proven to be smaller than the 15~30 nm grain length of the cross section which is perpendicular to the direction and the specimen crystallized by the conventional method. It was possible to control shape and orientation of crystal grain by the crystallization in the electrical field.

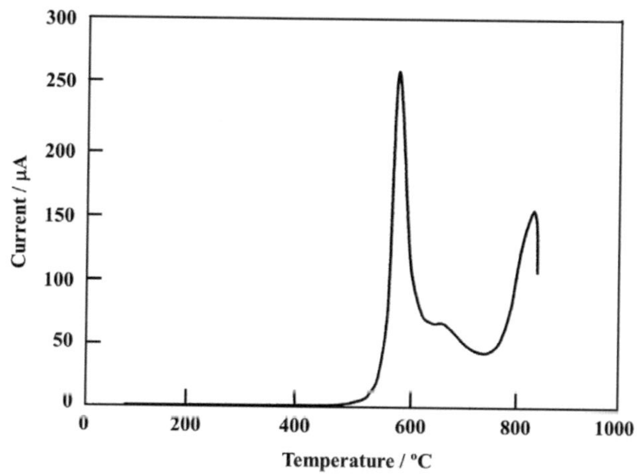

Fig. 18. Current profile in relation to temperature during crystallization process in electric field.

Due to the bias field an electric current in relation to temperature was measured during crystallization process. Fig. 18 shows current profile in relation to temperature during crystallization process in the electric field. The largest observed current was 250 μA. The current profile exhibits three peaks at about 600°C, 700°C and 850°C. These temperatures correspond to those of nucleation, phase transition from N3-phase to N5-phase, and crystallization of glass specimens determined by DTA analysis, respectively. An electric current in relation to temperature was measured newly by applying the bias voltage only in two limited temperature range, because two main peaks were observed in Fig. 18. One range is from right before of the first main peak (511°C to 652°C), and another range is from right before of the second main peak (790°C to 865°C). The resulting current profile is shown in Fig. 19. It was found that the mass transfer in the specimen is being generated even in the condition that does not apply the voltage.

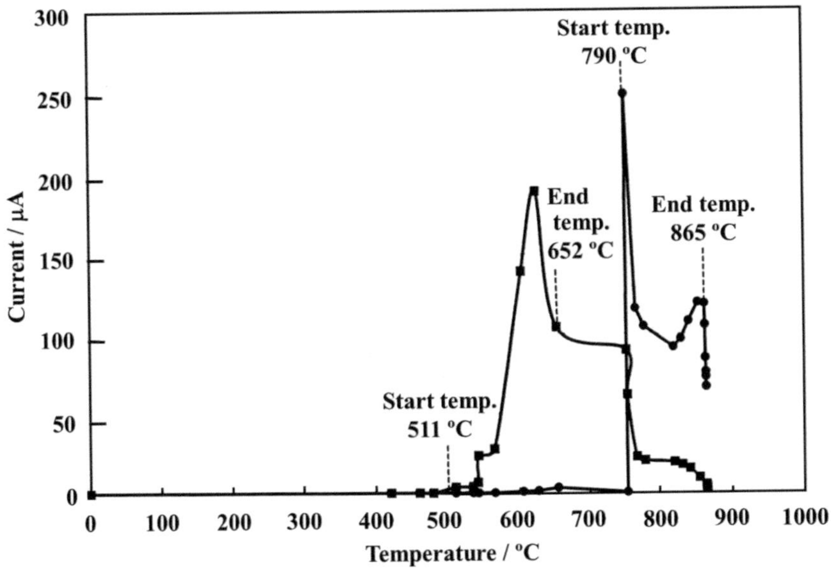

Fig. 19. Current profile in relation to temperature measured by applying voltage in two limited temperature ranges.

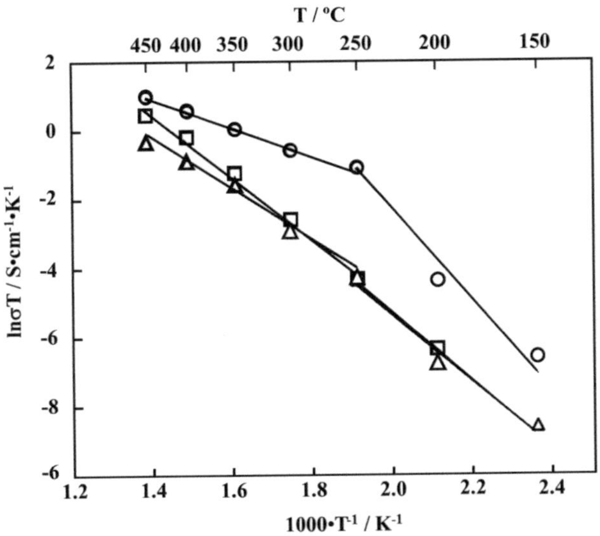

Fig. 20. Temperature dependence Arrhenius plots of the conductivities of the bias crystallized NaRPSi glasses. ○ Conventional △ Parallel □ Perpendicular

Crystalline phases were identified on the sample after the crystallization in the electric field by the X-ray diffraction (XRD) method in order to consider the possibility of structural change by the movement of Na^+ ion which is a carrier. In the several cut sections, no difference in the fundamental structure was observed. Judging from the patterns, the N5 single phase ionic conductors were successfully produced by bias crystallization of glasses.

	E_a / kJ•mol^{-1}		σ_{300} / mS•cm^{-1}
	Temperature		
	High	Low	
○ Conventional	33.0	99.9	0.9860
△ Parallel	66.6	73.5	0.0923
□ Perpendicular	76.4	82.2	0.1320

Table 6. Conduction properties of the bias-crystallized NaRPSi glasses.

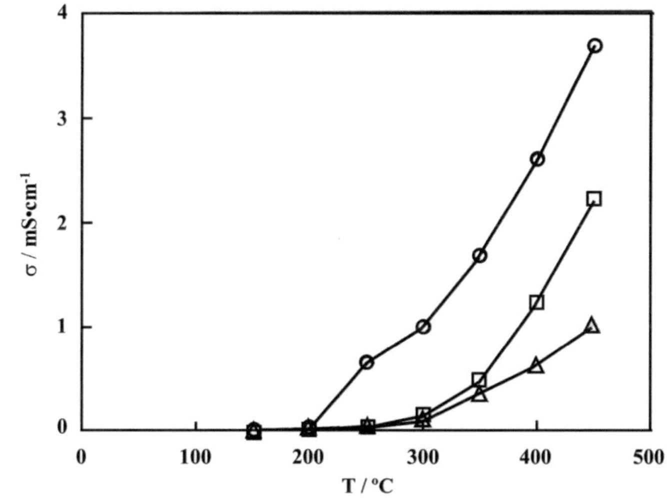

Fig. 21. Temperature dependence of conductivity of the bias crystallized NaRPSi glasses.

Fig. 20 shows the temperature dependence Arrhenius plots of the conductivities of various specimens. The complex admittances of the measured glass-ceramics consisted of two semicircles below 300°C. The two intercepting points on the real axis are interpreted as the resistance of crystallized grains (R_G) and the total resistance of grains and remaining glassy grain boundaries (R_{GB}). As R_{GB} decreases rapidly with increasing temperature because of high $(E_a)_{GB}$ to a comparable value with R_G at 300°C, the total conductivities (R_G+R_{GB}) are dominated by grain boundary conductivity. The effect of the grain boundary is greatly seen on the appearance at lower temperatures. Table 6 summarizes the conduction properties obtained from Fig. 20. The cross sections which are parallel and perpendicular to the electric field direction showed the ionic conductivities of 0.0923 and 0.132 mS/cm at 300°C, respectively. It was found that the bias crystallized specimens were less conductive than that crystallized by the conventional method. Fig. 21 shows the temperature dependence of conductivity of the bias crystallized specimen. In the temperatures over 300°C, anisotropy in the conductivity was observed. It was also found that the cross section which is perpendicular to the electric field direction was more conductive than that in parallel with the direction. The microstructure and the electric conductivity of the NaRPSi glass-ceramics perpendicular to the electric field direction were significantly different from those in parallel.

6. Conclusion

The Na^+-superionic conducting glass-ceramics with N5-type structure were successfully produced using the sodium rare earth silicophosphate composition of $Na_{3+3x-y}R_{1-x}P_ySi_{3-y}O_9$, in which the rare earth elements of Sc to Sm were applicable to R. The possible combinations of x and y became more limited for the crystallization of the superionic conducting phase as the ionic radius of R increased, while the Na^+ conduction properties were more enhanced in the glass-ceramics of larger R. The meaning of the composition formula can be signified in the thermodynamic and kinetic study of crystallization and phase transformation of metastable to stable phase in the production of N5-type glass-ceramics. It was demonstrated that the medium value of content product as [P]×[R] is important in the crystallization of N5 single phase. Conduction properties of these glass-ceramics were strongly dependent upon the crystallization conditions as well as compositions. Not only complex impedance analysis but also TEM observation confirmed that this dependence was attributed to the conduction properties of grain boundaries which were glasses condensed at triple points enclosed by grains.

Glass-ceramics of the N5-type superionic conductors in the system Na_2O-Sm_2O_3-P_2O_5-SiO_2 were prepared by crystallization of glasses with the composition $Na_{3+3x-y}Sm_{1-x}P_ySi_{3-y}O_9$. The optimum conditions for crystallization were discussed with reference to the conduction properties and the preparation of crack free N5-type glass-ceramics. The crystallization of the N5 single-phase glass-ceramics was dependent strongly on the concentrations of both [R] and [P] (or x and y in the composition parameters) and the temperature for crystallization of glass specimens. The ionic conductivity of the glass-ceramic $Na_{4.1}Sm_{0.5}P_{0.4}Si_{2.6}O_9$ was 4.78×10^{-2} S/cm at 300°C. The grain size of the specimen was about 3-5 μm. The state of grain growth is promoted with increase of heating temperature and heating time for crystallization. Although grain growth may cause high conductivity, it was difficult to prevent the sample heated for a long time from cracking during crystallization.

Glass-ceramics of the titanium-, germanium- or tellurium-containing N5-type superionic conductors were prepared by crystallization of glasses with the composition $Na_{3+3x}Y_{1-x}X_ySi_{3-y}O_9$ (X=Ti, Ge, Te), and the effects of X elements on the separation of the phase and the microstructural effects on the conduction properties of glass-ceramics were discussed. The combination of x and y was most varied in N5YGeS and more limited in the order of N5YTeS>N5YTiS. Their conductivities and activation energies are of the order of 10^{-2} S/cm at 300°C and of 15 to 24 kJ/mol, respectively. The conductivity of the glass-ceramic N5YXS decreases giving the order N5YGeS>N5YTeS>N5YTiS. It is considered that this order corresponds to the N5 single phase region. Large enhancement of electrical conductivity was observed in the glass-ceramics as the grain growth was promoted with increase of heating temperature and heating time for crystallization.

Glass-ceramics of the vanadium- or molybdenum-containing N5-type superionic conductors were prepared by crystallization of glasses with the compositions $Na_{3+3x-y}Y_{1-x}V_ySi_{3-y}O_9$ or $Na_{3+3x-2y}Y_{1-x}Mo_ySi_{3-y}O_9$. The combination of x and y was most varied in N5-type NYPS and more limited in N5-type NYVS and NYMS. The conductivities of the glass-ceramic specimens with the $Na_{3.9}Y_{0.6}V_{0.3}Si_{2.7}O_9$ (A) and $Na_{3.7}Y_{0.7}Mo_{0.1}Si_{2.9}O_9$ (B) compositions were 0.87×10^{-2} and 3.58×10^{-2} S/cm at 300°C, respectively. The conductivity decreases giving the order NYPS>NYMS>NYVS. It is considered that this order corresponds to the N5 single phase region. We assume that the effect of the substitution of Si with V or Mo should be to bring about the difference of homogeneity in the N5 ring structure. The Na^+ ionic transport numbers of these glass-ceramics determined by Wagner polarization method were nearly 0.9 for the specimen (A) and 1 for the specimen (B) at 300°C, respectively. It is considered that about 10% of total conduction is electronic conduction in the specimen (A). This result can explain following facts; the conductivity of the specimen (A) are lower than other N5 conductors, and it is seen in the temperature dependence Arrhenius plots for the specimen (A) that the lines drawn from the conductivity of grains are bending upwards.

We have successfully produced the N5-type glass-ceramic conductors by bias crystallization of the glasses with the composition $Na_{4.05}Y_{0.55}P_{0.3}Si_{2.7}O_9$ in an electric field. The microstructure and the conduction properties were dependent on the current direction in the process of crystallization. The cross sections which are parallel and perpendicular to the electric field direction showed the ionic conductivities of 0.0923 and 0.132 mS/cm at 300°C, respectively. The microstructure and the electric conductivity of the glass-ceramics perpendicular to the electric field direction were significantly different from those in parallel.

7. Acknowledgment

I would like to thank Prof. Kimihiro Yamashita (Tokyo Medical and Dental University, Japan) and Professor emeritus Hideki Monma (Kogakuin University, Japan) for their support and warm encouragement.

8. References

Alexander, M. G. (1987). Effect of modifier cations on Na^+ conductivity in sodium silicate glasses. *Solid State Ionics*, 22, 257.

Banks, E. & Kim. C. H. (1985). Ionic conductivity in glass and glass-ceramics of the $Na_3YSi_3O_9$ and $Na_5YSi_4O_{12}$ type materials. *J. Electrochem. Soc.*, 132, 2617-2621.

Beyeler, H. U. & Himba, T. (1978). The sodium conductivity paths in the superionic conductors $Na_5RESi_4O_{12}$. *Solid State Commun.*, 27, 641-643.

Beyeler, H. U.; Shannon, R. D. & Chen H. Y. (1980). Ionic conductivity of single-crystal $Na_5YSi_4O_{12}$. *Appl. Phys. Lett.*, 37, 934-936.

Cervantes, F.; Marr, L. J. & Glasser, F. P. (1981). Compounds in the Na_2O-Y_2O_3-SiO_2 system. *Ceram. Intl.*, 7(2), 43-47.

Hong, H. Y-P.; Kafalas, J. A. & Bayard, M. (1978). High Na^+-ion conductivity in $Na_5YSi_4O_{12}$. *Mater. Res. Bull.*, 13, 757-761.

Kim, C. H.; Qiu, B. & Banks, E. Ionic conductivity in solid solutions of the $Na_3YSi_3O_9$ type. (1985). *J. Electrochem. Soc.*, 132, 1340-1345.

Kingery, W. D.; Bowen, H. K. & D. R. Uhlmann. (1976). *Introduction to Ceramics, 2nd ed.*, John Wiley & Sons, Inc., New York, p. 368-369.

Maksimov, B. A.; Petrov, I. V.; Rabenau, A. & Schulz, H. (1982). X-ray investigations and possible mechanism for the ionic conductivity in the $Na_5RESi_4O_{12}$ (RE=Y, Sc) class of compounds. *Solid State Ionics*, 6, 195-200.

Morimoto, S. (1989). Ionic conductivity of Na_2O-ZrO_2-P_2O_5-SiO_2 system glass ceramic. *J. Ceram. Soc. Jpn.*, 97, 1097-1103.

Okura, T.; Tanaka, M.; Kanzawa, H. & Sudoh, G. (1996). Synthesis and conduction properties of Na^+ superionic conductors of sodium samarium silicophosphates. *Solid State Ionics*, 86-88, 511-516.

Okura, T.; Yamashita, K. & Umegaki, T. (1996). Na^+-ion conduction properties of glass-ceramic Narpsio in the Y-Sm mixed system. *Phosphorus Research Bulletin*, 6, 237-240.

Okura, T.; Tanaka, M. & Sudoh, G. (1997). Conduction properties of Na^+-ion implanted glass-ceramics in the system Na_2O-$R_2O_3P_2O_5$-SiO_2 (R = rare earth). *Mat. Res. Soc. Symp. Proc.*, Vol. 453, 611-616.

Okura, T. & Yamashita, K. (2000). Ionic conductivities of Na^+-ion implanted silico-phosphate glass-ceramics. *Solid State Ionics*, 136-137, 1049-1054.

Okura, T.; Inami, Y.; Monma, H.; Nakamura, S. & Yamashita. K. (2002). Structure and conduction properties of Na^+ superionic conductor Narpsio synthesized by bias crystallization of glass. *Solid State Ionics*, 154-155, 361-366.

Okura, T.; Tanaka, M.; Monma, H.; Yamashita, K. & Sudoh, G. (2003). New superionic conducting glass-ceramics in the system Na_2O-Y_2O_3-Sm_2O_3-P_2O_5-SiO_2: crystallization and ionic conductivity. *J. Ceram. Soc. Jpn.*, 111/4, 257-261.

Okura, T.; Monma, H. & K. Yamashita. (2004). Na^+-superionic conductors of glass-ceramics in the system Na_2O-Sm_2O_3-X_2O_3-P_2O_5-SiO_2 (X=Al, Ga). *Solid State Ionics*, 172, 561-564.

Okura, T.; Monma, H. & Yamashita, K. (2004). Effect of substitution of Si with Ge and Te on ionic conductivity of $Na_5SmSi_4O_{12}$–type glass-ceramics. *J. Ceram. Soc. Jpn.*, 112/ 5, S685-S689.

Okura, T.; Monma, H. & Yamashita, K. (2006). Superionic conducting $Na_5SmSi_4O_{12}$-type glass-ceramics: Crystallization condition and ionic conductivity. *J. Eur. Ceram. Soc.*, 26, 619.

Okura, T.; Takahashi, T; Monma, H. & Yamashita. K. Effect of substitution of Si with V and Mo on ionic conductivity of $Na_5YSi_4O_{12}$–type glass-ceramics. (2008). *Solid State Ionics*, 179, 1291-1295.

Okura, T.; Saimaru, M; Monma, H. & Yamashita. K. (2009). Ionic conductivities of Nasicon-type glass-ceramic superionic conductors in the system Na_2O-Y_2O_3-XO_2-SiO_2 (X=Ti, Ge, Te). *Solid State Ionics*, 180, 537-540.

Okura, T; Monma, H & Yamashita, K. (2010). Na^+-fast ionic conducting glass-ceramics of silicophosphates. *J. Electroceram.*, 24, 83-90.

Shannon, R. D.; Taylor, B. E.; Gier, T. E.; Chen, H. Y. & Berzins, T. (1978). Ionic conductivity in $Na_5YSi_4O_{12}$–type silicates. *Inorg. Chem.*, 17, 958-964.

Shannon, R. D.; Gier, T. E.; Foris, C. M.; Nelen, J. A. & Appleman, D. E. (1980). Crystal data for some sodium rare earth silicates. *Phys. Chem. Minerals*, 5, 245-253.

Yamashita, K.; Ohkura, S.; Umegaki, T. & Kanazawa, T. (1988). Synthesis and ionic conduction of C_3A-type Nasicon $Na_{3+3x-y}Y_{1-x}Si_{3-y}P_yO_9$. *Solid State Ionics*, 26, 279-286.

Yamashita, K.; Ohkura, S.; Umegaki, T. & Kanazawa, T. (1988). Synthesis, polymorphs and sodium ionic conductivity of sodium yttrium silicophosphates with the composition $Na_{3+3x-y}Y_{1-x}P_ySi_{3-y}O_9$. *J. Ceram. Soc. Jpn.*, 96, 967-972.

Yamashita, K.; Nojiri, T.; Umegaki, T. & Kanazawa, T. (1989). New fast sodium-ion conducting glass-ceramics of silicophosphates: Crystallization, microstructure and conduction properties. *Solid State Ionics*, 35, 299-306.

Yamashita, K.; Nojiri, T.; Umegaki, T. & Kanazawa, T. (1990). Na^+ superionic conductors of glass-ceramics in the system Na_2O-Re_2O_3-P_2O_5-SiO_2 (Re=rare-earth elements). *Solid State Ionics*, 40/41, 48-52.

Yamashita, K.; Tanaka, M. & Umegaki, T. (1992). Thermodynamic and kinetic study on the phase transformation of the glass-ceramic Na^+ superionic conductors $Na_{3+3x-y}Re_{1-x}P_ySi_{3-y}O_9$. *Solid State Ionics*, 58, 231-236.

Yamashita, K.; Tanaka, M.; Kakuta, T.; Matsuda, M. & Umegaki, T. (1993). Effects of rare earth elements on the crystallization of the glass-ceramic Na^+ superionic conductor Narpsio-V. *J. Alloys and Compounds*, 193, 283-285.

Yamashita, K.; Umegaki, T.; Tanaka, M.; Kakuta, T. & Nojiri, T. (1996). Microstructural effects on conduction properties of $Na_5YSi_4O_{12}$-type glass-ceramic Na^+-fast ionic conductors. *J. Electrochem. Soc.*, 143, 2180-2186.

5

Crystallization Kinetics of Metallic Glasses

Arun Pratap and Ashmi T. Patel
Condensed Matter Physics Laboratory, Applied Physics Department, Faculty of Technology & Engineering, The Maharaja Sayajirao University of Baroda, Vadodara, India

"Crystallization is still in many ways, more an art than a science."
–David Oxtoby, Nature, August 3, 2000.

1. Introduction

Metallic glasses are kinetically metastable materials. Metallic glass is defined as "A liquid, which has been cooled into a state of rigidity without crystallizing". Properties of metallic glasses differ form non metallic glasses. Ordinary glasses are made up of silica while metallic glasses are made of alloy metals. Ordinary glasses are transparent whereas metallic glasses are opaque. In ordinary glasses, covalent bond is observed while in metallic glasses metallic bond is observed. On the basis of internal arrangement of atoms or molecules and type of force acting between them, the material can be classified into the following two categories:

i. Crystalline solid: Those materials in which the constituent ions or atoms and molecules are arranged in regular pattern are called crystalline solids. Besides, crystalline solids have a definite external geometrical form.
   e.g. Quartz, Calcite, Diamond, Sugar, and Mica
ii. Amorphous or glassy solid: Those materials do not have definite geometric pattern are called amorphous solids. In amorphous solid atoms, ions or molecules are not arranged in definite pattern.
   e.g. Rubber, Glass, Plastic and Cement

Also, an amorphous solid is a solid in which there is no long range order of the positions of the atoms. Solids in which there is long-range atomic order are called crystalline solids.

At high cooling rate, any liquid can be made into an amorphous solid. Cooling reduces molecular mobility. If the cooling rate is faster, then molecules can not organize into a more thermodynamically favourable crystalline state and an amorphous solid will be formed. Materials in which such a disordered structure is produced directly from the liquid state during cooling are called "Glasses" and such amorphous metals are commonly referred to as "Metallic Glasses" or "Glassy Metals". The metallic glasses have a combination of amorphous structure and metallic bond. This combination provides a metallic glass a new and unique quality, which cannot be found in either pure metals or regular glass.

In the past, small batches of amorphous metals have been produced through a variety of quick-cooling methods. For instance, amorphous metal wires have been produced by sputtering molten metal onto a spinning metal disk. The rapid cooling, of the order of millions of degrees a second, is too fast for crystals to form and the material is "locked in" a glassy state. Now-a-days number of alloys with critical cooling rates low enough to allow formation of amorphous structure in thick layers (over 1 millimetre) have been produced; these are known as bulk metallic glasses (BMG).

However, there are various methods in which amorphous metals can be produced, preventing the crystallization. Sputtering, glow discharge sputtering, chemical vapour deposition (CVD), gel desiccation, electrolyte deposition, reaction amorphization, pressure–induced amorphization, solid state diffusion amorphization, laser glazing, ion implantation, thin-film deposition, melt quenching and melt spinning are some of them.

The study of the thermally-activated phase transformations is of great significance in the field of materials science as the properties of materials change due to the change in the composition and/or microstructure. The properties of fully or partly crystalline materials are usually different from their amorphous counterparts. From the viewpoint of a materials scientist, the crystallization of amorphous or non-crystalline materials involves the nucleation and growth processes. The processes driven by nucleation and growth have attracted a lot of interest for tailoring technological applications. For example, the recrystallization of the deformed metals, controlling the nucleation and growth of islands on terraces in order to get large scale arrays of nanostructures in the manufacturing of thin-film transistors (Castro, 2003). Thus, the knowledge of the kinetics of crystallization would help to attain products with the required crystallized fraction and microstructure (e.g. nanocrystalline or quasicrystalline) or to avoid the degradation of materials at high processing (& operating) temperatures.

The kinetics of the crystallization process can be studied with the help of thermo-analytical techniques namely, differential scanning calorimetry (DSC) and differential thermal analyzer (DTA). The DSC/DTA experiments can be carried out in isothermal as well as non-isothermal (linear heating) conditions (Ligero et al., 1990; Moharram et al., 2001; Rysava et al., 1987; Giridhar & Mahadevan, 1982; Afify, 1991). Efforts made by the researchers in this field so far, to analyze the data obtained from DSC and hence to determine the kinetic parameters of the crystallization processes (say, activation energy, rate constant etc.), raise two important issues: (i) the selection of the mode of experiment (isothermal or non-isothermal) and, (ii) the choice of a sound method for the analysis of the experimental data. However, we are more concerned with the later issue due to the fact that several methods for the kinetic analysis are available in the literature. These methods are generally based on either the isokinetic hypothesis or the isoconversional principle and they can be accordingly categorized as: (1) isokinetic methods where the transformation mechanism is assumed to be the same throughout the temperature/time range of interest and, the kinetic parameters are assumed to be constant with respect to time and temperature; (2) isoconversional methods, which are generally used for non-isothermal analysis, assume that the reaction (transformation) rate at a constant extent of conversion (degree of transformation) is only a function of temperature (Lad et al, 2008; Patel & Pratap, 2012). The kinetic parameters, in this case, are considered to be dependent on the degree of transformation at different temperature and time. The use of isoconversional methods is widespread in the physical chemistry for the determination of the kinetics of the thermally activated solid-state reactions. The physicochemical changes during an exothermic or endothermic event in DSC (or DTA) are complex and involve multi-step (serial or parallel) processes occurring simultaneously at different rates. Therefore, the activation energies for such processes can logically not be same and it may vary with the degree of conversion. This is contrary to the isokinetic view assuming all the constituents of the material to react simultaneously at the same rate. The activation energy, in the isokinetic case, is thus constant and independent of the degree of conversion. A strong difference of opinion persists among the researchers in the field of thermal analysis about the concept of variable activation energy (Galwey, 2003; Vyazovkin, 2003). In the metallurgical branch of materials science, most of the thermal phase transformations (like crystallization, recovery) are morphological and are considered to be governed by the nucleation and growth processes. The transformation mechanisms in these processes are also complex e.g. interface-controlled, diffusion-controlled growth. Notwithstanding this, the kinetic analysis of the transformation process like crystallization is done according to isokinetic hypothesis. The isoconversional methods are scarcely used for the study of the crystallization kinetics of metallic glasses.

2. Theory

To study the phase transformation, which involves nucleation and growth, many methods are developed. Most of the methods depend on the transformation rate equation given by Kolmogorov, Johnson, Mehl and Avrami (Lesz & Szewieczek, 2005; Szewieczek & Lesz, 2005; Szewieczek & Lesz, 2004; Jones et al., 1986; Minic & Adnadevic, 2008), popularly known as KJMA equation, basically derived from experiments carried out under isothermal conditions. The KJMA rate equation is given by

$$\frac{d\alpha}{dt} = nk(1-\alpha)[-ln(1-\alpha)]^{(n-1)/n} \quad (1)$$

where, $a \rightarrow$ degree of transformation at a given time t,
$n \rightarrow$ Avrami (growth) exponent
$k \rightarrow$ the rate constant

The Arrhenius form of the rate constant is given by

$$k(T) = k_0 \, exp\left(-\frac{E}{RT}\right) \quad (2)$$

where, $k_0 \rightarrow$ pre-exponential factor
E $\rightarrow$ activation energy, and
R $\rightarrow$ universal gas constant

KJMA rate equation is based on some important assumptions and it has been suggested that the KJMA kinetic equation is accurate for reactions with linear growth subject to several conditions (Minic et al., 2009).

The isoconversional methods are also known as model-free methods. Therefore, the kinetic analysis using these methods is more deterministic and gives reliable values of activation energy E, which depends on degree of transformation, a. However, only activation energy will not give a perfect picture of crystallization kinetics. The microstructural information (e.g. dimensionality of the growth) of the precipitating phase during the transformation is also very important for understanding the whole kinetics of crystallization. Microstructural information would be known to us when we take the isokinetic methods into account. Therefore, the complementary use of both the methods is more useful for understanding the kinetics of crystallization.

Differential Scanning Calorimetry (DSC) has become a convenient and widely used tool for studying the kinetics of phase transformations. The volume fraction (x) of the sample transformed in crystalline phase during the crystallization event has been obtained from the DSC curve as a function of temperature (T). The volume fraction of precipitated crystal can be obtained from the DSC curve by using

$$x = {}^{S}\!/_{S_0}$$

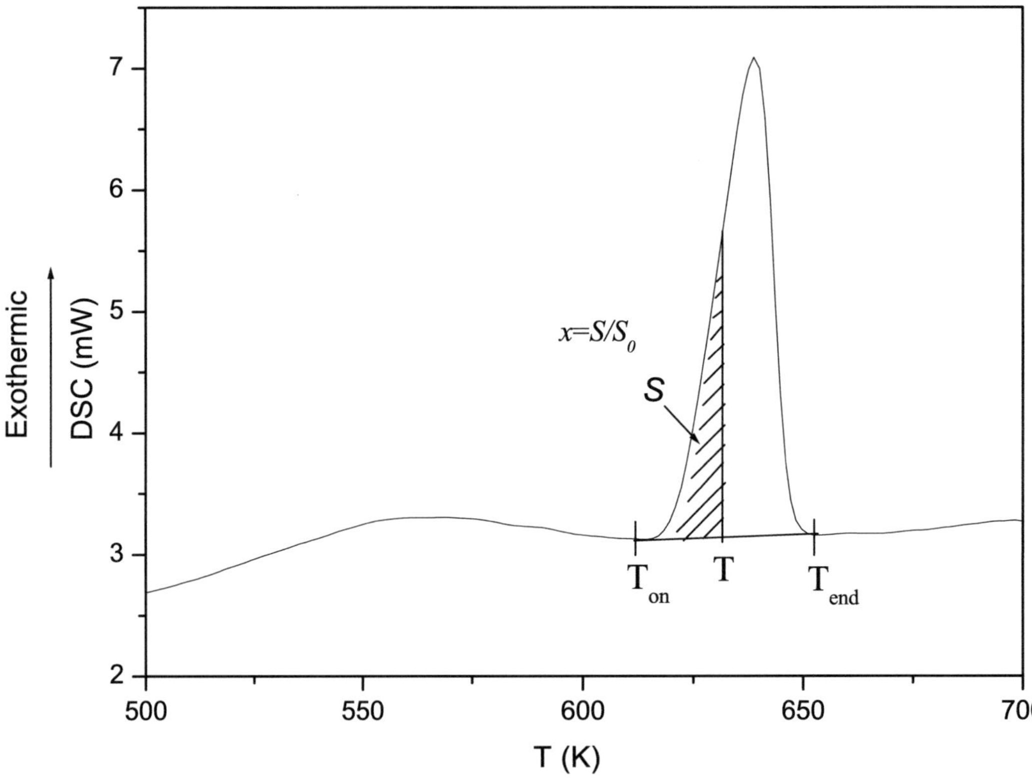

Where S_0 is the total area under the crystallization curve i.e. the area under the curve between the temperature at the onset of crystallization T_{on} and the end-set temperature T_{end} when the crystallization is completed. S is the area at any temperature T between T_{on} and T at which the fractional crystallization is required to be known.

There are three important modes of crystallization involving nucleation and growth processes, depending on the composition of a particular alloy: primary crystallization, polymorphous crystallization and eutectic crystallization (Hsiao et al., 2002). In primary crystallization the primary phase of the alloy constituents crystallizes first. The dispersed primary crystallized phase coexists with the amorphous matrix and may serve as the nucleation site for secondary or tertiary crystallization. In Fe-based alloys α-Fe crystallizes first, which is a kind of primary crystallization. Polymorphous crystallization is a transition of the amorphous phase to a crystalline one without any change in the composition of that phase. There is no concentration difference across the reaction front because the concentration does not change. Eutectic crystallization is simultaneous crystallization of two crystalline phases by a discontinuous reaction. This reaction takes longer than polymorphous crystallization to proceed because the two components have to separate by diffusion into two separate phases within the crystallized region (Minic, 2006).

3. Results and discussion

The DSC thermograms at four different heating rates are shown in Fig.1. The thermograms show three-stage crystallization. The first crystallization peak is evaluated for heating rates 4, 6, 8 and 10 deg/min. Glass transition becomes clear as we go for the higher heating rates, but the third crystallization peak becomes less prominent as we go to the higher heating rates. The onset and endset of first crystallization exotherms exhibit different levels of heat flow i.e. the crystallization ends at slightly higher level followed by the second and third crystallization peak. This difference of the level indicates that the phases at the start of crystallization and at the end of it are not same. The analysis of DSC data to evaluate the kinetic parameters can be obtained from non-isothermal rate laws by both isokinetic also known as model fitting methods and isoconversional methods.

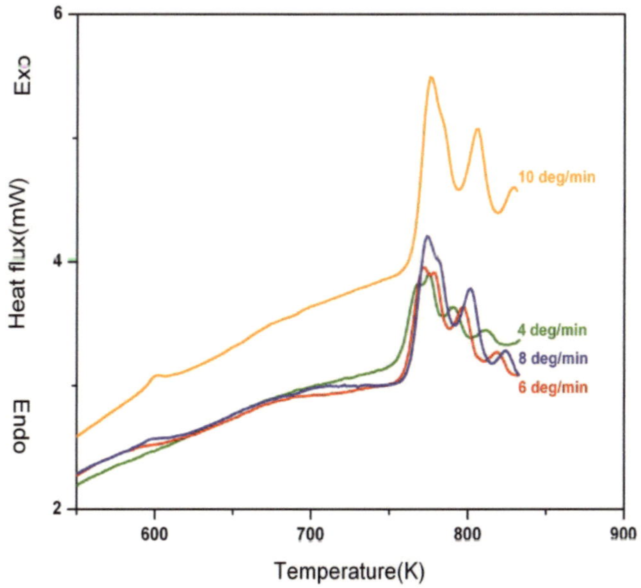

Fig. 1. DSC thermograms of the metallic glass $Co_{66}Si_{12}B_{16}Fe_4Mo_2$ at different heating rates

3.1 Isoconversional analysis

The isoconversional methods require the knowledge of temperatures $T_\alpha(\beta)$ at which an equivalent stage of reaction occurs for various heating rates. The equivalent stage is defined as the stage at which a fixed amount is transformed or at which a fixed fraction, α of the total amount is transformed (Starink, 1997). These methods are further categorized as linear and non-linear isoconversional methods. The linear integral isoconversional methods (Kissinger, 1957; Ozawa, 1965; Augis & Bennett, 1978; Boswell, 1980; Flynn & Wall, 1966; Akahira & Sunose, 1971; Li & Tang, 1999) depend on the approximation of the temperature integral and require the data on $T_\alpha(\beta)$. The differential isoconversional methods depend on the rate of transformation at $T_\alpha(\beta)$ and the data on $T_\alpha(\beta)$ (Gupta et al., 1988; Friedman, 1964; Gao & Wang, 1986). Vyazovkin (Vyazovkin & Wight, 1997) introduced a non-linear isoconversional method to increase the accuracy of evaluating the activation energy. The isoconversional methods are based on the basic kinetic equation (Paulik, 1995)

$$\frac{d\alpha}{dT} = \frac{1}{\beta}k(T)f(\alpha) = \frac{A}{\beta}exp\left(-\frac{E}{RT}\right)f(\alpha) \tag{3}$$

where $k(T)$ is the rate constant, β is the heating rate, α is the conversion fraction and $f(\alpha)$ is the reaction model which in case of KJMA formalism gives the Eq. (1). Eq. (3) can also be expressed in the integral form as

$$g(\alpha) = \int_0^\alpha [f(\alpha)]^{-1} d\alpha = \frac{A}{\beta}\int_0^T exp\left(-\frac{E}{RT}\right)dT \tag{4}$$

As mentioned earlier, exact solution of the temperature integral is not available and various approximations made for this has resulted into different methods. We have selected a few most commonly used methods. The accuracy of various isoconversional methods and, the experimental and analytical errors associated with these methods are discussed in detail by Starink (Starink, 2003). Roura and Farjas (Roura & Farjas, 2009) have proposed an analytical solution for the Kissinger equation. Rotaru and Gosa (Rotaru & Gosa, 2009) describe their recently developed software that implements a number of known techniques such as various isoconversional methods, a method of invariant kinetic parameters, master plots methods, etc. Cai and Chen (Cai & Chen, 2009) have proposed a new numerical routine for a linear integral isoconversional method that allows one to obtain accurate values of the activation energy in the cases when the latter varies strongly with the extent of conversion. Criado et al. (Criado et al., 2008) provide a critical overview of isoconversional methods, putting the focus on establishing whether the observed variations in the activation energy are real or apparent (Vyazovkin, 2010).

Linear integral isoconversional methods

a. Ozawa-Flynn-Wall (OFW) method

In this method (Ozawa, 1965; Augis & Bennett, 1978; Boswell, 1980; Flynn & Wall, 1966) the temperature integral in Eq. (4) is simplified by using the Doyle's approximation (Doyle, 1961, 1962, 1965) and hence we obtain the following equation:

$$ln\,\beta = -1.0516\frac{E(\alpha)}{RT_\alpha} + const \qquad (5)$$

The plot of $ln\beta$ vs $1000/T_\alpha$ (Fig.2) gives the slope $-1.0516\ E(\alpha)/R$ from which the activation energy has been evaluated (Table 1). At $T_\alpha = T_p$, (Ozawa method) the value of E determined using Eq. (5) is given in Table 2.

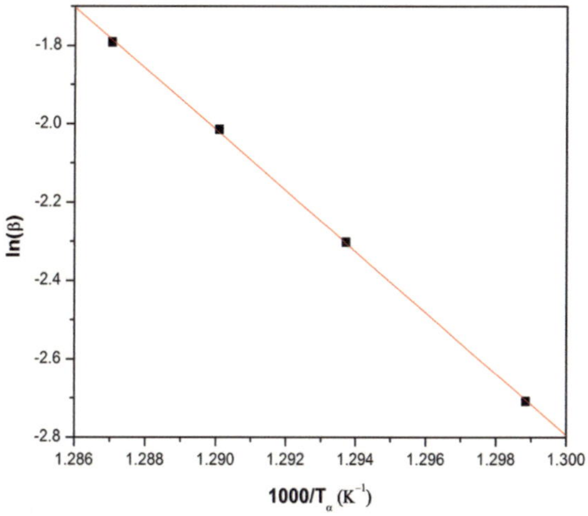

Fig. 2. OFW plot for peak-1

b. Kissinger-Akahira-Sunose (KAS) method

Kissinger, Akahira and Sunose (Kissinger, 1957; Ozawa, 1965; Augis & Bennett, 1978; Boswell, 1980; Flynn & Wall, 1966; Akahira & Sunose, 1971) used the approximation given by Coats & Redfern (Coats & Redfern, 1964) to evaluate the integral in the rate Eq. (4). KAS method is based on the expression

$$ln\left(\frac{\beta}{T^2}\right) = ln\left(\frac{AR}{Eg(\alpha)}\right) - \frac{E}{RT} \qquad (6)$$

The activation energy can be evaluated from the slope of plot $ln(\beta/T^2)$ vs $1000/T$ for constant conversion, α (Fig. 3) Values of E are given in Table 1. The discussion given ahead describes some of the methods available in the literature which are basically special cases of the KAS equation (6).

i) Kissinger method: This well-known method assumes that the reaction rate is maximum at the peak temperature (T_p). This assumption also implies a constant degree of conversion (α) at T_p. The equation used by Kissinger is

$$ln\left(\frac{\beta}{T_p^2}\right) = -\frac{E}{RT_p} + ln\left(\frac{AR}{E}\right) \qquad (7)$$

A plot of $ln(\beta/T_p^2)$ vs $1000/T_p$ gives an approximate straight line and the activation energy E is calculated using the slope (Table 2).

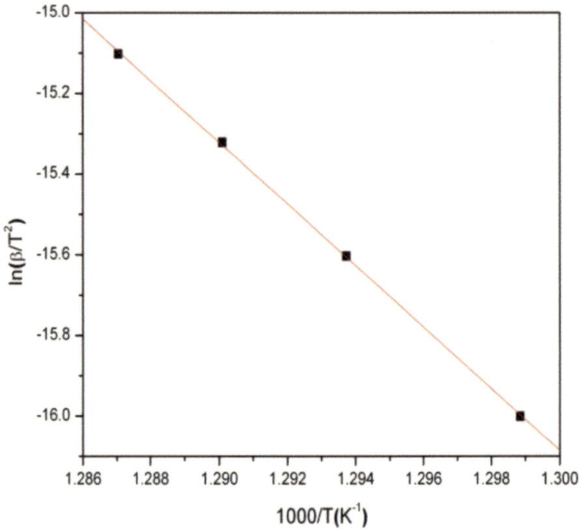

Fig. 3. KAS plot for peak-1

ii) Augis & Bennett's method: This method was suggested by Augis and Bennett (Augis & Bennett, 1978) and is an extension of Kissinger method showing its applicability to heterogeneous reaction described by Avrami expression. Apart from the peak crystallization temperature it also incorporates the onset temperature of crystallization, T_o and it is supposed to be a very accurate method of determining E through the equation

$$ln\left(\frac{\beta}{\left(T_p - T_o\right)}\right) = -\frac{E}{RT_p} + ln\,A \tag{8}$$

where T_p and T_o are the peak and the onset temperatures of crystallization respectively. The values of E obtained from the plot $(ln(\beta/(T_p\text{-}T_o))$ vs $1000/T_p$ is given in Table 2.

Further,

$$n = 2.5\frac{T_p^2}{\Delta T\left(\frac{E}{R}\right)} \tag{9}$$

where ΔT is the full width at half maximum of the DSC curve. n can be derived using Eq. (9).

iii) Boswell method: Boswell (Boswell, 1980) has found a limitation in the Augis & Bennett method that if

$$\frac{T_p - T_o}{T_p} \approx 1$$

then Augis & Bennett gives crude results. However, it may be noted that this condition may not apply to the present case.

Boswell method determines the activation energy at peak temperature (Table 2) using the following equation

$$ln\left(\frac{\beta}{T_p}\right) = -\frac{E}{RT_p} + const \tag{10}$$

c. Li-Tang Method

Li and Tang (Li & Tang, 1999) have developed an isoconversional integral method which does not make any assumption about the kinetic model and involves no approximation in the Eq. (3) as

$$\int_0^{\alpha}\left(ln\frac{d\alpha}{dt}\right)d\alpha = G(\alpha) - \frac{E_{\alpha}}{R}\int_0^{\alpha}\left(\frac{1}{T}\right)d\alpha \tag{11}$$

Where $G(\alpha) = \alpha \ln A + \int_0^{\alpha}(\ln f(\alpha))d\alpha$ has the same value for a given reaction under study and a given α irrespective of β. A plot of $\int_0^{\alpha} ln\left(\frac{d\alpha}{dt}\right)d\alpha$ vs $\int_0^{\alpha}\left(\frac{1}{T}\right)d\alpha$, for a set of βs at constant conversion α will have the slope *-E/R*.

Linear differential isoconversional method

The method suggested by Friedman (Friedman, 1964) sometimes known as transformation rate-isoconversional method, utilizes the differential of the transformed fraction and hence it is called differential isoconversional method. Substituting value of *k(T)* in Eq. (3) and taking logarithm, Friedman derived a linear differential isoconversional expression as

$$ln\left(\frac{d\alpha}{dt}\right)_{\alpha} = ln\,\beta\left(\frac{d\alpha}{dT}\right)_{\alpha} = ln\left(A\,f(\alpha)\right) - \frac{E_{\alpha}}{RT_{\alpha}} \tag{12}$$

by taking logarithm on both sides of Eq.(3). For a constant α, the plot of $ln\left(\beta\frac{d\alpha}{dT}\right) vs \left(\frac{1}{T}\right)$ should be a straight line (Fig. 4) whose slope gives us the value of *E* (Table 1).

Since this method does not take any mathematical approximation for the temperature integral, it is considered to give accurate estimate of *E*. Thus the method does not require any assumption on *f(α)*, i.e. it is a so-called model-free method. However, being a differential method, its accuracy is limited by the signal noise (Dhurandhar et al, 2010).

A method suggested by Gao and Wang (Gao & Wang, 1986) is a special case of the Friedman method. This method uses the following expression to determine the activation energy.

$$ln\left(\beta\frac{d\alpha}{dT_p}\right) = -\frac{E}{RT_p} + const \tag{13}$$

$$K_p = \frac{\beta E}{RT_p^2} \tag{14}$$

where,

$$K_p = A\,exp\left(\frac{-E}{RT_p}\right) and \left(\frac{d\alpha}{dt}\right)_p = 0.37nK_p$$

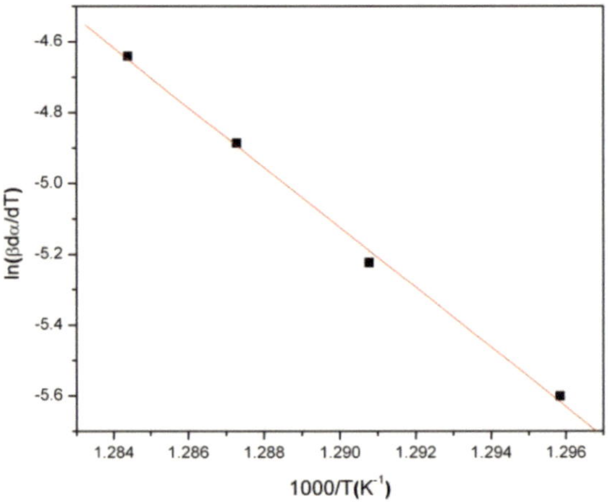

Fig. 4. Friedman plot for peak-1

α	E (kJ/mol)		
	KAS	OFW	Friedman
0.1	602 ± 2	584 ± 2	555 ± 1
0.2	597 ± 1	580 ± 1	626 ± 1
0.3	603 ± 1	586 ± 2	648 ± 1
0.4	615 ± 1	597 ± 1	687 ± 1
0.5	635 ± 1	616 ± 1	725 ± 1
0.6	654 ± 1	634 ± 1	702 ± 3
0.7	648 ± 1	629 ± 1	522 ± 5
0.8	606 ± 1	589 ± 1	398 ± 5
0.9	549 ± 1	534 ± 1	318 ± 2

Table 1. Local Activation energy (E) at different conversion for different methods.

Method	E (kJ/mol)
Kissinger	553 ± 2
Ozawa	546 ± 2
Augis & Bennett	532 ± 2
Boswell	443 ± 7

Table 2. Activation energy (E) using various methods.

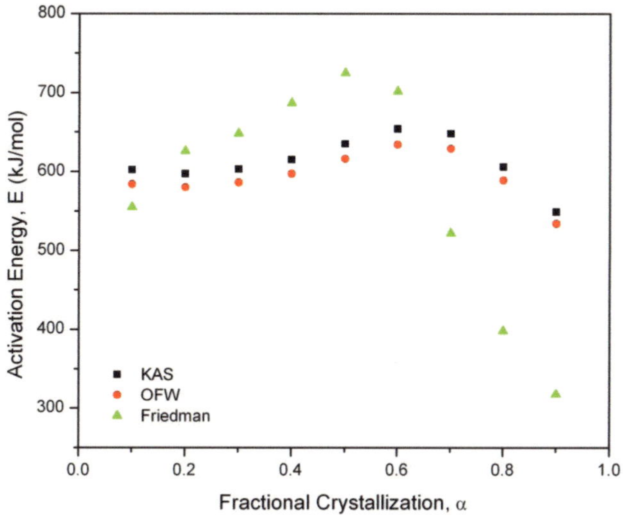

Fig. 5. Local Activation energy E at different α from different methods

Non-linear isoconversional method

Vyazovkin and Wight (Vyazovkin & Wight, 1997) described an advanced isoconversional method. Similar to other isoconversional methods, this method is also based on the assumption that the reaction model, $g(\alpha)$ is independent of the heating program. So, for any two experiments conducted at different heating rates the ratio of the temperature integral I (E, T_α) to the heating rate β is a constant. For a given conversion and a set of n experiments performed under different heating rates, the activation energy can be determined at any particular value of α by finding the value of E_α for which the function

$$\sum_{i \neq j}^{n}\sum^{n}\frac{\left[I(E_\alpha,T_{\alpha i})\beta_j\right]}{\left[I(E_\alpha,T_{\alpha j})\beta_i\right]} \tag{15}$$

is a minimum. The minimization procedure is repeated for each value of α to find the dependence of activation energy on the extent of conversion.

3.2 Isokinetic methods

a. Matusita and Sakka method

Matusita and Sakkka (Matusita & Sakka, 1979) suggested the following equation specifically for the non-isothermal data

$$\ln[\ \ln(1\ \ \alpha)] = \ n\ln\beta \ \ \frac{mE}{RT}\ \text{ı}\ Const \tag{16}$$

where m is an integer depends on the dimensionality of the crystal and the Avarami exponent n depends on the nucleation process. For a constant temperature, the plot of ln[-

ln(1-α)] versus lnβ gives a straight line (Fig.6) and the slope gives the value of *n*, which come out to be n = 1.33 and *n* = 1.36 for temperatures T = 775 K and *T* = 778 K respectively. The plot of ln[-ln(1-α)] versus 1/T at constant heating rate should be a straight line and the value of mE is obtained from the slope (Fig.7).

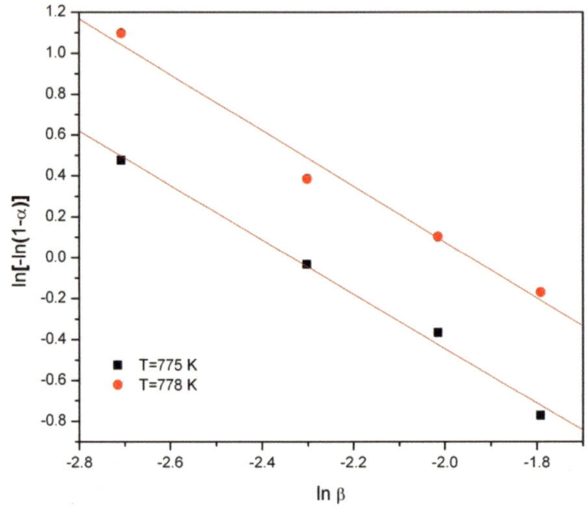

Fig. 6. Plot of ln[-ln(1- α)] Vs. lnβ for diff. Temp.

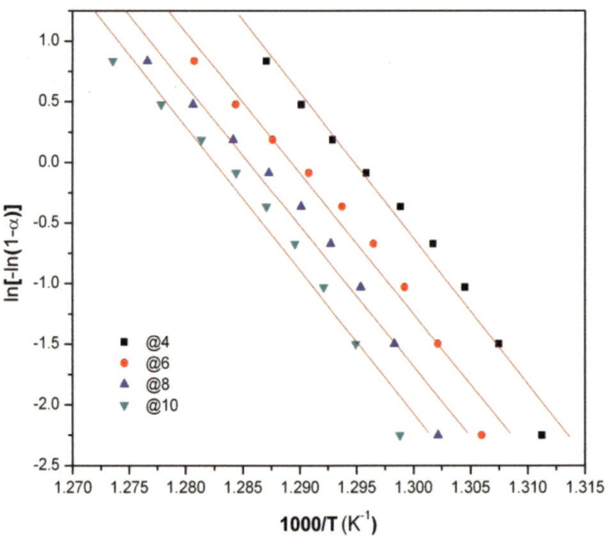

Fig. 7. Plot of ln[-ln(1- α)] Vs. 1000/T for diff. heating rates.

a) Modified Kissinger method

The modified Kissinger equation (Matusita & Sakka, 1980) given below can be utilized to derive the activation energy (*E*).

$$ln\left(\frac{\beta^n}{T_p^2}\right) = -\frac{mE}{RT_p} + Const \tag{17}$$

Where E is the activation energy for crystallization, T_p is the peak temperature and R is the universal gas constant. m is known as the dimensionality of growth and here $m = n$. In order to derive E from this equation, one must know the value of n. The n value can be obtained from the slope of the plot of ln[-ln(1-α)] Vs. lnβ at constant temperature. In order to evaluate E, the values of n are substituted in Eq. (17). Then plots of $ln\left(\frac{\beta^n}{T_p^2}\right)$ Vs. $\frac{1}{T_p}$ (Fig. 8) gives the values of activation energy E, and the average E obtained is 549.80 kJ/mol.

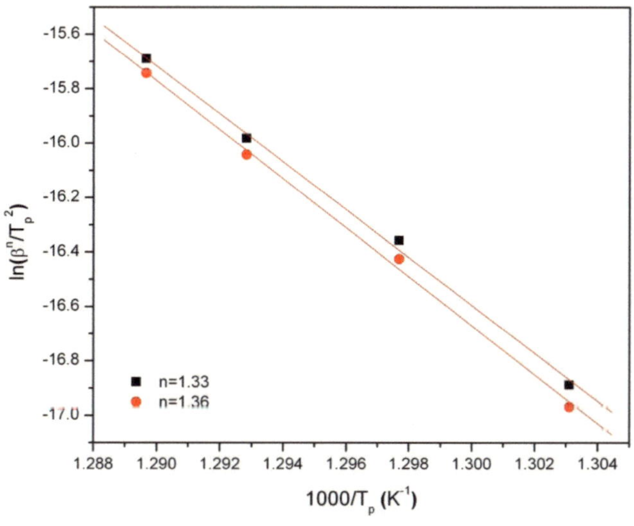

Fig. 8. Mod. Kissinger plot for n = 1.33 and n = 1.36

A general trend of decrease in the values of n with increasing heating rate can be observed. Such trend has been also seen by Matusita and Sakka (Matusita & Sakka, 1979) and in other Fe-based (Raval et al., 2005) metallic glasses.

Heating rate	*Matusita & Sakka*
4	*1.81*
6	*1.76*
8	*1.75*
10	*1.79*

Table 3. Values of Avrami exponent (n)

b. Coats & Redfern method

One of the most popular model-fitting methods is the Coats and Redfern method (Coats & Redfern, 1964). This method is based on the equation

$$ln\frac{g_i(\alpha)}{T^2} = ln\left[\frac{A_iR}{\beta E_i}\left(1-\frac{2RT}{E_i}\right)\right] - \frac{E_i}{RT} \tag{18}$$

$$\cong ln\frac{A_iR}{\beta E_i} - \frac{E_i}{RT}$$

The graph of $ln\frac{g_i(\alpha)}{T^2}$ Vs. $\frac{1}{T}$ gives a straight line whose slope and intercept allow us to calculate E and A for a particular reaction model. For the different kinetic models and for $0.1 \leq \alpha \leq 0.9$, the straight lines corresponding to CR method are characterized by correlation coefficients (r). The general practice in this method to determine E is to look for the model corresponding to maximum r. In some cases, the so-obtained value of E is significantly different from those obtained from other methods.

c. The invariant kinetic parameter (IKP) method

It has been observed that the same experimental curve $\alpha = \alpha(T)$ can be described by different function of conversion $(f(\alpha))$. Further, the values of the activation energy obtained for various $f(\alpha)$ for single non-isothermal curve are correlated through the compensation effect (Galwey, 2003). These observations form the basis of the IKP method. In order to apply this method, $\alpha = \alpha(T)$ curves are obtained at different heating rates (β_v = 4, 6, 8, 10) using DSC. For each heating rate the pairs (A_{vj}, E_{vj}), where j corresponds to a particular degree of conversion, are determined using the following equation:

$$ln\frac{g(\alpha)}{T^2} = ln\frac{AR}{\beta E} - \frac{E}{RT} \tag{19}$$

For constant β, a plot of $ln\frac{g(\alpha)}{T^2}$ Vs. $\frac{1}{T}$ is a straight line whose slope allows the evaluation of activation energy E_v and intercept, pre-exponential factor, A_v for different reaction models $g(\alpha)$. The same procedure is repeated to obtain the pairs (E_v, A_v) for different heating rates. Now, the calculation of invariant activation parameters is done using the compensation relation (Budrugeac, 2007)

$$ln A_v = \alpha^* + \beta^* E_v \tag{20}$$

The Eq. (20) represents a linear relationship between lnA and E; any increase in the magnitude of one parameter is offset, or compensated, by appropriate increase of the other. Plotting lnA_v Vs. E_v for different heating rates, the compensation effect parameters α^* and β^* are obtained. These parameters follow an equation

$$\alpha^* = \ln \boldsymbol{A} - \beta^* \boldsymbol{E} \tag{21}$$

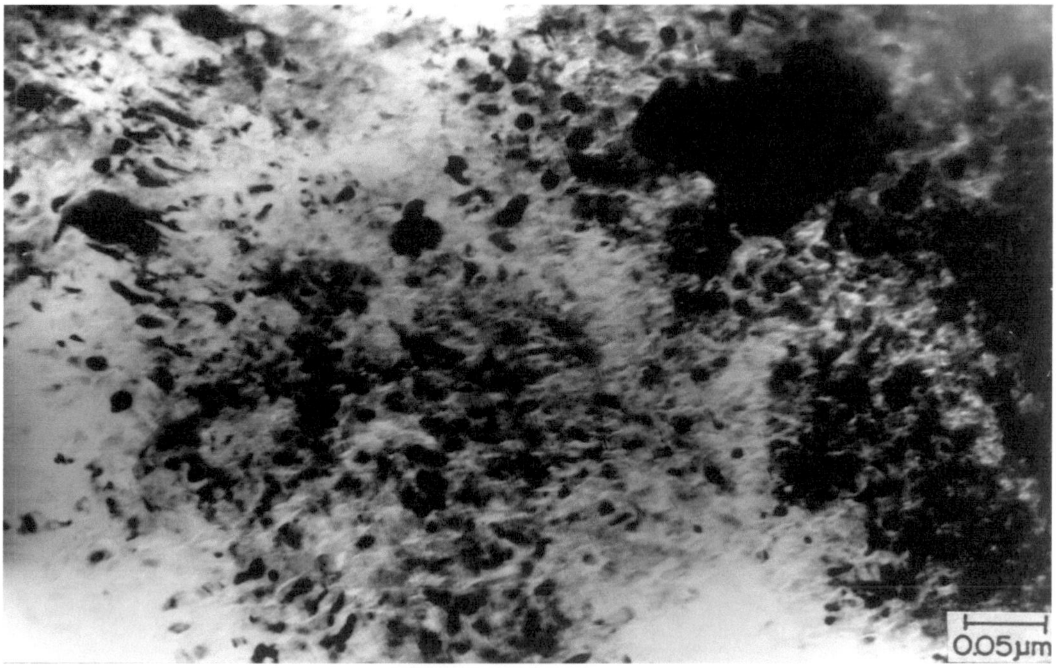

Fig. 9. Bright field TEM image of $Ti_{20}Zr_{20}Cu_{60}$ metallic glass after annealing at 673 K for 4 hours

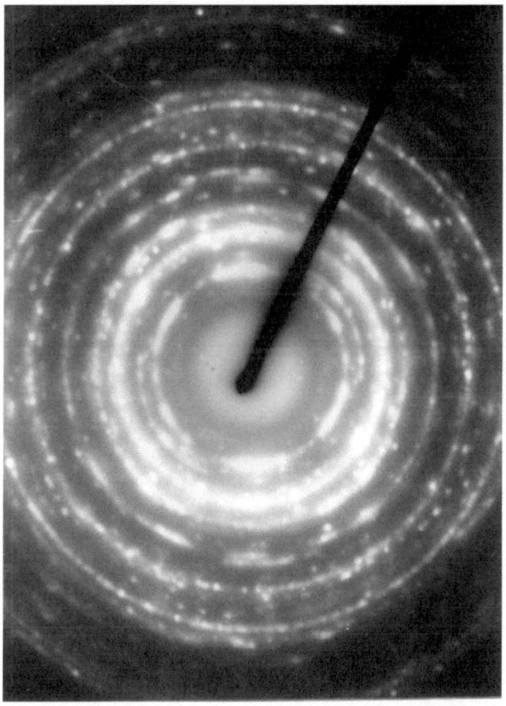

Fig. 10. SAD pattern of $Ti_{20}Zr_{20}Cu_{60}$ metallic glass after annealing at 673 K for 4 hours

The plot of α^* and β^* gives the true values of E and A.

Nano-structures can be synthesized by controlled crystallization of metallic glasses also known as de-vitrification method.

The selected area diffraction (SAD) pattern shows characteristic rings with discontinuity. The phases can also be idetified as seen from fig.11.

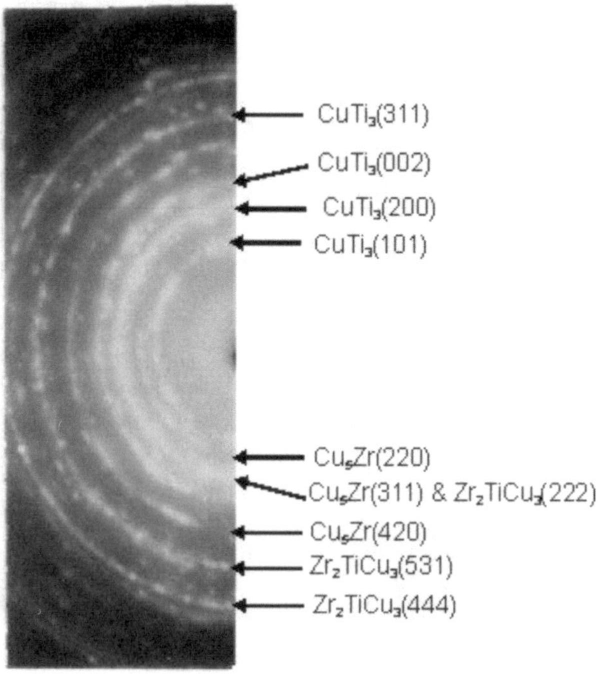

Fig. 11. Nano-phases present in $Ti_{20}Zr_{20}Cu_{60}$ metallic glass after annealing at 673 K for 4 hours

4. Conclusion

Based on the analysis of thermo-analytical data, the very obvious and straight forward question is: which is better, iso-conversional or iso-kinetic?

Iso-conversional methods provide activation energy values, E as a function of conversion, α. The iso-kinetic methods, on the other hand, are used considering the crystallization mechanism to be the same throughout the entire conversion (crystallization) and give single constant value of activation energy. For metallic glasses, the thermally activated phase transformations are more physical than chemical. In fact, crystallization is a complex process involving nucleation and growth and on rigorous grounds, it can not be considered to be a single-step process. The iso-kinetic analysis always leads to a single activation energy (rather say, apparent activation energy) giving an overall picture of the crystallization process. However, the difficulty (and hence uncertainty) in choosing the proper reaction model persists in isokinetic analysis. Therefore, the isoconversional methods are definitely superior to the isokinetic methods as far as the determination of *E* is concerned (Pratap et al, 2007). Nonetheless, accurate determination of E is not the only issue in the kinetic analysis of crystallization process in metallic glasses. The micro-structural evolution during the non-isothermal heating of the metallic glasses is also important. For the determination of the dimensionality of the growth and the grain size, one needs to know a precise reaction model that closely follows the crystallization process. A reaction model independently proposed by John-Mehl- Avrami-Kolmogorov (JMAK) is found to be the most suitable for describing the nucleation and growth process during the non- isothermal crystallization of metallic glasses. This model does help to determine of the kinetic parameters, like the dimensionality of growth (apart from *E* and *A*). The model-free isoconversional methods are definitely superior to the isokinetic methods for the accurate determination of kinetic parameters like E and A. However, the knowledge of accurate E and A is not sufficient for the detailed investigations of the dimensionality of the growth and the grain size using thermal analysis. A precise reaction model accounting for the phase transformations during the crystallization process is a prerequisite for deriving such micro-structural information. This could be a valid proposition if it is explicitly related to the phase transformations involving significant chemical changes. One can find numerous publications where JMAK formalism has been found to be the most appropriate for the description of kinetics of nucleation and growth processes in metallic glasses. Therefore, in our opinion, isokinetic methods (despite its limited applicability) are important and useful for the analysis of non-isothermal crystallization data. So, as far as the study of thermally activated phase transformation in metallic glasses is concerned, both the types of methods are complementary and provide not only useful data, but also pave way into the insight of the crystallization process.

5. References

Afify, S. (1991). Differential scanning calorimetric study of chalcogenide glass $Se_{0.7}Te_{0.3}$, *Journal of Non Crystalline Solids*, Vol. 128, pp. 279-284; ISSN: 0022-3093

Akahira, T. & Sunose, T. (1971). Joint convention of four electrical institutes, *Research Report, Chiba. Institute of Technology (Science and Technology)*, Vol. 16, pp. 22-31

Augis, J.A. & Bennett, J.E. (1978). Calculation of the Avrami parameters for heterogeneous solid state reactions using a modification of the Kissinger method, *Journal of Thermal Analysis and Calorimetry*, Vol. 13, pp. 283-292; ISSN:1388-6150 (Print), 1572-894 (electronic version)

Boswell, P.G. (1980). On the calculation of activation energies using modified Kissinger method, *Journal of Thermal Analysis and Calorimetry*, Vol. 18, pp. 353-358 ISSN:1388-6150 (Print), 1572-894 (electronic version)

Budrugeac, P. (2007). The Kissinger law and the IKP method for evaluating the non-isothermal kinetic parameters, *Journal of Thermal Analysis and Calorimetry*, Vol. 89, pp. 143-151 ISSN:1388-6150 (Print), 1572-894 (electronic version)

Cai, J.M. & Chen, S.Y. (2009).A new iterative linear integral isoconversional method for the determination of the activation energy varying with the conversion degree *Journal of Computational Chemistry*, Vol. 30, pp. 1986-1991; Online ISSN: 1096-987X

Castro, M. (2003). Phae-field approach to heterogeneous nucleation, *Physical Review B*, Vol. 67, pp. 035412-035419; ISSN: 0163-1829 (Print), 1095-3795 (electronic version)

Coats, A.W. & Redfern J.P. (1964). Kinetic Parameters from Thermogravimetric Data, *Nature (London)*, Vol. 201, pp. 68-69; ISSN: 0028-0836 (Print), EISSN: 1476-4687

Criado, J.M.; Sanchez-Jimenez, P.E. & Perez-Maqueda, L.A. (2008). *Journal of Thermal Analysis and Calorimetry*, Vol. 92, pp. 199-203; ISSN:1388-6150 (Print), 1572-894 (electronic version)

Dhurandhar, H.; Patel, A.T.; Shanker-Rao, T.L.; Lad, K.N. & Pratap, A. (2010). Kinetics of crystallization of Co-based multi-component amorphous alloy, *Journal of ASTM International*, Vol. 7, pp. 1-15; 978-0-8031-7516-7

Doyle, C.D. (1961). Kinetic Analysis of Thermogravimetric Data, *Journal of Applied Polymer Science*, Vol. 5, pp. 285–292; ISSN: 1097-4628

Doyle, C.D. (1962). Estimating Isothermal Life from Thermogravimetric Data, *Journal of Applied Polymer Science*, Vol. 6, pp. 693-642; ISSN:1097-4628

Doyle, C.D. (1965). Series Approximations to the Equation of Thermogravimetric Data, *Nature (London)*, Vol. 207, pp. 290-291; ISSN: 0028-0836 (Print), EISSN: 1476-4687

Flynn, J.H. & Wall, L.A. (1966). General Treatment of the Thermogravimetry of Polymers, *Journal of Research of the National Bureau of Standards. Section- A. Physics and Chemistry*, Vol. 70A, pp. 487-523; 0022-4332

Friedman, H.L. (1964). Kinetics of thermal degradation of char-forming plastics from thermogravimetry. Application to phenolic plastic, *Journal of Polymer Science Part C*, Vol. 6, pp. 183-195; ISSN: 0449-2994

Galwey, A.K. (2003). Eradicating erroneous Arrhenius arithmetic, *Thermochimica Acta*, Vol. 399, pp. 1-29; ISSN: 0040-6031

Gao, Y.Q. & Wang, W. (1986). On the activation energy of crystallization in metallic glasses, *Journal of Non-Crystalline Solids*, Vol. 81, pp. 129-134; ISSN: 0022-3093

Giridhar, A. & Mahadevan, S. (1982). Studies on the As-Sb-Se glass system, *Journal of Non Crystalline Solids*, Vol. 51, pp. 305-315; ISSN: 0022-3093

Gupta, A.K.; Jena, A.K. & Chaturvedi, M.C. (1988). A differential technique for the determination of the activation energy of precipitation reactions from Differential Scanning Calorimetric data, *Scripta Metallurgica*, Vol. 22, pp. 369-371; ISSN: 0036-9748

Hsiao, A.; McHenry, M.E.; Luaghlin, D.E.; Kramer, M.J.; Ashe, C. & Ohkubo, T. (2002). The Thermal, Magnetic, and Structural Characterization of the Crystallization Kinetics of $Fe_{88}Zr_7B_4Cu_1$, An Amorphous Soft Magnetic Ribbon, *IEEE Transactions on Magnetics*, Vol. 38, No. 5, pp. 3039-3044; ISSN:0018-9464

Jones, G.A.; Bonnett, P. & Parker, S.F.H. (1986). Crystallilzation kinetics of the amorphous magnetic material 2605 Co ($Fe_{67}Co_{18}B_{14}Si_1$), *Journal of Magnetism & Magnetic Materials*, Vol. 58, pp. 216-226; ISSN:0304-8843

Kissinger, H.E. (1957). Reaction kinetics in differential thermal analysis, *Analytical Chemistry*, Vol. 29, pp. 1702-1706; ISSN:0003-2700

Lad, K.N.; Savalia, R.T.; Pratap, A., Dey, G.K. & Banerjee, S. (2008). Isokinetic and isoconversional study of crystallization kinetics of a Zr-based metallic glass, *Thermochimica Acta*, Vol. 473, pp. 74-80; ISSN: 0040-6031

Lesz, S. & Szewieczek, D. (2005). *Proceedings of the Worldwide Congress on Materials and Manufacturing Engineering and Technology COMMENT'2005*, Poland, Gliwice-Wisla (CD-ROM), Vol. 637, May 16-19, 2005

Li, C.R. & Tang, T.B. (1999). A new method for analyzing non-isothermal thermoanalytical data from solid-state reactions, *Thermochimica Acta,* Vol. 325, pp. 43-46; ISSN:0040-6031

Ligero, R.A.; Vazquez, J.; Villares, P. & Jimenez-Garay, R. (1990). Crystallization kinetics in the As-Se-Te system, *Thermochimica Acta,* Vol. 162, pp. 427-434; ISSN: 0040-6031

Matusita, K. & Sakka, S. (1979). Kinetic study of crystallization of glass by differential scanning calorimetry, *Physical Chemistry of Glasses,* Vol. 20, pp. 81-84; 0031-9090

Matusita, K. & Sakka, S. (1980). Kinetic study on Crystallization of glass by differential thermal analysis - criterion on application of Kissinger plot, *Journal of Non-Crystlline Solids,* Vol. 38-39, pp. 741-746; ISSN: 0022-3093

Minic D.M. (2006). Synthesis, Characterization and Stability of Amorphous Alloys, *Science of Sintering,* Vol. 38, pp. 83-92; 0350-820x

Minic, D.M. & Adnadevic, B. (2008). Mechanism and kinetics of crystallization of α-Fe in amorphous $Fe_{81}B_{13}Si_4C_2$ alloy, *Thermochimica Acta.,* Vol. 474, pp. 41-46; ISSN: 0040-6031

Minic, D.M.; Gavrilovic, A.; Angerer, P.; Minic, D.G. & Mariclc, A. (2009). Thermal stability and crystallization of $Fe_{89.8}Ni_{1.5}Si_{5.2}B_3C_{0.5}$ amorphous alloy, *Journal of Alloys and Compound,* Vol. 482, pp. 502-507; ISSN: 0925-8388

Moharram, A.H.; El-Oyoun, M.A. & Abu-Sehly, A.A. (2001). Calorimetric study of the chalcogenide $Se_{72.5}Te_{20}Sb_{7.5}$ glass, *Journal of Physics D Applied Physics,* Vol. 34, pp. 2541-2546; ISSN 0022-3727 (Print) ISSN 1361-6463 (Online)

Ozawa, T. (1965). A New Method of Analyzing Thermogravimetric Data, *Bulletin of the Chemical Society of Japan,* Vol. 38, pp. 1881-1886; ISSN 0009-2673

Patel, A.T. & Pratap, A. (2012). Kinetics of crystallization of $Zr_{52}Cu_{18}Ni_{14}Al_{10}Ti_6$ glass, *Journal of Thermal Analysis and Calorimetry,* Vol. 107, pp. 159-165; ISSN:1388-6150 (Print), 1572-894 (electronic version)

Paulik, F. (1995). Ch. 10, *Special Trends in Thermal Analysis,* John Wiley & Sons, Chichester, UK

Pratap, A.; Shanker-Rao, T.L.; Lad, K.N. & Dhurandhar, H.D. (2007). Isoconversional vs. Model fitting methods A case study of crystallization kinetics of a Fe-based metallic glass, *Journal of Thermal Analysis and Calorimetry,* Vol. 89, pp. 399-405; ISSN:1388-6150 (Print), 1572-894 (electronic version)

Raval, K.G.; Lad, K.N.; Pratap, Arun; Awasthi, A.M. & Bhardwaj, S. (2005). Crystallization kinetics of a multicomponent Fe-based amorphous alloy using modulated differential scanning calorimetry, *Thermochimica Acta,* Vol. 425, pp. 47-57; ISSN: 0040-6031

Rotaru, A. & Gosa, M. (2009). Computational thermal and kinetic analysis Complete standard procedure to evaluate the kinetic triplet form non-isothermal data, *Journal of Thermal Analysis and Calorimetry,* Vol. 97, pp. 421-426; ISSN:1388-6150 (Print), 1572-894 (electronic version)

Roura, P. & Farjas, J. (2009).Analytical solution for the Kissinger equaton, *Journal of Materials Research,* Vol. 24, pp. 3095-3098;

Rysava, N.; Spasov, T. & Tichy, L. (1987). Isothermal DSC methods for evaluation of the kinetics of crystallization in the Ge-Sb-S glassy system, *Journal of Thermal Analysis,* Vol. 32, pp. 1015-1021; ISSN:1388-6150 (Print), 1572-894 (electronic version)

Starink, M.J. (2003). The determination of activation energy form linear heating rate Experiments: A comparison of the accuracy of isoconversion methods, *Thermochimica Acta*, Vol. 404, pp. 163-176; ISSN: 0040-6031

Starink, M.J. (1997). On the applicability of isoconversion methods for obtaining the activation energy of reactions within a temperature-dependent equilibrium state, *Journal of Materials Science*, Vol. 32, pp. 6505-6512; ISSN: 0022-2461

Szewieczek, D. & Lesz, S. (2004). The Structure and Selected Physical Properties of the Nanocrystalline $Fe_{92.4}Hf_{4.2}B_{3.4}$ Alloy, *Journal of Materials Processing Technology*, Vol. 157–158, pp. 771–775; ISSN:0924-0136

Szewieczek, D. & Lesz, S. (2005). Influence of Structure on Magnetic and Mechanical Properties ofAmorphous and Nanocrystalline $Fe_{85.4}Hf_{1.4}B_{13.2}$ Alloy, *Proceedings of the 13th International Scientific Conference, Achievements in Mechanical & Materials Engineering AMME'05*, Gliwice-Wisla, Vol. 637

Vyazovkin, S. & Wight, C.A. (1997). Isothermal and Nonisothermal Reaction Kinetics in Solids: In Search of Ways toward consensus,*The Journal of Physical Chemistry A*, Vol. 101, pp. 8279-8284; ISSN: 1089-5639

Vyazovkin, S. (2003). Reply to "What is meant by the term 'variable activation energy when applied in the kinetics analyses of solid state decompositions (crystolysis reactions)?", *Themochimica Acta*, Vol. 397, pp. 269-271; ISSN:0040-6031

Vyazovkin, S. (2010). Thermal Analysis, *Analytical Chemistry*, Vol. 82, pp. 4936-4949; ISSN 0003-2700

6

Crystallization Kinetics of Amorphous Materials

Miray Çelikbilek, Ali Erçin Ersundu and Süheyla Aydın
Istanbul Technical University
Turkey

1. Introduction

Amorphous (non-crystalline) materials have no crystal structure where the atoms appear to have a random distribution (Omar, 1993). There are different classes of amorphous materials. Glasses, inorganic materials which have no long-range order (<10 Å) and high viscosity greater than 1013 Poise, are the most typical amorphous materials (Doremus, 1973; Jackson, 2004; Park, 2009). The regular arrangement resulting from the distribution over long distances of a repeating atomic arrangement, which is characteristic of a crystal, is missing in glasses (see Fig. 1). However, there is often evidence of a short-range order in glasses, which corresponds to the atomic arrangement in the immediate vicinity of any selected atom (Carter & Norton, 2007).

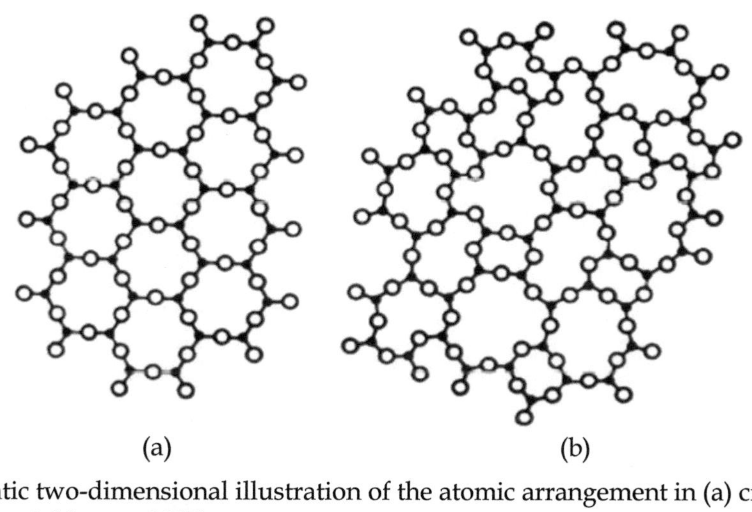

Fig. 1. Schematic two-dimensional illustration of the atomic arrangement in (a) crystal and (b) glass (Carter & Norton, 2007)

Glass can be formed by cooling from a liquid state without a change in its specific volume, which delays crystallization and assist to reach the glass transition temperature before the crystallization occurs (Jackson, 2004; Park, 2009). This is why amorphous materials, such as glasses, are sometimes referred to as supercooled liquids (Omar, 1993).

In a liquid, the atoms or molecules move around much more rapidly than in a crystal. They are constantly in motion, jiggling around relative to each other, unlike in a crystal, where the atoms are bound to specific lattice sites. When a liquid is cooled, the space for the atoms to move around decreases and on further cooling below the glass transition temperature, the atoms can no longer move around with respect to each other, and so the material becomes a solid. A measure of this is the specific volume, which can be measured as the difference between the density of the crystal and of the liquid. As a glass-forming material cools, this excess volume decreases, and finally the density of the glass approaches that of the crystal, as illustrated in Fig. 2. In practice, the formation of an amorphous or crystalline solid depends on how rapidly the liquid is cooled through the glass transition temperature (Jackson, 2004). Upon cooling the liquid, if there is a discontinuity in volume change or in rate of cooling the liquid crystallizes, however if the liquid passes into a supercooled state the volume decreases and no crystallization occurs (Carter & Norton, 2007).

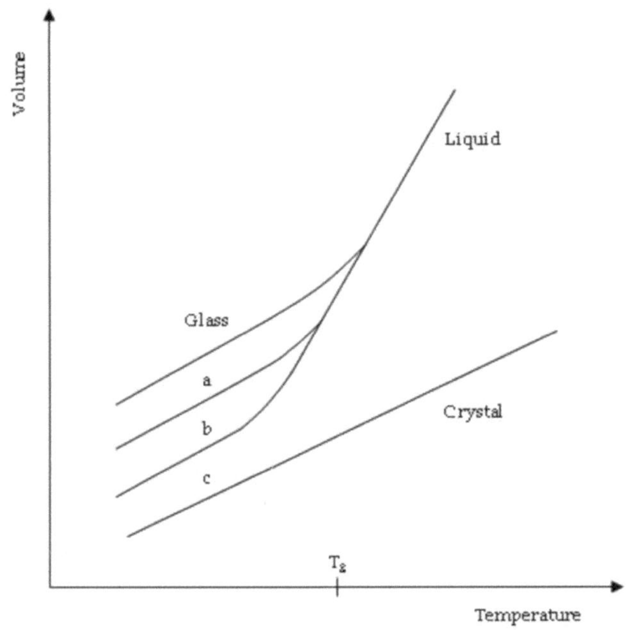

Fig. 2. The specific volume in a liquid decreases more rapidly with temperature than the crystal. The thermal expansion coefficient of glass is similar to that of the crystal. The final specific volume of the glass depends on the cooling rate: (a) fast cooling, (b) normal cooling, (c) slow cooling (Carter & Norton, 2007; Jackson, 2004)

It is believed that most materials can be prepared as glasses by sufficiently rapid quenching but there is a notable exception: no pure metal has been prepared in an amorphous state (Jackson, 2004). Materials which can form non-crystalline solids with the atomic arrangement shown in Fig. 1b and at an appreciable size are found in glass systems in oxides, halides, and chalcogenides. The three-dimensional random network of strong bonds is developed by the constituent called the "network former". In principle, glass formation is possible for a system of any composition provided if it contains sufficient of the network former. Network modifiers can also participate in glass formation by acting to modify the glass properties. These components do not form networks but occupy thermodynamically stable sites as illustrated schematically in Fig. 3 or act as a replacement for a part of network former. Among various oxides used in the industrial materials, SiO_2, GeO_2, B_2O_3, and P_2O_5 are known to be good network formers which can develop the three-dimensional random network and can form a glass by themselves (Yamane & Ashara, 2000).

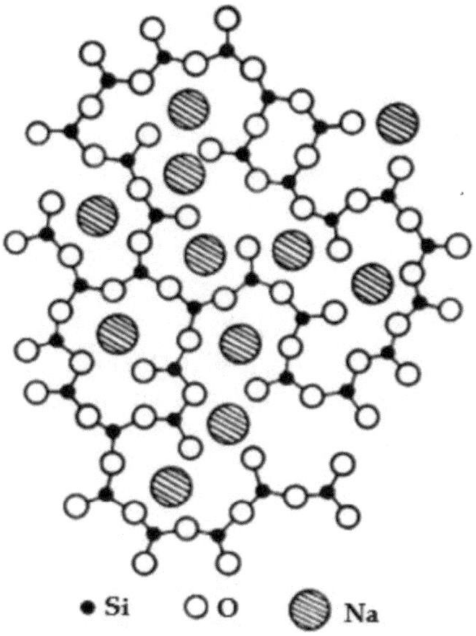

Fig. 3. Schematic two-dimensional illustration of the structure of a binary sodium silicate glass (Yamane & Ashara, 2000)

The special glasses used as key components of various devices in the fields of optics, electronics, and opto-electronics are not always silicates but are often non-silicate glasses of the phosphate, borate, germanate, vanadate or tellurite systems. Although non-silicate glasses are not generally applied to mass production due to the high cost of raw materials and their rather inferior chemical durability, they do show unique properties that cannot be obtained for silicate glasses (Yamane & Ashara, 2000). Comparing with silicate, borate and phosphate glasses, tellurite glasses have drawn considerable attention because of their various promising properties, such as relatively low-phonon energy, high linear and non-linear refractive index, high dielectric constant, good infrared transmission, good corrosion resistance, thermal and chemical stability, low crystallization ability and easy fabrication at low temperatures. Therefore, tellurite glasses are of scientifically and technologically important for their potential use in fiber optics, laser hosts, infrared and infrared to visible upconversion applications in optical data storage, sensors, and spectroscopic applications. TeO_2 is a conditional glass former which does not transform to the glassy state under normal quenching conditions. Therefore, addition of a secondary component, such as heavy metal oxides, alkalis or halogens increases its glass forming ability (Çelikbilek et al., 2010, 2011a, 2011b; El-Mallawany, 2002; Ersundu et al., 2010a, 2010b, 2011; Karaduman et al., 2011).

In order to use special glasses in the fields of opto-electronics at high optical intensities without exposure to thermal damage it is important to recognize their physical and thermal characteristics and also to control the crystallization processes. The investigation of crystallization in terms of kinetics is an important informational tool for the audience and therefore this chapter "Crystallization Kinetics of Amorphous Materials" is crucial to have a complete understanding of the crystallization phenomenon.

2. Crystallization

Crystallization of a liquid or an amorphous solid is a complex process involving simultaneous nucleation and growth of crystallites (Yinnon & Uhlmann, 1983). Crystallization is initiated by crystal nucleation. Nucleation may occur spontaneously or it may be induced artificially. It is not always possible, however, to decide whether a system has nucleated of its own accord or whether it has done so under the influence of some external stimulus (Mullin, 2001). The nucleation either occurs without the involvement of a foreign substance in the interior of the parent phase which is called "homogeneous or primary nucleation" or with the contact of the parent phase with a foreign substance that acts as a preferred nucleation site which is called "heterogeneous or secondary nucleation". The nucleation process is followed by the growth of the crystal nuclei to macroscopic dimensions, which is called "crystal growth" (Kalb, 2009).

2.1 Homogeneous nucleation

Homogeneous nucleation occurs in the interior of the parent phase without the involvement of a foreign substance. At temperatures below a material's melting point (T_m), the driving force for solidification is the difference in Gibbs free energy (ΔG) between the liquid and the solid. If we assume that the heat capacities of the liquid and solid are equal, then the molar enthalpy and molar entropy of solidification will each remain constant as a function of temperature, and ΔG can be calculated as follows:

$$\text{liquid} \rightarrow \text{solid}$$

$$\Delta G = \Delta H - T\Delta S \tag{1}$$

Note that $\Delta H = -L$, where L is the latent heat of fusion.

$$\Delta G = -L + T\frac{L}{T_m} \tag{2}$$

$$\Delta G = \frac{L}{T_m}(T - T_m) \tag{3}$$

When a spherical particle of solid of radius r is formed, the change in Gibbs free energy is the volume of the particle multiplied by the volumetric Gibbs free energy change, ΔG_v.

$$\Delta G_{vol} = \frac{4}{3}\pi r^3 \Delta G_v \tag{4}$$

where ΔG_v is the Gibbs free energy per unit volume,

$$\Delta G_v = \frac{1}{V}\frac{L}{T_m}(T - T_m) \tag{5}$$

$$\Delta G_{vol} = \frac{4}{3}\pi r^3 \frac{1}{V}\frac{L}{T_m}(T - T_m) \tag{6}$$

But when the particle of radius r is formed, there is another energy term to be considered, the surface energy. The surface energy of the particle is:

$$\Delta G_s = 4\pi r^2 \gamma \tag{7}$$

where $\gamma = \gamma_{s\text{-}l}$ the surface energy between solid and liquid.

The sum of the two energy term is:

$$\Delta G_r = 4\pi r^2 \gamma + \frac{4}{3}\pi r^3 \Delta G_v \tag{8}$$

The first of these terms involves the increase in energy required to form a new surface. The second term is negative and represents the decrease in Gibbs free energy upon solidification. Because the first is a function of the second power of the radius, and the second a function of the third power of the radius, the sum of the two increases, goes through a maximum, and then decreases (Fig. 4).

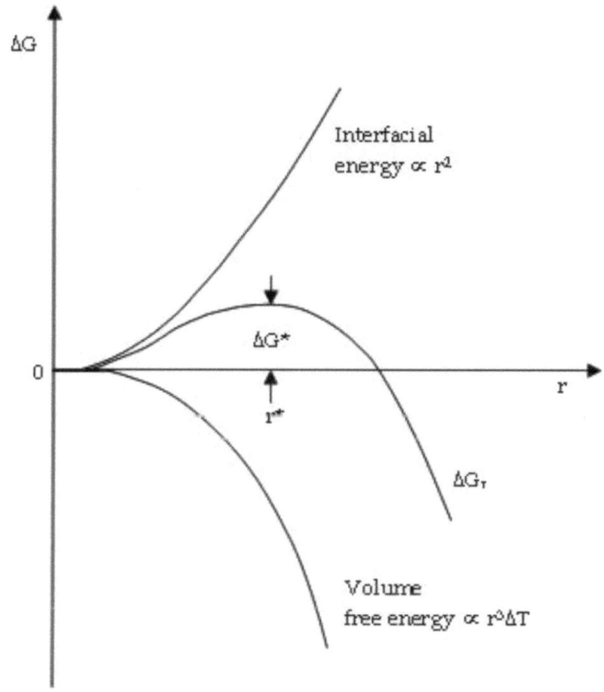

Fig. 4. The free energy change associated with homogeneous nucleation of a sphere of radius r (Ragone, 1994)

The radius at which the Gibbs free energy curve is at maximum is called the critical radius r^*, for a nucleus of solid in liquid. The driving force of the Gibbs free energy will tend to cause a particle with a smaller radius than r^* to decrease in size. This is a particle of subcritical size for nucleation. A viable nucleus is one with radius greater than or equal to r^*. The critical Gibbs free energy corresponding to the radius r^* is ΔG^*. These terms can be shown to be:

$$r^* \text{ when } \left(\frac{\partial \Delta G_r}{\partial r}\right)_T = 0 = 8\pi r\gamma + 4\pi r^2 \Delta G_v \tag{9}$$

$$r^* = -\frac{2\gamma}{\Delta G_v} \tag{10}$$

$$\Delta G^* = \frac{16}{3} \frac{\pi\gamma^3}{\Delta G_v^2} \tag{11}$$

In practice homogeneous nucleation is rarely encountered in solidification. Instead heterogeneous nucleation occurs at crevices in mould walls, or at impurity particles suspended in the liquid.

2.2 Heterogeneous nucleation

Usually, foreign phases like container walls and impurities aid in the nucleation process and thereby increase the nucleation rate. In this case, nucleation is called heterogeneous (Kalb, 2009).

Under practical solidification conditions, supercooling of only a few degrees is observed because nuclei of the solid can be formed on surfaces that catalyze solidification, such as inclusions in the material being solidified, the walls of the container in which it is being held, or the surfaces of the casting molds. To catalyze solidification, the nucleus of solid must wet the catalyst to some extent. A nucleus catalyzed on a surface is shown schematically in Fig. 5.

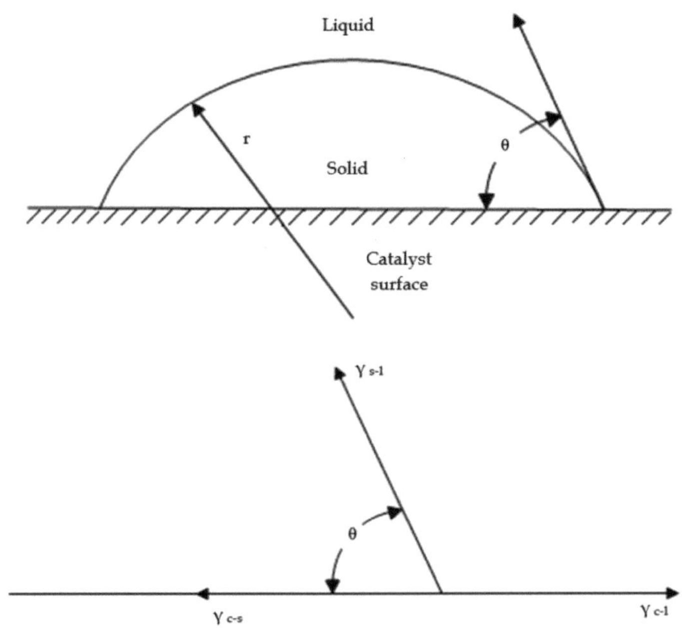

Fig. 5. Catalysis of a nucleus on a surface (Porter & Easterling, 1992)

The analysis of the energies involved in heterogeneous nucleation follows the same method as the one used for homogeneous nucleation. In the case of a heterogeneous nucleus in the form of a spherical cap (Fig. 5), the surface energy term involves the surface energy of the catalyst surface as it is coated by the nucleus.

The surface energy term is derived as follows:

solid-liquid surface = $2\pi r^2(1 - \cos\theta)$

catalyst-solid surface = $\pi r^2(1 - \cos^2\theta)$

where r is the radius of the curvature of the nucleus. Then we write

$$\Delta G_{surface} = 2\pi r^2 (1-\cos\theta)\gamma_{s-l} + \pi r^2 (1-\cos^2\theta)(\gamma_{c-s} - \gamma_{c-l}) \tag{12}$$

where γ_{s-l} = solid-liquid interfacial energy

γ_{c-s} = solid-catalyst interfacial energy

γ_{c-l} = liquid-catalyst interfacial energy

The terms involving the interactions between the catalyst surface and the liquid and solid can be expressed in terms of the solid-liquid interfacial energy by noting the relationships among them.

$$\gamma_{c-l} = \gamma_{c-s} + \gamma_{s-l}(\cos\theta) \tag{13}$$

The volumetric Gibbs free energy change is the product of the volume of the cap and ΔG_v, the specific Gibbs free energy change. That volume, in terms of its radius of the curvature and contact angle is:

$$V = \frac{4}{3}\pi r^3 \left\{\frac{(2+\cos\theta)(1-\cos\theta)^2}{4}\right\} \tag{14}$$

or

$$\Delta G_{vol} = \frac{4}{3}\pi r^3 \Delta G_v f(\theta) \tag{15}$$

where

$$f(\theta) = \left\{\frac{(2+\cos\theta)(1-\cos\theta)^2}{4}\right\} \tag{16}$$

$$r^* = -\frac{2\gamma_{s-l}}{\Delta G_v} \tag{17}$$

$$\Delta G^* = \frac{16}{3}\frac{\pi\gamma_{s-l}^3}{\Delta G_v^2} f(\theta) \tag{18}$$

It is particularly important to note that the critical radius of curvature, r^*, does not change when the nucleation becomes heterogeneous. The critical Gibbs free energy, ΔG^*, however, is strongly influenced by the wetting that occurs at the surface of the material that catalyzes the nucleation. A lower values of ΔG^* means a lower activation energy to be overcome in nucleation; that is, nucleation takes place more easily. The magnitude of the effect can be appreciated by considering values of $f(\theta)$. $f(\theta)$ is the ratio of the volume of the heterogeneous nucleus (the cap) to the volume of the sphere with the same radius of curvature (Ragone, 1994).

Fig. 6, a graph of ΔG as a function of radius of curvature of the nucleus, shows the effect of wetting on the critical Gibbs free energy to be overcome for the nucleus to form.

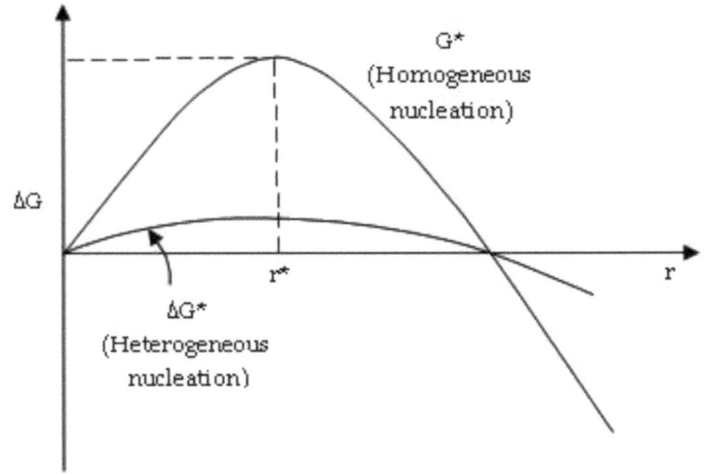

Fig. 6. Plot of ΔG versus r for homogeneous nucleation and an example of heterogeneous nucleation (Ragone, 1994)

The critical Gibbs free energy for nucleation depends on the nucleus volume. This can be demonstrated by considering a nucleus having the shape of spherical cap with radius of curvature r. The Gibbs free energy of the nucleus depends on the interfacial energy and the volumetric Gibbs free energy change as follows:

$$\Delta G_r = \alpha r^2 \gamma + \beta r^3 \Delta G_v \quad (19)$$

The parameters α and β are determined by the particular geometry of the nucleus. The surface energy term, γ, is an average surface energy for the nucleus determined according to the geometrical factors.

The volume of the nucleus is βr^3. To determine r^*,

$$\left(\frac{\partial \Delta G_r}{\partial r}\right)_T = 0 \quad (20)$$

$$2\alpha\gamma r^* + 3\beta r^{*2} \Delta G_V = 0 \quad (21)$$

$$r^* = -\frac{2\alpha}{3\beta\Delta G_V}\gamma \text{ or } \alpha = -\frac{3\beta\Delta G_V}{2\gamma} r^* \quad (22)$$

Substituting in Equation 19, we have

$$\Delta G_r^* = -\frac{3\beta\Delta G_V r^*}{2}(r^*)^2 + \Delta G_V \beta (r^*)^3 \quad (23)$$

$$\Delta G_r^* = -\frac{1}{2}\beta (r^*)^3 \Delta G_V \quad (24)$$

$$\Delta G_r^* = -\frac{1}{2} V^* \Delta G_V \quad (25)$$

where V^* is the volume of the critical nucleus.

It is apparent that any factors that reduce the volume of the nucleus reduce the critical Gibbs free energy of formation of that nucleus, making nucleation more probable (Ragone, 1994).

The activation energy barrier against heterogeneous nucleation (ΔG^*_{het}) is smaller than ΔG^*_{hom} by the shape factor $f(\theta)$. In addition, the critical nucleus radius (r^*) is unaffected by the mould wall and only depends on the undercooling. This result was to be expected since equilibrium across the curved interface is unaffected by the presence of the mould wall.

The effect of undercooling on ΔG^*_{het} and ΔG^*_{hom} is shown schematically in Fig. 7a. If there are n atoms in contact with the mould wall the number of nuclei should be given by:

$$n^* = n_1 \exp\left(-\frac{\Delta G^*_{het}}{kT}\right) \tag{26}$$

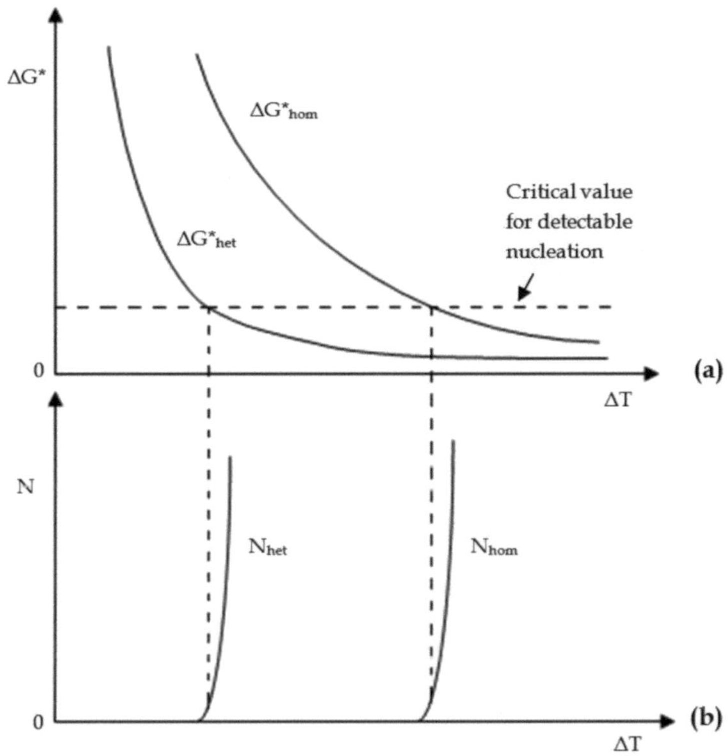

Fig. 7. (a) Variation of ΔG^* with undercooling (ΔT) for homogeneous and heterogeneous nucleation, (b) The corresponding nucleation rates assuming the same critical value of ΔG^* (Porter & Easterling, 1992)

Therefore, heterogeneous nucleation should become feasible when ΔG^*_{het} becomes sufficiently small. The critical value for ΔG^*_{het} should not be very different from the critical value for homogeneous nucleation. It will mainly depend on the magnitude of n_1, in Equation 26. It can be seen from Fig. 7b that heterogeneous nucleation will be possible at much lower undercoolings than are necessary for homogeneous nucleation.

2.3 Growth

Nucleation is the birth of new crystal nuclei either spontaneously from solution or in the presence of existing crystals. Crystal growth is the increase in size of crystals as solute is deposited from solution. These often competing mechanisms ultimately determine the final crystal size distribution. The rate of growth of a transformation product is determined by the driving force for the transformation and the frequency with which molecules successfully make the transition from the reactant phase to the product phase. To use solidification as an example, the driving force is the negative of the ΔG of solidification:

$$-\Delta G = \frac{L}{T_m}(T_m - T) \quad (27)$$

The jump frequency across the liquid-solid interface has a temperature dependence of the form:

$$f = f_0 \exp\left(-\frac{\Delta G_M}{kT}\right) \quad (28)$$

where ΔG_M is the activation energy for movement across the liquid-solid interface. The product of the two is:

$$\text{solidification rate} = f_0 \exp\left(-\frac{\Delta G_M}{kT}\right)\left(\frac{L}{T_m}\right)(T_m - T) \quad (29)$$

As the temperature decreases, the driving force increases but the jump frequency decreases. These two opposing dependencies can produce a maximum in the rate of growth as a function of temperature, as illustrated in Fig. 8.

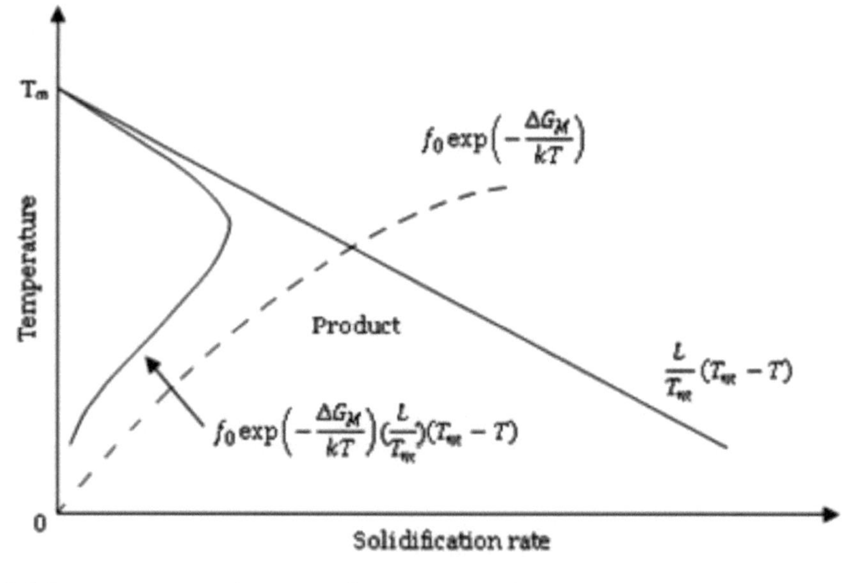

Fig. 8. Solidification rate as a function of temperature (Ragone, 1994)

The temperature dependencies of both nucleation of a new phase and its rate of growth result in a strong temperature dependence of transformation rate.

2.4 Supersaturation

Supersaturation is the difference between the actual concentration and the solubility concentration at a given temperature which is the driving force for all solution crystallization processes. The figure below (Fig. 9) illustrates the concept of supersaturation and introduces the metastable zone width (MSZW), the kinetic boundary at which crystallization occurs (Porter & Easterling, 1992).

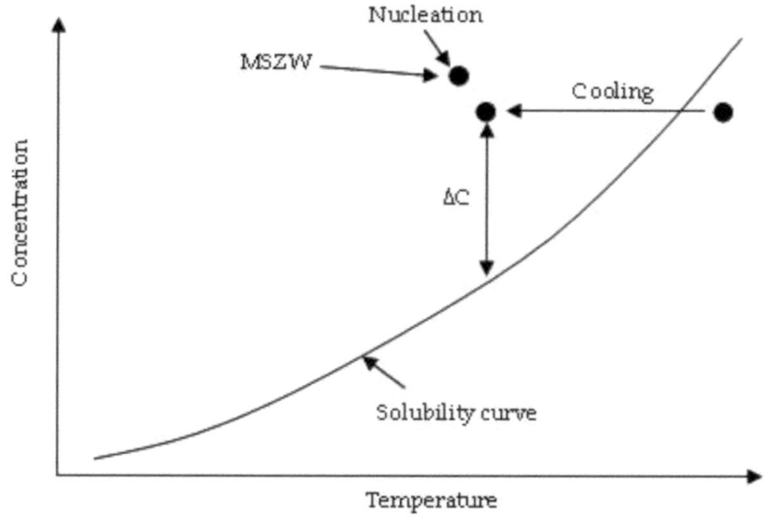

Fig. 9. The excess free energy of solid clusters for homogeneous and heterogeneous nucleation. Note r^* is independent of nucleation site

Supersaturation is critical because it is the driving force for crystal nucleation and growth. The relationship between supersaturation and nucleation and growth is defined by the following equations.

$$G = k_g \Delta C^g \tag{30}$$

$$B = k_b \Delta C^b \tag{31}$$

where, G is growth rate, k_g is growth constant, g is growth order, B is nucleation rate, k_b is nucleation constant, b is nucleation order, ΔC is supersaturation.

At low supersaturation, crystals can grow faster than they nucleate resulting in a larger crystal size distribution. However, at higher supersaturation, crystal nucleation dominates crystal growth, ultimately resulting in smaller crystals. This diagram, relating supersaturation to nucleation, growth and crystal size clearly illustrates how controlling supersaturation is vitally important when it comes to creating crystals of the desired size and specification (see Fig. 10) (Porter & Easterling, 1992).

The rate of crystal nucleation in glasses reaches its maximum at a temperature somewhat higher than the glass transition temperature and then decreases rapidly with increasing temperature, while the rate of crystal growth reaches its maximum at a temperature much higher than the temperature at which the nucleation rate is highest. Therefore, when a glass is heated at a constant rate, crystal nuclei are formed at lower temperature and grow in size at higher temperatures without any increase in number (Matusita & Sakka, 1981).

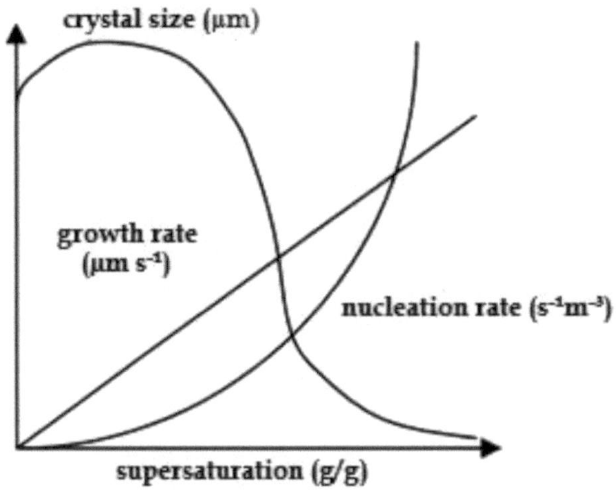

Fig. 10. The relationship between supersaturation and nucleation and growth

3. Kinetic investigation of crystallization

The theory of crystallization in amorphous materials can be explained by considering the structure and the kinetics of the crystallization. Therefore, the investigation of crystallization kinetics is important since it quantifies the effect of the nucleation and growth rate of the resulting crystallites (Carter & Norton, 2007; Kashchiev, 2000). In this chapter, crystallization kinetics of amorphous materials was investigated by explaining the crystallization mechanism and the crystallization activation energy in terms of isothermal and non-isothermal methods with different approaches. Different thermal analysis techniques used in crystallization kinetic studies were presented and a correlation between kinetic and structural investigations was made to determine the crystallization mechanism (Araújo & Idalgo, 2009; Malek, 2000; Prasad & Varma, 2005).

The crystallization kinetics of amorphous materials can be investigated either isothermally or non-isothermally by using thermal analysis techniques. In the isothermal method, the sample is heated above the glass transition temperature and the heat absorbed during the crystallization process is measured as a function of time. On the other hand, in the non-isothermal method, the sample is heated at a fixed rate and then the change in enthalpy is recorded as a function of temperature. Thermal analysis techniques such as differential thermal analysis (DTA) and differential scanning calorimetry (DSC) are quite popular for kinetic analysis of crystallization processes in amorphous solids (Araújo & Idalgo, 2009; Malek, 2000; Prasad & Varma, 2005).

Many of the reactions of interest to materials scientists involve transformations in the solid state, reactions such as recrystallization of a cold-worked material, the precipitation of a crystalline polymer from an amorphous phase, or the growth of an equilibrium phase from a non-equilibrium structure, the driving force for which is brought about by cooling from one temperature to another. Consider the initial phase to be α and the resulting phase to be β; the reaction is written:

$$\alpha \rightarrow \beta$$

The total volume of the sample is the sum of the volumes of α and β:

$$V = V^{\alpha} + V^{\beta} \tag{32}$$

The fraction transformed can be represented in the literature as x, α or F:

$$x = \frac{V^{\beta}}{V} \tag{33}$$

Assume that the transformation from α to β is controlled by nucleation and growth, that is, the nucleation of phase β within α and then the rate of growth of β.

If we consider N is the nucleation rate per unit volume and G is the growth rate in one direction = dr/dt (assuming spherical form of β). Consider the time line from zero to a time, t (see Fig. 11). We will consider another measure of time (τ), which starts when a nucleus is formed. The number of nuclei formed in differential time $d\tau$ is equal to

$$NV^{\alpha}\, d\tau \tag{34}$$

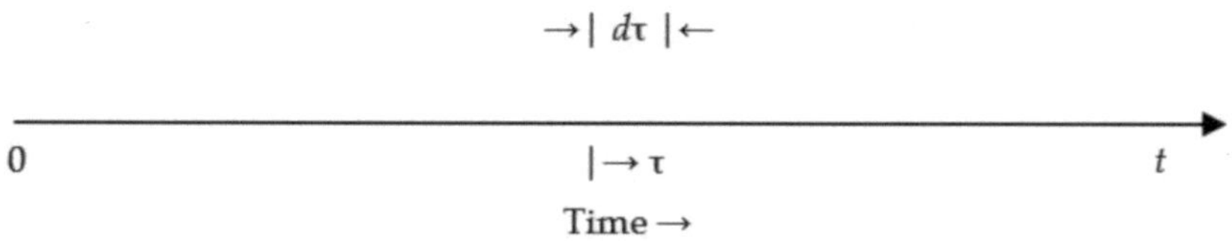

Fig. 11. Definition of t and τ for derivation of solid state transformation equations (Ragone, 1994)

Assuming that the particles nucleated in this time $d\tau$ grow as spheres, the radius of the particles formed during $d\tau$, after they have grown to time t, is:

$$\int_0^r dr = \int_{\tau}^{t} G\, dt \tag{35}$$

$$r = G(t - \tau) \tag{36}$$

The volume of the particle nucleated during $d\tau$ at time t is:

$$dV^{\beta} = \frac{4}{3}\pi G^3 (t - \tau)^3 (NV^{\alpha}) d\tau \tag{37}$$

$$dV^{\beta} = \frac{4}{3}\pi G^3 N(V - V^{\beta})(t - \tau)^3 d\tau \tag{38}$$

Early in the transformations, when V^{β} is small, V^{β} can be considered negligible with respect to V. In this case, the fraction transformed may be calculated as follows:

$$\int_0^{V^{\beta}} dV^{\beta} = \int_0^t \frac{4}{3}\pi G^3 NV(t - \tau)^3\, d\tau \tag{39}$$

$$V^{\beta} = V\frac{\pi}{3}G^3Nt^4 \tag{40}$$

$$x = \frac{V^{\beta}}{V} = \frac{\pi}{3}G^3Nt^4 \tag{41}$$

To treat the regime beyond the early transformation the extended volume concept is adopted. In this case, the nucleation and growth rates are separated from geometrical considerations such as impingement. The extended volume (V_e) is the volume that would have been formed if the entire volume had participated in nucleation and growth, even that portion transformed (V^{β}). In this case,

$$dV_e^{\beta} = V\frac{4}{3}\pi G^3N(t-\tau)^3\,d\tau \tag{42}$$

$$V_e^{\beta} = \frac{4}{3}\pi V\int_0^t G^3\,N(t-\tau)^3\,d\tau \tag{43}$$

But the total volume is equal to the sum of the volumes of α and β:

$$V = V^{\alpha} + V^{\beta} \tag{44}$$

$$\frac{V^{\alpha}}{V} = 1 - \frac{V^{\beta}}{V} = 1\text{-}x \tag{45}$$

where $x = V^{\beta}/V$

The amount of β formed, dV^{β}, is the fraction of α times dV_e^{β}

$$dV^{\beta} = \left(1\text{-}\frac{V^{\beta}}{V}\right)dV_e^{\beta} \tag{46}$$

Integrating Equation 46,

$$V_e^{\beta} = -V\ln\left(1-\frac{V^{\beta}}{V}\right) = -V\ln(1-x) \tag{47}$$

Combining Equations 43 and 47 yields,

$$-\ln(1-x) = \frac{4}{3}\pi\int_0^t G^3\,N(t-\tau)^3\,d\tau \tag{48}$$

If G and N are constant,

$$-\ln(1-x) = \frac{4}{3}\pi G^3N\int_0^t(t-\tau)^3\,d\tau = \frac{\pi}{3}G^3Nt^4 \tag{49}$$

$$x = 1\text{-}\exp\left(-\frac{\pi}{3}G^3Nt^4\right) \tag{50}$$

The resulting equation relating the fraction transformed to nucleation rate, growth rate and time is called Johnson-Mehl equation.

A similar treatment of the subject is given by Avrami. In general he expresses the fraction transformed as

$$x = 1 - \exp(-kt^{n}) \tag{51}$$

where n is called "the Avrami n". To determine the value of Avrami n from Equation 50, the following mathematical manipulation is performed:

$$x = 1 - \exp\ (-kt^{n}) \tag{52}$$

$$1 - x = \exp\ (-kt^{n}) \tag{53}$$

$$\ln(1 - x) = -kt^{n} \tag{54}$$

$$\ln\ [\ln(1 - x)] = \ln k - n \ln t \tag{55}$$

Thus the Avrami n is the slope of the plot of the logarithm of the logarithm of $(1 - x)$ versus the negative of the logarithm of t (Ragone, 1994).

3.1 Thermal analysis techniques

The study of crystallization kinetics in glass-forming liquids has often been limited by the elaborate nature of the experimental procedures which are employed. The increasing use of thermoanalytical techniques such as differential thermal analysis (DTA) or differential scanning calorimetry (DSC) has, however, offered the promise of obtaining useful data with simple methods (Yinnon & Uhlmann, 1983).

When a reaction occurs in thermal analysis, the change in heat content and in the thermal properties of the sample is indicated by a deflection (Kissinger, 1957). It is conventional to represent an endothermic effect by a negative deflection and an exothermic effect by a positive deflection (see Fig. 12). The deflections, whether positive or negative, are called peaks (Kissinger, 1956). If the reaction proceeds at a rate varying with temperature, possesses an activation energy the position of the peak varies with the heating rate if other experimental conditions are maintained fixed. This variation in peak temperature could be used to determine the energy of activation for first order reactions (Kissinger, 1957).

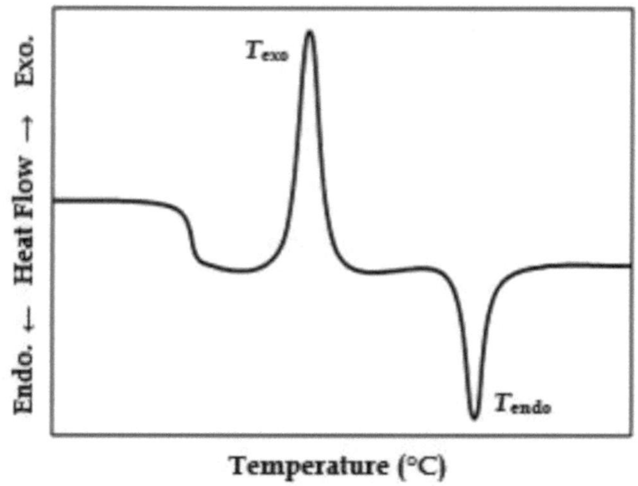

Fig. 12. A representative thermal analysis thermogram

Differential thermal analysis (DTA) has been extensively used as a rapid and convenient means for detecting the reaction process. The rate of chemical reaction was analyzed quantitatively by DTA and the activation energies were obtained. Furthermore, this method was used to obtain the activation energy for the crystallization of glass, assuming that the process of crystallization is a first order reaction (Matusita & Sakka, 1981). DSC measurements are useful in obtaining kinetic parameters related to the glass crystallization process especially in non-isothermal method due to the rapidity of this thermoanalytical technique (Araújo & Idalgo, 2009; Cheng et al., 2007).

3.2 Thermal analysis methods

3.2.1 Isothermal method

In the isothermal method, kinetic parameters of amorphous materials crystallization are obtained by monitoring the shift in the crystallization peak as a function of time (Prasad & Varma, 2005). Crystallization peak temperatures, T_p and crystallized volume fractions, x, are determined from the thermal analysis curves with respect to time in isothermal method.

In isothermal analysis, the volume fraction crystallized, x, at any time t is given as $x = S_t / S$ where S is the total area of the exothermic peak between the time t_i at which the crystallization begins and the time t_f at which the crystallization is completed. S_t is the area between t_i and t (see Fig. 13) (Ray et al., 1991).

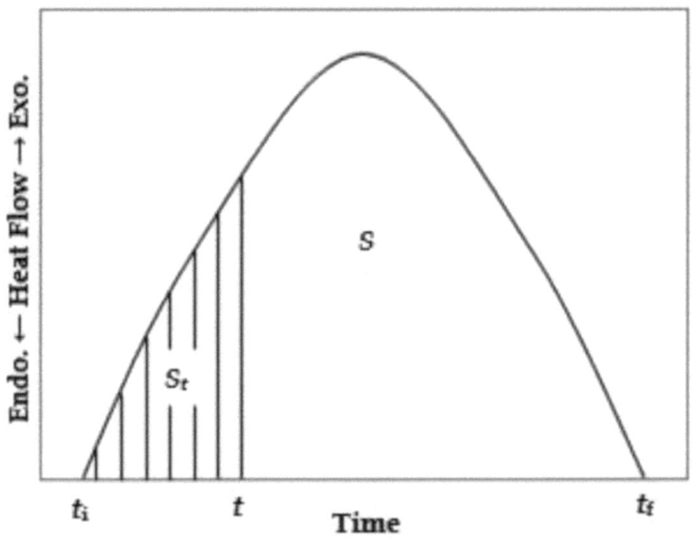

Fig. 13. Computation of the volume fraction crystallized, x, in isothermal method (Prasad & Varma, 2005)

Isothermal investigation of crystallization in amorphous materials can be described by the Johnson-Mehl-Avrami equation as given in Equation 55 (Araújo & Idalgo, 2009; Avrami, 1939, 1940; Çelikbilek et al., 2011; Prasad & Varma, 2005).

From the slopes of the linear fits to the experimental data from a plot of $\ln[-\ln(1-x)]$ versus $\ln t$, k and n values are calculated for different isothermal hold temperatures (see Fig. 14).

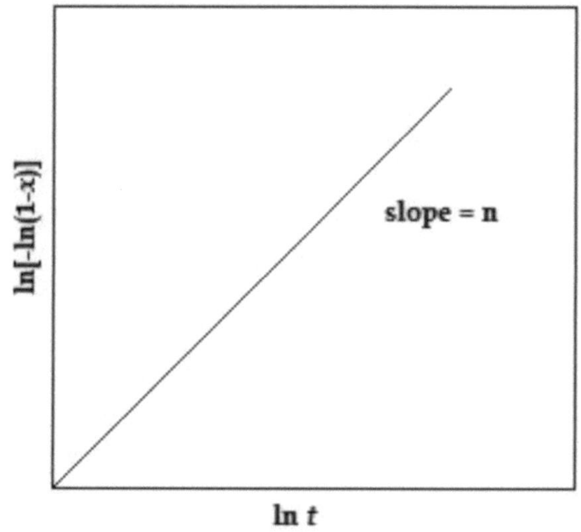

Fig. 14. Plot of ln[-ln(1- x)] against lnt for determining the Avrami parameter, n

The activation energy can be evaluated by the Arrhenius-type equation (Araújo & Idalgo, 2009; Çelikbilek et al., 2011; Prasad & Varma, 2005):

$$\ln(k) = \ln k_0 - E_A/RT \tag{56}$$

where k is the reaction rate constant, k_0 is the frequency factor, E_A is the activation energy, R is the gas constant. The activation energy for several isothermal temperatures is calculated from the slopes of the linear fits to the experimental data from a plot of ln$k$ versus $1/T$ (see Fig. 15).

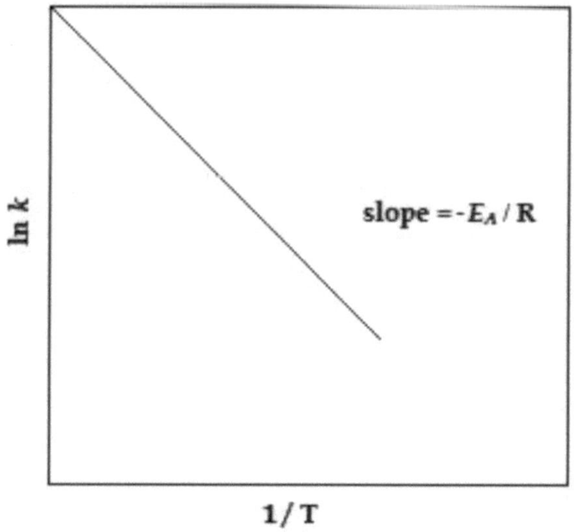

Fig. 15. Plot of ln$k$ against 1/T for determining the activation energy, E_A

In crystallization kinetic studies, as-cast (non-nucleated) and pre-nucleated samples can be used to recognize the effect of increasing and constant number of nuclei on the crystallization mechanism, respectively (Çelikbilek et al., 2011; Prasad & Varma, 2005). For the as-cast (non-nucleated) samples, when the nucleation takes place during thermal analysis, the number of nuclei of the as-cast sample is proportional to the heating rate. For the pre-nucleated samples, the number of nuclei of the pre-nucleated sample does not depend on the heating rate (Çelikbilek et al., 2011).

Crystallization mechanism of amorphous materials can be detected regarding to the following approaches; when the nucleation rate is zero during the thermal analysis experiment, $n = m$, when nucleation takes place during thermal analysis, $n = m + 1$ and when surface crystallization is the predominant mechanism, $n = m = 1$, where the parameters of n and m represent the values of the growth morphology depending on the crystallization mechanism (Çelikbilek et al., 2011; Matusita & Sakka, 1981; Prasad & Varma, 2005). As seen in Table 1, different crystallization mechanisms appear for different numerical factors, such as rod-like for one dimensional growth or surface crystallization (m = 1), disk-like for two-dimensional growth (m = 2) and spherical for three-dimensional growth (m = 3) (Çelikbilek et al., 2011; Matusita & Sakka, 1981; Matusita et al., 1984; Prasad & Varma, 2005; Ray et al. 1991).

Crystallization mechanism	n	m
Bulk crystallization with a constant number of nuclei		
Three-dimensional growth of crystals	3	3
Two-dimensional growth of crystals	2	2
One-dimensional growth of crystals	1	1
Bulk crystallization with an increasing number of nuclei		
Three-dimensional growth of crystals	4	3
Two-dimensional growth of crystals	3	2
One-dimensional growth of crystals	2	1
Surface crystallization	1	1

Table 1. Values of n and m for different crystallization mechanisms (Matusita & Sakka, 1981)

3.2.2 Non-isothermal method

Non-isothermal measurements offer some advantages if compared with isothermal studies. The kinetics of crystallization of several amorphous materials has been extensively obtained from thermal analysis techniques in isothermal mode. However, non-isothermal measurements, using a constant heating rate until the complete crystallization, are usually applied to study the devitrification on different glasses since the rapidity with which this thermoanalytical technique can be performed (Araújo & Idalgo, 2009). In the non-isothermal method, crystallization peak temperatures, T_p and crystallized volume fractions, x, are determined from the thermal analysis curves with respect to temperature. The volume fraction crystallized, x, at any temperature T is given as $x = S_T / S$, where S is the total area of the exothermic peak between the temperature, T_i, at which the crystallization begins and the temperature, T_f, at which the crystallization is completed and S_T is the partial area of the exothermic peak up to the temperature T (see Fig. 16) (Ray et al., 1991).

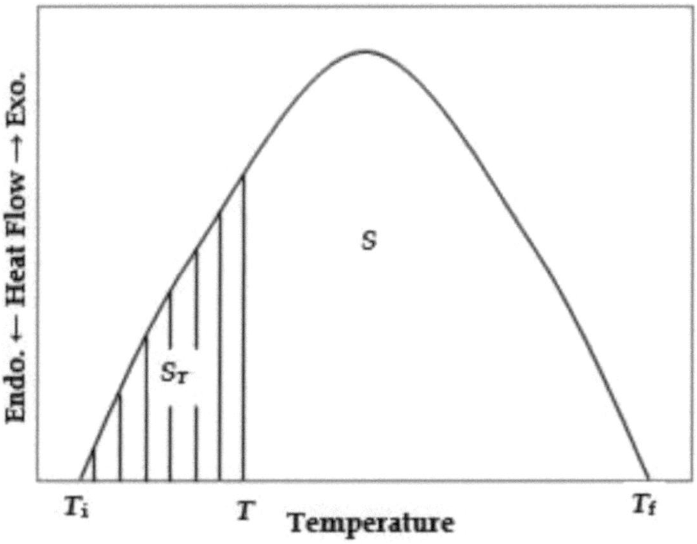

Fig. 16. Computation of the volume fraction crystallized, *x*, in non-isothermal method (Çelikbilek et al., 2011; Prasad & Varma, 2005)

Some authors have applied the Johnson-Mehl-Avrami equation to the non-isothermal crystallization process, although it is not appropriate because the Johnson-Mehl-Avrami equation was derived for isothermal crystallization. Table 2 shows different approaches for the interpretation of the kinetic data obtained from thermal analysis measurements. The Kissinger equation was basically developed for studying the variation of the peak crystallization temperature with heating rate. According to Kissinger's method, the transformation under non-isothermal condition is represented by a first-order reaction. Moreover, the concept of nucleation and growth has not been included in Kissinger equation. Matusita et al. have developed a method on the basis of the fact that crystallization does not advance by an nth-order reaction but by a nucleation and growth process. They emphasized that crystallization mechanisms such as bulk crystallization or surface crystallization should be taken into account for obtaining avtivation energy. In addition to activation energy, Matusita's method provides information about the Avrami exponent and

Method	Approach	
Ozawa	$\ln[-\ln(1-x)] = -n\ln\beta + \text{const}$	(Equation 57)
Kissinger modified by Matusita et al.	$\ln\left(\frac{T_p^2}{\beta^n}\right) = \frac{mE_A}{RT_P} + \text{const}$	(Equation 58)
Ozawa modified by Matusita et al.	$\ln\beta = -1.052\left(\frac{mE_A}{nRT_p}\right) - \left\{\frac{\ln[-\ln(1-x)]}{n}\right\} + \text{const}$	(Equation 59)
Augis and Bennett	$\ln(\beta/T_p) = -E_A/RT_p + \ln k_o$	(Equation 60)
Afify	$\ln(\beta/T_p^2) = -(m/n)(E_A/RT_p) + \text{constant}$ $\ln(\beta) = -(m/n)(E_A/RT_p) + \text{constant}$	(Equation 61) (Equation 62)

Table 2. Different methods for interpretation of non-isothermal kinetic data

dimensionality of growth. Augis and Bennett method is helpful in obtaining kinetic parameters such as frequency factor (k_o), rate constant (k) along with activation energy of crystallization and therefore preferred for the calculation of the kinetics over the other models (Deepika et al., 2009).

Ozawa and Kissinger plots are the most commonly used equations to calculate non-isothermal kinetic data, such as Avrami constant, n and crystallization activation energy, E_A, respectively (Çelikbilek et al., 2011; Kissinger, 1956; Ozawa, 1971; Prasad & Varma, 2005).

In the non-isothermal method, the values of the Avrami parameter, n, are determined from the Ozawa equation (Çelikbilek et al., 2011; Ozawa, 1971; Prasad & Varma, 2005):

$$\ln[-\ln(1-x)] = -n\ln\beta + \text{const} \tag{57}$$

where x is the crystallized volume fraction at T for the heating rate of β. From the slopes of the linear fits to the experimental data from a plot of $\ln[-\ln(1-x)]$ versus $\ln \beta$, n values are calculated (see Fig. 17). Crystallization mechanism, m, of the glass samples can be detected from Matusita et al.'s approach (Matusita & Sakka, 1981) as shown in Table 1.

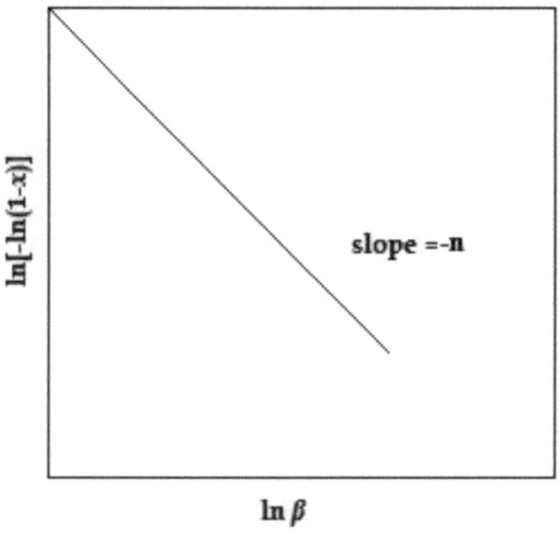

Fig. 17. Plot of $\ln[-\ln(1- x)]$ against $\ln\beta$ for determining the Avrami parameter, n

The activation energy can be evaluated by the modified Kissinger equation by Matusita et al. (Araújo & Idalgo, 2009; Çelikbilek et al., 2011; Kissinger, 1956; Matusita & Sakka, 1981; Prasad & Varma, 2005):

$$\ln\left(\frac{T_P^2}{\beta^n}\right) = \frac{mE_A}{RT_P} + \text{const} \tag{58}$$

where T_p is the crystallization peak temperature for a given heating rate β, E_A is the activation energy, R is the gas constant, n is the Avrami parameter and m is the numerical factor of crystallization mechanism. The activation energy is calculated from the slopes of the linear fits to the experimental data from a plot of $\ln(T_p^2/ \beta^n)$ versus $1/T_p$ (see Fig. 18).

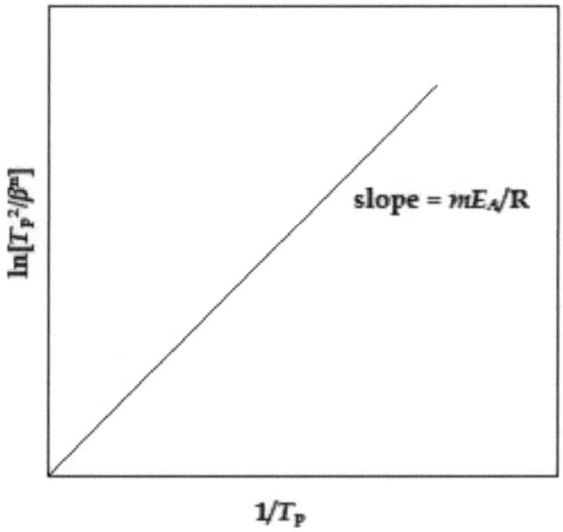

Fig. 18. Plot of $\ln[T_p^2/\beta^n]$ against $1/T_p$ for determining the activation energy, E_A

There exist also different approaches for the interpretation of the activation energy in the literature, such as the modified Ozawa equation by Matusita et al. (Equation 59) (Matusita & Sakka, 1981):

$$\ln\beta = -1.052\left(\frac{mE_A}{nRT_p}\right) - \left\{\frac{\ln[-\ln(1-x)]}{n}\right\} + \text{const} \tag{59}$$

where x is the crystallized volume fraction, β is the heating rate, T_p is the peak temperature, E_A is the activation energy, R is the gas constant, n is the Avrami parameter and m is the crystallization mechanism.

The activation energy of crystallization can also be determined by an approximation proposed by Augis and Bennett (Augis & Bennett, 1978). The relation used by them is of the form:

$$ln(\beta / T_p) = -E_A / RT_p + ln\, k_o \tag{60}$$

where, k_o is the frequency factor and R is gas constant. The plot of ln (β/T_p) against $1/T_p$ gives activation energy of crystallization (E_A).

This method has an extra advantage over the modifed Ozawa method employed in the literature for the determination of activation energy of crystallization that the intercept of ln (β/T_p) against $1000/T_p$ gives the value of pre-exponential

factor k_o of Arrhenius equation (Eq. 56), which is defined as the number of attempts made by the nuclei per second to overcome the energy barrier. This also provides information for the calculation of number of nucleation sites, present in the material for crystal growth (Deepika et al., 2009).

The value of activation energy can also be calculated also by using the variation of T_p with the heating rate β for both crystallization phases (Afify, 1990; Afify et al., 1991). By using similar relations to Kissinger equation, the relations can be written in the form:

$$\ln(\beta/T_p^2) = -(m/n)(E_A/RT_p) + \text{const} \quad (61)$$

$$\ln(\beta) = -(m/n)(E_A/RT_p) + \text{const} \quad (62)$$

where, R is gas constant, T_p is the peak temperature, n is the Avrami parameter and m is the crystallization mechanism.

3.3 Crystallization kinetic studies in amorphous materials

Numerous studies exist on crytallization kinetics of amorphous materials, such as glasses (Araújo, & Idalgo, 2009; Araújo, 2009; Cheng, 2007; Çelikbilek, 2010; Çelikbilek, 2011; El-Mallawany, 1997; Idalgo, 2006; Jeong, 2007; Prasad & Varma 2005; Ray, C.S., Huang, W.H. & Day, 1991; Shaaban, 2009; Yukimitu, 2005), amorphous alloys (Abu El-Oyoun, 2009; Afify, 1990; Afify, 1991; Al-Ghamdi, 2010; Al-Ghamdi, 2011; Aly, 2009; Dahshan, 2010; Deepika, 2009; Elabbar, 2008; Huang, 2008; Mehta, 2004; Yahia, 2011, Zhang, 2008), amorphous thin films (Abdel-Wahaba, 2005; Bhargava, 2010; Chen & Wu, 1999; Hajiyev, 2009; Lei, 2010; Liu & Duh, 2007; Seeger & Ryder, 1994), amorphous nanomaterials (Ahmadi, 2011; Gridnev, 2008; Qin, 2004; Tomasz, 2010), etc. In this section, selected studies reported in the literature on crystallization kinetics of different types of amorphous materials were given to make the theory of kinetics more understandable for the reader.

3.3.1 Glasses

Isothermal crystallization kinetic studies in the glass system (100-x) $LiBO_2$-$x Nb_2O_5$ ($5 \le x \le 20$, in molar ratio) have been realized using differential thermal analyses by Prasad et al. (Prasad & Varma, 2005). The isothermal experiments were carried out by heating the samples to the desired temperature at a rate of 50 °C/min. After attaining the required temperature, the run was held for a period of about 30 seconds to reach the equilibrium. The temperature range of 527-547 °C with an interval of 5 °C was selected for isothermal experiments because the glass shows reasonable peak shapes in this range, which is recommended for accurate data analyses.

The crystallized fraction x as a function of time at all the holding temperatures is shown in Fig. 19. It reveals that the time taken to complete the crystallization peak is indirectly proportional to the isothermal holding temperature. A typical isothermal DTA trace obtained at 537 °C (holding temperature) is shown in the inset of Fig. 19.

Plots of ln[-ln(1-x)] against ln t are shown in Fig. 20. Values of Avrami exponent n and the reaction rate constants k were determined by least square fits of the experimental data. The average value of n is 2.62. Since n takes only integer values from 1 to 4, the n (close to 3) value observed in the present study indicates the near three-dimensional growth of $LiNbO_3$ (see Table 1). The values of ln k are determined for all the temperatures from the plots of ln[-ln(1-x)] against ln t (Fig. 20).

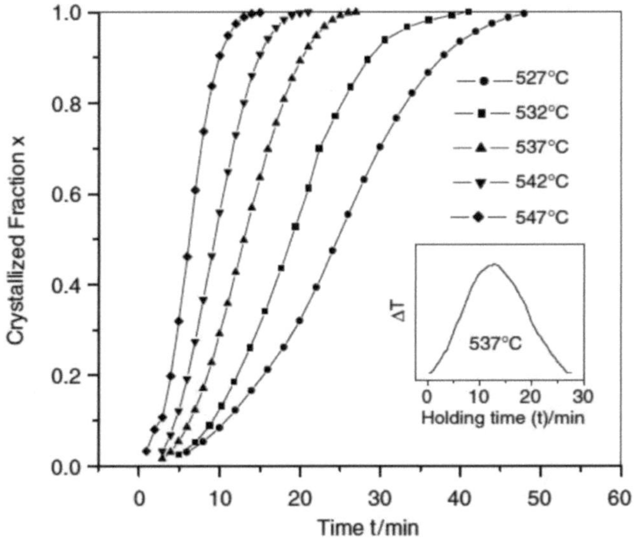

Fig. 19. The crystallized fraction x as a function of isothermal time. Inset shows the isothermal differential thermal analyses trace obtained at 537 °C (Prasad & Varma, 2005)

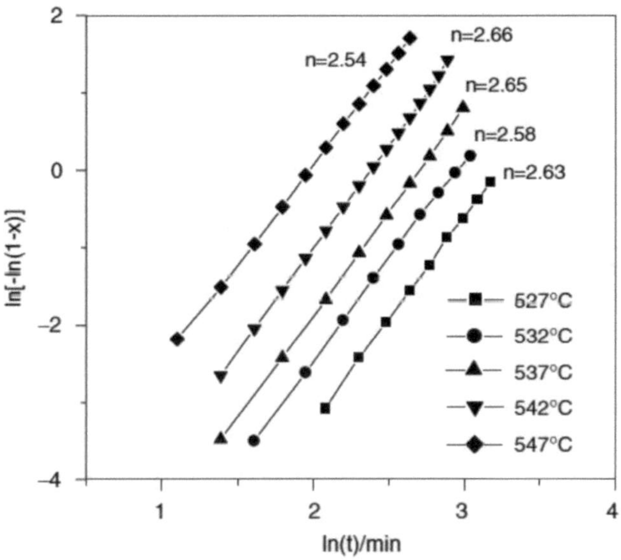

Fig. 20. Plot of ln[-ln(1-x)] against ln t for determining the n (Prasad & Varma, 2005)

A plot of ln k versus $1/T$ shown in Fig. 21 yielded the activation energy, E_A, to be 293 kJ/mol (Prasad & Varma, 2005).

Non-isothermal crystallization kinetics of the $(1-x)TeO_2-xWO_3$ (where x=0.10, 0.15 and 0.20, in molar ratio) glass system was studied by Çelikbilek et al. and DSC curves of the as-cast (non-nucleated) sample recorded at different heating rates, β, are shown in Fig. 22. The T value for the calculation of the volume fraction crystallized of $0.90TeO_2$–$0.10WO_3$ glass was determined at 420 °C (Çelikbilek et al., 2011).

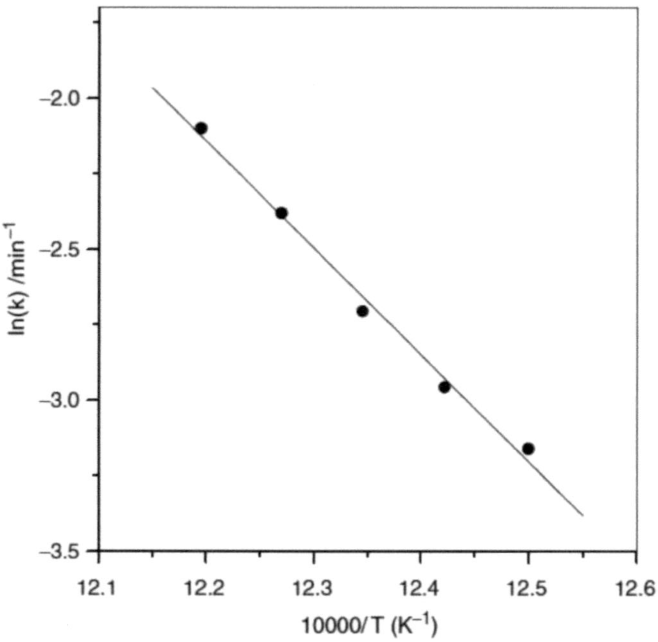

Fig. 21. Plot of ln*k* versus 1/*T* from which the values of crystallization activation energy are obtained (Prasad & Varma, 2005)

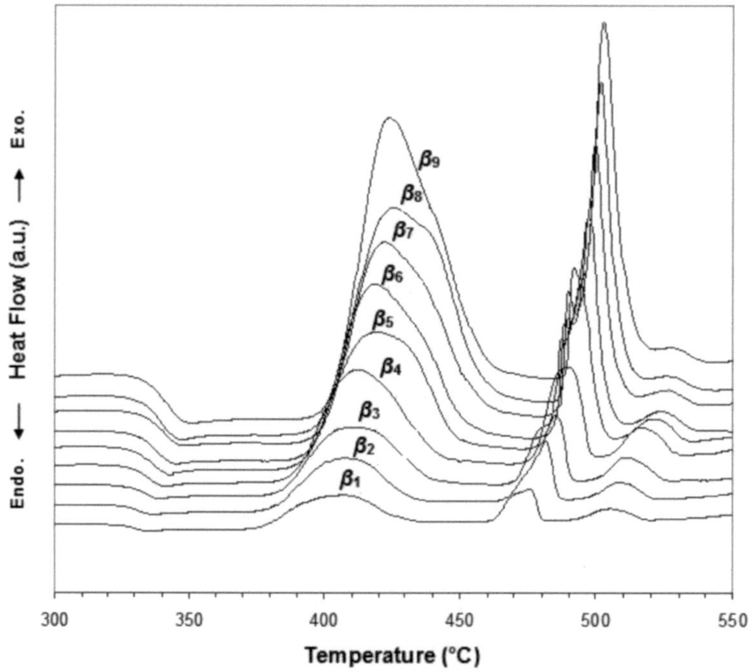

Fig. 22. DSC curves with heating rate β (5, 7.5, 10, 15, 20, 25, 30, 35 and 40 °C/min) for $0.90TeO_2$–$0.10WO_3$ sample (Çelikbilek et al., 2011)

The values of the Avrami parameter, *n*, were calculated from the linear fits to the experimental data based on the Ozawa equation (Equation 57), as shown in Fig. 23. The *n* value was determined as 1.14 for $0.90TeO_2$–$0.10WO_3$ glasses. On the basis of the determination about the non-integer value of the Avrami parameter, in this study the *n* value was determined as 1, indicating the formation of surface crystallization during the crystallization process (see Table 1).

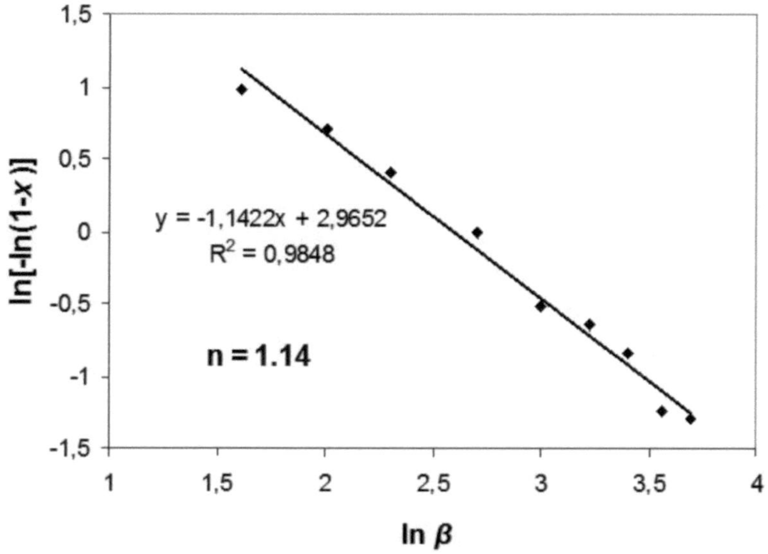

Fig. 23. The Ozawa plot for determining *n* associated with the first exotherm of the $0.90TeO_2$–$0.10WO_3$ sample (Çelikbilek et al., 2011)

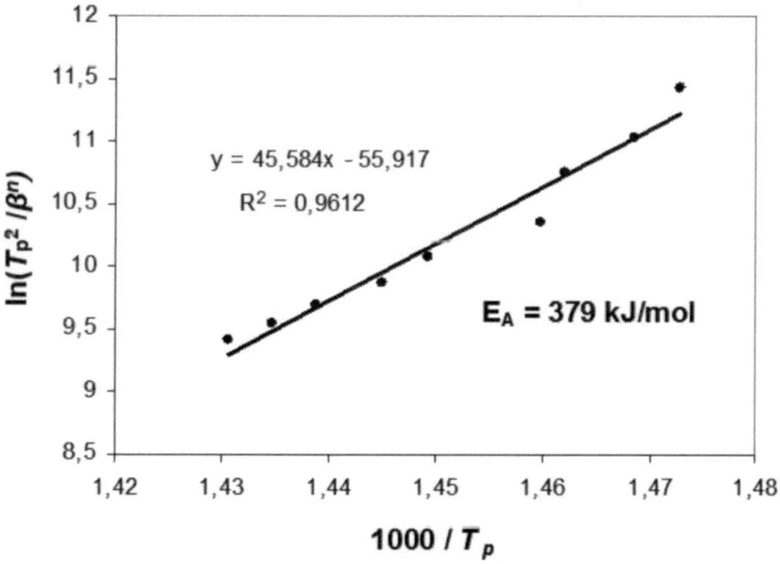

Fig. 24. The Kissinger plots for determining E_A associated with the first exotherm of the pre-nucleated $0.90TeO_2$–$0.10WO_3$ sample (Çelikbilek et al., 2011)

Using the modified Kissinger equation (Equation 58), activation energy, E_A, for the first crystallization reaction of $0.90TeO_2$–$0.10WO_3$ glass was determined from the linear fits of $\ln(T_p^2/\beta^n)$ versus $1/T_p$ plots, as shown in Fig. 24. The activation energy of the first exotherm was calculated as 379 kJ/mol. Comparing with this result, in another study realized by Çelikbilek et al., the E_A value was determined as 372 kJ/mol by applying the modified Ozawa method from the slopes of the linear fits to the experimental data from a plot of ln β versus $1/T_p$ for $0.90TeO_2$–$0.10WO_3$ sample.

To determine the crystallization mechanism of the glasses with fixed nuclei number, $0.90TeO_2$–$0.10WO_3$ sample was pre-nucleated by heat-treating for 2 hours at 350 °C. The pre-nucleation temperature was determined by nucleating the as-cast glass sample for 2 hours at three different temperatures between T_g and T_p. The nucleation temperature which corresponds to the maximum peak temperature was selected as the pre-nucleation temperature (350 °C) (Çelikbilek et al., 2011).

The value of the Avrami constant of $0.90TeO_2$–$0.10WO_3$ glass pre-nucleated at 350 °C is shown in Fig. 25. The n value of the pre-nucleated sample was calculated as 1 from the Ozawa equation (Equation 57) and while the number of nuclei do not depend on the heating rate for pre-nucleated samples, according to the approach $n = m$, the mechanism was determined as one-dimensional growth of the crystals (see Table 1). The n value calculated for the as-cast $0.90TeO_2$–$0.10WO_3$ glass was also calculated as 1, indicating the surface crystallization.

The activation energy, E_A, of the first crystallization reaction of pre-nucleated $0.90TeO_2$–$0.10WO_3$ glass was calculated using the Kissinger equation, as shown in Fig. 26. The E_A value of the pre-nucleated sample was determined as 382 kJ/mol, very close to the calculated E_A value of the as-cast sample (379 kJ/mol) and from the obtained data it was concluded that constant or increasing number of nuclei does not have a significant effect on crystallization activation energy.

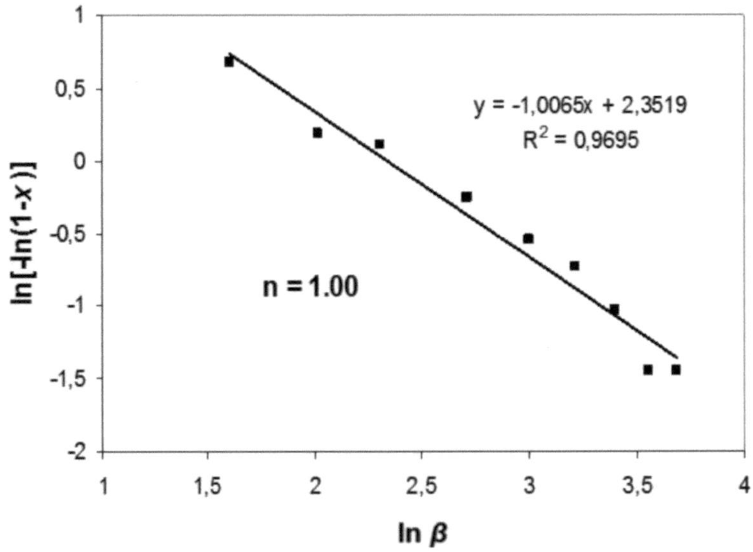

Fig. 25. The Ozawa plot for determining n associated with the first exotherm of the $0.90TeO_2$–$0.10WO_3$ sample pre-nucleated at 350 °C for 2 hours (Çelikbilek et al., 2011)

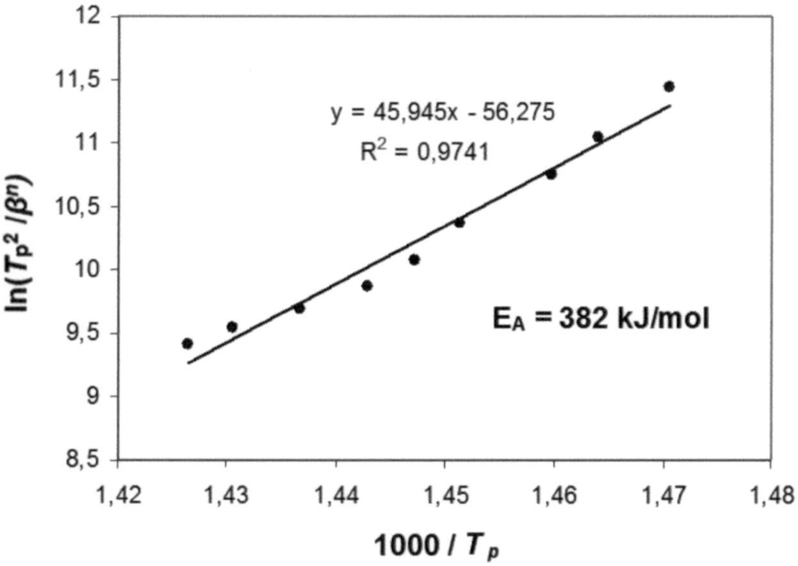

Fig. 26. The Kissinger plot for determining E_A associated with the first exotherm of the $0.90TeO_2$–$0.10WO_3$ sample pre-nucleated at 350 °C for 2 hours (Çelikbilek et al., 2011)

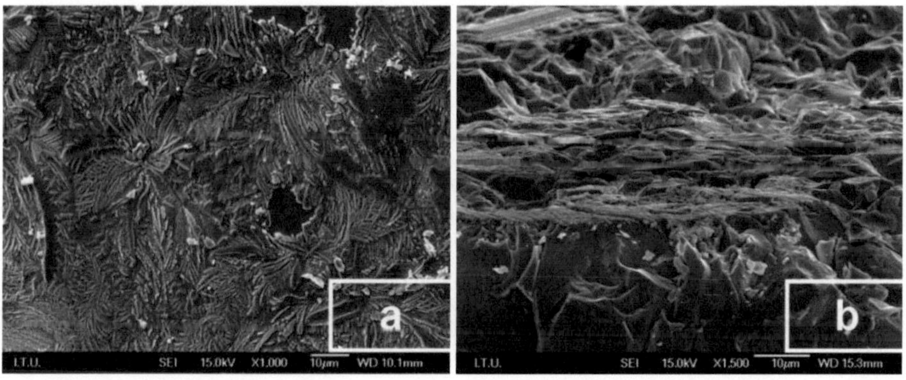

Fig. 27. SEM micrographs of the $0.90TeO_2$–$0.10WO_3$ sample heat-treated at 410 °C (a) surface, (b) cross-section (Çelikbilek et al., 2011)

Fig. 27a-b represents the SEM micrographs taken from the surface and the cross-section of the $0.90TeO_2$–$0.10WO_3$ sample heat-treated at 410 °C, above the first crystallization onset temperature, respectively. Fig. 27a exhibits the presence of dendritic leaf-like crystallites differently oriented on the surface. However, in the cross-sectional micrograph (see Fig. 27b), a typical amorphous structure without any crystallization on bulk structure can be clearly observed following the crystallites on the surface. Based on the SEM investigations, it was determined that the crystallites formed on the surface and did not diffuse into the bulk structure proving the surface crystallization mechanism (Çelikbilek et al., 2011).

3.3.2 Amorphous alloys

Non-isothermal crystallization kinetic investigation of $Se_{58}Ge_{42-x}Pb_x$ (x = 9, 12) alloy studied by Deepika et al. is given here as an example of the crystallization kinetics in amorphous alloys. In their study, glassy alloys of $Se_{58}Ge_{33}Pb_9$ and $Se_{58}Ge_{30}Pb_{12}$ were prepared by melt-quenching technique and after realizing thermal analyses, the samples were annealed at 633 and 643 K, which lie between the first and second crystallization, respectively. The activation energy of crystallization for the first and second crystallization stages of $Se_{58}Ge_{33}Pb_9$ and $Se_{58}Ge_{30}Pb_{12}$ glassy systems was derived using the approximation methods developed by the Kissinger (Equation 58), Matusita (Equation 59), Augis and Bennett (Equation 60). The values of activation energy of crystallization of as-prepared and annealed samples using different theoretical models are given in Table 3 (Deepika et al., 2009).

Models applied	Activation energy of crystallization (kJ/mol)					
	$Se_{58}Ge_{33}Pb_9$			$Se_{58}Ge_{30}Pb_{12}$		
	Before annealing		After annealing	Before annealing		After annealing
	I peak	II peak		I peak	II peak	
Kissinger Model	167.06 ± 0.88	165.29 ± 0.45	144.97 ± 0.88	171.46 ± 1.19	146.15 ± 2.03	139.71 ± 1.13
Augis–Bennett Method	155.65 ± 1.37	168.20 ± 2.27	66.35 ± 0.77	195.66 ± 0.53	186.43 ± 3.12	134.22 ± 2.94
Matusita Model	330.34 ± 7.46	316.28 ± 2.0	237.58 ± 1.33	384.52 ± 3.79	412.39 ± 2.61	195.60 ± 1.96

Table 3. Activation energy of crystallization of as-prepared and annealed samples of $Se_{58}Ge_{33}Pb_9$ and $Se_{58}Ge_{30}Pb_{12}$ glassy systems using different theoritical models (Deepika et al., 2009)

From Table 3, it is observed that activation energy of crystallization decreases after annealing. This means that group of atoms in the glassy state requires less amount of energy to jump to crystalline state hence, making the sample less stable and more prone to crystallization. This is again an indication of the fact that annealing of glass leads it to a quicker crystallization. The crystallization mechanism of crystals decreases to one dimension from two and three dimensions after annealing, suggesting a decrease from bulk nucleation to surface nucleation in annealed samples.

It is also observed that activation energies of amorphous alloys calculated by means of the different theoretical models differ substantially from each other. This difference in the activation energy as calculated with the different models may be attributed to the different approximations used in the models.

3.3.3 Amorphous thin films

The crystallization kinetics in $Ni_{50.54}Ti_{49.46}$ film was studied by Liu et al. using differeantial scanning calorimetry through non-isothermal and isothermal techniques. $Ni_{50.54}Ti_{49.46}$ thin films were prepared by a mixed NiTi target using DC magnetron sputtering. The NiTi thin film was deposited on Si wafer and the substrate was unheated to achieve the amorphous structure. For non-isothermal analyses, a set of DSC scans was recorded at heating rates ranging from 5 to 50 °C/min. For isothermal analyses, the amorphous samples were first heated to a fixed temperature with 250 °C/min (between 793 and 823 K), and then held for a certain period of time until fully crystalline state was achieved.

The activation energy for crystallization was determined to be 411 and 315 kJ/mol by Kissinger (Equation 58) and Augis & Bennett (Equation 60) method, respectively (Fig. 28a-b) (Liu & Duh, 2007). Comparing with this study, previous works on near equiatomic NiTi films showed that the activation energy was 370–419 kJ/mol (Chen & Wu 1999; Seeger & Ryder, 1994).

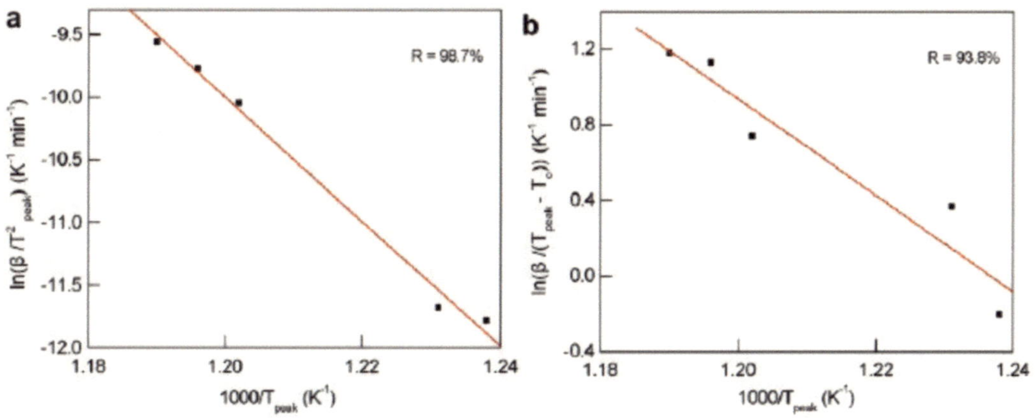

Fig. 28. Plot of the (a) Kissinger and (b) Augis & Bennett equations for the crystallization in $Ni_{50.54}Ti_{49.46}$ thin films (Liu & Duh, 2007)

The isothermal crystallization kinetics of amorphous materials is described by the Johnson-Mehl-Avrami (JMA) equation. The Avrami exponent n for different temperature ranges from 2.63 to 3.12 between 793 and 823 K (Fig. 29a), which indicates that the isothermal annealing was governed by diffusion-controlled three-dimensional growth for $Ni_{50.54}Ti_{49.46}$

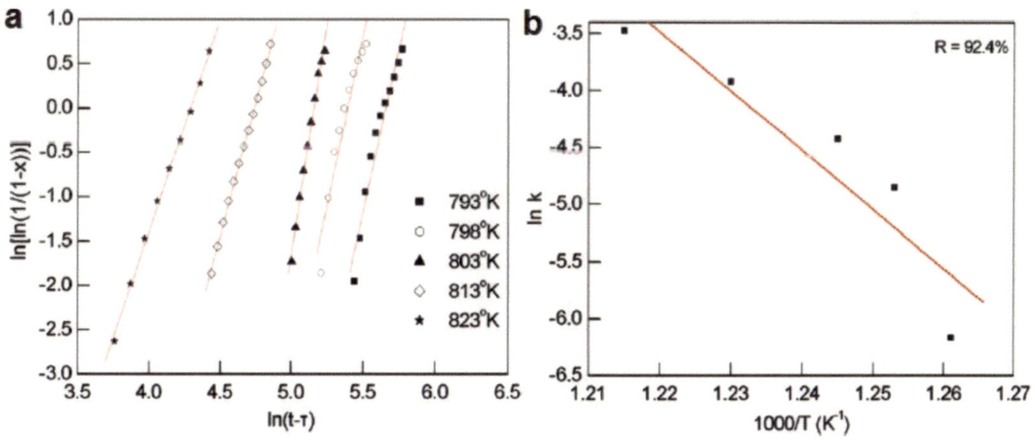

Fig. 29. Plots of the (a) Avrami and (b) Arrhenius equations for the isothermal crystallization of $Ni_{50.54}Ti_{49.46}$ thin films(Liu & Duh, 2007)

thin films. The E_A value was calculated from the Arrhenius equation and activation energy for crystallization was determined as 424 kJ/mol (Fig. 29b). This value is very similar to that from non-isothermal method as determined by the Kissinger analysis (411 kJ/mol). It isdemonstrated that the crystallization on both non-isothermal and isothermal methods induces a similar phase transformation mechanism (Liu & Duh, 2007).

3.3.4 Amorphous nanomaterials

The nanocrystallization kinetics of $Ni_{45}Ti_{23}Zr_{15}Si_5Pd_{12}$ alloy has been investigated using differential scanning calorimetry by means of non-isothermal and isothermal techiques and the products of crystallization have been analyzed by transmission electron microscopy and X-ray diffraction. The bulk amorphous alloy has been prepared by copper mold casting. In the non-isothermal experiments, a set of DSC scans were obtained at a heating rate ranging from 10 to 40 K/min. For the isothermal analysis, the amorphous samples were first heated to a fixed temperature between 820 and 835 K at a rate of 200 K/min, then maintained for a certain period of time until the completion of exothermic process (Qin, 2004).

By using the values of glass transition, T_g, crystallization onset, T_x, and crystallization peak, T_p, temperatures a Kissinger plot yields approximate straight lines as shown in Fig. 30. From

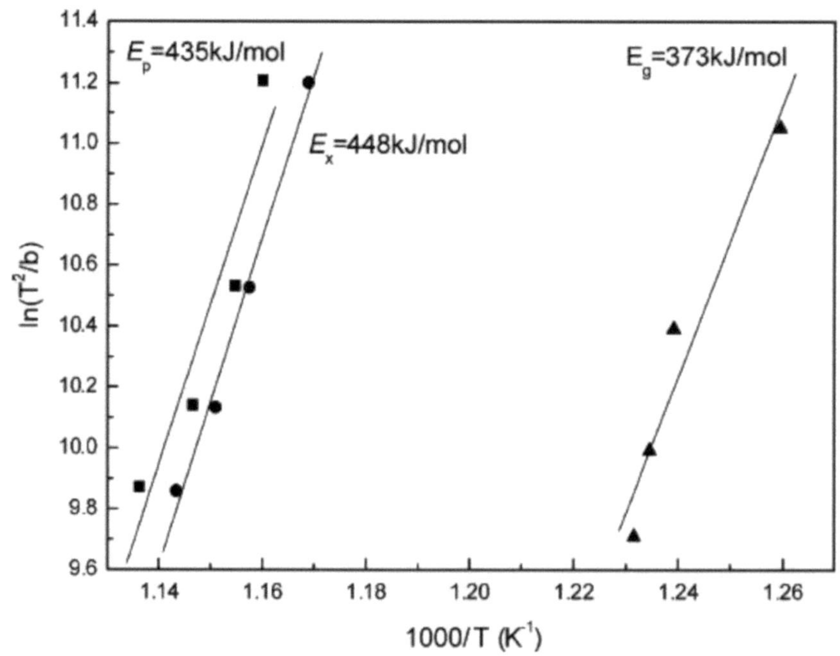

Fig. 30. Kissinger plots of the $Ni_{45}Ti_{23}Zr_{15}Si_5Pd_{12}$ bulk amourphus alloy (Qin, 2004)

the slopes of these lines the activation energies were determined as 373, 448 and 435 kJ/mol for T_g, T_x, T_p, respectively. The apparent activation energy E_x deduced from the Kissinger equation for onset crystallization is higher than that for crystallization peak, meaning that nucleation process is more difficult than growth (Qin, 2004).

The bright field TEM image and a corresponding selected area diffraction pattern (SADP) of the bulk specimen after annealing at 840 K for 900 seconds are shown in Fig. 31. The bright field image shows crystalline nanoparticles embedded in the amorphous matrix. The SADP consists of several ring patterns superimposed on a diffuse halo patterns, also indicating a mixture of nanocrystalline and residual amorphous phase. All the continuous rings can be analyzed into the ordered bcc structure phase with lattice parameter little larger than that of NiTi, which is in agreement with the realized XRD results (Qin, 2004).

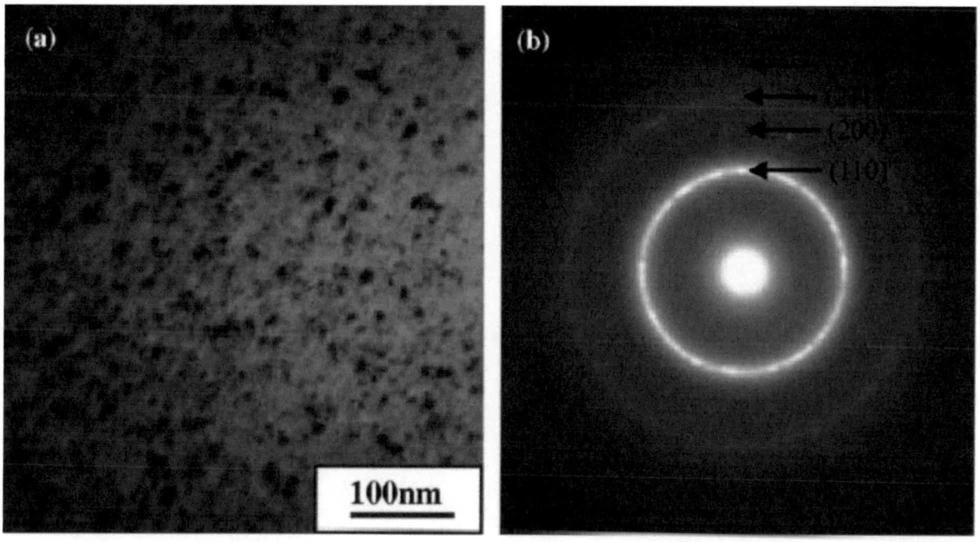

Fig. 31. Bright-field TEM image (a) and selected-are eletron diffraction pattern, (b) of the $Ni_{45}Ti_{23}Zr_{15}Si_5Pd_{12}$ bulk amourphus alloy annealed for 900 seconds at 840K (Qin, 2004)

4. Conclusion

In conclusion, crystallization of an amorphous material is a complex phenomenon involving nucleation and growth processes and it can be investigated by taking into account the structure and the kinetics of the crystallization reaction. Crystallization kinetics is crucial since it studies the effect of nucleation and growth rate of the resulting crystalline phase.

Therefore, this chapter covers the investigation of the crystallization kinetics of amorphous materials by studying the crystallization mechanism in terms of isothermal and non-isothermal methods and describing different thermal analysis techniques used in crystallization kinetic studies and explaining the structural characterization techniques used to determine the crystallization mechanism.

5. References

Abdel-Wahaba, F.A., El-Hakim, S.A. & Kotkat, M.F. (2005). Electrical conductivity and crystallization kinetics of amorphous $Se_{0.81}In_{0.19}$ films. *Physica B*, 366, pp. 38-43, 0921-4526

Abu El-Oyoun, M. (2009). Determination of the crystallization kinetic parameters of $Ge_{22.5}Te_{77.5}$ glass using model-free and model-fitting methods. *Journal of Alloys and Compounds*, 486, pp. 1-8, 0925-8835

Afify, N. (1990). Crystallization kinetics of overlapping phases in $Se_{0.6}Ge_{0.2}Sb_{0.2}$ chalcogenide glass. *Journal of Non-Crystalline Solids*, 126, pp. 130-140, 0022-3093

Afify, N., Abdel-Rahim, M.A., Abd El-Halim, S. & Hafiz, M.M. (1991). Kinetics study of non-isothermal crystallization in $Se_{0.7}Ge_{0.2}Sb_{0.1}$ chalcogenide glass. *Journal of Non-Crystalline Solids*, 128, pp. 269-278, 0022-3093

Ahmadi, S., Shahverdi, H.R. & Saremi, S.S. (2011). Effects of Nb Alloying on nano-crystallization kinetics of $Fe_{(55-x)}Cr_{(18)}Mo_{(7)}B_{(16)}C_{(4)}Nb_{(x)}$ (x=0, 3) bulk amorphous alloys. *Journal of Materials Science & Technology*, 27, pp. 735-740, 1005-0302

Al-Ghamdi, A.A., Alvi, M.A., Khan, S.A. (2010). Glass transition and crystallization kinetics of $In_x(Se_{0.75}Te_{0.25})_{100-x}$ chalcogenide glasses. *Journal of Alloys and Compounds*, 491, pp. 85-91, 0925-8838

Al-Ghamdi, A.A., Alvi, M.A., Khan, S.A. (2011). Non-isothermal crystallization kinetic study on $Ga_{15}Se_{85-x}Ag_x$ chalcogenide glasses by using differential scanning calorimetry. *Journal of Alloys and Compounds*, 509, pp. 2087-2093, 0925-8835

Aly, K.A., Othman, A.A., Abousehly, A.M. (2009). Effect of Te additions on the glass transition and crystallization kinetics of $(Sb_{15}As_{30}Se_{55})_{100-x}Te_x$ amorphous solids. *Journal of Alloys and Compounds*, 467, pp. 417-423, 0925-8838

Araújo, E.B. & Idalgo, E. (2009). Non-isothermal studies on crystallization kinetics of tellurite $20Li_2O$-$80TeO_2$ glass. *Journal of Thermal Analysis and Calorimetry*, 95, pp. 37-42, 1388-6150

Araujo, E.B., Idalgo, E., Moraes, A.P.A., Souza Filho, A.G. & Mendes Filho, J. (2009). Crystallization kinetics and thermal properties of $20Li_2O$–$80TeO_2$ glass. *Materials Research Bulletin*, 44, pp. 1596-1600, 0025-5408

Avrami, M. (1939). Kinetics of phase change I. *Journal of Chemical Physics*, 7, pp. 1103-1012, 0021-9606

Avrami, M. (1940). Kinetics of phase change II. *Journal of Chemical Physics*, 8, pp. 212-224, 0021-9606

Augis, A.J. & Bennett, J.E. (1978). Calculation of the Avrami parameters for heterogeneous solid state reactions using a modification of the Kissinger method. *Journal of Thermal Analysis and Calorimetry*, 13, pp. 283-292, 1388-6150

Bhargava, A., Kalla, J. & Suthar, B. (2010). Crystallization process in amorphous Sn-Te-Se thin films. *Chalcogenide Letters*, 7, pp. 175-180, 1584-8663

Carter, C.B. & Norton, M.G. (2007). *Ceramic Materials Science and Engineering*, Springer, ISBN-10: 0387462708, New York

Chen, J.Z. & Wu, S.K. (1999). Crystallization behavior of r.f.-sputtered TiNi thin films. *Thin Solid Films*, 339, pp. 194-199, 0040-6090

Cheng, Y., Xiao, H. & Guo, W. (2007). Structure and crystallization kinetics of PbO–B_2O_3 glasses. *Ceramics International*, 33, pp. 1341-1347, 0272-8842

Çelikbilek, M., Ersundu, A.E., Solak, N. & Aydin, S. (2011). Crystallization kinetics of TeO_2–WO_3 glasses. *Journal of Non-Crystalline Solids*, 357, pp. 88-95, 0022-3093

Çelikbilek, M., Ersundu, A.E., Solak, N. & Aydin, S. (2011). Investigation on thermal and microstructural characterization of the TeO_2–WO_3 system. *Journal of Alloys and Compounds*, 509, pp. 5646–5654, 0925-8835

Çelikbilek, M., Ersundu, A.E., Solak, N. & Aydin, S. (2010). Kinetic studies on the tungsten tellurite glasses, *ACerS 112th Annual Meeting combined with Materials Science & Technology 2010 Conference & Exhibition,* Houston, Texas, U.S.A., October 2010

Dahshan, A., Amer, H.H. & Aly, K.A. (2010). Thermal stability and crystallization kinetics of Ge-Se-Cd glasses. *Philosophical Magazine*, 90, pp. 1435-1449, 1478-6435

Deepika, Jain, P.K., Rathore, K.S. & Saxena, N.S. (2009). Structural characterization and phase transformation kinetics of $Se_{58}Ge_{42-x}Pb_x$ (x = 9, 12) chalcogenide glasses. *Journal of Non-Crystalline Solids*, 355, pp. 1274-1280, 0022-3093

Doremus, R. H. (1973). *Glass science*, Wiley, ISBN-10: 0471891746, New York

Elabbar, A.A., Abu El-Oyoun, M., Abu-Sehly, A.A. & Alamri, S.N. (2008). Crystallization kinetics study of $Pb_{4.3}Se_{95.7}$ chalcogenide glass using DSC technique, *Journal of Physics and Chemistry of Solids*, 69, pp. 2527-2530, 0022-3697

El-Mallawany, R.A.H. (2002). *Tellurite Glasses Handbook*, CRC Press, ISBN-10: 0849303680, Boca Raton London New York Washington, D.C.

El-Mallawany, R., Abousehly, A., El-Rahamani, A. & Yousef, E. (1997). Calorimetric study on tellurite glasses. *Physica Status Solidi*, 163, pp. 377-386, 1862-6319

Ersundu, A.E., Karaduman, G., Çelikbilek, M., Solak, N. & Aydin, S. (2010). Stability of the δ-TeO_2 phase in the binary and ternary TeO_2 glasses. *Journal of the European Ceramic Society*, 30, pp. 3087–3092, 0955-2219

Ersundu, A.E., Karaduman, G., Çelikbilek, M., Solak, N. & Aydin, S. (2010). Effect of rare-earth dopants on the thermal behavior of tungsten–tellurite glasses. *Journal of Alloys and Compounds,* 508, pp. 266–272, 0925-8838

Ersundu, A.E., Çelikbilek, M., Solak, N. & Aydin, S. (2011). Glass formation area and characterization studies in the CdO–WO_3–TeO_2 ternary system. *Journal of the European Ceramic Society*, 31, pp. 2775–2781, 0955-2219

Gridnev, S.A., Gorshkov, A.G., Sitnikov & A.V. (2008). The Study of the Isothermal Crystallization and Electrical Properties of Heterogeneous Metal-Ferroelectric Nano-Composites. *Ferroelectrics*, 374, pp. 194-201, 0015-0193

Hajiyev, E.S., Madadzadeh, A.I. & Ismayilov, J.I. (2009). Structure and Kinetics of Crystallization of Thin Amorphous Films of $Yb_{(1-x)}Sm_{(x)}As_{(4)}S_{(7)}$. *Semiconductors,* 43, pp. 1492-1495, 1090-6479

Huang, L.J., Li, L., Liang, G.Y., Guo, Y. & Wu, D.C. (2008). Crystallization kinetics of $Mg_{65}Cu_{25}Nd_{10}$ amorphous alloy. *Journal of Non-Crystalline Solids*, 354, 1048-1053, 0022-3093

Idalgo, E., Araujo, E.B., Yukimitu, K., Moraes, J.C.S., Reynoso, V.C.S. & Carvalho, C.L. (2006). Effects of the particle size and nucleation temperature on tellurite $20Li_2O$–$80TeO_2$ glass crystallization. *Materials Science and Engineering A*, 434, pp. 13-18, 0921-5093

Jackson, K.A. (2004). *Kinetic Processes: Crystal Growth, Diffusion, and Phase Transitions in Materials*, Wiley-VCH, ISBN-10: 3527306943, Weinheim

Jeong, E.D., Bae, J.S., Hong, T.E., Lee, K.T., Ryu, B.K., Komatsu T., & Kim, H.G. (2007). Thermal properties and crystallization kinetics of tellurium oxide based glasses. *Journal of Ceramic Processing Research*, 8, pp. 417-420, 1229-9162

Kashchiev, D. (2000). *Nucleation*, Butterworth Heinemann, ISBN 0750646829, Burlington.

Kalb, J. A. (2009). Crystallization Kinetics, In: Phase Change Materials: Science and Applications, Raoux S., Wuttig M., pp. 125-148, Springer, ISBN: 0387848738, New York

Karaduman, G., Ersundu, A.E., Çelikbilek, M., Solak, N. & Aydin, S. (2011). Phase equilibria and glass formation studies in the $(1-x)TeO_2$–$xCdO$ ($0.05 \leq x \leq 0.33$ mol) system. *Journal of the European Ceramic Society*, 10.1016/j.jeurceramsoc.2011.09.027, 0955-2219

Kissinger, H.E. (1956). Variation of peak temperature with heating rate in differential thermal analysis. *Journal of Research of the National Bureau of Standards*, 57, 217-221, 1044-677X

Kissinger, H.E. (1957). Reaction Kinetics in Differential Thermal Analysis. *Analitical Chemistry*, 29, pp. 1702-1706, 0003-2700

Lei, Y., Zhao, H., Cai, W., An, X. & Gao, L. (2010). Crystallization kinetics of Ni_xTi_{1-x} alloy thin films. *Physica B-Condensed Matter*, 405, pp. 947-950, 0921-4526

Liu, K.T. & Duh, J.G. (2007). Kinetics of the crystallization in amorphous NiTi thin films. *Journal of Non-Crystalline Solids*, 353, 1060–1064, 0022-3093

Malek, J. (2000). Kinetic analysis of crystallization processes in amorphous materials. *Thermochimica Acta*, 355, pp. 239-253, 0040-6031

Marotta, A., Saiello, S. & Buri, A. (1983). Remarks on determination of the Avrami exponent by non-isothermal analysis. *Journal of Non-Crystalline Solids*, 57, pp. 473-475, 0022-3093

Matusita, K. & Sakka, S. (1981). Kinetic study on non-isothermal crystallization of glass by thermal analysis. *Bulletin of the Institute for Chemical Research*, 59, 3, pp. 159-171, 1747-5198

Matusita, K., Komatsu, T. & Yokota, R. (1984). Kinetics of non-isothermal crystallization process and activation energy for crystal growth in amorphous materials. *Journal of Materials Science*, 19, pp. 291-296, 0022-2461

Mehta, N., Kumar, D. & Kumar, A. (2004). Calorimetric studies of the crystallization growth process in glassy $Se_{70}Te_{30-x}Ag_x$ alloys. *Turkish Journal of Physics*, 28, pp. 396-406, 1300-0101

Mullin, J.W. (2001). *Crystallization*, Butterworth- Heinemann, ISBN-10: 0750648333, London

Omar, M.A. (1993). *Elementary Solid State Physics*, Addison-Wesley, ISBN-10: 0201607336,

Ozawa, T. (1971). Kinetics of non-isothermal crystallization. *Polymer*, 12, pp. 150-158, 0032-3861

Park, J. (2009). *Bioceramics: Properties, Characterizations, and Applications*, Spinger, ISBN-10: 0387095446, New York

Porter, D.A. & Easterling, K.E. (1981). *Phase Transformations in Metals and Alloys*, Chapman & Hall, ISBN: 0412450305, London, Weinheim, New York, Tokyo, Melbourne, Madras

Prasad, S. & Varma, K.B.R. (2005). Crystallization kinetics of the $LiBO_2$–Nb_2O_5 glass using differential thermal analysis. *Journal of the American Ceramic Society*, 88, pp. 357-361, 0002-7820

Qin, F.X., Zhang, H.F., Ding, B.Z. & Hu, Z.Q. (2004). Nanocrystallization kinetics of Ni-based bulk amorphous alloy. *Intermetallics*, 12, pp. 1197-1203, 0966-9795

Ragone, D.V. (1994). *Thermodynamics of Materials*, John Wiley & Sons.Inc, ISBN-10: 0471308854 New York, Chichester, Brisbane, Toronto, Singapore

Ray, C.S., Huang, W.H. & Day, D.E. (1991). Crystallization kinetics of a lithia-silica glass: Effect of sample characteristics and thermal analysis measurement techniques. *Journal of American Ceramic Society*, 74, pp. 60-66, 0002-7820

Seeger, C. & Ryder, P.L. (1994). Kinetics of the crystallization of amorphous Ti-Ni and Ti-Ni-Si alloys, *Materials Science and Engineering, A179/A 180,* pp. 641-644, 0921-5093

Shaaban, E.R., Saddeek, Y.B. & Abdel Rafea, M. (2009). Crystallization kinetics of the TeO_2-BaO glass system. *Philosophical Magazine*, 89,pp. 27-39, 1478-6435

Tomasz, C. (2010). Mechanism and kinetics of nano-crystallization of the thermally stable NiNb(ZrTi)Al metallic glasses. *Journal of Thermal Analysis and Calorimetry*, 101, pp. 615-622, 1388-6150

Yahia, I.S., Shakra, A.M., Fadel, M., Hegab, N.A., Salem, A.M. & Farid, A.S. (2011). Kinetics of non isothermal crystallization of ternary $Se_{85}Te_{15-x}Sb_x$ glassy alloys. *Chalcogenide Letters*, 8, pp. 453-467, 1584-8663

Yamane, M. & Ashara, Y. (2000). *Glasses for photonics*, Cambridge University Press, ISBN 0-511-03862-3, Cambridge, New York, Port Melbourne, Madrid, Cape Town

Yinnon, H. & Uhlmann, D.R. (1983). Applications of thermoanalytical techniques to the study of crystallization kinetics in glass-forming liquids, Part I: Theory. *Journal of Non-Crystalline Solids*, 54, pp. 253-275, 0022-3093

Yukimitu, K., Oliveira, R.C., Araujo, E.B., Moraes, J.C.S. & Avanci, L.H. (2005). DSC studies on crystallization mechanisms of tellurite glasses. *Thermochimica Acta*, 426, pp. 157-161, 0040-6031

Zhang, S.N., Zhu, T.J. & Zhao, X.B. (2008). Crystallization Kinetics of $Si_{15}Te_{85}$ and $Si_{20}Te_{80}$ Chalcogenide Glasses. *Physica B-Condensed Matter*, 403, pp. 3459-3463, 0921-4526

Thermodynamics of Enthalpy Relaxation and Hole Formation of Polymer Glasses

Nobuyuki Tanaka
Gunma University, Tsutsumicho, Kiryu, Japan

1. Introduction

The enthalpy relaxation of the glassy materials has been investigated rheologically for years with a view to approaching the ideal glass[1 - 5]. The imaginary liquid at Kauzmann temperature[2, 3], T_K, at which the extrapolation line of enthalpy or entropy as a function of temperature for the liquid intersected the enthalpy or entropy line of the crystal, had been considered once to be the ideal glass. However because at T_K, the enthalpy or entropy for the liquid was same as that of crystal, the liquid like this was hard to take thermodynamically, bringing the entropy crisis. Fig. 1 depicts the change of the enthalpy difference, ΔH, between the liquid and the crystal upon cooling for a polymer. For stable liquids, ΔH should be almost constant from near T_g upon cooling as described below. Therefore, the liquid line can never intersect that of the crystal. The transition from liquid to crystal or vice versa means the emission or absorption of the latent heat accompanying the enthalpy jump. Thus for polymers, T_K is merely a temperature parameter.

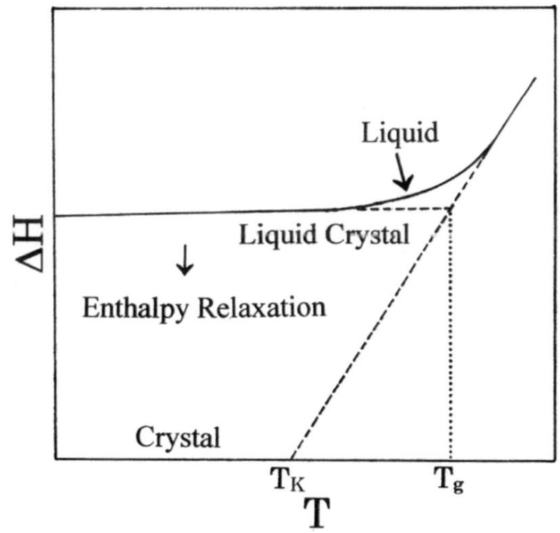

Fig. 1. The change of ΔH for a polymer liquid or liquid crystal (solid line) upon cooling. The small arrow mark shows the direction of enthalpy relaxation. The base line is that of crystal. The dashed lines are extrapolated to T_g and T_K, and the dot line is complementary.

For the glass transition of polymeric or other matter glasses, whether it is the phase transition or not is nowadays yet controversial[6 - 11]. However for polymers, the criterion of the glass transition temperature, T_g, presented by us is decisive, supporting the first order hole phase transition[12 - 14] and the broad heat capacity jump at the glass transition without the enthalpy relaxation has been understood successfully[15]. The cooled polymers unidentified by the criterion should belong to the liquid on the way of enthalpy relaxation even if they were glassy. When without the enthalpy relaxation at temperatures below T_g, the super-cooled liquid could be stable thermodynamically. Glassy pitch[4] might be the stable liquid, which is flowing still at a drop per 9 years from 1927. Recently, for poly(ethylene terephthalate) (PET) [16], polystyrene (PS), isotactic polypropylene (iPP), polyethylene (PE) and nylon-6 (N6), it was predicted that during the enthalpy relaxation at temperatures below T_g, the "ordered part / hole" pairs should generate and consequently, bringing a reduction in the relaxation time to the goal of the ideal glass, e.g., the quasi-crystals lacked periodicity but with symmetry[17], the alternative and stable glass parts should be formed between their pairs. At T_g, the unfreezing of glass parts is caused by the first order hole phase transition. Then the holes should play the free volume jump as an affair of dynamic equilibrium in the disappearance and generation of "ordered part / hole" pairs. The free volume jump at the glass transition, accompanying the enthalpy and entropy jumps, is the characteristic of the first order hole phase transition. In this chapter, introducing the constant volume heat capacity of photons to the holes[16, 18, 19], the generation of "ordered part / hole" pairs during the enthalpy relaxation at temperatures below T_g and the subsequent disappearance at the glass transition, accompanied by the jumps of free volume, enthalpy and entropy, were discussed for PET, PS, iPP, PE and N6 on the basis of thermodynamics. IPP, PE and N6 were investigated as the peculiar cases for the comparison with PET and PS, which are the glassy polymers with the almost same values of the constant, c_2, of WLF equation[20], i.e., 55.3 K[21] and 56.6 K[20], respectively. The holes taken in the helix structure of iPP should hold the interaction of partnership with the helical ordered parts. For PE, the glasses with T_g = 135 K and 237 K depending on the structure of the ordered parts[15] were discussed. It seems likely that from near 130 K, the growth of open space in the PE glasses occurs[22]. DSC (Differential Scanning Calorimetry) on PE films[18] revealed that ~38 % of the crystal lamella was constituted by the inter-grain aggregates containing the glass with a secondary T_g. The generation of "crystal / anti-crystal hole" pairs from the secondary glass was discussed. For N6, the ordered parts were the stretched sequences of -$(CH_2)_5$- between amido groups. The hole energy of "ordered part / hole" pairs was concerned with the frequency of absorption bands in the infrared spectrum.

2. Enthalpy relaxation and hole formation of PET, PS, iPP, PE and N6 glasses

2.1 Thermodynamics of glass transition

The glass transition temperature, T_g, of polymer glasses could be identified as that of the first order hole phase transition by satisfying the criterion consisted of Eqs. (1) and (2), in which f_x has been added under the operational definition concluding that the stable glasses could not be formed easily without the generation of "ordered part / hole" pairs during the enthalpy relaxation at temperatures below T_g[13]:

$$f_x = f_{flow} (= h_{flow} - T_g s_{flow}) = 0 \tag{1}$$

$$s_{flow} = 0\ (\therefore\ h_{flow} = 0) \tag{2}$$

where f_x (= $h_x - T_g s_x$), h_x and s_x are the free energy, enthalpy and entropy per molar structural unit for ordered parts, f_{flow}, h_{flow} and s_{flow} are the free energy, enthalpy and entropy per molar structural unit for flow parts. The molar free energy of holes held photons at the temperature, T, is generally given by $f^h = -RT\ln(v_f/v_0)$ and then the molar entropy of holes, s^h, at a constant pressure, p, is derived:

$$s^h = -(\partial f^h/\partial T)_p = R\ln(v_f/v_0) + RT\{\partial \ln(v_f/v_0)/\partial T\}_p \tag{3}$$

where v_f and v_0 are the molar free and core volumes of holes and R is the gas constant. When v_0 is almost constant, from $f^h = h^h - Ts^h$ and Eq. (3), the molar enthalpy of holes, h^h, is derived:

$$h^h = RT^2(\partial \ln v_f/\partial T)_p \tag{4}$$

When the ordered parts at T_g are in equilibrium with the holes; f_x (= 0) = f^h, from $f_x = 0$ and $h_x = h_g + \Delta h$ (see Eq. (13)), s_x at T_g is derived:

$$s_x\ (= h_x/T_g) = \Delta s_g + \Delta h/T_g \tag{5}$$

with $\Delta s_g = h_g/T_g$, defining the entropy of unfreezing for the glass parts at T_g, where h_g is the molar glass transition enthalpy, Δh is the heat per molar structural unit required still to melt ordered parts, relating to the jump of v_f (see Eq. (8)). Further from Eq. (2), the h_g and the molar glass transition entropy, s_g, are derived[13]:

$$h_g = RT_g^2(\partial \ln v_f/\partial T)_p \tag{6}$$

$$s_g\ (= s_g^{conf} + s_g^{int}) = R\ln(v_f/v_0) + RT_g(\partial \ln v_f/\partial T)_p \tag{7}$$

with $s_g^{int} = (3R/2)\ln(2\pi mkT_g/h^2) - (1/x)(R/N)\ln N! + R\ln q - R\ln v_0$

where s_g^{conf} and s_g^{int} are the conformational and cohesive entropies per molar structural unit for glass parts at T_g (see Eq. (16) for s_g^{conf}), m is the mass of a structural unit, k is Boltzmann constant, h is Planck constant, N is the number of chains, x is the degree of polymerization and q (≤ 1) is the packing factor. When $v_f = v_0$, from Eqs. (6) and (7), $h_g = 0$ and $s_g = 0$ are derived. On the other hand, from $f^h = 0$ (= f_x) at T_g, h^h is given as:

$$h^h = RT_g^2(\partial \ln v_f/\partial T)_p + RT_g\ln(v_f/v_0) \tag{8}$$

In the case of $RT_g\ln(v_f/v_0) = \Delta h$, the relation of $h_x = h^h$ is derived from Eq. (8), because of $RT_g^2(\partial \ln v_f/\partial T)_p = h_g$. However, when the length distribution by lengthening of ordered parts occurred during the enthalpy relaxation at temperatures below T_g, as longer the length of ordered parts, the melting temperature, T_x, for ordered parts should be elevated from T_g to the higher temperature[23] (see Eq. (29)). Therefore the shortage of Δh (= $RT_g\ln(v_f/v_0)$) corresponding to the latent heat of disappearance for the holes at T_g is made up by the supply of the heat required to melt all ordered parts:

$$\Delta h = \int_{T_g}^{T_\ell} \Delta C_p dT \tag{9}$$

where T_ℓ is the end temperature of melting for ordered parts, ΔC_p is the difference between the observed isobaric heat capacity, C_p^ℓ, for the equilibrium liquid and C_p^g for the hypothesized super-heated glass at the glass transition from T_g to T_ℓ. In the equilibrium liquid, the isobaric heat capacities of ordered parts and flow parts, C_p^x and C_p^{flow}, are equal to C_p^ℓ, respectively[13]:

$$C_p^\ell = C_p^x = C_p^{flow} \tag{10}$$

In the flow parts, the tube-like space exists between a chain and the neighboring chains, behaving as if it is the counterpart of a chain[24]. Therefore when the hole energy at T (> T_g) is given by ε (= $3C_v^{ph}T$), C_p^{flow} is represented as[16]:

$$C_p^{flow} = 3C_v^{ph}(1 + T\mathrm{dln}J/\mathrm{d}T) \tag{11}$$

where C_v^{ph} (= 2.701R)[16, 18, 19] is the constant volume heat capacity for photons, J is the number of holes lost by T and 3 is the degree of freedom for photons.

Fig. 2 shows the schematic curves of the molar entropy, s, and the v_f around T_g upon cooling and heating for polymers. Upon cooling in Fig. 2 (upper), the dashed line is the s curve for the liquid glass frozen partially from the super-cooled liquid and the solid line upon heating

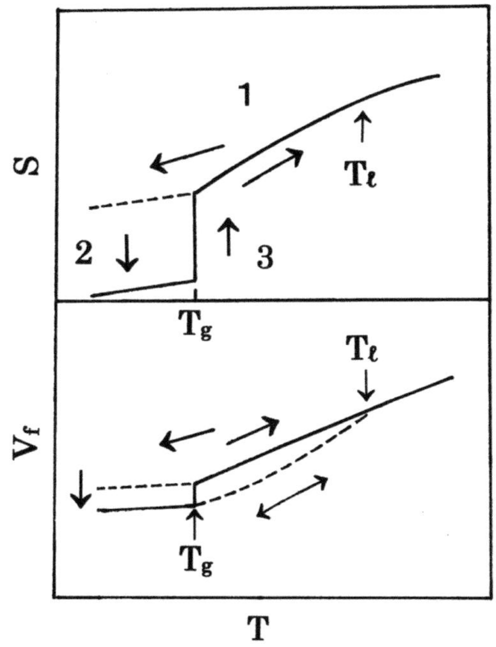

Fig. 2. The schematic curves of the entropy, s, and the free volume, v_f, around T_g for polymers. Upper; **1**: the change of s for the liquid glass frozen partially from the super-cooled liquid, shown by the dashed line, **2**: the entropy relaxation and **3**: the change of s with a jump at T_g upon heating. Lower; The dashed line upon cooling is the v_f curve for the same liquid glass, the solid line upon heating after relaxation shows the v_f curve with a jump at T_g and the dashed line shows a reversible jump of v_f between T_g and T_ℓ.

after relaxation shows the subsequent s curve with a jump at T_g. Upon cooling in Fig. 2 (lower), the dashed line is the v_f curve for the same liquid glass. Upon heating after relaxation, the solid line shows the v_f curve with a jump at T_g and the dashed line shows a reversible jump of v_f between T_g and T_ℓ.

2.2 "Ordered part / hole" pairs

Next whether h_x agrees to h^h at T_g or not is investigated for PET, iPP, PS, PE and N6 glasses. The agreement provides one of evidences for the generation of "ordered part / hole" pairs during the enthalpy relaxation at T (< T_g). h^h at T_g is given by[16]:

$$h^h = 3C_v^{ph}T_g \quad (12)$$

where $C_v^{ph} = 2.701R$. For PET with $T_g = 342$ K, $h^h = 23.0$ kJ/mol was derived. While h_x at T_g is given by[13, 25]:

$$h_x = h_g + \Delta h \quad (13)$$

In Eq. (13), h_g is given approximately by three expressions[13, 25]; (1) RT_g^2/c_2 (c_2 is the constant of WLF equation[20]), (2) the molar enthalpy difference between the super-cooled liquid and the crystal at T_g, $H_g^a - H_g^c$, and (3) the sum of the conformational and cohesive enthalpies per molar structural unit at T_g, $h_g^{conf} + h_g^{int}$. Δh is given by either Eq. (14) or (15)[25, 26]:

$$\Delta h = \Delta H - Q \quad (14)$$

with $\Delta H = H_m^a - H_c^a$, where H_m^a is the enthalpy per molar structural unit for the liquid at the equilibrium melting temperature, T_m^∞, H_c^a is the enthalpy per molar structural unit for the super-cooled liquid at the onset temperature, T_c, of a DSC crystallization peak upon cooling and Q is the heat per molar structural unit corresponding to the total area of the DSC endothermic peak upon heating. Or, rewriting Eq. (13),

$$\Delta h = (h_x^{conf} - h_g^{conf}) + \Delta h^{int} \quad (15)$$

where h_x^{conf} is the conformational enthalpy per molar structural unit for ordered parts, Δh^{int} is the molar cohesive enthalpy difference between the ordered parts and the glass parts. Thus when $h_x^{conf} = h_g^{conf}$ at T_g, $\Delta h = \Delta h^{int} = (RT_g \ln Z_t)/x$ (see Table 3 for N6) and when $h_x^{conf} \neq h_g^{conf} = 0$ at T_g, the another Δh is derived[26, 27].

$$\Delta h = T_g\{s_g^{conf} - (R\ln Z_0)/x\} \quad (16)$$

with $s_g^{conf} = (R\ln Z + RT_g d\ln Z/dT)/x$

where Z is the conformational partition function for a chain, Z_0 (= Z/Z_t) and Z_t are the component conformational partition function for a chain regardless of temperature and as a function of temperature, respectively. The differential of Eq. (15) by temperature represents the heat capacity jump at the glass transition[15].

Table 1 shows the values of T_g, Δs_g, h_g, Δh, h_x, h^h and h^h/h_x for PET, iPP and PS. PET showed the good agreement between h_x, i.e., 22.3 ~ 24.1 kJ/mol, and h^h, i.e., 23.0 kJ/mol. The values of them also agreed with the heat of fusion, h_u = 23.0 kJ/mol, for the smectic crystals of mesophase with the conformational disorder between the phenylene groups but along

Polymer	T_g K	Δs_g J/(K mol)	h_g kJ/mol	Δh kJ/mol	h_x kJ/mol	h^h kJ/mol	h^h/h_x
PET	342	47.1	16.1*1	6.5*4	22.6	23.0	1.0
		51.5	17.6*2	6.2*5	23.8		1.0
		51.2	17.5*3	6.2*5	23.7		1.0
iPP	270	23.0	6.2*1	1.1*4	7.3	18.2	2.5 (1.0*7)
		23.7	6.4*3	1.0*5	7.4		2.5 (1.0*7)
PS	359	52.6	18.9*2	5.3	24.2*6	24.2	1.0
		56.5	20.3*3	3.9	24.2*6		1.0

*1: $H_g^a - H_g^c$, *2: RT_g^2/c_2, *3: $h_g^{conf} + h_g^{int}$, *4: Eq. (14), *5: Eq. (16), *6: $h_x = h^h$ and *7: $(h^h/2.5)/h_x$. The data of iPP used to calculate Δh in Eq. (14) are as follows: T_c = 403.6 K, T_m^∞ = 449 K for the α form crystal[28], ΔH ($=H_m^a - H_c^a$) = 4.89 kJ/mol[29] and Q = 3.76 kJ/mol for the sample annealed at 461.0 K for 1 hour.

Table 1. The values of T_g, Δs_g, h_g, Δh, h_x, h^h and h^h/h_x for PET, iPP and PS.

a chain axis[25, 30]. For the smectic-c crystals with stretched sequences, h_u is 28.5 kJ/mol. Further, DSC revealed[25] that for the crystalline films of smectic-c crystals, the ordered parts in the amorphous regions were like smectic crystals and for the crystalline films of smectic crystals, the ordered parts were like the smectic-c crystals. Fig. 3 shows the sequence models of smectic crystal (A) and smectic-c crystal (B), together with the four conformations (a, b, c and d) that an isolated chain can take preferentially below 10 K. An arrow mark shows the direction of ordering or crystallization for a, b, c and d. From these results, the ordered parts are like the smectic crystal and the hole of a pair should have the free volume coming from the difference of conformation between A and a, b, c or d in Fig. 3. For iPP, h^h was 2.5 times as much as h_x. This result suggested that the hole of a pair was the inside space of a 3/1 helical ordered part composed of 3 structural units, holding 3 photons, but each photon was concerned in the potential energy of 2.5 structural units in a helical sequence, and that, $(h^h/2.5)/h_x = 1$. This was comparable to $h^h/h_x = 1$ for PET. The value of

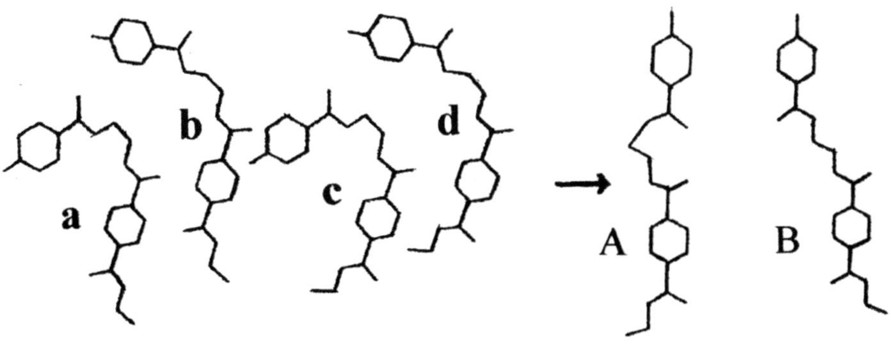

Fig. 3. Right: The sequence models of the smectic crystal of mesophase with a conformational disorder (A) and the smectic-c crystal with a stretched conformation (B) for PET[30]. Left: Four conformations taken preferentially below 10 K for an isolated chain; a: TTTTG'T, b: TTTTGT, c: TCTTG'T and d: TCTTGT, by Flory's theory[31]. T, G and G' are the trans, gauche and gauche' isomers, respectively. **T** and **C** are the trans and cis isomers of phenylene groups (lower groups). An arrow mark shows the direction of ordering or crystallization.

h_x was almost equal to h_u (= 7.6 kJ/mol for α form). Fig. 4 shows the photon sites in the helix structure and the helical conformation of an isolated sequence with an inversion defect isomer *TT*, taking preferentially at temperatures below 70 K[26].

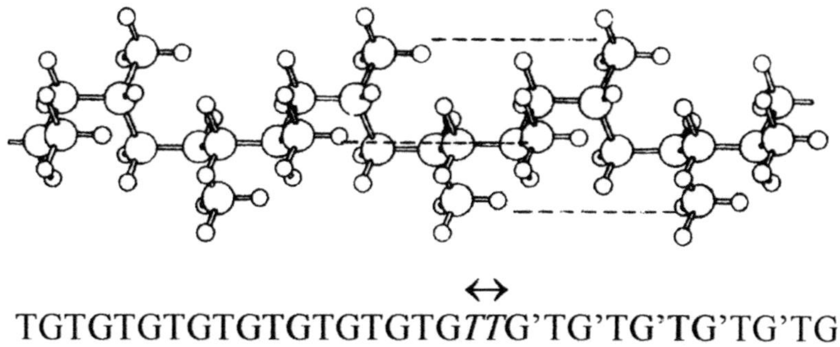

Fig. 4. Upper: The photon sites (the dashed line parts) in the helix structure with TGTG or TG'TG' conformation at temperatures below T_g for iPP. Large circle: C and small circle: H. Lower: The helical conformation of an isolated sequence with an inversion defect isomer *TT*, taking preferentially at temperatures below 70 K. The allow mark shows the shift of *TT* on a helical sequence.

For PS, supposing $h_x = h^h$, Δh was evaluated. From the value of Δh to be near that of PET, the v_f jump at T_g should be due to the release between phenyl groups. The T_g of PE [15], producing the entropy of unfreezing for the glass parts; $\Delta s_g = h_g/T_g$ (see Eq. (5)), was almost dependent on h_g^{int}. When a value of h_g^{int} was that of the glass with T_g = 237 K, $h_g^{int}/2$ gave T_g = 135 K. Table 2 shows the values of T_g, Δs_g, h_g, Δh, h_x, h^h and h^h/h_x for PE glasses with T_g = 135 K (h_g^{int} = 2.8/2 kJ/mol) and 237 K (h_g^{int} = 2.8 kJ/mol). The above relation in T_g and h_g^{int} for the glass parts was linked to the ordered parts. For both glasses, h^h was about 5 times as much as h_g. Thus from Eq. (13), the common relations of $h_x = h^h/4$ and $\Delta h = h^h/4 - h_g$ were predicted for the ordered parts in both glasses and shown in Table 2. Fig. 5 depicts the sequence models of ordered parts (A and B) and the schematic transition from the glassy state (C: left) to that of the "ordered part / hole" pair (C: right).

For the glass with T_g = 135 K, the coarse 4/1 helical ordered parts with GG or G'G' conformation in Fig. 5A and as the hole of a pair, the inside space holding four photons per a helical segmental unit, $-(CH_2)_4-$, were predicted. Further the length distribution of helical ordered parts in the glass and as the end temperature of melting for the ordered parts, ~237 K were predicted. In this case, the same value of Δh for both glasses enabled the scheme as depicted in Fig. 6. For the glass with T_g = 237 K, the ordered part of fringe-type formed by

Polymer	T_g K	Δs_g J/(K mol)	h_g kJ/mol	Δh kJ/mol	h_x kJ/mol	h^h kJ/mol	h^h/h_x
PE	135	13.8	1.8*1	0.5	2.3*2	9.1	4 (1*3)
	237	14.8	3.5*1	0.5	4.0*2	16.0	4 (1*3)

*1: $h_g^{conf} + h_g^{int}$, *2: $h_x = h^h/4$ and *3: $(h^h/4)/h_x$.

Table 2. The values of T_g, Δs_g, h_g, Δh, h_x, h^h and h^h/h_x for PE.

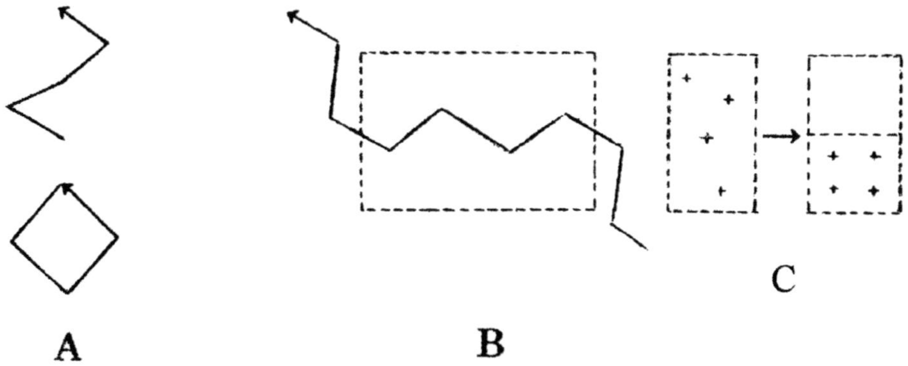

Fig. 5. The sequence models of ordered parts; A: the 4/1 helix structure and B: the stretched structure in the region surrounded by the dashed line. C (left): the glassy state and C (right): the state of a "ordered part / hole" pair. The plus mark (+) shows the cross section of a stretched segmental unit, $-(CH_2)_4-$. An arrow mark shows the direction of ordering.

bundling TTT parts of four sequences at least, and that, the smallest crystal of PE and the neighboring hole were predicted (see Fig. 5B and C: right), since the value of h_x was almost equal to h_u = 4.1 kJ/mol. Fig. 6 shows the bar graph of h_g at T_g = 135 K and 237 K, together with Δh (= 0.5 kJ/mol) at 237 K, for PE. The difference in two complementary lines suggests the supply schedule of heat over the temperature range from 135 K to ~237 K in order to make up the shortage of Δh required to melt all ordered parts in the glass with T_g = 135 K. For the glass with T_g = 237 K, the glass transition only at T_g is shown. The enthalpy relaxation, accompanied by the generation of stretched segments, at temperatures over 135 K to ~237 K

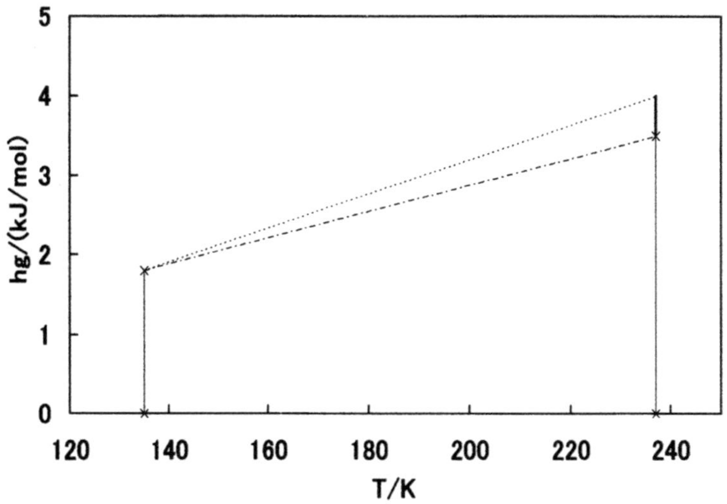

Fig. 6. The bar graph of h_g at T_g = 135 K and 237 K (the vertical thin lines between the cross marks) and Δh (= 0.5 kJ/mol) at 237 K (the thick line part) for PE. The dot and dot-dashed lines are drawn complementarily.

should vitrify or order the melted helical ordered parts and confine the larger helical ordered parts, which were not yet melted even over T_g = 135 K, in the glass. The liquid on the way of enthalpy relaxation like this should reach step by step to the glass with T_g = 237 K (h_g^{int} = 2.8 kJ/mol) at temperatures below T_g. Table 3 shows the values of T_g, Δs_g, h_g, Δh, h_x, h_u and h^h for N6. For N6, $h_g^{conf} = h_x^{conf}$ in Eq. (15), i.e., $\Delta h = (RT_g \ln Z_t)/x$ was predicted, because of the strong interaction between amido groups. h^h was 4.7 times as much as h_x in the parenthesis. Accordingly a photon was concerned in the potential energy of a stretched segmental unit, $-(CH_2)_5-$. Further Δh = 2.5 kJ/mol was 0.5 kJ per molar methylene unit, $-CH_2-$, which agreed with Δh = 0.5 kJ/mol for the PE glasses with T_g = 135 K and 237 K (see Table 2). In addition, the value of h^h was almost equal to h_u of the heat of fusion. This agreement suggested that the ordered parts were the stretched segmental units in the smallest crystals of N6. Fig. 7 shows the structural unit of N6 and the photon site in the structural unit.

Polymer	T_g K	Δs_g J/(K mol)	h_g kJ/mol	Δh*2 kJ/mol	h_x kJ/mol	h_u kJ/mol	h^h kJ/mol
N6 [27]	313	120.1 (6.4)	37.6*1 (2.0)	2.5, 0.5*3	40.1 (4.5)	21.3	21.1

*1: $h_g = h_g^{conf} + h_g^{int}$, *2: $\Delta h = (RT_g \ln Z_t)/x$ and *3: the value of Δh per molar methylene unit, $-CH_2-$. The values of Δs_g, h_g and h_x in the parentheses are those without the cohesive energy of amido group, 35.6 kJ/mol.

Table 3. The values of T_g, Δs_g, h_g, Δh, h_x, h_u and h^h for N6.

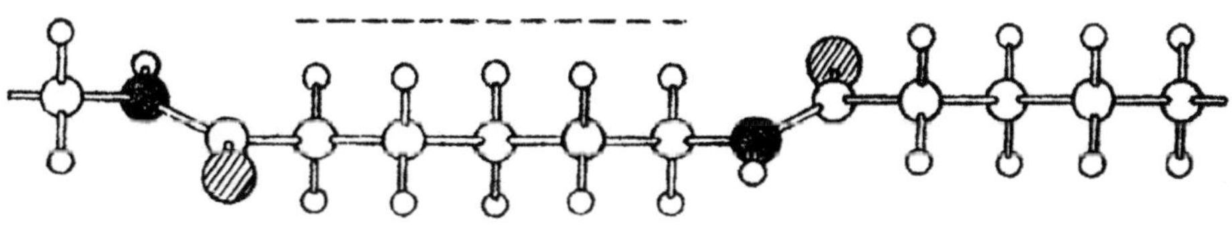

Fig. 7. The structure of N6 structural unit and the photon site (the dashed line part) in the unit. The filled circle: N, the shaded circle: O, the large circle: C, and the small circle: H.

3. Hole energy and a photon

A photon has the property as a boson or a wave. Therefore h^h is also represented as the vibrational energy of a wave with the quantum number n = 1 (meaning one photon) and the frequency per second (sec), ν:

$$h^h = N_A(3/2)h\nu \tag{17}$$

where N_A is Avogadro constant. Thus from Eqs. (12) and (17), the zero-point energy, ε_0 (= $(1/2)h\nu$), is derived:

$$\varepsilon_0 = C_v^{ph} T_g / N_A \tag{18}$$

Table 4 shows the values of T_g, h^h, ν, λ and 1/λ for PE, iPP, PS and PET, where λ is the wavelength and 1/λ is the wavenumber. According to the infrared spectroscopy, for PE, 1/λ = 510 cm^{-1} and 893 cm^{-1} might be concerned with 720 cm^{-1} and 731 cm^{-1} bands assigned to the rocking of $-CH_2-$[32)]. For iPP, 1/λ = 1022 cm^{-1} almost agreed with 1045 cm^{-1} relating to the crystallinity[33)]. Also for PS, 1/λ = 1359 cm^{-1} almost agreed with one of conformation sensitive bands[34)], i.e., 1365 cm^{-1} band. For PET, 1/λ = 1292 cm^{-1} was near 1339 cm^{-1} and 1371 cm^{-1} bands assigned to the wagging of $-CH_2-$ with trans and gauche conformations, respectively[35)].

Polymer	T_g K	h^h kJ/mol	ν sec^{-1}	λ μm	1/λ cm^{-1}
PE	135	9.1	15.3×10^{12}	19.6	510
	237	16.0	26.9×10^{12}	11.2	893
iPP	270	18.2	30.6×10^{12}	9.78	1022
PS	359	24.2	40.6×10^{12}	7.36	1359
PET	342	23.0	38.6×10^{12}	7.74	1292

Table 4. The values of T_g, h^h, ν, λ and 1/λ for PE, iPP, PS and PET.

4. Conclusions and introduction to next section

For PET, PS, iPP, PE and N6 glasses, the generation of "ordered part / hole" pairs during the enthalpy relaxation at temperatures below T_g and the subsequent disappearance at the glass transition were discussed under the operational definition leading the criterion of T_g. Thus it was concluded that the unfreezing of the glass parts at T_g was caused by the first order hole phase transition, accompanied by the jumps of free volume, enthalpy and entropy. h^h was concerned with the frequency of the absorption bands in the infrared spectrum. In particularly, 1/λ for iPP and PS coincided closely with the respective sensitive bands with physical meaning. In the next section, the generation of "crystal / anti-crystal hole" pairs from the secondary glass in PE crystal lamella was discussed on the DSC curves. The secondary glass was distinguished from the primary glass discussed in the above sections. The C_v^{ph} was also available in the discussion of the cavity radiation from the anti-crystal holes filled by photons. The anti-crystal holes were regarded as the lattice crystal made up of photons without the mass.

5. Crystallization of secondary glass in PE lamella

5.1 Thermal analysis

DSC is capable of quantitatively determining by way of standard and dynamical measurements[36)] the common thermal phenomena in polymers, e.g., the melting, the crystallization and the glass transition. Such analyses are carried out on the basis of thermodynamics, mathematics and molecular dynamics simulation[36, 37)]. This section describes an attempt to understand the peculiar DSC curves of PE films containing orthorhombic crystals. DSC demonstrated two indications of the secondary glass in the crystal lamella. One of the underlying reasons was the much larger heat of melting as opposed to the heat of crystallization upon cooling and the other was the fact that the glass transition enthalpy was larger than the molar enthalpy of the ordered parts in the amorphous regions; $\Delta h < 0$ in Eq. (14). At temperatures above its T_g, the generation and disappearance of the "crystal / anti-crystal hole" pairs from the secondary glass were predicted as the simultaneous phenomena in the crystallization and the melting. Hexagonal and monoclinic forms of PE crystals are also well known. However, the hexagonal crystals should not be related to the melting of the orthorhombic crystals since the DSC melting peak of the hexagonal crystals generally cannot be observed for the samples without restraints such as high pressure[38]. Moreover, the DSC melting peak of monoclinic crystals disappears before the melting of the orthorhombic crystals[39, 40]. Thus, when the monoclinic crystals are in the bulk state, the heat due to their melting should contribute to the activation heat required to release the secondary glass state in the orthorhombic crystal lamella.

5.2 Secondary glass

Fig. 8 depicts the DSC crystallization peak upon cooling and the two peaks divided from a DSC endothermic peak upon heating for the PE film annealed at 416.6 K (near $T_m^{\infty} = 415$ K) for 1 hour. The thin line in T_b^* and T_e^* is the curve before division. T_c (= 391.5 K) is the onset temperature of crystallization, T_b^* ($\approx T_c$) is the intersection between the base line and the extrapolation line from the line segment with the highest slope on the lower temperature side of the melting peak, and that, the onset temperature of the higher temperature side peak and T_e^* is the end temperature of the lower temperature side peak, and that, the origin of the extrapolation line, respectively. Q_m is the heat per molar structural unit corresponding to the endothermic peak area of crystal lamella that starts to melt at T_b^* and h_c (= 0.89 kJ/mol) is the heat of crystallization per molar structural unit corresponding to the area surrounded by the dashed line and the exothermic curve. ΔQ_m (= $Q_m - h_c$) corresponds to the area between T_b^* and T_e^* of the higher temperature side

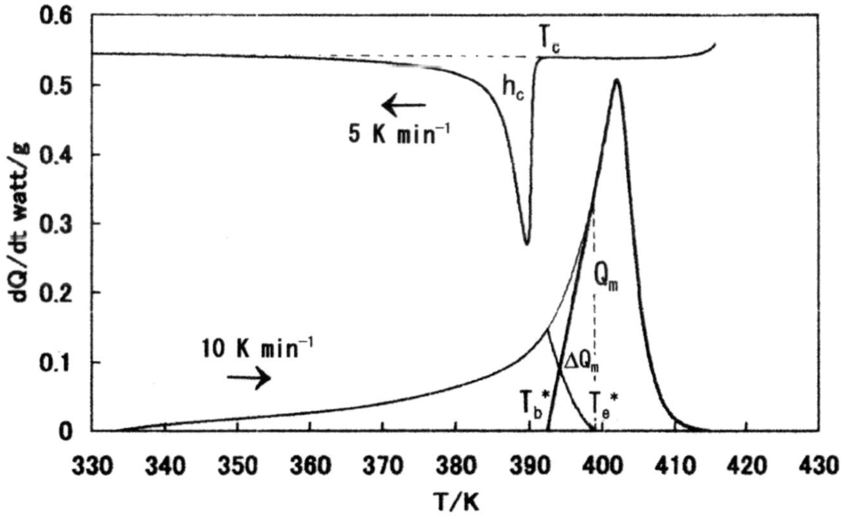

Fig. 8. The DSC crystallization peak upon cooling and the two peaks divided from a DSC endothermic peak upon heating for the PE film annealed at 416.6 K for 1 hour. dQ/dt is the heat flow rate. The cooling and heating rates are 5 K/min and 10 K/min, respectively.

peak, which is equal to the area surrounded by the thin line and the lower temperature side peak curve between T_b^* and T_e^*. The endothermic peak on the lower temperature side is due to the melting of small crystals around the crystal lamella[41]. The decrease of heat flow rate from T_b^* to T_e^* for the peak on the lower temperature side is believed to be due to the crystallization of secondary glass in the inter-grain aggregates belonging to the crystal lamella (see Fig. 9). This precedes the increase of heat flow rate due to the melting of newly crystallized parts from T_b^* to T_e^* in the peak on the higher temperature side. The equilibrium melting of the ordered parts in the amorphous regions does not show any peak. Its enthalpy, h_x, has been represented as Eq. (13), in which Δh given by Eq. (14) is usually positive; 6.5 kJ/mol and 11.5 kJ/mol for PET with two values of T_m^∞ (535 K and 549 K)[25], respectively. Also for iPP, Δh (= 1.1 kJ/mol) of Eq. (14) was positive, as shown in Table 1. Nevertheless, it was found to be negative for PE (see Table 5). In order to satisfy $\Delta h < 0$ ($h_g > h_x$), the glass with a secondary T_g, which formed near T_c upon cooling and disappeared near T_c after melting of the ordered parts in the amorphous regions upon heating, must exist in the crystal lamella. When the secondary T_g is approximated to T_c, h_g^* is given by:

$$h_g^* \approx H_c^a - H_c^c \tag{19}$$

where h_g^* is h_g at the secondary T_g, H_c^a and H_c^c are the enthalpy per molar structural unit for the super-cooled liquid and the crystal at T_c. h_x is given by $h_g^* + \Delta h$ ($\Delta h < 0$). Here, ΔH (= $H_m^a - H_c^a$) in Eq. (14) is regarded as the heat emitted when only one single crystal lamella without deformation is formed. According to the ATHAS databank[29], ΔH is 0.83 kJ/mol, which is close to the value of h_c = 0.89 kJ/mol as observed in Fig. 8. The difference in h_c and ΔH, 0.06 kJ/mol, might be the additive heat of emission due to the release of lamellar deformation. The spherulites observed in the films are substantially like disks[42]. The twist deformation energy of ribbon-like lamella is believed to originate from the irregular growth of lamella. Fig. 9 shows a schematic structure of the crystal lamella after release of the twist deformation from the ribbon-like lamella. The dark parts between the rectangular parallelepiped blocks correspond to the inter-grain aggregates described above. The crystal lamellae for the samples used here are described in the section 5.5 of "*Crystal length distribution function*".

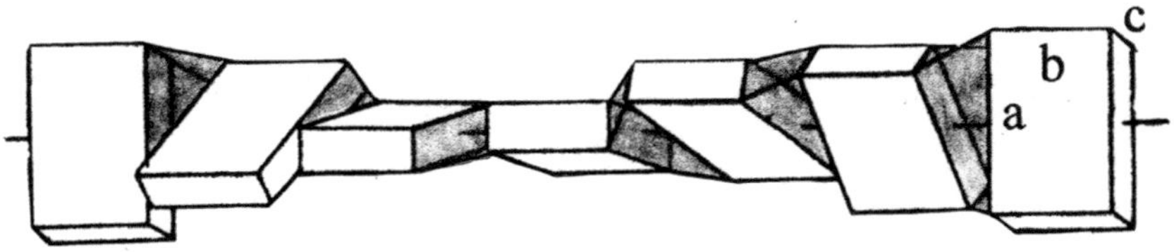

Fig. 9. A schematic structure of the crystal lamella after release of the twist deformation from the ribbon-like lamella. a, b and c (chain axis) correspond to the three cell axes of the orthorhombic crystal and the dark parts represent the inter-grain aggregates. The cell lengths of a, b and c axes for PE are 0.74 nm, 0.49 nm and 0.25 nm, respectively.

Table 5 shows the values of T_c ($\approx T_g^*$), T_b^*, Q, ΔH ($=H_m^a - H_c^a$), Δh, h_g^* and h_x for the samples annealed at T_a = 376.6 K, 416.6 K and 426.6 K for 1 hour, where T_g^* is the secondary T_g and T_a is the annealing temperature. h_x was found to decrease with an increasing T_a. The values of h_x for "T_a = 416.6 K and 426.6 K"-samples were near h_x = 2.3 kJ/mol for the glass with T_g =

135 K in Table 2. In the amorphous region of these samples, the ordered parts with the coarse 4/1 helix structure might be formed[5]. The ordered parts in the inter-grain aggregates, being like the crystals of fold-type with h_x^i instead of h_x, could also have the holes as the pair (see Eq. (28)).

Sample T_a/K	T_c K	T_b^* K	Q kJ/mol	ΔH kJ/mol	Δh kJ/mol	h_g^* kJ/mol	h_x kJ/mol
376.6	391.5	392.4	1.79	0.83	–0.96	3.99	3.03
416.6	391.5	392.5	2.38	0.83	–1.55	3.99	2.44
426.6	391.5	393.3	2.85	0.83	–2.02	3.99	1.97

Table 5. The values of T_c (≈ T_g^*), T_b^*, Q, ΔH, Δh, h_g^* and h_x for PE films annealed at 376.6 K, 416.6 K and 426.6 K for 1 hour.

5.3 "Crystal / anti-crystal hole" pairs from secondary glass

The crystal length, ζ, as a function of the melting temperature, T_m, is according to Gibbs-Thomson given by:

$$\zeta = \{T_m^{\infty}/(T_m^{\infty} - T_m)\}\{2\sigma_e/(\mu h_u)\} \tag{20}$$

where σ_e is the end-surface free energy per unit area for crystals, μ is the conversion coefficient. For PE, μ = (10⁶/14) mol/m³, h_u = 4.11 kJ/mol[36] and T_m^{∞} = 415 K[18, 36]. σ_e is given as[25, 43]:

$$\sigma_e = \mu h_u c^*[\{RT_m^2 + (H_x - h_x)(T_m^{\infty} - T_m)\}/\{2(H_x - h_x)T_m^{\infty}\}] \tag{21}$$

with $H_x = 2h_u - Q_m$, where c^* is the cell length of c-axis. The term of the square bracket in Eq. (21) is dimensionless. Table 6 shows the values of σ_e for "T_a = 376.6 K, 416.6 K and 426.6 K"-samples, together with the values of T_p, T_e^*, h_x, Q_m, ΔQ_m (= $Q_m - h_c$) and ΔQ (= Q – Q_m) used in the calculation of σ_e, where T_p is the melting peak temperature and ΔQ is the heat per molar structural unit corresponding to the area of the lower temperature side peak in Fig. 8, contributing to the activation heat required to release the secondary glass state. ΔQ_m is given by:

$$\Delta Q_m = \int_{T_b^*}^{T_e^*}(dQ/dt)(1/\alpha_s)dT \tag{22}$$

where α_s (= dT/dt) is the heating rate. T_e^* was derived from Eq. (22) using ΔQ_m in Table 6. The fact that T_e^* was almost equal to that by observation, as shown in Table 6, supported that h_c was due only to the formation of the crystal lamella, thus giving rise to the melting peak on the higher temperature side. T_e^* is also the end temperature of melting for ordered parts of fold-type in the inter-grain aggregates. Upon heating over T_e^*, the flow parts in the amorphous regions should start to participate directly in the melting of crystals. With an increasing T_a, σ_e and h_x decreased, whereas Q_m, ΔQ_m and ΔQ increased. ΔQ should influence h_x. For all samples, the value of σ_e at T_p was 1.3 times larger than that at T_m^{∞} (Q_m = 0). The experimental values of σ_e from Eq. (20), 30 ~ 90 mJ/m², differed significantly from those in Table 6, probably due to the length of the lamellae after annealing and the use of cooling as the substitute of ζ at T_m[23, 44].

Sample T_a/K	T_p K	T_e^* K	h_x kJ/mol	Q_m kJ/mol	ΔQ_m kJ/mol	ΔQ kJ/mol	σ_e at T_p mJ/m^2
376.6	401.2	399.2 (397.7)	3.03	1.37	0.48	0.42	16.3 (12.4)
416.6	401.8	400.1 (399.0)	2.44	1.42	0.53	0.96	14.3 (11.1)
426.6	401.9	400.7 (389.7)	1.97	1.52	0.63	1.33	12.9 (10.1)

The values in the parentheses of T_e^* and σ_e columns are the apparent T_e^* by observation and σ_e at T_m^∞ = 415 K (Q_m = 0), respectively. T_p is corrected by 0.1 K to the lower temperature side.

Table 6. The values of T_p, T_e^*, h_x, Q_m, ΔQ_m, ΔQ and σ_e for PE films annealed at 376.6 K, 416.6 K and 426.6 K for 1 hour.

On the other hand, according to Flory's theory[45)] on the melting of the fringe-type crystals with a finite ζ, σ_e at λ and $(df_u/d\zeta)_\lambda = 0$ is given by:

$$\sigma_e = \mu(RT\zeta/2)[1/(x - \zeta + 1) + (1/\zeta)\ln\{(x - \zeta + 1)/x\}] \tag{23}$$

where λ is the amorphous fraction, f_u is the free energy per molar structural unit for the crystals and x is the degree of polymerization. In this context:

$$2\sigma_e/\zeta = \mu(f_x - f_u) \tag{24}$$

$$f_x' = RT[(1/\zeta)\ln\{(x - \zeta + 1)/x\} - \ln P_c] \tag{25}$$

where P_c, given by $\{(x - \zeta + 1)/x\}^{1/\zeta}$ for fringe-type crystals, is the probability that a sequence occupies the lattice sites of a crystalline sequence. Moreover:

$$f_u - (f_x - f_x') = 0 \tag{26}$$

Eq. (23) is obtained when $\ln P_c = -1/(x - \zeta + 1)$. From Eq. (26), the relations are derived based on f_u and f_x at $f_x' \geq 0$ and those can be grouped into four equilibrium classes (A ~ D) and one non-equilibrium class (X) as shown in Table 7. Class A of $f_x = f_u$ at $f_x' = 0$ shows the dynamic equilibrium relation between the ordered parts and the crystal parts of equivalent fringe-types, leading to $\sigma_e = 0$ in Eq. (24) as expected for highly oriented fibers. For class B, $f_u = -f_x'$ from Eq. (26) with $f_x = 0$ refers to the anti-crystal holes and $f_x = 0$ is assigned to the ordered parts of $\zeta = \infty$. For class C, $f_x = f_x'$ from Eq. (26) with $f_u = 0$ is assigned to the ordered parts of $\zeta \neq \infty$ (i.e., a kebab structure) and $f_u = 0$ to the crystals of $\zeta = \infty$ (i.e., a shish structure). Class D of f_u (= f_x') = $f_x/2$ is related to the equilibrium in crystal and ordered parts. For those with folded chains, the reversible change from crystal or ordered parts to other parts is expected to take place automatically. The relations in class X do not satisfy Eq. (26), suggesting that the holes of class B can not be replaced by the crystals with $\zeta \neq \infty$.

When $f_u = -f_x'$ for class B, the temperature at which the anti-crystal holes disappear (melt), i.e., T_m^h, is given by:

$$T_m^h = T_m^\infty\{1 + 2\sigma_e/(\mu\zeta h_u)\} \tag{27}$$

where h_u, σ_e and ζ are imaged for the anti-crystal holes, and that, the photonic crystals made up only of photons without the mass. According to Eq. (27), $T_m{}^h$ approaches $T_m{}^\infty$ with an increasing ζ. However, the interface between the anti-crystal holes and the ordered parts, which satisfy $f_x = 0$ (described below), should work as the reflector of photons attached to the anti-crystal holes. In this case, the even interface made of the folded chain segments should be avoided through the random reflection. At such an interface, from Eq. (24), the following relation of energy balance is derived:

$$h_x - h_u = T_m{}^h(s_u - s_x) = \sigma_e/(\mu\zeta) \quad (28)$$

where s_u is the entropy per molar unit for the anti-crystal holes. As well as h_u, one mole of units (photons) corresponds to three moles of the oscillators, since three oscillators can be coordinated to each point of the crystal lattice. According to Eq. (28) or (24), the respective interface energies of the hole and the ordered part are compensated each other at the common interface, leading to $f_x = 0$ of class B and further, $T_m{}^h$ approaches 0 K with an increasing ζ. From the interface at $\zeta = \infty$, the photons are not reflected, and that, do not exist in the holes. This is exactly the real "dark hole". Therefore when the anti-crystal hole of $\zeta = \infty$ is pairing with the neighboring crystal as shown in Fig. 11C, the crystal is set at 0 K. As opposed to Eq. (27), from $f_u = f_x'$ of class D, the T_m of the crystals is derived:

$$T_m = T_m{}^\infty\{1 - 2\sigma_e/(\mu\zeta h_u)\} \quad (29)$$

Eq. (29) is the same as Eq. (20). In Eq. (29), T_m is $T_m{}^\infty$ at $\zeta = \infty$ and from Eqs. (27) and (29), $T_m{}^\infty = (T_m{}^h + T_m)/2$ is derived. According to the pair relation, the emission of heat from the anti-crystal holes after crystallization is necessarily linked to the supply of heat of the same quantity to the newly crystallized parts. However $T_m{}^h$ should be depressed down to T_m by the emission of heat to the outside. For a model of the inter-grain aggregates shown in Fig. 9, the interaction in the inter-grain aggregates and the a - c face of the crystals must be neglected and the ordered parts in the inter-grain aggregates must satisfy Eqs. (1) and (2) at $T_g{}^*$. It is thus presumed that over $T_g{}^*$, the chains that cross the glass regions give rise to the newly crystallized parts of fringe-type, whereas the folded chain segments around the glass excluded from the ordered parts give rise to the two end-surfaces of the anti-crystal hole with the same ζ as the new crystal. The T_m of the crystals from the secondary glass, being equal to $T_m{}^h$, was found to change from $T_b{}^*$ to $T_e{}^*$ as a function of ζ in Eq. (29). The time spent from $T_b{}^*$ to $T_e{}^*$ was 0.73 s, in which the probability of observing a spontaneous generation of crystallization or melting should be 1/2, according to the uncertainty principle.

f_x'	f_x	f_u	Class
$f_x' = 0$	$f_x = f_u$	$f_u = f_x$	A
$f_x' > 0$	$f_x = 0$	$f_u = -f_x'$	B
	$f_x = f_x'$	$f_u = 0$	C
	$f_x = 2f_u$	$f_u = f_x/2 = f_x'$	D
$f_x' > 0$	$f_x = 0$	$f_u = f_x'$	X

Table 7. Relations of equilibrium (A ~ D) and non-equilibrium (X) in f_x and f_u at $f_x' \geq 0$ for crystalline polymers[46].

Fig. 10 shows the schematic behaviors of sequences and photons on the way of crystallization and melting. The heat of emission, ΔU_h, corresponds to ΔQ_m of the area between the observed melting curve (thin line) and the lower temperature side peak curve from T_b^* to T_e^* and the heat of absorption, ΔU_m (= ΔU_h), corresponds to ΔQ_m of the under area of the higher temperature side peak curve from T_b^* to T_e^* in Fig. 8. Fig. 11 shows the cross sections of the glass (A) and the "crystal / anti-crystal hole" pair (C). The two end-surfaces of the anti-crystal hole in Fig. 11C contact in equilibrium those of the ordered parts. Supposing that this model of aggregates was valid for the "T_a = 416.6 K"-sample shown in Fig. 8, a derivation of f_x' (= $-f_u$) = 0.13 kJ/mol was obtained from Eq. (24) using the values of σ_e (see Table 6) and ζ (= 3.1 nm) at T_p = 401.8K (see Table 9). Here, f_x is rewritten as f_x^i for the ordered parts in the inter-grain aggregates. Furthermore, under the assumption that the strain energy in the glass should be spent to build the interface between the anti-crystal hole and the ordered part, by substituting $h_c - \Delta H$ (= 0.06 kJ/mol) for $h_x^i - h_u$, it was possible to derive $s_u - s_x^i$ = 0.15 J/(K mol) at T_c (= 391.5 K), where h_x^i and s_x^i are the enthalpy and entropy per molar structural unit for the ordered parts in the inter-grain aggregates. The relation of $h_x^i - h_u \approx h_c - \Delta H$ (= 0.06 kJ/mol) could be supported by determining $\sigma_e/(\mu\zeta)$ = 0.06 ~ 0.07 kJ /mol in Eq. (28), which was obtained for all samples using ζ and σ_e at T_p. From f_x^i = 0, h_x^i = 4.17 kJ/mol and s_x^i = 10.7 J/(K mol) at T_c were obtained. Moreover, using h_u = 4.11 kJ/mol, s_u was 10.9 J/(K mol) at T_c.

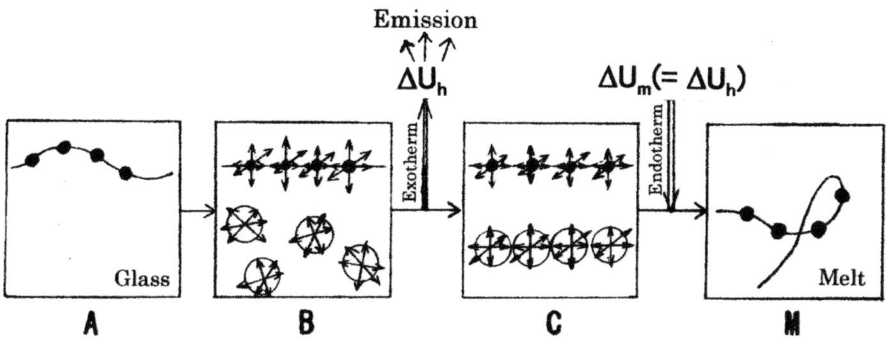

Fig. 10. The schematic process from the glass (A) to the generation of "crystal / anti-crystal hole" pairs (B → C) by emission of ΔU_h and then the disappearance of them (M) by absorption of ΔU_m (= ΔU_h). The filled circle (•) is the segmental unit, the arrow mark (↔) is the oscillator and the large circle is the photon.

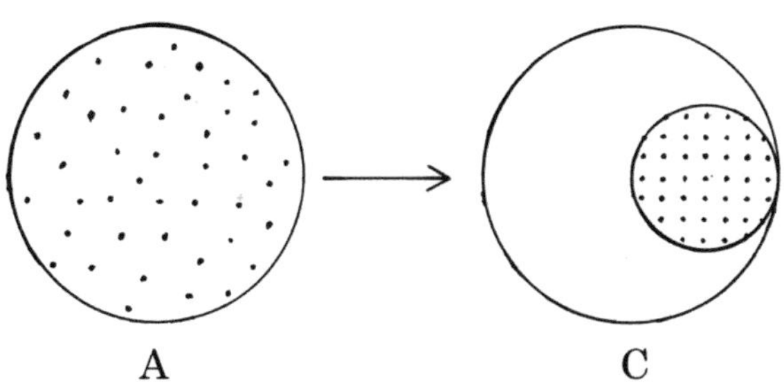

Fig. 11. The cross sections of the glass (A) and the "crystal / anti-crystal hole" pair (C). The dot in large and small circles is the cross section of a segment and the blank in C is the hole. The arrow mark shows the crystallization from A to C.

5.4 Fraction of secondary glass

The anti-crystal holes should be permeated by the photons obeying the frequency distribution function with an upper limit. This is due to the interface between the anti-crystal hole and the ordered part be able to act as a filter for the photons. The molar photon energy loss of the anti-crystal holes, ΔU_h, due to the cavity radiation from T_b^* to T_e^* is given by[18]:

$$\Delta U_h = 3C_v^{ph}(T_e^* - T_b^*) \tag{30}$$

On the other hand, the heat change per molar structural unit, ΔU_m, due to the melting of the newly crystallized parts from T_b^* to T_e^* is given by:

$$\Delta U_m = \Gamma \Delta Q_m \tag{31}$$

where Γ is the fraction of the secondary glass, contributed to the generation of "crystal / anti-crystal hole" pairs, in the inter-grain aggregates at temperatures below T_g^*. From $\Delta U_h = \Delta U_m$ at T_e^*, Γ is given by[18]:

$$\Gamma = 3C_v^{ph}(T_e^* - T_b^*)/\Delta Q_m \tag{32}$$

Table 8 shows the values of $T_e^* - T_b^*$, Q_m, ΔQ_m, ΔU_h (= ΔU_m), $\Delta Q_m/Q_m$ and Γ (= $\Delta U_h/\Delta Q_m$) for PE films annealed at 376.6 K, 416.6 K and 426.6 K for 1 hour. From $\Delta Q_m/Q_m$ and Γ, 35 ~ 41 % of the lamella was constituted of the inter-grain aggregates and approximately 79 ~ 96 % of this was the glass at $T \leq T_g^*$. The difference in ΔQ_m and ΔU_m was believed to represent the irreversible heat change due to the melting of the ordered parts of fold-type, since the folded segments have the excess defect energy. The values of ΔU_h were almost same as Δh (= 0.5 kJ/mol) for the glasses with T_g = 135 K and 237 K in Table 2, giving the latent heat required to disappear the holes.

Sample T_a/K	$T_e^* - T_b^*$ K	Q_m kJ/mol	ΔQ_m kJ/mol	ΔU_h kJ/mol	$\Delta Q_m/Q_m$	Γ
376.6	6.8	1.37	0.48	0.46	0.35	0.95
416.6	7.6	1.42	0.53	0.51	0.37	0.96
426.6	7.4	1.52	0.63	0.50	0.41	0.79

Table 8. The values of $T_e^* - T_b^* \rightarrow T_e^* - T_b^*$, Q_m, ΔQ_m, ΔU_h (= ΔU_m), $\Delta Q_m/Q_m$ and Γ (= $\Delta U_h/\Delta Q_m$) for PE films annealed at 376.6 K, 416.6 K and 426.6 K for 1 hour.

5.5 ζ distribution function, F(ζ)

The occurrence of ζ distribution by crystallization is one of the characteristics of bulk polymers. The conversion of a DSC melting peak into the ζ distribution by Eq. (20) needs the values of σ_e and T_m^∞. The σ_e can be evaluated by Eq. (21) using only the DSC data. On T_m^∞, the inherent temperature of the crystal form should be selected. The $F(\zeta)$ is defined as[25]:

$$F(\zeta) = (\delta Q_m/Q_m)/\ \zeta = n_\zeta/\{N_c(T_e - T_b)\} \tag{33}$$

where δQ_m ($=\zeta n_\zeta Q_m/\{N_c(T_e - T_b)\}$) is the heat change per molar structural unit per K, n_ζ is the number of crystal sequences with ζ and N_c is the number of structural units of crystals melted in the temperature range from T_b (= T_b^* here) to T_e. $\delta Q_m/Q_m$ is given by:

$$\delta Q_m/Q_m = (dQ/dt)/\int_{T_b}^{T_e}(dQ/dt)dT \tag{34}$$

where dQ/dt is the heat flow rate of the melting curve. Fig. 12 shows $F(\zeta)$ of each melting curve from T_b^* for "T_a = 376.6 K, 416.6 K and 426.6 K"-samples. Table 9 lists the values of ζ-range, ζ_c and ζ_p of $F(\zeta)$ curve for each sample, where ζ_c is ζ at T_e^* and ζ_p is ζ at T_p. For "T_a = 376.6 K"-sample, ζ_p was slightly larger than for other samples. The small value of ζ ($< \zeta_c$) was believed to be caused by the crystallization from the secondary glass in the restricted space of inter-grain aggregates. The large ζ value at the maximum for "T_a = 416.6 K and 426.6 K"-samples might be related to the long period change of lamellae at the higher temperature upon heating[36]. Whereas, the very narrow ζ-range for "T_a = 376.6 K"-sample might be due to the effective annealing.

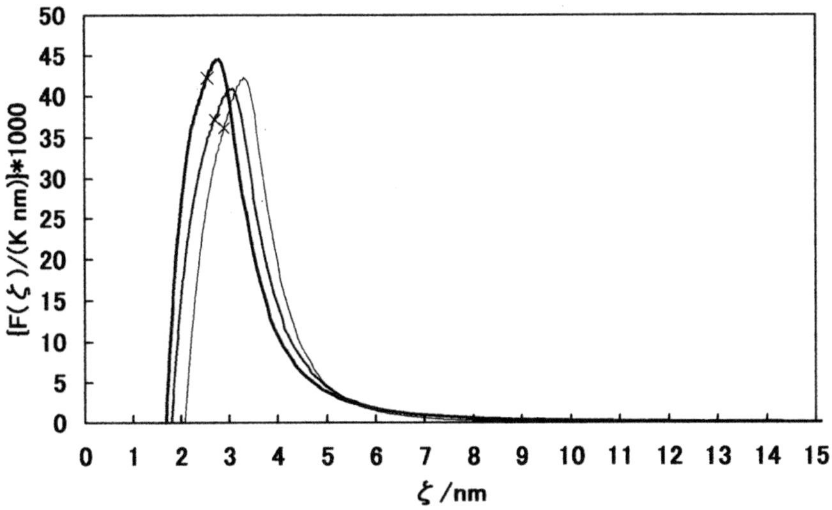

Fig. 12. $F(\zeta)$ for PE films annealed at 376.6 K (right), 416.6 K (middle) and 426.6 K (left) for 1 hour. ×; $F(\zeta_c)$.

Sample T_a/K	ζ-range nm	ζ_c nm	ζ_p nm
376.6	2.1 - 14	2.9 (2.7)	3.3
416.6	1.8 – 810 ~	2.7 (2.6)	3.1
426.6	1.7 – 730 ~	2.5 (2.2)	2.8

The values in the parentheses are ζ at the apparent T_e^*.

Table 9. The values of ζ-range, ζ_c and ζ_p in F(ζ) for PE films annealed at 376.6 K, 416.6 K and 426.6 K for 1 hour.

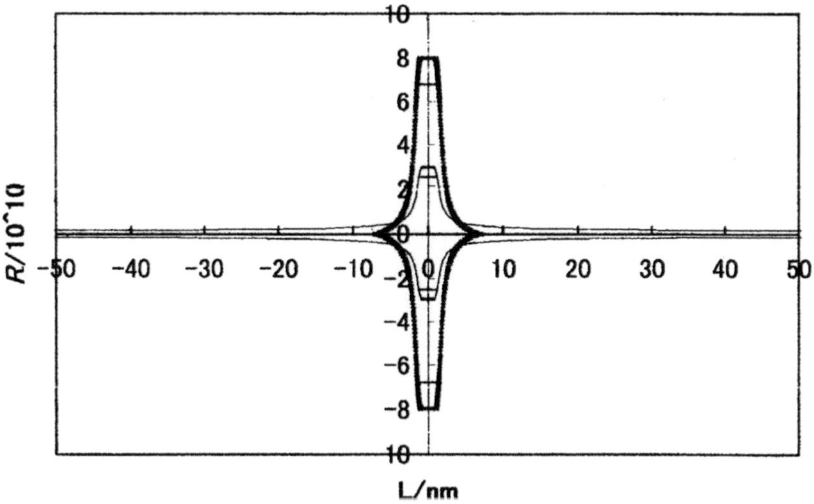

Fig. 13. The relationship between R (= $\pm R_n$) and L (= $\pm\zeta/2$) for PE films (1 g) annealed at 376.6 K (thick line) and 426.6 K (thin line) for 1 hour. The horizontal lines show R of the crystal melting from ζ_n or ζ_c to ζ = 0.

In the last stage, a single crystal-like image was drawn from F(ζ). Rewriting Eq. (33), n_ζ is given by:

$$n_\zeta = F(\zeta)N_c(T_e - T_b) \quad (35)$$

The number of the crystal sequences from ζ_n to ζ, N_ζ, is as follows:

$$N_\zeta = N_c(T_e - T_b) \int_{\zeta n}^{\zeta} F(\zeta)d\zeta \quad (36)$$

Accordingly, the number of the crystal sequences from ~ζ (> ζ) to ζ_x, ΔN, is given by:

$$\Delta N = N_c(T_e - T_b)(\int_{\zeta n}^{\zeta x} F(\zeta)d\zeta - \int_{\zeta n}^{\zeta} F(\zeta)d\zeta) \quad (37)$$

where ζ_x and ζ_n are the maximum and minimum of ζ, respectively. When the crystal sequences are bundled, like a fringe in a circle, the number of crystal sequences in a radius direction, R_n, as a function of ζ is given by:

$$R_n = (\Delta N/\pi)^{1/2} \tag{38}$$

Fig. 13 shows the relationship between R (= $\pm R_n$) and L (= $\pm\zeta/2$) for the samples (1 g) annealed at 376.6 K and 426.6 K for 1 hour. In the ζ-range of 0 ~ $\pm\zeta_n/2$, R at $\zeta_n/2$ is represented by a solid line, which leads to the supposition of a melting process from the end-surfaces of the crystal with ζ_n at T_b^*. The horizontal line of R at $\zeta_c/2$ depicts the same imaginable melting process of the crystal with ζ_c at T_e^*. The distinct difference of R or L between both types of crystals should be available in order to evaluate the annealing effects. The single crystal image from R and L for "T_a = 376.6 K"-sample (thick line) was very similar to the electron microscope (EM) image of self-seeded PE crystals[36].

6. Conclusions

The generation and disappearance of the "crystal / anti-crystal hole" pairs from the secondary glass in PE crystal lamella were discussed on the DSC curves. Thus the fraction of the secondary glass in the lamella was derived from the molar photon energy loss of the anti-crystal holes upon heating, which agreed with the latent heat of disappearance for the holes at the primary T_g of the first order hole phase transition.

7. Acknowledgements

The author wishes to thank Professor em. B. Wunderlich of the University of Tennessee and Rensseler Polytechnic Institute for the encouragement during this work.

8. References

[1] C. A. Angell, Science, 267, 1924(1995).
[2] F. H. Stillinger, Science, 267, 1935(1995).
[3] P. G. Debenedetti, F. H. Stillinger, Nature, 410, 259(2001).
[4] J. Mainstone, T. Parnell, Ig Nobel Prize in Physics, 2005.
[5] M. R. Tant and A. J. Hill, Ed: "Structure and Properties of Glassy Polymers", Oxford Univ. Press, Oxford, p37, 53(1998).
[6] T. Hecksher, A. I. Nielsen, N. B. Olsen, J. C. Dyre, Nature Physics, 4, 737(2008).
[7] S. Wei, I. Gallino, R. Busch, C. A. Angell, Nature Physics, 7, 178(2011).
[8] W. L. Johnson, G. Kaltenboeck, M. D. Demetriou, J. P. Schramm, X. Liu, K. Samwer, C. P. Kim, D. C. Hofmann, Science, 332, 828(2011).
[9] F. Zamponi, Nature Physics, 7, 99(2011).
[10] C. A. Angell, I. S. Klein, Nature Physics, 7, 750(2011).
[11] B. Wunderlich, J. Appl. Polym. Sci., 105, 49(2007).
[12] N. Tanaka, Polymer, 19, 770(1978).
[13] N. Tanaka, Polymer, 33, 623(1992).

[14] N. Tanaka, G. Wypych Ed: "Handbook of Solvents", ChemTech Publishing, Tronto, p253(2001).
[15] N. Tanaka, Thermochimica Acta, 374, 1(2001).
[16] N. Tanaka, Preprints of 46th Japanese Conference on Calorimetry and Thermal Analysis (Tsu), p27(2010).
[17] ACS, C & EN, October 10, 7(2011).
[18] N. Tanaka, Preprints of 45th Japanese Conference on Calorimetry and Thermal Analysis (Hachioji), p20(2009).
[19] L. H. Hill, "Introduction to Statistical Thermodynamics", Addison-Wisley, Massachusetts, p456(1960).
[20] J. D. Ferry, "Viscoelastic Properties of Polymers", Reinhold, New York, (1961).
[21] H. Sasabe, K. Sawamura, S. Sawamura, S. Saito, K. Yoda, Polymer J., 2, 518(1971).
[22] A. Uedono, S. Tanigawa, Japanese J. Polym.Sci. and Tech., 53, 563(1996).
[23] G. Kanig, Colloid & Polymer Sci., 260, 356(1982).
[24] T. McLeish, Physics Today, 61, No. 8, (2008).
[25] N. Tanaka, Polymer, 49, 5353(2008).
[26] N. Tanaka, Polymer, 34, 4941(1993).
[27] N. Tanaka, Polymer, 35, 5748(1994).
[28] N. Tanaka, 56th SPSJ Symposium on Macromolecules, Polymer Preprints (CD), Nagoya, 56, (2007).
[29] B. Wunderlich, ATHAS databank (1992 Recommended Data).
[30] L. Lew, W. H. de Jeu, "Interfaces and Mesophases in Polymer, Crystallization 2", In: G. Allegra, Ed: Adv. Polym. Sci., 181, Springer, Berlin, p88(2005).
[31] P. J. Flory, "Statistical Mechanics of Chain Molecules", Wiley & Sons, Inc., New York, (1969).
[32] Spectroscopical Soc. of Japan, "Infrared and Raman Spectroscopy", Kodansha Sci., Tokyo, (2011).
[33] J. P. Luongo, J. Appl. Polym. Sci., 3, 302(1960).
[34] M. Kobayashi, S. Hanafusa, T. Yoshioka, S Koizumi, Japanese J. Polym.Sci. and Tech., 53, 575(1996).
[35] Y. Yamashita, K. Monobe, Japanese J. Fiber Sci. and Tech., 37, T-53(1981).
[36] B. Wunderlich, "Thermal Analysis of Polymeric Materials", Springer, Berlin, p141, 247, 659, 780(2005).
[37] B. Wunderlich, Y. Jin, A. Boller, Thermochimica Acta, 237, 277(1994).
[38] S. Rastigi, A. E. Terry, "Interphases and Mesophases in Polymer, Crystallization 1", In: G. Allegra, Ed: Adv. Polym. Sci., 180, Springer, Berlin, p161(2005).
[39] Y. K. Kwon, A. Boller, M. Pyda, B. Wunderlich, Polymer, 41, 6237(2000).
[40] J. Pak, B. Wunderlich, Thermocimica Acta, 421, 203(2004).
[41] B. Wunderlich, Thermochimica Acta, 446, 128(2006).
[42] S. Okamura, A. Nakajima, S. Onogi, H. Kawai, Y. Nishijima, T. Higashimura, N. Ise, Polymer Chemistry, Kagakudojin, Kyoto, p177(1970).
[43] N. Tanaka, H. Fujii, J. Macromol. Sci., Phys., B42, 621(2003).
[44] B. Wunderlich, G. Czornyj, Macromolecules, 10, 906(1977).

[45] P. J. Flory, J. Chem. Phys. 17, 223(1949).
[46] N. Tanaka, 58th SPSJ Annual Meeting, Polymer Preprint (CD), Kobe, 58, (2009).

8

Crystallization Behavior and Control of Amorphous Alloys

Lai-Chang Zhang
The University of Western Australia
Australia

1. Introduction

Since the discovery in 1960 by Duwez (Klement et al., 1960), considerable effort has been devoted to form amorphous (or glassy) alloys either by rapid solidification techniques or by solid-state amorphization techniques (Inoue, 2000; Johnson, 1999; Suryanarayana & Inoue, 2011; Wang et al., 2004). However, the geometry of the amorphous samples has long time been limited in the form of ribbons or wires. The first "bulk" amorphous alloys, arbitrarily defined as the amorphous alloys with a dimension no less than 1 mm in all directions, was discovered by Chen and Turnbull (Chen & Turnbull, 1969) in ternary Pd-Cu-Si alloys. These ternary bulk glass-forming alloys have a critical cooling rate of about 10^2 K s^{-1} and can be obtained in amorphous state with a thickness up to 1 mm and more. Since then, especially after the presence of new bulk metallic glasses (BMGs) in $La_{55}Al_{25}Ni_{20}$ (Inoue et al., 1989) and $Zr_{41.2}Ti_{13.8}Cu_{12.5}Ni_{10.0}Be_{22.5}$ (Peker & Johnson, 1993), multicomponent BMGs, which could be prepared by direct casting from molten liquid at low cooling rates, have been drawing increasing attention in the scientific community. A great deal of effort has been devoted to developing and characterizing BMGs with a section thickness or diameter of a few millimetres to a few centimetres (Suryanarayana & Inoue, 2011). A large variety of multicomponent BMGs in a number of alloy systems, such as Pd-, Zr-, Mg-, Ln-, Ti-, Fe-, and Ni-based BMGs, have been developed via direct casting method with low cooling rates of the order of 1 – 10^2 K s^{-1} (Inoue, 2000; Johnson, 1999; Suryanarayana & Inoue, 2011; Wang, et al., 2004). In this method, the alloy compositions were carefully designed to have large glass-forming ability (GFA) so that "bulk" amorphous alloys can be formed at a low cooling rate to frustrate crystallization from molted liquid state. A number of parameters/indicators have been proposed to evaluate the GFA of multicomponent alloy systems to search for BMGs with larger dimensions (Suryanarayana & Inoue, 2011). So far, the "record" size of the BMGs is 72 mm diameter for a $Pd_{40}Cu_{30}Ni_{10}P_{20}$ bulk metallic glass (Inoue et al., 1997). The discovery of amorphous alloys has attracted widespread research interests because of their technological promise for practical applications and scientific importance in understanding glass formation and glass phenomena.

Arising from their disordered atomic structure and unique glass-to-supercooled liquid transition, amorphous alloys represent a new class of structural and functional materials with excellent properties (Eckert et al., 2007; Inoue, 2000; Johnson, 1999; Suryanarayana & Inoue, 2011; Wang, 2009; Xu et al., 2010), e.g. high strength about 2–3 times of their crystalline counterparts, large elastic limit about 2% which is very near to some polymer materials, including extreme strength at low temperature and high flexibility at high temperature, high corrosion resistance, high wear resistance, superior chemical and physical properties, etc. These properties, which can be rarely found in crystalline materials, are attractive for the practical applications as a new class of structural and functional materials. Fig. 1 summarizes the relationship between fracture strength and Young's modulus for typical engineering materials in amorphous and crystalline states. There is a clear tendency for fracture strength to increase with increasing Young's modulus, but the slope of the linear relation corresponding to elastic elongation is significantly different between the bulk amorphous and crystalline alloys and the elastic elongation of the amorphous alloys is ~3 times larger than those for the crystalline alloys. The amorphous alloys also exhibit high strength which is ~3 times higher than those for crystalline alloys, when the comparison is made at the same Young's modulus level. Currently, amoprhous alloys have a variety of uses for sports and luxury goods, microelectromechanical systems (MEMS), biomedicine and nanotechnology.

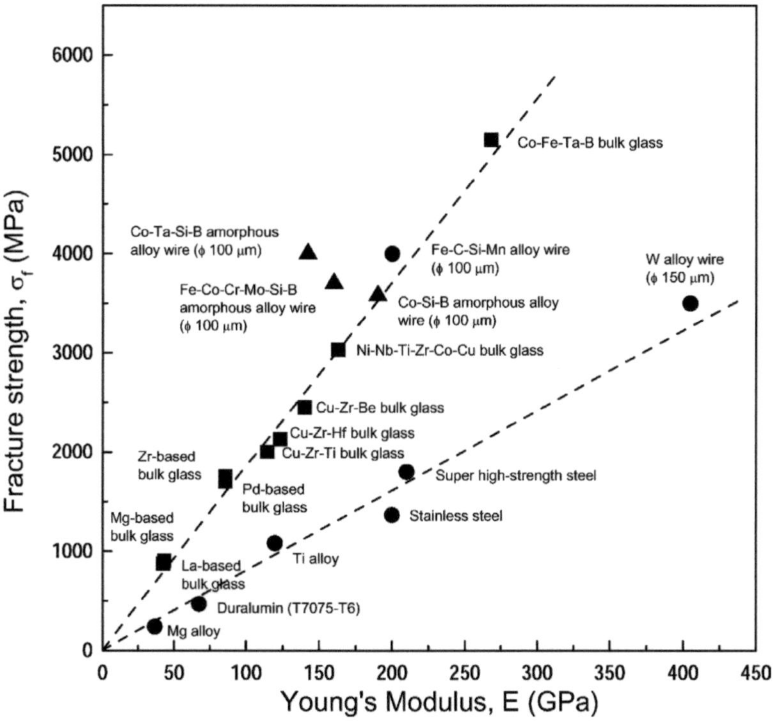

Fig. 1. Relation between strength and Young's modulus for bulk alloys in amorphous and crystalline states. Reprinted from (Inoue et al., 2004b), with permission from Elsevier.

2. Crystallization mechanisms

In general, the best practice way to describe a microstructure is in terms of its thermodynamic state before configurationally freezing set in (Turnbull, 1981). In this way, an amorphous alloy in configurationally frozen state as an undercooled liquid would be considered metastable. By considering the local potential wells between which atoms must make diffusional jumps, even states which are thermodynamically unstable may be regarded as kinetically metastable. Such kinetic metastability can exist only where thermal activation of atomic jumps is required. Regardless of the processing route used for the formation of amorphous state, the amorphous alloys are in thermodynamically metastable state and are susceptible to transform into more stable states under appropriate circumstances. Crystallization is such a transformation during which an amorphous phase devitrified into one or more metastable or stable crystalline phases. The driving force for the crystallization is the Gibbs free energy difference between the amorphous and the crystalline states. Crystallization could happen when an amorphous solid is subject to heat treatment (Calin et al., 2007; Suryanarayana & Inoue, 2011; Zhang et al., 2002; Zhang et al., 2003; Zhang & Xu, 2004; Zhang et al., 2005a; 2005b; Zhang et al., 2006a; Zhang et al., 2007a; Zhang et al., 2007b), mechanical deformation (Fornell et al., 2010; Lohwongwatana et al., 2006; Setyawan et al., 2010), pressure (Jiang et al., 2000; Jiang et al., 2002; Jiang et al., 2003b; Yang et al., 2006; Ye & Lu, 1999), and/or irradiations (Azam et al., 1979). Amongst these processing techniques, conventionally thermal annealing is the most commonly used in the investigation on crystallization of amorphous alloys. The dimensions and morphologies of the crystallization products strongly depend on the transformation mechanism, which is closely related to the chemical composition of the amorphous phase and to the thermodynamic properties of the corresponding crystalline phase. The crystallization products could include crystalline solids (solid solution, intermetallics, and/or compounds) (Foley et al., 1997; Kelton et al., 2003; Lu, 1996; Sahu et al., 2010; Zhang, et al., 2002; Zhang, et al., 2003) or quasicrystalline (Murty et al., 2000). As the crystallization process upon annealing of an amorphous phase is much slower than during solidification of liquids, it is relatively easier to fundamentally investigate crystallization in amorphous phases than in liquids on the processes of nucleation and growth, in particular of nucleation kinetics difficult to study quantitatively in the liquid state. The study of crystallization behaviors on amorphous alloys is of primary importance not only to characterize the thermal stability of amorphous alloys against crystallization but also to investigate the fundamental aspect of the processes of nucleation and growth, which are of relevance for the understanding glass formation.

Three types of crystallization reactions that may occur during devitrification can be classified, depending on their chemical compositions (Köster & Herold, 1981; Lu, 1996): *polymorphous*, *eutectic* and *primary* crystallization. Fig. 2 shows a hypothetical free energy diagram to illustrate the crystallization reactions during crystallization. This schematic is essentially a representation of the variation of free energy with the chemical compositions of the amorphous phase and various crystalline phases (in this case, two crystalline phases, a solid solution α and a compound β, are included) at a chosen annealing temperature.

2.1 Polymorphous crystallization

In *polymorphous* crystallization, an amorphous solid crystallizes into a single crystalline phase with different structure but with same chemical composition as the amorphous phase. This reaction can only occur in concentration ranges near to those of stable compounds (C_1 in Fig. 2) or pure elements (C_2) and needs only single jumps of atoms across the crystallization front. The polymorphous crystallization reaction (reaction (1) or (2)) may produce a single compound phase (β) or a supersaturated solid solution phase (α), as shown in Fig. 2. The crystallization mechanisms of $Fe_{33}Zr_{67}$ (Spassov & Koster, 1993), $Ni_{33}Zr_{67}$ (Lu et al., 1996), $Co_{33}Zr_{67}$ (Nicolaus et al., 1992) and $Zr_{50}Co_{50}$ (Köster & Meinhardt, 1994) amorphous alloys are the typical polymorphous crystallization.

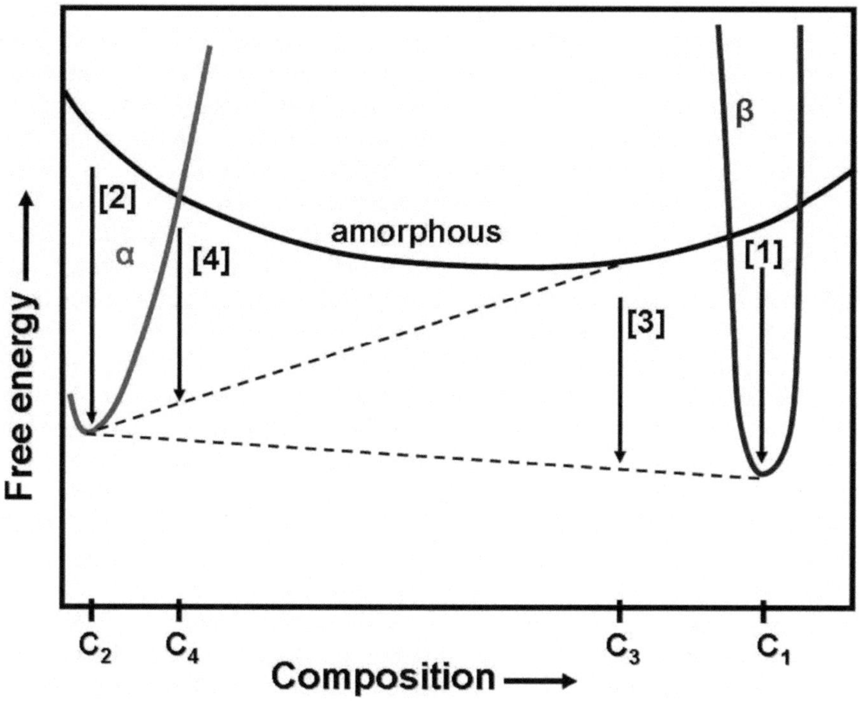

Fig. 2. Hypothetical free energy diagram to illustrate the crystallization of amorphous alloys. Reprinted from (Lu, 1996), with permission from Elsevier.

2.2 Eutectic crystallization

In case of *eutectic* crystallization, amorphous phase crystallizes into two crystalline phases simultaneously (e.g. reaction (3) in Fig. 2, α + β), during which two phases grow in a coupled fashion. This is similar to the eutectic crystallization of liquids. The reaction has the largest driving force and the overall composition of the two phases remains the same as that of the amorphous matrix. The eutectic crystallization can occur within a concentration range around the equilibrium eutectic composition rather than a specific eutectic composition as observed in conventional crystallization. A possible reason might be that the crystalline material contains a large amount of interface that may have higher energetic configurations and thus allows a relatively wide composition range (Lu, 1996). For example, e.g. in the Ni-P binary system eutectic crystallization occurs within 18.2–20.0 at.% P (i.e. amorphous → Ni + Ni_3P), where the equilibrium eutectic composition is 19.0 at.% (Dong et al., 1994).

2.3 Primary crystallization

In *primary* crystallization, amorphous phase crystallizes into a phase with different composition (C_4 in Fig. 2) in the first step (this can be either a supersaturated solid solution or an intermetallic compound) embedded in an amorphous matrix (amorphous phase (C_4) = α + amorphous phase' (C_3)). During this reaction, a concentration gradient occurs at the interface between the precipitate and the matrix until the reaction reaches the metastable equilibrium. The residual amorphous phase (with the new concentration C_3) crystallizes, in the second step, into crystalline phases through the mechanism of either the *eutectic* or *polymorphous* crystallization. The crystallization mechanisms of most of Al-based, e.g. $Al_{88}Ni_4Y_8$ (Jiang et al., 1997), and Fe-based amorphous alloys, e.g. $Fe_{73.5}Si_{13.5}B_9Nb_3Cu_1$ (Finemet) (Hono et al., 1992), are typically primary crystallization (Foley, et al., 1997; Kelton, et al., 2003). The control of primary crystallization behaviors could lead to nanocrytalline-amorphous composites with special mechanical or functional properties (see *Section 5.1*).

3. Influences on crystallization of amorphous alloys

The mechanisms and products of crystallization of amorphous alloys are influenced by both inherent (e.g. composition, oxygen) and extraneous (e.g. preparation method, pressure, etc) factors.

3.1 Effect of chemical composition

During the searching for strong glass-forming alloys, the effect of composition on the crystallization behavior has been extensively studied in a variety of amorphous alloys, despite of the preparation methods (Suryanarayana & Inoue, 2011). Two examples are listed in this section to show how the chemical compositions of amorphous alloys influence the crystallization mechanism and crystallization products.

Zhang et al. (Zhang, et al., 2002) has investigated the addition of Al on the glass formation and crystalliztion in the ball-milled amorphous $Ti_{50}(Cu_{0.45}Ni_{0.55})_{44-x}Al_xSi_4B_2$ (x=0, 4, 8 and 12) alloys. Al additions were introduced to simultaneously replace part of the Cu and Ni in $Ti_{50}Cu_{20}Ni_{24}Si_4B_2$ (Zhang & Xu, 2002) to further reduce the density of the resulting alloys and improve the thermal stability of the supercooled liquid. The Ti-based amorphous alloy powders prepared through this solid-state process exhibit a well-defined glass transition and a supercooled liquid region. Al addition has changed the crystallization mechanims and crystallizaiton products of the amoprhous $Ti_{50}Cu_{20}Ni_{24}Si_4B_2$ alloy. Fig. 3 (a) displays the differential scanning calorimetry (DSC) scans for the as-milled samples with different Al contents. In all cases, an endothermic signal associated with the glass transition is evident. As see from Fig. 3 (a), the onset of glass transition temperature (T_g) is apparently insensitive to the change in the overall alloy composition. With increasing Al substitution, the exothermic reaction due to crystallization occurs at higher temperatures and the single-step crystallization event changes to a two-step process. X-ray diffraction (XRD) has been used to identify the structural changes associated with the exothermal events at several different temperatures, as marked by dots in the DSC traces in Fig. 3 (b). For x = 0, the XRD pattern at 777 K crystallization peak and after the crystallization event (810 K) showed that the amorphous phase transformed into the cubic NiTi phase and an unknown phase. The same products were found for x = 4 after crystallization, as shown in the XRD pattern at 820 K. Such a transition can be regarded as a eutectic crystallization, by which the amorphous phase simultaneously transforms into more than two phases in one step (as stated in *Section 2.2*). For x = 8 and x = 12, on the other hand, crystallization is completed in two steps. Fig. 3

(b) indicates that in addition to the NiTi phase precipitated in the first stage of the crystallization, the second crystallization peak in the DSC traces arises from the appearance of the Ti_2Ni intermetallic compounds in the final crystallization products.

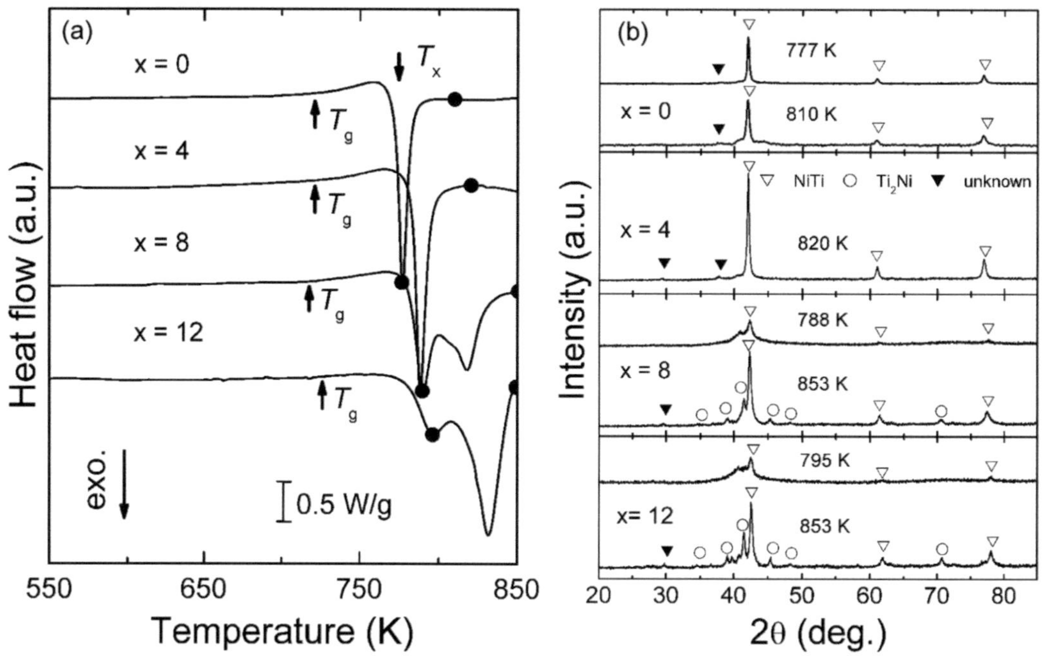

Fig. 3. (a) DSC scans and (b) the corresponding XRD patterns measured at room temperature after heating to different temperatures in DSC at a heating rate of 40 K/min for the mechanically alloyed $Ti_{50}(Cu_{0.45}Ni_{0.55})_{44-x}Al_xSi_4B_2$ (x=0, 4, 8, 12) powders. Reprinted from (Zhang, et al., 2002), with permission from Cambridge University Press.

The glass formation and crystallization behavior in multicomponent Zr-based alloys have been intensively investigated. In general, multicomponent Zr-based alloys can be used for the production of fully amorphous bulk samples with dimensions up to centimeter order, or for the formation of bulk nanostructured materials. However, the phase selection upon crystallization is strongly affected by the chemical composition of the amorphous phase. In order to obtain nanostructured materials from amorphous precursors (see *Section 5.1*), amorphous specimens are typically annealed at temperatures within the supercooled liquid region (the temperature region between onset glass transition temperature, T_g, and the onset of crystallization, T_x,) or close to T_x. Eckert et al (Eckert et al., 2001) has investigate the crystallization behaviors of Zr-based BMGs and produce bulk nanostructured alloys by partial crystallization of the Zr-based BMGs precursors. Fig. 4 (a) displays the DSC scans for as-cast $Zr_{62-x}Ti_xCu_{20}Al_{10}Ni_8$ glassy alloys (x = 0, 3, 5, and 7.5). $Zr_{62}Cu_{20}Al_{10}Ni_8$ crystallizes via one sharp exothermic peak to form several intermetallic compounds. Upon Ti addition, the crystallization mode changes toward a double-step process. With increasing Ti content, the first DSC peak shifts to lower temperatures and the enthalpy related to the second exothermic peak decreases. The samples were isothermally annealed for different times below T_x for further study of the crystallization process. The crystallization products after annealing were investigated by XRD (Fig. 4(b)). $Zr_{62}Cu_{20}Al_{10}Ni_8$ transforms into cubic $NiZr_2$-type and tetragonal CuZr2-type compounds. Annealing the alloy with $x = 3$ leads to primary precipitation of an icosahedral quasicrystalline (QC) phase with spherical morphology and a size of about 50 to 100 nm. For $x = 5$, the diffraction peaks are weaker in intensity and broader because the precipitates are as small as about 5 nm. For $x = 7.5$, the precipitates are about 3 nm in size. At first glance the XRD pattern (Fig. 4 (b)) after annealing displays no obvious reflections but only broad amorphous-like maxima. However, careful examinations of the annealed state by high intensity synchrotron radiation and/or by transmission electron microscopy (TEM) (Eckert, et al., 2001) clearly shows differences in scattering intensity compared to the as-cast state indicates the precipitation of a metastable cubic phase with a grain size of ~2 nm coexisting with a residual amorphous phase.

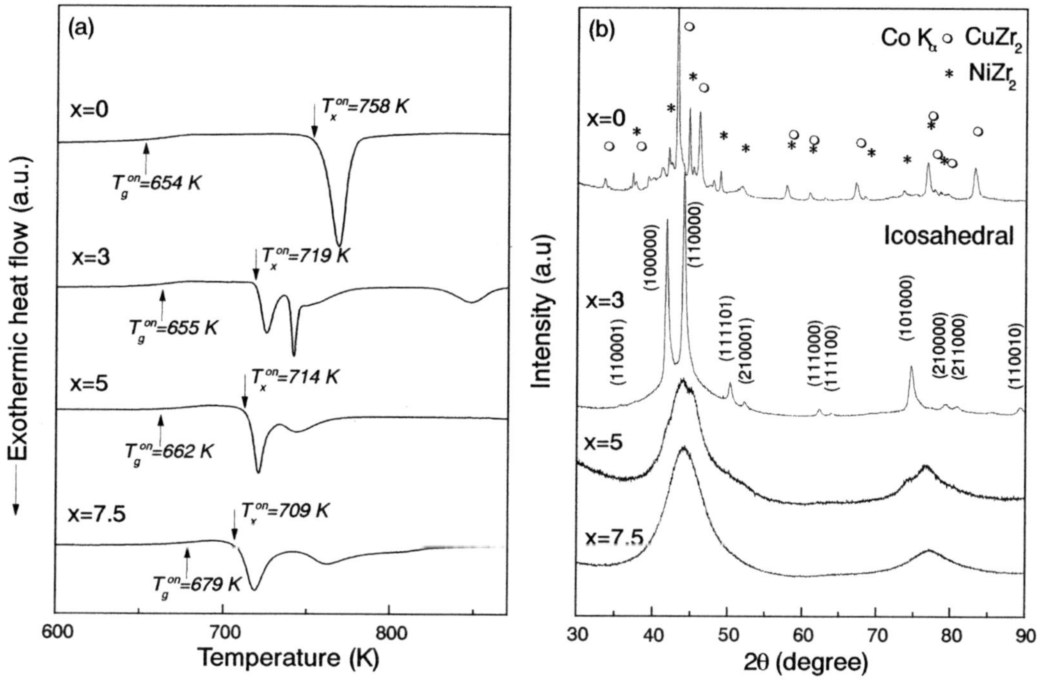

Fig. 4. (a) DSC scans and (b) corresponding XRD patterns after isothermal annealing for the $Zr_{62-x}Ti_xCu_{20}Al_{10}Ni_8$ (x = 0, 3, 5, and 7.5) amorphous alloys: $x = 0$, annealed at 723 K for 30 min; $x = 3$, annealed at 703 K for 5 min; $x = 5$, annealed at 683 K for 30 min and $x = 7.5$ annealed at 688 K for 40 min. Reprinted from (Eckert, et al., 2001), with permission from Elsevier.

3.2 Effect of oxygen

Although Zr-based BMGs have shown high glass-forming ability, high thermal stability and excellent mechanical properties, the glass-forming ability of these BMGs appears to be significantly affected by the contamination of oxygen either from the raw materials or from the processing (Inoue et al., 1995a; Kubler et al., 1998; Lin et al., 1997). The investigation of the influence of oxygen on the crystallization behavior of Zr-based amorphous alloys

(Altounian et al., 1987) showed that the oxygen induces the formation of metastable face-centered cubic (fcc) $NiZr_2$, thereby reducing the thermal stability of the Zr-Ni amorphous alloys. Extensive studies have proved that oxygen enhances the crystallization reaction in Zr-based amorphous alloys. For example, Lin et al (Lin, et al., 1997) reported for undercooled Zr–Ti-Cu-Ni-Al molten liquids that oxygen addition strongly affects crystal nucleation and can dramatically increase the necessary critical cooling rate for glass formation, thus limiting bulk glass formation and reducing the maximum attainable sample thickness. Over the range of oxygen content studied (300 – 5000 at. ppm), the time–temperature-transformation curves vary roughly by two orders of magnitude along the time axis. In other words, oxygen contamination ranging up to 0.5 at.% can increase the necessary cooling rate for glass formation by two orders of magnitude (Lin, et al., 1997). Köster et al. (Köster et al., 1996; Köster et al., 1997) reported the formation of an icosahedral phase during primary crystallization in $Zr_{65}Cu_{17.5}Ni_{10}Al_{7.5}$ amorphous alloys, whereas such a crystallization process was not reported in the same alloy composition by Zhang et al. (Zhang et al., 1991), indicating that the formation of quasicrystals is induced by the oxygen contamination in the alloy. Eckert and his co-workers (Eckert et al., 1998; Gebert et al., 1998) also reported the strong influence of the oxygen contamination on the crystallization kinetics and products in $Zr_{65}Cu_{17.5}Ni_{10}Al_{7.5}$ amorphous alloy, where supercooled liquid region decreases with increasing oxygen content due to the change in crystallization sequence from a single- to a double-step process. It was also shown that an fcc $NiZr_2$ phase is formed at a higher oxygen level in the Zr–Cu–Ni–Al system. Therefore, oxygen contamination is of primary importance for the glass formation and crystallization behavior of Zr-based amorphous alloys.

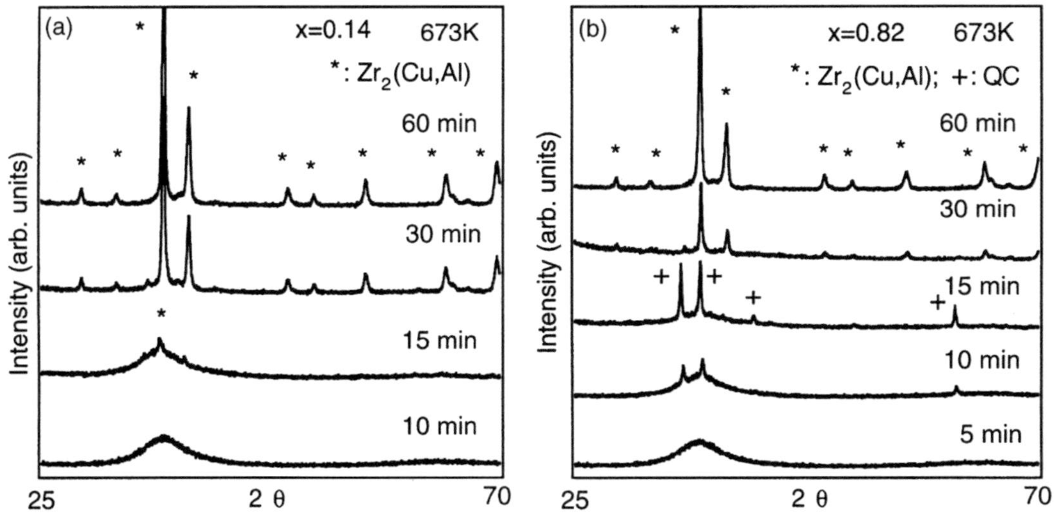

Fig. 5. XRD patterns of the $Zr_{65-x}Cu_{27.5}Al_{7.5}O_x$ bulk amorphous alloys with (a) x = 0.14 and (b) x = 0.82 after annealing at 673 K for different durations. Reprinted from (Murty, et al., 2000), with permission from Elsevier.

Murty et al. (Murty, et al., 2000) investigated the influence of oxygen on the crystallization behavior of melt-spun amorphous $Zr_{65-x}Cu_{27.5}Al_{7.5}O_x$ (x = 0.14, 0.43 and 0.82) ribbons. DSC results showed that the T_g increases with the addition of oxygen. The base alloy containing the lowest amount of oxygen (x = 0.14) crystallizes in a single step. The addition of oxygen significantly decreases the width of supercooled liquid region (ΔT_x) from 85 K for x = 0.14 to 58 K for x = 0.82. The decrease in ΔT_x is partly due to the increase in T_g with increasing oxygen and also due to the appearance of a pre-crystallization peak in the oxygen-containing alloys before the main crystallization event. The base alloy heated to 673 K (in supercooled liquid region) shows an amorphous nature. In the x = 0.82 alloy, precipitation of spherical icosahedral particles in nanocrystalline state was observed within 10 min annealing at 673 K. Then it crystallizes to $Zr_2(Cu,Al)$ when heated to 723 K. No other phase is present in the alloy heated up to 753 K. The x = 0.82 alloy heated to 673 K has a similar trend to that of the base alloy. However, the alloy heated near the pre-crystallization peak (708 K) led to the presence of an icosahedral phase along with a small amount of $Zr_2(Cu,Al)$. Only $Zr_2(Cu,Al)$ is present in the alloy heated to 723 and 753 K. The base alloy with x = 0.14 remains amorphous for up to 10 min at 673 K, after which the formation of $Zr_2(Cu,Al)$ was observed. The XRD patterns in Fig. 5 (a) clearly indicate the formation of $Zr_2(Cu,Al)$ beyond 10 min of annealing at 673 K. No other phase was identified even after a longer annealing for 60 min (Fig. 5 (a)). XRD patterns of the alloy heat treated at 673 K (Fig. 5 (b)) show that the icosahedral phase starts forming after 10 min and persists for up to 15 min, beyond which the icosahedral phase transforms to $Zr_2(Cu,Al)$.

The mechanism of the oxygen-induced precipitation of metastable fcc $Zr_2(Cu,Al)$ and icosahedral quasicrystalline phases is rationalized by considering the effect of oxygen on the nucleation process. The high thermal stability of multicomponent Zr-based amorphous alloys is generally attributed to the difficulty of precipitation of crystalline compounds from the undercooled liquid. The combination of elements with significantly different atomic sizes and negative enthalpies of mixing leads to a homogeneously mixed dense random packed structure of the liquid resulting in a large liquid–solid interface energy (Inoue, 2000). If the nucleating phase has a different composition with respect to the homogeneous undercooled liquid, then the nucleation of the phase requires substantial atomic rearrangement (Eckert, et al., 1998). The driving force for the polymorphous crystallization is ΔG_{total}. However, if the icosahedral phase is stabilized by oxygen addition, the driving force for the primary crystallization of the icosahedral phase can be comparable to or higher than that for the polymorphous crystallization, ΔG_{total}. In such a case, icosahedral phase would initially precipitate from the amorphous matrix by the primary crystallization. The free energy reduction is accompanied with this crystallization and there is still a driving force to form $Zr_2(Cu,Al)$ from the icosahedral phase and/or the remaining amorphous phase in the second stage. The formation of the icosahedral phase would be preferable if the driving forces for the polymorphous crystallization and the primary crystallization are comparable, because it is believed that icosahedral clusters are present in the amorphous phase, and these would act as nuclei for the icosahedral phase primary crystals. If such icosahedral clusters are stabilized by the presence of oxygen, the oxygen-enriched alloy would be favour to form an icosahedral phase by primary crystallization. The differences in the sequence of the phase formation in these alloys with x = 0.14 and 0.82 are illustrated schematically in Fig. 6, in which the darkness of the gray scale corresponds to the concentration of Zr. In the base Zr–Cu–Al and Zr–Cu amorphous alloys, crystallization proceeds by polymorphous reaction without change in composition. On the other hand, in the oxygen-containing ternary alloys, the first stage of crystallization occurs by primary crystallization of the icosahedral phase. The icosahedral phase is enriched in Zr and O and depleted in Cu and Al. When $Zr_2(Cu,Al)$ precipitates peritectically, the concentration of the $Zr_2(Cu,Al)$ particles becomes the same as the initial alloy composition. When this reaction is complete, only the grains of single-phase $Zr_2(Cu,Al)$ remain.

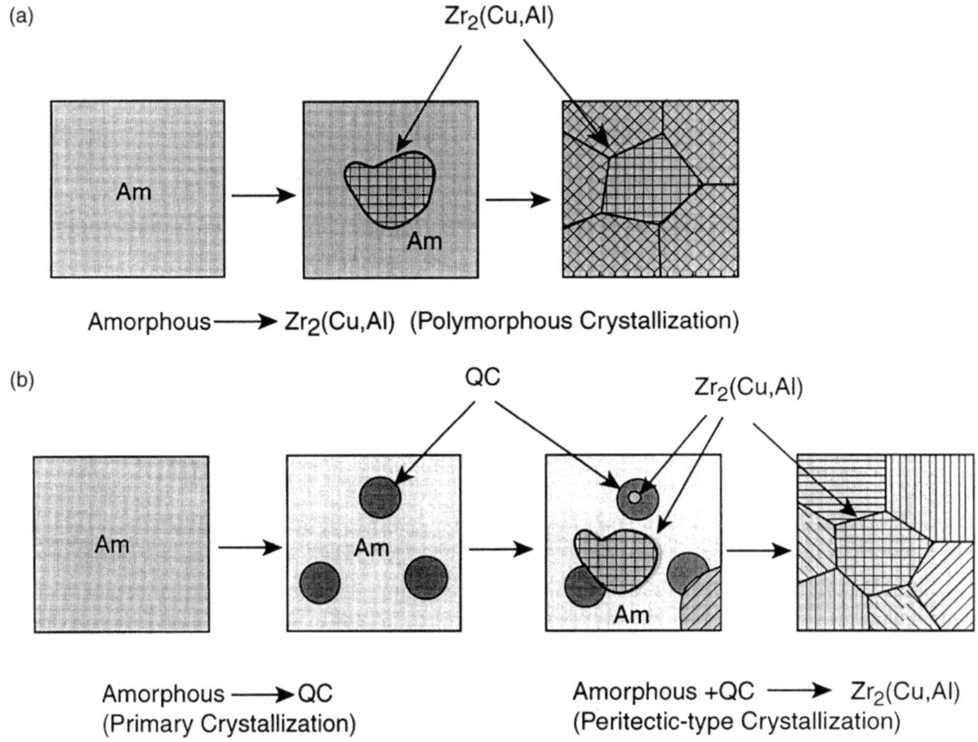

Fig. 6. Schematic diagrams showing the evolution of microstructure during crystallization of $Zr_{65-x}Cu_{27.5}Al_{7.5}O_x$ amorphous with (a) $x = 0.14$ and (b) $x = 0.82$. Reprinted from (Murty, et al., 2000), with permission from Elsevier.

3.3 Effect of sample preparation method

In general, when the transformation temperatures (e.g. T_g and T_x, etc) of an amorphous alloy are measured by DSC, there is no appreciable difference between in the amorphous samples prepared by direct melt cooling from molten liquid (e.g. by melt spinning, casting, water quenching, etc). Table 1 summarizes the transformation temperatures determined from DSC for some typical amorphous alloys prepared by different routes. Furthermore, there is no difference in the transformation temperatures of the amorphous rods with different sizes. As seen from Table 1, same transformation temperatures are obtained in the $Mg_{65}Cu_{15}Y_{10}Ag_{10}$ amorphous rods in 6 mm diameter prepared by injection casting and in 10 mm diameter prepared by squeeze casting.

Composition	Synthesis route	T_g (K)	T_x (K)	ΔT_x (K)	Reference
$Cu_{50}Zr_{50}$	Cu-mold casting	675	732	57	(Inoue et al., 2005)
	Melt spinning	686	744	58	
$Cu_{60}Zr_{30}Ti_{10}$	Cu-mold casting (2.5 mm∅ rod)	714	758	44	(Jiang et al., 2003a)
	Melt spinning	711	754	43	
$Pd_{40}Ni_{40}P_{20}$ (fluxed)†	Water quenching (7 mm∅ rod)	576	678	102	(He et al., 1996)
	Melt spinning	590	671	91	(Inoue, et al., 1997)
$Pd_{81}Si_{19}$ (fluxed)†	Air cooling	638	696	58	(Yao & Ruan, 2005)
	Melt spinning	633	675	42	
$Mg_{65}Cu_{15}Y_{10}Ag_{10}$	Melt spinning	428	469	41	(Kang et al., 2000)
	Injection casting (6 mm∅ rod)	428	469	41	
	Squeeze casting (10 mm∅ rod)	428	469	41	
$Pd_{40}Cu_{30}Ni_{10}P_{20}$ (fluxed)	Melt spinning	572	670	98	(Inoue, et al., 1997)
$Pd_{40}Cu_{30}Ni_{10}P_{20}$ (unfluxed)	Melt spinning	572	663	91	
$Zr_{65}Al_{7.5}Ni_{10}Cu_{17.5}$	Water quenching (16 mm∅ rod)	625	750	125	(Inoue et al., 1993b)
	Melt spinning	622	749	127	
$Zr_{55}Ni_{25}Al_{20}$	Planar flow casting (30 μm thick ribbon)	805	820	15	(Illeková et al., 1997)
	Water quenching (9 mm∅ rod)	738	795	57	
$La_{55}Al_{25}Ni_{10}Cu_{10}$	High-pressure die casting (9 mm∅ rod)	460	527	67	(Inoue et al., 1993a)
	Melt spinning	480	550	90	
$Ti_{50}Cu_{35}Ni_{12}Sn_3$	Ball Milling	652	717	65	(Zhang, et al., 2005a)
	Melt spinning	675	739	64	
$Ti_{50}Cu_{18}Ni_{22}Al_4Sn_6$	Ball Milling	705	771	66	(Zhang, et al., 2005a)
	Melt spinning	721	789	68	

† heating rate is 0.33 K/s.

Table 1. Comparison of the transformation temperatures determined from DSC at heating rate is at 0.67 K/s (if not indicated) for some typical amorphous alloys prepared by different methods. T_g: glass transition temperature; T_x: onset crystallization temperature; ΔT_x: the width of supercooled liquid region, which is equal to $T_x - T_g$.

Fig. 7 compares the DSC curves for the $[(Fe_{0.8}Co_{0.2})_{0.75}B_{0.2}Si_{0.05}]_{96}Nb_4$ bulk amorphous alloy rods with different diameters up to 2.5 mm with the data for melt-spun ribbon of the same composition. No appreciable difference is recognized in the transformation temperatures or crystallization process between the melt-spun ribbon and cast rod samples, in spite of an increase in Curie temperature (T_c) with the increase of diameter. All samples exhibit a distinct glass transition at 830 K, followed by crystallization at 880 K, resulting in a large supercooled liquid region of 50 K. Similar results have been obtained in a number of bulk glass-forming alloy systems (Suryanarayana & Inoue, 2011).

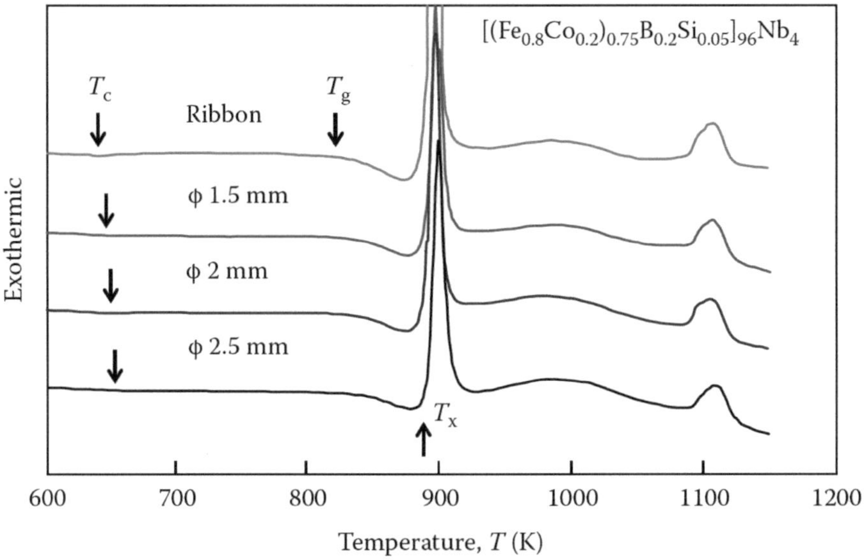

Fig. 7. DSC curves at a heating of 0.67 K/s for $[(Fe_{0.8}Co_{0.2})_{0.75}B_{0.2}Si_{0.05}]_{96}Nb_4$ bulk amorphous alloy rods (1.5, 2 and 2.5 mm∅) as well as the melt-spun amorphous alloy ribbon of the same composition. Reprinted from (Inoue et al., 2004a) and (Suryanarayana & Inoue, 2011), with permission from Elsevier.

Note that, although no appreciable difference in transformation temperatures has been observed in the amorphous alloys prepared by direct melt cooling from molten liquid, some alloys do have shown some differences in the transformation temperatures in the amorphous ribbon and rod samples, even though they have an identical chemical composition. As seen from Table 1, the $Zr_{55}Ni_{25}Al_{20}$ glassy alloys prepared by two different solidification methods (one is planar flow casting with cooling rate of about 10^5 K s^{-1} and the other is water quenching with a solidification rate of about 10^2 K s^{-1}) showed a significant difference in the transformation temperatures, i.e. 69 K difference in T_g and 15 K in T_x (Illeková, et al., 1997). By comparing the enthalpy of structural relaxation in DSC curves and the full width at half maximum (FWHM) of the first diffuse peak in XRD patterns, it is concluded that the samples produced from both methods represent the same amorphous state, but the amorphous ribbon sample contains a higher degree of short-range order (SRO) (Illeková, et al., 1997).

A number of investigations have reported a distinct difference in the transformation temperatures between the amorphous alloys prepared by melt cooling and that formed by solid-state amorphization techniques (e.g. ball milling or mechanical alloying). Fig. 8 compares the structural feature and transformation temperatures for $Ti_{50}Cu_{35}Ni_{12}Sn_3$ and

$Ti_{50}Cu_{18}Ni_{22}Al_4Sn_6$ alloys prepard by ball-milling (BM) and melt-spinning (MS). The broad diffuse maximum for the amorphous phase formed by BM is determined to be 26.89 nm^{-1} for $Ti_{50}Cu_{35}Ni_{12}Sn_3$ and 26.64 nm^{-1} for $Ti_{50}Cu_{18}Ni_{22}Al_4Sn_6$, respectively. They are well in agreement with the values of the amorphous alloys prepared using MS method, Q_p = 26.81 nm^{-1} for $Ti_{50}Cu_{35}Ni_{12}Sn_3$ and 26.67 nm^{-1} for $Ti_{50}Cu_{18}Ni_{22}Al_4Sn_6$, respectively. It implies that for a given alloy, the amorphous phase obtained using the different preparation methods is very similar in the all cases. However, both T_g and T_x of the BM alloy shift towards a lower temperature, with respect to the MS alloys, by about 20 K for $Ti_{50}Cu_{35}Ni_{12}Sn_3$ and 16-18 K for $Ti_{50}Cu_{18}Ni_{22}Al_4Sn_6$, respectively, even though a very close ΔT_x is obtained in the BM and MS amorphous alloys for each phases. Furthermore, the heat of crystallization in the BM amophous state is slightly lower than that in MS one for both alloys. The difference in the transformation temperaturees betwen the BM and MS amorphous phase is likely caused by the minor difference in the composition, oxygen content, and/or short-range order in the amorphous phases formed by different processing route.

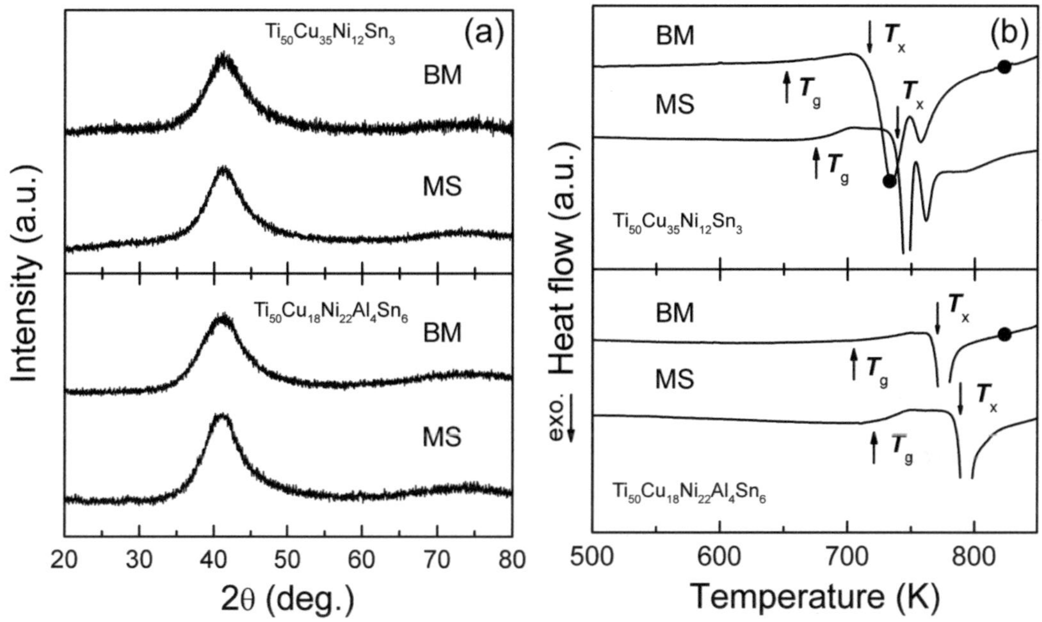

Fig. 8. (a) XRD patterns and (b) DSC curves at a heating of 0.67 K/s for $Ti_{50}Cu_{35}Ni_{12}Sn_3$ and $Ti_{50}Cu_{18}Ni_{22}Al_4Sn_6$ alloys prepard by ball-milling (BM) and melt-spinning (MS). Reprinted from (Zhang, et al., 2005a), with permission from Elsevier.

3.4 Effect of pressure

A few work has investigated the effect of high pressure on the crystallization of amorphous alloys, e.g. see the references (Jiang, et al., 2000; Jiang, et al., 2002; Jiang, et al., 2003b; Ye & Lu, 1999; Zhuang et al., 2000). In general, the crystallization temperature of an amorphous alloy increases with increasing pressure. However, the rate and the range of such temperature increase are closely related to the alloy systems. Fig. 9 shows the pressure dependence of the crystallization temperatures (i.e. T_{x1} and T_{x2}) for the $Al_{89}La_6Ni_5$

amorphous alloy. Both T_{x1} and T_{x2} firstly decrease with the increase in pressure in the range of 0–1 GPa and then increase with pressure increasing up to 4 GPa. Such changes in crystallization temperature with pressure is related to the competing process between the thermodynamic potential barrier and the diffusion activation energy under pressure (Zhuang, et al., 2000).

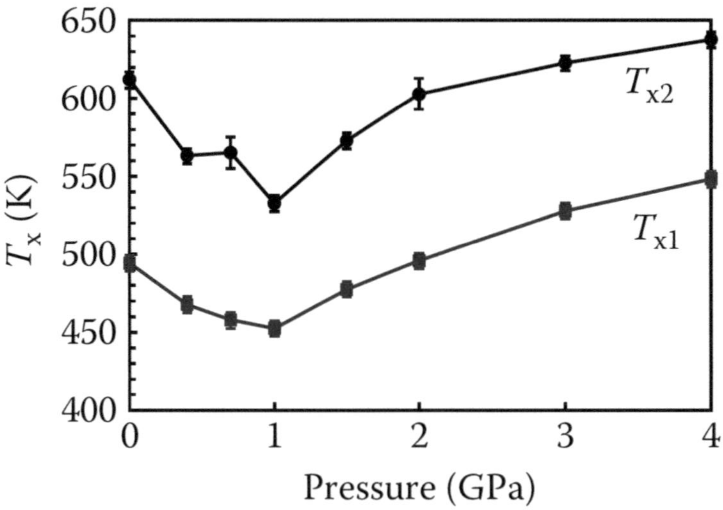

Fig. 9. Pressure dependence of the crystallization temperatures for $Al_{89}La_6Ni_5$ amorphous alloy. Reprinted from (Zhuang, et al., 2000), with permission from American Institute of Physics.

Crystallization of an amorphous alloy is normally regarded as a process proceeding by nucleation and subsequent growth of crystals. During the initial stage of nucleation of crystals in the amorphous phase, the effect of pressure on the crystallization kinetics is associated with the atomic diffusion process and the volume change effect. The crystallization temperature(s) of an amorphous alloy may be governed by the thermodynamic potential barrier of nucleation and diffusion activation energy. According to crystallization kinetics theory, the nucleation rate I can be written as,

$$I = I_0 \exp[-(\Delta G^* + Q_n) / kT] \tag{1}$$

where I_0 is a constant, ΔG^* is the free energy required to form a nucleus of the critical size, i.e., the thermodynamic potential barrier of nucleation, Q_n is the activation energy for the transport of an atom across the interface of an embryo, and k is the Boltzmann's constant. The sum $\Delta G^* + Q_n$ is called the nucleation work.

In the $Al_{89}La_6Ni_5$ alloy, ΔG^* is much larger than Q_n and the dominant factor at low pressures (0–1 GPa). Thus, the nucleation work decreases with increasing pressure, leading to an enhancement of nucleation rate I and a reduction of the crystallization temperature with increasing pressure, as shown in Fig. 9. With increasing pressure, ΔG^* rapidly decreases while Q_n increases, resulting in atomic diffusion a dominant factor in the nucleation process. Hence, the nucleation work $\Delta G^* + Q_n$ increases with increasing pressure. Consequently, nucleation rate I decreases with the increase in pressure and an enhancement of crystallization temperature is expected, as observed for the $Al_{89}La_6Ni_5$ amorphous alloy in the pressure range of 1–4 GPa in Fig. 9.

In addition to the aforementioned factors, the crystallization temperature(s) of an amorphous phase significantly depend on the heating rate used in DSC measurement (Kissinger, 1957). In contrast, the heating rate has a slight influence on the glass transition temperature. All these transformation temperatures of amorphous alloys increase with increasing the heating rate that is used in DSC. Therefore, the heating rate is usually indicated when describing the transformation temperatures of an amorphous phase.

4. Kinetics of crystallization

The kinetics of crystallization of amorphous alloys has been extensively studied by using differential scanning calorimetry (DSC) or differential thermal analysis (DTA), as the structural change in a material upon heating or cooling is indicated by a defection or peak in the DSC/DTA curve. The kinetic behavior associated with a structural change leading to an alternative metastable state in an amorphous alloy above its glass transition is a key subject since it provides new opportunities for structural control by innovative design and processing strategies. *Section 5* will show some application examples by controlling crystallization from amorphous precursors in order to tailor microstructure for excellent properties. Such crystallization control requires fundamental understanding of the specific mechanisms influencing structural transformations.

In general, crystallization is a thermally activated reaction, either by isothermal or isochronal heating. The transformation rate during a reaction could be described as

$$d\alpha / dt = f(\alpha)k(T) \tag{2}$$

where a is the fraction transformed. The temperature dependent function is generally assumed to follow an Arrehnius type dependency

$$k = k_0 \exp(-E / RT) \tag{3}$$

where k_0 is the reaction constant, R is the gas constant and E is the activation energy. In general, the reaction function $f(\alpha)$ is unknown. From the above equations it follows that for transformation studies by performing studies at a constant temperature T, E can be obtained as below:

$$ln\left(t_f\right) = E / RT + c_i \tag{4}$$

where t_f is the time needed to reach a certain fraction transformed, and c_i is a constant, which depends on the reaction stage and on the kinetic model. Thus, E can be obtained from two or more experiments at different T. For isothermal experiments $k(T)$ is constant, the determination of $f(\alpha)$ is relatively straightforward, and is independent of E. For non-isothermal experiments, the reaction rate at all times depends on both $f(\alpha)$ and $k(T)$, and the determination of $f(\alpha)$, k_0 and E (the so-called kinetic triplet) is an interlinked problem. A deviation in the determination of any of the three parameters will cause a deviation in the other parameters of the triplet. Over the past decades a variety of non-isothermal methods have been proposed. Among them, the Kissinger method (Kissinger, 1957) is widely used in the isochronal method for the calculation of the activation energy for the crystallization. A higher value of the activation energy is generally interpreted as a measure of the high stability and resistance of the amorphous phase towards crystallization. The activation energy for crystallization could be determined using

$$ln\left(\beta / T_p^2\right) = -E / RT_p + C \quad (5)$$

where β is the heat rates that used to heating amorphous samples in DSC, T_p is the temperature corresponding to the peak of the crystallization event (exothermic peak), R is the gas constant and C is a constant. Thus, by plotting $ln\left(\beta / T_p^2\right)$ against $1/T_p$, one could obtain a straight line whose slope is $-E/R$, from which the activation energy for the transformation, E, can be calculated.

On the other hand, kinetic data on first-order transformations are often obtained by isothermal analysis. The isothermal crystallization kinetics of the amorphous phase can be usually analyzed in terms of the generalized theory of the well-known Kolmogorov-Johnson-Mehl-Avrami (JMA) equation (Christian, 2002) for a phase transition:

$$x_c(t,T) = 1 - \exp[-k(t-\tau)^n] \quad (6)$$

or

$$ln\left[-ln\left(1 - x_c(t,T)\right)\right] = n\,lnk + n\,ln(t-\tau) \quad (7)$$

where $x_c(t,T)$ is the volume fraction of crystallized phases after annealing time t, τ is the incubation period of transient nucleation, which is the time period that must elapse prior to formation of nuclei, k is a temperature-dependent kinetic parameter and n is the Avrami exponent, which is a significant parameter to describe the crystallization mechanism, such as nucleation and growth behavior, and varies from 1 to 4 (Doherty, 1996). For diffusion-controlled growth, one may have the following cases: $1 < n < 1.5$ indicates growth of particles with an appreciable initial volume; $n = 1.5$ indicates growth of particles with a nucleation rate close to zero; $1.5 < n < 2.5$ reflects growth of particles with decreasing nucleation rate; $n = 2.5$ reflects growth of particles with constant nucleation rate, and $n > 2.5$ pertains to the growth of small particles with an increasing nucleation rate (Doherty, 1996). A JMA plot of $\ln[-\ln(1- x_c(t,T))]$ vs. $\ln(t-\tau)$ yields a straight line with slop n and intercept $n\ln k$. Using a DSC, operated under isothermal mode, phase transformations can be distinguished unambiguously in terms of those occurring only by growth of existing nuclei or those occurring by nucleation and growth. For a transformation resulting in grain growth or structural relaxation results in a monotonically decreasing signal, a "bell-shape" exothermic curve is the classical signature for a nucleation-and growth transformation (Chen & Spaepen, 1991).

The transformed volume fraction, x, during the isothermal process at a particular temperature, T, can then be determined by measuring the area under the exothermic curve. It is assumed that the volume fraction transformed, x, up to any time, t, is proportional to the fractional area of the exothermic peak or the integrated enthalpy. Therefore, in the isothermal DSC scans, the transformed volume fraction, $x_c(t,T)$, up to any time t is proportional to the fractional areas of the exothermic peak. Hence, the crystallized volume fraction during the isothermal annealing process can be accurately determined by measuring the area of the exothermic peak. The crystallized fractions $x_c(t,T)$ after time t at a certain temperature T for amorphous phase could be derived from the isothermal DSC curves by assuming that $x_c(t,T)$ is proportional to the integrated enthalpy

$$x_c(t,T) = \int_0^t h(t,T)dt \,/\, \int_0^\infty h(t,T)dt \tag{8}$$

where $h(t,T)$ is the enthalpy release. Using the JMA equation, the reaction rate as well as the parameters governing the nucleation rate and/or the growth morphology can be obtained. As shown before, it is inappropriate to describe the crystallization mechanism by using the average Avrami exponent derived from the non-linear JMA plot in the whole range of volume fraction. An alternative method of examining the isothermal DSC results is to evaluate the local value of the Avrami exponent, N_{loc}, which is defined as (Calka & Radlinski, 1988)

$$N_{loc} = \partial ln\left[-ln\left(1 - x_c(t,T)\right)\right] / \partial \ln(t-\tau) \tag{9}$$

as a function of crystallized volume fraction $x_c(t,T)$. Such a differential Avrami plot can highlight changes in reaction kinetics during the progress of crystallization.

The isothermal activation energy for the crystallization process can also be determined in terms of the incubation period τ at different temperatures during isothermal annealing, using the Arrhenius equation for a thermally activated process (Luborsky, 1977):

$$\tau = \tau_0(-E_{iso} / RT) \tag{10}$$

where τ_0 is a constant and E_{iso} is the activation energy for crystallization. The plot of $\ln\tau$ vs. $1/T$ yields a straight line. From the slope, the activation energy E_{iso} for crystallization of an amorphous phase is calculated.

Samples	Temperature range (K)	n	$x_c(t,T)$ range	E_{iso} (kJ/mol)	E_x (kJ/mol)
$Ti_{50}Cu_{18}Ni_{22}Al_4Sn_6$	735-755	2.5-3.3	0.05-0.60	399±55	392±17
with 10 vol.% TiC	723-750	2.1-2.8	0.05-0.60[a]	384±10	382±22

[a] 0.05-0.40 was used for the composite at 723 K.

Table 2. Avrami exponent (n) and activation energy of crystallization (E_{xT}) in terms of incubation time during isothermal annealing and the activation energy of crystallization (E_{iso}) determined from a Kissinger plot for the ball-milled amorphous $Ti_{50}Cu_{18}Ni_{22}Al_4Sn_6$ alloy and its composite containing 10 vol.% TiC. Reprinted from (Zhang, et al., 2006a), with permission from American Institute of Physics.

Table 2 compares the active energy of crstallization estimated by the aforementioned two methods for the ball-milled amorphous $Ti_{50}Cu_{18}Ni_{22}Al_4Sn_6$ alloy and its composite containing 10 vol.% TiC (Zhang, et al., 2006a). As seen from Table 2, there are no essential differences in the activation energies between those evaluated using the Arrhenius equation in isothermal annealing and those obtained by isochronal annealing as revealed by Kissinger analysis for the Ti-based amorphous alloy with and without TiC particles. The activation energy of crystallization determined from the Kissinger analysis and the Arrhenius equationfor both powders show that the composite has slightly lower activation energy. The addition of 10 vol.% TiC particles into the Ti-based amorphous alloy may slightly affect the crystallization kinetics of the amorphous phase and the TiC particles may act as potential heterogeneous nucleation sites.

5. Crystallization control for applications

5.1 Nanocrystalline alloy created from amorphous precursor via partial crystallization

The subject of preparation of nanostructured alloys by nanocrystallization of amorphous solid precursors has been reviewed by Lu (Lu, 1996) and by McHenry et al. (McHenry et al., 1999). The formation of nanocrystalline structures during crystallization of amorphous alloys is of a great interest from both the fundamental and the technical point of view. Fundamental studies of the mechanisms of crystal nucleation and growth as well as kinetics of transformation will to a certain degree aid in tailoring the structure for excellent physical (e.g. magnetic properties) and mechanical properties of nanostructured materials attractive for practical applications. In general, this method has extensively applied for those amorphous alloys where ductile solid solution phase(s) or functional phase(s) is formed through primary crystallization. Amorphous alloys of appropriate chemical compositions, crystallized at temperatures above their primary crystallization temperature but below the secondary crystallization temperature, can yield nanocrystalline grains dispersed in an amorphous matrix. Three important groups of nanocrystalline materials produced by primary crytallization from amorphous alloy precursors can be distinguished: constructional Al-based alloys (Kim et al., 1990; Latuch et al., 1997; Zhong et al., 1997), magnetically soft (Lachowicz & Slawskawaniewska, 1994; Makino et al., 1997; Suzuki et al., 1990; Suzuki et al., 1993; Willard et al., 1998) and magnetically hard (Inoue et al., 1995b; Manaf et al., 1993; Takeuchi et al., 1997; Withanawasam et al., 1994) Fe-based alloys. Examples of the alloys compositions and main aspects of their structure are presented in Table 3. There are two basic parameters characterizing structure of these materials: crystallite diameter, D, and volume fraction, V_{cr}, of nanocrystals. The optimum amount of nanocrystalline phase differs from each group. In the case of magnetically hard nanocrystalline materials, full (Manaf, et al., 1993; Takeuchi, et al., 1997) or almost full (Inoue, et al., 1995b) crystallization is required. For constructional and magnetically soft nanocrystalline materials the optimum mechanical and magnetic properties, respectively, are obtained after partial crystallization of their amorphous precursors (Inoue et al., 1988), which means that they are dual-phase materials composed of nanocrystals and an amorphous matrix. To preserve ductility in Al-based nanocrystalline alloys, V_{cr} should not exceed 20% in ternary Al–Y–Ni (Inoue, et al., 1988) and 40% in quaternary Al–Y–Ni–Cu (Latuch, et al., 1997) alloys. Mechanical properties of these materials can be explained and predicted using mixture model based on the volume fractions of amorphous matrix and nanocrystals, proposed by Kim et al. (Kim et al., 1999).

Inoue and Kimura (Inoue & Kimura, 2000) have summarized the microstructure and mechanical properties of aluminum based alloys produced by controlling the crystallization of amorphous alloy precursors, as shown in Fig. 10. A high mechanical strength exceeding

1000 MPa is achieved by the formation of an amorphous phase. The bulk nanocrystalline alloys, which contain a mixed structure of intermetallic compounds embedded fcc-Al matrix by the crystallization of Al-based amorphous phase, exhibit high mechanical strength of 700–1000 MPa and have been commercialized as a commercial name of GIGAS. By controlling the crystallization of Al-based amorphous alloys, the tensile strength of the Al-based amorphous alloys increases to 1560 MPa by the homogeneous precipitation of nanoscale fcc-Al particles into an amorphous phase, which is higher than the strength of 1260 MPa by the formation of an amorphous single phase.

Nanocrystalline materials	Magnetically soft (Fe-based)	Constructional (Al-based)	Magnetically hard (Fe-based)
Alloys	Finemet® ($Fe_{73.5}Cu_1Nb_3Si_{13.5}B_9$) Nanoperm® ($Fe_{84}Zr_{3.5}Nb_{3.5}B_8Cu_1$) Hitperm ($Fe_{44}Co_{44}Zr_7B_4Cu_1$)	Al–RE–TM (RE=Y, Ce, Nd, Sm; TM=Ni, Co, Fe, Cu) GIGAS®	Fe–RE–B e.g. $Fe_{82.3}Nd_{11.8}B_{5.9}$ $Fe_{88}Nb_2Pr_5B_5$
Structure	Nanocrystals (bcc-Fe) + Amorphous matrix	Amorphous matrix + Nanocrystals (fcc-Al)	Nanocrystals $Fe_{14}Nd_2B$ (+Fe_3B, bcc-Fe, Am)
Sructural parameters			
V_{cr}	70–75% $\Rightarrow \lambda_s \approx 0$	$\leq 40\% \Rightarrow$ ductility	$\leq 100\%$
D	≤ 15 nm $\Rightarrow \langle K \rangle \approx 0$	$V_{cr} \uparrow, D \downarrow \Rightarrow \sigma_f \uparrow$	< 25 nm
Properties	High permeability, low magnetic losses	High specific strength at high temperatures	High coercivity, high remanence

Table 3. General characteristics of the three main groups of nanocrystalline materials produced by devitrification of amorphous alloys (V_{cr} – volume fraction of crystalline phase, D – diameter of nanocrystals, λ_s – saturation magnetostriction constant, $\langle K \rangle$ – averaged magnetocrystalline anisotropy, σ_f – fracture strength). Reprinted from (Kulik, 2001), with permission from Elsevier.

5.2 Net-shape (micro-)forming in supercooled liquid region

Although amorphous alloys have exhibited unique properties compared the conventional polycrystalline materials, the metastable nature of amorphous phase has imposed a barrier to broad commercial adoption, particularly where the processing requirements of these alloys conflict with conventional metal processing methods. In general, amorphous alloys are super-strong with compressive yield strengths of approximately 2 GPa and even up to 5 GPa for some exotic bulk glass-forming alloys, as has already shown in Fig. 1. However, amorphous alloys suffer from a strong tendency toward shear localization upon yielding, which results in macroscopically brittle failure at ambient temperatures. Therefore, processing of amorphous alloys at ambient temperatures is extremely hard. When an amorphous solid is continuously heated in the supercooled liquid region the viscosity decreases dramatically as the alloy relaxes into the metastable equilibrium state of the supercooled liquid and the large viscous flowability is obtained (Bakke et al., 1995; Volkert

& Spaepen, 1989). Larger values of ΔT_x (T_x - T_g) indicate higher metastability of the liquid with respect to crystallization. The considerable softening of an amorphous alloy (viscous flowability) in its supercooled liquid region can be used for net-shape micro-forming of bulk amorphous alloy components and creation of bulk amorphous alloy samples via powder processing of amorphous powder precursors (see *Section 5.3*). In order to maintain their unique properties, processing of an amorphous alloy requires special attention. The main challenge is to avoid crystallization during the processing of amorphous alloy. By utilizing the low viscosity and large viscous flowability, bulk amorphous alloys could be deformed (Inoue & Takeuchi, 2002; Nishiyama & Inoue, 1999) to various complicated shapes in the maintenance of good mechanical properties. For example, from a bulk amorphous $Pd_{40}Cu_{30}Ni_{10}P_{20}$ alloy rod in 6 mm diameter, the die-forging into a three-stage die with pitch circle diameters of 4, 5 and 6 mm and a module of 0.3 was made for 120 s at 610 K under a compressive stress of 10 kPa and a three-stage gear was formed. The shape and dimension

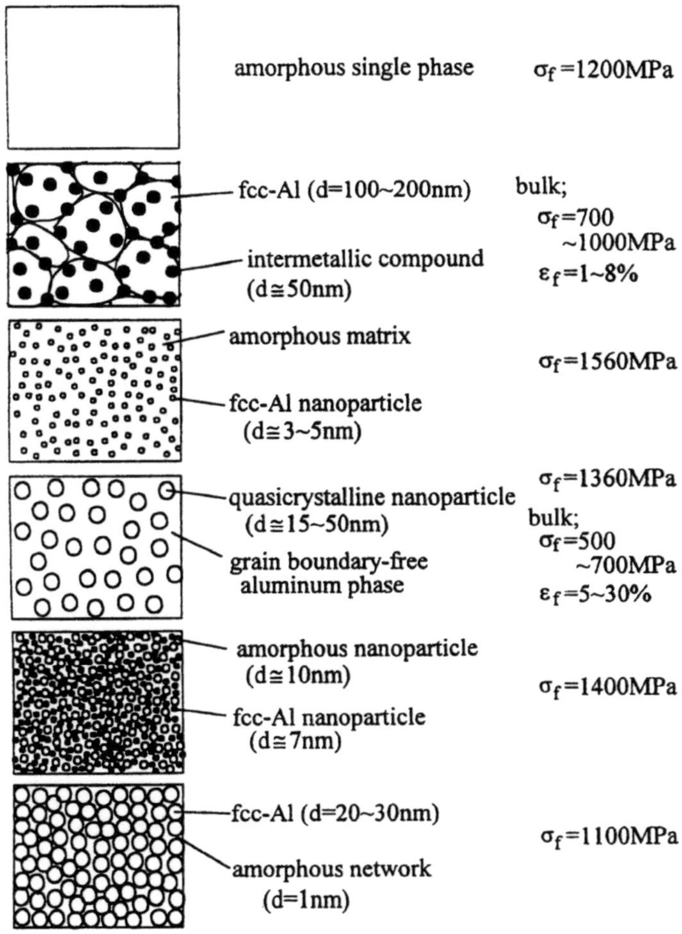

Fig. 10. Summary of the microstructure and mechanical properties of aluminum based alloys produced from amorphous alloy precursors. Reprinted from (Inoue & Kimura, 2000), with permission from Elsevier.

of the gear agree with the inner size of the die within a scattering of ±1% (Nishiyama & Inoue, 1999). The utilization of viscous flow of supercooled liquid is useful for secondary working of the bulk amorphous alloys which can produce a final product with near-net shape. In addition, in the supercooled liquid region, successful joining of the $Pd_{40}Ni_{40}P_{20}$ bulk amorphous components has been achieved by the friction-welding method utilizing the viscous flowability of the supercooled liquid (Kawamura & Ohno, 2001).

Recently, Schroers and his co-workers (Kumar et al., 2009; Schroers et al., 2007; Schroers, 2008; 2010; Schroers et al., 2011) have used a developed novel thermoplastic forming (TPF)-based processing to fabricate complex amorphous components. The process of TPF is also known as hot forming, hot pressing, super plastic forming, viscous flow working, and viscous flow forming. TPF takes advantage of the drastic softening of amorphous alloys upon heating above glass transition temperature and its thermal stability of supercooled liquid, which is quantified by the width of the supercooled liquid region. During TPF, the amorphous

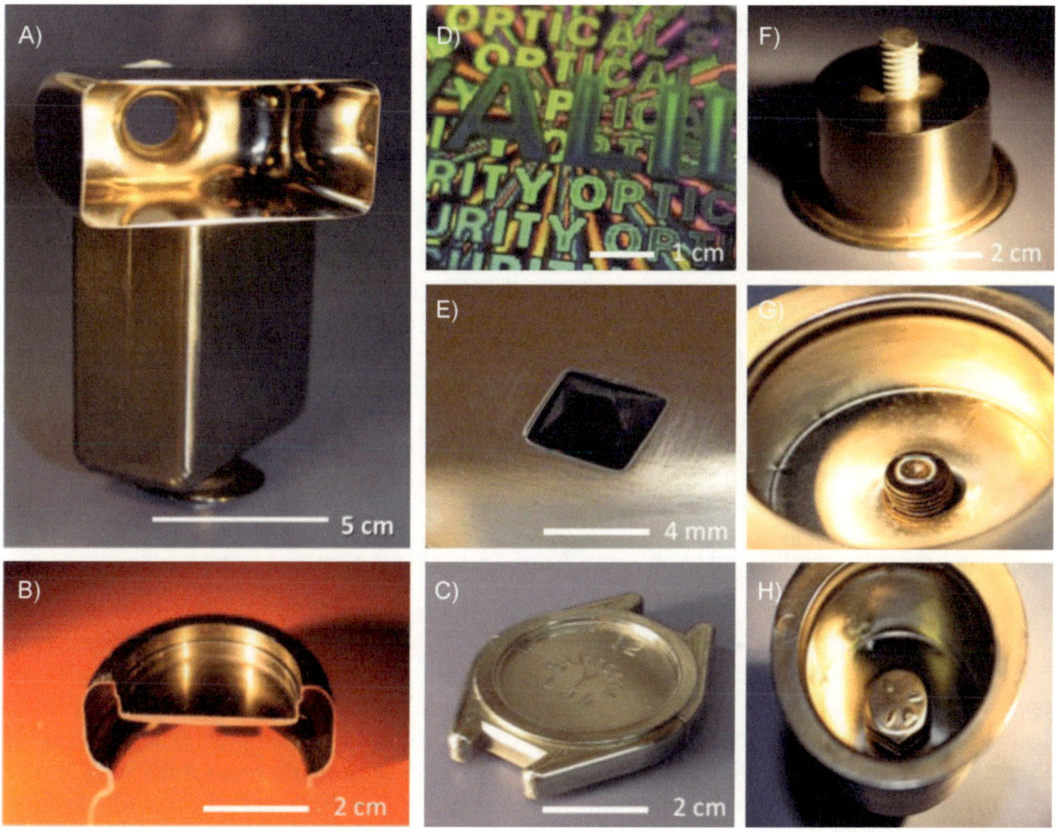

Fig. 11. Through TPF-based blow molding Blow molding with bulk metallic glasses (BMGs) permits hollow, thin, seamless shapes, which can include undercuts. These shapes were previously unachievable with any other metal processing method (A–C). The surface can be patterned, e.g., to reveal a hologram (D), joints can be created such as threads (F,H), and a second material can be joined to the BMG (E) in the same processing step than the blow molding. Reprinted from (Schroers, 2010), with permission from John Wiley and Sons.

solid is reheated into the supercooled liquid region, where it relaxes into a supercooled and metastable liquid before it eventually crystallizes. For a variety of BMG formers, a large processing window exists, which permits access to temperatures in this region on a practical experimental time scale in order to avoid crystallization. In general, at low temperatures a long processing time is available accompanied by a high viscosity. In contrast, at high temperatures, the viscosity is significantly reduced but, at the same time, the processing time is shortened. Currently, for a wide range of alloys, viscosities of 10^6 Pa s and lower can be accessed in the supercooled liquid region on a practical time scale (Schroers, 2010). For the highest formability of the BMG former in supercooled liquid region, optimum processing such as low viscosity and long processing time are required. The formability is a material property that reflects the maximum strain a metastable material can undergo before crystallization under given geometry and processing parameters.

As a novel technique with integration of shaping, joining, and finishing into one processing step, TPF-based blow molding allows one to net shape complex geometries in an economical and precise manner, including shapes, which can not be produced with any other metal processing method. In particular when pre-shaped parisons are used, BMGs can be blow molded into shapes that were previously not achievable with any metal processing method. Examples of such shapes are given in Fig. 11. They include hollow seamless shapes, which can comprise of complex undercuts, and very large thin sections. Due to the low forming pressure, together with the ability to replicate smallest features, as shown in Fig. 11D, the dimensional accuracy that can be achieved with this process is even superior over other TPF-based processes. In addition, this method is capable to combine the three processing steps typically required for metal processing—shaping, joining, and finishing—into one step (Schroers, 2010). For example potential joints such as threads, as shown in Fig. 11E–H can be formed in the BMG during the expansion process. Surface finishes that can be achieved with blow molding of BMGs include mirror finish. The superior properties of BMGs relative to plastics and typical structural metals, combined with the ease, economy, and precision of blow molding, have the potential to impact society in a manner similar to the development of synthetic plastics and their associated processing methods.

5.3 Bulk amorphous alloy consolidated from amorphous powder precursor

Synthesis of three-dimensional bulk amorphous materials has been an attractive object for several decades, not only for its significance in basic studies of the intrinsic properties of bulk amorphous materials (instead of the form of powders, fibers, or ribbons), but also for technological applications of these advanced materials with many novel properties. In principle, there are two approaches to obtain bulk amorphous samples. The first one is direct casting of alloy melts into bulk form in amorphous state (Suryanarayana & Inoue, 2011). An alternative approach that can potentially lead to bulk amorphous alloys is to exploit the viscous flow resulting from the significant decrease of the viscosity in supercooled liquid region. This is an especially attractive route to bulk amorphous alloys, especially to obtain bulk samples for the alloy systems with insufficient or limited glass-forming ability. A number of amorphous alloys with a sizable supercooled liquid region have been reported (Inoue, 2000; Johnson, 1999). This opens up the possibility of preparing truly bulk samples through powder consolidation in supercooled liquid region. In the Zr-, Cu-, Fe- and Ni-based alloy systems, some successful consolidation of amorphous powders with wide supercooled liquid region has been achieved by warm extrusion, spark plasma sintering and equal channel angular pressing (ECAP) (Choi et al., 2007; Ishihará et al., 2002; Itoi et al., 2001; Karaman et al., 2004; Kawamura et al., 1997; Kim et al., 2004; Kim et al., 2009; Lee et al., 2003; Mear et al., 2009; Robertson et al., 2003; Senkov et al., 2004; Senkov et al., 2005; Sordelet et al., 2002; Zhang et al., 2006b; Zhang, et al., 2007a). The consolidated samples show almost the same thermal properties, mechanical properties, and/or soft magnetic properties as those of the BMGs prepared by direct melt casting from molted

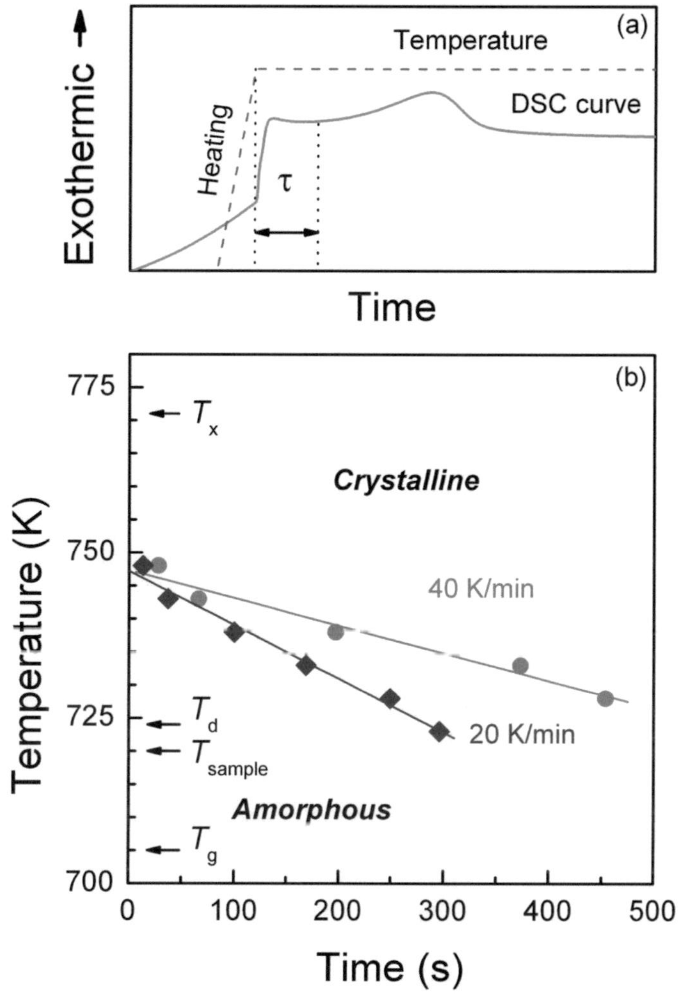

Fig. 12. (a) A representative DSC curve to determine the holding time (τ) up to the initial crystallization, and (b) TTT diagram for the onset of crystallization of the amorphous $Ti_{50}Cu_{18}Ni_{22}Al_4Sn_6$ powders heated to set temperatures at heating rate of 0.33 or 0.67 K s^{-1}. The data of the onset temperature of crystallization (T_x) and the glass transition temperature (T_g) at the heating rate of 0.67 K s^{-1} are also shown. Reprinted from (Zhang, et al., 2006b), with permission from Elsevier.

liquid. Among the aforementioned consolidation methods, it has recently been shown that ECAP is a particular effective and novel approach used for the consolidation of amorphous powders. ECAP is a method for subjecting a volume fraction of materials to severe shear deformation by forcing them around a mold corner (Karaman, et al., 2004; Robertson, et al., 2003; Zhang, et al., 2006b). The advantages of ECAP have allowed to fabricate bulk materials with large cross-sections.

In order to utilize the viscous flow of amorphous phase, the crystallization of an amorphous alloys must be well controlled. Therefore, the temperature-time-transformation (TTT) diagram should be determined for the selected amorphous powders by measuring the onset time of the exothermic reaction due to crystallization on the DSC curves during isothermal annealing (e.g. see Fig. 12 (a)), where the sample was heated to the selected annealing temperature(s) in the supercooled liquid region, and the time that the sample began to crystallize (the onset of an exothermic reaction) was recorded. Fig. 12 (b) shows an example of the TTT diagram for the amorphous $Ti_{50}Cu_{18}Ni_{22}Al_4Sn_6$ powders, which provides a window for processing in supercooled liquid state. The temperature and the time before crystallization (or the time to remain in the fully amorphous state at a certain temperature) exhibits approximately a linear relationship. At a given heating rate, the lower the temperature is, the longer the time is for the supercooled liquid to remain stable without crystallization. For the same temperature, the time window is longer at a faster heating rate. Therefore, for the ECAP processing at a given length of the can, it is necessary to select a suitable extrusion temperature (T_e) and extrusion rate (v_e). Two extrusion temperatures (700 and 705 K) near the calorimetric glass transition temperature (T_g) were used in when with extrusion rate of 0.40 mm s^{-1} (Zhang, et al., 2006b). By using ECAP with these processing parameters, bulk nanocrystal-amorphous composites with a relative density about 97% have been achieved from the amorphous $Ti_{50}Cu_{18}Ni_{22}Al_4Sn_6$ powders. Full densification was not reached, mainly owing to that the powders experienced insufficient shear deformation and that partial crystallization occurred during ECAP processing (Zhang, et al., 2006b).

Karaman et al (Karaman, et al., 2004) has optimized the ECAP process to consolidate the gas-atomized Vitreloy 106a ($Zr_{58.5}Nb_{2.8}$Cu15.6$Ni_{12.8}Al_{10.3}$) powder in supercooled liquid region at different strain rates and temperatures. The microstructure of all consolidates shows significant particle deformation. The increase in aspect ratio of particles due to shear strain is correlated with the extrusion temperature. Extrusions processed close to glass transition temepraure showed significant porosity. There is an increase in the consolidate hardness, depending on the extrusion temperature. Compression experiments on the consolidated V106a shows that good consolidate samples have strength levels of 1500 - 1700 MPa, which are comparable to that of cast V106 ($Zr_{57}Nb_5Al_{10}Cu_{15.4}Ni_{12.6}$). In spite of some nanocrystallization and short-range order formation upon processing, most of the fracture surfaces of the consolidates show shear banding and well-developed vein patterns, typical fracture characteristics of amorphous alloys with good ductility.

6. Conclusions

The amorphous alloys have attracted widespread research interests because of their technological promise for practical applications due to execellent properties and scientific importance in understanding glass formation and glass phenomena. Due to the nature of metastability, amorphous phase tends to crystalize to more stable crystalline state through *polymorphous*, *eutectic* and/or *primary* crystallization mechanisms. The crystallization mechanisms and crystallization products are influenced by both inherent (e.g. chemical composition of amorphous phase, oxygen) and extraneous (e.g. preparation method, pressure, etc.) factors. The study of kinetic behavior associated with a structural change in amorphous alloys above glass transition temperature could provide opportunities for structure control by innovative design and processing strategies. By controlling the crystallization of amorphous alloys, bulk nanocrystalline alloys and/or nanocrystalline-amorphous composites with excellent properties could be achieved from amorphous alloys precursors. By utilizing the viscous flowability of amorphous alloys in supercooled liquid region, net-shaped microforming could be realized for bulk amorphous alloys and bulk amorphous components with "true" bulk size might be produced from amorphous powder precursors.

7. Acknowledgments

The author is grateful to J. Xu, E. Ma, J. Eckert, H.B. Lu and M. Calin for their stimulating discussions. Financial support provided by Research Services of The University of Western Australia (through UWA Research Development Award Scheme) is gratefully acknowledged.

8. References

Altounian, Z.; Batalla, E.; Stromolsen, J.O. & Walter, J.L. (1987). The influence of oxygen and other impurities on the crystallization of NiZr2 and related metallic glasses. *Journal of Applied Physics,* Vol. 61, No. 1, (January 1987), pp. 149-155, ISSN 0021-8979

Azam, N.; Lenaour, L.; Rivera, C.; Grosjean, P.; Sacovy, P. & Delaplace, J. (1979). Crystallization and irradiation effects of amorphous Fe40Ni38Mo4B18 alloy. *Journal of Nuclear Materials,* Vol. 83, No. 2, (Febuary 1979), pp. 298-304, ISSN 0022-3115

Bakke, E.; Busch, R. & Johnson, W.L. (1995). The viscosity of the Zr46.75Ti8.25Cu7.5Ni10Be27.5 bulk metallic-glass forming alloy in the supercooled liquid. *Applied Physics Letters,* Vol. 67, No. 22, (November 1995), pp. 3260-3262, ISSN 0003-6951

Calın, M.; Zhang, L.C. & Eckert, J. (2007). Tailoring of microstructure and mechanical properties of a Ti-based bulk metallic glass-forming alloy. *Scripta Materialia,* Vol. 57, No. 12, (December 2007), pp. 1101-1104, ISSN 1359-6462

Calka, A. & Radlinski, A.P. (1988). The local value of the avrami exponent - a new approach to devitrification of glassy metallic ribbons. *Materials Science and Engineering,* Vol. 97, No. 1, (January 1988), pp. 241-246, ISSN 0025-5416

Chen, H.S. & Turnbull, D. (1969). Formation, stability and structure of Palladium-Silicon based alloy glasses. *Acta Metallurgica,* Vol. 17, No. 8, (August 1969), pp. 1021-1031, ISSN 0001-6160

Chen, L.C. & Spaepen, F. (1991). Analysis of calorimetric measurements of grain-growth. *Journal of Applied Physics,* Vol. 69, No. 2, (January 1991), pp. 679-688, ISSN 0021-8979

Choi, P.P.; Kim, J.S.; Nguyen, O.T.H.; Kwon, D.H.; Kwon, Y.S. & Kim, J.C. (2007). A1-La-Ni-Fe bulk metallic glasses produced by mechanical alloying and spark-plasma sintering. *Materials Science and Engineering A*, Vol. 449, No. 1-2, (March 2007), pp. 1119-1122, ISSN 0921-5093

Christian, J.W. (2002). *The Theory of Transformations in Metals and Alloys* (3rd ed.). Elsevier Science, ISBN 9780080440194, Oxford

Doherty, R.D. (1996). Diffusive phase transformations In the solid state. In: Cahn, R.W. & Haasen, P. (Eds.), *Physical metallurgy* (Fourth, revised and enhanced edition ed., Vol. II, pp. 1363-1506). Amsterdam: North-Hollaiw.

Dong, Z.F.; Ma, Y.H. & Lu, K. (1994). Crystallization process and thermal stabilities of the melt-spun amorphous Ni100-xPx (x=16.0-20.0 at-percent) alloys. *Scripta Metallurgica et Materialia,* Vol. 31, No. 1, (July 1994), pp. 81-86, ISSN 0956-716X

Eckert, J.; Mattern, N.; Zinkevitch, M. & Seidel, M. (1998). Crystallization behavior and phase formation in Zr-Al-Cu-Ni metallic glass containing oxygen. *Materials Transactions JIM,* Vol. 39, No. 6, (June 1998), pp. 623-632, ISSN 0916-1821

Eckert, J.; Kuhn, U.; Mattern, N.; Reger-Leonhard, A. & Heilmaier, M. (2001). Bulk nanostructured Zr-based multiphase alloys with high strength and good ductility. *Scripta Materialia,* Vol. 44, No. 8-9, (May 2001), pp. 1587-1590, ISSN 1359-6462

Eckert, J.; Das, J.; Pauly, S. & Duhamel, C. (2007). Mechanical properties of bulk metallic glasses and composites. *Journal of Materials Research,* Vol. 22, No. 2, (Febuary 2007), pp. 285-301, ISSN 0884-2914

Foley, J.C.; Allen, D.R. & Perepezko, J.H. (1997). Strategies for the development of nanocrystalline materials through devitrification. *Materials Science and Engineering A,* Vol. 226, No. 1-2, (June 1997), pp. 569-573, ISSN 0921-5093

Fornell, J.; Rossinyol, E.; Surinach, S.; Baro, M.D.; Li, W.H. & Sort, J. (2010). Enhanced mechanical properties in a Zr-based metallic glass caused by deformation-induced nanocrystallization. *Scripta Materialia,* Vol. 62, No. 1, (January 2010), pp. 13-16, ISSN 1359-6462

Gebert, A.; Eckert, J. & Schultz, L. (1998). Effect of oxygen on phase formation and thermal stability of slowly cooled Zr65Al7.5Cu7.5Ni10 metallic glass. *Acta Materialia,* Vol. 46, No. 15, (September 1998), pp. 5475-5482, ISSN 1359-6454

He, Y.; Schwarz, R.B. & Archuleta, J.I. (1996). Bulk glass formation in the Pd-Ni-P system. *Applied Physics Letters,* Vol. 69, No. 13, (September 1996), pp. 1861-1863, ISSN 0003-6951

Hono, K.; Hiraga, K.; Wang, Q.; Inoue, A. & Sakurai, T. (1992). The microstructure evolution of a Fe73.5Si13.5B9Nb3Cu1 nanocrystalline soft magnetic material. *Acta Metallurgica Et Materialia,* Vol. 40, No. 9, (September 1992), pp. 2137-2147, ISSN 0956-7151

Illeková, E.; Jergel, M.; Duhaj, P. & Inoue, A. (1997). The relation between the bulk and ribbon Zr55Ni25Al20 metallic glasses. *Materials Science and Engineering A,* Vol. 226-228, No. 1-2, (June 1997), pp. 388-392, ISSN 0921-5093

Inoue, A.; Ohtera, K.; Tsai, A.P. & Masumoto, T. (1988). New amorphous-alloys with good ductility in Al-Y-M and Al-La-M (M=Fe, Co, Ni or Cu) systems. *Japanese Journal of Applied Physics,* Vol. 27, No. 3, (March 1988), pp. L280-L282, ISSN 0021-4922

Inoue, A.; Kita, K.; Zhang, T. & Masumoto, T. (1989). An Amorphous La55Al25Ni20 Alloy Prepared by Water Quenching. *Materials Transactions JIM,* Vol. 30, No. 9, (September 1989), pp. 722-725, ISSN 0916-1821

Inoue, A.; Nakamura, N.; Sugita, T.; Zhang, T. & Masumoto, T. (1993a). Bulky La–Al–TM (TM=Transition Metal) amorphous alloys with high tensile strength produced by a

high-pressure die casting method. *Materials Transactions JIM,* Vol. 34, No. 4, (April 1993), pp. 351-358, ISSN 0916-1821
Inoue, A.; Zhang, T.; Nishiyama, N.; Ohba, K. & Masumoto, T. (1993b). Preparation of 16 mm diameter rod of amorphous Zr65Al7.5Ni10Cu17.5 alloy. *Materials Transactions JIM,* Vol. 34, No. 12, (December 1993), pp. 1234-1237, ISSN 0916-1821
Inoue, A.; Shinohara, Y.; Yokoyama, Y. & Masumoto, T. (1995a). Solidification analyses of bulky Zr60Al10Ni10Cu15Pd5 glass produced by casting into wedge-shape copper mold. *Materials Transactions JIM,* Vol. 36, No. 10, (October 1995), pp. 1276-1281, ISSN 0916-1821
Inoue, A.; Takeuchi, A.; Makino, A. & Masumoto, T. (1995b). Hard magnetic-properties of nanocrystalline Fe-rich Fe-Nd-B alloys prepared by partial crystallization of amorphous phase. *Materials Transactions JIM,* Vol. 36, No. 7, (July 1995), pp. 962-971, ISSN 0916-1821
Inoue, A.; Nishiyama, N. & Kimura, H. (1997). Preparation and Thermal Stability of Bulk Amorphous Pd40Cu30Ni10P20 Alloy Cylinder of 72 mm in Diameter. *Materials Transactions JIM,* Vol. 38, No. 2, (February 1997), pp. 179-183, ISSN 0916-1821
Inoue, A. (2000). Stabilization of metallic supercooled liquid and bulk amorphous alloys. *Acta Materialia,* Vol. 48, No. 1, (January 2000), pp. 279-306, ISSN 1359-6454
Inoue, A. & Kimura, H. (2000). High-strength aluminum alloys containing nanoquasicrystalline particles. *Materials Science and Engineering A,* Vol. 286, No. 1, (June 2000), pp. 1-10, ISSN 0921-5093
Inoue, A. & Takeuchi, A. (2002). Recent progress in bulk glassy alloys. *Materials Transactions,* Vol. 43, No. 8, (August 2002), pp. 1892-1906, ISSN 1345-9678
Inoue, A.; Shen, B.L. & Chang, C.T. (2004a). Super-high strength of over 4000 MPa for Fe-based bulk glassy alloys in [(Fe1−xCox)0.75B0.2Si0.05]96Nb4 system. *Acta Materialia,* Vol. 52, No. 14, (August 2004), pp. 4093-4099, ISSN 1359-6454
Inoue, A.; Shen, B.L.; Koshiba, H.; Kato, H. & Yavari, A.R. (2004b). Ultra-high strength above 5000 MPa and soft magnetic properties of Co-Fe-Ta-B bulk glassy alloys. *Acta Materialia,* Vol. 52, No. 6, (April 2004), pp. 1631-1637, ISSN 1359-6454
Inoue, A.; Zhang, W.; Tsurui, T.; Yavari, A.R. & Greer, A.L. (2005). Unusual room-temperature compressive plasticity in nanocrystal-toughened bulk copper-zirconium glass. *Philosophical Magazine Letters,* Vol. 85, No. 5, (May 2005), pp. 221-229, ISSN 0950-0839
Ishihará, S.; Zhang, W. & Inoue, A. (2002). Hot pressing of Fe-Co-Nd-Dy-B glassy powders in supercooled liquid state and hard magnetic properties of the consolidated alloys. *Scripta Materialia,* Vol. 47, No. 4, (August 2002), pp. 231-235, ISSN 1359-6462
Itoi, T.; Takamizawa, T.; Kawamura, Y. & Inoue, A. (2001). Fabrication of Co40Fe22Nb8B30 bulk metallic glasses by consolidation of gas-atomized powders and their soft-magnetic properties. *Scripta Materialia,* Vol. 45, No. 10, (November 2001), pp. 1131-1137, ISSN 1359-6462
Jiang, J.Z.; Olsen, J.S.; Gerward, L.; Abdali, S.; Eckert, J.; Schlorke-de Boer, N.; Schultz, L.; Truckenbrodt, J. & Shi, P.X. (2000). Pressure effect on crystallization of metallic glass Fe72P11C6Al5B4Ga2 alloy with wide supercooled liquid region. *Journal of Applied Physics,* Vol. 87, No. 5, (March 2000), pp. 2664-2666, ISSN 0021-8979
Jiang, J.Z.; Gerward, L. & Xu, Y.S. (2002). Pressure effect on crystallization kinetics in Zr46.8Ti8.2Cu7.5Ni10Be27.5 bulk glass. *Applied Physics Letters,* Vol. 81, No. 23, (December 2002), pp. 4347-4349, ISSN 0003-6951

Jiang, J.Z.; Saida, J.; Kato, H.; Ohsuna, T. & Inoue, A. (2003a). Is Cu60Ti10Zr30 a bulk glass-forming alloy? *Applied Physics Letters,* Vol. 82, No. 23, (June 2003), pp. 4041-4043, ISSN 0003-6951

Jiang, J.Z.; Yang, B.; Saksl, K.; Franz, H. & Pryds, N. (2003b). Crystallization of Cu60Ti20Zr20 metallic glass with and without pressure. *Journal of Materials Research,* Vol. 18, No. 4, (April 2003), pp. 895-898, ISSN 0884-2914

Jiang, X.Y.; Zhong, Z.C. & Greer, A.L. (1997). Primary crystallization in an amorphous Al88Ni4Y8 alloy. *Philosophical Magazine B,* Vol. 76, No. 4, (October 1997), pp. 419-423, ISSN 0141-8637

Johnson, W.L. (1999). Bulk glass-forming metallic alloys: Science and technology. *MRS Bulletin,* Vol. 24, No. 10, (October 1999), pp. 42-56, ISSN 0883-7694

Kang, H.G.; Park, E.S.; Kim, W.T.; Kim, D.H. & Cho, H.K. (2000). Fabrication of bulk Mg-Cu-Ag-Y glassy alloy by squeeze casting. *Materials Transactions Jim,* Vol. 41, No. 7, (July 2000), pp. 846-849, ISSN 0916-1821

Karaman, I.; Robertson, J.; Im, J.T.; Mathaudhu, S.N.; Luo, Z.P. & Hartwig, K.T. (2004). The effect of temperature and extrusion speed on the consolidation of zirconium-based metallic glass powder using equal-channel angular extrusion. *Metallurgical and Materials Transactions A,* Vol. 35A, No. 1, (January 2004), pp. 247-256, ISSN 1073-5623

Kawamura, Y.; Kato, H.; Inoue, A. & Masumoto, T. (1997). Fabrication of bulk amorphous alloys by powder consolidation. *International Journal of Powder Metallurgy,* Vol. 33, No. 2, (March 1997), pp. 50-61, ISSN 0888-7462

Kawamura, Y. & Ohno, Y. (2001). Superplastic bonding of bulk metallic glasses using friction. *Scripta Materialia,* Vol. 45, No. 3, pp. 279-285, ISSN 1359-6462

Kelton, K.F.; Croat, T.K.; Gangopadhyay, A.K.; Xing, L.Q.; Greer, A.L.; Weyland, M.; Li, X. & Rajan, K. (2003). Mechanisms for nanocrystal formations in metallic glasses. *Journal of Non-Crystalline Solids,* Vol. 317, No. 1-2, (March 2003), pp. 71-77, ISSN 0022-3093

Kim, H.J.; Lee, J.K.; Shin, S.Y.; Jeong, H.G.; Kim, D.H. & Bae, J.C. (2004). Cu-based bulk amorphous alloys prepared by consolidation of amorphous powders in the supercooled liquid region. *Intermetallics,* Vol. 12, No. 10-11, (October-November 2004), pp. 1109-1113, ISSN 0966-9795

Kim, H.S.; Warren, P.J.; Cantor, B. & Lee, H.R. (1999). Mechanical properties of partially crystallized aluminum based amorphous alloys. *Nanostructured Materials,* Vol. 11, No. 2, (March 1999), pp. 241-247, ISSN 0965-9773

Kim, J.S.; Povstugar, I.V.; Choi, P.P.; Yelsukov, E.P. & Kwon, Y.S. (2009). Synthesis of Al-Y-Ni-Co composites by mechanical alloying and consecutive spark-plasma sintering. *Journal of Alloys and Compounds,* Vol. 486, No. 1-2, (November 2009), pp. 511-514, ISSN 0925-8388

Kim, Y.H.; Inoue, A. & Masumoto, T. (1990). Ultrahigh tensile strengths of Al88Y2Ni9Mn1 or Al88Y2Ni9Fe1 amorphous-alloys containing finely dispersed fcc-Al particles. *Materials Transactions JIM,* Vol. 31, No. 8, (August 1990), pp. 747-749, ISSN 0916-1821

Kissinger, H.E. (1957). Reaction kinetics in differential thermal analysis. *Analytical Chemistry,* Vol. 29, No. 11, (November 1957), pp. 1702-1706, ISSN 0003-2700

Klement, W.; Willens, R.H. & Duwez, P. (1960). Non-crystalline structure in solidified Gold-Silicon alloys. *Nature,* Vol. 187, No. 4740, (September 1960), pp. 869-870, ISSN 0028-0836

Köster, U. & Herold, U. (1981). Crystallization of metallic glasses. In: Guntherodt, H.-J. & Beck, H. (Eds.), *Glassy Metals I, Ionic Structure, Electronic Transport, and Crystallization* (pp. 225-259). Berlin: Springer-Verlag.

Köster, U. & Meinhardt, J. (1994). Crystallization of highly undercooled metallic melts and metallic glasses around the glass-transition temperature. *Materials Science and Engineering A,* Vol. 178, No. 1-2, (April 1994), pp. 271-278, ISSN 0921-5093

Köster, U.; Meinhardt, J.; Roos, S. & Liebertz, H. (1996). Formation of quasicrystals in bulk glass forming Zr-Cu-Ni-Al alloys. *Applied Physics Letters,* Vol. 69, No. 2, (July 1996), pp. 179-181, ISSN 0003-6951

Köster, U.; Meinhardt, J.; Roos, S. & Busch, R. (1997). Formation of quasicrystals in bulk glass forming Zr-Cu-Ni-Al alloys. *Materials Science and Engineering A,* Vol. 226, No. 1-2, (June 1997), pp. 995-998, ISSN 0921-5093

Kubler, A.; Eckert, J.; Gebert, A. & Schultz, L. (1998). Influence of oxygen on the viscosity of Zr-Al-Cu-Ni metallic glasses in the undercooled liquid region. *Journal of Applied Physics,* Vol. 83, No. 6, (March 1998), pp. 3438-3440, ISSN 0021-8979

Kulik, T. (2001). Nanocrystallization of metallic glasses. *Journal of Non-Crystalline Solids,* Vol. 287, No. 1-3, (July 2001), pp. 145-161, ISSN 0022-3093

Kumar, G.; Tang, H.X. & Schroers, J. (2009). Nanomoulding with amorphous metals. *Nature,* Vol. 457, No. 7231, (Febuary 2009), pp. 868-872, ISSN 0028-0836

Lachowicz, H.K. & Slawskawaniewska, A. (1994). Coexistence of various magnetic phases in nanocrystalline Fe-based metallic glasses. *Journal of Magnetism and Magnetic Materials,* Vol. 133, No. 1-3, (May 1994), pp. 238-242, ISSN 0304-8853

Latuch, J.; Kokoszkiewicz, A. & Matyja, H. (1997). The effect of Cu addition on the formation of fcc-Al phase in rapidly quenched Al-Y-Ni alloys. *Materials Science and Engineering A,* Vol. 226-228, No. 1-2, (June 1997), pp. 809-812, ISSN 0921-5093

Lee, M.H.; Bae, D.H.; Kim, W.T.; Kim, D.H.; Rozhkova, E.; Wheelock, P.B. & Sordelet, D.J. (2003). Synthesis of Ni-based bulk amorphous alloys by warm extrusion of amorphous powders. *Journal of Non-Crystalline Solids,* Vol. 315, No. 1-2, (January 2003), pp. 89-96, ISSN 0022-3093

Lin, X.H.; Johnson, W.L. & Rhim, W.K. (1997). Effect of oxygen impurity on crystallization of an undercooled bulk glass forming Zr-Ti-Cu-Ni-Al alloy. *Materials Transactions JIM,* Vol. 38, No. 5, (May 1997), pp. 473-477, ISSN 0916-1821

Lohwongwatana, B.; Schroers, J. & Johnson, W.L. (2006). Strain rate induced crystallization in bulk metallic glass-forming liquid. *Physical Review Letters,* Vol. 96, No. 7, (Febuary 2006), pp. 075503, ISSN 0031-9007

Lu, K. (1996). Nanocrystalline metals crystallized from amorphous solids: Nanocrystallization, structure, and properties. *Materials Science and Engineering R,* Vol. 16, No. 4, (April 1996), pp. 161-221, ISSN 0927-796X

Lu, K.; Liu, X.D. & Yuan, F.H. (1996). Synthesis of the NiZr2 intermetallic compound nanophase materials. *Physica B,* Vol. 217, No. 1-2, (January 1996), pp. 153-159, ISSN 0921-4526

Luborsky, F.E. (1977). Crystallization of some Fe-Ni metallic glasses. *Materials Science and Engineering,* Vol. 28, No. 1, (April 1977), pp. 139-144, ISSN 0025-5416

Makino, A.; Bitoh, T.; Inoue, A. & Masumoto, T. (1997). Nanocrystalline Fe-M-B-Cu (M=Zr,Nb) alloys with improved soft magnetic properties. *Journal of Applied Physics,* Vol. 81, No. 6, (March 1997), pp. 2736-2739, ISSN 0021-8979

Manaf, A.; Buckley, R.A. & Davies, H.A. (1993). New nanocrystalline high-remanence Nd-Fe-B alloys by rapid solidification. *Journal of Magnetism and Magnetic Materials,* Vol. 128, No. 3, (December 1993), pp. 302-306, ISSN 0304-8853

McHenry, M.E.; Willard, M.A. & Laughlin, D.E. (1999). Amorphous and nanocrystalline materials for applications as soft magnets. *Progress in Materials Science,* Vol. 44, No. 4, (October 1999), pp. 291-433, ISSN 0079-6425

Mear, F.O.; Xie, G.Q.; Louzguine-Luzgin, D.V. & Inoue, A. (2009). Spark plasma sintering of Mg-based amorphous ball-milled powders. *Materials Transactions,* Vol. 50, No. 3, (March 2009), pp. 588-591, ISSN 1345-9678

Murty, B.S.; Ping, D.H.; Hono, K. & Inoue, A. (2000). Influence of oxygen on the crystallization behavior of Zr65Cu27.5Al7.5 and Zr66.7Cu33.3 metallic glasses. *Acta Materialia,* Vol. 48, No. 15, (September 2000), pp. 3985-3996, ISSN 1359-6454

Nicolaus, M.M.; Sinning, H.R. & Haessner, F. (1992). Crystallization behavior and generation of a nanocrystalline state from amorphous Co33Zr67. *Materials Science and Engineering A,* Vol. 150, No. 1, (Febuary 1992), pp. 101-112, ISSN 0921-5093

Nishiyama, N. & Inoue, A. (1999). Glass transition behavior and viscous flow working of Pd40Cu30Ni10P20 amorphous alloy. *Materials Transactions JIM,* Vol. 40, No. 1, (January 1999), pp. 64-71, ISSN 0916-1821

Peker, A. & Johnson, W.L. (1993). A hghly processable metallic glass: Zr41.2Ti13.8Cu12.5-Ni10.0Be22.5. *Applied Physics Letters,* Vol. 63, No. 17, (October 1993), pp. 2342-2344, ISSN 0003-6951

Robertson, J.; Im, J.T.; Karaman, I.; Hartwig, K.T. & Anderson, I.E. (2003). Consolidation of amorphous copper based powder by equal channel angular extrusion. *Journal of Non-Crystalline Solids,* Vol. 317, No. 1-2, (March 2003), pp. 144-151, ISSN 0022-3093

Sahu, K.K.; Mauro, N.A.; Longstreth-Spoor, L.; Saha, D.; Nussinov, Z.; Miller, M.K. & Kelton, K.F. (2010). Phase separation mediated devitrification of Al88Y7Fe5 glasses. *Acta Materialia,* Vol. 58, No. 12, (July 2010), pp. 4199-4206, ISSN 1359-6454

Schroers, J.; Nguyen, T.; O'Keeffe, S. & Desai, A. (2007). Thermoplastic forming of bulk metallic glass - applications for MEMS and microstructure fabrication. *Materials Science and Engineering A,* Vol. 449, No. 1-2, (March 2007), pp. 898-902, ISSN 0921-5093

Schroers, J. (2008). On the formability of bulk metallic glass in its supercooled liquid state. *Acta Materialia,* Vol. 56, No. 3, (Febuary 2008), pp. 471-478, ISSN 1359-6454

Schroers, J. (2010). Processing of Bulk Metallic Glass. *Advanced Materials,* Vol. 22, No. 14, (April 2010), pp. 1566-1597, ISSN 0935-9648

Schroers, J.; Hodges, T.M.; Kumar, G.; Raman, H.; Barnes, A.J.; Quoc, P. & Waniuk, T.A. (2011). Thermoplastic blow molding of metals. *Materials Today,* Vol. 14, No. 1-2, (January-Febuary 2011), pp. 14-19, ISSN 1369-7021

Senkov, O.N.; Miracle, D.B.; Scott, J.M. & Senkova, S.V. (2004). Equal channel angular extrusion compaction of semi-amorphous Al85Ni10Y2.5La2.5 alloy powder. *Journal of Alloys and Compounds,* Vol. 365, No. 1-2, (Febuary 2004), pp. 126-133, ISSN 0925-8388

Senkov, O.N.; Senkova, S.V.; Scott, J.M. & Miracle, D.B. (2005). Compaction of amorphous aluminum alloy powder by direct extrusion and equal channel angular extrusion. *Materials Science and Engineering A,* Vol. 393, No. 1-2, (Febuary 2005), pp. 12-21, ISSN 0921-5093

Setyawan, A.D.; Saida, J.; Kato, H.; Matsushita, M. & Inoue, A. (2010). Deformation-induced structural transformation leading to compressive plasticity in Zr65Al7.5Ni10Cu12.5M5 (M = Nb, Pd) glassy alloys. *Journal of Materials Research*, Vol. 25, No. 6, (June 2010), pp. 1149-1158, ISSN 0884-2914

Sordelet, D.J.; Rozhkova, E.; Huang, P.; Wheelock, P.B.; Besser, M.F.; Kramer, M.J.; Calvo-Dahlborg, M. & Dahlborg, U. (2002). Synthesis of Cu47Ti34Zr11Ni8 bulk metallic glass by warm extrusion of gas atomized powders. *Journal of Materials Research*, Vol. 17, No. 1, (January 2002), pp. 186-198, ISSN 0884-2914

Spassov, T. & Koster, U. (1993). Grain-growth in nanocrystalline zirconium-based alloys. *Journal of Materials Science*, Vol. 28, No. 10, (May 1993), pp. 2789-2794, ISSN 0022-2461

Suryanarayana, C. & Inoue, A. (2011). *Bulk metallic glasses*. CRC Press, ISBN 978-1-4200-8597-6, Boca Raton London New York

Suzuki, K.; Kataoka, N.; Inoue, A.; Makino, A. & Masumoto, T. (1990). High saturation magnetization and soft magnetic-properties of bcc Fe-Zr-B alloys with ultrafine grain-structure. *Materials Transactions JIM*, Vol. 31, No. 8, (August 1990), pp. 743-746, ISSN 0916-1821

Suzuki, K.; Makino, A.; Inoue, A. & Masumoto, T. (1993). Low core losses of nanocrystalline Fe-M-B (M=Zr, Hf, or Nb) alloys. *Journal of Applied Physics*, Vol. 74, No. 5, (September 1993), pp. 3316-3322, ISSN 0021-8979

Takeuchi, A.; Inoue, A. & Makino, A. (1997). Improvement of hard magnetic properties of Fe90Nd7B3 alloys by two-stage crystallization treatment. *Materials Science and Engineering A*, Vol. 226-228, No. 1-2, (June 1997), pp. 636-640, ISSN 0921-5093

Turnbull, D. (1981). Metastable structures in metallurgy. *Metallurgical Transactions B*, Vol. 12, No. 2, (June 1981), pp. 217-230, ISSN 0360-2141

Volkert, C.A. & Spaepen, F. (1989). Crossover relaxation of the viscosity of Pd40Ni40P19Si1 near the glass-transition. *Acta Metallurgica*, Vol. 37, No. 5, (May 1989), pp. 1355-1362, ISSN 0001-6160

Wang, W.H.; Dong, C. & Shek, C.H. (2004). Bulk metallic glasses. *Materials Science and Engineering R*, Vol. 44, No. 2-3, (June 2004), pp. 45-89, ISSN 0927-796X

Wang, W.H. (2009). Bulk metallic glasses with functional physical properties. *Advanced Materials*, Vol. 21, No. 45, (December 2009), pp. 4524-4544, ISSN 0935-9648

Willard, M.A.; Laughlin, D.E.; McHenry, M.E.; Thoma, D.; Sickafus, K.; Cross, J.O. & Harris, V.G. (1998). Structure and magnetic properties of (Fe0.5Co0.5)(88)Zr7B4Cu1 nanocrystalline alloys. *Journal of Applied Physics*, Vol. 84, No. 12, (December 1998), pp. 6773-6777, ISSN 0021-8979

Withanawasam, L.; Murphy, A.S.; Hadjipanayis, G.C. & Krause, R.F. (1994). Nanocomposite R(2)Fe(14)B/Fe exchange-coupled magnets. *Journal of Applied Physics*, Vol. 76, No. 10, (November 1994), pp. 7065-7067, ISSN 0021-8979

Xu, J.; Ramamurty, U. & Ma, E. (2010). The fracture toughness of bulk metallic glasses. *JOM*, Vol. 62, No. 4, (April 2000), pp. 10-18, ISSN 1047-4838

Yang, C.; Wang, W.K.; Liu, R.P.; Zhan, Z.J.; Sun, L.L.; Zhang, J.; Jiang, J.Z.; Yang, L. & Lathe, C. (2006). Crystallization of Zr41Ti14Cu12.5Ni10Be22.5 bulk metallic glass under high pressure examined by in situ synchrotron radiation x-ray diffraction. *Journal of Applied Physics*, Vol. 99, No. 2, (January 2006), pp. 023525, ISSN 0021-8979

Yao, K.F. & Ruan, F. (2005). Pd-Si binary bulk metallic glass prepared at low cooling rate. *Chinese Physics Letters*, Vol. 22, No. 6, (June 2005), pp. 1481-1483, ISSN 0256-307X

Ye, F. & Lu, K. (1999). Pressure effect on crystallization kinetics of an Al-La-Ni amorphous alloy. *Acta Materialia,* Vol. 47, No. 8, (June 1999), pp. 2449-2454, ISSN 1359-6454

Zhang, L.C. & Xu, J. (2002). Formation of glassy Ti50Cu20Ni24Si4B2 alloy by high-energy ball milling. *Materials Letters,* Vol. 56, No. 5, (November 2002), pp. 615-619, ISSN 0167-577X

Zhang, L.C.; Xu, J. & Ma, E. (2002). Mechanically alloyed amorphous Ti50(Cu0.45Ni0.55)(44-x)AlxSi4B2 alloys with supercooled liquid region. *Journal of Materials Research,* Vol. 17, No. 7, (July 2002), pp. 1743-1749, ISSN 0884-2914

Zhang, L.C.; Shen, Z.Q. & Xu, J. (2003). Glass formation in a (Ti,Zr,Hf)-(Cu,Ni,Ag)-Al high-order alloy system by mechanical alloying. *Journal of Materials Research,* Vol. 18, No. 9, (September 2003), pp. 2141-2149, ISSN 0884-2914

Zhang, L.C. & Xu, J. (2004). Glass-forming ability of melt-spun multicomponent (Ti, Zr, Hf)-(Cu, Ni, Co)-Al alloys with equiatomic substitution. *Journal of Non-Crystalline Solids,* Vol. 347, No. 1-3, (November 2004), pp. 166-172, ISSN 0022-3093

Zhang, L.C.; Shen, Z.Q. & Xu, J. (2005a). Mechanically milling-induced amorphization in Sn-containing Ti-based multicomponent alloy systems. *Materials Science and Engineering A,* Vol. 394, No. 1-2, (March 2005), pp. 204-209, ISSN 0921-5093

Zhang, L.C.; Shen, Z.Q. & Xu, J. (2005b). Thermal stability of mechanically alloyed boride/Ti50Cu18Ni22Al4Sn6 glassy alloy composites. *Journal of Non-Crystalline Solids,* Vol. 351, No. 27-29, (August 2005), pp. 2277-2286, ISSN 0022-3093

Zhang, L.C.; Xu, J. & Eckert, J. (2006a). Thermal stability and crystallization kinetics of mechanically alloyed TiC/Ti-based metallic glass matrix composite. *Journal of Applied Physics,* Vol. 100, No. 3, (August 2006), pp. 033514, ISSN 0021-8979

Zhang, L.C.; Xu, J. & Ma, E. (2006b). Consolidation and properties of ball-milled Ti50Cu18Ni22Al4Sn6 glassy alloy by equal channel angular extrusion. *Materials Science and Engineering A,* Vol. 434, No. 1-2, (October 2006), pp. 280-288, ISSN 0921-5093

Zhang, L.C.; Calin, M.; Branzei, M.; Schultz, L. & Eckert, J. (2007a). Phase stability and consolidation of glassy/nanostructured Al85Ni9Nd4Co2 alloys. *Journal of Materials Research,* Vol. 22, No. 5, (May 2007), pp. 1145-1155, ISSN 0884-2914

Zhang, L.C.; Kim, K.B.; Yu, P.; Zhang, W.Y.; Kunz, U. & Eckert, J. (2007b). Amorphization in mechanically alloyed (Ti, Zr, Nb)-(Cu, Ni)-Al equiatomic alloys. *Journal of Alloys and Compounds,* Vol. 428, No. 1-2, (January 2007), pp. 157-163, ISSN 0925-8388

Zhang, T.; Inoue, A. & Masumoto, T. (1991). Amorphous Zr-Al-TM (TM = Co, Ni, Cu) alloys with significant supercooled liquid region of over 100 K. *Materials Transactions JIM,* Vol. 32, No. 11, (November 1991), pp. 1005-1010, ISSN 0916-1821

Zhong, Z.C.; Jiang, X.Y. & Greer, A.L. (1997). Nanocrystallization in Al-based amorphous alloys. *Philosophical Magazine B,* Vol. 76, No. 4, (October 1997), pp. 505-510, ISSN 0141-8637

Zhuang, Y.X.; Jiang, J.Z.; Zhou, T.J.; Rasmussen, H.; Gerward, L.; Mezouar, M.; Crichton, W. & Inoue, A. (2000). Pressure effects on Al89La6Ni5 amorphous alloy crystallization. *Applied Physics Letters,* Vol. 77, No. 25, (December 2000), pp. 4133-4135, ISSN 0003-6951

Section 3

Crystallization of Nanomaterials

9

Influence of Crystallization on the Properties of SnO_2 Thin Films

Daniya M. Mukhamedshina and Nurzhan B. Beisenkhanov
Institute of Physics and Technology, Kazakhstan

1. Introduction

The interest in the surface structures with their special properties has increased considerably due to extensive applications in micro- and optoelectronics. It is known that the properties of films of submicron size can be different from those of structures having macroscopic dimensions. The parameters that change the properties of films, are the thickness, number of layers, uniformity of the films, the size of clusters and nanocrystals. The presence of small particles and nano-sized elements leads to changes in material properties such as electrical conductivity, refractive index, band gap, magnetic properties, strength, and others (Suzdalev, 2006; Kobayashi, 2005).

One of the most promising materials in this regard is tin dioxide. Such advantages as high transparency in a visible range of wavelengths and high conductivity make SnO_2 very suitable as transparent conductive electrodes in such devices as solar sells, flat panel displays, etc (Rembeza et al., 2001; Jarzebski et al., 1976; Das and Banerjee, 1987; Song, 1999). Wide-gap semiconductor SnO_x films exhibit quantum confinement effect with decreasing of crystallite sizes, i.e. the band gap becomes size dependent and is increased from 3.6 to 4.2 eV.

Semiconductor gas sensors on the base of nanoscale SnO_2 films are manufactured (Evdokimov et al., 1983; Buturlin et al., 1983b; Watson et al., 1993). The possibility of tin dioxide layers to change their electrical conductivity upon adsorption of gases due to the reactions of reduction and oxidation, is used (Bakin et al., 1997; Srivastava, R. et al., 1998a; Jiang et.al., 2002; Vigleb, 1989). An increase in adsorption possibilities of SnO_2 films during the transition from single-crystal to nanocrystalline system (Srivastava, R. et al., 1998b) is one of the main directions of work to improve the sensitivity and reduce of response time (Bakin et al., 1997; Ramamoorthy et al., 2003; Karapatnitski et al., 2000; Xu et al., 1991; McDonagh et al., 2002).

As is known, tin dioxide (SnO_2) is a crystal of white color, the density is 7.0096 g/cm^3, melting point is about 2000°C (Knuniants, 1964). The SnO_2 films have predominantly amorphous or polycrystalline structure with a tetragonal lattice of rutile with parameters a = b = 0.4737 nm, c = 0.3185 nm, with two tin atoms and four oxygen atoms in the unit cell (Dibbern et al., 1986; Shanthi et al., 1981; Weigtens et al., 1991). Depending on the method of film synthesis, the band gap varies in the range 3.35–4.3 eV and the refractive index - in the range 1.8–2.0 (Martin et al., 1986; Melsheimer et al., 1986; Nagamoto Takao et al., 1990).

Under normal conditions in air, at the surface of tin dioxide, a layer of the adsorbed oxygen molecules is formed. Oxygen molecules carry out the electron trapping from the surface layer (Kaur et al., 2007). In the simpler schematization, in clean air the conductivity of SnO_2 is low because the conduction electrons are bound to surface oxygen, whereas in the presence of a reducing gas, electrons are no longer bound to the surface states and the conductivity increases. The transfer of electrons from SnO_2 to oxygen leads to the creation near the surface of a crystal a space charge layer (depletion layer), the concentration of electrons in which is less than in bulk. The net charge at the surface generates an electric field, which causes a bending of the energy bands in the SnO_2. A negative surface charge bends the bands upward, i.e. pushes the Fermi level into the band gap of the SnO_2, effectively reducing the charge carrier concentration and resulting in an electron depletion zone. In other words, this is an electron trapping process at the surface, which leads to an increase in the electrical resistance (or decrease in conductivity) of the surface layer. When a reducing gas, e.g. CO (or H_2, CO, H_2S, NH_3, etc.), reacts with the adsorbed oxygen to form CO_2, the concentration of oxygen in layer is decreased. The electrons released in this reaction are injected into the conduction band of the SnO_2, which results in a decrease in the electrical resistance (or increase in conductivity) of materials of n-type of conductivity (Kaur et al., 2007; Bakin et al., 1997; Andryeeva et al., 1993). According to barrier model, reaction of adsorbed oxygen species with reducing gases decreases the potential barrier heights, resulting in a huge decrease in resistance.

Reduced tin dioxide is characterized by a deficiency of oxygen – $SnO_{2-\delta}$, where $10^{-5} < \delta < 10^{-3}$ - deviation from stoichiometry (Rumyantseva et al., 2003; Kaur et al., 2007). Vacancies of ionized oxygen are the main intrinsic defects and define the electrical properties of the material - n-type of conductivity and free carrier concentration. Energy levels of oxygen vacancies V_O^+ and V_O^{2+} lie at a depth of 30–40 and 140–150 meV below the conduction band edge, respectively (Ryabtsev et al., 2008). The concentration of oxygen vacancies can be reduced by annealing of the material in an oxygen atmosphere.

In the transition from single crystal to a polycrystalline system, each crystal grain is considered as a closed volume, near the surface of which is located a depletion layer. Reducing of the charge carrier concentration at the grain boundaries leads to a formation of intergranular energy barriers, the magnitudes of which determines the electrical conductivity of polycrystalline material as a whole. The greatest influence of the surface state on the electrical properties of the material occurs when the condition: $l/2 \le L$, where l is the size of the crystallites; L – length of the depletion layer, which for different oxides varies from 3 to 10 nm: $L = L_D\sqrt{\frac{eV_s^2}{kT}} = \sqrt{\frac{\varepsilon kT}{2\pi e^2 N_d}} \cdot \sqrt{\frac{eV_s^2}{kT}} = \sqrt{\frac{\varepsilon V_s^2}{2\pi e N_d}}$, where ε is the dielectric constant, V_S - surface/Schottky potential, N_d is the concentration of the donor impurity, L_D is the Debye length, k is the Boltzmann constant, T is the temperature. Wide range of sizes of the SnO_2 crystallites (from several to several hundred nanometers) affects the mobility of charge carriers, which may vary from a few to 100 $cm^3/V \cdot s$. For example, Xu Chaonan et al. (1991) investigated the dependence of gas sensitivity of film on the diameter D of SnO_2 crystallites (in the range 5-32 nm) and the Debye length L_D. For undoped SnO_2 D increased from 4 nm with increasing annealing temperature to 13 nm at 400°C and 27 nm at 900°C. The dependences of the electrical resistance of the films in air (R_a) and a mixture of hydrogen-air (R_g) at 300°C on the crystallite sizes were presented. The values of R_a and R_g increase with decreasing D for D ≤ 6 nm, as well as increase with increasing D for D ≥ 10 nm. The value of D at the boundary between two regions is equal to 2L and the optimum value for L is obtained by 3 nm. The gas sensitivity of film was determined as R_a/R_g. For the both H_2 and CO the sensitivity increases if the value D ≤ 2L (~6 нм). It follows, that the gas-sensitive properties of the tin dioxide films are determined by their nanostructure, formed during crystallization process. This makes it extremely important to study the processes of crystallization in tin dioxide films, depending on the conditions of their preparation and processing.

Good quality SnO_2 films have been grown by several techniques (Buturlin et al., 1983a). As is known, the method of preparation of SnO_x films significantly influences on their characteristics (Bakin et al., 1997; Srivastava R. et al, 1998; Jiang et al., 2002; Srivastava R. et al., 1998b; Minami et al, 1989). Not all techniques of deposition are used for production of functional metal oxide films (Buturlin et al., 1983b; Jarzebski, 1982; Khol'kin and Patrusheva, 2006). The physical methods of obtaining films of metal oxides include the method of thermal evaporation of metal in vacuum with deposition on the dielectric substrate and oxidation in air (Buturlin et al., 1983b) at 720–770K. The particles of deposited material one can obtain also by evaporation in a gas. In an atmosphere of reactive gas material is heated to evaporation, then the atoms of substance reacts with residual gas atoms, resulting in formation of ultrafine particles of size in range 1–100 nm. The resulting film has an extremely developed surface and requires no further heat treatment (Buturlin et al., 1983a). During thermal sputtering of an oxide target (Das and Banerjee, 1987; Feng et al., 1979) an annealing is not required, too. The method is based on the sputtering of target made of pressed powder of desired oxide (Jarzebski, 1982) or in plasma at high frequency in a three-electrode circuit, or by an ion beam. The spraying process is carried out in an plasma of argon of high purity with the addition of oxygen. Das and Banerjee (1987) show the relation between a structure and other physical properties of the films. For the deposited by electron-beam evaporation SnO_x films a change of x value by varying the substrate temperature T_S was revealed. The films deposited at T_S = 150°C were amorphous, but after annealing in air (T_A = 550°C, 2h) the presence of SnO_2 by X-ray diffraction was revealed. Films deposited at T_S = 225°C were also amorphous, but after annealing under identical conditions consisted of a mixture of phases SnO and β-SnO. These amorphous films are characterized by high electrical resistivity (10^{10} ohm·cm at room temperature), which sharply decreased after annealing in air (550°C, 2 h) to the values ρ = 9,5×10^{-2} ohm·cm at room temperature, seemingly, due to the formation of 0.015 eV donor level (T_S = 150°C) and 0.26 eV (T_S = 250°C) below the bottom of the conduction band of tin oxide. For T_S = 150°C the band gap Eg = 3.46 eV corresponds to SnO_2 and decreases with increasing T_S up to 225°C due to the presence of different oxide phases, and increases again at T_S>225°C. Annealing in air (550°C, 2h) leads to deterioration of the transmission T in the wavelength range of 400–850 nm for the films deposited at T_S = 150°C and to the increase of transmission for films deposited at T_S ≥ 250°C. At T_A = 650°C the full transformation into the SnO_2 phase with the dominant lines (101) and (200) is completed. The increase of the substrate temperature (T_S = 350°C) causes a decrease in the resistance of the film, indicating that activation occurred because of donors at a depth of 0.26 eV.

In the method of ion-stimulated deposition (Song, 1999) flow of neutral Sn atoms was obtained in a high vacuum using partially ionizing radiation source, the flow of oxygen ions with energies in the range 0-1000 eV was obtained using ion gun with hollow cathode. The properties of SnO_2 films on substrates of glass, amorphous SiO_2/Si and Si (100) were studied. In films deposited on glass, with increasing of the energy of the ion flux both the surface grain size up to 7–10 nm and roughness are increased. The increase of energy of the oxygen ions led to the increase of crystal perfectness, stoichiometry, porosity of the films and their transmission in the shortwave region. All tin oxide films on the glass after deposition had very low optical absorption in the visible wavelength region and high absorption in the shortwave region.

Method for production of SnO_2 films by cathode sputtering of tin provides the creation of plasma in an inert gas (argon) at a pressure in the vacuum chamber from 50 to 10^{-2} Pa. A target is fixed to the cathode , substrate – to the anode. When the diode sputtering is used, between the anode and cathode DC voltage is applied, the growth rate of SnO_2 films is small (~ 1.7 nm/min) and only conductive targets can be used. In the high-frequency sputtering, between the cathode and the anode an alternating RF voltage is applied. As a result, high-speed deposition of the layer up to 800 nm/min is provided, depending on the voltage and the partial pressure of O_2 and Ar. One can use a non-conductive target. Conductivity of SnO_2 films varies in the range 10^{-9}–10^{2} $ohm^{-1}\cdot cm^{-1}$.

One of the most promising physical methods of production of the oxide films is the reactive magnetron sputtering of a metal target (Geoffroy et al., 1991; Kisin et al., 1999; Lewin et al., 1986; Semancik and Cavicchi , 1991). As a result of the deposition process, a mixture of two phases - SnO and SnO_2 - is formed. By changing the parameters of deposition (substrate temperature, chamber pressure, the rate of film growth) both amorphous and nanocrystalline films can be prepared. For example, Semancik S. and Cavicchi R.E. (1991) by this method have deposited the SnO_2 films on substrates of sapphire and TiO_2. It is shown that the use of substrates with perfect structure and an increase in their temperature up to 500°C leads to the epitaxial growth of the film and the formation of a single crystal structure. Gas sensitivity of the films was close to sensors based on bulk single-crystal SnO_2. Using a VUP-4 installation, Rembeza et al. (2001) have deposited polycrystalline SnO_2 films with thickness of 1–5 microns on glass substrate by magnetron sputtering of tin target doped by antimony (3% vol.) in a gas mixture of Ar (25%) and O_2 (75%). In the films has been revealed only a well-crystallized tetragonal phase of SnO_2 with average grain size 11–19 nm after annealing. After annealing of the films (600°C, 4h) the concentration of electrons decreases with the increase of temperature up to 400°C in the range ~ 2×10^{18} - 3×10^{17} cm^{-3}, and the mobility of free charge carriers increases from 70 to 150 $cm^2/(V\cdot s)$ at 130°C and further up to 400°C does not change. Stjerna and Granqvist (1990) studied the optical and electrical properties of SnO_x films, obtained by magnetron sputtering of tin in a mixture of argon and oxygen. The optical properties show a transparency window in the wavelength range of ~ 0.4–2 microns. It is noted that the behaviors of the curves of transmission and reflection strongly depend on the ratio of values of partial pressure of O_2 and Ar during the growth of the films, and is particularly strong changed near ratio of ~ 4%. Near this point (from 4.1 to 4.2% vol. O_2) a sharp minimum of the surface resistance is observed. If the power discharge is increased from 10 to 60 watts, the position of minimum is shifted toward higher concentrations of oxygen - up to 11.6–11.8%. Kissin et al. (2000) gave a description of dependence of the concentration of free electrons on the partial pressure of oxygen for the film deposited by rf magnetron sputtering of target, prepared from powder of SnO_2 doped with CuO, followed by a two-hour annealing in a flowing oxygen at a temperature of 700°C. Polycrystalline SnO_2 films with an average grain size of 100 nm and a thickness of 1 micron were monophasic and had a texture in the (110). In the range of partial pressure of oxygen in the chamber from 0.013 to 1.3 Pa a power dependence of the conductivity of the films on the pressure with an exponent of 0.6–0.8 is observed. Rumyantseva et al. (2001) showed that the effect of porosity on gas sensitivity of the film is more significant than the influence of the partial pressure of oxygen. Saturation by vacancies and the formation of pores in the gas-sensitive layers was carried out by doping the initial loading of tin by volatile impurities - iodine and tellurium. The porosity of the material with a grain size comparable with the wavelength of the radiation manifests itself in the form of reducing the magnitude of the refractive index, averaged over the volume.

Chemical methods of preparation of SnO_2 films can be divided as obtained from the gas phase and from solution. Hydrolysis of chemical compounds in the gas phase occurs at relatively low temperatures. The chlorides and other compounds easily hydrolyzed in the vapor phase and are deposited as thin films of hydroxides (Suikovskaya, 1971). Obtaining of the corresponding oxides is carried out by annealing or by deposition on the heated substrate. Depending on the temperature of the substrate, the amorphous or polycrystalline films can be produced (Melsheimer and Teshe, 1986). In chemical vapor deposition (Kern and Ban, 1978), the laminar flow of an inert carrier gas completely covers the substrate in the reaction chamber, forming over it motionless boundary layer. The reagents are fed as a vapor-gas mixture, the deposition rate of which depends on the speed of their diffusion through the boundary layer. Growth of the SnO_2 films (Kim and Chun, 1986) can occur in several modes. The optimal regime is the nucleation on the substrate surface and their growth by surface diffusion of the incoming material. This leads to the texturing of the film due to the formation of oriented nuclei on the surface.

To obtain homogeneous films of SnO_2 from solutions one must take into account the properties of precursors and solvents. As a precursor the tin chlorides are often used. Suitable solvents include ethanol and acetone containing a small amount of water (Suikovskaya, 1971). Aerosol methods of producing of tin oxide films are based on the decomposition at high temperature (720–970 K) of chloride or organometallic tin compounds on the surface of the substrate. At spraying under the influence of the inert gas, the solution was transferred to the substrate surface in the form of small droplets. The deposition rate of films is 100 nm/min. Dimitrov et.al. (1999) by the hydrothermal method (t_{sint} = 130–250°C, τ_{sint} = 2–6 h) synthesized nanocrystalline samples of SnO_2. Crystallite size of SnO_2, synthesized by hydrothermal treatment of amorphous gel of tin acid is 4–5 nm and is virtually independent of temperature and duration of treatment.

An actively developing chemical method of producing of SnO_2 thin films is a sol-gel method. It is based on transfer of a substance in the sol and gel states. The method allows obtaining the multilayer films with a thickness of one layer of 2-3 nm, and it does not require a vacuum and complex installations. Deposition of the film can occur on a cold substrate. The main methods of film deposition are dip-coating, centrifugation (spin-on-films) and spray pyrolysis (Kim and Chun, 1986; Aranowich et al.,1979; Sanz Maudes and

Rodriguez, 1980; Fantini and Torriani, 1986; Jitianu et al., 2003; Chatelon et al., 1997). It is one of the simplest and economic among known techniques. The advantages of the sol-gel process are the possibility of doping and its potential to fabricate films of large area substrates. The flexibility of the technological cycle management provides opportunities to obtain the necessary surface morphology and sizes of structural formations. One can obtain the films with a high degree of homogeneity and controlled stoichiometry by sol-gel method at a sufficiently low temperature of synthesis . Often (Suykovskaya NV, 1971; Zang and Liu, 1999) salt volatile acids, such as chlorides, as film-forming substances are used. As a solvent mainly ethanol is used. Chlorine can be included in the formed SnO_2 film or by replacing the oxygen, or by interstitial mechanism (Bosnell and Waghorne, 1973; Aboaf et al., 1973). The admixture of chlorine causes an increase in carrier concentration, leads to the stabilization of the parameters of the oxide layer, to a low density of surface states, and to high values of the lifetime of the major carriers. Growth of SnO_2 crystallite size after annealing is noted by many authors (Asakuma et al., 2003; Brito et al., 1997). However, the introduction of additional components in the composition of the film-forming mixture (Torkhov et al., 2003) leads to a decrease in the rate of growth of crystallites of the both SnO_2 and substance injected with increasing temperature. Torkhov et al. (2003) synthesized the nanocrystalline SnO_2 and WO_3 and nanocomposite Sn:W = 1:9, 1:1, 9:1 by α-$SnO_2 \cdot n \cdot H_2O$ and $WO_3 \cdot H_2O$ (series X) gel deposition and krizol-method (series K). For nanocomposites of Sn (X, K) and Sn_9W_1 (X, K) after annealing at T ≥ 150°C a phase of tin dioxide SnO_2 (cassiterite) was detected, the degree of crystallinity of which increases with increasing temperature. An increase in the crystallite size of SnO_2 (2–15 nm) and WO_3 (10–50 nm) is also observed. For SnO_2 films the change of grain size in range from 10 to 6 nm is accompanied by increase of resistance from 2×10^4 up to 6×10^4 ohms. Anishchik et al. (1995) annealed the films obtained by centrifugation from an aqueous solution at temperatures of 400 or 500°C for 30 minutes with further rapid cooling in air ("hardening of soft"). In SnO_2 films, subjected to cooling from 400°C, the SnO phase also found. Rapid cooling from 500°C leads to an appearance of Sn_3O_4 phase, while at slow cooling this phase is not detected. Rapid cooling of the samples from temperatures of 400 and 500°C leads to "freezing" on the surface of films of elemental tin and its oxides, as they do not have time to oxidize up to SnO_2.

The films to ensure maximum sensitivity are heated during operation of sensors. The temperature effect causes an increase in the crystallite sizes of SnO_2. On the one hand, there is a reduction of the effective activation energy of the interaction of SnO_2 with oxygen, and, on the other hand, increasing the concentration of electrons with sufficient energy to overcome the barrier created by the negatively charged surface (Rumyantseva et al., 2008). The degree of surface coverage by chemisorbed oxygen $O^-_{(ads)}$, $O^{2-}_{(ads)}$ is increased. The films with a fixed size of the crystallites SnO_2 are required to ensure a long-term operation of the sensors. Kukuev and Popov (1989) used a SnO_2 film in the manufacture of microelectronics devices in installations for heating of ultrapure deionized water, as well as in commercially available installations for infrared heat treatment of photo resist as heating elements. The most suitable as heaters are the films with grain size ~ 0.2-0.3 μm, in which such processes as crystallization, condensation and oxidation do not occur. Adamyan et.al. (2006) obtained high-quality films using sodium stannate as a precursor. 1 M of H_3PO_4 was added into Na_2SnO_3 solution with constant stirring to neutralize the solution (pH = 7). The transparent, stable sol, indicating the small size of its particles and the stability of their surface, was obtained.

One of the ways to improve the selectivity of sensors based on SnO_2 and increase the contribution of molecules of this type in the gas phase to the total electrical signal is the introduction of alloying elements into highly dispersed oxide matrix (Rumyantseva et al., 2003; Fantini and Torriani, 1986; Okunara et al., 1983 ; Bestaev et al., 1998). Dopants are usually divided into two groups: catalytic (Pt, Pb, Ru, Rh) and electroactive (In, Sb, Cu, Ni, Mn). Of considerable interest is also the effect of processing by plasma on the properties of films. Minami et al. (1989) placed the transparent conductive films of tin oxide (TCO) in a quartz tube filled with hydrogen gas. Films at high temperature treatment are completely painted in dark gray. Transmittance of ITO indium doped and undoped TO films decreased rapidly during annealing above 300°C for 30 minutes. Decrease the transparency and color of the films are attributed to the formation of the oxygen-depleted surface level due to chemical reduction of these films by hydrogen. The upper surface of the films passed into the metallic state, and then indium and tin were thermally evaporated at a higher temperature. Hydrogen glow discharge plasma was obtained at a pressure of 400 Pa in the installation with a capacitive circuit (frequency of 13.56 MHz or 2.45 GHz) and power of 300 watts. Transmittance and film thickness are greatly reduced when exposed by hydrogen plasma at 250°C. ECR hydrogen plasma was obtained in the volumetric resonator cavity at a pressure of 6.5×10^{-2} Pa, microwave power (2.45 GHz) 300 watts when a magnetic field was 8.75×10^{-2} T. Indium containing (ITO) film and undoped (TO) SnO_2 were stained even when treated for 2 and 5 minutes.

An analysis of published data shows that in papers a great attention is paid to influence of the structure of the synthesized tin dioxide films on their sensitivity to different gases and transparency in a wide range of wavelengths. Usage of different methods of preparation, as well as modification of the films by different types of treatments leads to a change of phase composition and microstructure of the films causing a change in their optical, electrical and gas-sensitive properties. It is of considerable interest to study the effect of the concentration component, nanoclusters, the composition of films and their heat or plasma treatment on the crystallization processes and clustering, the size of the nanocrystals and, consequently, the physical properties of the films.

In this paper we consider the influence of composition and structure on the optical and electric characteristics of SnO_x thin films deposited on glass substrates by sol-gel technique. For comparison, the data for films prepared by magnetron sputtering, are presented. Using various techniques the effect of both thermal and plasma treatments on the structure and properties of the SnO_x films were studied.

2. Experimental

A solution to obtain the nanostructured films by sol-gel technique (the method of spreading) was prepared from crystalline hydrate of tin tetrachloride ($SnCl_4{\cdot}5H_2O$) by dissolution in pure ethanol. SnO_2 sol of desired concentration was obtained under strong stirring to obtain a colorless and transparent solution. Freshly prepared solution has a neutral pH value of ~ 7. After maturation of the solution, which lasted more than six hours, pH was equal about 0.88, indicating the release of HCl during the dissolution of $SnCl_4$. To study the effect of concentration of components in the solution on the film properties were obtained the solutions with a concentration of tin atoms: 0.82, 0.41, 0.30 and 0.14 mol/L. Aliquots of these solutions with volumes of 0.04 ml, 0.08 ml, 0.11 ml, 0.23 ml, respectively, using micropipettes were deposited on the cleaned surface of the microscope glass slides. It was assumed that the number of tin atoms in the films will be identical (~ 3.25×10^{-5} mole), and the film thickness is ~ 350 nm. After the deposition, the films were dried for 1 hour at 80ºC. Then the samples were annealed at 100, 200 or 400ºC for 1 hour. The thickness of the films, estimated from the change in mass of the sample, was 360±40 nm.

The SnO_2 films with thickness of 300 nm were also fabricated by method of centrifugation. A solution was obtained by dissolving of anhydrous $SnCl_4$ in 97% ethanol. The solution was deposited onto a glass substrate located on a table of centrifuge rotor, the rotation speed of which was ~ 3800 rpm. Centrifugation time was 3–5 s. Substrate with a layer was dried using an infrared radiation at 80°C for 3–5 minutes, and then in a muffle furnace at 400°C for 15 minutes. After cooling, the cycle repeats. Number of deposited SnO_2 layers ranged from 12 to 15.

The SnO_2 films with thickness of 300 nm were also deposited on cleaned microscopy glass slides by magnetron sputtering. The magnetron sputtering mode parameters were: cathode voltage U_c = 470 V, the ion beam current I_{ion} = 35 mA, the argon-oxygen mixture pressure inside the chamber ~ 1–2.7 Pa, the oxygen concentration in the Ar-O_2 mixture ~10%, deposition rate of films ~ 0.05 nm/c, the temperature of the substrate ~ 200°C.

The SnO_2 film's structure was investigated by X-ray diffraction using a narrow collimated (0.05×1.5 mm^2) monochromatic (CuK_{α}) X-ray beam directed at an angle of 5º to the sample surface. The average crystallite size was estimated from the width of X-rays lines by Jones method. The surface of the layers was analyzed by Atomic force microscopy (JSPM 5200, Jeol, Japan) using AFM AC technique. The optical transmittance spectra of SnO_x films were measured in the wavelength range from 190 to 1100 nm by means of the SF-256 UVI and from 1100 to 2500 nm by means of the SF-256 NIR spectrophotometers (LOMO, Russia). Electrical resistance of the films was measured by a four probe technique at room temperature. For the measurements of electrical characteristics and parameters of gas sensitivity of the thin films in a wide temperature range was used a specialized experimental setup.

The glow discharge hydrogen plasma was generated at a pressure of 6.5 Pa with a capacitive coupled radio frequency (r.f.) power (27.12 MHz) of about 12.5 W. The temperature of processing did not exceed 100°C. The processing time was 5 min.

3. Results and discussion

3.1 Physical properties of tin oxide thin films obtained by sol-gel technique (the method of spreading)

Fig. 1 shows the transmission spectra of SnO_x films deposited on a glass substrate by spreading method. Transparency of films lies in the range of 80-90% and increases over the entire range of wavelengths with the increase of temperature up to 400°C.

X-ray diffraction studies of the SnO_x films, obtained at various concentrations of tin ions in the colloidal solutions, led to the following results.

a. In the case of high tin concentration 0.83 mol/L (~ 0.04 ml aliquot) immediately after deposition and drying at 80°C, the film structure is similar to the amorphous (Fig. 2a) and remains practically unchanged after annealing for 1 hour at 100°C (Fig. 2b). Separation of the broad band into two SnO_2 lines, indicating the formation of crystallites, visually observed only after annealing at 200°C (Fig. 2c). Crystallite size in the planes (110) and (101) of SnO_2 is ~ 1.5 nm, i.e. the crystallites are small and have an imperfect structure.

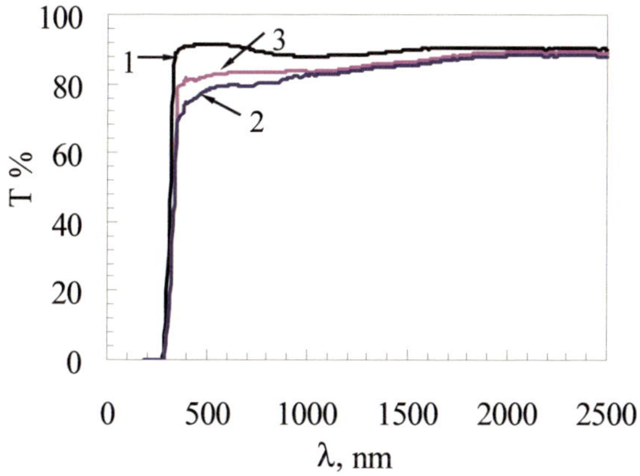

Fig. 1. The transmission spectra of tin oxide film synthesized by sol-gel technique: 1 – glass substrate, 2 – SnO_2 film after annealing at 100°C, 3 – SnO_2 film after annealing at 400°C.

b. A similar pattern after the deposition and annealing at 100°C is observed for films of SnO_2, produced at lower tin concentrations of 0.41 mol/L (~ 0.08 ml aliquot) and 0.30 mol/L (~ 0.11 ml aliquot). After annealing at 200°C the differences are appeared, which manifest themselves in increasing intensities of SnO_2 lines with decreasing of tin concentration in the colloidal solution.

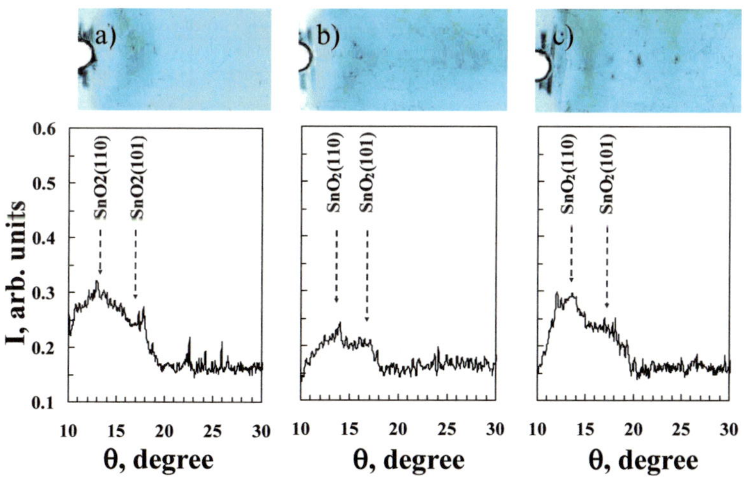

Fig. 2. X-ray diffraction patterns and intensity curves for the SnO_x films on glass substrates obtained by the sol-gel technique (0.04 ml solution with tin concentration of 0.83 mol/L, d = 320 nm) after: a) deposition and drying at a temperature of 80°C, and b) annealing at 100°C; c) annealing at 200°C.

c. In case of the minimum concentration of 0.14 mol/L (~ 0.23 ml aliquot), the formation of a SnO_2 polycrystalline film (Fig. 3a) with a crystallite size of 2.5–3.5 nm (Table 1) was observed immediately after deposition and drying. Annealing at 400°C led to the improvement of the crystallite perfection (4 lines) and an increase in their size up to ~ 5 nm (Fig. 3b).

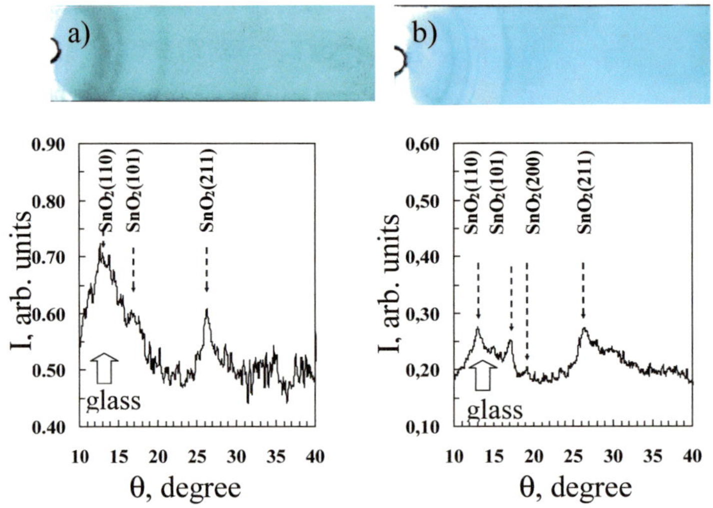

Fig. 3. X-ray diffraction patterns and intensity curves for thin SnO_x films on glass substrates obtained by the sol-gel technique (0.23 ml solution with tin concentration of 0.14 mol/L, d = 400 nm): a) after deposition and drying at a temperature of 100°C, b) after annealing at 400°C.

SnO_2 plane	Crystallite size after annealing, nm	
	100°C	400°C
SnO_2 (110)	2.5	5.5
SnO_2 (101)	3.0	5.0
SnO_2 (211)	3.5	3.0

Table 1. Results of measurements of SnO_2 crystallite sizes by X-ray diffraction

Thus, the SnO_2 polycrystalline film immediately after deposition and drying without additional annealing at elevated temperatures by the sol-gel technique was derived. This may be due to better conditions for the processes of SnO_2 crystallization and evaporation of HCl in the case of a lower concentration of tin atoms (0.14 mol/L) in solution. Additional annealing at 400°C leads to the formation of SnO_2 crystallite with size of about 5 nm, ie close to optimal (6 nm) for high gas-sensitive films (Xu et al., 1991).

Gas sensitivity was determined from the expression $\gamma = \frac{R_0 - R_g}{R_0} \cdot 100\% = \frac{\Delta R}{R_0} \cdot 100\%$, where R_0 is the resistance of gas-sensitive layer in the clean air, R_g is the resistance of the layer in the mixture of air with detectable reducing gas. Fig.4 shows the temperature dependence of the sensitivity of SnO_2 film for a given concentration of ethanol vapor in the atmosphere. Since the temperature range of sensitivity to different gases is different, it is necessary to determine the temperature at which a maximum sensitivity of the film to the test gas is observed.

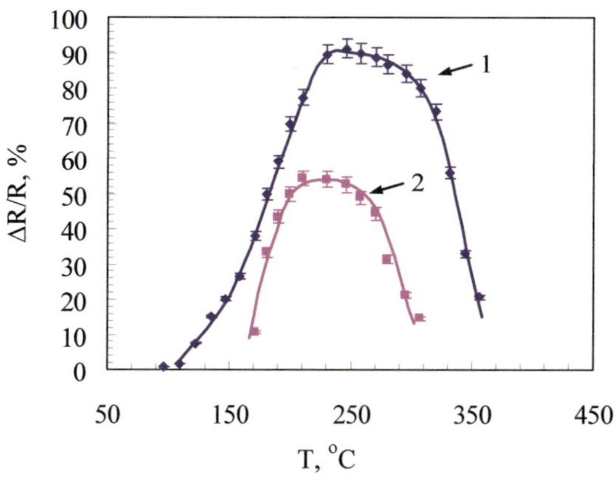

Fig. 4. Temperature dependence of the sensitivity of the SnO_2 films (obtained from a colloidal solution with concentration of tin ions 0.14 mol/L) to the ethanol vapors of various concentration: 1 – 15 mg/L of ethanol in the air; 2 – 0.7 mg/L of ethanol in the air.

The greatest sensitivity of films to ethanol is observed at a temperature of 230°C. Adsorbed oxygen ions create a space charge region near the surface of SnO_2 grains by extracting electrons from the material. Ethanol, being by nature a reducing gas, reacts with adsorbed O^- ions and removes them from the surface of the grains, re-injecting electrons back into the material and thus lowering the resistance of the film. Peak sensitivity at 230°C shows that at this temperature, the amount of chemisorbed oxygen ions, which react with the molecules of ethanol, is maximal.

In Fig. 5 the sensitivity of the SnO_2 films at 230°C as a function of the concentration of ethanol vapor is presented. Figs.4 and 5 show that sensitivity increases significantly with increasing of ethanol concentration. At low concentrations of ethanol vapor (0.1–1 mg/L), the dependence of the film sensitivity on the concentration is linear (Fig. 5), as is likely, there is a sufficient number of oxygen ions that react with molecules of ethanol. Linearity of sensitivity at low concentrations of ethanol can be used to create sensors basing on this film, which are sensitive to ethanol vapor in exhaled human. A visible sensitivity of the film is evident even when the concentration of ethanol vapor is about 0.05 mg/L (25 ppm).

The films obtained by deposition of a colloidal solution with a concentration of tin atoms 0.83 mol/L (Fig. 5, curve 1), have higher gas sensitivity in comparison with films deposited from a solution with a concentration of tin atoms 0.14 mol/L (Fig. 5, curve 4). This may be due to the cluster structure of the film synthesized from a solution with a concentration of tin atoms 0.83 mol/L. Indeed, the average size of the crystallites in these films is much smaller (1.5 nm).

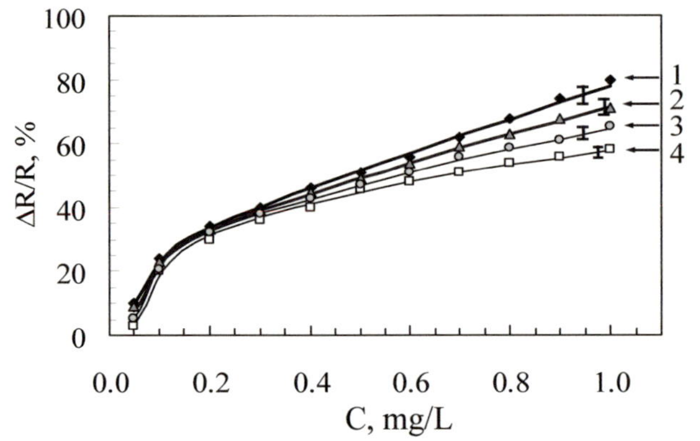

Fig. 5. The dependence of the sensitivity of the SnO_2 films on the concentration of ethanol vapor at 230°C: 1 – film synthesized from a solution with a concentration of tin ions 0.83 mol/L, 2 – 0.41 mol/L, 3 – 0.30 mol/L; 4 – 0.14 mol/L.

To measure the response time, the changes of resistance are recorded as a function of time beginning from the moment when ethanol vapors are introduced into the chamber. The concentration of ethanol wondered equal to 1 mg/L (500 ppm) at T = 230°C. As a response time of the sensor was taken the time duration for which the sensor resistance dropped by 90%. Response time and recovery time of resistance to the initial value for the film deposited from a colloidal solution with a concentration of tin atoms 0.83 mol/L (Fig. 6, curve 1) are ~ 3 and 30 seconds, respectively, in contrast to the second film (0.14 mol/L), where response time and recovery time were twice as much – ~ 6 and 65 seconds, respectively.

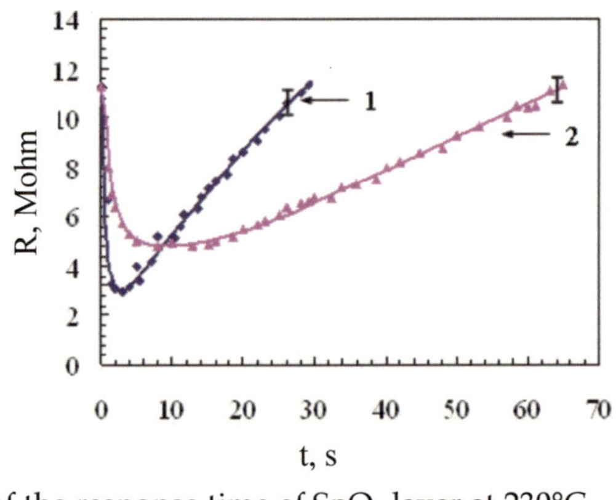

Fig. 6. The dependence of the response time of SnO_2 layer at 230°C (concentration of ethanol vapors 1 mg/L) on concentration of tin atoms in colloidal solution: 1 - 0.83 mol/L, 2 - 0.14 mol/L.

Figs 7a and 7b show the atomic force microscopy and scanning electron microscopy images, respectively, of the surface of tin oxide film, obtained by deposition of a colloidal solution with a tin concentration 0.83 mol/L on a glass substrate. Areas, protruding above the surface, have a light color, areas below the surface have a dark color (Fig.7a), and counting the height begins from the bottom point. In the process of drying, the SnO_2 film, obtained by the method of spreading, is transformed into a structure with an undulating surface, the wave amplitude of which is comparable to the thickness of the film. The heterogeneity of the thickness of the films may be the main reason for the lack of interference in the transmission spectra. Consequently, in order to obtain uniform thickness of the films with relatively smooth surface the centrifugation method was used, when under the centrifugal force during the sample rotation the alignment of the surface is reached.

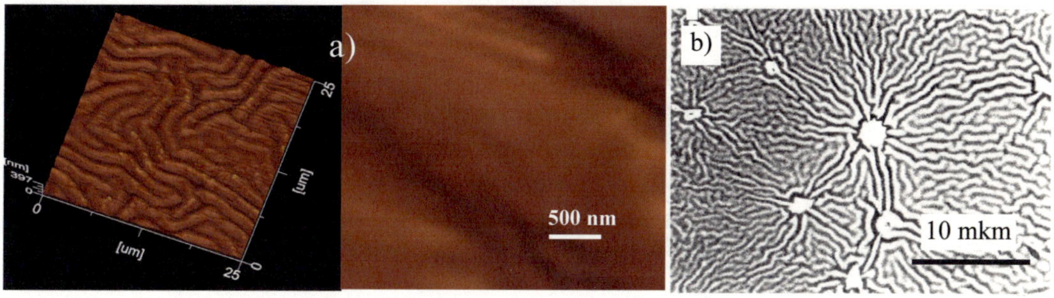

Fig. 7. Surface topography of tin dioxide film synthesized by sol-gel technique (a method of spreading, a colloidal solution with a tin concentration 0.83 mol/L): a) atomic force microscopy; b) scanning electron microscopy

3.2 Optical, electrical, structural and sensory properties of the SnO_x films, prepared by the sol-gel technique (centrifugation method)

The films were made by centrifugation in order to reduce their electrical resistance. Scheme of the synthesis and study of the SnO_2 films is shown in Figure 8. A solution of the desired tin concentration for producing of SnO_2 film with thickness ~ 300 nm was obtained by dissolving of anhydrous $SnCl_4$ in 97% ethanol. Kinematic viscosity of the solution was ~ 1.9 mm^2/s. The radius of sol particles in solution, determined by the turbidimetric method according to formula $r = \sqrt[3]{\frac{3V}{4\pi}}$, was 5.4 nm. The solution was deposited on a glass substrate, located on a table of centrifuge rotor. Rotational speed of the centrifuge was ~ 3800 rpm. Centrifugation time was 3–5 seconds. The deposited films were dried by an infrared emitter at 80°C for 3–5 minutes and also in a muffle furnace at 400°C for 15 minutes.

As shown above, the films obtained by spreading, in the case of high concentration of tin atoms (0.83 mol/L) had higher gas sensitivity and lower response time and recovery time compared to films with a low concentration of tin atoms (0.14 mol/L). However, the films obtained by centrifugation from solution with a high concentration of tin atoms (0.83 mol/L) were susceptible to detachment due to poor adhesion of the film and had a greater thickness. Therefore, preference was given to films obtained from solution with low concentration of tin atoms (0.14 mol/L). Thin layers were obtained. By subsequent deposition of additional layers the multilayered films were obtained with good adhesion to glass substrate, acceptable optical properties and resistance of about 200 ohms.

Characteristics of tin oxide films deposited by magnetron sputtering (pressure Ar-O_2 mixture in the chamber –1 Pa) and sol-gel technique were compared. It was found that the relative resistance of films prepared by sol-gel technique, much faster decrease with increasing temperature (Fig. 9), and the increase of their resistance associated with a decrease in carrier mobility due to scattering by impurities occurs at higher temperatures (> 270°C).

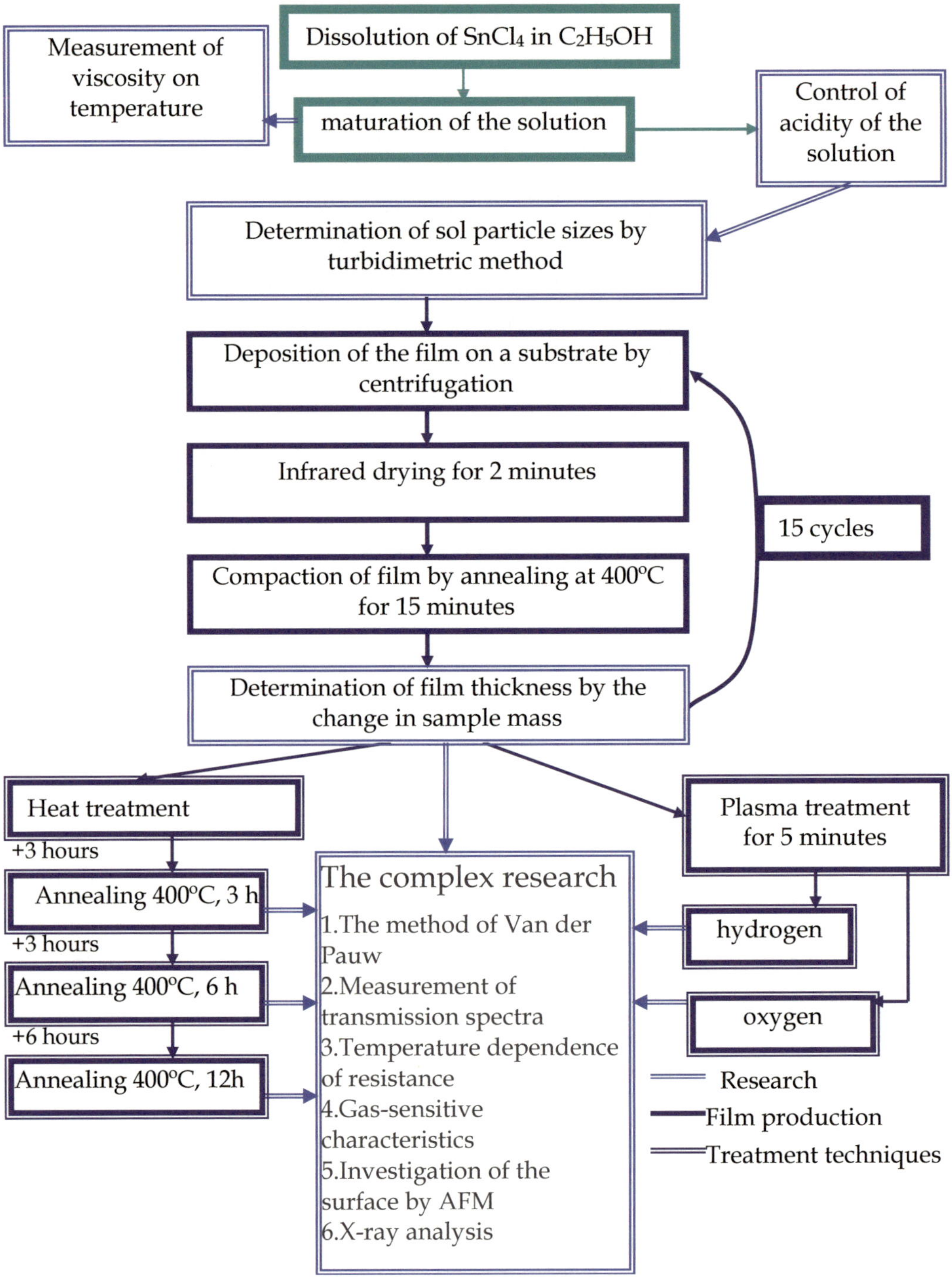

Fig. 8. Scheme of the synthesis and study of nanostructured SnO_2 films (sol-gel technology)

The absolute values of the resistance of films, prepared by magnetron sputtering, changed in the temperature range 20-230°C from 69 to 11 kohm, while for the sol-gel films from 192 to 2.3 kohm. Low resistance of films obtained by magnetron sputtering at room temperature may be due to the presence of particles of tin, whose resistance increases with increasing temperature. On the other hand, the alleged occurrence of dielectric inclusions of tin monoxide may be the cause of the increased resistance of these films at 230°C in comparison with the sol-gel films.

Fig. 10 shows the optical transmission spectra of films obtained in two ways. Transparency of films was about 90%. At long wavelengths there is a reduction of the transmission coefficient for the film SnO_x obtained by magnetron sputtering. The absence of a similar reduction for the films prepared by sol-gel technique indicates a lack of tin particles in films and, consequently, a better stoichiometry of the films of tin dioxide.

Table 2 shows the optical parameters determined from the transmission spectra by ways described in papers (Mishra et al., 2002; Ryzhikov et al., 2002). Films prepared by the sol-gel technique have high porosity, which is considered a positive factor contributing to increase of their gas sensitivity.

The method of film deposition	n, refractive index	D, film thickness, nm	E_g, band gap, eV	ρ, density, g/cm^3	k, absorption coefficient, 1/cm	V, porosity, %
Magnetron sputtering	1.81	282	4.05	5.76	1.67×10^3	17
The sol-gel technique	1.73	318	4.10	5.35	2.49×10^3	23

Table 2. Optical parameters of SnO_x film

SnO_x film prepared by magnetron sputtering (pressure of Ar-O_2 mixture in the chamber 1 Pa), after the deposition had an amorphous structure. Annealing in the temperature range 400–550°C led to the formation of polycrystalline phases of SnO_2, SnO, Sn_2O_3 (Fig. 11a). The results help to explain the slow decrease in the resistance of these films with increasing annealing temperature (Fig. 9) by the dielectric properties of the SnO crystallites. The SnO_x film obtained by sol-gel technology had a polycrystalline structure after deposition and short-term annealing at 400°C for 15 minutes, included in the process of drying the film. X-ray diffraction patterns and intensity curves of these films indicate the presence of only polycrystalline SnO_2 phase (Fig. 11b). The crystallites sizes were of the order of 3–5 nm (Table 3), that is more optimal to obtain the gas-sensitive films.

The method of film deposition	SnO_2 plane	Crystallite sizes, nm
Magnetron sputtering	SnO_2 (101)	11.4
	Sn_2O_3 (021)	8.5
	SnO (101)	10.0
The sol-gel technique	SnO_2 (110)	5.2
	SnO_2 (101)	5.1
	SnO_2 (211)	3.0

Table 3. Average crystallite sizes of SnO_2 films

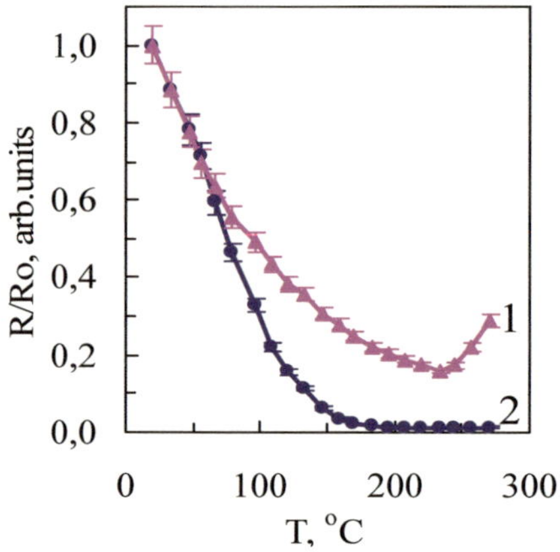

Fig. 9. Temperature dependences of resistivity of the SnO_x films: 1 – film deposited by magnetron sputtering, 2 – by sol-gel method.

Research of gas sensitivity of the deposited films was conducted in two stages. In the first stage, the temperature dependences of film resistance in both pure air, $R_0(T)$, and at a given concentration of the test gas, $R_g(T)$, were measured. It was determined the temperature at which a maximum sensitivity of film to test gas observed. In the second stage, the effects of different concentrations of the gas at this temperature are measured.

Fig. 12 shows curves of temperature dependences of the sensitivity for films synthesized by both sol-gel method and magnetron sputtering, at an ethanol concentration of 1 mg/L. Maximum sensitivity of the sol-gel films to the ethanol vapors is shifted to lower temperatures (230°C), relative to films prepared by magnetron sputtering (270°C). This may be due to smaller crystallites, higher porosity and homogeneity of the phase composition of the sol-gel SnO_2 films.

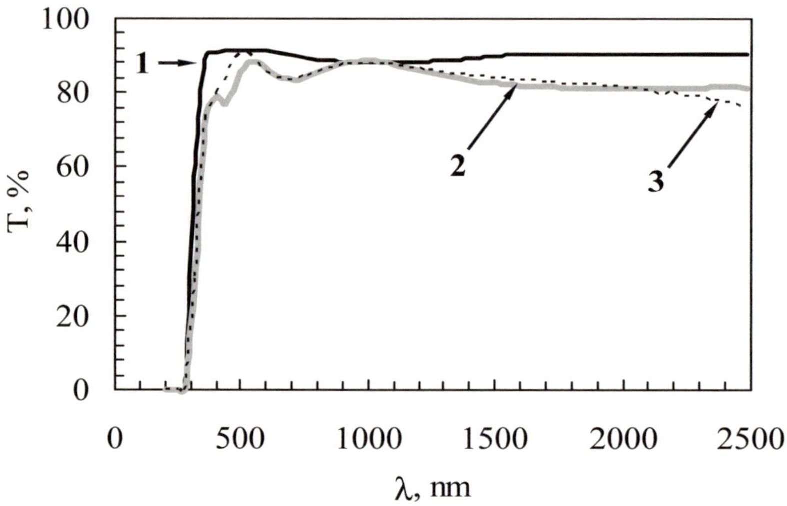

Fig. 10. Optical transmission spectra of SnO_x films: 1 – glass, 2 – film, deposited by the sol-gel technique, 3 – film deposited by magnetron sputtering.

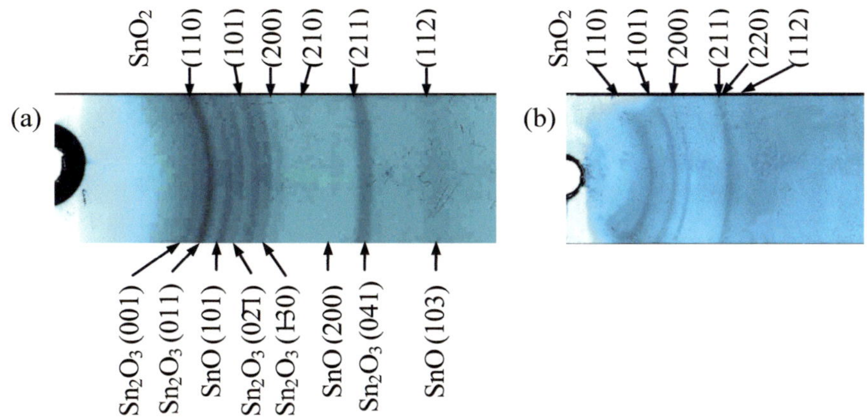

Fig. 11. X-ray diffraction patterns for thin SnO_x films on glass substrates: a) films obtained by magnetron sputtering, and b) by the sol-gel technique.

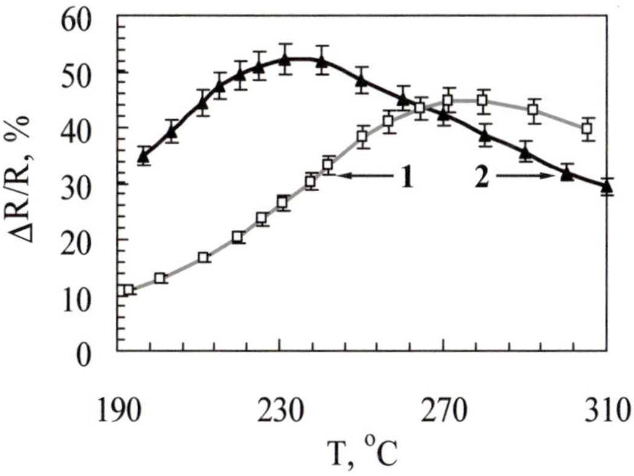

Fig. 12. Temperature dependence of the sensitivity of tin dioxide films to ethanol (concentration 1 mg/L): 1 – film deposited by magnetron sputtering, 2 – by sol-gel technique.

The listed above factors also affect the dynamic performance. To measure the response time of gas-sensitive film, in the volume of the reactor a certain number of identified gas for the time of the order of tenths of a second was introduced. Fig. 13 shows the response time for the presence of ethanol vapor with a concentration of 1 mg/L (500 ppm), for which the film resistance dropped by 90%. For the films obtained by sol-gel method and magnetron sputtering, the response time was ~ 3 and 16 s, recovery time of resistance to the initial value was 70 and 90 s, respectively.

Fig. 14 shows the dependence of sensitivity of the SnO_2 films synthesized by both sol-gel technique and magnetron sputtering, on the concentration of ethanol vapor at a temperature corresponding to the maximum of sensitivity. There is a sensitivity to trace amounts of ethanol vapor of films synthesized by sol-gel technique. In addition, the films have a linear sensitivity to small concentrations of ethanol.

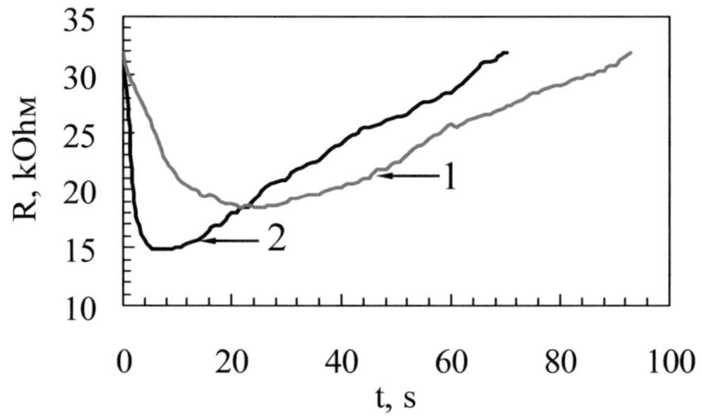

Fig. 13. The dependence of the response time of SnO_2 layer on ethanol vapors (concentration 1 mg/L): 1 – film deposited by magnetron sputtering, 2 – by sol-gel technique.

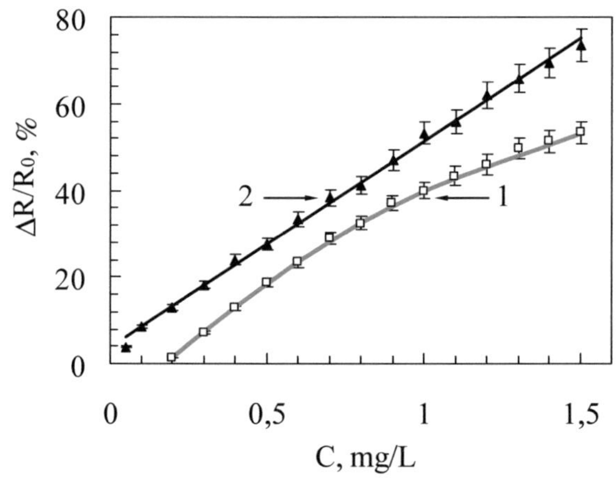

Fig. 14. The dependence of sensitivity of SnO_2 films to the concentration of ethanol vapors: 1 – film deposited by magnetron sputtering, 2 – by sol-gel technique.

Fig. 15 shows the atomic-force microscope images of the surface microstructure of the glass substrate and SnO_x film deposited by magnetron sputtering and sol-gel technique. The film deposited by magnetron sputtering, consists of large agglomerates, and the film synthesized by sol-gel technique has a fine-grained structure of the surface.

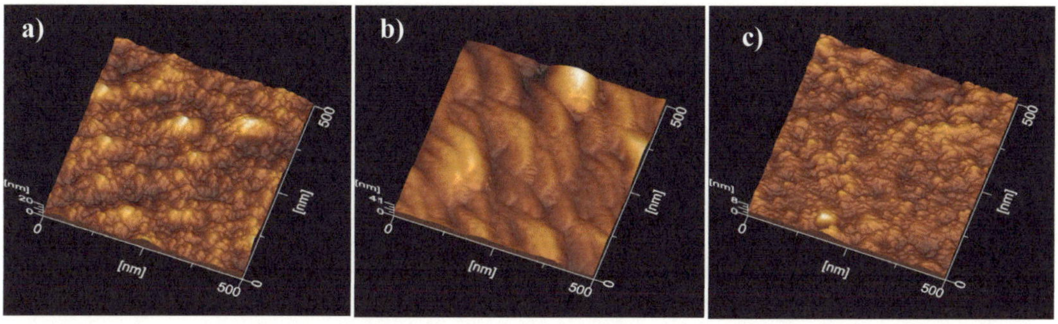

Fig. 15. Surface topography of the SnO_x films on glass substrates: a) glass, b) SnO_x film, deposited by magnetron sputtering, c) by sol-gel technique.

Table 4 presents data on the film surface structure obtained using the processing software to an atomic force microscope, where: R_a is average roughness; R_{zjis} – average roughness of 10 points; S – area of the image; R_q – RMS roughness; R_z – the difference between maximum and minimum height of an image; S_{ratio} – the ratio of the image area S to the area of flat surface S_0.

Parameters	Microscope slides (glass)	SnO_x films	
		Magnetron sputtering	The sol-gel method
R_a, nm	2.31	3.58	0.66
R_{zjis}, nm	17.7	27.1	5.67
S, nm^2	264632.0	259867.1	251484.2
R_q, nm	2.92	4.69	0.842
R_z, nm	19.0	35.6	7.02
S_{ratio}	1,06	1.04	1.01

Table 4. Parameters of the surface structure of SnO_2 films on glass

3.3 Influences of thermal and plasma treatments on phase composition, microstructure and physical properties of the SnO_x films deposited by sol-gel technique and magnetron sputtering

To modify the properties of SnO_2 films, different ways of their treatment are widely used. Of particular interest is the study on the influence of thermal and plasma treatments on optical and structural properties of the SnO_x films, deposited by the sol-gel techniques.

As is known, the crystallization of SnO_2 films on glass substrate takes place most intense at temperatures of 400–600°C. Heat treatment for 1 hour at temperatures of 550–600°C on energy costs can be equivalent to several hours of treatment at 400°C. However, the changes of the structural and sensory properties of the films may be non-equivalent in these conditions. Analysis of the influence of the duration of isothermal annealing on the properties of the films is of a separate scientific and practical interest.

In recent years, the plasma treatment for modifying the properties of SnO_2 films is used (Karapatnitski et al., 2000). It has been reported about the influence of plasma on the sensitivity of the sensors. An analysis of changes in the optical parameters determined from transmission spectra in combination with changes of structural characteristics on the basis of X-ray diffraction and atomic force microscopy, may help to better understand the dynamics of the physical and structural properties of thin films of tin dioxide.

This section presents the results of the study of the effect of isothermal annealing (15 min, 3 hours, 6 hours and 12 hours at T = 400°C) and processing by hydrogen and oxygen glow discharge plasma on the microstructure, optical and electrical properties, thickness, porosity and gas sensitivity of SnO_2 films, deposited by the sol-gel technique (the method of centrifugation) on a glass substrate. The results are interpreted by comparing with recent data on the effect of plasma treatment on the properties of films obtained by magnetron sputtering or ion implantation.

3.3.1 Influence of thermal and plasma treatment on phase composition, microstructure and physical properties of the SnO_x films deposited by magnetron sputtering

Fig. 16 shows the dependence of sensitivity on the concentration of ethanol vapor for thin SnO_2 film, obtained by magnetron sputtering, after treatment by oxygen plasma (27.12 MHz, 12.5 W, 6.5 Pa, 100°C, 5 or 20 min). The SnO_2 film acquires a high sensitivity (> 50%) to ethanol vapor with concentrations below 0.2 mg/L after treatment by oxygen plasma for 5 and 20 min. The short-term treatment for 5 min was more effective for increasing the sensitivity of the SnO_2 films (~70%). This significant effect of plasma treatment on the gas-sensitive properties of the SnO_2 films, probably, is caused by changes in the structure of the films during processing.

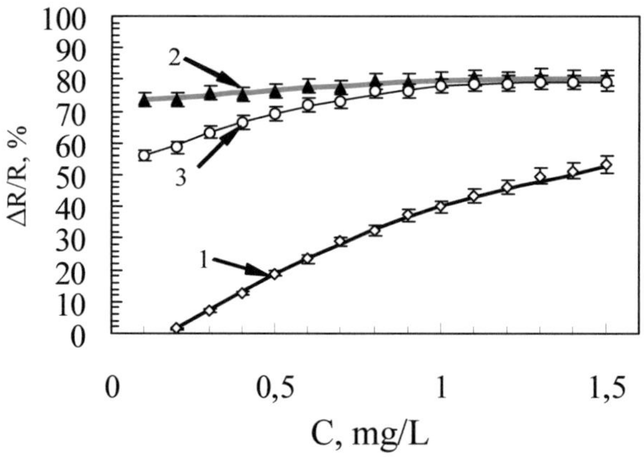

Fig. 16. Dependence of the sensitivity on the concentration of ethanol vapor for SnO_2 film: 1 – after deposition, 2 – after treatment by oxygen plasma for 5 minutes, 3 – for 20 minutes.

As shown in Fig. 17, immediately after the deposition by magnetron sputtering (cathode voltage is 470 V, discharge current – 35 mA, the pressure in a $Ar\text{-}O_2$ mixture in the chamber – 2.7 Pa, oxygen concentration 10%, the rate of deposition of films ~ 0.05 nm/s, substrate temperature 200°C) the films have a nanocrystalline structure and contain SnO_2 crystallites with sizes about 4 nm. In this case there are clear reflections from the systems of planes of SnO_2 with Miller indices (110), (101), (211) and (112). After oxygen plasma treatment for 5 minutes (Fig. 17b) and 20 minutes (Fig. 17c) the intensities of SnO_2 lines are greatly reduced. This may occur due to destruction of crystallites of tin oxide under the influence of oxygen plasma to form a cluster structure of the film. Clustering is not accompanied by appreciable sputtering of the film, since after annealing at 550°C for 1 h the structure of SnO_2 crystallites is restored and the intensity of lines on X-ray Debye patterns almost restored (Fig. 17d). Table 5 shows the optical parameters of the films, calculated by standard methods (Song, 1999). It is seen that the thickness of the film after oxygen plasma treatment decreases slightly.

If the deposition of SnO_x film is carried out in condition of a lack of oxygen at reduced pressure ~ 1 Pa in a Ar-O_2 mixture in the chamber, an increase of the concentration of excess tin in the film results the formation of tin (β-Sn) crystallites with an average size of ~ 30 nm (Fig. 18a) during treatment by hydrogen plasma. So, that demonstrates a segregating effect by hydrogen plasma on the film structure. After annealing of these films in air at 550°C for 1 h, both as-treated and not treated by H-plasma, the appearance of X-ray lines of SnO_2 (6 lines), Sn_2O_3 (5) and SnO (1) phases (Fig. 18b), is observed. Crystallite sizes are in the range of 8–18 nm. Processing of annealed samples by hydrogen plasma led to a blurring of line sections in the angle range of $15^o < \theta < 20^o$ (in the region between arrows 1 in Figs. 18c and 19), corresponding to reflection from the plane systems of SnO(101), SnO_2 (101) and (200), Sn_2O_3 $(0\bar{2}1)$ and $(0\bar{3}0)$. Since the X-ray lines on the Debye photograph are a consequence of a number of specular reflections from the set of crystallites which are in reflecting position in accordance with the equation of Bragg $2d \cdot \sin\theta = \lambda$, the blurring of line sections can occur when selective destruction (amorphization and clusterization) only those crystallites which are oriented to reflect in the area of blurring (between arrows 1). This assumes such orientation of SnO_x film towards the movement of the plasma particles, that the symmetry and arrangement of atoms in the systems of planes of the crystallites SnO(101), SnO_2(101) and (200), Sn_2O_3 $(0\bar{2}1)$ and $(0\bar{3}0)$ (plane A in Fig. 19) are disordered. The estimation of the angles α between the projection onto a plane perpendicular to the incident X-ray beam, of normals to the sample surface and to the planes A of crystallites which were in the reflecting position before clustering by hydrogen plasma treatment. Angle α for the plane systems SnO(101), SnO_2(101) and (200), Sn_2O_3 $(0\bar{2}1)$ and $(0\bar{3}0)$ lies in the range of 15–45°, and the angle θ in the range of 15–20°.

Sample	n, refractive index	D, film thickness, nm	E_g, band gap, eV	k, absorption coefficient, 1/cm	V, porosity, %	ρ, density, g/cm^3
Magnetron sputtering	1.830	280	4.05	$1.65 \cdot 10^3$	15.5	5.89
O-plasma	1.750	276	4.05	$1.68 \cdot 10^3$	21.0	5.45
H- plasma	1.805	294	4.05	$2.25 \cdot 10^3$	17.0	5.75

Table 5. The parameters of SnO_x films deposited by magnetron sputtering at a pressure of Ar-O_2 mixture inside the camera – 2.7 Pa, after deposition and treatment by a glow discharge plasma

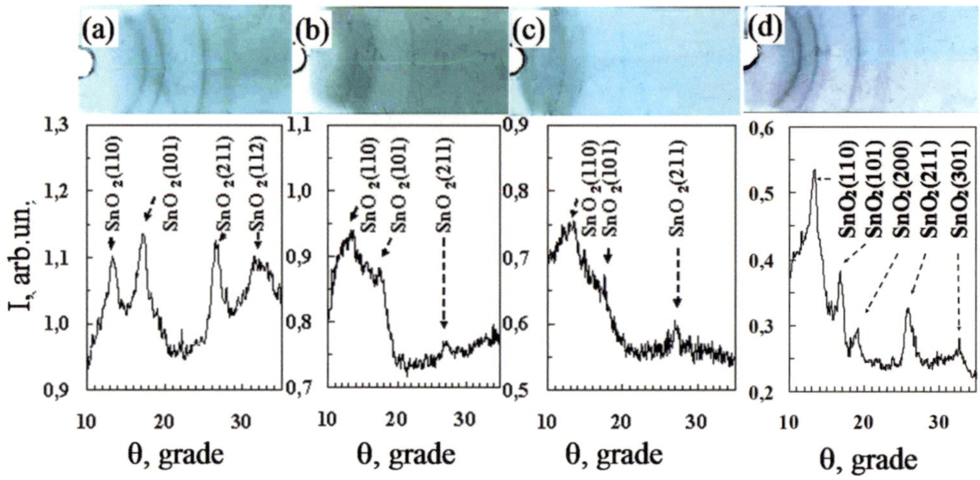

Fig. 17. X-ray diffraction patterns and intensity curves for the thin SnO_2 film after deposition by magnetron sputtering (pressure 2.7 Pa in Ar-O_2 mixture) on a glass substrate (a), after treatment by a glow discharge oxygen plasma for 5 min (b), 20 min (c) and annealing at 550°C for 1 h (d).

In other directions, areas of X-ray lines not subjected to blur (in the region between the arrows 2 in Figs. 18c and 19), the crystallites were solid and even increased their size, demonstrating segregate effects of hydrogen plasma. The order of arrangement of atoms in the systems of planes of SnO(101), SnO_2 (101) and (200), Sn_2O_3 in these crystallites (plane B in Fig. 19) is intact. Angle α for these planes lies in the range of 15-(-45)°, and the angle θ in the same range of 15-20°. Therefore, the possibility of obtaining by treatment in glow discharge hydrogen plasma of crystal-amorphous nanostructures in which high-quality nanocrystals of tin oxide alternate with nanosized clusters of tin oxides, is shown.

Repeated annealing for 1 h at 550°C (Fig. 20, curve 7) causes the restoration of the integrity of the X-ray lines of SnO(101), SnO_2(101) and (200), Sn_2O_3 (021) and (130) (Fig. 18d).

It follows from the optical transmission spectra (Fig. 20, curve 2), low pressure of Ar-O_2 mixture in the chamber (~ 1 Pa), which leads to an increase in the concentration of excess tin in the film, leads to a significant deterioration of its transparency. The presence of significant amounts of tin clusters after deposition is probably the reason for the lack of transparency of the films at low wavelengths (<500 nm) and low transparency in the region above 500 nm, as β-Sn crystallites yet not observed on X-ray patterns. The formation of tin (β-Sn) crystallites with an average size of ~ 30 nm (Fig. 18a) after treatment by hydrogen plasma leads to a further reduction in the transparency of the film (Fig. 20, curve 3) over the entire considered range of wavelengths (500–1100 nm). The decomposition and oxidation of tin crystallites at the subsequent annealing in air at 550°C for 1 h (Fig. 18b) leads to significant increase of film transparency (Fig. 20, curve 4) both in range of > 500 nm, and at small length waves ~ 350–500 nm ($T(\lambda)$ from 0 to ~ 80%), for which the film was not transparent. This may be caused not only by oxidation of the β-Sn crystallites, but also tin clusters. At the same time the transparency of these films is higher than the transparency of untreated by H-plasma films after annealing under the same conditions (Fig. 20, curve 5). This may be due to more intense oxidation processes in the plasma-treated films in the annealing process due to their higher porosity (11.8 instead of 7.7 according to Table 6). As seen from Fig. 20 (curve 6), the processing by H-plasma of the annealed polycrystalline films, leads to a marked deterioration in transmittance in the wavelength range 300–1100 nm. This may be due to disordering of the structure and the formation of crystal-amorphous structure containing opaque inclusions of SnO, in accordance with (1):

$$SnO_2 \xrightarrow[-H_2O]{+H_2} SnO \xrightarrow[-H_2O]{+H_2} Sn \qquad (1)$$

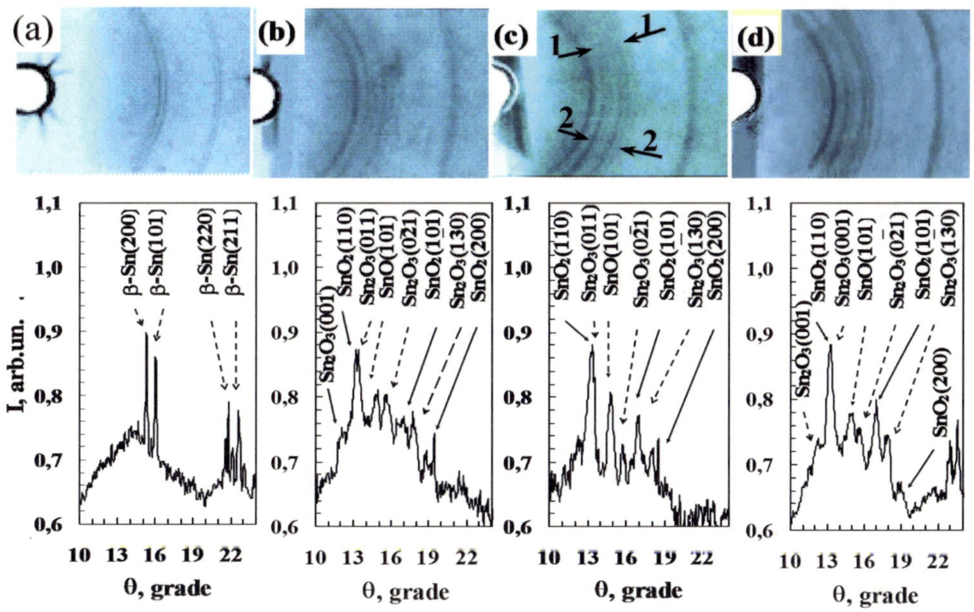

Fig. 18. X-ray diffraction patterns and intensity curves for the SnO_x films after deposition by magnetron sputtering (pressure of 1 Pa in Ar-O_2 mixture) and H-plasma treatment (a), deposition and annealing at 550°C for 1 h (b), the subsequent processing of H-plasma (c) and re annealing at 550°C (d).

Further annealing at 550°C (1 h) permits to maximize the transparency of the films produced under low pressure in Ar-O_2 mixture in the chamber (~ 1 Pa). This is due to the restoration of the structure of microregions, disordered during plasma treatment. It is assumed that the transformation of these areas during the oxidation into SnO_2 crystallites with a higher density than Sn_3O_4, increases the porosity of the films (Table 6), which is important for the gas sensitivity of films.

The obtained data allows us to interpret the transmission spectra of SnO_x films, obtained by magnetron sputtering under higher pressure in Ar-O_2 mixture in the chamber ~ 2.7 Pa (Fig. 21). X-ray data show the presence of polycrystalline SnO_2 phase immediately after the deposition and the absence of β-Sn crystallites (Fig. 17a). As is seen in the figure 21 (curves 3 and 4), a significant drop in $T(\lambda)$ in range of 1200–2500 nm after processing of the SnO_x films for 5 min by O-plasma or H-plasma is observed. The decrease of the transmission of these films, consisting mainly of SnO_2 crystallites (Fig. 17a), can occur as a result of increase of the concentration of free charge carriers due to segregation of excess atoms of tin and the formation of tin clusters with sizes of tenths of a nanometer. It should be noted that the phenomenon of reducing of the transparency due to the presence of nanoparticles in the films is less pronounced in the case of oxygen plasma treatment, because the process of segregation of tin nanoparticles is accompanied by a process of oxidation of some part of them up to SnO_2.

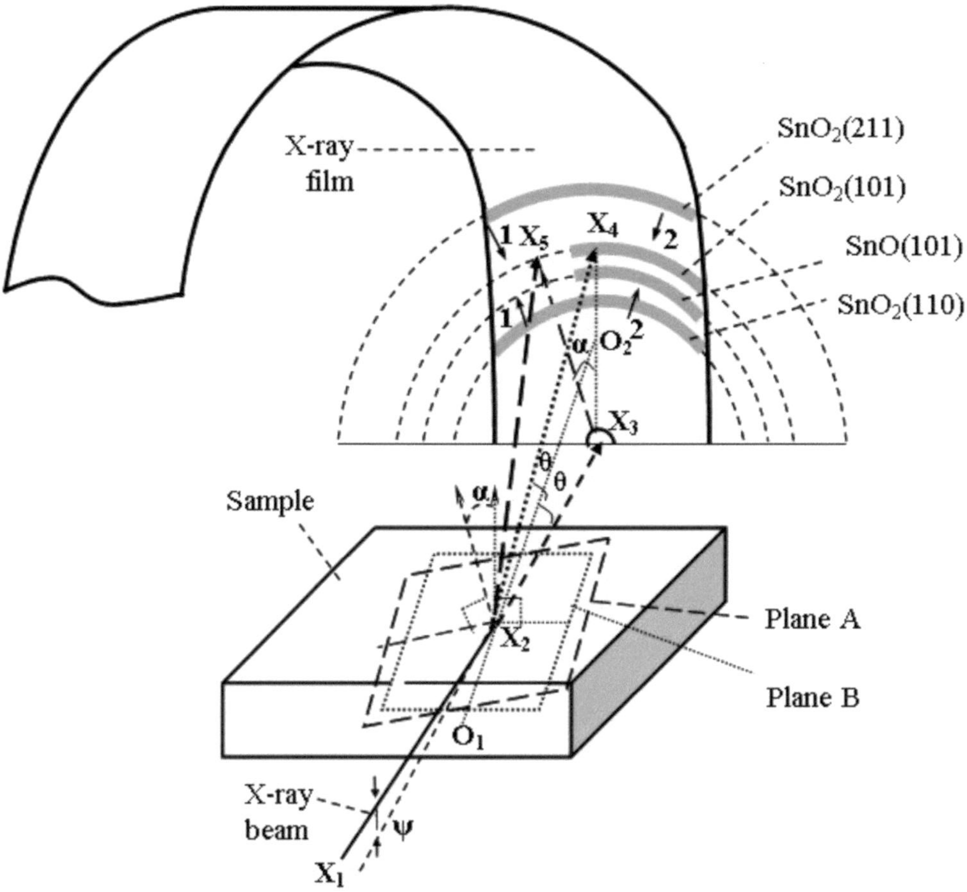

Fig. 19. Illustration of the blurring of X-ray line sections (using $SnO_2(101)$ and $SnO(101)$ as an example) in the angle ranges $15^o < \theta < 20^o$ and $15^o < \alpha < 45^o$ (in the region between arrows 1) after treatment of the annealed samples by H-plasma. The X-ray beam paths as follows: X_1X_2 is the X-ray beam incident at an angle of $\psi = 5^o$ to the sample surface, X_2X_3 is the direction of the transmitted beam, X_2X_4 is the X-ray beam reflected from the SnO_2 (101) plane system of crystallite B (Plane B), X_2X_5 is the beam reflected from the plane system SnO_2 (101) of crystallite A before its destruction by plasma treatment (Plane A). O_1O_2 is the straight line lying in the A plane and making an angle θ with X_2X_3 and X_2X_4. α is the angle between the projections onto the plane (perpendicular to the incident X-ray beam) of the normals to the sample surface and to the A plane.

Thus established that the short-term treatment (5 min) by O- or H-plasma on the deposited at a pressure of 2.7 Pa SnO_2 film results in the formation of Sn clusters, which reduces the transparency of the film from 80% to 50% and 40%, respectively, in the near infrared region (1200-2500 nm).

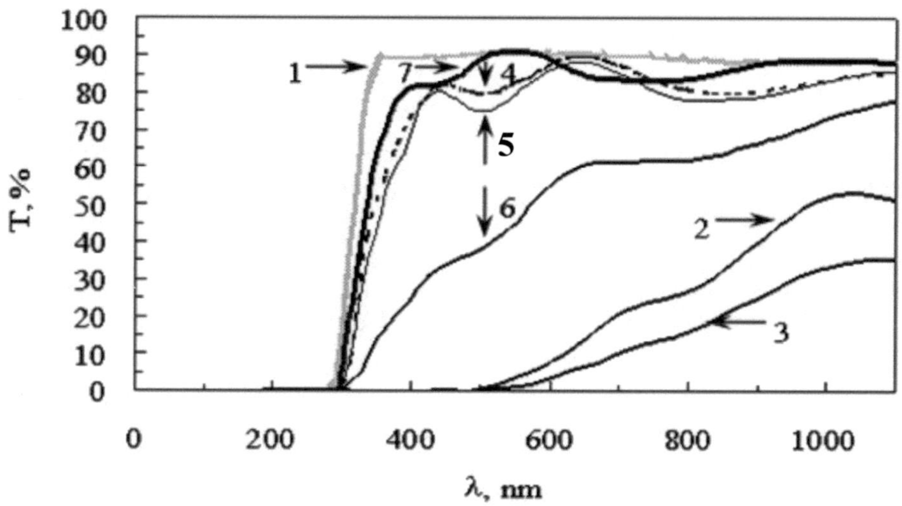

Fig. 20. Optical transmission spectra of glass substrate and thin SnO_x films, deposited by magnetron sputtering (pressure 1 Pa of Ar-O_2 mixture) and treated by H-plasma and annealing: 1 is a spectrum of substrate (glass), 2 – after deposition of SnO_x film on glass, 3 – deposition + H-plasma; 4 – deposition + H-plasma + annealing at 550°C (1 h); 5 – deposition + annealing at 550°C; 6 – deposition + annealing at 550°C + H-plasma; 7 – deposition + annealing at 550°C + H-plasma + annealing at 550°C.

Sequence of operations	n, refractive index	D, film thickness, nm	E_g, band gap, eV	ρ, density, g/cm^3	V, porosity, %
Annealing	1.94	330	4.00	6.42	7.70
Plasma - annealing	1.88	341	4.01	6.13	11.8
Annealing - plasma - annealing	1.82	296	4.05	5.82	16.2

Table 6. Optical parameters of SnO_x film deposited by magnetron sputtering at a pressure of Ar-O_2 mixture in the chamber – 1 Pa

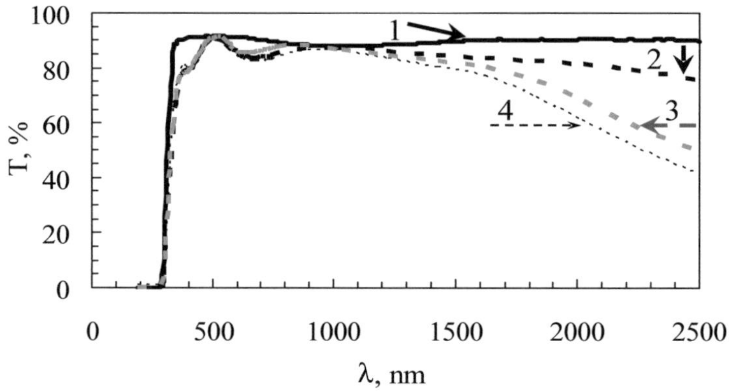

Fig. 21. Optical transmission spectra: the glass substrate (1); thin SnO_x films on a glass substrate after deposition by magnetron sputtering (2) and treatment by oxygen (3) and hydrogen (4) plasma.

3.3.2 Influence of isothermal annealing and plasma treatment on phase composition, microstructure and physical properties of the SnO_x films deposited by sol-gel technique

Colloidal solution for preparation of tin dioxide films was deposited on glass substrate. Rotational speed of the centrifuge was 3800 rpm. Centrifugation time was 3−5 s. The substrates with the deposited films were dried by an infrared emitter at 80°C for 3−5 min. Low temperature annealing was maintained to prevent the occurrence of cracks on the SnO_2 films. Then the sample is placed in a muffle furnace and dried at 400°C for 15 min. The number of deposited layers of SnO_2 was 15. The thickness of the deposited film was estimated from the weight of the film and was about 300 nm.

As shown earlier (Fig. 11b), X-ray diffraction study of SnO_2 films after drying at 400ºC for 15 minutes showed that the films have a polycrystalline structure. The structure of SnO_2 crystallites is sufficiently advanced, it contributed to recording X-ray reflections from 6 plane systems with Miller indices SnO_2 (110), (101), (200), (211), (220), (112). The increase of the duration of isothermal annealing leads to an increase in the average size of SnO_2 crystallites in the films: 15 minutes – 6 nm, 3 and 6 hours – little more than 6 nm, 12 hours – more than 10 nm.

Fig. 22 shows the spectra of optical transmission of tin dioxide films after annealing at 400ºC for 15 min, 3, 6 and 12 hours. Fig. 23 shows the spectra of optical transmission after treatment by oxygen and hydrogen plasma.

The prepared films have high transparency (~90%). With increasing of annealing time the transparency at short wavelengths is increased, which indicates the improvement of the stoichiometry of tin dioxide films and the removal of residual solvent (Fig. 22). As shown in Table 7, the increase of isothermal annealing time to 6 and 12 h leads to compaction of the film, reducing its thickness, porosity and absorption coefficient.

The treatment by glow discharge hydrogen plasma resulted in a decrease in the transparency on 3–15% in the short- wave range (Fig. 23, curve 4), while the oxygen plasma treatment resulted in a slight decrease of 1–5% (Fig. 23, curve 3). Noticeable decrease in the transparency of the thin film after hydrogen plasma treatment can be attributed to the formation of opaque compounds such as SnO or Sn, formed by reducing properties of hydrogen in accordance with the reaction (1).

However, X-ray diffraction data do not confirm the presence of crystallites of SnO and Sn in the films treated by hydrogen plasma (Table 7). Nevertheless, the presence of SnO inclusions in the amorphous state or in the form of clusters is possible that leads to an increase in the absorption coefficient from $2.5 \cdot 10^3$ up to $5.91 \cdot 10^3$ cm^{-1} (Table 7). The presence of metal clusters of tin in appreciable amounts is unlikely, since there is no reduction of the transmission coefficient $T(\lambda)$ in the range 1200-2500 nm, associated with an increase in the concentration of free charge carriers.

The treatment by oxygen plasma does not result the formation of SnO, and the slight change in the transparency of the film is caused by the damaging effects of plasma on the structural perfection and crystallite size. Reducing the size of the SnO_2 crystallites is caused by the destructive influence of the massive oxygen ions on the structure of the crystallites (Table 7). The increased density of the film in proportion to the reduction of its thickness is presumably caused by the filling of intergranular voids by SnO_2 clusters. The increase of the SnO_2 crystallites sizes during processing by hydrogen plasma is caused by the segregating effect of light hydrogen ions of plasma, under the influence of which the alteration, destruction, and merging of the crystallites are taken place. Reducing the thickness by 10% while maintaining the density of the film can be occur through the formation and desorption of H_2O molecules and the formation of oxygen vacancies. The formation of SnO molecules on the surface of SnO_2 crystallites is taken place; thereby the stoichiometry and transparency of the film are reduced.

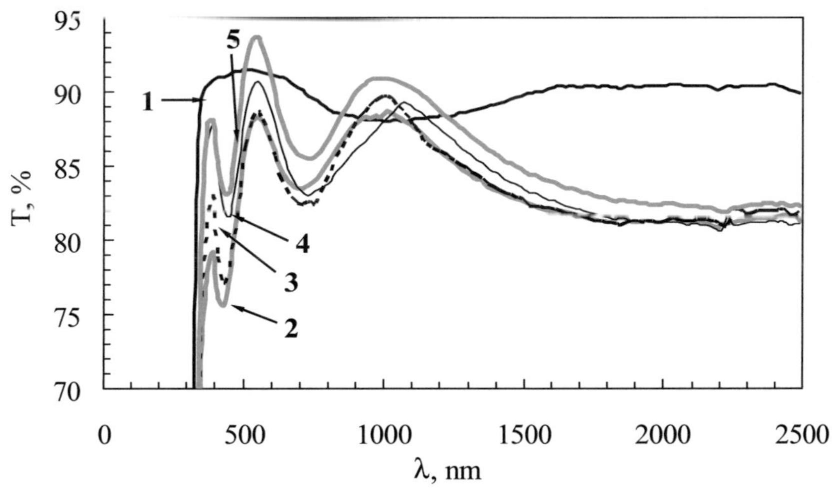

Fig. 22. Optical transmission spectra of SnO_2 films after isothermal annealing at 400°C:
1 – glass substrate, 2 – the film after deposition and annealing for 15 minutes,
3 – 3 hours of annealing; 4 – 6 hours of annealing, 5 – 12 hours of annealing.

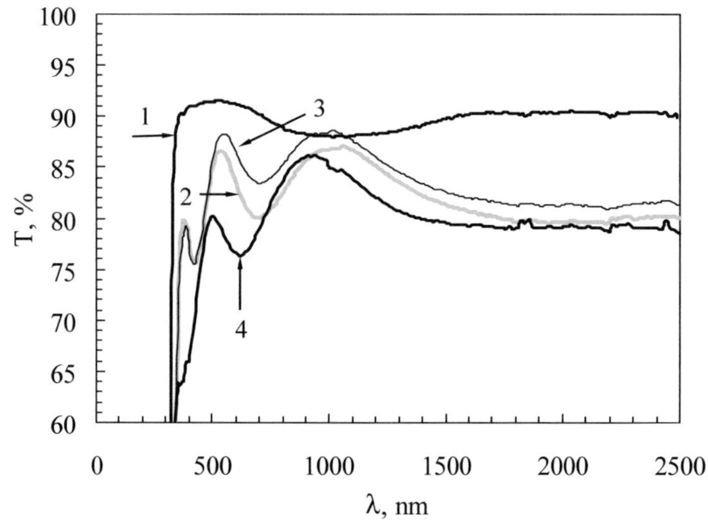

Fig. 23. Optical transmission spectra of SnO_2 films after plasma treatment: 1 – substrate, 2 – after film deposition, 3 – after treatment by oxygen plasma, 4 – after treatment by hydrogen plasma.

Optical and structural parameters		Isothermal annealing				Plasma treatment		
		400°C, 15 min	400°C, 3 h	400°C, 6 h	400°C, 12 h	400°C, 15 min	O-plasma	H-plasma
n, refractive index		1.72	1.76	1.81	1.81	1.74	1.81	1.73
D, film thickness, nm		319	312	304	304	316	297	288
E_g, band gap, eV		4.1	4.1	4.1	4.1	4.1	4.1	4.0
P, density, g/cm^3		5.29	5.53	5.76	5.76	5.40	5.80	5.36
k, absorption coefficient, 1/cm		$2.4 \cdot 10^3$	$2.3 \cdot 10^3$	$1.6 \cdot 10^3$	$0.6 \cdot 10^3$	$2.5 \cdot 10^3$	$3.05 \cdot 10^3$	$5.91 \cdot 10^3$
V, porosity, %		23.9	20.5	17.2	17.1	22.3	16.5	22.8
Average size ε of crystallites, nm	SnO_2 (110)	5	7.5	6	8	5	5.5	6.5
	SnO_2 (101)	6.5	7	6.5	10	6.5	6	9.5
	SnO_2 (211)	6	6	6	14.5	6	4.5	10

Table 7. Optical and structural parameters of SnO_2 films

The surface resistance of SnO_2 films after deposition and annealing at 400°C for 15 minutes by four-probe method is measured, and it was 16 ohm · cm (Fig. 24). The surface resistance of the SnO_2 film increases linearly with increasing duration of isothermal annealing (400°C). Resistance measurements were performed at room temperature (22°C). It is seen that each hour of annealing leads to an increase in surface resistance of ~ 24 kOhm · cm (Fig. 24). This may be due to a decrease in the number of oxygen vacancies as a result of improving of the crystallite structure and stoichiometry during the long process of annealing in air.

After treatment in oxygen plasma, the film sheet resistance increased from 16 to 133 kohm·cm, apparently due to reducing the number of oxygen vacancies. In contrast, after treatment by hydrogen plasma, it decreased from 16 to 0.7 kohm·cm, presumably due to increase in the number of oxygen vacancies.

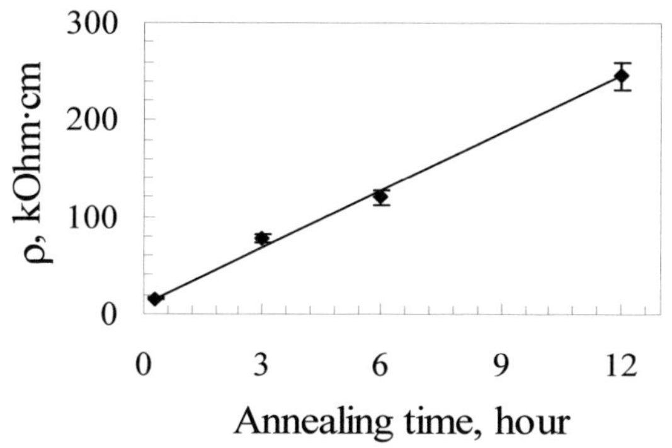

Fig. 24. The dependence of sheet resistance (22°C) of the SnO_2 film on the duration of isothermal annealing at 400°C.

Fig. 25 shows the temperature dependence of the resistance of the films after isothermal annealing, and Fig. 26 - after treatment by hydrogen and oxygen plasmas. The substrate with the film placed on a heated table, located in a cylindrical chamber. The resistance of the deposited films decreases rapidly with increasing temperature (Figs. 25 and 26, curves 1). Increase of the duration of isothermal annealing has little influence on the temperature dependence of resistance (Fig. 25, curves 2–4).

The processing of the film by hydrogen plasma leads to an increase in the number of oxygen vacancies and, correspondingly, a decrease in resistance from 192 kohm to 3.1, resulting in a decrease to 1.6 kohm of resistance with increasing temperature is smooth and the relative resistance decreases slowly (Fig. 26, curve 3). Treatment by oxygen plasma leads to occurrence of excess oxygen in the film and, accordingly, to increase resistance from 192 to 640 kohm, resulting in a decrease to 1.9 kohm of resistance with increasing temperature is faster due to the desorption of excess oxygen (Fig. 26, curve 2).

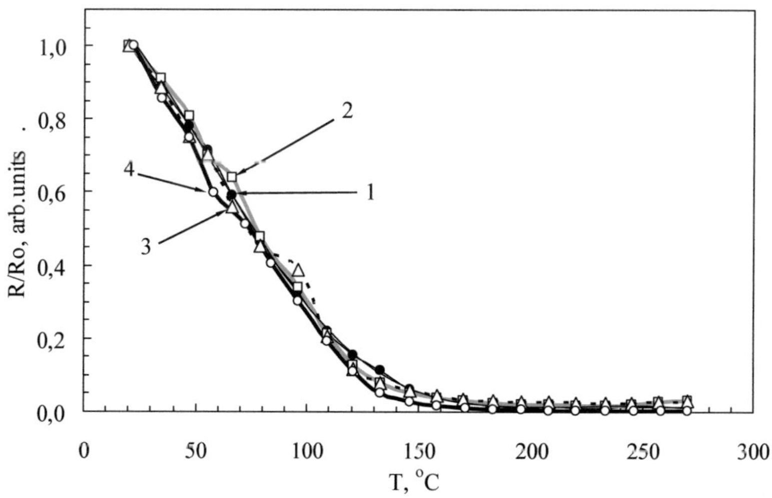

Fig. 25. Temperature dependences of the resistance of the SnO_2 film after annealing at 400°C for 15 minutes (1 - •), 3 hours (2 - □), 6 hours (3 - Δ) and 12 hours (4 - ○).

Thin films of SnO_2, subjected to isothermal annealing and plasma treatment, were investigated for sensitivity to ethanol vapor. Gas sensitivity in this case defined as the ratio

$$\gamma = \frac{R_0}{R_g}, \quad (2)$$

where R_0 is resistance of gas-sensitive layer in the clean air, Rg is the resistance of the layer in the mixture of air with detectable gas.

Fig. 27 shows the temperature dependence of the sensitivity for films synthesized by sol-gel technique at concentration of ethanol 1 mg/L after isothermal annealing (a) and treatment by plasma (b). Maximum sensitivity of the sol-gel films to the vapors of ethanol takes place at a temperature of 235°C (Fig. 27, curves 1). There is a significant increase in sensitivity of the films both after annealing and after treatment by plasmas. The maximum sensitivity of the films is observed after annealing at 400°C for 6 hours. A significant decrease in sensitivity after annealing for 12 hours in comparison with annealing for 3 and 6 hours may be due to a significant increase in the crystallite size from ~ 6.5 to ~ 11 nm. As shown by Xu et al. (1991), optimal crystallite size for good gas sensitivity should be ~ 6 nm.

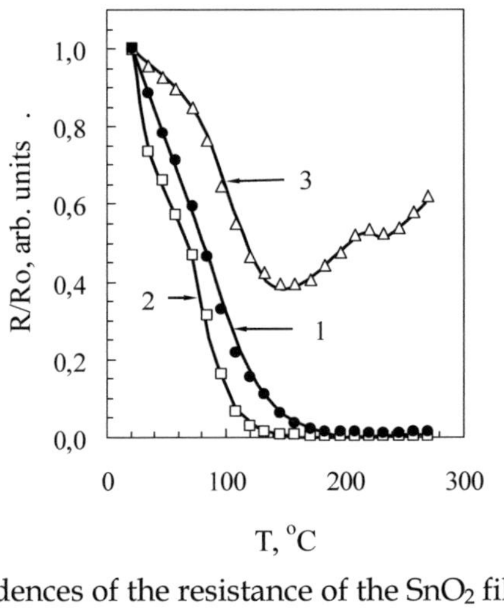

Fig. 26. Temperature dependences of the resistance of the SnO_2 film after deposition (1) and glow discharge oxygen (2) and hydrogen (3) plasma treatment for 5 min.

After processing by plasma, the sensitivity increases without increase of the operating temperature. The observed increase in sensitivity can be caused by an increase in the size of pores and cracks during the bombardment by ions of oxygen or hydrogen. The sensitivity of the films increases linearly (Fig. 28), if the concentration of ethanol vapor increases from 0.05 to 0.8 mg/L and above, after which saturation occurs. The growth of sensitivity is more intense after annealing or plasma treatment. The most intensive growth was observed after annealing for 3–6 hours. A marked sensitivity of the film is evident even when the concentration of ethanol vapor is less than 0.05 mg/L. Response time of sensor to ethanol with concentration of 1 mg/L is decreased from 5 to 2 sec both with the increase of annealing time and when exposed to plasma (Table 8). The recovery time of resistance to the initial value increases from 90 up to 140 s, except the 12-hour annealing (75 s).

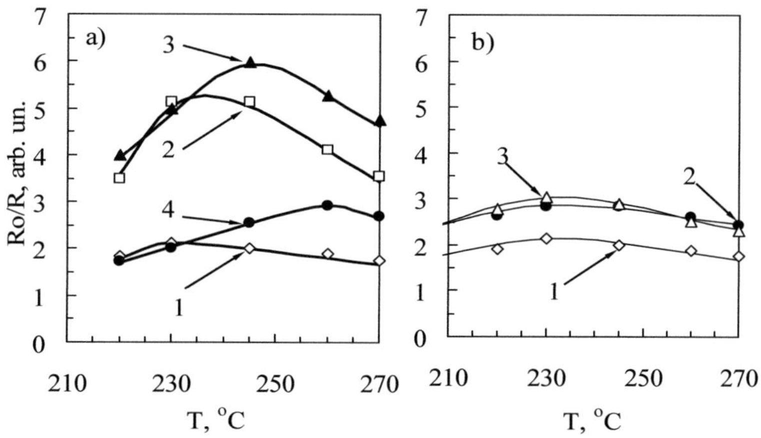

Fig. 27. Temperature dependence of sensitivity to the vapors of ethanol (1 mg/L) of the SnO_2 films: a) after annealing at 400°C for 15 minutes (1), 3 hours (2), 6 hours (3) and 12 hours (4); b) after annealing at 400°C for 15 minutes (1) and treatment by oxygen (2) and hydrogen (3) plasmas for 5 minutes.

The parameters of sensitivity of SnO_2 films	Isothermal annealing (400°C)				Plasma treatment		
	15 min	3 h	6 h	12 h	400°C, 15 min	O-plasma	H-plasma
Response time, s	5	3	2	2	5	2	3
Recovery time, s	90	140	130	75	90	110	100

Table 8. Response time and recovery time of SnO_2 films

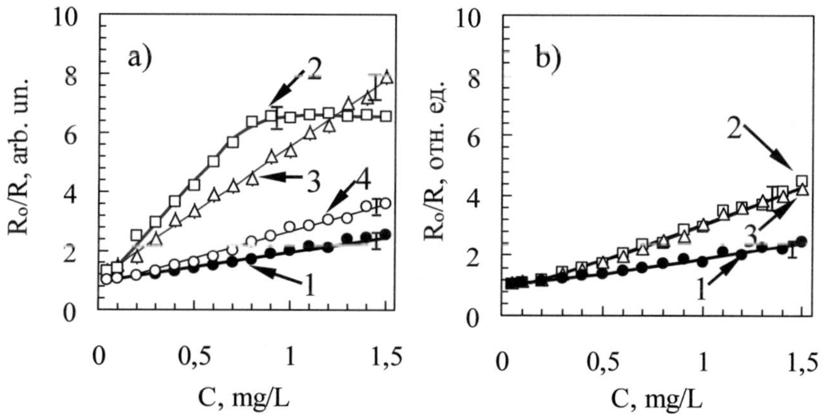

Fig. 28. The dependence of sensitivity on the concentration of ethanol vapor of SnO_2 film: a) after isothermal annealing at 400°C for 15 min (1), 3 h (2), 6 h (3) and 12 hours (4); b) after annealing at 400°C for 15 min (1) and treatment by oxygen (2) and hydrogen (3) plasmas for 5 min.

AFM images of the surface of tin dioxide multilayer film of size 500 × 500 nm (Fig. 29), annealed at 400°C for 15 minutes, 3, 6 and 12 hours, suggest a fine-grained structure of the film surface. It turned out that after annealing at 400°C for 15 minutes about 80% of the areas of film surface lies at elevations of 1.8–3.7 nm, after annealing for 3 hours - in the range 2.4–4.9 nm, for 6 hours - in the range of 2.7–4.9 nm and 12 hours - in the range of 3.2–6.4 nm. After annealing for 3 and 6 hours on the film surface are observed pronounced granules (Fig. 29b, c). After annealing for 12 hours the surface contains a large number of projections.

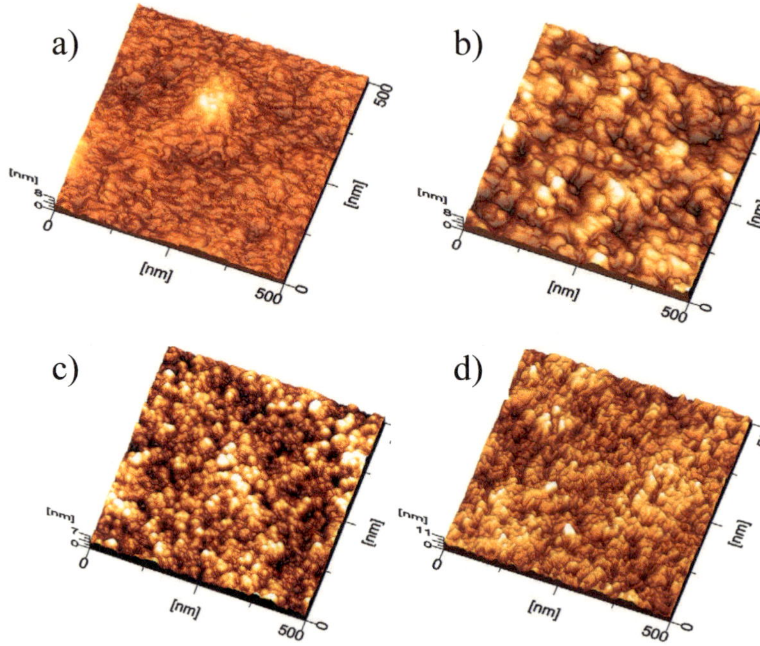

Fig. 29. Surface topography of the films of tin dioxide of 500×500 nm after annealing at 400°C for 15 min (a), 3 h (b), 6 h (c) and 12 h (d).

Treatment by oxygen plasma (Fig. 30b) leads to the disintegration of the granular structure of the film, confirming the assumption about the clustering of structure. Treatment by hydrogen plasma (Fig. 30c) leads to the formation of agglomerates of sizes up to 200 nm. The dimensions of the agglomerates is much greater than the dimensions of the SnO_2 crystallites (~ 9 nm), ie the agglomerates are composed of SnO_2 crystallites.

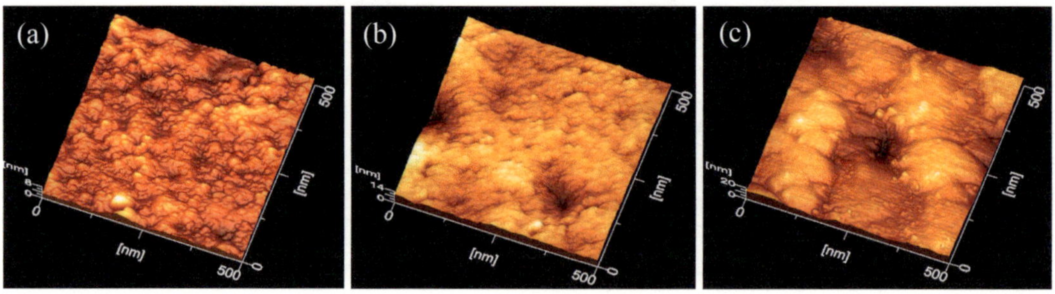

Fig. 30. Surface topography of SnO_2 film after after deposition and annealing (400°C, 15 min) (a) and treatment by O-plasma (b) and H-plasma (c) for 5 min.

Table 9 presents data on the effect of isothermal annealing and treatment by plasma on the roughness of the SnO_2 films, where R_a is average roughness; R_{zjis} - average roughness of 10 points; R_q - RMS roughness; R_z - the difference between the maximum and minimum height of an image. The roughness of the SnO_2 films is increased both after isothermal annealing and after the plasma treatment. This increases the gas sensitivity of film (Figs. 27b and 28b, curves 2 and 3) because the growth of roughness increases the surface area of the film. However, after isothermal annealing roughness varies slightly, but sensitivity is increased more significantly.

Sample	R_a, nm	R_{zjis}, nm	R_z, nm	R_q, nm
Glass substrate	2.31	17.7	19.0	2.92
400°C, 15 min	0.66	5.67	7.02	0.84
400°C, 3 h	0.79	6.32	6.93	0.98
400°C, 6 h	0.74	6.49	7.14	0.93
400°C, 12 h	1.02	9.06	10.0	1.27
400°C, 15 min	0.66	5.67	7.02	0.84
O- plasma	1.15	12.3	13.6	1.58
H- plasma	2.04	16.5	19.7	2.61

Table 9. Analysis of the surface topography of SnO_2 films (500 × 500 nm)

The essential differences in the films deposited by magnetron sputtering and sol-gel technique, include the composition, structure and stoichiometry influencing the properties of the films after deposition as well as after treatment. The comparison of the effect of treatment by hydrogen or oxygen plasma on the individual properties of the SnO_2 films, obtained by these methods, is of particular scientific and practical interest.

Sensitivity to ethanol vapor of films prepared by sol-gel technique, exceeds the sensitivity of the films obtained by magnetron sputtering (Fig. 31, curves 4 and 1). The maximum sensitivity of the sol-gel films at temperature ~ 235°C reaches up to 53%. Treatment by both H-and O-plasma (Fig. 31, curves 5 and 6) increases the sensitivity of the film up to 66% at

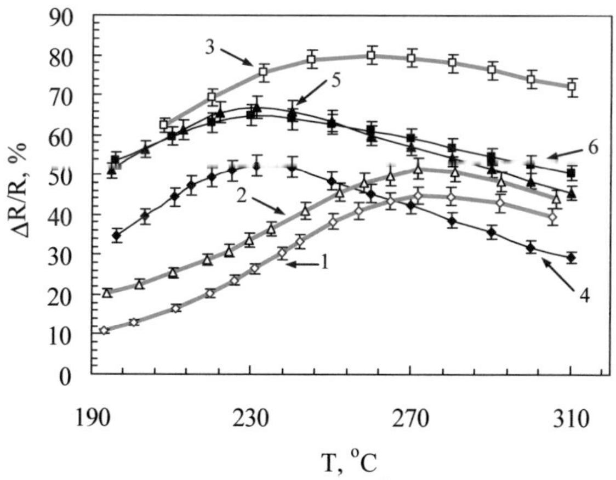

Fig. 31. Temperature dependence of sensitivity to the vapors of ethanol (1 mg/L) of the SnO_x films after deposition by magnetron sputtering (1), H-plasma treatment (2) and O-plasma (3); deposition by sol-gel method (4), H-plasma (5) and O-plasma (6) treatment.

the same operating temperature. For the films synthesized by magnetron sputtering, at an operating temperature of 270°C sensitivity is about 45% (Fig. 31, curve 1). Treatment of H-plasma increases the sensitivity to 52% at the same operating temperature, the same treatment on oxygen plasma increases the sensitivity up to 78% with a decrease in operating temperature to 255°C (Fig. 31, curves 2 and 3).

Thus, the regularities of the influence of treatment by H- or O-plasma on growth of the gas sensitivity of SnO_2 films were established. The plasma treatment of SnO_2 films, obtained by the sol-gel technique, does not alter their phase composition, resulting in gas sensitivity growth does not depend on the type of plasma. Treatment by O-plasma of films prepared by magnetron sputtering, leads to a more significant increase in gas sensitivity than by H-plasma, due to the oxidation of SnO_x film.

4. Conclusion

The problem of modifying the structural properties of the SnO_x films by treatment in glow discharge plasma is studied. Treatments by hydrogen and oxygen plasma permit significantly improve the transparency and gas sensitivity of the SnO_x films. The conditions of film synthesis were revealed, which allow by sol-gel technique (400°C, 6 h) to synthesize the SnO_2 films with high gas sensitivity, the minimum response time (2 seconds) and the optimum crystallite size (6 nm); by magnetron sputtering (10% of O_2 in ArO_2 mixture, 2.7 Pa) to deposit the SnO_2 films with optimum properties (average grain size ~ 4 nm, the transparency ~ 90%, band gap ~ 4.0 eV, the refractive index ~ 1.8) without annealing.

1. A method of formation of the cluster structure of film deposited by a way of spreading, and decrease the average SnO_2 crystallite size up to 1.5 nm by increasing the concentration of tin ions in the film-forming solution from 0.14 to 0.83 mol/L, was developed. This leads to an increase in film sensitivity to the ethanol vapors on 20% (for 1 mg/L), decrease of response time in half (up to 3 seconds) and to appearance of sensitivity to the presence of micro-amounts of ethanol vapor (~ 0.05 mg/L). If the concentration of tin atoms in solution is low (0.14 mol/L), the SnO_2 nanocrystals in film after deposition, drying (100°C, 2.5 nm) and annealing (400°C, 5 nm) were formed.
2. The surfaces of films produced by sol-gel technique (centrifugation method) and magnetron sputtering are fine-grained and in the form of large agglomerates, have an average roughness of 0.6 and 3.6 nm, respectively. The maximum sensitivity of films to the vapors of ethanol is observed at temperatures of 230 and 270°C, the response time is ~ 3 and 16 s, recovery time ~ 70 and ~ 90 s, respectively. The differences may be due to smaller crystallites, higher porosity and the absence of SnO crystallites in the films prepared by sol-gel technique. The treatment by hydrogen plasma of these fine-grained films leads to the formation of large agglomerates. The same treatment by oxygen plasma leads to the disintegration of the granular surface structure and a partial clustering of the film. Isothermal annealing at 400°C for 3–6 h leads to the formation of a pronounced granular surface structure. Treatment by hydrogen plasma of coarse grained (~ 60–120 nm) films, synthesized by magnetron sputtering, leads to a slight decrease in the size of the agglomerates, whereas the treatment by oxygen plasma leads to the destruction of the grains and the film clusterization.

3. The lower resistance at room temperature of films deposited by magnetron sputtering is caused by the presence of tin particles, and a smoother decrease of resistance with increasing temperature is caused by the inclusions of SnO. Films synthesized by the sol-gel technique are composed entirely of SnO_2 phase, have optimal crystallite sizes (3-5 nm) and high values of porosity for higher gas sensitivity. There is no decrease in transmittance at long wavelengths due to absence of tin particles. After treatment by oxygen plasma the excess oxygen atoms increases the film resistance, and its rapid decrease is caused by the desorption of excess oxygen with increasing temperature and absence of SnO inclusions. The treatment by hydrogen plasma results in formation of oxygen vacancies and a decrease of film resistance.
4. It is shown that tin dioxide films synthesized by sol-gel method, after isothermal annealing at 400°C for 15 min, 3, 6 and 12 h have a high transparency (~ 80–90%). The optimum mode of thermal annealing (400°C, 6 h) of SnO_2 films is identified, which permits to achieve the maximum of gas sensitivity and minimum of response time (2 seconds) due to the better structure and optimal sizes (6 nm) of the crystallites.
5. Treatment by glow discharge hydrogen plasma of tin dioxide films synthesized by sol-gel technique resulted in a decrease in the transparency of 3–15% in the visible wavelength range due to the formation of opaque inclusions of SnO. Treatment by oxygen plasma resulted in a slight decrease in the transparency of 1–5% due to the damaging effects of plasma on the structural perfection of the SnO_2 crystallites.
6. For nonstoichiometric SnO_x films ($x<2$), deposited by magnetron sputtering in condition of a lack of oxygen (pressure ArO_2 mixture 1–2 Pa), the formation of polycrystalline β-Sn phase after treatment by H-plasma is shown. Both the segregation of Sn particles at temperatures of 150–200°C and the formation of oxides SnO, Sn_2O_3 and SnO_2 at temperatures of 50–550°C, are taken place. The short-term treatment (5 min) by O- or H-plasma of deposited at a pressure of 2.7 Pa SnO_2 film leads the formation of Sn clusters, which reduces the transparency in the near infrared region (1200–2500 nm) from 80% to 50% and 40%, respectively.
7. The possibility of obtaining of the crystal-amorphous nanostructures in which crystallites of tin oxides alternate with clusters obtained by selective amorphization of SnO, SnO_2 and Sn_2O_3 crystallites during treatment by glow discharge H-plasma, is shown. In this case, the order of the atoms in those (hkl) planes of crystallites is violated, along which there are variations of the plasma particles.
8. The regularities of the influence of treatment by H- or O-plasma on growth of gas sensitivity of SnO_2 films were studied. The plasma treatment of SnO_2 films, obtained by the sol-gel technique, does not alter their phase composition, and gas sensitivity growth does not depend on the type of plasma. Treatment by O-plasma of films prepared by magnetron sputtering, leads to a more significant increase in gas sensitivity than by H-plasma, due to the oxidation of SnO_x film.
9. The phenomena of self-organization of matter in SnO_x layers were identified, consisting in the intensive formation of SnO_2 crystallites with sizes of 4 nm in the process of film deposition by magnetron sputtering (pressure in the $Ar-O_2$ mixture is 2.7 Pa), and SnO_2 crystallites with sizes of 3 nm at the deposition of films by the sol-gel technique (concentration of Sn 0.14 mol/L in a solution).

5. Acknowledgement

The authors are very grateful to Dmitriyeva E.A. for preparation of the films and Mit' K.A. for AFM measurements.

6. References

Aboaf, J.A., Marcotte, V.C., Chou, N.J. (1973). Doping of tin oxide films and analysis of oxygen content. J. Electrochem. Soc., Vol. 120, p. 701.

Adamyan, A.Z., Adamyan, Z.N., Arotyunyan, V.M. (2006). Zol'-gel' tehnologiya polucheniya chuvstvitel'nykh k vodorodu tonkikh plenok (Sol-gel technology for producing hydrogen sensitive thin films). Al'ternativnaya energetika i ekologiya (Alternative Energy and Ecology), Vol. 40, №8, pp. 50–55. In Russian.

Andryeeva, Ye.V., Zilberman, A.B., Il'in, Yu.L., Makhin, A.V., Moshnikov, V.A., Yaskov, D.A. (1993). Vliyanie etanola na elektrofizicheskie svoĭstva dioksida olova (Effect of ethanol on the electrophysical properties of tin dioxide). Fizika i tehnika poluprovodnikov, T. 27, №7, pp. 1095–1100. In Russian.

Anishchik, V.M., Konyushko, L.I., Yarmolovich, V.A., Gobachevskiĭ, D.A., Gerasimova, T.G. (1995). Struktura i svoistva plenok dioksida olova (Structure and properties of tin dioxide films). Nyeorganicheskie materialy (Inorganic Materials), Vol. 31, №3, pp. 337-341. In Russian.

Aranowich, J., Ortiz, A., Bube, R.H. (1979). Thin oxide films prepared by spray hydrolysis. J. Vacuum Sci. Technol., Vol. 16, pp. 994.

Asakuma N., Fukui T., Toki M., Imai H. (2003) Low-Temperature synthesis of ITO thin films using an ultraviolet laser for conductive coating on organic polymer substrates. Journal of Sol-Gel Science and Technology, Vol. 27, pp. 91–95.

Bakin, A.S., Bestaev, M.V., Dimitrov, D.Tz., Moshnikov, V.A., Tairov, Yu.M. (1997). SnO_2 based gas sensitive sensor. Thin Solid Films, Vol. 296, pp. 168–171.

Bestaev, M. V., Dimitrov, D.Ts., Il'in, A.Yu., Moshnikov, V.A., Träger, F., Steitz, F. (1998). Study of the surface structure of tin dioxide layers for gas sensors by atomic-force microscopy. Semiconductors, Vol. 32, iss. 6, pp. 587-589.

Bosnell, J.R. and Waghorne, R. (1973). The method of growths tin films. Thin Solid Films, Vol. 15, pp. 141.

Brito G.E.S., Ribeiro S.J.L., Briois V., Dexpert-Ghys J., Santilli C.V., Pulcinelli S.H. (1997). Short range order evolution in the preparation of SnO_2 based materials. Journal of Sol-Gel Science and Technology, Vol.8, pp. 261–268.

Buturlin, A.I., Gabuzyan, T.A., Golovanov, N.A., Baranenkov, I.V., Yevdokimov, A.V., Murshudli, M.N., Fadin, V.G., Chistyakov, Yu.D. (1983a). Gazochuvstvitel'nye datchiki na osnove metallookisnykh poluprovodnikov (Gas-sensitive sensors based on metal oxide semiconductors). Zarubezhnaya elektronnaya tehnika. 10, 269, pp. 3–39. In Russian.

Buturlin, A.I., Obrezkova, M.V., Gabuzyan, T.A., Fadin, V.G., Chistyakov, Yu.D. (1983b). Elektronnye datchiki dlya kontrolya kontsentratsii etanola v vydykhaemom vozdukhe (Electronic sensors for monitoring the concentration of ethanol in the exhaled air). Zarubezhsnaya elektronnaya tehnika, № 11, pp.67-87. In Russian.

Chatelon J.P., Terrier C., Roger J.A. (1997). Consequence of the pulling solution ageing on the properties of tin oxide layers elaborated by the sol-gel dip-coating technique. Journal of Sol-Gel Science and Technology, Vol. 10, pp. 185–192.

Das, Debajyoti and Banerjee, Ratnabali (1987). Properties of electron-beam-evaporated tin oxide films. Thin Solid Films. Vol. 147. pp.321–331.

Dibbern, U., Kuersten, G., Willich, P. (1986). Gas sensitivity, sputter conditions stoichiometry of pure tin oxide layers. Proc. 2nd Int. Meeting on Chem. Sensors. Bordeaux, 1986. p. 142.

Dimitrov, D.Ts., Luchinin, V.V., Moshnikov, V.A., Panov, M.V. (1999). Ellipsometry as a rapid method of establishing a correlation between the porosity and the gas sensitivity of tin dioxide layers. Technical Physics, Vol. 44, iss. 4, pp.468-469.

Evdokimov, A.V., Murshudlu, M.N., Podlepetsky, B.I., Rzhanov, A.E., Fomenko, S.V., Filippov, V.I., Yakimov S.S. (1983). Mikroelektronnye datchiki himicheskogo sostava gasa (Microelectronic sensors of the gas chemical composition). Zarubezhnaya elektronnaya tehnika, № 10, pp. 3-39. In Russian.

Fantini, M., Torriani, I. (1986). The compositional and structural properties of sprayed SnO_2:F thin films. Thin Solid Films, Vol. 138, pp. 255–265.

Geoffroy C., Campet G., Portier J. (1991). Preparation and characterization of fluorinated indium tin oxide films prepared by R.F. sputtering. Thin Solid Films, Vol. 202, pp. 77–82.

Jarzebski, Z.M. (1982). Preparation and physical properties of transparent conducting oxide films. Phys. Stat. Sol., Vol. 71 (a), pp.13–41.

Jarzebski, Z.M., and Marton, J.P. (1976). Physical properties of SnO_2. Materials. J. of the Electrochemical Society. Vol. 123, № 7. pp. 199C–205C; № 9. pp. 299C–310C; № 10, pp. 333C–346C.

Jiang, J.C., Lian, K., Meletis, E.I. (2002). Influence of oxygen plasma treatment on the microstructure of SnO_x thin films. Thin Solid Films, Vol.411, pp. 203–210.

Jitianu, A., Altindag,Y., Zaharescu, M., Wark, M. (2003). New SnO_2 nanoclusters obtained by sol-gel route, structural characterization and their gas sensing applications. Journal of Sol-Gel Science and Technology, Vol. 26, pp. 483–488.

Karapatnitski, I.A., Mit', K.A., Mukhamedshina, D.M., Baikov, G.G. (2000). Effect of hydrogen plasma processing on the structure and properties of tin oxide thin film produced by magnetron sputtering. Proc. of the 4th Int. Conf. on Thin Film Physics and Applications, 8-11 May, 2000. Shanghai, China. Vol. 4086, p. 323.

Kaur, M., Aswal, D.K., Yakhmi J.V. (2007). Chemiresistor gas sensor: materials, mechanisms and fabrication. In: Science and Technology of Chemiresistor Gas Sensors. Editors: Aswal D.K., Gupta S.K. Chapter 2, pp. 33-93.

Kern, W., Ban, V.S. (1978). In: Thin Film Processes. New-York: Academic Press. Ed. J.L. Vossen and W. Kern. P. 257–331.

Khol'kin, A.I., Patrusheva, T.N. (2006) Ekstraktsionno-piroliticheskiĭ metod: Poluchenie funktsional'nykh oksidnykh materialov. (Extraction-pyrolytic method: Preparation of functional oxide materials) – M.: Kom Kniga, 288 s. In Russian.

Kim, K.H., Chun, J.S. (1986). X-ray studies of SnO_2 prepared by chemical vapour deposition. Thin Solid Films, Vol. 141, pp. 287–295.

Kissin V. V., Sysoev V. V., Voroshilov S. A. (1999). Discrimination of acetone and ammonia vapour using an array of thin-film sensors of the same type. Technical Physics Letters, V. 25, Iss.8, pp. 657-659.

Kissin, V. V., Sysoev, V. V., Voroshilov, S. A., Simakov, V. V. (2000). Effect of oxygen adsorption on the conductivity of thin SnO_2 films. Semiconductors, Vol. 34, № 3, pp. 308-311.

Knuniants, I.L. (1964). In: Kratkaya khimicheskaya entsiklopediya (Brief Chemical Encyclopedia). Moscow: Sovetskaya entsiklopediya, 1964. 1112 p. with illustrations. In Russian.

Kobayashi, N. (2005). In: Vvedenie v nanotekhnologiyu (Introduction to Nanotechnology). Moscow: Binom, Laboratoriya Znanii, 134 p. In Russian (translated from Japanese).

Kukuev, V.I., Popov, G.P. (1989). Povyshenie nadezhnosti plenochnykh nagrevatel'nykh elementov na osnove dvuokisi olova (Improving the reliability of film heating elements based on tin dioxide). Elektronnaya promyshlennost' (Electron industry), №3, pp. 33–35.

Lewin, R., Howson, R.P., Bishop, C.A., Ridge, M.I. (1986). Transparent conducting oxides of metals and alloys made by reactive magnetron sputtering from elemental targets. Vacuum, Vol. 36, pp. 95–98.

Martin, P.J., Netterfield, R.P. (1986). Properties of indium tin oxide films prepared by ion-assisted deposition. Thin Solid Films, Vol. 137, pp. 207–214.

McDonagh, C., Bowe, P., Mongey, K., MacCraith, B.D. (2002). Characterization of porosity and sensor response times of sol-gel-derived thin films for oxygen sensor applications. Journal of Non-Crystalline Solids, Vol. 306, pp. 138–148.

Melsheimer, J., Teshe, B. (1986). Electron microscopy studies of sprayed thin tin dioxide films. Thin Solid Films, Vol. 138, pp. 71–78.

Minami, T., Sato, H., Nanto, H., Takata, S. (1989). Heat treatment in hydrogen gas and plasma for transparent conducting oxide films such as ZnO, SnO_2 and indium tin oxide. Thin Solid Films, Vol. 176, pp. 277–282.

Mishra, S., Ghanshyam, C., Ram, N., Singh, S., Bajpai, R.P., Bedi, R.K. (2002). Alcohol sensing of tin oxide thin film prepared by sol-gel process. Bull. Mater. Sci., Vol. 25, № 3, pp. 231-234.

Nagamoto Takao, Maruta Yukihiro, Omoto Osami (1990). Electrical and optical properties of vacuum-evaporated indium-tin oxide films with high electron mobility. Thin Solid Films, Vol. 193/194, pp. 704–711.

Okunara, T., Kasai, A., Hayakova, N., Voneda, G., Misono, M. (1983). Catalysis by eteropolicompounds. Journal of catalysis, Vol. 83, pp. 121–130.

Ramamoorthy, R., Kennedy, M.K., Nienhaus, H., Lorke, A., Kruis, F.E., Fissan, H. (2003). Surface oxidation of monodisperse SnO_x nanoparticles. Sensors and Actuators B, Vol. 88, pp. 281–285.

Rembeza, S. I. , Svistova, T. V., Rembeza, E. S., Borsyakova, O. I. (2001). The microstructure and physical properties of thin SnO_2 films. Semiconductors, Vol. 35, № 7, pp. 762-765.

Rumyantseva, M.N., Bulova, M.N., Charyeev, D.A., Ryabova, L.I., Akimov, B.A., Arhangel'skiĭ, I.V., Gas'kov, A.M. (2001). Sintez i issledovanie nanokompozitov na osnove poluprovodnikovykh oksidov SnO_2 i WO_3 (Synthesis and study of nanocomposites based on semiconducting oxides, SnO_2 and WO_3). Vestn. Mosk. Un-ta. Ser.2: Khimiya, T. 42, №5, pp. 348–355. In Russian.

Rumyantseva, M.N., Makyeeva, Ye.A., Gas'kov, A.M. (2008). Vliyanie mikrostruktury poluprovodnikovykh sensornykh materialov na khemosorbtsiyu kisloroda na ikh poverhnosti (Influence of microstructure of semiconductor sensor materials on chemisorptions of oxygen on their surface). Ros. khim. zhurn. (Zh. Ros. Khim. ob-va im. D.I.Mendelyeeva) (Ros. Chem. Journal) - Vol. LII, №2, pp. 122–129. In Russian.

Rumyantseva, M.N., Safonova, O.V., Bulova, M.N., Ryabova, L.I., Gas'kova, A.M. (2003). Gazochuvstvitel'nye materialy na osnove dioksida olova (Gas-sensitive materials based on tin dioxide). Sensor, № 2(8), pp. 8–33. In Russian.

Rumyantseva, M.N., Safonova, O.V., Bulova, M.N., Ryabova, L.I., Gas'kova, A.M. (2003). Legiruyushchie primesi v nanokristallicheskom diokside olova (Dopants in nanocrystalline tin dioxide). Izvestiya Akademii nauk. Seriya khimicheskaya (Proceedings of the Academy of Sciences. Chemical Bulletin), №6, pp. 1151–1171.

Ryabtsev, S.V., Yukish, A.V., Khango, S.I., Yurakov, Yu.A., Shaposhnik, A.V.; Domashevskaya, E.P. (2008). Kinetics of resistive response of SnO_{2-x} thin films in gas environment. Semiconductors, vol. 42, issue 4, pp. 481-485.

Ryzhikov, A.S., Vasiliev, R.B., Rumyantseva, M.N., Ryabova, L.I., Dosovitsky, G.A., Gilmutdinov, A.M., Kozlovsky, V.F., Gaskov, A.M. (2002). Microstructure and electrophysical properties of SnO_2, ZnO and In_2O_3 nanocrystalline films prepared by reactive magnetron. Materials Science and Engineering. B, Vol. 96, pp. 268–274.

Sanz Maudes, J., Rodriguez, T. (1980). Pyrolysis method for preparing of transparent conducting films. Thin Solid Films, Vol. 69, p. 183.

Semancik S., Cavicchi R.E. (1991). The growth of thin, epitaxial SnO_2 films for gas sensing applications. Thin Solid Films, Vol. 206, pp. 81–87.

Seok-Kyun Song. (1999). Characteristics of SnO_x films deposited by reactive ion-assisted deposition. Phys. Rev., Vol. 60, pp. 11137–11148.

Shanthi, E., Dutta, V., Banerjee, A., Chopra, K.L. (1981). SnO_2 Based Gas Sensitive Sensor. J. Appl. Phys., Vol. 51, pp. 6243–6249.

Srivastava, R., Dwivedi, R., Srivastava, S.K. (1998a). Effect of oxygen, nitrogen and hydrogen plasma processing on palladium doped tin oxide thick film gas sensors. Physics of Semiconductor Devices. - India, New Delhi: Narosa Publishing House, pp. 526–528.

Srivastava, R., Dwivedi, R., Srivastava, S.K. (1998b). Effect of oxygen and hydrogen plasma treatment on the room temperature sensitivity of SnO_2 gas sensors. Microelectronics Journal, Vol. 29, pp. 833–838.

Stjerna B. and Granqvist C.G. (1990). Optical and electrical properties of SnO_x thin films made by reactive R.F. magnetron sputtering. Thin Solid Films, 193/194, pp. 704–711.

Suikovskaya, N.V. (1971). Khimicheskie metody polucheniya tonkikh prozrachnykh plenok (Chemical methods of thin transparent films). - Leningrad. Khimiya (Chemistry), 200 s. In Russian.

Suzdalev, I.P. (2006). In: Nanotekhnologiya: phiziko-khimiya nanoklasterov, nanostruktur i nanomaterialov (Nanotechnology: physics and chemistry of nanoclusters, nanostructures and nanomaterials), M.: KomKniga, 592 p. In Russian.

Torkhov, D.S., Burukhin, A.A., Churagulov, B.R., Rumyantseva, M.N., Maksimov, V.D. (2003). Hydrothermal synthesis of nanocrystalline SnO_2 for gas sensors. Inorganic Materials. Vol. 39, № 11, pp. 1158-1162.

Vigleb, G. (1989). In: Datchiki (Sensors). - Moscow: Mir, 196 p. In Russian.

Watson, J., Ihokura, K. and Coles, S. (1993). The tin dioxide gas sensor. Meas. Sci. Technol., Vol. 4, pp. 713–719.

Weigtens, C.H.L., Loon, P.A.C. (1991). Influence of annealing on the optical properties of indium tin oxide. Thin Solid Films, Electronics and Optics, Vol. 196, pp. 17–25.

Xu, Ch., Jun, T., Norio, M., Nobory, Y. (1991). Grain size effects on gas sensitivity of porous SnO_2 based elements. Sensors and Actuators B., Vol. 3, pp. 147–155.

Zang, G. and Liu, M. (1999). Preparation of nanostructured tin oxide using a sol-gel process based on tin tetrachloride and ethylene glycol. Journal of Material Science, Vol. 34, pp. 3213–3219.

10

Crystallization of Sub-Micrometer Sized ZSM-5 Zeolites in SDA-Free Systems

Nan Ren[1,*], Boris Subotić[2] and Josip Bronić[2]
[1]*Department of Chemistry, Shanghai Key Laboratory of Molecular Catalysis and Innovative Materials and Laboratory of Advanced Materials, Fudan University, Shanghai*
[2]*Ruđer Bošković Institute, Division of Materials Chemistry, Laboratory for the Synthesis of New Materials Zagreb*
[1]*People's Republic of China*
[2]*Croatia*

1. Introduction

In recent years, the threat of the potential shortage of oil becomes one of the difficult problems which not only influences the quality of human life but also triggers the regional conflicts or wars. Thus, the oil management becomes the hottest topic in the contemporary economical and political world. To well solve that problem, there are two different approaches, 1) to discover/develop the alternative energy sources such as: bio-energy (biogas, biofuel), sunlight energy (or power plants), energy of wind (or power plants), nuclear energy (or power plants), etc. and 2) to increase the efficiency of crude oil (fossil fuels) processing and quality of final product. Although the first one seems interesting and ambitious, the existing drawbacks such as the low (but increasing) efficiency of sunlight transformation and the potential risk of the leakage of radioactive materials from nuclear facilities, become the great obstacle for the fast and promising development of these types of 'New plants Energetic Strategy'. Compare with the first approach, the latter seems more mild, reliable and realizable. To achieve such goal, the increase of the efficiency of catalysts is the key. Most frequently, different types of zeolites are used as catalysts for crude oil processing - among them, zeolite ZSM-5 has most expressive role.

Zeolite ZSM-5, as a member of the family of pentasil zeolites, has aroused tremendous interest after its first discovery by the research group of Mobile Company in the year 1972 [1]. With its adjustable framework Al content (from 0 to about 8Al per unit cell), two dimensional micropore channels (0.55 nm × 0.54 nm; Fig. 1a), sinusoidal pore geometry along c axis (Fig.1b) and easy insertion of hetero-T atoms, this material plays an important role in many of crucial catalytic processes such as hydro-cracking, de-waxing, alkylation, etc., [2-5] as well as in separation of organic compounds with different sizes and shapes [6]. In the case when zeolite ZSM-5 was used as catalyst, most of reactions are 'diffusion-controlled' [7]. This means that the product distribution largely depend on the nature and location of active sites in the crystalline framework of catalyst. Thus, the increase of the

* Corresponding Author

catalyst reactivity and thus, its efficiency can be achieved by increase of easily accessible active sites and by decrease of the length of diffusion path of reactants/intermediates/products and eliminating the probability of the occurrence of coke-formation side-reactions. Upon decreasing the crystal size, the diffusion paths of the reactant and product molecules inside the pores becomes shorter, and thus this can result in the reduction or elimination of undesired diffusion limitations of the reaction rate [8-10].

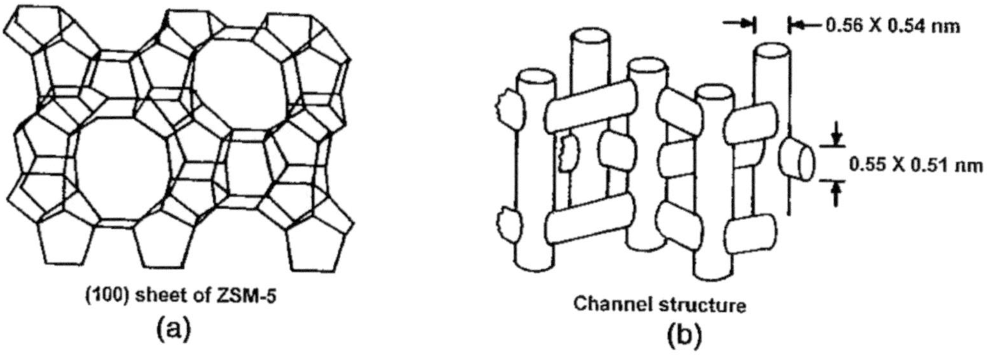

Fig. 1. The topological view of ZSM-5 crystals, (a) skeletal diagram of the (100) face and (b) channel structure.

To synthesize small-sized ZSM-5 zeolites and achieve the precise control of crystal size, the synthesis mechanism of ZSM-5 and the critical processes occurring during its crystallization should be understood thoroughly. From the well documented literature data, it is evident that, besides the framework constituents such as silica and alumina in different forms [1,11-18] and different alkaline bearing cations [13,19-24], the presence of organic structure directing agent (SDA) is vital for the synthesis of zeolite ZSM-5 [1,11-13,15,18,21,25] and other high-silica types of zeolites.

Zeolite ZSM-5 is conventionally synthesized by hydrothermal treatment of the reactive gel containing aluminosilicate as well as the tetrapropylammonium ions (TPA^+) as structure directing agent [1,11-13,15,18,25-27]. For the sake of industrial scale production, the synthesis of ZSM-5 zeolites could also be performed by using cheaper silica sources (fumed or precipitated silica) with reduced amount of SDAs. In such cases, the starting reaction mixture appears in the form of dense gel and the gelation phenomena can be observed at very short period of the mixing of the reactants; such system is also denoted as heterogeneous crystallization system [28,29]. However, the size of the products obtained from this route is normally in the range of several micrometers to tens micrometers, and the crystal size seems very difficult to be adjusted.

Using the known gel compositions for the synthesis of ZSM-5 zeolites, recently, a method called 'clear solution' synthesis has been proposed for the successful preparation of small sized ZSM-5 zeolites [30]. Since the 'clear solution' approach generally uses silica species in molecular form and thus a transparent solution appears at the initial stage of crystallization, such synthesis system is also called homogeneous crystallization system [28,29]. Although the crystal size could be facilely controlled by this method, such approach has obvious drawbacks such as low product yield (less than 5 wt. %), use of large amount of both SDAs

(SDA/SiO_2 ratio normally exceeds 0.2) and expensive silica sources (normally TEOS), which inevitably increases the production cost [31].

Thus, an expected ideal approach for the synthesis of small sized ZSM-5 zeolites is a compromise between heterogeneous and homogeneous crystallization systems. However, even in the case of successful compromise between heterogeneous and homogeneous crystallization systems, the problem of use of SDAs is still persisting [27, 32-36]. On the other hand, many attempts of the synthesis of zeolite ZSM-5 without organic templates [6,25,27,34-49] were encountered with another group of problems such as that zeolite ZSM-5 can be synthesized only in narrow ranges of SiO_2/Al_2O_2 (~ 40-70) and Na_2O/SiO_2 (~ 0.13-0.20) [36], formation of impurity phases such as un-reacted amorphous solids [47], quartz [41], mordenite [42] and analcime [44], uncontrollable crystal size [25], long crystallization time [40] and low yield [40].

Most of the mentioned problems can, however, be overcome by addition of small amount of seed crystals (ZSM-5, silicalite-1) in the TPA^+-free reaction mixture [6, 27, 35, 38, 40, 45, 50]. Seed induction synthesis is a well developed strategy which could not only shorten the duration of synthesis, but also control the product properties [51]; addition of seed crystals results in the formation of zeolite ZSM-5 with high degree of crystallinity and a narrow size distribution at short synthesis times [38,40]. Such method has been used for the synthesis of zeolites with various framework topologies [52]. Recently, small sized zeolites were obtained fastly, using this approach [53]. This method, although old, is still under developing.

Taking into consideration the background of ZSM-5 synthesis conditions, and accompanying this with the basic knowledge of seed induced synthesis, we have developed a seed surface crystallization (SSC) approach, in which the sub-micrometer sized ZSM-5 zeolites can be obtained in a controllable manner [50, 54, 55]. More importantly, the growth of crystalline end products is achieved in a heterogeneous, SDA-free crystallization system, which is good basis for their further industrial-scale production and use in various applications.

In this chapter, SDA-free, SSC approach for the crystallization of sub-micrometer sized ZSM-5 zeolites including the influence of various synthesis parameters on the product properties and the crystallization mechanism will be discussed in detail. The relevant content is divided into five parts: (1) Controllable synthesis, (2) Influence of batch alkalinity, (3) Influence of sodium ions and gel ageing, (4) Crystallization mechanism, and (5) Modeling approach.

2. Controllable synthesis of sub-micrometer sized ZSM-5 zeolites

Batch oxide molar chemical composition of the reaction mixture (hydrogel) for the synthesis was 1.0 Al_2O_3/100 SiO_2/28 Na_2O/4000 H_2O. A series of silicalite-1 nanocrystals having different mean diameters (90, 180, 220, 260 and 690 nm; Fig. 2) were prepared by synthesis from clear solution and used as seeds for the further growth of ZSM-5 nanocrystals.

Fig. 3 shows that the well crystalline ZSM-5 crystals are obtained after 2 h of hydrothermal treatment at 483K with addition of 4 wt.% of seeds (with respect to the total amount of silica in the reaction mixture). The crystal size of product increases with increasing size of seed crystals as can be seen in the corresponding SEM images (Fig. 4).

On the other hand, using different amount of the same seed crystals (260 nm in this case), the crystal size of the product decreases with increasing amount of added seed crystals (Fig. 5).

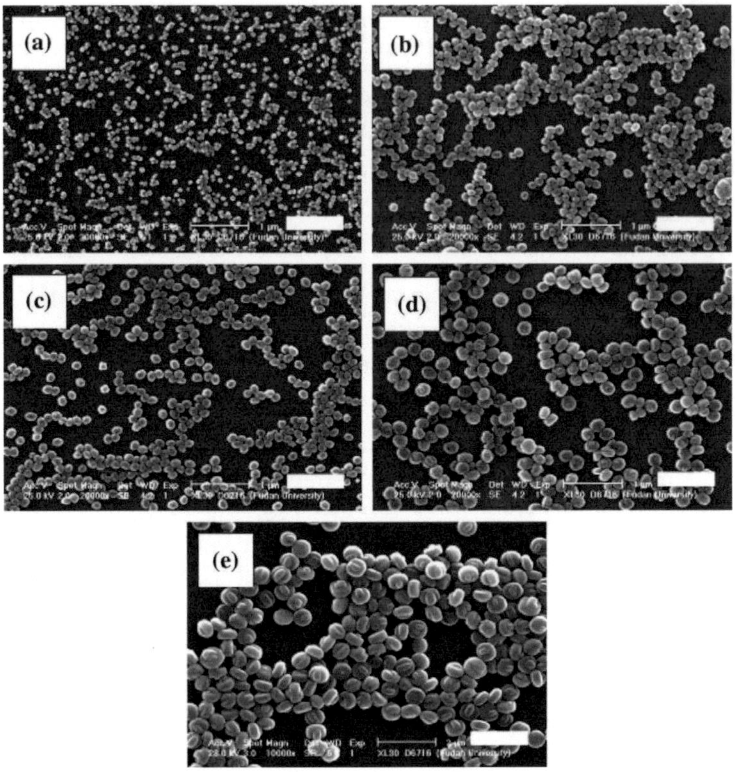

Fig. 2. SEM images of silicalite-1 nanocrystals with diameters of 90 (a), 180 (b), 220 (c), 260 (d), and 690 nm (e). The scale bars in a-d are of 1 μm, and of 2 μm in e. (Adopted from Ref. [50] with permission of Publisher.)

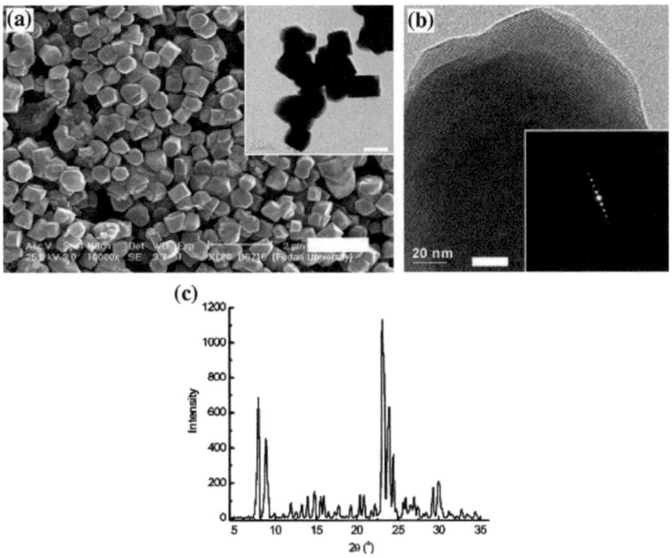

Fig. 3. The SEM (a), TEM (inset of a), High magnification TEM images (b), electron diffraction (ED) pattern (inset of b) and XRD pattern of ZSM-5, synthesized using 4 wt. % of

seed (260 nmsilicalite-1 crystals). The scale bars are of 2 μm in (a) and 20 nm in (b). (Adopted from Ref. [50] with permission of Publisher.)

Fig. 4. SEM images of ZSM-5 samples synthesized with using 4 wt. % of: 90 nm (a), 180 nm (b), 220 nm (c) and 690 nm (d) silicalite-1 seed crystals. The scale bars in all the images are of 2 μm. (Adopted from Ref. [50] with permission of Publisher.)

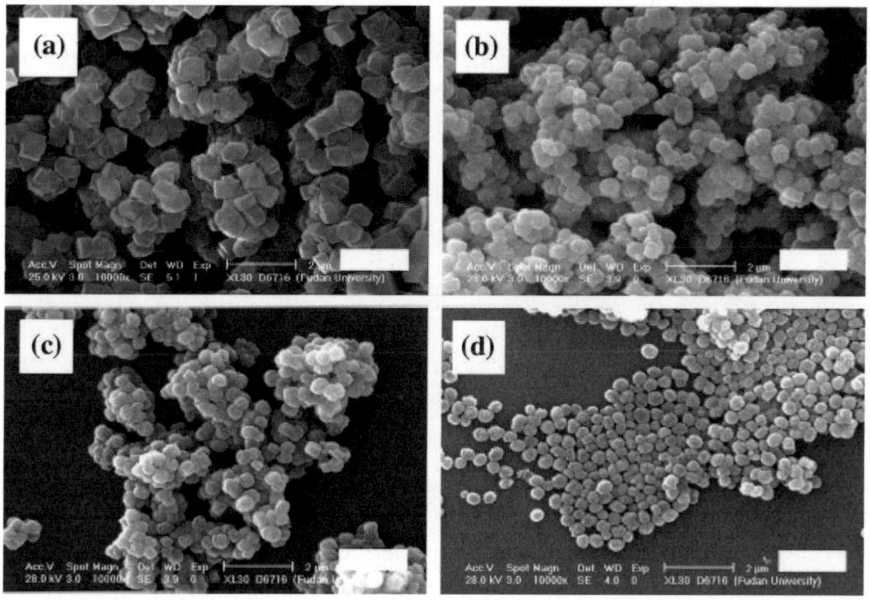

Fig. 5. SEM images of ZSM-5 samples synthesized by using 4 (a), 8 (b), 16 (c) and 32 (d) wt. % of 260 nm silicalite-1 seed crystals. The scale bars in all the images are of 2 μm. (Adopted from Ref. [50] with permission of Publisher.)

All the products are entirely crystalline and have adjustable size in the range from 270 nm to 1100 nm. This indicates that the SSC approach is a powerful tool for controlling the crystal size of the ZSM-5 zeolites. Besides the crystal size, the morphology of the product could also be controlled via changing the morphology of the seed crystals. The surface-stacking morphology of the silicalite-1 seed crystals (synthesized via microwave heating method) could be fairly well 'cloned' to the final ZSM-5 products (Figure 6). Based on the above phenomena, it could be found that the product properties could be adjusted via a 'seed-dependent' manner.

Fig. 6. The silicalite-1 seed crystals with surface-stacking morphology (a) and the corresponding ZSM-5 product (b). The scale bars is of 1 μm. (Adopted from Ref. [50] with permission of Publisher.)

3. Influence of batch alkalinity

Although the controllability of SSC approach has been proved by successful syntheses of sub-micrometer sized ZSM-5 zeolites by using different kinds of silicalite-1 seed crystals, another interesting aspect of this investigation is a study of the critical processes which occur during the crystallization. The materials with more controllable functionalities may be synthesized only with the thorough understanding of the crystallization mechanisms. In order to achieve this goal, the parameters influencing the crystallization pathway and properties of the products are adjusted. Since it is well known that the batch alkalinity play the key role in zeolite synthesis [13, 22, 30, 56-58], the influence of batch alkalinity, $A=[Na_2O/SiO_2]_b$, of this SDA-free system, is in detail studied in this part. In this part of study, the reaction mixtures (hydrogels) having different batch alkalinities, A, were seeded with 4 wt. % of 260 nm silicalite-1 seed crystals. The samples were characterized by SEM, XRD and PSD analyzes.

Fig. 7 shows the SEM images of the crystalline end products (zeolite ZSM-5) synthesized at different batch alkalinities. It can be clearly observed that at low batch alkalinities ($A \leq 0.003$), the ZSM-5 crystals are either surrounded or covered with amorphous species, indicating that the crystallization process is not completed. Increase of the batch alkalinity causes the morphological change of the samples from irregular particles ($0.004 \leq A \leq 0.005$) to cubic crystals ($0.006 \leq A \leq 0.010$). With further increase of the batch alkalinity ($0.011 \leq A \leq 0.012$), the morphologies of the samples again return to irregular shapes.

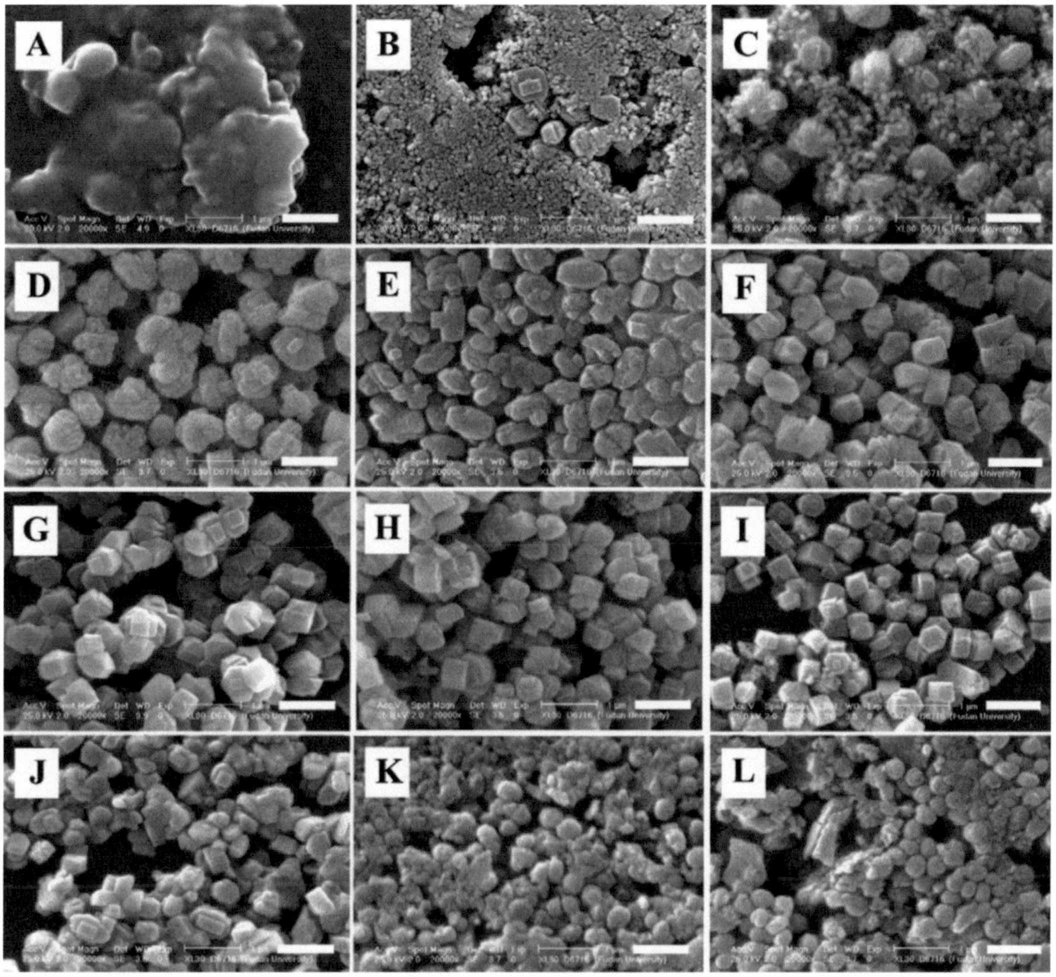

Fig. 7. SEM images of the products hydrothermally synthesized (at 438 K for 2 h) from the reaction mixtures (hydrogels) having: A = 0.001 (A), A= 0.002 (B), A = 0.003 (C), A = 0.004 (D), A = 0.005 (E), A = 0.006 (F), A = 0.007 (G), A = 0.008 (H), A = 0.009 (I), A = 0.010 (J), A = 0.011 (K), and A = 0.012 (L). The scale bars in all figures are of 1 μm.
(Adapted from Ref. [54] with permission of Publisher)

XRD analysis of the products revealed that fully crystalline zeolite ZSM-5 are obtained within the alkalinity range of $0.006 \leq A \leq 0.010$, which is in accordance with the observation in the corresponding SEM images (Figs. 7F - 7L).

In order to find more details of particulate properties, the Laser Light Scattering analysis of the obtained samples has been carried out. The particle size distribution (PSD) curves of the samples, measured by Mastersizer 2000 (Malvern) particle size analyzer, are shown in Fig. 8. The crystalline end products obtained under optimal alkalinity range ($0.006 \leq A \leq 0.01$) appear as a mixture of single crystals (50 - 60 % by number) having the size between 400 and 600 nm and aggregates of the single particles (40 – 50 % by number) having the size higher than 600 nm.

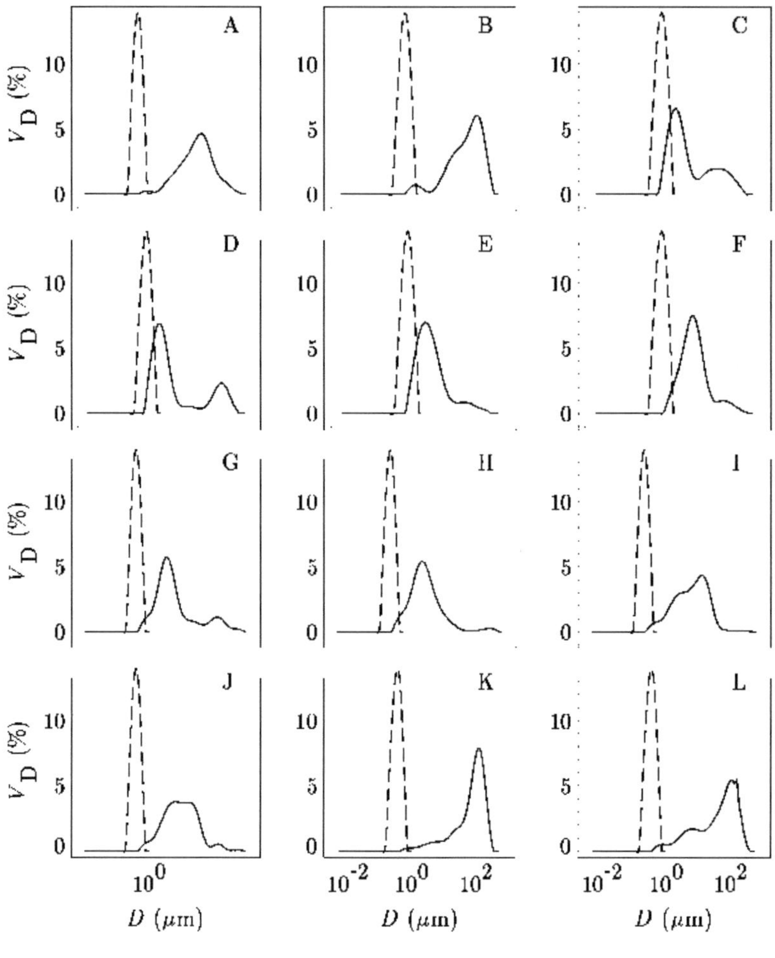

Fig. 8. Particle size distribution (by volume) of the products hydrothermally synthesized (at 438 K for 2 h) from the reaction mixtures (hydrogels) having: $A = 0.001$ (A), $A = 0.002$ (B), $A = 0.003$ (C), $A = 0.004$ (D), $A = 0.005$ (E), $A = 0.006$ (F), $A = 0.007$ (G), $A = 0.008$ (H), $A = 0.009$ (I), $A = 0.010$ (J), $A = 0.011$ (K), and $A = 0.012$ (L). Dashed curves represent the particle size distribution of silicalite-1 nanocrystals. V_D is volume percentage of crystals having the sphere equivalent diameter D. (Adopted from Ref. [54] with permission of Publisher)

Besides the phase purity and particulate properties, the Si/Al ratio of final products also correlates very well with the batch alkalinity. Table 1 summarizes the influence of alkalinity on the product properties. It could be concluded that the well crystallized ZSM-5 zeolites with adjustable Si/Al ratio from 10 - 18 can be obtained in the alkalinity range of $0.006 \leq A \leq 0.010$.

Batch alkalinity[a]	Phase[b]	Si/Al ratio[c]	Particle shape[d]
0.001	Amor[e]+ZSM-5	47	*f*
0.002	Amor[e]+ZSM-5	43	*f*
0.003	Amor[e]+ZSM-5	32	irregular
0.004	ZSM-5	28	irregular
0.005	ZSM-5	23	irregular+cubic
0.006	ZSM-5	18	cubic
0.007	ZSM-5	15	cubic
0.008	ZSM-5	12	cubic
0.009	ZSM-5	10	cubic
0.010	ZSM-5	9	cubic
0.011	Phillipsite+ZSM-5 +Amor[e]	11	Spherical+irregular
0.012	Phillipsite+ZSM-5 +Amor[e]	11	Spherical+irregular

a. $A=[Na_2O/H_2O]_b$
b. Determined from corresponding XRD patterns
c. Calculated from XRF results
d. Obtained from SEM observations
e. Amor, abbreviation of amorphous
f. Exact information cannot be provided because of the existence of large portion of amorphous impurities
(Adopted from Ref. [54] with permission of Publisher)

Table 1. Summary of structural, chemical and morphological properties of the products obtained at different alkalinities

4. Influence of sodium ions and gel ageing

It is expected that, besides the influence of alkalinity, sodium ions [13, 22, 56-58] and duration of room-temperature gel ageing [59] also influence the crystallization pathway and properties of zeolite ZSM-5. The sodium ion is recognized as the potential template ions which make the influences on the nucleation process of MFI zeolites especially in the SDA-free system [13, 22, 57, 58].

On the other hand, room temperature ageing of the reaction mixture shortens not only the duration of 'induction period' and the entire crystallization process, but also diminishes the differences in crystal size distributions of the final product [59, 60]. Moreover, in some cases, the ageing of the reaction mixture (hydrogel) influences also the morphology [61] and even phase composition of the final product [62]. From the above reasons, the influences of the mentioned parameters on the properties of crystallized zeolite ZSM-5 are described in this part. Since the increase of batch alkalinity ($A=[Na_2O/H_2O]_b$) is accompanied with the increase of sodium content, the related studies are performed at different batch alkalinities with the addition of sodium sulphate as the source of sodium ions excess. The batch content of sodium ions is expressed as $B = [Na^+/SiO_2]_b$. The batch molar composition of the reaction mixture for the synthesis of zeolite ZSM-5 is expressed as, $1.0Al_2O_3/100SiO_2/xNa_2O/4000H_2O/yNa_2SO_4$, where the values x and y are changed to adjust the batch alkalinity and batch content of sodium ions, respectively. The samples synthesized at different alkalinity and different sodium ion content are denoted as A/B (e.g., 0.003/0.24 for A = 0.003 and B = 0.24).

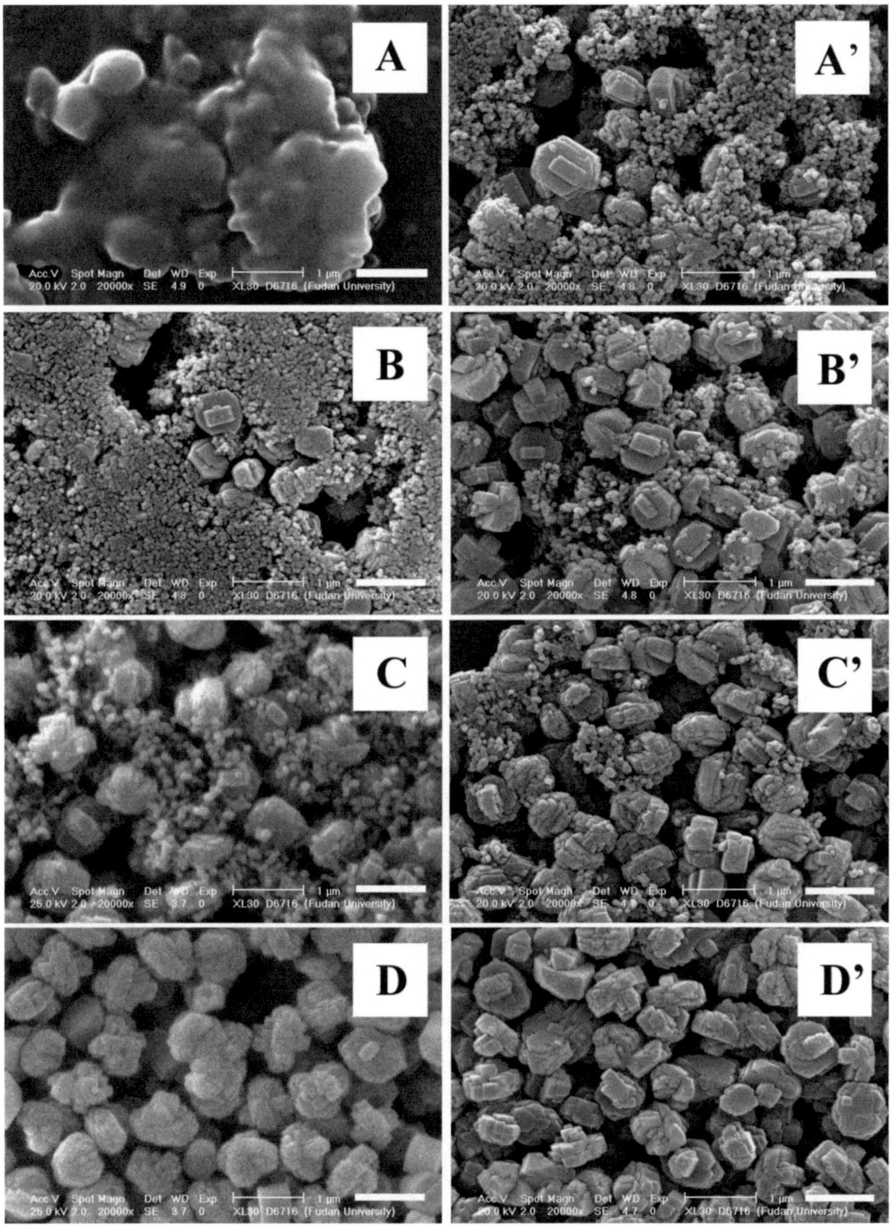

Fig. 9. SEM images of the products obtained by hydrothermal treatment (at 483 K for 2 h) of the reaction mixtures 0.001/0.08 (A), 0.001/0.64 (A'), 0.002/0.16 (B), 0.002/0.64 (B'), 0.003/0.24 (C), 0.003/0.64 (C'), 0.004/0.32 (D) and 0.004/0.64 (D'). The scale bars are of 1 μm. (Adopted and reproduced from Ref. [55] with permission of Publisher)

Fig. 9 shows the SEM images of the product obtained by hydrothermal treatment (at 483K for 2h) of the reaction mixtures having low alkalinities (A = 0.001 - 0.004) and different batch contents of sodium ions (B = 0.08 and 0.64). It can be observed that the increase of the batch content, B, of sodium ions at the same batch alkalinity, A, does not significantly influence the morphology of the final product, as can be seen comparing of the samples prepared without (Figs. 9A, 9B, 9C and 9D) and with addition of sodium sulphate (Figs. 9A', 9B', 9C' and 9D'). Also, the addition of sodium sulphate in the reaction mixture does not affect the phase composition (crystallinity) of products as revealed by XRD analysis of the product samples.

Knowing the templating role of sodium ions in the TPA-free synthesis of MFI-type zeolites [13, 22, 56-58], these findings indicate that the rate-determining factor is concentration of "free" low-molecular weight silicate species, in the short synthesis duration of this seed-induction system

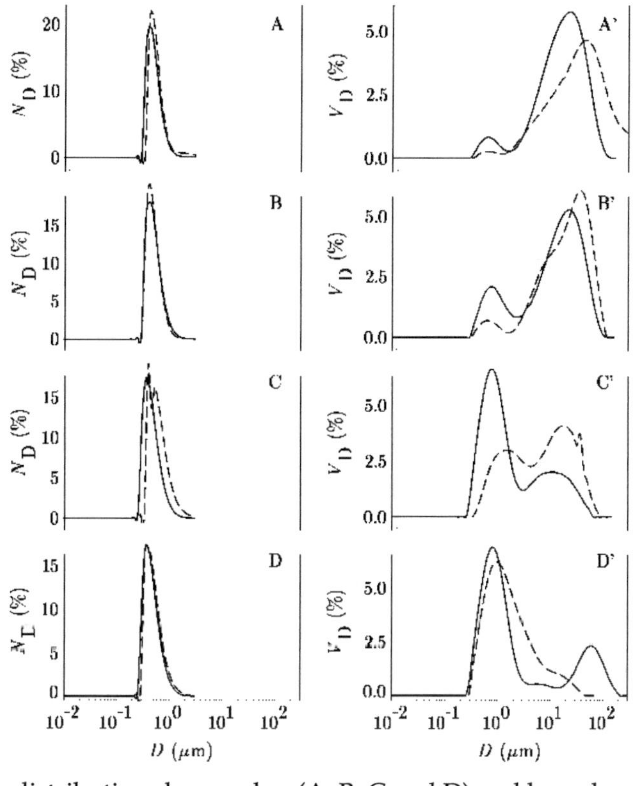

Fig. 10. Particle size distributions by number (A, B, C and D) and by volume (A', B', C' and D') of the products obtained from reaction mixtures having two batch content ($B = [Na^+/SiO_2]_b$) of Na ions 0.08 (dashed curves) 0.64 (solid curves). Alkalinity ($A = [Na_2O/H_2O]_b$) were: 0.001 for A and A', 0.002 for B and B', 0.002 for B and B', 0.003 for C and C', 0.003 for C and C', 0.004 for D and D'. N_D is number percentage and V_D is volume percentage of crystals having the sphere equivalent diameter D. (Adopted from Ref. [55] with permission of Publisher)

Since the generation of these species is mainly caused by hydrolysis of silica precursors which depend on the concentration of "free" OH^- ions and thus, on the batch alkalinity, A [54], it is evident that the alkalinity of system, A, determines the rate of crystallization and not the concentration of "free" Na^+ ions. In the other words, when $A < 0.004$, fully crystalline product cannot be obtained for $t_c \leq 2$ h regardless to the content of sodium ions in the reaction mixture [54,55].

Fig. 10 shows that the fraction of the small single particles (ZSM-5 crystals formed by growth of silicalite-1 seed crystals) having the size 400 – 600 nm increases with increasing alkalinity, *A*, of the reaction mixture and that, for a given batch alkalinity, the fraction of the single 400 – 600 nm particles increases with increasing value of *B*.

According to the principals of aggregation of zeolite particles, the tendency to aggregation decreases with the increase of the (negative) crystal surface charge (repulsion effect), and increases with the increase of the concentration of the "compensating" Na^+ ions [55]. It is evident that at the simultaneous increase of OH^- ions and Na^+ ions with increasing alkalinity, *A*, the repulsive force prevails the "compensating" effect of Na^+ ions. In this context, addition of sodium sulfate additionally increases the negative charge of the crystal surface by oxy-anion effect of SO_4^{2-} ions [63] and, at the same time, additionally reduces the "compensating" effect of Na^+ ions. The result is that the aggregation is considerably reduced in the presence of sodium sulphate.

When the batch alkalinity further increases to $A = 0.007$, the perfect small-sized ZSM-5 crystals can be obtained, as described in previous section. An increase of the batch amount of sodium ions does not influence the phase purity of products as it was revealed by XRD analysis of the samples. However, the surface of the crystals become rough as it can be observed in the corresponding SEM images (Fig. 11). This phenomenon could be attributed to the effect of sodium ions on the cross-linking and polymerization of the silicate species in solution [64]; the species formed by cross-linking and polymerization are further deposited onto the crystalline surface of the final crystals, causing the roughing of the surface morphology of the product. This assumption is further supported by the increase of Si/Al ratio of the crystalline end product from 15 for $[Na^+/SiO_2]_b = 0.56$ to 20 for $[Na^+/SiO_2]_b = 1.0$ and simultaneous decrease of Si concentration in the liquid phase of the reaction mixture from 20.23 mg/ml for $[Na^+/SiO_2]_b = 0.56$ to 14.00 mg/ml for $[Na^+/SiO_2]_b = 1.0$, accompanying with the increase of the solid recovery yield from 42% to 50%. The surface roughing phenomena caused by the deposition of active species from liquid to solid phase (crystals) can also be evidenced by the increase of the average size of the crystalline end products (see PSD curves in Fig. 12). More interestingly, the particle size distribution also becomes narrower with the increased amount of sodium ions. This indicates that the addition of sodium sulphate into reaction mixture also prevents aggregation of the ZSM-5 crystals formed during hydrothermal treatment as was already shown in the cases of low batch alkalinities.

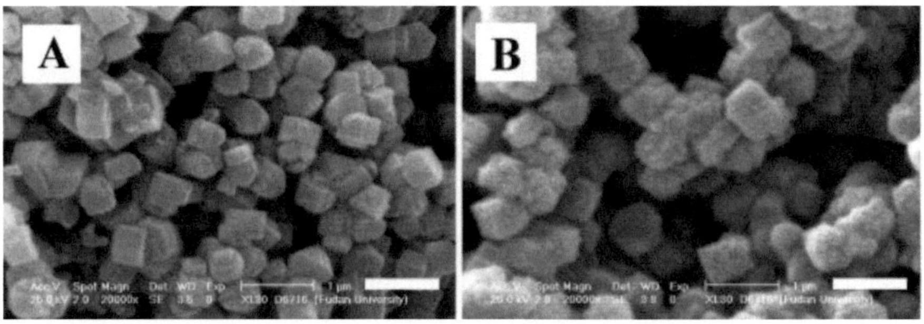

Fig. 11. SEM images of the products obtained by hydrothermal treatment (at 483 K for 2 h) of the reaction mixtures 0.007/0.56 (A) and 0.007/1.0 (B). The scale bars are of 1 μm. (Adopted from Ref. [55] with permission of Publisher)

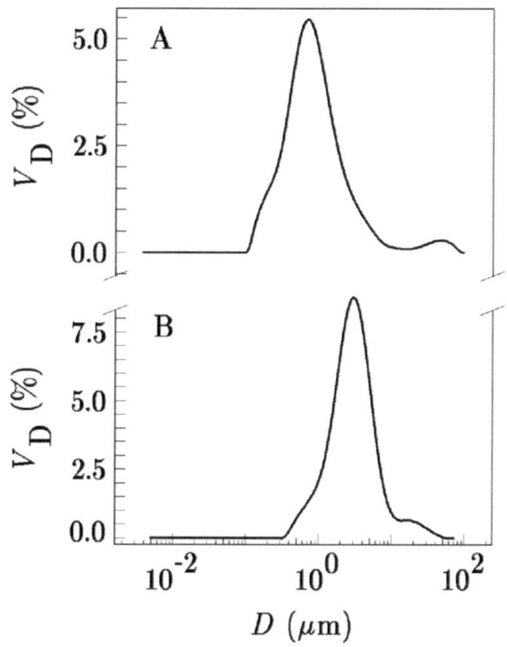

Fig. 12. Particle size distributions by volume of the products obtained by hydrothermal treatment (at 483 K for 2 h) of the reaction mixtures 0.007/0.56 (A) and 0.007/1.0 (B). (Adopted from Ref. [55] with permission of Publisher)

From the above results, it could be concluded that the addition of excess amount of sodium ions into the crystallization system has apparent effect on the particulate properties of the product. At low batch alkalinity, the additional sodium ions causes de-aggregation of the final products, rendering the particles with more uniform size distributions. At high batch alkalinity, the excess amount of sodium ions triggers the surface condensation reactions on the crystalline end products. However, the crystallization rate is not enhanced by the increase of batch sodium ion content, indicating that the determining factor of crystallization of ZSM-5 zeolites in SDA-free system is concentration of low molecular weight silicate species, determined by batch alkalinity.

Fig. 13. SEM images of the products obtained by hydrothermal treatment (at 483 K for 2 h) of the reaction mixtures aged at room temperature for 0 (A), 3 (B) and 48 h (C) in the presence of silicalite-1 seed crystals. The scale bars are of 1 μm. (Adopted from Ref. [55] with permission of Publisher)

In addition, the influence of room temperature ageing of the reaction mixture on the product properties has also been investigated. Since in the investigated crystallization system, the seed crystals are added as one of the gradients of the gel precursors, the relevant studies are separated as ageing in the presence or absence of seed crystals.

Figs. 13 and 14 show the SEM images and PSD curves of the ZSM-5 zeolites crystallized from the reaction mixtures aged for different times in the presence of silicalite-1 seed crystals. Although the ageing does not change morphology of the crystals, the particle size distribution becomes narrower with increasing the ageing duration. Moreover, it can also be observed that the particle size of the product slightly decreases with ageing.

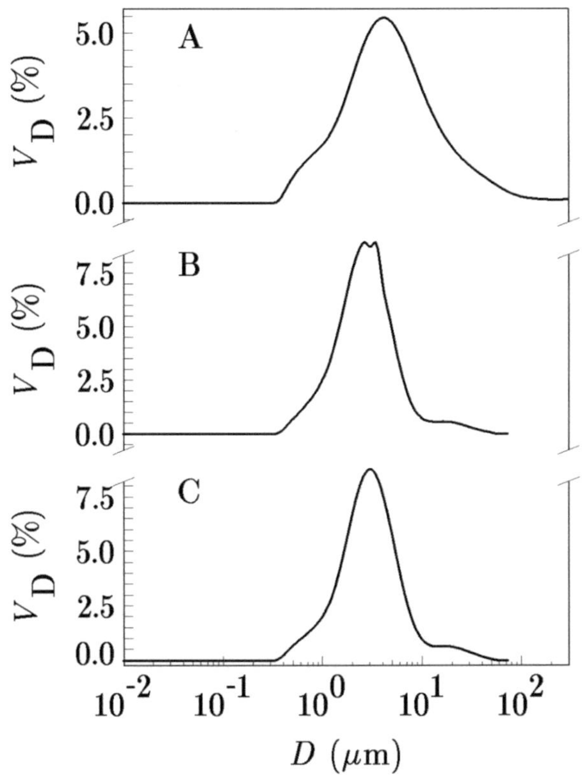

Fig. 14. Particle size distributions by volume of the products obtained by hydrothermal treatment (at 483 K for 2 h) of the reaction mixtures having aged at room temperature for 0 (A), 3 (B) and 48 h (C) in the presence of silicalite-1 seed crystals. (Adopted from Ref. [55] with permission of Publisher)

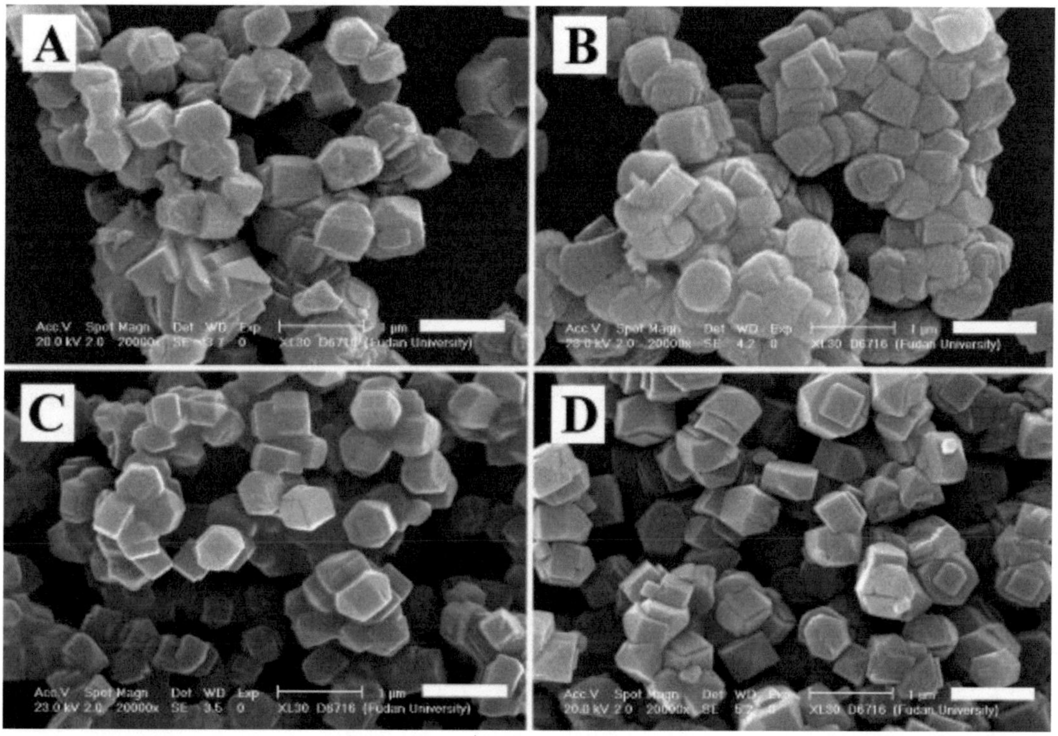

Fig. 15. SEM images of the products obtained by hydrothermal treatment (at 483 K for 2 h) of the reaction mixture (A = 0.008; $[Na^+/SiO_2]_b$ = 0.64) aged at room temperature for 5 (A), 20 (B), 48 (C) and 120 h (D) before addition of seeds and heating. The scale bars are of 1 μm. (Adopted from Ref. [55] with permission of Publisher)

Similar like in the cases of the ageing in the presence of seeds, the results from the ageing of the seed-free reaction mixture also show the same trend (Figs. 15 and 16). The crystallinity of the product does not markedly change with ageing. On the other hand, the PSD curves in Fig. 16 show a gradual decrease of the fraction of large-size "particles" (10 - 500 μm) and simultaneous increase of the fraction of discrete particles (crystals) having the size in the range from 0.45 - 10 μm, with the increase of the ageing duration. This undoubtedly shows that the ageing of the seed-free reaction mixture prevents the agglomeration of growing crystals of zeolite ZSM-5.

The independence of the properties of crystalline end products on the duration of room temperature ageing of reaction mixtures indicates that the formation of precursors for the subsequent growth of ZSM-5 zeolites does not occur during the room-temperature ageing in the current system and, at the same time, this confirms the hypothesis that, under such conditions, the active growth precursors can only be formed at reaction temperature. Since in SDA-free crystallization systems, the formation of new nuclei is greatly suppressed in the presence of seed crystals [51], the size of discrete crystals in the product depend only on the amount and size of the seeds present in the reaction mixture.

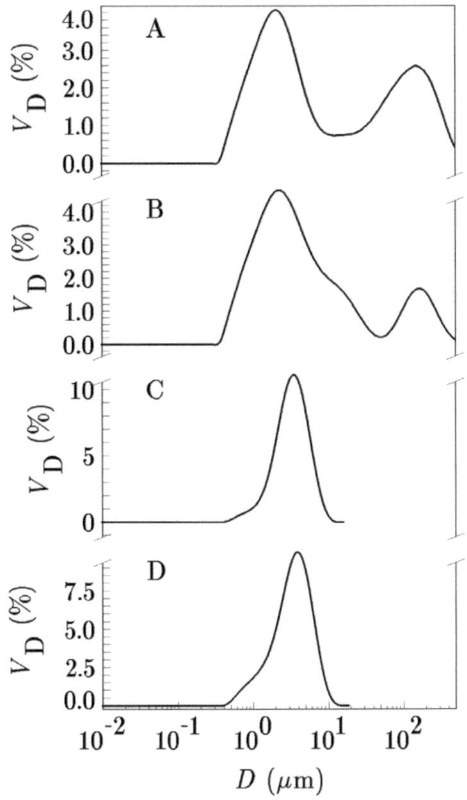

Fig. 16. PSDs by volume of the products obtained by hydrothermal treatment (at 483 K for 2 h) of the reaction mixture (A = 0.008; $[Na^+/SiO_2]_b$ = 0.64) aged at room temperature for 5 (A), 20 (B), 48 (C) and 120 h (D) before addition of seeds and heating. (Adopted from Ref. [55] with permission of Publisher)

On the other hand, the influence of the hydrogel ageing on the PSDs of the crystalline end products is probably connected with the change (increase) of the concentration of Si (most probably in the form of low-molecular weight silicate species) in the liquid phase of the reaction mixture with prolonging ageing duration (Fig. 17). Although the increase of the concentration of Si in the liquid phase during ageing (Fig. 17) does not cause the formation of the growth precursor species at the ageing (room) temperature, it certainly facilitates the formation of the growth precursor species during heating of the reaction mixture, i.e., higher starting concentration of low-molecular weight silicate species causes higher rate of formation of the growth precursor species and thus, higher rate of growth of zeolite ZSM-5 on the silicalite-1 crystals. From this reason, particle (crystal) size at given crystallization time increases with increasing time of hydrogel ageing (Fig. 16). Since, on the other hand (i) crystal growth and agglomeration take place simultaneously and (ii), the rate of agglomeration increases with decreasing particle size [65], tendency of agglomeration decreases with increasing time of hydrogel ageing (Fig. 16).

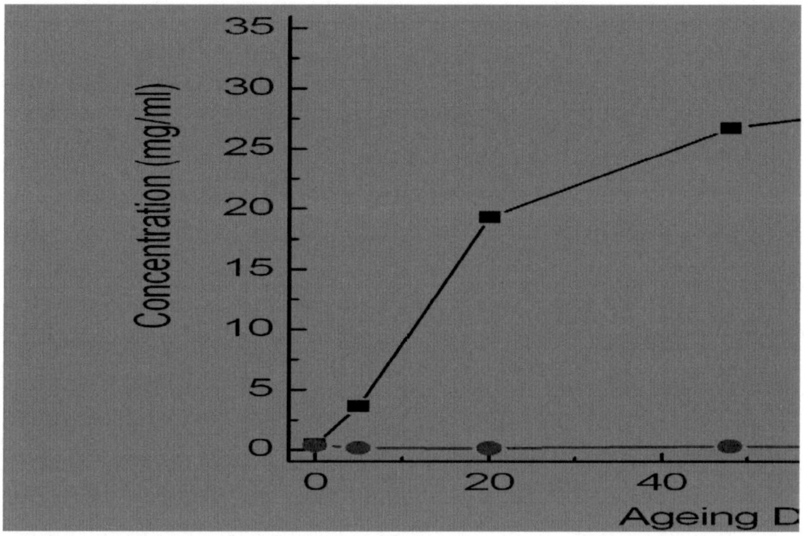

Fig. 17. Influence of the room-temperature ageing of the seed-free reaction mixture (A = 0.008; $[Na+/SiO2]_b = 0.64$) on: (A) changes in the concentrations of silicon (■) and aluminum (●) in the liquid phase. (Adopted and reproduced from Ref. [55] with permission of Publisher)

5. Crystallization mechanism

With the above studies, the influence of various parameters on the three most important properties (phase purity, particulate properties and chemical composition) of ZSM-5 zeolites can be clearly concluded in Table 2. From this summary, the most pronounced influence is attributed to the batch alkalinity, which changes all the relevant properties of the crystalline-end product. In the large-scale synthesis of ZSM-5 zeolites for application, the batch alkalinity should be strictly controlled to obtain the qualified product. On the other hand, among the mentioned three properties of ZSM-5 zeolites, the particulate properties are most sensitive to the change of the environment of crystallization. Thus, the synthesis of ZSM-5 zeolites with controllable uniform crystal sizes is always a hot, attractive and enduring topic in the field of zeolites.

	Seed size & morphology	Seed amount	Batch alkalinity	Batch sodium ion content	Gel ageing effect
Phase purity (structural property)	-	-	+	-	-
Particulate property (size and distribution)	+	+	+	+	+
Chemical composition (Si/Al in framework)	-	+	+	+/-	-

a. The influences are expressed as pronounced influence (+), less influence (-), and influence depends on the detailed condition

Table 2. Summary of the influences of synthetic parameters on the properties of ZSM-5 product. [a]

The systematic analysis of the parameters governing the crystallization of zeolite ZSM-5 also makes a chance to define the crystallization mechanism of ZSM-5 zeolites in SDA-free system. To achieve the controllable synthesis of ZSM-5 zeolites with designed particulate and chemical properties, the thorough understanding of the critical processes during the crystallization is the only way. Such goal is achieved by both efforts, the kinetic analysis of crystallization processes and the alkaline, post-treatment of final products.

Fig. 18 shows the crystallization curves of the systems containing various amounts (expressed as the weight percentage of silica amount in the batch) of silicalite-1 seed crystals of different sizes. It can be clearly observed that the rate of crystallization considerably depends on both the size and the amount of the added silicalite-1 seed crystals. The time for the completion of crystallization decreases with both decreasing seed size (at constant amount) and increasing amount of seeds (at constant seed size). Furthermore, the crystallization process of the system with 4 wt.% of 260 nm seed crystals was followed by TEM observations (Fig. 19). At the initial stage of crystallization, the seed crystals and amorphous precursors (existing as small particles) can be identified clearly (Fig. 19a). At prolonged hydrothermal treatment, the amorphous precursors start to stack onto the surface of seed crystals and agglomerate together (Figs. 19b and c), followed by depleting of the amorphous precursors and complete ordering of the structure into ZSM-5 crystals (Fig. 19 d). High magnification TEM image (Fig. 20) shows a "core-shell" transformation process of amorphous to crystalline phase (zeolite ZSM-5), which happens at the seed-amorphous interface.

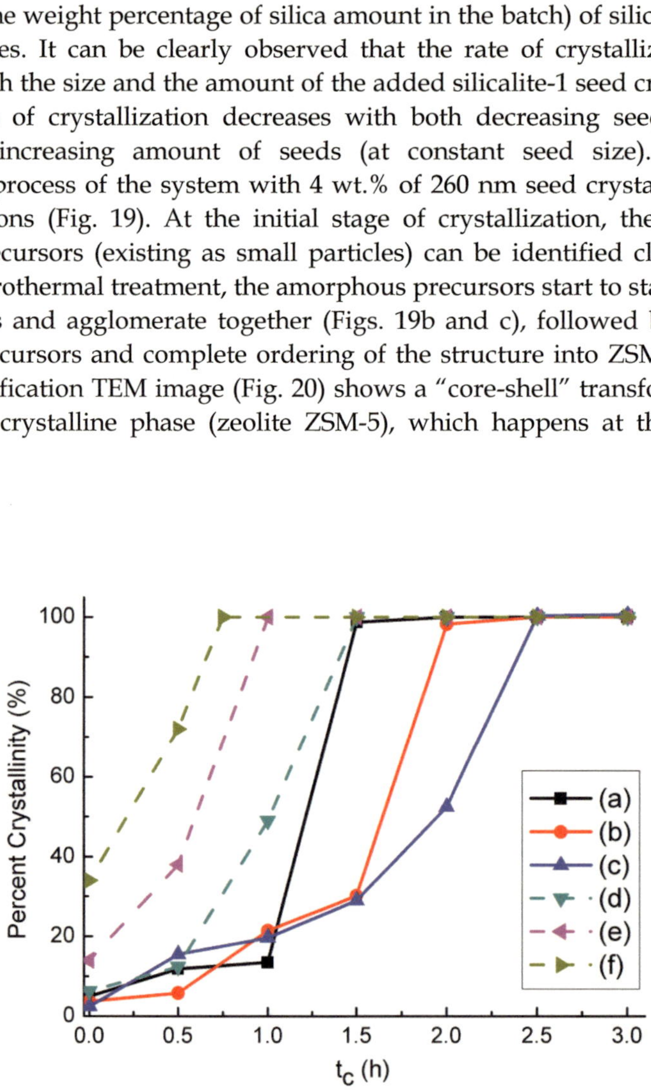

Fig. 18. Crystallization curves of systems with using silicalite-1 seeds of 90 nm-4wt.% (a), 260 nm-4wt.% (b), 690 nm-4wt.% (c), 260 nm-8wt.% (d), 260 nm-16wt.% (e), and 260 nm-32wt.%(f). t_c is the time of crystallization. The percent crystallinity was calculated from the corresponding XRD patterns. (Adopted from Ref. [50] with permission of Publisher)

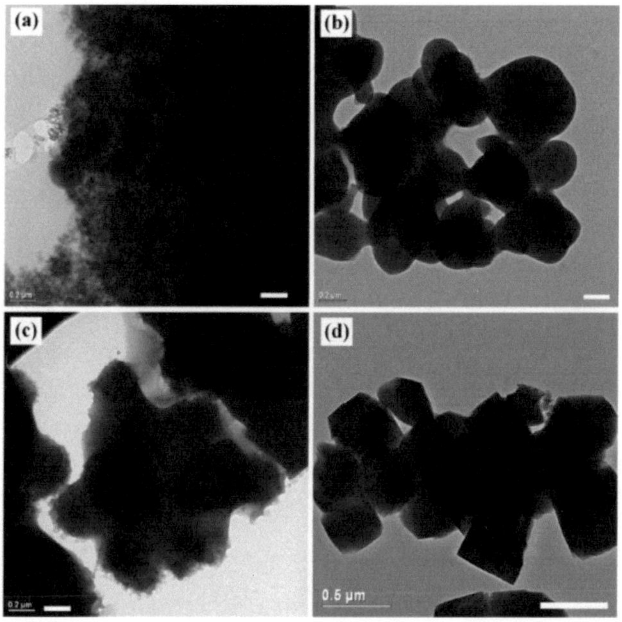

Fig. 19. TEM images of the samples obtained by using 4 wt. % of 260 nm silicalite-1 seed crystals at t_c = 0.5 h (a), t_c = 1.0 h (b), t_c = 1.5 h (c), and t_c = 2.0 h (d). The scale bars in (a), (b), (c), and (d) are of 0.2 μm, 0.2 μm, 0.2 μm, and 0.5 μm, respectively. (Adopted from Ref. [50] with permission of Publisher)

Fig. 21 displays the PSD curves (by number) of the 260 nm silicalite-1 seed crystals (Fig. 21A) and that of the ZSM-5 zeolites obtained using same type of seeds (Fig. 21B). It can be observed that both curves possess almost the same shape and trend. The only difference is the size of crystals. Such phenomenon, although simple, clearly revealed the fact that the total number of crystals during the process of crystallization remains unchanged. It can be deduced that the nucleation process is completely suppressed under the current synthesis condition (SDA-free system with the presence of seed crystals), further proves the only occurrence of the growth of ZSM-5 zeolites on the surface of silicalite-1 seeds.

In addition, the post-synthesis alkaline treatment was carried out on the obtained samples. Since the tetrahedral aluminium centres are relatively inert to hydroxide attack due to the negative charges associated with these centers, and the aluminum atom protects not only the adjacent silicon atoms but also those in the positions of next nearest neighbours [17, 57, 58], it could be expected that, after the alkaline treatment of the final products, the all-silica seed crystals would be dissolved away, leaving behind the aluminium-containing framework of ZSM-5. To check this assumption, the samples are treated with 0.8 M sodium carbonate solution at 80 °C for 36 h under stirring. Fig. 22 shows the TEM images of the alkaline treated samples. By comparison of untreated and alkaline-treated samples, it can be found that both the size and crystal structure remains unchanged, but the hollow core appears in the centres of particles (crystals) after alkaline treatment. This finding indicates that the crystallization takes place on the surface of seed crystals, which is in accordance with the analysis of the results of kinetic study.

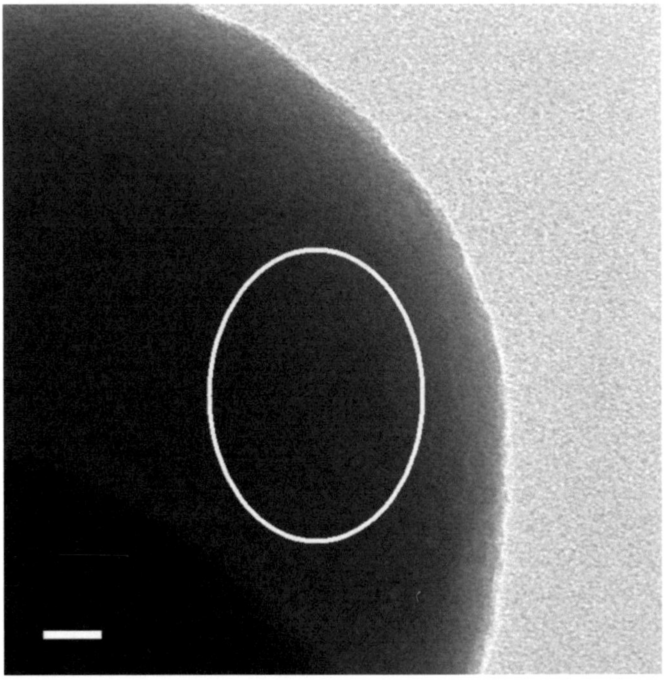

Fig. 20. High magnification TEM image of the sample separated from the reaction mixture containing 4 wt. % of 260 nm silicalite-1 seed crystals at $t_c = 1$ h. The scale bar is of 20 nm and the crystalline lattice can be identified from the indicated area. (Adopted from Ref. [50] with permission of Publisher)

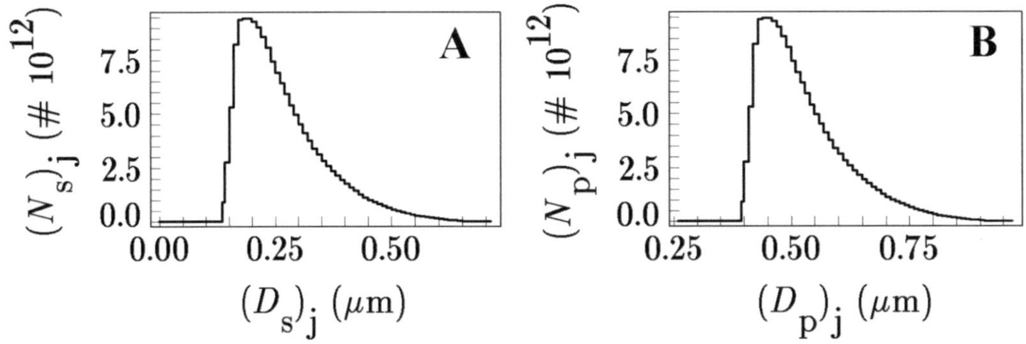

Fig. 21. The PSD curves (by number) of silicalite-1 seed crystals (A) and ZSM-5 product (B) synthesized using 4 wt.%% of the same seed crystals.

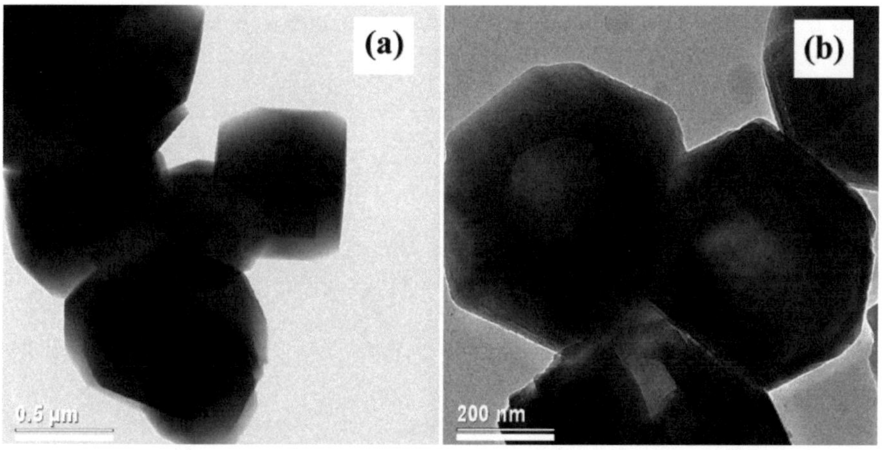

Fig. 22. TEM images of ZSM-5 samples, synthesized from the reaction mixture containing 4 wt. % of 260 nm seed crystals, before (a) and after (b) alkaline treatment. The scale bars in (a) and (b) are of 0.5 μm and 0.2 μm, respectively. (Adopted from Ref. [50] with permission of Publisher)

Taking into consideration of all the relevant mechanistic studies, the crystallization process of SDA-free seed-induced approach can be described as the growth of active species on the surface of seed crystals without formation of new nuclei. All the synthetic parameters influence the properties of the final products by changing either the relative rate or the environment of the growth step during crystallization process.

6. Modeling approach

After the relevant mechanistic studies, the mathematical analysis of the growth step of ZSM-5 zeolites on the surface of silicalite-1 seed crystals becomes possible. Compared with the phenomenological description, the quantitative explanation of the crystallization step makes the further control of the process much easier and convenient. On the basis of above presented data, it can be assumed that the crystallization process is a typical seed-induced one which can be mathematically expressed by a cubic function [50, 59, 60], i.e.

$$m_t = m_0 + K_1 t + K_2 t^2 + K_3 t^3 \tag{1}$$

where m_t is the mass of zeolite crystallized up to the time t and m_0 is the mass of the seed crystals added into the reaction mixture. From the above observations and general knowledge on the crystal growth of zeolites [61], it can be assumed that the crystallization proceeds by a linear, size-independent growth of seed crystals and thus, that the crystal size, d_t, at the crystallization time t, can be expressed as:

$$d_t = d_0 + K_g t \tag{2}$$

where, d_0 represents the size of the seed crystals and K_g is the growth rate constant. Taking into consideration that the formation of nuclei in the crystallizing system is depressed by both by the absence of SDA [50] and presence of seed crystals [50, 59, 60] it is reliable to assume that the number of growing crystals, N_t, is constant and equal to the number, N_0, of the added seed crystals. Then, the change of the mass, m_t, of the crystallized zeolite ZSM-5 with the crystallization time t, can be expressed as,

$$m_t = G\rho N_t (d_t)^3 = G\rho N_0 (d_0 + K_g t)^3 = G\rho N_0 (d_0)^3 (1 + \frac{K_g}{d_0} t)^3 \tag{3}$$

where, G is the geometrical shape factor of the growing crystals, ρ is the density of crystalline phase (zeolite), and the mass, m_o, of seed crystals can be expressed as:

$$m_0 = G\rho N_0 (d_0)^3 \tag{4}$$

The Eq. (3) can be transformed into

$$m_t = m_0 (1 + \frac{K_g}{d_0} t)^3 = m_0 [1 + 3(\frac{K_g}{d_0})t + 3(\frac{K_g}{d_0})^2 t^2 + (\frac{K_g}{d_0})^3 t^3] \tag{5}$$

where,

$$K_1 = 3(\frac{K_g}{d_0}) m_0, \; K_2 = 3(\frac{K_g}{d_0})^2 m_0 \text{ and } K_3 = (\frac{K_g}{d_0})^3 m_0 .$$

Since the total silica-alumina source in the reaction mixture is constant during the synthesis, the values of m_t and m_0 can also be expressed as the fractions x_t and x_0 of aluminosilicate material in the reaction mixture at different crystallization time t. Thus, the Eq. (5) can also be expressed as [59]:

$$x_t = x_0 (1 + \frac{K_g}{d_0} t)^3 \tag{6}$$

where, x_0 is fraction of aluminosilicate material contained in seed crystals at $t = 0$, and x_t is the fraction of aluminosilicate material in the crystalline phase (seeds + newly crystallized zeolite) at $t > 0$.

Then, the crystal growth rate constant, K_g, can be obtained by solving the Eq. (6), e.g.,

$$K_g = (\sqrt[3]{\frac{x}{x_0}} - 1) \frac{d_0}{t} \tag{7}$$

Thus, the value of K_g can be calculated by Eq. (7) using d_0 = seed size and x/x_0 = (crystallinity at t/crystallinity at t = 0) (see Fig. 18); K_g = 0.15 μm/h for all investigated system. Then, inserting the calculated value of K_g = 0.15 μm/h and the appropriate values of d_0 = 0.26 μm and x_0 = 1 wt. % into Eq. (5), the kinetics of crystallization of zeolite ZSM-5 from the reaction mixture containing 1 wt. % of 260 nm silicalite-1 seed crystals is calculated (curve in Fig. 23) and compared with the measured kinetics (points in Fig.23). Almost perfect agreement between calculated and measured kinetics indicates that the crystallization process proceeds in expected way, namely by a linear, size-independent growth of silicalite-1 seed crystals.

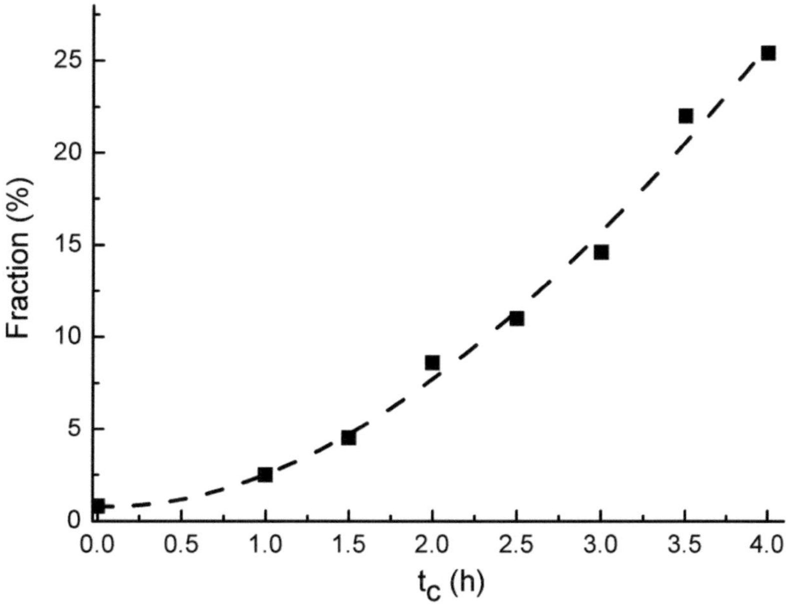

Fig. 23. Correlation between measured kinetics of crystallization of zeolite ZSM-5 from the reaction mixture containing 1 wt. % of 260 nm silicalite-1 seed crystals (points) and the kinetics of crystallization calculated by Eq. (5) (dashed curve). (Adopted from Ref. [50] with permission of Publisher)

On the other hand, an almost perfect correlation can be observed between measured sizes, d_t(det), of the crystalline end products and the corresponding sizes, d_t(cal), calculated by Eq. (2), using the value K_g = 0.15 μm/h, and the corresponding seed sizes, d_0 (Table 3). This phenomenon is an additional evidence to prove that the process of crystallization takes place by the mechanism mathematically described by Eqs. (1) – (6).

d_0 (nm)	Seed addition amount (wt.%)	t_c[a] (h)	d_t(cal)[b] (nm)	d_t(det)[c] (nm)
90	4.0	1.5	315	270
690	4.0	2.5	1065	1100
260	4.0	2.0	560	520
260	8.0	1.5	485	440
260	16.0	1.0	410	410
260	32.0	0.75	373	350

a. determined from the corresponding crystallization curve in Fig. 18 by choosing the first 100% crystallinity data point
b. calculated from Eq. (2) using crystal growth rate of 0.15 μm/h
c. determined from the corresponding SEM images
(Adopted from Ref. [50] with permission of Publisher)

Table 3. The comparison of measured, d_t(det), and calculated, d_t(cal), final crystal sizes of zeolite ZSM-5 synthesized from the reaction mixtures containing different amounts of silicalite-1 seed crystals having different sizes.

Based on the analysis of experimental data in both phenomenological and mathematical way, the crystallization mechanism and the corresponding critical processes occurring during crystallization are depicted in Fig. 24. These processes are: (i) dissolution of amorphous silica and/or alumina sources, (ii) formation of alumino(silicate) gel by poly-condensation reactions and the partial deposition of gel onto the surface of seed crystals, (iii) formation of growth precursor species in the gel matrix at high temperature, (iv) deposition of growth species onto the surface of seed crystals, and (v) final ordering of growth species to form the crystalline end products.

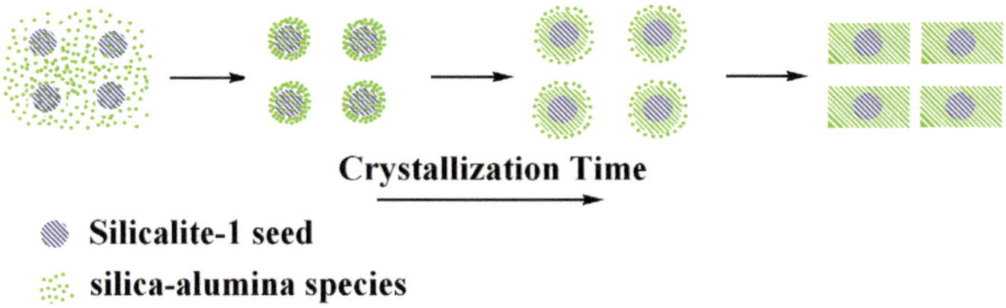

Fig. 24. Schematic presentation of the crystallization process of seed-induced SDA-free approach for the synthesis of sub-micrometer sized ZSM-5 zeolites. (Adopted from [50] with permission of Publisher)

7. Conclusion

In this chapter, the seed-induced, SDA-free crystallization system for the growth of sub-micrometer sized ZSM-5 zeolites is carried out, which clearly shows that:

- The crystallization process is predominantly determined by the growth steps in which the product size and morphologies could be well tuned by the variations of size and amount of silicalite-1 seed crystals;
- The batch alkalinity, expressed as $A = [Na_2O/SiO_2]_b$ plays the key role on the crystallization processes, which influences the phase purity, particulate and chemical composition (Si/Al ratio) of the final product. The optimal alkalinity for growth of pure phase ZSM-5 zeolite is $0.006 \leq A \leq 0.010$;
- The excess batch content of sodium ions and prolonged duration of hydrogel ageing do not enhance the rate of crystallization. Instead, the aggregation behaviour/size uniformity of the products is influenced by these two parameters. This indicates that the precursor species for the growth of ZSM-5 are generated at high temperature controlled by batch alkalinity, rather than by the concentration of sodium ions at room temperature;
- The crystallization mechanism of the seed-induction SDA-free approach is revealed as the typical size-independent, linear growth process on the seed surface.

With the understanding of the critical processes occurring during the crystallization of ZSM-5 zeolites, it is expected that the ZSM-5 zeolites possessing more interesting properties could be obtained by the co-ordinately variation of multi-synthesis parameters at the same time. This work will also make corresponding contributions on the rational design of ZSM-5 zeolites with desired functionalities in the domain of both materials and industrial catalysis.

8. Acknowledgement

This work is realized in the frame of the projects: NSFC (20803010), "Chen Guang" project supported by Shanghai Municipal Education Commission and Shanghai Education Development Foundation (09CG02), 'Brain Gain' Post-Doc project (I-668-2011) supported by Croatian Science Foundation and the project 098-0982904-2953, financially supported by the Ministry of Science, Education and Sport of the Republic of Croatia.

9. References

[1] Argauer , R. J.; Landolt, G. R., Mobile Oil, US Patent 3, 702, 886, 1972.
[2] Singh, R.; Dutta, P. K., 'MFI: A Case Study of Zeolite Synthesis', in Auerbach, S.M.; Carrado, K.A.; Dutta, P.K. Eds. *Handbook of Zeolite Science and Technology*, Chapter 2, Marcel Dekker Inc, New York-Basel, 2003, p21.
[3] Adewuyi, Y.G.; Klocke, D.J.; Buchnan, J.S. *Appl. Catal. A-General* 1995, *131*, 121.
[4] Chen, N.Y.; Garwood, W.E.; Dwyer, F.G. *Shape Selective Catalysis in Industrial Applications*, 2nd edition, Marcel Dekker Inc., New York, 1996.
[5] Kim, S.D.; Noh, S.H.; Park, J.W.; Kim, W.j. *Microporous Mesoporous Mater*. 2006, *92*, 181.
[6] Narita, E.; Sato, K.; Okabe, T. *Chem. Lett.* 1984, 1055.
[7] Jacobs, P. A.; Martens, J. A., *Stud. Surf. Sci. Catal.* 1987, *33*, 1.
[8] Weitkamp, J. *Solid State Ionics* 2000, *131*, 175.
[9] Vedrine, J.C. *ACS Symp. Ser.* 1985, *297*, 257.
[10] Renzo, F.D. *Catal. Today* 1998, *41*, 37.
[11] Chao, K.-J.; Tsai, T.C.; Chen, M.-S. *J. Chem. Soc. Faraday Trans.* 1981, *77*, 547.
[12] Romannikov, V.N.; Mastikhin, V.M.; Hočevar, S.; Držaj, B. *Zeolites* 1983, *3*, 310.
[13] Gabelica, Z.; Derouane, E.G.; Blom, N. *Adv. Chem. Ser.* 1984, *248*, 219.
[14] Padovan, M.; Leofanti, G.; Solari, M.; Moretti, E. *Zeolites* 1984, *4*, 295.
[15] Scholle, K.F.M.G.J.; Veeman, W.S.; Frenken, P.; van der Velden, G.P.M. *Catal. Today* 1985, *17*, 233.
[16] Gabelica, Z., Nagy, J.B.; Debras, G.; Derouane, E.G. *Acta Chimica Hungarica* 1985, *119*, 275.
[17] Čižmek, A.; Subotić, B.; Aiello, R.; Crea, F.; Nastro, A.; Tuoto, C. *Microporous Mater.* 1995, *4*, 159.
[18] Cundy, C.S.; Henty, M.S.; Plaisted, R.J. *Zeolites* 1995, *15*, 353.
[19] Erdem, A.; Sand, L.B. *J. Catal.* 1979, *60*, 241.
[20] Derouane, E.G.; Detremmerie, S.; Gabelica. Z.; Blom, N. *Appl. Catal.* 1981, *1*, 201.
[21] Mostowicz, R.; Sand, L.B. *Zeolites* 1983, *3*, 219.
[22] Gabelica, Z.; Blom, N.; Derouane, E.G. *Appl. Catal.* 1983, *5*, 227.
[23] Nastro, A.; Sand, L.B. *Zeolites* 1983, *3*, 57.
[24] Nastro, A.; Aiello, R.; Colella, C. *Ann. Chim.* 1984, *74*, 579.
[25] Ghamami, M.; Sand, L.B. *Zeolites* 1983, *3*, 155.
[26] Chang, C.D.; Lang, W.H.; Silvestri, A.J. US Patent 3 894 106, 1975.
[27] Pan, M.; Lin, J.S. *Microporous Mesoporous Mater.* 2001, *43*, 319.
[28] Dokter, W.H.; van Garderen, H.F.; Beleen, T.P.M.; van Santen, R.A.; Bras, V. *Angew. Chem. Int. Ed. Engl.* 1995, *34*, 73.
[29] Subotic, B.; Bronic, J.; Jelic, T. A. in *Ordered Porous Solids*, Elsevier B.V. 2009, Chapter 6.
[30] Persson, A.E.; Schoeman, B. J.; Sterte, J.; Otterstedt, J.E. *Zeolites* 1995, *15*, 611.
[31] Hsu, C.Y.; Chiang, A.S.T.; Selvin, R.; Thompson, R.W. *J. Phys. Chem. B* 2005, *109*, 18804.
[32] Geus, E.R.; van Bekkum, H. *Zeolites* 1995, *15*, 333.
[33] Lin, X.; Falconer, J.L. Noble, R.D. *Chem. Mater.* 1998, *10*, 3716.

[34] Lai, R. ; Gavalas, G.R. *Microporous Mesoporous Mater.* 2000, *38*, 239.
[35] Lassinatti, M.; Jereman, F.; Hedlund, J.; Creaser, D.; Sterte, J. *Catal. Today* 2001, *67*, 109.
[36] Kalipcilar, H.; Culfaz, A. *Cryst. Res. Technol.* 2001, *36*, 1197.
[37] Grose, R.W.; Flanigen, E.M US Patent 4 257 885, 1981.
[38] Narita, E.; Sato, K.; Yatabe, N.; Okabe, T. *Ind. Eng. Chem. Prod. Res. Dev.* 1985, *24*, 507.
[39] Berak, J.M.; Mostowicz, R. *Stud. Surf. Sci. Catal.* 1985, *24*, 47.
[40] Batista, J.; Kaučič, V. Vestn. *Slov. Kem. Drus.* 1987, *34*, 289.
[41] Aiello, R.; Crea, F.; Nastro, A.; Pellegrino, A. *Zeolites* 1987, *7*, 549.
[42] Dai, F.-Y.; Suzuki, M.; Takahashi, H.; Y. Saito, *ACS Sym. Ser.* 1989, *398*, 224.
[43] Schweiger, W.; Bergk, K.-H.; Freude, D.; Hunger, M.; Pfeifer, H. *ACS Sym. Ser.* 1989, *398*, 274.
[44] Nastro, A.; Crea, F.; Hayhurst, D.T.; Testa, F.; Aiello, R.; Toniolo, L. *Stud. Surf. Sci. Catal.* 1989, *49A*, 321.
[45] Mravec, D.; Riečanova, D.; Ilavsky, J.; Majing, *J. Chem. Papers* 1991, *45*, 27.
[46] Lowe, B. M.; Nee, J.R.D.; Casci, J.L. *Zeolites* 1994, *14*, 610.
[47] Otake, M. *Zeolites* 1994, *14*, 42.
[48] Cheng, Y.; Wang, L.J.; Li, J.S.; Yang, Y.C.; Sun, X.Y. *Mater. Lett.* 2005, *59*, 3427.
[49] Cheng, Y.; Liao, R.H.; Li, J.S.; Sun, X.Y.; Wang, L.J. J. *Mater. Process. Technol.* 2008, *206*, 445.
[50] Ren, N.; Yang, Z.J.; Lv, X.C.; Shi, J.; Zhang, Y.H.; Tang, Y. *Micropor. Mesopor. Mater.* 2010, *131*, 103.
[51] Gonthier, S.; Thompson, R.W. *Stud. Surf. Sci. Catal.* 1994, *85*, 43.
[52] Majano, G.; Delmotte, L.; Valtchev, V.; Mintova, S. *Chem. Mater.* 2009, *21*, 4184.
[53] Cundy, C.S.; Plaisted, R.J.; Zhao, J.P. *Chem. Commun.* 1998, 1465.
[54] Ren, N.; Bronić J.; Subotić, B.; Lv, X.C.; Yang, Z.J.; Tang, Y. *Microporous Mesoporous Mater.* 2011, *139*, 197.
[55] Ren, N.; Bronić J.; Subotić, B.; Song, Y.M.; Lv, X.C.; Tang, Y. *Microporous Mesoporous Mater.* 2012, *147*, 229.
[56] Derouane, E.G.; Gabelica, Z. *J. Solid State Chem.* 1986, *64*, 296.
[57] Nastro, A.; Gabelica, Z.; Bodart, P.; Nagy J. B. in: Kaliaguine, S.; Mahay, A. (Eds.), *Catalysis on the Energy Scene*, Elsevier Science Publishers B.V., Amstardam, 1984, p. 131.
[58] Nagy, J.B.; Bodart, P.; Collette, H.; Fernandez, C.; Gabelica, Z.; Nastro, A.; Aiello, R. *J. Chem. Soc. Faraday Trans. I* 1989, *85*, 2749.
[59] Čižmek, A.; Subotić, B.; Kralj, D.; Babić-Ivančić, V.; Tonejc, A. *Microporous Mater.* 1997, *12*, 267.
[60] Zhdanov, S.P.; Samulevich, N.N. in: L.V. Rees (Ed.), Proceedings of the Fifth International Conference on Zeolites, Heyden, London, Philadelphia, Rheine, 1980, p. 75.
[61] Larlus, O.; Valthev, V.P. *Chem. Mater.* 2004, *16*, 3381.
[62] Zhu, G. ; Li, Y. ; Chen, H. ; Liu, J. ; Yang, W. *J. Mater. Sci.* 2008, *43*, 3279.
[63] Kang, N.Y. ; Song, B.S. ; Lee, C.W. ; Choi, W.C. ; Yoon, K.B. ; Park, Y.K. *Microporous Mesoporous Mater.* 2009, *118*, 361.
[64] Iller, R.K. *The Chemistry of Silica*, Willey, New York, 1979.
[65] Kawashima, Y.; Capes, C.E. *Powder Technology*, 1976, *13*, 279.
[57] Lechert, H.; Kacirek, H. *Zeolites* 1991, *11*, 720.
[58] Dessau, R. M.; Valyocsik, E. W.; Goeke, N. H. *Zeolites* 1992, *12*, 776.
[59] Kacirek, H.;. Lechert, H. *J. Phys. Chem.* 1975, *79*, 1589.
[60] Grujic, E.; Subotic, B.; Despotovic, L.J.A. *Stud. Surf. Sci. Catal.* 1989, *49A*, 261.
[61] Subotić, B.; Bronić, J. in: S.M. Auerbach, K.A. Carrado and P.K. Dutta (Eds.), *Handbook of Zeolite Science and Technology*, Marcel Dekker Inc., New York - Basel, 2003, p. 129.

11

The Growth of Chalcedony (Nanocrystalline Silica) in Electric Organs from Living Marine Fish

María Prado Figueroa
Instituto de Investigaciones Bioquímicas (INIBIBB)
CONICET, Universidad Nacional del Sur (UNS)
Departamento de Biología, Bioquímica y Farmacia
Bahía Blanca,
Argentina

1. Introduction

The Rajidae Family are weakly electric fish (Fessard, 1958). *Psammobatis extenta* (Rajidae Family) is a South American electric fish. Electric organs of the electric fish have constituted the choice system for studying the biochemistry, morphology, physiology and cell biology of the nervous cholinergic system (Barrantes *et al.*, 1983; Changeux, 1981; 2010; Fox *et al.*, 1990; Prado Figueroa *et al.*, 1995; Wittaker, 1977). Electric organs derive embryologically from myoblasts and are constituted by cells called electrocytes.

The electric organs (EO) of *Psammobatis extenta* produce weak electrical discharges to the surrounding environment. Electrocytes have evolved and differential independently, losing the contractile ability. In previous works we employed a microanalysis (EDS-SEM) and documented the presence of aluminium and silicon in significant concentrations. Zinc, oxygen and cupper were also localized.

Silicon (Si) is an essential nutrient for animal biology (Carlisle 1982). It has been shown that silicon is required for bone, cartilage and connective tissue formation (Bissé *et al.* 2005). Silicon may function as a biological cross-linking agent and may contribute to the architecture and resilience of connective tissue (Schwarz 1973). Aluminium (Al) and Si accumulations have been detected in electric organs by a combination of scanning electron microscopy and X-ray spectrometry (EDS-SEM) (Prado Figueroa *et al.*, 2008). Al and Si have been also detected in the human cerebral cortex from elderly people by using EDS/SEM (Perl and Brody 1980, Candy *et al.* 1985). These inorganic elements are related to pathological changes in the human brain (Candy *et al.* 1985, Tokutake *et al.* 1995).

Biomineralization is the process by which living organisms produce minerals, often to harden or stiffen existing tissues. Chalcedony is a microcrystalline fibrous form of silica (SiO_2) and it is the product of biomineralization by silica (Erhlich *et al.*, 2010). We have identified chalcedony in living fish electric organs by using a standard polarized light microscope (Prado Figueroa *et al.*, 2008). In plane-polarized light, chalcedony is rounded in shape, 12–15 micron in size, translucent, with a low refraction index. The crossed-polarizer image shows first order birefringence colour (grey–white) and radial extinction.

In this chapter, we document the visualization and identification of chalcedony crystals in electric organs, by using a Leica TCS - SP2 Laser Scanning Confocal Microscope (LSCM). Three ion lasers were used i.e.: argon with emission band in 458 nm (cyan), 476 nm (blue-green), 488 nm (green) and 514 nm (yellow); He/Ne in 543 nm with emission band in red, and He/Ne with emission band in 633 nm (blue). The autofluorescent character of chalcedony (a mineral) allowed us to obtain images of the crystals together with a topographic study. Chalcedony consists of nanoscale intergrowths of silica polymorphs: quartz and moganite (Heaney & Post, 1992). These silica polymorphs, with their two different nanocrystal structures, are described in the present chapter. Quartz and moganite are both identified in three-dimensional (3-D) images using a LSCM and Leica software. 3-D images were generated as "surface": the blank spaces between the pixels are filled. 3-D images were also generated as "wireframe": all pixels are linked with lines, while the blank spaces remain free. Many images generated by differential interference contrast (DIC) are also shown in this chapter.

2. Biomineralization by silica in electric fish

Biomineralization by silica is a complicated process observed in living organisms. Of the intriguing topics that are receiving renewed attention, the study of biomineral formation based on organic templates is one of the most fascinating topics today (Ehrlich, 2010; Ehrlich *et al.*, 2010). Biosilicification is an evolutionarily old and widespread type of biomineralization both in unicellular and multicellular organisms, including sponges, diatoms, radiolarians, choanoflagellates, and higher plants (Schroeder *et al.*, 2008).

We have studied *Psammobatis extenta* electric organs from the Rajidae family, a group of elasmobranch electric fish. Electric organs are structures specialized in the production of electric discharges (Fessard, 1958). Their major cell components, called electrocytes, are highly polarized. We could detect biomineralization by microcrystalline silica in *P. extenta* electric organs (Prado Figueroa *et al.*, 2008).

2.1 *Psammobatis extenta*: A Rajidae family fish

Adult female and male *P. extenta* were collected from the Bahía Blanca Estuary (38° 40'S and 39°30'S, 62°16'W and 63°26'W) in the Buenos Aires Province, Argentina and transported to the laboratory in sealed polyethylene bags containing oxygen-saturated seawater.

The fish were anesthetized by immersion in ice cold seawater for 10 min and then killed by pithing. Immediately after dissection of the ray, the electric tissue was frozen in liquid nitrogen at -198°C.

Electrocytes from *P. extenta* are cup-shaped cells, multinucleated and polarized. They have an anterior, concave, innervated face and a posterior, convex, non-innervated face, that shows a very large system of caveolae (Prado Figueroa *et al.*, 1995). These cup-shaped electrocytes are plesiomorphic, phylogenies based on morphological data (Jacob *et al.*, 1994). Neuronal cell death and synaptic terminal degeneration have been noted in the adult electric organs of fish from the Rajidae family (Fox *et al.*, 1990 and our observations).

Understanding cellular and molecular mechanisms participating in neurodegenerative process is thus an important field of research.

2.2 Silicon in electric organs

Silicon is an essential element for animals (Carlisle, 1982). Silicon may function as a biological cross-linking agent and may contribute to the architecture and resilience of connective tissue (Schwarz, 1973). It has also been documented that silicon is present as a silanolate, i.e., an ester-like derivative of silicic acid and plays a role in the structural organization of glycosaminoglycans and polyuronides (Schwarz, 1973).

We could observe aluminium and silicon in the cytoplasmic extracts of *P. extenta* electric tissue (Prado Figueroa *et al.*, 2008) using a combination of scanning electron microscopy and X-ray spectrometry (EDS/SEM) (see Fig. 1). The result of this microanalysis is an energy-dispersive spectrum in which the peaks are localized at energy lines characteristic for each element

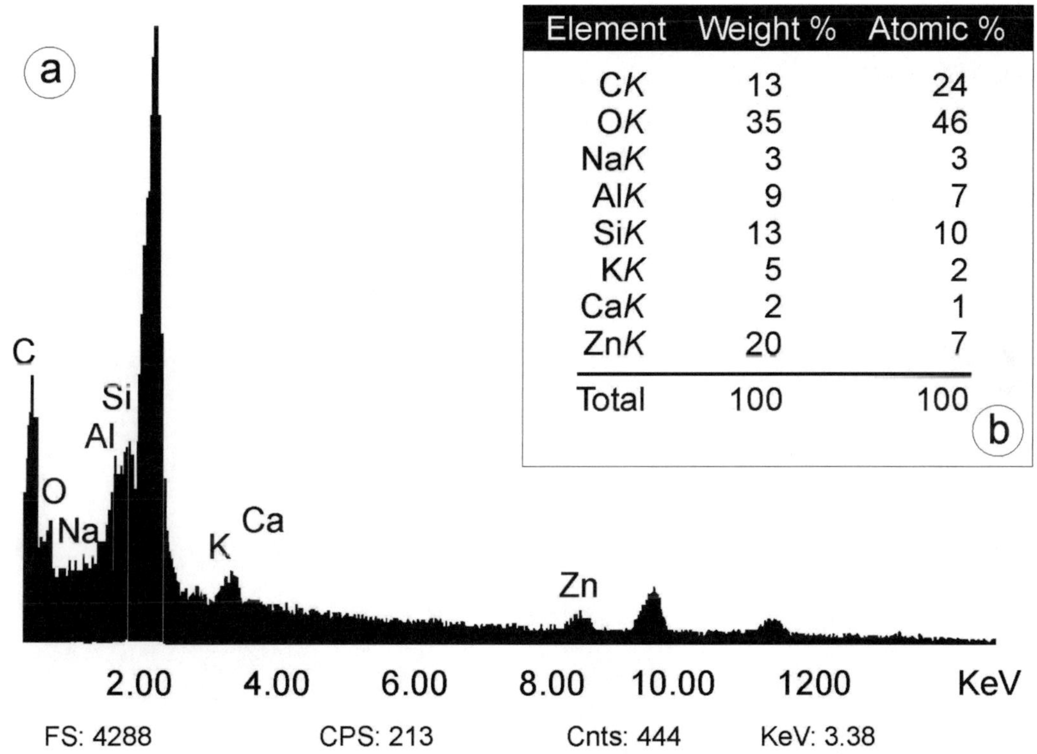

Element	Weight %	Atomic %
C*K*	13	24
O*K*	35	46
Na*K*	3	3
Al*K*	9	7
Si*K*	13	10
K*K*	5	2
Ca*K*	2	1
Zn*K*	20	7
Total	100	100

Fig. 1. Electric organ cytoplasmic extracts from *Psammobatis extenta* on lyophilised paper and metalized with gold were microanalyzed by using EDS/SEM. This energy dispersive spectrum (a) shows high peaks localized at energy lines characteristic for oxygen, silicon and aluminium. Sodium, potassium, calcium and zinc are also observed. Weight and atomic percents for these elements are indicated next to the spectrum (b).

An electric organ cytoplasmic extract (50 µl) on lyophilised paper (Labconco Corp., USA) was fixed in 2.5% glutaraldehyde in a 0.05 M sodium phosphate buffer (pH 7.2) for 60 min at 4 °C. Samples were washed with buffer and bi-distilled water for 2 h, then, dehydrated. (For more detail of this method, see Prado Figueroa *et al.*, 2008).

This method (EDS/SEM) has also been used for studying photosensitizers in electric tissue (Prado Figueroa & Santiago, 2004).

2.3 Microcrystalline silica in electric organs

Based on the evidence of aluminum and silicium accumulation in *P. extenta,* we documented the presence of silica minerals in *P. extenta* electric tissue by means of mineralogical techniques (Prado Figueroa *et al.*, 2008). It was thought that these compounds could form minerals (i.e., solid inorganic substances with a defined chemical composition and determined crystallography).

Fractionation of electric tissue homogenates by differential centrifugation was carried out as described for other tissues (Beaufay and Amar-Costesec, 1976; Amar-Costesec *et al.*, 1985)

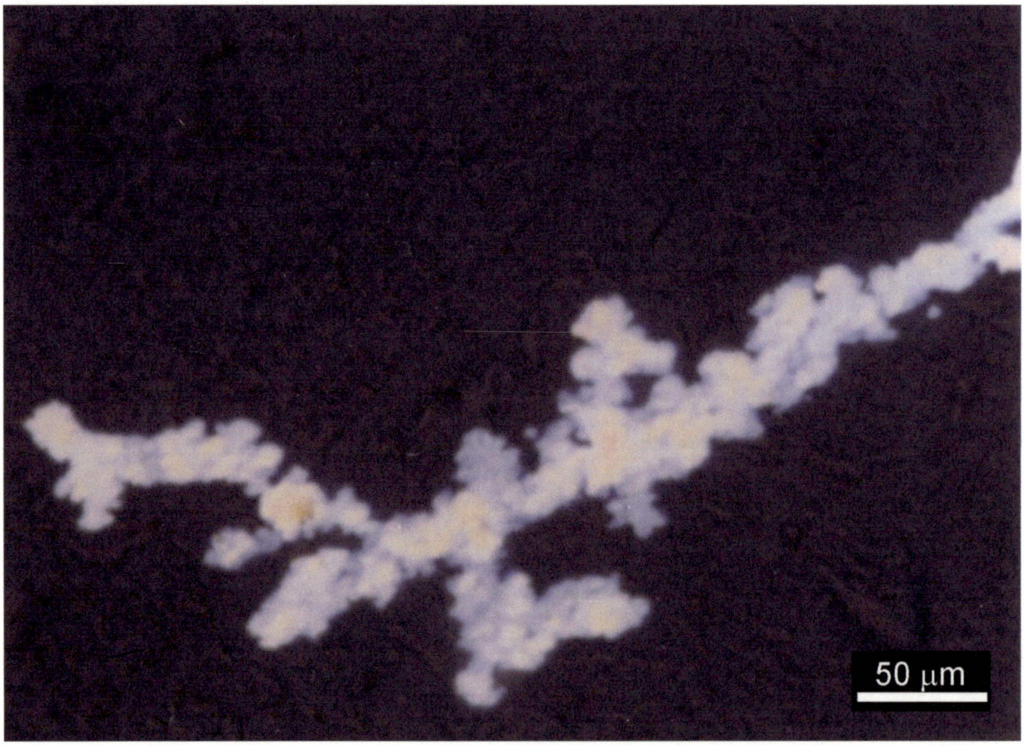

Fig. 2. Photomicrograph of a cytoplasmic extract of the electric organ from *P. extenta* in crossed-polarizers, by using a standard polarized light microscope.

The crossed-polarizer image of a cytoplasmic extract shows SiO_2 replacements in grey and white arrangements, with undulatory extinction.

using isotonic 3 mM imidazole–HCl-buffered 0.25 M sucrose (pH 7.4). The following fractions were obtained: cytoplasmic extract (E), nuclear fraction (N), large granules (ML), microsomes (P) and supernatant (S). Drops of the fractions were collected on glass slides, dried and mounted in PBS/glycerol. These fractions were inspected using a Leica polarized light microscope (DMLP). This microscope has a polarizer and a switchable analyzer. In mineralogical microscopy, when the light enters an anisotropic mineral, one which transmits light at different rates in different orientations, it is decomposed in two rays, oscillating in two orthogonal planes. This phenomenon is known as birefringence and allows for the identification of each mineral. In this microscope, with a circular graduated stage capable of a 360° rotation, minerals in different positions display their optical properties, such as birefringence colour and extinction position, with crossed polarizers.

All electric organ fractions in crossed-polarizers show SiO_2 replacements in grey and white arrangement, with undulatory extinction. The crossed-polarizer image of a cytoplasmic extract shows SiO_2 replacements in grey and white arrangements (chalcedony), Fig. 2.

Electric organs without any treatment were also used for X-ray diffraction analysis. Different peaks were obtained from diffractometric analysis; specifically peaks belonged to a quartz (low quartz; Moore and Reynolds, 1997). (See: Prado Figueroa *et al.*, 2008).

3. Autofluorescent crystalline silica detected by using a LSCM

Autofluorescence characteristics of minerals have been described by Henkel (1989). In this chapter, we document the visualization and identification of chalcedony crystals in electric organs, by using a Leica TCS - SP2 Laser Scanning Confocal Microscope (LSCM). This microscope has three ion lasers i.e.: argon with emission band in 458 nm (cyan), 476 nm (blue-green), 488 nm (green) and 514 nm (yellow); He/Ne with emission band in 543 nm (red) and He/Ne with emission band in 633 nm (blue). Since chalcedony is characteristically an autofluorescent mineral, we were allowed to obtain images of crystals.

3.1 Autofluorescent microcrystalline silica (chalcedony)

Adult female and male *P. extenta* were collected from Bahía Blanca Estuary in Buenos Aires Province, Argentina. Fractionation of electric tissue homogenates by differential centrifugation was carried out as described in Section 2.1.

Fractions (nuclear fraction "N"; microsomes "P" and supernatant "S") were used and observed with a LSCM. A nuclear fraction shows many autofluorescent crystals and also little electrocytes, Fig. 3. Electrocytes from the electric organ of the Patagonian ray *Psammobatis extenta* are very unusual cells: semicircular in shape, multinucleated and highly polarized. Their anterior face is concave and innervated by numerous nerve-endings.

Fig. 3 shows an unbroken electrocyte with many autofluorescent crystals, these were observed with an argon ion laser with emission band at 488 nm (green). This image was merged with (DIC).

Normally, the microsomal fraction contains membranes from the synaptic region. The microsomal fraction has many autofluorescent crystals. A crystal of chalcedony from the microsomal fraction, its dimensions and autofluorescent intensity (I, arbitrary units) are

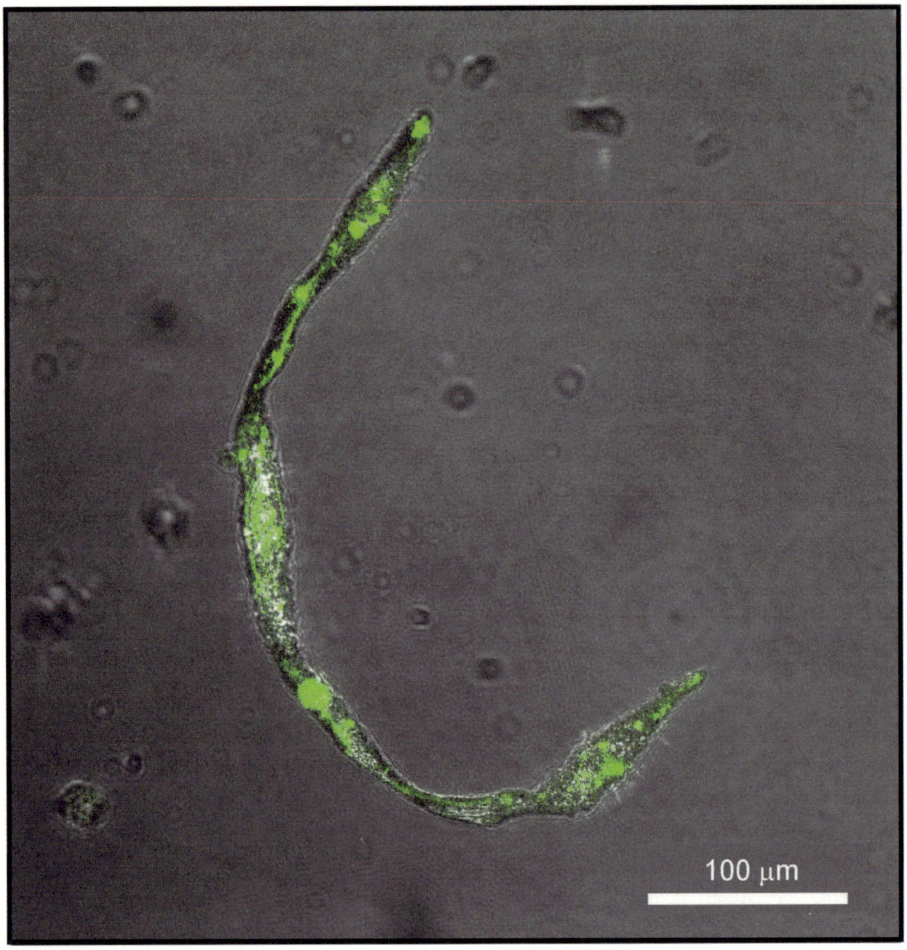

Fig. 3. Photomicrograph of a nuclear fraction of the electric organ from *P. extenta* in LSCM. This fraction shows a little electrocyte, which is unbroken. Many autofluorescent crystals were observed in the electrocyte with an Ar ion laser with emission band at 488 nm (green). This image was merged with DIC.

shown in Fig. 4. This image was obtained with an argon ion laser with emission band at 488 nm (green). This crystal is about 20 micron.

The crystal dimension and autofluorescent intensity (I) are shown in this image. This crystal was observed by using an argon ion laser with emission band at 488 nm (green).

Images of a chalcedony crystal from the microsomal fraction in LSCM are shown in Fig. 5.

Crystals from this fraction are rhombohedral in shape and they are in large quantities. This crystal is about 20 micron. The chalcedony crystal has autofluorescence with different ion lasers and is about 20 micron. A He/Ne ion laser with emission band in 543 nm (red), a He/Ne ion laser with emission band in 633 nm (blue) and an Ar ion laser with emission bands in 458 nm (cyan), and 514 nm (yellow) were used. An image DIC of the crystal is shown (top, right side) and also all the images merged (botton, right side).

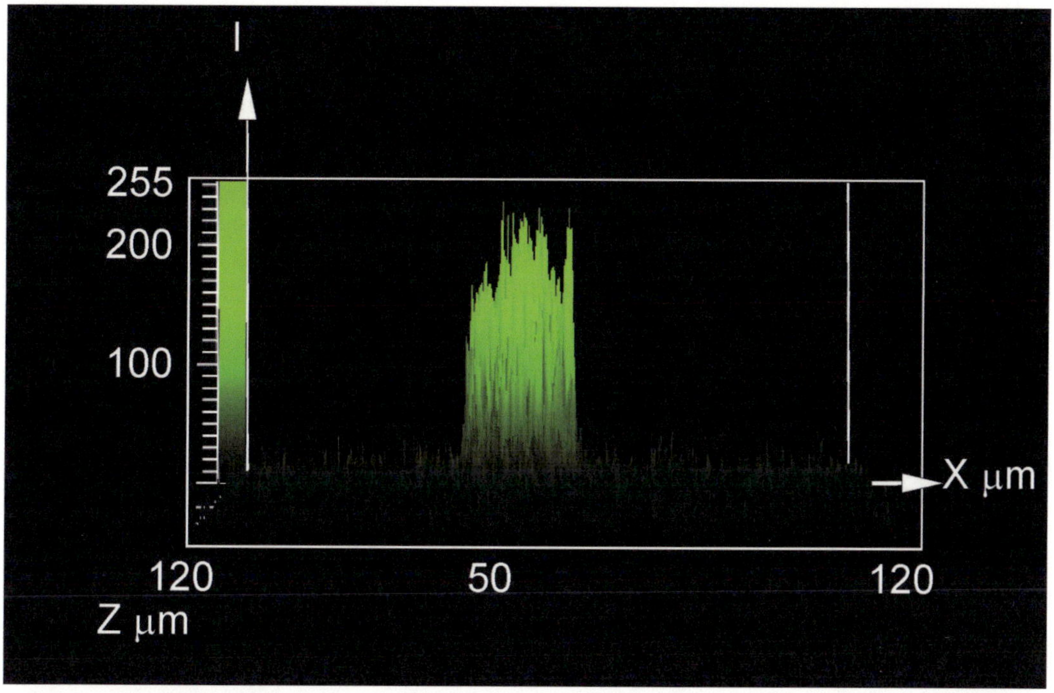

Fig. 4. An autofluorescent crystal of chalcedony from the microsomal fraction in LSCM.

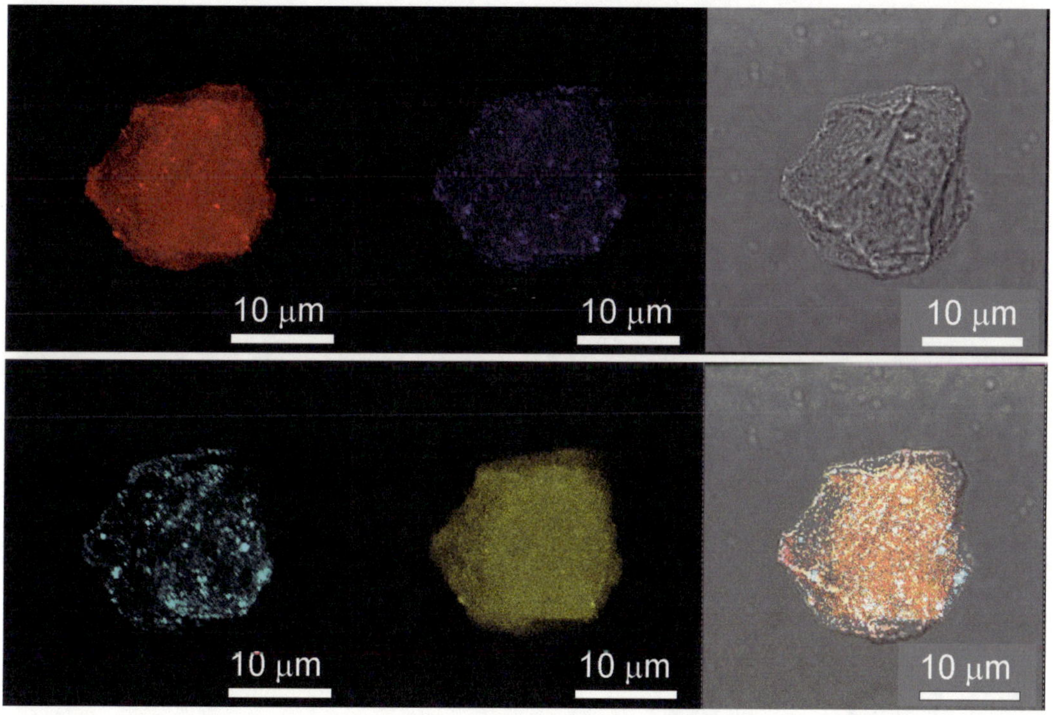

Fig. 5. Images of a chalcedony crystal from the microsomal fraction in LSCM.

The chalcedony crystal has autofluorescence with different ion lasers: a He/Ne ion laser with emission band in 543 nm (red); a He/Ne ion laser with emission band in 633 nm (blue); an Ar ion laser with emission bands in 458 nm (cyan), and 514 nm (yellow) were used. An image DIC of the crystal is shown and also all the images merged (right side, top and botton). This crystal is about 20 μm in size.

An image of a chalcedony crystal from the supernatant fraction in LSCM is shown in Fig. 6, it was observed by using an argon ion laser with emission band at 488 nm (green). The chalcedony crystal has autofluorescence with different ion lasers and is about 10 micron in size. The supernatant fraction has many crystals.

3.2 Autofluorescent nanocrystalline silica (silica polymorphs)

Chalcedony is a microcrystalline fibrous form of silica which actually consists of nanoscale intergrowths of quartz and the optically length-slow fibrous silica polymorph moganite (Conrad *et al.*, 2007; Deer *et al.*, 1966; Heaney & Post, 1992; Heaney, 1993; Heaney *et al.*, 1994, 2007). Quartz and moganite are both silica minerals, but they differ in that quartz has a trigonal crystal structure (α, β, γ different to 90º), whilst moganite has a monoclinic crystal structure (α different to 90º and β, γ = 90º). An image of silica polymorphs were obtained by using a LSCM with different emission bands, Fig. 7. This Fig. 7 shows a crystal of chalcedony from the microsomal fraction using an argon ion laser with emission bands in 458 nm (cyan) and 514 (yellow) merged and contrasted with DIC.

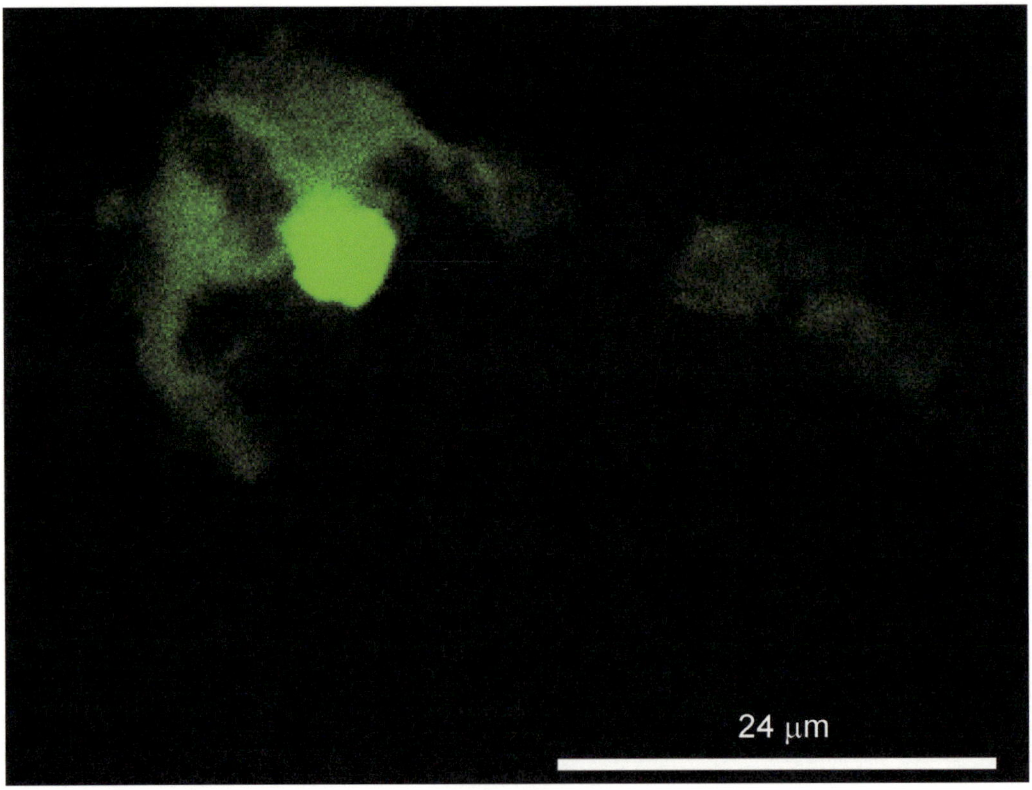

Fig. 6. A crystal of chalcedony from a supernatant fraction in LSCM.

This chalcedony crystal was observed by using an argon ion laser with emission band at 488 nm (green). This crystal is about 10 µm. This fraction has many little autofluorescent crystals.

The same crystal of chalcedony (from Fig. 7) is shown in 3D images, Fig. 8. An argon ion

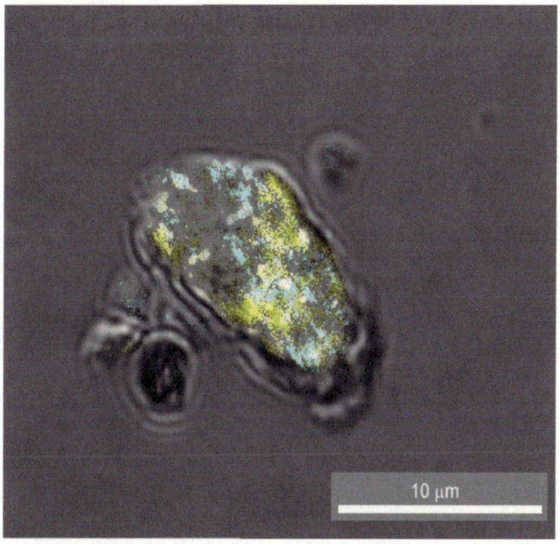

Fig. 7. A crystal of chalcedony from the microsomal fraction in LSCM. An argon ion laser was used, with emission bands in 458 nm (cyan) and 514 nm (yellow). Emissions were merged and contrasted with DIC.

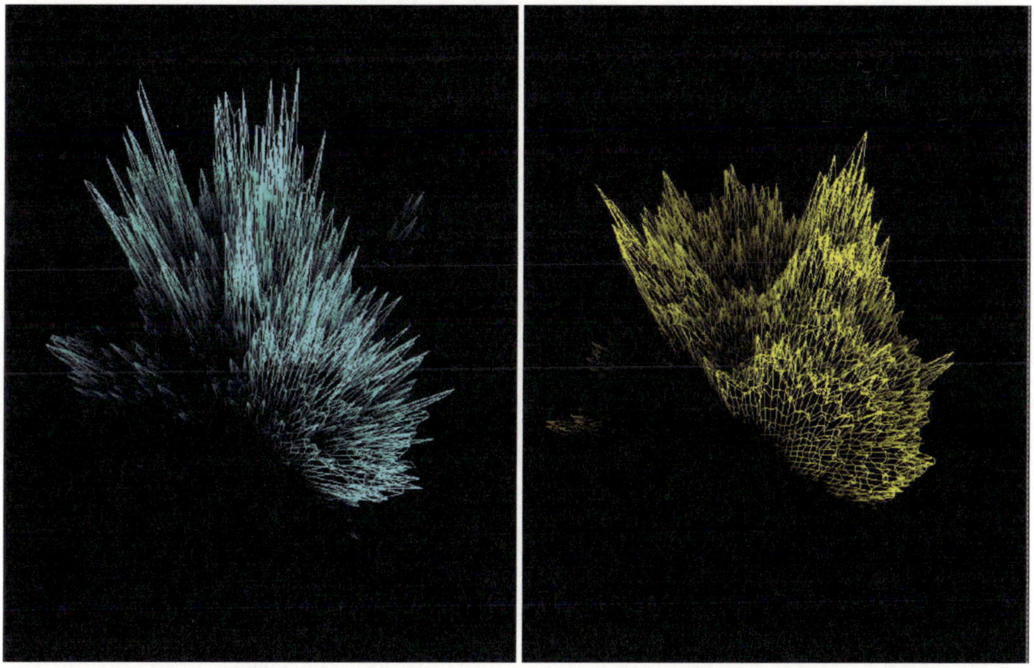

Fig. 8. 3-D images of the silica polymorphs from the microsomal fraction in LSCM .

These images were obtained of the same crystal from Fig. 7. Cyan colour nanocrystals are acicular (pinacoid) in shape and nanocrystals in yellow are trapezohedral in shape.

laser was used with emission band in 458 nm (cyan) and 514 nm (yellow). Both nanocrystals are very different. Nanocrystals in cyan colour are acicular in shape. Nanocrystals in yellow are trapezohedral in shape. Quartz and moganite were detected by using ion lasers with different emission bands.

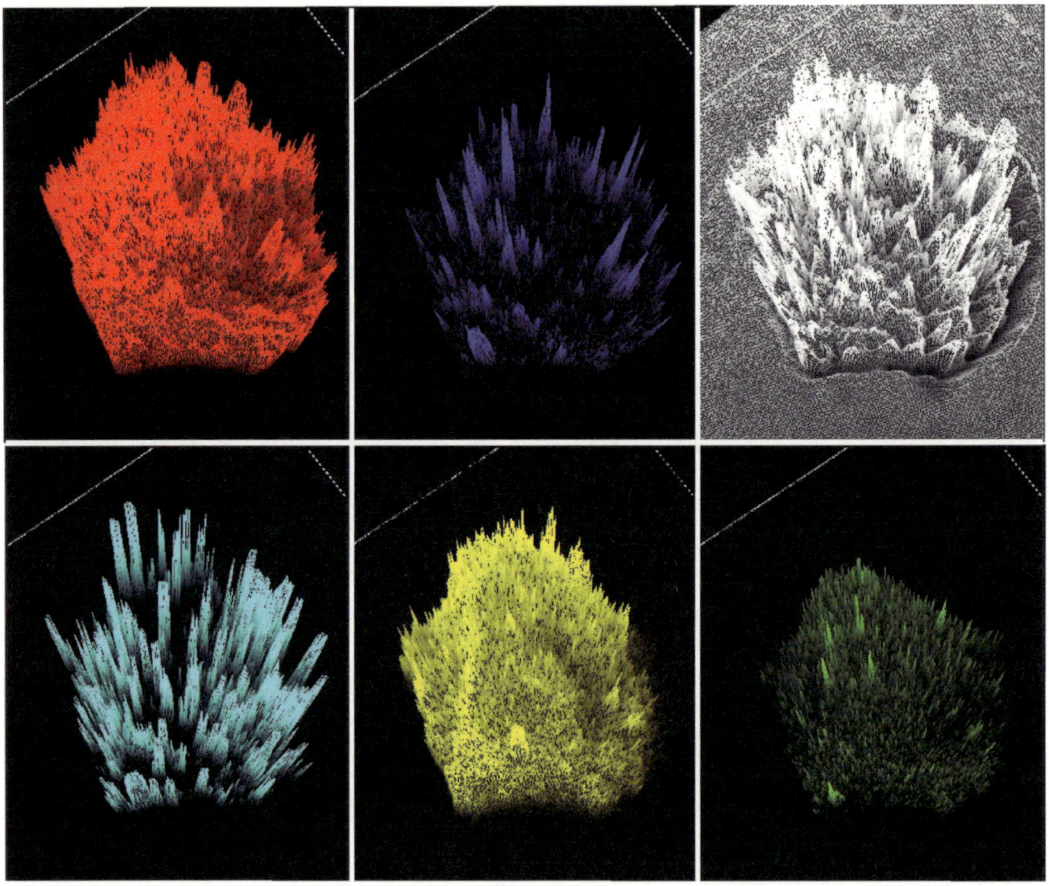

Fig. 9. 3D images of silica polymorphs nanocrystals from the microsomal fraction in LSCM.

The chalcedony crystal has autofluorescence with different ion lasers: a He/Ne ion laser with emission band in 543 nm (red); a He/Ne ion laser with emission band in 633 nm (blue);

an Ar ion laser with emission bands in 458 nm (cyan), 514 nm (yellow) and 488 nm (green). A DIC image of the crystal is shown (Top, right). This crystal is about 20 µm in size.

Moganite nanocrystals are pinacoid (acicular) in shape and about 1 micron in size. Quartz nanocrystals are trapezohedral in shape, about 2 or 3 micron in size, and they are present in large quantities in chalcedony crystal.

3D images of nanocrystals from the microsomal fraction, shows silica polymorphs, Fig. 9.

The chalcedony crystal has autofluorescence with different ion lasers: a He/Ne ion laser with emission band in 543 nm (red); a He/Ne ion laser with emission band in 476 nm (blue); an argon ion laser with emission bands in 458 nm (cyan), 514 nm (yellow) and 488 nm (green). A DIC image of the crystal is shown (Fig. 9, top, right). This chalcedony crystal is about 20 micron in size and shows very different autofluorescent images. May be, this crystal image shows different degree of crystallization. This mineralization of the electrocytes implies the death of the cell and the nerves, revealing that these conditions of pH and Eh are necessary for this process to occur.

4. Conclusion

The origin of chalcedony (SiO2) has been widely discussed in the literature. Biomineralization by silica can occur under a wide variety of circumstances (Heaney, 1993; Fernández López, 2000; Nash & Hopkinson, 2004). A slight oversaturation of silicon is necessary for allowing chalcedony formation from the solution.

A crystal is a solid material whose constituent atoms, molecules, or ions are arranged in an orderly repeating pattern extending in all three spatial dimensions (Hahn, 2002). Crystal habit depends on two main factors: the inner, crystalline structure determines the faces the crystal can present; growth conditions, however, determine the relative size of each face, and hence also the overall shape (Rasmuson, 2009).

The conditions at which silica formation occur is at a pH 7 or near a pH of 8 and an Eh (oxidation potentials) of 0.0 to -0.2 (Fig. 10; Krumbein & Garrels, 1952). It was documented oxidative stress in electric organs from Rajidae family fish (Prado Figueroa, 2005). In such oxidative conditions, the presence of iron could contributed to silica formation.

Crystal growth is a major stage in the crystallization process, and consists in the addition of new atoms, ions, or polymer. Details of the early stages of chalcedony genesis are not fully understood but could involve either the direct precipitation of amorphous silica from a hydrous fluid, which then evolves into chalcedony, or the direct growth of crystalline chalcedony (Moxon & Carpenter, 2009).

The proportion of moganite decreases with age (Moxon, 2004; Moxon & Carpenter, 2009). There is a correlation of crystallite growth with moganite content. The recrystallization of moganite to alfa-quartz clearly occurs at the same time as crystallite growth. Water has an important role: In the absence of water vapour, crystallite growth and transformation of moganite to quartz ceases (Moxon & Carpenter, 2009). Moganite is abundant in arid environment, this is probably due to the lack of water available to support the dissolution of moganite and the simultaneous precipitation of quartz (Bustillo, 1992).

The change in composition of internal water is proposed as a method for approximate dating of agates (a variety of chalcedony) (Moxon, 2004). It has been demonstrated in geological environments that less stable silica polymorphs appear to have transformed over time to chalcedony and microquartz (Nash & Hopkinson, 2004).

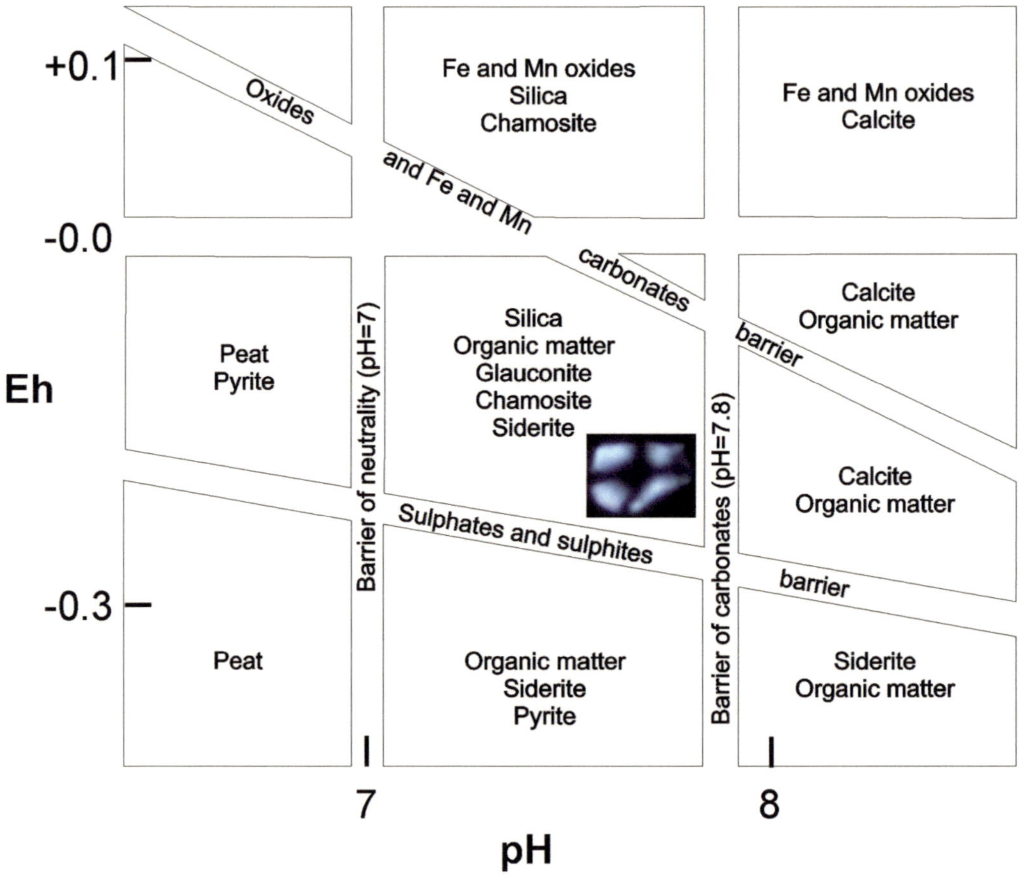

Fig. 10. Chalcedony formation. Krumbein, W.C., Garrels, R.M., (1952).

It is possible to identify the silica polymorph components of chalcedony: quartz and moganite in 3-D images of nanocrystals, by using a Laser Scanning Confocal Microscope and Leica software, see Fig. 7. These quartz and moganite nanocrystals can be differentiated by using an argon ion laser with two emission bands: 514 nm (yellow emission) and 458 nm (cyan emission). These are shown at the Figs. 7 to 9. These figures perhaps show the growth of chalcedony in the electric organ from marine fish.

Quartz is estimated to occupy circa 12 % of the earth's crust, so it is not surprising that research into its diverse applications has been of major interest for over 100 years. Important uses are made of quartz minerals that range from piezoelectric devices to the literal down-to earth quartz aggregates required by the construction industry. During the past (1950's and 60'0s, there were commercial demands for quality quartz crystal required in the manufacture of medical and aerospace sensors (Moxon, 2004; Moxon & Carpenter, 2009).

Of the intriguing topics that are receiving renewed attention nowadays, the study of the "triangle" biomineralization/demineralization/remineralization is among the most fascinating (Ehrlich *et al.*, 2010).

Electric organs of the electric fish have constituted the choice model for studying the nervous cholinergic system (see: Changeux, 1981; 2010). This chapter shows the autofluorescent chalcedony in electric organs. Autofluorescent chalcedony was also documented in human brains from elderly patients (Prado Figueroa & Sánchez Lihón, 2010).

There are many similarities in the occurrence of biosilicification in both systems. One of the major neurochemical features of Alzheimer's disease is the marked reduction of nicotinic acetylcholine receptor in relevant diseased brain regions such as the cerebral cortex and hippocampus (Oddo & LaFerle, 2006). An important use of chalcedony crystals from electric organs is, maybe, in relation with the human medicine.

This paper has demonstrated, using different samples of electric organs, the way Laser Scanning Confocal Microscope with three dimensional (3D) images obtained by using a Leica Confocal Software can be employed to identify chalcedony and nanocrystalline silica polymorphs (quartz and moganite) in electric organs from marine fish.

This communication provides the first experimental evidence of biologically produced crystalline silica mineral phase (i.e., chalcedony) and its growth (crystallinity) in electric organs from living electric fish.

5. Acknowledgment

This research was partially supported by grants to María Prado Figueroa from Secretaría General de Ciencia y Técnica, Universidad Nacional del Sur (UNS), Bahía Blanca, Argentina. MPF is grateful to Dr. T. Moxon, 55 Common Lane, Auckley, Doncaster DN93HX, UK, for many interesting suggestions about chalcedony formation and isolation.

MPF is grateful to Prof. F.J. Barrantes, ex-Director of the Centro Científico Tecnológico Bahía Blanca, Argentina (CCT-BBca, CONICET - UNS) for his permanent interest in this study.

MPF appreciate Lic. E. Buzzi and Mr. M. Diestefano from the CCT-BBca, CONICET - UNS, for technical assistance. Finally, the author is very grateful to Lic. M. Salaberry, English Translator, for her excellent work.

6. References

Amar-Costesec, A., Prado Figueroa, M., Beaufay, H., Nagelkerke, J.F., van Berkel, T.J.C., 1985. Analytical study of microsomes and isolated subcellular membranes from rat liver. IX. Nicotinamide adenine dinucleotide glycohydrolase: a plasma membrane enzyme prominently found in Kupffer cells. *Journal of Cell Biology* 100, 189–197.

Beaufay, H., Amar-Costesec, A., 1976. Cell fractionation techniques. In: Korn, E.D. (Ed.), *Methods in Membrane Biology*, vol. 6. Plenum Press, New York, London, pp. 1-100.

Barrantes, F.J., Mieskes, G. & Wallimann T., (1983). Creatine kinase activity in the Torpedo electrocyte and in the nonreceptor v-proteins from acetylcholine receptor-rich membranes. *Proceeding National Academic of Sciences U.S.A*, 80, 5440-5444.

Bustillo, M.A., 2002. Aparición y significado de la moganita en las rocas de la sílice: una revisión. *Journal of Iberian Geology* 28,157-166.

Carlisle, EM (1982) The nutritional essentiality of silicon. *Nutritional Review*. 40, 193-198.

Changeaux, J.P., (1981) The acetylcholine receptor: an "allosteric" membrane protein. *Harvey Lecture* 75, 85-254.

Changeaux, J.P., (2010) Allosteric receptors: from electric organ to cognition. *Annu Rev Pharmacol Toxicol* 50, 1-38.

Conrad, C.F., Yasuhara, H., Bandstra, J.Z., Icopini, G.A., Brantley, S.L., Heaney, P.J., (2007). Modeling the kinetics of silica nanocolloid formation and growth in aqueous solutions as a function of pH and ionic strength. *Geochimic Cosmochimical. Acta* 71, 531-542.

Deer, W.A., Howie, R.A., Zussman, J. (1966). An introduction to the rock-forming minerals. In: Longman Scientific and Technical, Longmans, Green and Co. Ltd., W. Clowes and Sons Ltd., London, 696 pp.

Erhlich, H. (2010). Chitin and collagen as universal and alternative templates in biomineralization, *International Geology Review* 52, 7, 661 – 699

Erhlich, H., Demadis, K.D. Pokrovsky, O.S. Koutsoukos, P.G. (2010). Modern views on desilicification: biosilica and abiotic silica dissolution in natural and artificial environments. *Chemical Reviews* 110, 4656-4689.

Fessard, A. (1958). Les organs éléctriques. In: P. Grassée (Ed.), *Traité de Zoologie*, vol. 13, 1143-1238. París, Masson.

Fernández López, S. (2000). *Temas de tafonomía*. Departamento de Paleontología. Fac. Cs. Geol. Universidad Complutense de Madrid. 167 pp. Madrid, Spain.

Fox, G.Q., Kriebel, M.E., Pappas, G.D. (1990). Morphological, physiological and biochemical observations on skate electric organ. *Anatomy and Embryology* 181, 305-315.

Hahn, Theo, (2002). *International Tables for Crystallography*, Vol. A: Space Group Symmetry, A. (5th ed.), Berlin, New York: Springer-Verlag.

Heaney, P.J. (1993). A proposed mechanism for the growth of chalcedony. *Contributions of Mineralogical Petrology* 114, 66–74.

Heaney, P.J., Post, J.E., (1992). The widespread distribution of a novel silica polymorph in microcrystalline quartz varieties. *Science* 255, 441–443.

Heaney, P.J., Veblen, D.R., Post, J.E., (1994). Structural disparities between chalcedony and macrocrystalline quartz. *American Mineralogist* 79, 452–460.

Heaney, P.J., Mc Keown, D.A., Post, J.E., (2007). Anomalous behavior at the I2/a to Imab phase transition in SiO2-moganite: an analysis using hard-mode Raman spectroscopy. *American Mineralogist* 92, 631–639.

Henkel, G. (1989). The Henkel glossary of fluorescent minerals. *Journal of the Fluorescent Mineral Society* 15, 1 - 91.

Jacob, B.A., Mc Eachran J.D., Lyons P.L. (1994) Electric organ in skate: variation and phylogenetic significance (Chondrichthyes: Rajoidei). *J. Morphology* 221, 45-63.

Krumbein, W.C., Garrels, R.M., (1952). The origin and classification of chemical sediments in terms of pH and oxidation–reduction potentials. *Journal of Geology* 60, 1–33.

Moore, D.M., & Reynolds, R.C., (1997). X-ray Diffraction and the Identification and Analysis of Clay Minerals. In Oxford University Press, New York, 378 pp.

Moxon, T. (2004) Moganite and water content as a function of age in agate. *European Journal of Mineralogy* 16, 269-278.

Moxon, T. & Carpenter, M.A. (2009). Crystallite growth kinetics in nanocrystalline quartz (agate and chalcedony). *Mineralogical Magazine*, 73(4), 551-568.

Nash, D.J., Hopkinson, L. (2004) A reconnaissance laser Raman and Fourier transform infrared survey of silcretes from the Kalahari desert, Botswana. *Earth Surf. Process. Landforms* 29, 1541-1558.

Oddo, S. & LaFerle, F.M. (2006). The role of nicotinic acetylcholine receptors in Alzeimer's disease. *Journal of Physiology* (Paris) 99, 172-179.

Prado Figueroa, M. (2005). Distribución cuantitativa del malondialdehido entre las fracciones subcelulares obtenidas por centrifugación diferencial de órganos eléctricos de peces de la familia Rajidae y topografía del nAChR. *III Jornadas de Bioquímica y Biología Molecular de Lípidos y Lipoproteínas.* Bahía Blanca, Argentina, p. 101.

Prado Figueroa, M., Santiago, J. (2004). Intracellular localization of a long alkyl chain tetraphenylporphyrin and chloride channel activation in Psammobatis extenta electrocytes. *Photochem. Photobiol. Sci.* 3, 33–35.

Prado Figueroa, M. & Sánchez Lihón, J (2010). Autofluorescent chalcedony in human brains from elderly patients. *Biotechnic and Histochemistry* 85, 171-176.

Prado Figueroa, M., Vidal, A. & Barrantes, F. J. (1995). Ultrastructure of Psammobatis extent (Rajidae) electrocytes and cytochemical localization of acetylcholinesterase, acetylcholine receptor and F-actin. *BIOCELL* 19, 113-123.

Prado Figueroa, M., Barrera, F., Cesaretti, N.N., (2008). Chalcedony (a crystalline variety of silica): biogenic origin in electric organs from living Psammobatis extenta (family Rajidae). *Micron* 39, 1027–1035.

Rasmuson, A.C. (2009). Introduction to crystallization of fine chemicals and pharmaceuticals. In: *Molecules: nucleation, aggregation and crystallization.* Ed. J. Sedzik & P. Riccio. World Scientific Publishing Co. Pte. Ltd., Singapore; New Jersey, USA; London, UK, pp. 145-172.

Schroder, H.C., Wang X., Tremel, W., Ushijima, H., Muller W.E. (2008). Biofabrication of biosilica-glass by living organisms. *Natural Product Reports* 25, 455-474.

Schwarz, K. (1973). A bound form of silicon in glycosaminoglycans and polyuronides. *Proceeding National Academic of Science USA*, 70, 1608-1612.

Whittaker, V.P. (1977). The electromotor system of Torpedo as a model cholinergic system. *Naturwissenschaften* 64, 606-611.

12

Synthesis and Characterization of Crystalline Zirconium Titanate Obtained by Sol-Gel

Venina dos Santos[1] and C.P. Bergmann[2]

[1]*Center of Exact Science and Technology, University of Caxias do Sul, Caxias do Sul, RS,*
[2]*Laboratory of Ceramic Materials (LACER), Departament of Materials, Engineering School, Federal University of Rio Grande do Sul, Osvaldo Aranha, Porto Alegre, RS, Brazil*

1. Introduction

Zirconium titanate (ZT)-based ceramic materials, ZrTiO4, have many attractive properties: high resistivity, high dielectric constant, high permittivity at microwave frequencies and excellent temperature stability for microwave properties (Bianco *et al*, 1999; Leoni *et al*, 2001). These materials have an extremely wide range of technological application, such as in microwave telecommunications (as capacitors, dielectric resonators in filters, and oscillators) (Navio *et al*, 1993; Navio *et al*, 1994; Azough *et al* 1996), in the manufacture of high-temperature pigments (Hund and Anorg, 1985; Dondi *et al*, 2006) in catalysis (as effective acid-base bifunctional catalysts and photocatalysts) (Tanabe, 1970; Araka and Tanabe, 1980), as structural ceramics (Parker, 1990) and, more recently, as biomaterial coated on 316L SS for biomedical applications (Devi *et al*, 2011).

The formation of $ZrTiO_4$ has been studied by several authors (McHale *et al*, 1989; Christoffersen *et al*, 1992a; Bhattacharya *et al*, 1996; Hom *et al*, 2001; Ananta *et al*, 2003; Kim *et al*, 2004; Troitzsch *et al*, 2005; Vittayakorn *et al*, 2006; Licina *et al*, 2008).

Zirconium titanate is normally synthesized by solid-state reaction of oxide mixtures between ZrO_2 and TiO_2 at elevated temperatures (1200-1600°C) and long heating times and also requires post treatment such as energy-intensive grinding/milling procedures for powder formation and this process usually leads to inhomogeneous, coarse, and multiphase powders of poor purity (Lynch, 1972; Swartz, 1982; McHale *et al*, 1983; McHale *et al*, 1986; Parker, 1990; Christoffersen *et al*, 1992b; Park et al, 1996; Stubicar *et al*, 2001; Troitzsch *et al*, 2004; Dondi *et al*, 2006). $ZrTiO_4$ ceramics have the orthorhombic structure of α-PbO_2 (Blasse, 1966; Newnham, 1967) and belongs to the space group Pbcn. Above 1200 °C the Zr ion and Ti ion of the high temperature normal phase are randomly distributed on octahedral site in α-PbO_2 type structure. (McHale *et al*, 1983; Christoffersen *et al*, 1992b; Azough *et al*, 1993; Ul'yanova *et al*, 1995). In contrast to the displaced transitions which take place at discrete temperature, $ZrTiO_4$ ceramics undergoes a continuous phase transition of the normal to incommensurate over the temperatures of 1200°C to 1100°C by increasing order in the Zr-Ti distributions.

The chemical preparation of reactive precursors offers advantages over traditional processing techniques because of the higher purity and better homogeneity obtained the lower processing temperatures and improved material properties. Table 1 summarizes the procedures currently used to prepare pure $ZrTiO_4$ materials by non-conventional or chemical routes (Navio et al, 1992a; Pol *et al*, 2007).

Precursors	Preparation procedure	References
$Zr(SO_4)_2$, $Ti(SO_4)_2$	1.Coprecipitation of mixed oxides using concentrated ammonia water 2.Calcined precipitates shaped into pellets, pressed at 100 MPa, sintered at 1600 °C for 10 h and post-treated at 1500 °C for 5 h, quenched to room-temperature air and annealed at 700 °C for 3days	Ikawa *et al*, 1988; Ikawa *et al*, 1991
Zirconium isopropoxide titanium isopropoxide	1. Using 2-propanol as solvent, mixing for 5 h at 82 °C, hydrolysis with H_20 at room temperature and increased to 75 °C 2. Solid filtered, washed with hot water and dried at 120 °C under reduced pressure	Yamagushi *et al*, 1989
Zirconium propoxide titanium isopropoxide	Partial and controlled hydrolysis using CH_3COOH-propanol, and a ratio $h = [CH_3COOH]/[metal]\ 1 \leq h \leq 8$ If $h \leq 3$, homogeneous gels were obtained If $h > 3$, white precipitate $M(OR)_x(CH_3COO)_y$	Muñoz *et al*, 1990
Zirconium alcoxide, titanium alcoxide	Classical hydrolytic sol-gel, where metal alcoxides are dissolved in alcohol, after o gel is dried and calcined.	Hirano *et al*, 1991; Komarneni *et al*, 1999; Karakchiev *et al*, 2001
Zirconyl chloride	1. Mixing and precipitation with methanol, hydrogen peroxide solution and aqueous ammonia solution at pH 8-9 2. The gelled mass was washed with acetone, filtered and dried slowly for several days	Navio *et al*, 1992b
Zirconium oxychloride and titanyl chloride	Process involves hydrolysis of the starting salts, nucleation of zirconia and titania, nucleus growth, and the formation of $ZrTiO_4$ above 1150 °C.	Gavrilov *et al*, 1996
$Ti(OBut)_4$, $ZrOC1_2.8H_2O$ (Polymeric precursor)	Mixing with ethylene glycol, $Ti(OBut)_4$, $ZrOC1_2.8H_20$ and citric acid, at 110-120 °C	Bianco *et al*, 1998
$TiCl_4$ and $ZrOCl_2$	Chemical precipitation using ammonia, produced ZT nanopowders	Leoni *et al*, 2001
Titanium sulfate, $ZrOC1_2.8H_2O$	Sol-gel process produced long $ZrTiO_4$ fibers	Lu *et al*, 2003

Table 1. Wet-chemical routes for synthesizing $ZrTiO_4$ materials

Actually the sol-gel processes were developed both in academic research and in industry, producing glasses by straightforward polymerization of molecular precursors in solution. Basically, sol-gel process carried out in a liquid medium. This process involves the evolution of inorganic networks through the formation of a colloidal suspension which is called sol and gelation of the sol to form a network in a continuous liquid phase which is denoted as gel. Three reactions generally describe the sol-gel process: (1) hydrolysis reaction, (2) alcohol condensation process, and (3) water condensation process. The sol-gel approach also provides an alternative and usual way for synthesis of nanomaterials. Combined with chemical nanotechnologies, remarkable progress has been achieved and sol-gel techniques have taken their place as a fundamental approach to the development of new nanomaterials. The sol-gel synthesis method has been used for the production of metal, metal oxide and ceramic nanoparticles with high purity and good homogeneity. If an organic surfactant is added to the sol as the structuring agent, it is even possible to obtain an ordered porous structure in two dimensions or three dimensions. The sol-gel process is considered as a low-temperature synthesis method that gives pure, homogeneous nanoparticles with good size distribution in the design of complex nanoarchitectures. Furthermore, many kinds of nanoparticles including oxides, sulfides, metals, and semiconductors with nanoporous structures can be synthesized through a precise heat treatment. The versatility of the process is largely due to the rich and varied chemistry of organometallic precursors, combined with the low processing temperature (Qiao *et al*, 2011).

Samples containing multi-phases are important from the technological perspective and are strongly superposed (Sham *et al*, 1998). Consentino *et al.* prepared ceramic powders from the mixture of zirconium oxychloride and titanium chloride in stoichiometric quantities in the presence of citric acid (60 °C) and ethylene glycol. By using this technique, the authors observed that after treated at 600 °C for 1 hour still amorphous. In 730 °C, had the crystalline phase of orthorhombic $ZrTiO_4$, contrary to reported by Karakchiev *et al* which obtained zirconium titanate by sols hydrated in 1:1 ratio of Zr:Ti with the presence of TiO_2 as anatase, below 600 °C. At 600 °C this form disappears and gives way to the $ZrTiO_4$ peaks. The preparation of $ZrTiO_4$ and $Zr_{0.8}Sn_{0.2}TiO_4$ by pulsed laser deposition was reported by Viticoli *et al.* Films of $ZrTiO_4$ were prepared, deposited between 450 and 550 °C. At 450 °C, a weak intensity peak at 2θ = 30.48° indicates the presence of crystalline zirconium titanate, peaks at 2θ = 32-35° also suggest the presence of phases for Ti_2O_3, TiO_2 and ZrO_2. Raising the temperature to 550 °C the intensity of reflections (1 1 1) characteristic of $ZrTiO_4$ and the reflections (0 2 0), (2 0 0) and (2 2 2) at 2θ = 32.6°, 35.7° and 63.3 can be identified. Under these conditions only the contributions of $ZrTiO_4$ can be observed, indicating the formation of a single crystalline phase of $ZrTiO_4$. The crystallographic structure for the films deposited, containing tin presents an intense peak at 450°C at 2θ = 33.06° associated with reflection (1 0 4) of Ti_2O_3. Peaks at 2θ = 32-35° suggest the presence of phases for SnO_2, TiO_2 and ZrO_2, while a weak intensity peak indicates the crystallization of $Zr_{0.8}Sn_{0.2}TiO_4$ (0 0 2). (Bhattacharya *et al*, 1996; Troitzsch *et al*, 2005).

In this study, ZT powders were obtained from hydrolysis reactions of zirconium n-propoxide (NPZ) and titanium isopropoxide (tetra-isopropyl titanate - TPT). These reactions were performed by hydrolyzing the alkoxides separately, at the molar ratio of Zr:Ti of 1:1, in n-propanol, using the sol-gel process at ambient temperature (20°C) and influence nitric acid, in order to verify the form of crystallization of $ZrTiO_4$ under these conditions. The results were compared to the previous work (Santos *et al*, 2010).

2. Materials and methods

The hydrolysis reaction to obtain zirconium titanate (ZT), $ZrTiO_4$, was performed with the alkoxides zirconium n-propoxide (NPZ) (Dupont), and titanium isoproxide (TPT) Dupont, deionized water in the presence of nitric acid (Merck). The solvent used was *n*-propanol (Merck).

2.1 Preparation of the zirconium titanate powders

The preparation of zirconium titanate powder from hydrolysis for alkoxides precursors, which was carried out separately, is illustrated in the block diagram reported in Figure 1. The synthesis performed with the mixture of alkoxides is described in Santos *et al.*

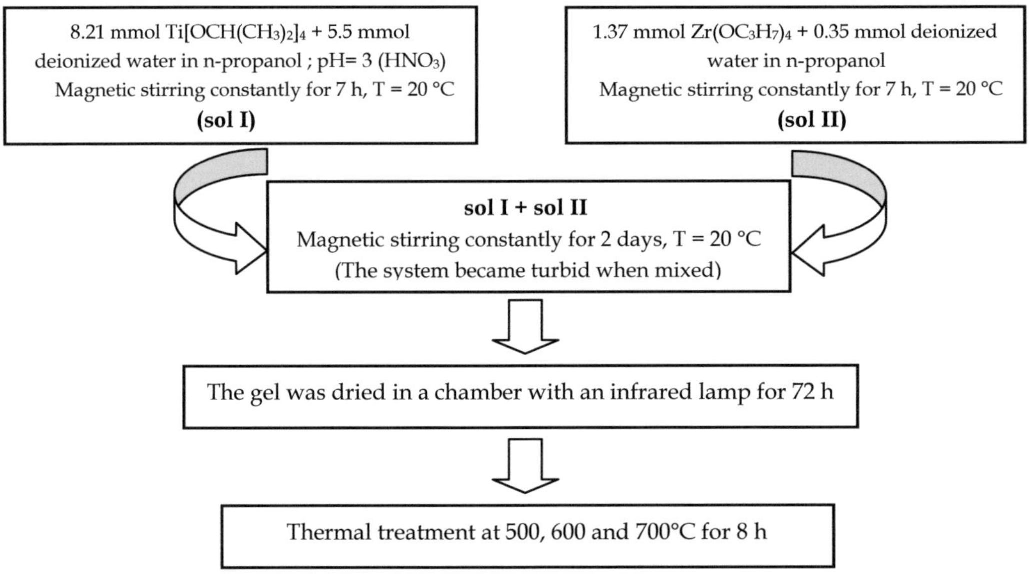

Fig. 1. Process of preparation of zirconium titanate powder from hydrolysis for alkoxides precursors.

2.2 Characterization methods

2.2.1 Thermal analyses

Thermogravimetric analyses (TGA) and thermodifferential analyses (DTA) of the powders were performed with an Harrop thermal analyzer, model STA 736, at a heating rate of 10 K· min^{-1} in air, with a 10 L·min^{-1} flow.

2.2.2 X-Ray diffraction

In order to determine the mineralogical phases by X-Ray Diffraction (XRD), a Philips X-Ray Diffractometer was used (model X'Pert MPD) equipped with a graphite monochromator and rotational anode, operated at 40 kV and 40 mV. The data were collected via Cu-K_{α} radiation at a step of 0.01° and time per step of 2 s, in order to determine the phases present in the samples.

2.2.3 Fourier transform infrared spectroscopy (FT-IR)

Fourier transform infrared spectroscopy (FT-IR) analysis of dried and annealed powders were carried out in an Impact 400, Nicolet spectrometer in the wavenumber range 400–4000 cm^{-1} at resolution of 4 cm^{-1} for studying the chemical groups. For this analysis, KBr pellets were pressed to hold the samples to be analyzed.

2.2.4 Particle size distribution by LASER diffraction spectrometry

The particle size distribution of agglomerates and particles of sintered powder was determined by a laser diffraction spectrometer Cilas, model 1180. The detection range of this equipment is between 0.04 and 2500 µm.

3. Results and discussion

Thermogravimetric and thermodifferential analyses are shown in Figures 2a for hydrolysis of alkoxides together and 2b for alkoxide hydrolysis separately. The curves obtained are very similar, initially, with a great loss of mass between approximately 40 °C and 200 °C, which is due, probably, to dehydration, accompanied by an endothermal peak due to the energy consumed to release volatiles. An exothermal mass can be seen in Figure 2a, at 703 °C. It is ascribed to crystallization of the orthorhombic phase of zirconium titanate. According to Khairulla and Phule (1992), the peaks by differential thermal analysis (DTA) at 350 and 550 °C are related to the removal of organic compounds and an exothermal event without loss of mass at approximately 710 °C is caused by the formation of the crystalline phase of $ZrTiO_4$. Macan *et al*[12], describe that a first loss of mass seen in the TGA and DSC curve as accompanied by an endothermal peak at 83 °C due to the evaporation of adsorbed water. Loss of mass diminishes constantly as temperature increases as a function of the slow degradation of residual organic matter, and the exothermal peak at 703°C is due to the crystallization of amorphous $ZrTiO_4$, confirmed in the preliminary research (Consentino *et al*, 2003).

In Figure 2b two exothermal peaks were observed, 261°C and 460 °C, accompanied by loss of mass, due to the burning of the remaining organic matter. Bhattacharya *et al* suggest that this inflexion is due to the removal of structural anionic species. As the temperature increases, decomposition reactions occur and a broad exothermic peak at 460 °C. The weight loss continues and stabilizes at about 500 °C. At 703°C no exothermal peak that could be attributed to the crystallization of zirconium titanate was observed. During the cooling process endothermal events occurred, with hardly any loss of mass at 1260, 948 and 897 °C, which may be related to crystalline phase transitions, as described by Park *et al* (1996) in the solid-state reaction of oxide mixtures between ZrO_2 and TiO_2 at elevated temperatures.

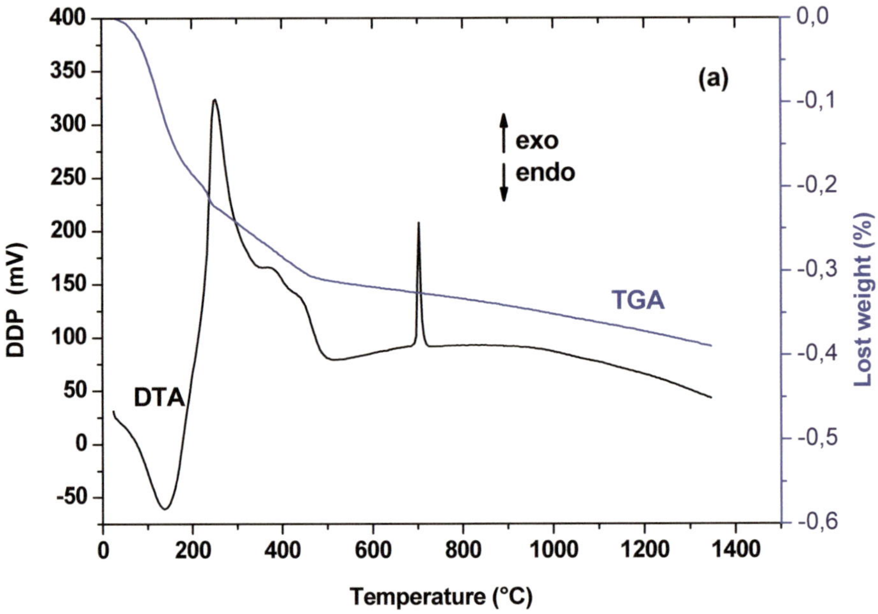

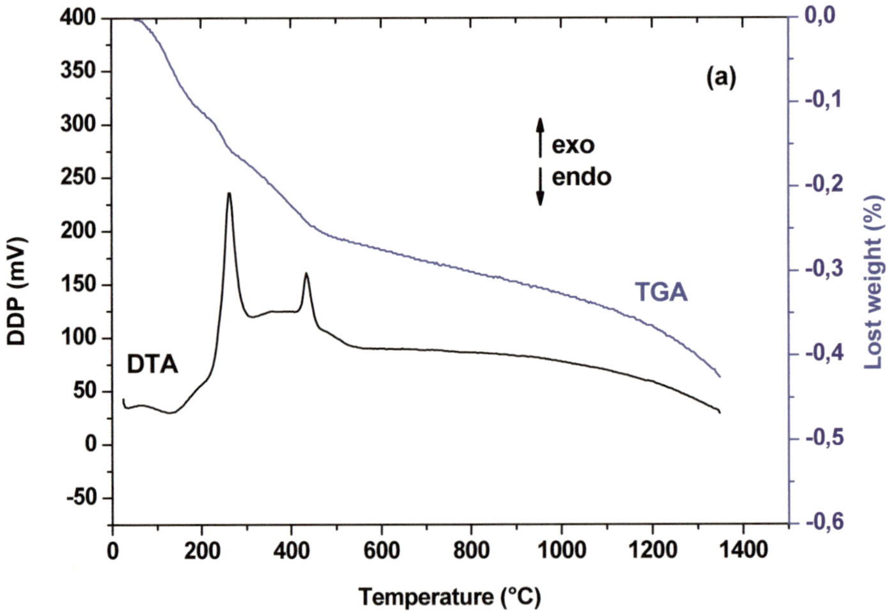

Fig. 2. Thermal analyses of powders after drying obtained by hydrolysis of the alkoxides, NTZ and TPT: (a) together (Santos *et al*, 2010) and (b) separately.

According to Navio *et al*, Figure 3 shows different structures for $ZrTiO_4$. The structure I could be postulated as the amorphous and it is expected to change to the structure II, III and IV (crystalline structures) as the temperature increases.

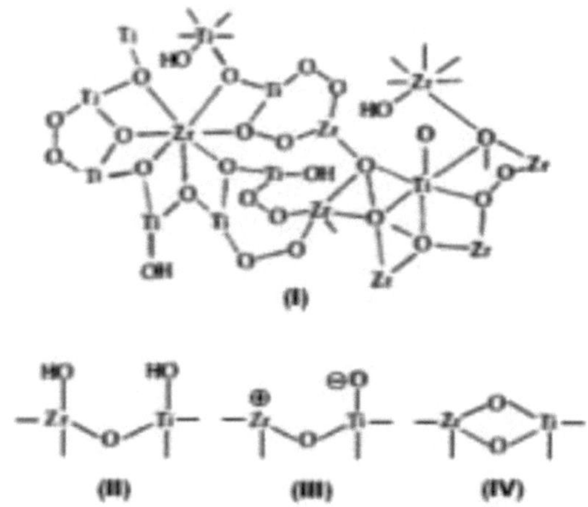

Fig. 3. Structures for $ZrTiO_4$. The structure I could be postulated as the amorphous and it is expected to change to the structure II, III and IV as the temperature increases (Navio *et al*, 1992b).

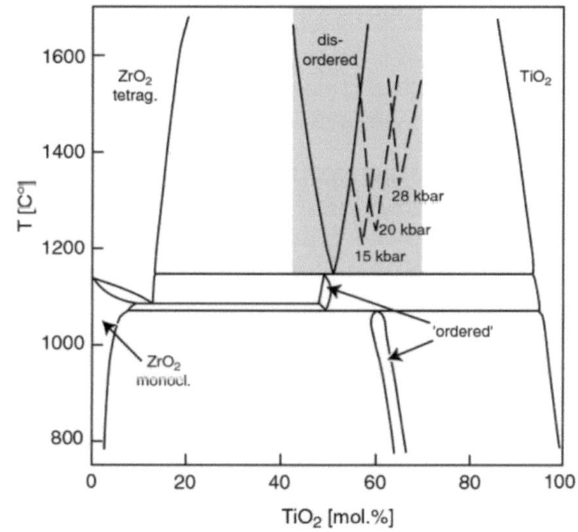

Fig. 4. ZrO_2-TiO_2 phase diagram (Troitzsch *et al*, 2005).

The ZrO_2-TiO_2 phase diagram, Figure 4, has been investigated by McHale and Roth (1986). Above 1100 °C there exists a stable solid solution, $Zr_{1+x}Ti_{1-x}O_4$, with approximate limits –0.17 < x < 0.1 (x ~ – 0.17 corresponds to Zr_5-Ti_7O_{24}). At lower temperatures $ZrTiO_4$ is not thermodynamically stable, but a solid solution centered around the composition $ZrTi_2O_6$ is stable. However, a wide range of compositions in the solid solution $Zr_{1+x}Ti_{1-x}O_4$ can be obtained in metastable form either by wet chemical methods (35-75 mol % Ti) or by quenching from high temperatures as the decomposition kinetics are very slow. The structure of $ZrTiO_4$ was first investigated in 1967. It was reported to be a close structural relative of columbite, (Fe,Mn)Nb_2O_6, a compound that has an α-PbO_2 substructure (Newnham, 1967).

The results of XRD of powders treated thermally at 500, 600 and 700 °C for 8 h are shown in Figure 5a (joint hydrolysis) (Santos *et al*, 2010) and 5b (separate hydrolysis). Figure 5a shows that below 700°C, the samples are amorphous. No presence of crystalline phases related to titania could be seen, probably because it presents in an amorphous form. The peak of orthorhombic $ZrTiO_4$ (2θ=30.595°) was observed in the diffractogram of the sample at 700 °C (López-López *et al*, 2010; Santos *et al*, 2010). López *et al.* prepared $ZrTiO_4$ films, deposited between 450 and 550 °C. At 450 °C, a weak intensity peak at 2θ = 30.48° indicates the presence of crystalline zirconium titanate, peaks at 2θ = 32-35° also suggest the presence of phases for Ti_2O_3, TiO_2 and ZrO_2. Raising the temperature to 550 °C the intensity of reflections (1 1 1) characteristic of $ZrTiO_4$ and the reflections (0 2 0), (2 0 0) and (2 2 2) at 2θ = 32.6°, 35.7° and 63.3 can be identified. Under these conditions only the contributions of $ZrTiO_4$ can be observed, indicating the formation of a single crystalline phase of $ZrTiO_4$. The crystallographic structure for the films deposited, containing tin presents an intense peak at 450°C at 2θ = 33.06° associated with reflection (1 0 4) of Ti_2O_3. Peaks at 2θ = 32-35° suggest the presence of phases for SnO_2, TiO_2 and ZrO_2, while a weak intensity peak indicates the crystallization of $Zr_{0.8}Sn_{0.2}TiO_4$ (0 0 2) (López-López *et al*, 2008). Figure 5b shows that baddeleyite, JCPDS data for baddeleyite (13-0307), is formed at 2θ = 28.22°, at 600 °C and 700 °C from powder treated thermally at 500 °C. Besides the crystalline phase related to zirconium, the peak of titania, JCPDS data for anatase (04-0477), was observed in the anatase crystalline form and the peak of orthorhombic zirconium titanate, JCPDS data for zirconium titanate (34-415), at angles 25.2° and 30.6°, respectively.

Rodrigues et al (2010) reported the synthesis and characterization of nanostructures of sodium titanate/zirconium oxide obtained from the hydrothermal treatment of mixed oxide. Based on these results, we showed that the morphology and crystal structure of the productsobtained via hydrothermal treatment depend of the x value of the precursor $Ti_{1-x}Zr_xO_2$-nH_2O (Ti/Zr molar ratio). For example, for sample with x equal to 0, only the presence of sodium titanate nanotubes were observed, while for small x values (less than 0.50) the nanoparticles showed morphology of nanoribbons and the presence of the sodium titanate phase and tetragonal ZrO_2. For large values of x (greater than 0.50, high amount of zirconium) no morphological changes occurred and the tetragonal ZrO_2 phase was observed for samples. Furthermore, only for the product obtained from x equal to 0.15, we observed the presence of three-dimensional flower-like arrangements. Thus, the influence that the Ti/Zr molar ratio of the precursor plays on the phase and morphology of the hydrothermal products obtained is significant. Second Rodrigues, the preferential coordination of Zr^{4+} ion is between 7 and 8, while for Ti^{4+} it is always 6. The ionic radius of Zr^{4+} in ZrO_2 is 0.84 , assuming the coordination number equal to 8, and for the Ti^{4+} in TiO_2 it is 0.61 Å , assuming the coordination number equal to 6. This significant difference in the ionic radius results in the differences in behavior of incorporation of the Ti^{4+} into the lattice of the ZrO_2 and of the Zr^{4+} into the titanate lattice. When Zr^{4+} was introduced into the titanate structure this ion tends to interrupt the crystal arrangement of the titanate. On the other hand, the Ti^{4+} ion can

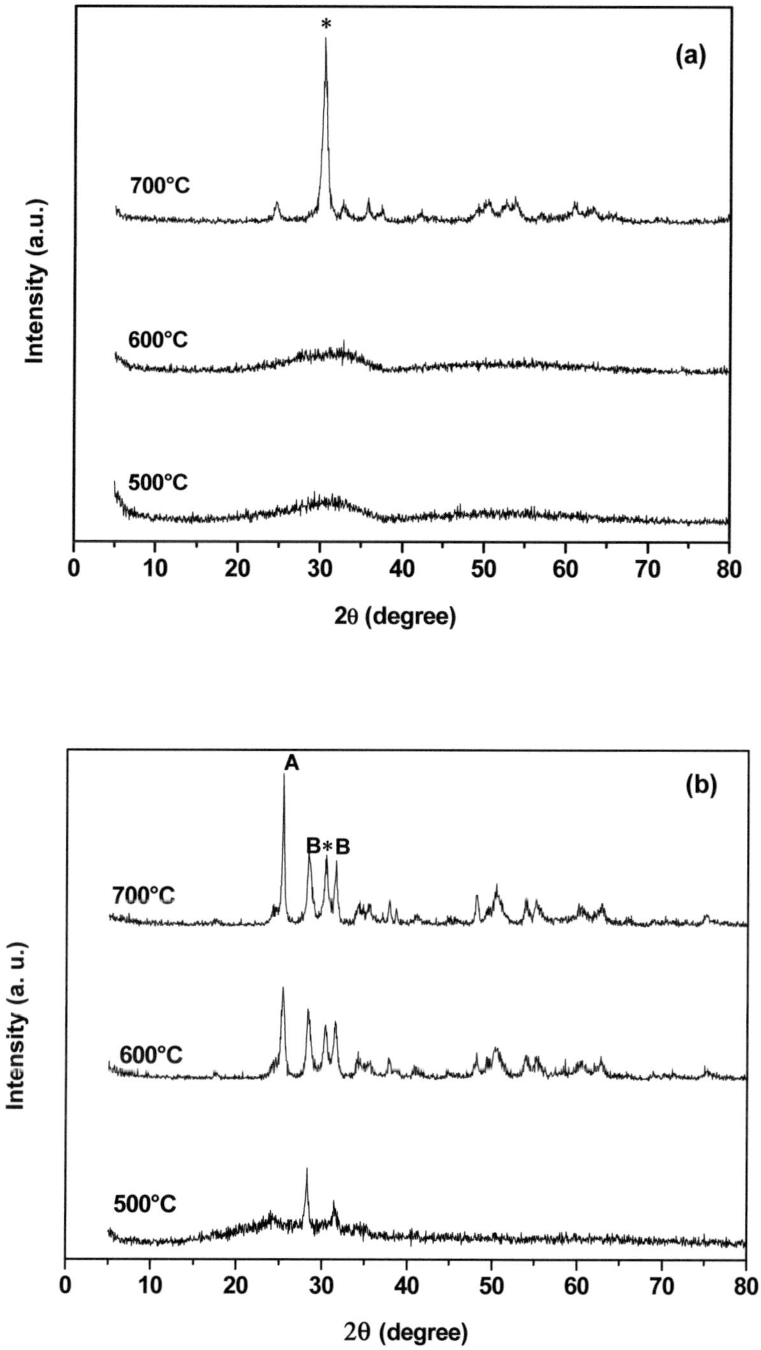

Fig. 5. Diffractograms of synthetized powders by hydrolysis of the alkoxides, after thermal treatment at 500, 600 and 700 °C for 8 h, NTZ and TPT: (a) together (*= ZT) (Santos et al, 2010) and (b) separately (A=anatase, B= baddeleyite, * = ZT).

be easily introduced into the ZrO_2 structure, because of its smaller radius and lower coordination number. However, because of its greater radius and coordination number Zr^{4+} ions could not be easily introduced into the titanate structure.

Figure 6 and 7 shows spectra obtained by means Fourier transform infrared spectroscopy (FT-IR) analysis of the dried and annealed ZT powders at different temperatures ranging from 100 °C to 800 °C with and without addition of nitric acid, respectively. The bands at 3380 cm^{-1} and 1565 cm^{-1} correspond to the vibration of stretching and deformation of the O–H bond due to the absorption of water and coordination water, respectively. As the annealing temperature increased, the formation of these bands gradually decreased, eventually disappearing. The absorption band at 466 cm^{-1} can not be observed, this band is related to the vibration of the Zr–O bond (Hao *et al*, 2004). According to Devi *et al* the bands between 3.500 and 3.300 cm^{-1} was assigned to fundamental stretching vibration of hydroxyl groups. Another peak related to hydroxyl group was found at 1.650 cm^{-1}. The set of overlapping peaks in the range of 810–520 cm^{-1} are related to Zr-O and Zr-O-Ti groups, respectively. From the above results it is indicated the presence of zirconium titanate on 316L SS (AISI stainless steel) substrate was confirmed. Similar results were also observed by Zhu *et al* (2009). The peaks between 500 and 710 cm^{-1} are related to Ti–O and Zr–O–Ti vibrations.

From the evaluation of the infrared spectra can be observed that the presence of nitric acid influences the absorption bands of OH, Zr–O and Zr–O–Ti. Hoebbel *et al* (1997) suggest that a low hydrolytic stability of metal oxide without functionality result in an additional, mostly indefinite number of H^+ groups, of the HNO_3 at the metal centres which causes a higher degree of hydrolysis reactions compared to stable complexes with a defined organic functionality.

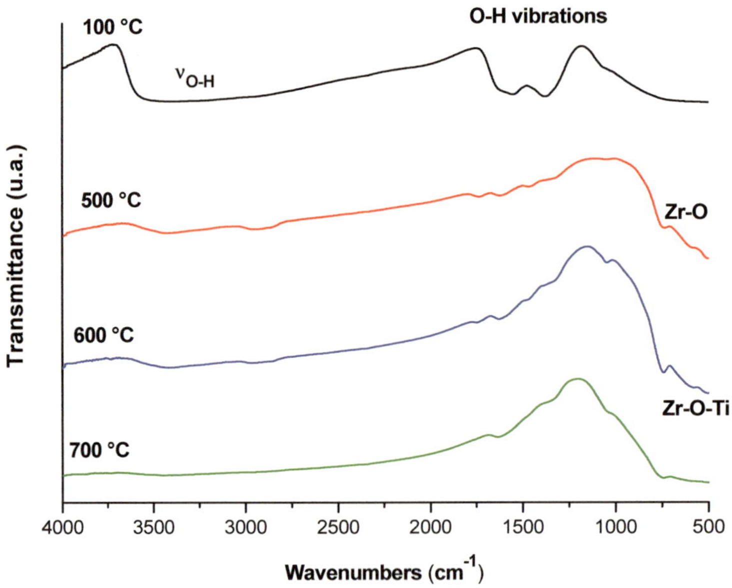

Fig. 6. FT-IR of the dried and annealed ZT powders at different temperatures ranging from 100, 500, 600 and 700 °C, with addition of nitric acid.

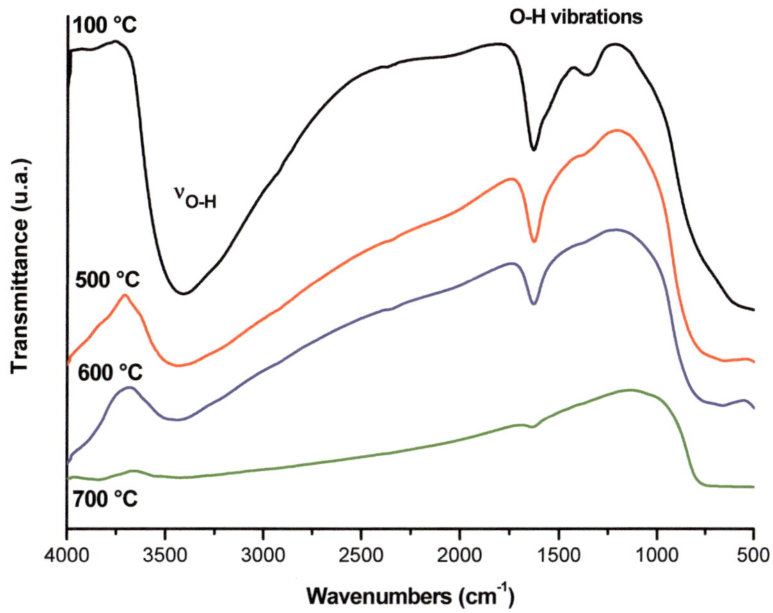

Fig. 7. FT-IR of the dried and annealed ZT powders at different temperatures ranging from 100, 500, 600 and 700 °C, without addition HNO_3.

With the thermal treatment at the temperatures evaluated in this study, the particle size increased. The mean size of the particle (D50) was 23 µm, determined by LASER diffraction spectrometry for powders after thermal treatment at 600°C. Generally in the sol-gel process, the particles are polydispersed and sometimes they can be multimodal, and/or nonspherical. In the sol-gel particles generally exhibit nanometer size, but in this work, the particle sizes presented in the order of micrometer size due to aggregation between the particles. Research on oxide-based materials have shown that dissolution reactions of titanium are initiated by the surface coordination of the material with H^+ and ligands that polarize, weaken and tend to break the metal-oxygen bonds of the surface (Blackwood *et al*, 2002). Therefore, durability of titanium oxide in acidic solutions can be envisaged as occurring by a parallel dissolution mechanism involving H^+ and ligands existing in solution. Unlike preferred complexing anions such as SO_4^{2-} or Cl^-, NO_3^- is a much weaker complexing agent, with very dilute HNO_3 known to be non-oxidising (Housecroft, 2005). The pronounced durability of zirconium oxide in nitric acid is well known for nanoparticles and films and can be extended here to zirconium titanate mesoporous materials from the results obtained (Gao *et al*, 1996; Andreeva *et al*, 1961). Moreover, Hoebbel *et al* suggest that a low hydrolytic stability of metal oxide complexes without functionality result in an additional, mostly indefinite number of H^+ groups at the metal centres which causes a higher degree of hydrolysis reactions compared to stable complexes with a defined organic functionality (Hoebbel *et al*, 1997).

It was possible to prepare $ZrTiO_4$ powders from zirconium *n*-propoxide and titanium isopropoxide, at a Zr:Ti molar ratio of 1:1, in propanol, with and without addition of nitric acid using the sol-gel process at ambient temperature (20 °C) and low thermal treatment.

Synthesis together with the alkoxides was obtained crystalline orthorhombic phase, confirmed by XRD and an exothermal peak without loss of mass by DTA analysis. When the synthesis was performed with separate alkoxides was not possible to obtain only zirconium titanate, probably due to the fact formed crystalline forms anatase, and baddeleyite observed by XRD, which competes with the formation of $ZrTiO_4$, which can be detected only at 700 °C. The $ZrTiO_4$ powders treated thermally at 700°C present. The thermal analysis showed also a great loss of mass, between 40 and 200 °C approximately, probably due to dehydration. The peaks in DTA at 350 and 550 °C are related to the removal of organic compounds. The water loss was confirmed in the spectrum FTIR. It was observed that the formation of the bands related to the presence of hydroxyl group showed a decrease with the increase of annealing temperatures, and at 700 °C they disappeared, indicating that the material structure no longer contains zirconium hydroxide. The sol-gel process was efficient in the preparation of $ZrTiO_4$ using the two routes compared, but should be studied the addition of other acids, as nitric acid caused the aggregation of the powers.

4. References

Ananta S., Tipakontitikul R., Tunkasiri T. (2003) Synthesis, formation and characterization of zirconium titanate (ZT) powders. *Mater Lett*; 57: 2637–42.

Andreeva V. V., Glukhova A. I. (1961) *J. Appl. Chem.* 11 390-397.

Araka, K.; Tanabe K. Bull. Chem. Soc. Japan (1980) 53, 299.

Azough F., Freer R. (1993) A Raman spectral characterization of ceramics in the system ZrO2-TiO2. *Journal of Materials Science*, 28 2273-2276.

Azough, F.; Freer, R.; Wang, C.-L.; Lorimer, G. W. (1996) The relationship between the microstructure and microwave dielectric properties of zirconium titanate ceramics. Journal of Materials Science 31 2539- 2549.

Baltrusaitis J., Schuttlefield J., Jensen J. H., Grassian V. H. (2007) *Phys. Chem. Chem.* Phys. 9, 4970 -4980.

Bhattacharya A. K., Mallick K. K., Hartridge A., Woodhead J. L. (1996) Sol gel preparation, structure and thermal stability of crystalline zirconium titanate. microspheres. *J Mater Sci*; 31:267–71.

Bianco, A.; Gusmano, G.; Freer, R.; Smith, P. (1998) *J. Eur. Ceram. Soc. 18*, 1235-1243.

Blackwood D. J., Chooi S. K. M. (2002) *Corros. Sci.* 44 395-405.

Blasse, G. (1966) J. Anorg. Chem. 345, 222.

Christoffersen R., Davies P.K. (1992a) Extended defect intergrowths in Zr1−*x*Ti1+*x*O4. *Solid State Ionics*; 57:59–69.

Christoffersen R., Davies P.K. (1992b) *J. Am Ceram. Soc.* 75, 563-569.

Consentino, I. C., Muccillo, E. N. S., Muccillo, R. (2003) Development of zirconia-titania porous ceramics for humidity sensors. *Sensors and Actuators B*, v. 96, p. 677-683.

Devi, K. B.; Singh, K.; Rajendran, N. (2011) Sol-gel synthesis and characterization of nanoporous zirconium titanate coated on 316L SS for biomedical applications. J. Sol-Gel Sci Technol 59:513-520).

Dondi, M.; Matteucci, F.; Cruciani, G. (2006) Zirconium titanate ceramic pigments: crystal structure, optical spectroscopy and technological properties. *J. of Solid State Chemistry* 179, 233-246.

Gao W., L. Dickinson, C. Grozinger, Morin F. G., Reven L. (1996) *Langmuir*. 12, 6429-6435.

Gavrilov, V.Yu. and Zenkovets, G.A. (1996) Pore Structure of Zirconia and Zirconia–Titania Materials Prepared via Precipitation from Solution, *Kinet. Katal.*, vol. 37, no. 4, pp. 617–621.

Hallett F. R. (1999) *Scattering and particle sizing, applications*, University of Guelph, Ontario, Canada, Academic Press, 2067-2071.

Hao Y., Li J., Yang X., Wang X., Lu L. (2004) *Materials Science and Engineering A* 367 243.

Hirano, S.; Hayashi, T.; Hattori, A. (1991) *J. Am. Ceram. Soc. 74*, 1320.

Hoebbel D., Reinert T., Schmidt H., Arpac E., (1997) *J. Sol-Gel Sci. Technol.* 10, 115-126.

Hom B. K., Stevens R., Woodfield B. F., Boerio-Goates J., Putnam L, Helean K. B., *et al.* (2001) The thermodynamics of formation, molar heat capacity, and thermodynamic functions of $ZrTiO_4$. *J Chem Thermodyn*; 33:165–78.

Housecroft C. E., Sharpe A. G. (2005) *Inorganic chemistry*. Pearson Prentice Hall.

Hund, F.; Anorg, Z. (1985) *Allg. Chem.* 525, 221.

Ikawa, H.; Iwai, A.; Hiruta, K.; Shimojima, H.; Urabe, K.; Udagawa, S. (1988) *J. Amer. Ceram. Soc.* 71 120.

Ikawa, H.; Yamada, T.; Kojima, K.; Matsumoto,S. (1991) *ibid.* 74, 1459.

Karakchiev, L. G.; Zima, T. M.; Lyakhov, N. Z. (2001) *Inorg. Mater. 37,* 386.

Khairulla F., Phule, P.P. (1992) Chemical synthesis and structural evolution of zirconium titanate, *Mater Sci Eng B*, v. 12, p. 327-336.

Kim I. J., Kim H. C.(2004) Zero level thermal expansion materials based on $ZrTiO_4$–Al_2TiO_5 ceramics synthesized by reaction sintering. *J Ceram Proc Res*; 5:308–12.

Komarneni, S.; Abothu, I. R.; Rao, A. V. P. (1999) *J. Sol-Gel Sci. Technol. 15*, 263.

Leoni, M.; Viviani, M.; Battilana, G.; Fiprello, A. M.; Vitticoli, M. (2001) *J. Eur. Ceram. Soc. 21*, 1739.

Licina V., Gajovic A., Mogus-Milankovic A., Djerdj I., Tomasic N., Su D. (2008) Correlation between the microstructure and the electrical properties of $ZrTiO_4$ ceramics. *J Am Ceram Soc*; 91:178–86.

López-López, E.; Baudín, C.; Moreno. (2008) R. Synthesis of zirconium titanate-based materials by colloidal filtration and reaction sintering. *Int. J. Appl. Ceram. Technol.* 5 394.

López-López E., Sanjuán M L, Moreno R, Baudín,C. (2010) Phase evolution in reaction sintered zirconium titanate based materials. *J Eur Ceram Soc* 30 981–991.

Lu, Q.; Chen, D.; Jiao, X. (2003) *J. Mater. Chem. 13*, 1127.

Lynch, R.W.; Morosin, B. (1972) *J. Amer. Ceram. Soc.* 55, 409.

Macan J., Gajovi´c A., Ivankovi H. (2009) Porous zirconium titanate ceramics synthesized by sol–gel process. J *Eur Ceram Soc* 29 691–696.

McHale, A.E.; Roth, R, S. (1983) *J. Amer. Ceram. Soc.* 66, 18-20.

McHale AE, Roth RS. (1986) Low-temperature phase relationships in the system ZrO2–TiO2. *J Am Ceram Soc*; 69:827–32.

Muiñoz, M.J.; Larbot, A. (1990) in Proceedings of the 3rd National Meeting on Materials Science, Sevilla, Spain, edited by I. Carrizosa, J. A. Odriozola, R. Alvero and M. A. Centeno (Instituto de Ciencia de Materiales, Universidad de Sevilla - CSIC, 1990) p. 260.

Navio, J. A.; Marchena, F. J.; Macias, M.; Sanchez-Soto. (1992a) *J. Mater. Sci. Let.* 11, 1570-1572.

Navio, J. A.; Marchena, F. J.; Macias, M.; Sanchez-Soto. (1992b) *J. Mater. Sci.* 27, 2463.

Navio, J. A.; Marchena, F. J.; Macias, M.; Sanchez-Soto, P. J. Pichat, P. (1993) *J. Therm. Anal.*, *40*, 1095.

Navio, J. A.; Colon, G. (1994) *New DeV. SelectiVe Oxid. II*, *82*, 721.

Newnham, R.E. (1967) J. Am. Cer. Soc. 50, 216.

Parker, F. J. J. Amer. Ceram. Soc. (1990) 73 929.

Park, Y.; Kim, Y. H.; Kim, H. G. (1996) Ferroelectric Letters. 21, 65-70.

Pol, S. V. *et al.* (2007) *J. Phys. Chem. C*, 111, 2484-2489.

Qiao S. Z., Liu J., Qing G. (2011) *Modern Inorganic Synthetic Chemistry, 2011, Pages 479-506*, Chapter 21 - Synthetic Chemistry of Nanomaterials.

Rodrigues C. M., Ferreira O. P., Alves O. L. (2010) Nanostructures of sodium titanate/zirconium oxide. *J Nanopart Res* 12:2355–2361.

Santos, V. dos; Zeni, M; Hohemberger, J. M; Bergmann, C. P. (2010) Preparation of crystalline $ZrTiO_4$ at low thermal treatment temperatures. *Rev Adv. Mater. Sci.* n° 1/2, Vol. 24, p. 44-47.

Sham E. L., Aranda M. A. G., Farfan-Torres E. M., Gottifredi J. C., Martinez-Lara M., Bruque S.. (1998) *Journal of Solid State Chemistry* 139.

Stubicar, M.; Bamanec, V.; Stubicar, N.; Kudrnovski, D.; Krumes, D. (2001) *J. Alloys Compd.* *316*, 316.

Stumm W. (1987) *Aquatic Surface Chemistry: Chemical processes at the particle-water interface.* John Wiley.

Swartz, S.L.; Shrout, T. R. (1982) *Mater. Res. Bull.* 17, 1245.

Tanabe, K. Solid acids and bases, their catalytic applications. Academic press, New York, 1970.

Troitzsch U., Ellis D.J. (2004) *Eur. J. Miner. 16*:577-584.

Troitzsch U., Ellis D.J. (2005). The ZrO2–TiO2 phase diagram. *J Mater Sci*; 40:4571–7.

Ul'yanova, T.M., Krut'ko, N.P., But'ko, Z.T., and Zonov, Yu.G. (1995) Formation of zirconium titanate in oxide fibers, *Neorg. Mater.* v. 31, n° 1, pp. 91–95 , *Inorg. Mater.* (Engl. Transl.), v. 31, n° 1, pp. 83–87.

Vittayakorn N. (2006) Synthesis and a crystal structural study of microwave dielectric zirconium titanate ($ZrTiO_4$) powders via a mixed oxide synthesis route. *J Ceram Proc Res*; 7:288–91.

Yamagushi, O.; Mogi, H. (1989) *ibid.* 72, 1065.

Zhu LY, Xu D, Yu G, Wang XQ (2009) *J Sol-Gel Techn* 49:341–346.

Characterization of Sol-Gel-Derived and Crystallized HfO_2, ZrO_2, ZrO_2-Y_2O_3 Thin Films on Si(001) Wafers with High Dielectric Constant

Hirofumi Shimizu and Toshikazu Nishide
College of Engineering, Nihon University
Tamura-machi, Koriyama, Fukushima,
Japan

1. Introduction

A very great number of metal oxide thin films are produced by sol-gel methods. Metallic compounds dissolved in organic solvents are hydrolyzed and polymerized by adding H_2O with an acid or a base and heating to obtain metal oxide sols. Metal oxide thin films are prepared by coating the sols on substrates followed by firing (Kozuka, 2005).

The sol-gel method produces amorphous or crystalline thin gel films of metallic solid compounds by solidifying a sol formed by hydrolyzing and polymerizing a solution containing metallic compounds. Sol-gel processes are widely employed in the field of chemistry to prepare ceramic powders and thin films of hafnium oxide (HfO_2) (Nishide et al., 2000) and zirconium oxide (ZrO_2) (Liu et al., 2002) for obtaining high-quality ceramics and insulators, offering the advantages of low cost, relative simplicity, and easy control of the composition of the layers formed. This chapter describes the characterization of sol-gel-derived and crystallized HfO_2 and ZrO_2 thin films intended for use as gate insulators with high dielectric constants in electronic devices.

In the electronic device field, the continuing miniaturization of silicon (Si) ultra-large-scale-integration (ULSI) devices has required an ultrathin gate Si dioxide (SiO_2) and oxynitride film; upon scaling down to 32-22 nm technology nodes and beyond, thinner SiO_2 gate oxide films have been required. At these thicknesses, gate leakage currents due to direct tunneling become comparable to the off-currents of metal-oxide-semiconductor (MOS) field-effect transistors (FETs), increasing the off-state power consumption of the devices. In further scaled-down advanced Si complementary MOS (CMOS) devices, scaling trends have required the substitution of gate SiO_2 by insulators with higher dielectric constants (high-k) (Huff & Gilmer, 2004). The aim of using high-k materials is to increase the film thickness, thus reducing the tunneling leakage current, while scaling the capacitance of the equivalent oxide thickness (EOT) below the direct tunneling limit of SiO_2 (Huff & Gilmer, 2004). Several high-k material candidates, such as HfO_2 (Blanchin et al., 2008), ZrO_2 (Niinisto et al., 2004), Al_2O_3, ZrO_2-Y_2O_3 (YANG, 1996), Y_2O_3 (Nishide & Shibata, 2001), La_2O_3 (Ng et al., 2005), and gate stack structures have been proposed and some materials have been put into practical use. All of them are either oxides or silicates of 4d or 5d transition metals or rare earth elements.

HfO_2 and ZrO_2 thin films are the most promising candidates as alternative high permittivity (high-*k*) oxides for replacing the SiO_2 gate dielectric material used in CMOS devices (Gusev, 2005, Wilk et al., 2001). Because of the higher permittivity, the dielectric gate insulator thickness can be increased for a given capacitance, resulting in reduced tunneling leakage current. HfO_2 has promising properties such as high permittivity (25~40) (Oniki et al. 2009, Wilk et al., 2000, 2001), a conduction band offset as high as 1.5 eV (Lucovsky, 2002), and a wide band gap (~5.68 eV) (Robertson & Chen, 1999, Robertson, 2000).

In device fabrication processes, HfO_2 and ZrO_2 thin film layers are deposited by chemical vapor deposition (CVD) or physical vapor deposition (PVD) or sputtered onto Si substrates (Gao et al., 2000, Wang et al., 2005) using argon (Ar) and O_2 mixed gases. A sol-gel process offers various advantages for fabricating ZrO_2, HfO_2, ZrO_2–Y_2O_3 (YANG, 1996) and HfO_2 – Y_2O_3 (Nishide et al., 2000) thin films. The properties of a sol-gel-derived thin film depend on the composition of the sol solution, and residual H_2O may affect the performance of the film. Investigations of the basic structural and optical properties of sol-gel-derived HfO_2 films have shown that HfO_2 films formed on quartz substrates begin to crystallize at a firing temperature of 550 °C as determined from X-ray diffraction (XRD) patterns (Nishide et al., 2000). From the interplanar spacing they derived from the XRD patterns and a comparison of their data with data from a Joint Committee on Powder Diffraction Standards (JCPDS) card, they determined the crystalline phase of the sol-gel-derived HfO_2 film to be monoclinic. Recently, on the basis of high-resolution transmission electron microscopy (HRTEM) measurements in combination with results of electron beam nanodiffraction analyses, sol-gel-derived HfO_2 thin films on Si(001) wafers were found to crystallize in a monoclinic face-centered cubic (fcc) structure (Shimizu et al., 2004). Sol-gel-derived ZrO_2 thin films fired in air at 350 and 450 °C on Si(001) wafers are reported to be amorphous and around 9-10 nm in thickness. Crystallization occurs first at 550 °C as amorphous/tetragonal (011), and finally at 700°C, the ZrO_2 film crystallizes into tetragonal (011)/monoclinic $(\bar{1}11)$ and (111) structures (Shimizu et al., 2009). Electrical characteristics have been evaluated using capacitors with an Al/ZrO_2 and/or HfO_2/Si sandwich structure. The leakage current and dielectric constant of the films have been examined using current-voltage (*I-V*) and capacitance-voltage (*C-V*) methods. On the basis of *C-V* characteristics, the dielectric constant (relative permittivity: ε_{ZrO2} and ε_{HfO2}) of sol-gel derived ZrO_2 and HfO_2 thin films fired in air were shown to be far higher than that of silicon dioxide (SiO_2: 3.9) (Shimizu et al., 2009, 2010). This chapter summarizes the characterizations of sol-gel-derived HfO_2, ZrO_2 and ZrO_2-Y_2O_3 thin films on Si(001) wafers with the aim of showing their suitability as alternative gate insulator materials in advanced CMOS devices.

2. Formation of sol-gel-derived HfO_2 on Si(001) wafers fired in air

A Hitachi HF-3000 transmission electron microscope (HRTEM) equipped with a cold type field-emission-gun (C-FEG) was used to obtain cross-sectional views of HfO_2 films on Si(001) wafers fired at 450 °C (amorphous state) and 700 °C (crystallized) [Figure 1 (a) and 1(b)]. The film fired at 700 °C did not show Moire patterns, indicating that the film consisted of single-crystal grains. Some subgrain boundaries or defects such as dislocations and twins were present, so overall the HfO_2 film fired at 700 °C was poly-crystalline.

Sol-gel-derived HfO_2 films on Si(001) wafers were evaluated by X-ray photoelectron spectroscopy (XPS). The XPS spectra of the Hf 4f [Figure 2(a)] and O 1s emissions [Figure 2(b)]

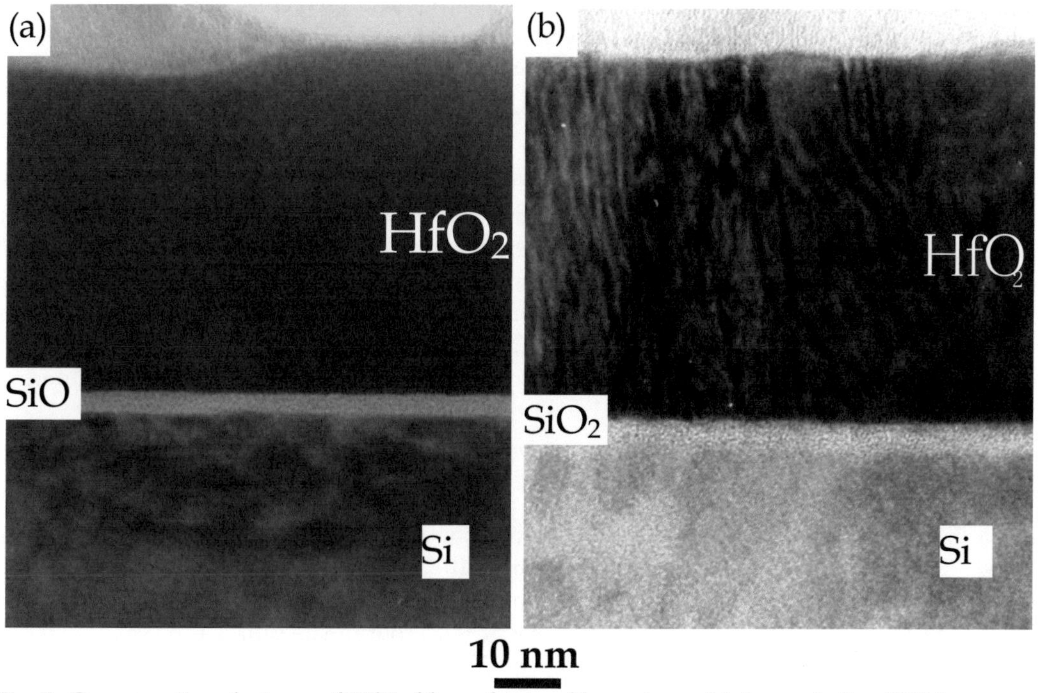

Fig. 1. Cross-sectional views of HfO_2 films obtained by using a high-resolution TEM: (a) a HfO_2 film fired at 450 °C and (b) a HfO_2 film fired at 700 °C (Shimizu et al., 2004).

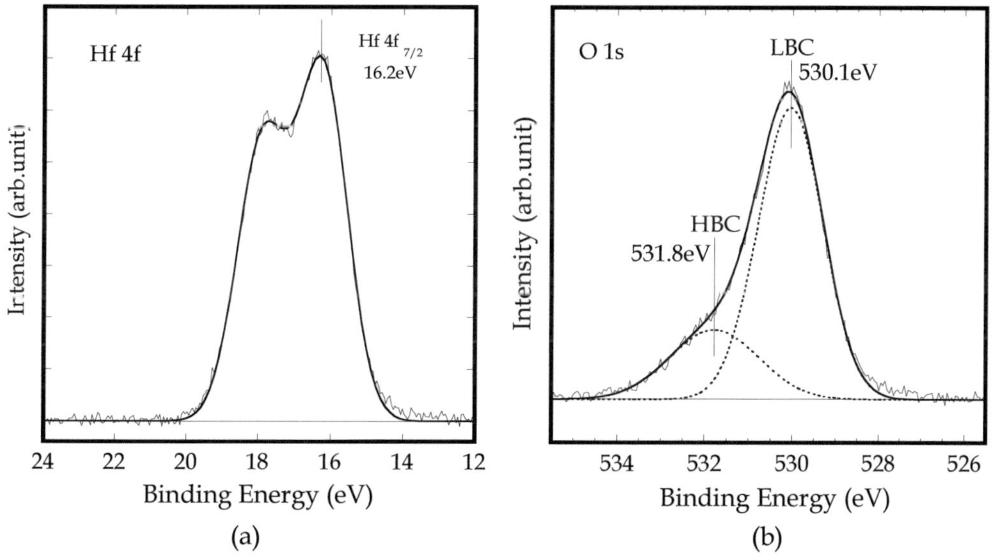

Fig. 2. XPS spectra of sol-gel-derived HfO_2 films. Solid lines are observed spectra and those fitted by the nonlinear least-squares algorithm. Dashed lines for O 1s spectra have two Gaussian peaks corresponding to Hf-OH (531.8 eV) and Hf-O (530.1 eV) (Shimizu et al., 2007).

from the sol-gel-derived HfO_2 film fired at 450 °C indicated that the HfO_2 film was amorphous. The Hf 4f $_{7/2}$ line was determined to be at 16.2±0.1 eV, which is in good agreement with that of the bulk HfO_2 (Chiou et al., 2007, Moulder et al., 1995).

Crystallized HfO_2 films fired at temperatures of 550 and 700 °C showed similar XPS spectra regardless of whether they were amorphous or crystalline. The crystallization of sol-gel-derived HfO_2 films will be discussed later. The O 1s spectrum at 450 °C [Figure 2(b)] was separated into two Gaussian-Lorentzian features corresponding to two chemical states by using the nonlinear least-squares method. One large peak at 530.1 eV (designated as the low-binding-energy component: LBC) was from Hf-O bonds and the other low peak at 531.8 eV (designated as the high-binding-energy component: HBC) was from Hf-OH bonds near the bulk at the surface area. However, since the binding energy of H_2O was slightly larger (533.2 eV) than that of OH, the peak due to physisorbed H_2O may have been included in HBC in the present XPS measurements.

3. Formation of sol-gel-derived ZrO_2 on Si(001) wafers fired in air

The XPS spectra of the Zr 3d and O 1s emissions from sol-gel-derived ZrO_2 thin films fired at 350 °C are shown in [Figure 3(a)] and [Figure 3(b)], respectively. ZrO_2 thin films fired at 450, 550, and 700 °C were also evaluated and similar results were obtained. The Zr 3d $_{5/2}$ line was at 182.1±0.1 eV, which is in good agreement with that of the bulk ZrO_2 (182~182.5 eV) (Moulder et al., 1995). The O 1s spectrum at 350 °C was separated into two Gaussian-Lorentzian features corresponding to two chemical states by the nonlinear least-squares method [Figure 3(b)].

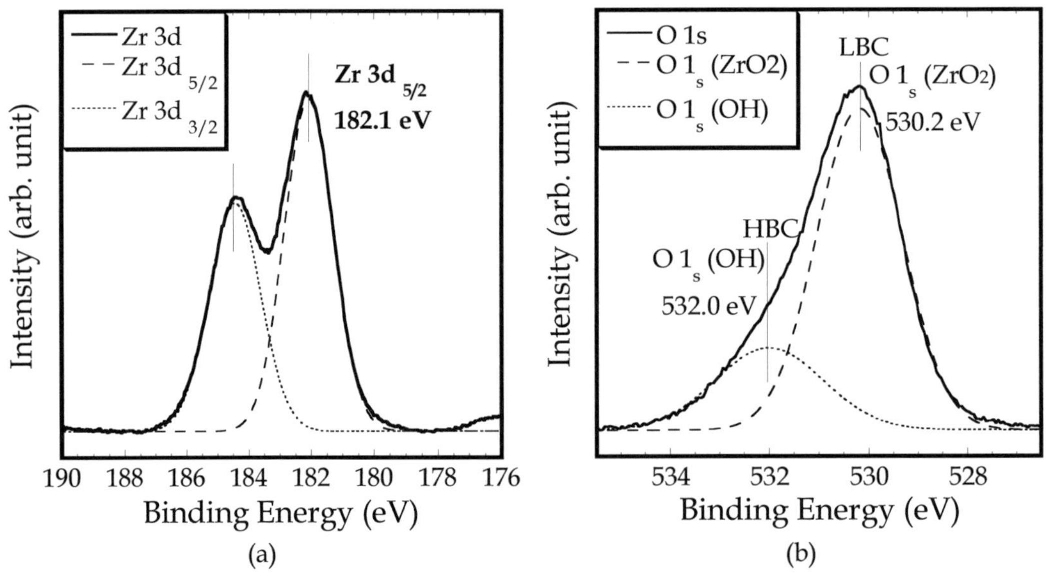

Fig. 3. XPS spectra of sol-gel-derived ZrO_2 thin films: (a) XPS Zr 3d spectrum and (b) XPS O 1s spectrum. Solid lines are the observed XPS spectra and dashed lines are for Zr 3d and O 1s spectra, which have two Gaussian peaks fitted by the nonlinear least-squares algorithm (Shimizu et al., 2009).

One large peak at 530.2 eV (designated as the low-binding-energy component: LBC) was from Zr-O bonds and the other low peak at 532.0 eV (designated as the high-binding-energy component: HBC) was from Zr-OH bonds near the bulk at the surface area. However, since the binding energy of H_2O was slightly higher (533.2 eV) than that of OH, the peak due to physisorbed H_2O may [have been included in HBC?] in the present XPS measurements.

4. Crystallinity of sol-gel-derived HfO_2 thin films on Si(001) wafers

The XRD patterns for HfO_2 films on Si(001) wafers fired at 450, 550 and 700 °C were found to be in good agreement with previously reported results (Nishide et al., 2000) by using a spectrometer (Rigaku RAD-2 XRD) with CuKα radiation (Figure 4). Specifically, the film was still amorphous at 450 °C, and at 550 °C, new peaks appeared at 2θ = 28.4 and 31.8°, as well as small peaks in the region from 18 to 41°; these have been assigned to monoclinic crystalline HfO_2 components (Nishide et al., 2000). At 700 °C, the HfO_2 film was completely crystallized.

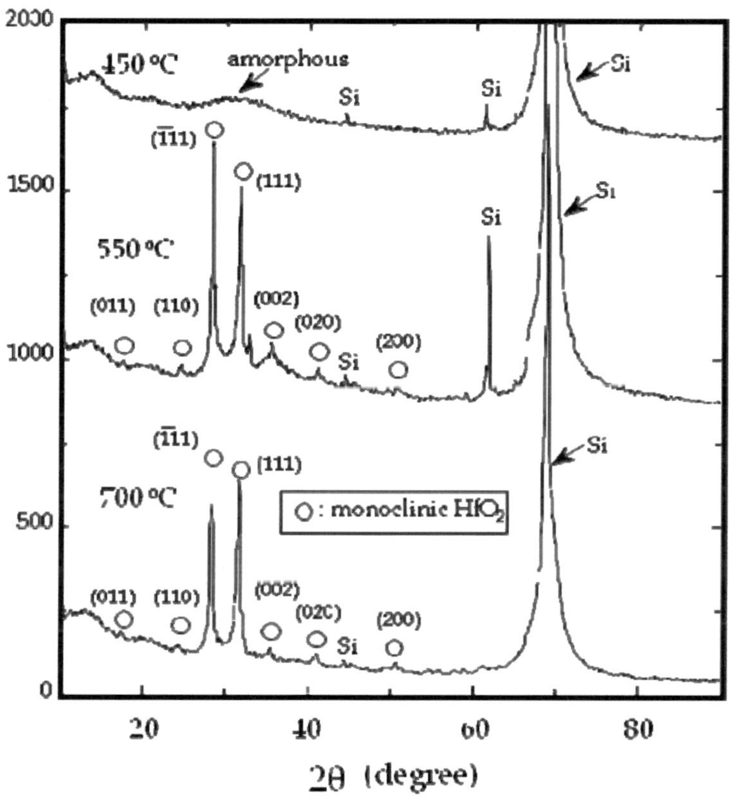

Fig. 4. XRD patterns obtained for HfO_2 films on Si(001) wafers fired at 450, 550 and 700 °C. Open circles indicate monoclinic HfO_2 (Shimizu et al., 2004).

In the electron beam (EB) nanodiffraction pattern for a cross section of the HfO_2 film fired at 550 °C, the Debye ring indicates the beginning of crystallization [Figure 5(a) and 5(b)].

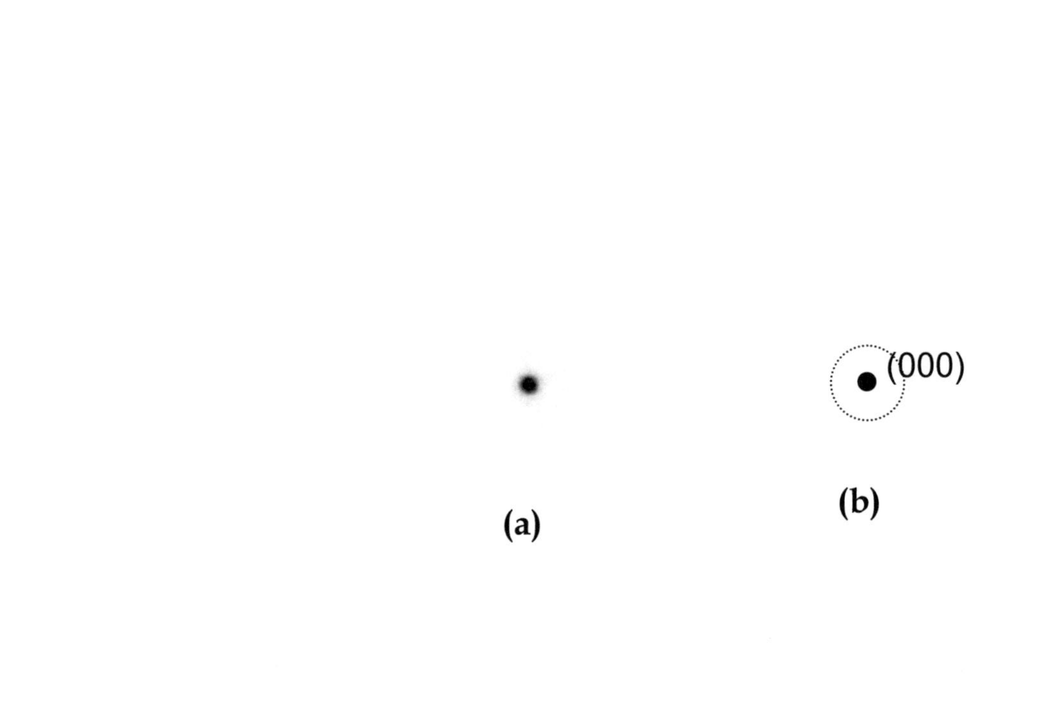

Le=542 mm, λ=0.001969 nm at 300 KV, spot size=1.0 nm
(Debye ring: r_1=3.7 mm d_1=0.288 nm)

Fig. 5. Electron beam nanodiffraction pattern for the (110) plane of the HfO_2 film fired at 550 ^{o}C. r is the distance between diffracted spots in (111) in the HfO_2 film and d is the corresponding interplanar spacing (Shimizu et al., 2004).

The crystalline structure of the sample fired at 700 ^{o}C was determined by analyzing the EB nanodiffraction patterns as follows. First, the camera length (Le) of the HRTEM was determined on the basis of the EB nanodiffraction pattern for a cross section of the Si (110) substrate [Figure 6(a)] and the assignment of the diffraction spots [Figure 6(b)]. Using the data for Si from the International Centre for Diffraction Data, the camera length of the HRTEM was determined to be 542 mm (Shimizu et al., 2004).

Based on the EB nanodiffraction pattern for the sample fired at 700 ^{o}C [Figure 7(a)], the distances (r) between spots in the electron diffraction pattern appearing on the microscopic film were r_1 = 3.7 mm, r_2 = 3.4 mm and r_3 = 6.9 mm for the (111), (111) and (220) planes, respectively [Figure 7(b)]. The corresponding interplanar spacings were determined to be d_1 = 0.288 nm, d_2 = 0.314 nm and d_3 = 0.181 nm using the camera length. A detailed analysis of the alignment of the nanodiffraction spots, with the (000) spot at the center surrounded by 4 (111) spots and 2 (220) spots, together with the interplanar spacings, revealed that the HfO_2 film sintered at 700 ^{o}C had a monoclinic fcc (face centered cubic) structure. One of the measured interplanar spacings of the crystalline HfO_2 was 0.314 nm, which is in good agreement with the spacing of the Si (111) planes. This implies the possibility of the epitaxial growth of HfO_2 films on the Si (111) surface (Shimizu et al., 2004).

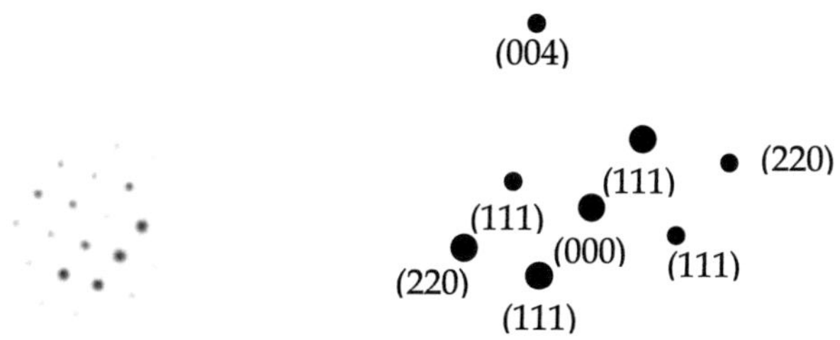

Le=542 mm、 λ=0.001969 nm、 spot size=1.0 nm

r_1= 3.4 mm d_1=0.314 nm (111)

r_2= 5.7 mm d_2=0.191 nm (220)

r_3= 8.2 mm d_3=0.131 nm (004)

Diamond

Fig. 6. Electron beam nanodiffraction patterns: (a) diffracted pattern on (110) plane of Si crystal and (b) schematic diffraction spots. *r* is the distance between diffracted spots in (111), (220) and (004) in the diamond structure and *d* is the corresponding interplanar spacing (Shimizu et al., 2004).

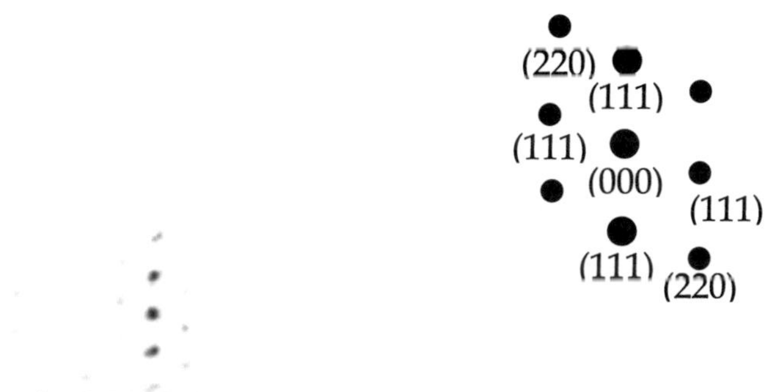

Le=542 mm, λ=0.001969 nm at 300 KV, spot size=1.0 nm

r_1=3.7 mm d_1=0.288 nm (111)

r_2=3.4 mm d_2=0.314 nm (111)

r_3=6.9 mm d_3=0.181 nm (220)

Monoclinic
FCC

Fig. 7. Electron beam nanodiffraction patterns: (a) diffracted pattern from a cross section of the HfO_2 film sintered at 700 °C and (b) schematic of the diffraction spots. *r* is the distance between diffracted spots in (111) and (220) in the HfO_2 film and *d* is the corresponding interplanar spacing (Shimizu et al., 2004).

5. Desorption of H_2O from sol-gel-derived HfO_2 and ZrO_2 thin films on Si(001) wafers during firing

During the firing of hafnia gel films, H_2O is not vaporized completely. Even after HfO_2 films are crystallized on the Si(001) surface, Hf-OH bonds and/or H_2O may be trapped between nanopores in HfO_2 films. Thus, the thermal properties, especially the desorption of H_2O from HfO_2 films, must be clarified after firing hafnia gel films. The electrical properties of sol-gel-derived HfO_2 films should also be characterized, in view of their possible application as gate insulators in next-generation CMOS devices.

Temperature-programmed desorption (TPD) is an excellent technique, not only for analyzing adsorbed gases on the surfaces of bulk sol-gel-derived HfO_2 films, but also for analyzing the species that evolve from the films.

5.1 Basic principles of TPD

TPD, also called thermal desorption spectroscopy, is essentially a method of analyzing desorped gases from samples heated under vacuum conditions using quadruple mass spectroscopy (QMS). The sample is heated by infrared light at a linear rate and evolved gases are introduced into a quadruple mass spectroscope that indicates the intensity of the signal according to the mass (*m*) and electric charge (*z*). TPD is now widely used to investigate the surfaces of ceramics and also semiconductors. In typical TPD spectra, the vertical axis shows the variation of the ion intensity of QMS (in amperes) for a specific m/z and the horizontal axis is the desorption temperature (Figure 8) (Nishide et al., 2004).

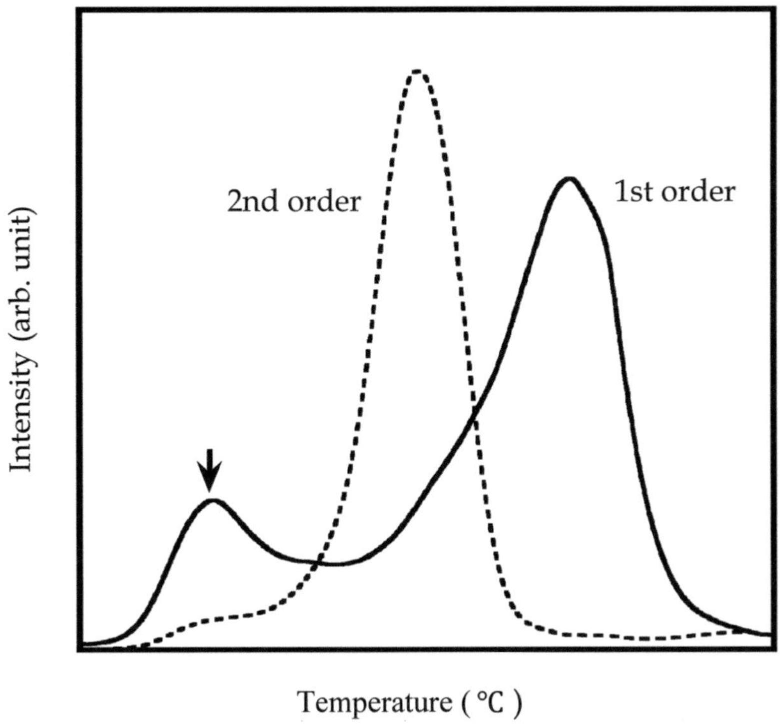

Fig. 8. Typical TPD curves plotted as a function of temperature (Shimizu et al., 2007).

TPD curves can be obtained for various *m/z*'s with increasing temperature, thereby enabling quantitative identification of species desorbed from materials and films. Simultaneously, the desorped species can be physically and chemically analyzed. In addition, reaction rate analyses of desorped gases can be performed. Figure 8 shows examples of (a) a nonsymmetrical TPD curve indicated by the solid line (the first-order reaction) and (b) a symmetrical TPD curve indicated by the dashed line (the second-order reaction) as a function of temperature. The arrow on the nonsymmetrical TPD curve corresponds to the evolution of physisorbed and chemisorbed H_2O, which is specified to be a liquid such as water and water molecules hydrogen-bonded to Si-OH bonds at nanopore sites in the films (Hirashita et al., 1993).

For chemical-vapor-deposited SiO_2 films, three distinct H_2O desorption states have been defined (Hirashita et al., 1993). They are physisorbed H_2O evolved at temperatures between 100 and 200 °C and chemisorbed H_2O evolved at temperatures between 150 and 300 °C in a TPD measurement. The higher desorption peak between 350 and 650 °C is ascribed to Si-OH bonds formed during film growth. Thus, TPD is a useful technique for evaluating surfaces and thin films of ceramics and semiconductors.

5.2 TPD spectral analyses of sol-gel-derived HfO_2 thin films

The desorption of H_2O (m/z = 18) that evolved from sol-gel-derived HfO_2 films on Si was analyzed by TPD (Figure 9). The HfO_2 films were fired at 450, 550, and 700 °C for 30 min. The vertical axis indicates the current value of QMS. The small peaks below 200 °C are due to the physisorbed H_2O (mere adsorption of H_2O) on the surface of the HfO_2 films and/or chemisorbed Hf-OH in the bulk at the surface area. Based on experiments, the small peaks are attributed to the adsorption of

H_2O immediately after the samples were taken out of the furnace and the amount of desorbed water (i.e., adsorbed water) saturated. The desorption states of physisorbed H_2O and/or chemisorbed Hf-OH bonds originate from liquid-like water, water molecules, and Hf-OH bonds at nanopore sites in the HfO_2 films.

Regarding the major peaks in the TPD spectra, two types of desorption curves are observed at temperatures higher than 200 °C. One has the form of a symmetrically shaped peak (Lorentzian distribution shoulder as shown by the dashed line) at around 320 °C (fired at 450 °C), which is reaction-controlled (Soraru, 2002) (the second-order reaction) (Nishide et al. 2004). The other consists of nonsymmetrical peaks at approximately 420 °C (fired at 550 °C) and 480 °C (fired at 700 °C) which are diffusion-controlled (the first order reaction) (Nishide et al., 2004). When the curve is symmetrical in shape, the peak at around 320 °C (fired at 450 °C) is not caused by physisorbed H_2O from the nanopores at the surface area, but can be ascribed to the associated desorption of chemisorbed water (Hf-OH) from the gel film, resulting in the formation of H_2O during firing, which is specified to be the second-order reaction (Nishide et al., 2004).

For the samples fired at 450 °C, Hf-OH bonds in the HfO_2 film bulk convert to HfO_2 and/or H_2O during heating and the resulting H_2O contributes to the major desorption peak in the TPD curve [Figure 10(a) and 10(b)], as a result of the following reaction

$$\equiv\text{Hf-OH HO-Hf}\equiv \;\rightarrow\; \equiv\text{Hf-O-Hf}\equiv + H_2O. \quad (1)$$

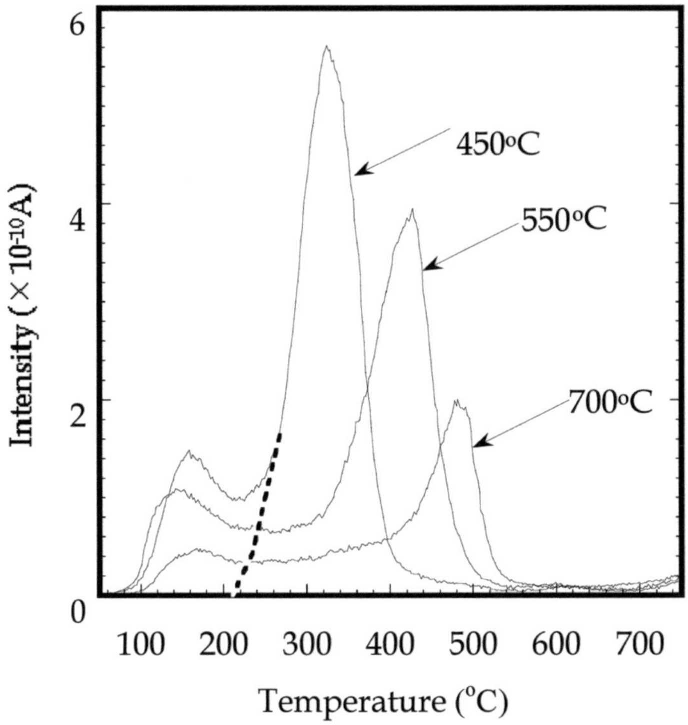

Fig. 9. TPD curves of H_2O (m/z = 18) that evolved from sol-gel-derived HfO_2 films on Si when fired at 450, 550, and 700 °C for 30 min (Shimizu et al., 2007).

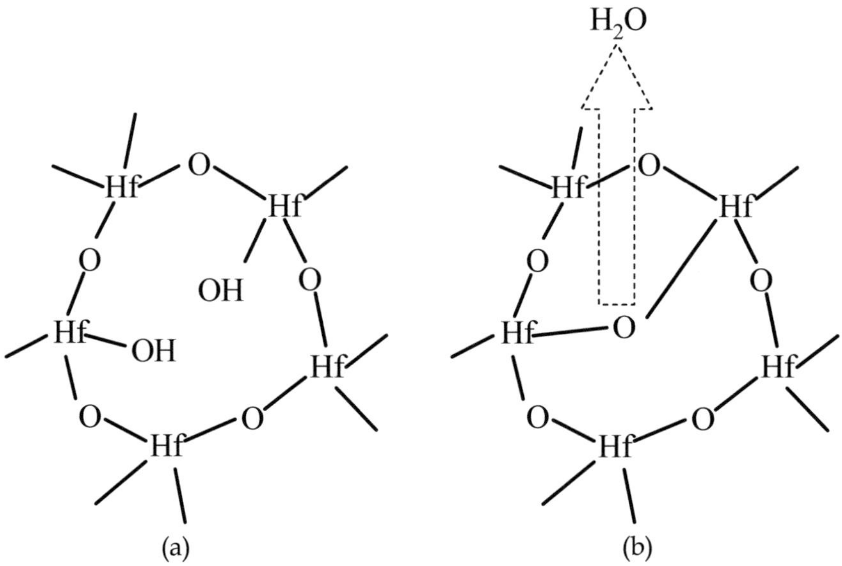

Fig. 10. A schematic speculation of H_2O desorption from HfO_2 films during TPD measurements for both amorphous and crystalline states. (a) Hf-OH bonds in the sol-gel derived HfO_2 films and (b) the formation of desorbed H_2O as a result of the following reaction: ≡ Hf-OH HO-Hf≡ → ≡ Hf-O-Hf≡ + H_2O (Shimizu et al., 2007).

In contrast, most of the HfO_2 films fired at 700 °C crystallized, where Hf-OH bonds in the films are conjectured to be tightly locked between crystallized grains and incorporated H_2O needs greater energy to percolate by the diffusion control mechanism through small gaps between grains. Therefore, in the TPD curve, the H_2O desorption peak shifts to higher temperatures and decreases steadily as the firing temperature increases.

5.3 Adsorption mechanism of physisorbed H_2O clarified by TPD using sol-gel-derived ZrO_2 thin films

To clarify whether the small peaks (small protrusions) between 100 and 200 °C in the TPD curves were due to physisorbed H_2O (mere adsorption of H_2O) on the surface of the ZrO_2 thin films and/or chemisorbed Zr-OH in the bulk at the surface area (Figure 11), the following three experiments were carried out (Shimizu et al., 2009). First (process ①), a sample immediately after firing at 350 °C for 30 min was measured by TPD until 350 °C and then (process ②) successively measured by TPD again from room temperature to 350 °C; finally (process ③), the sample was exposed to air for 59 h and then measured by TPD from room temperature to 700 °C. The TPD curve of H_2O in process ① was in good agreement with the typical curve of a ZrO_2 thin film fired at 350 °C for 30 min. No peaks were observed in process ②, indicating that the small protrusions and major peak vanished during heating

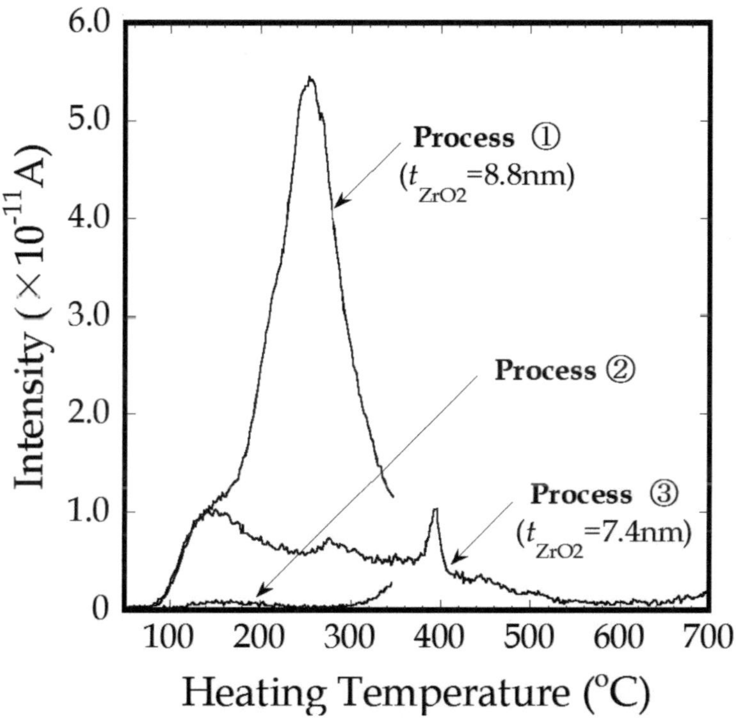

Fig. 11. TPD curves for processes (1), (2), and (3). In process (1), a sample immediately after firing at 350 °C for 30 min was measured by TPD until 350 °C, followed by process (2) in which the sample was successively measured by TPD again from room temperature to 350 °C; finally, process (3) in which the sample was exposed to air for 59 h and then measured by TPD from room temperature to 700 °C (Shimizu et al., 2009).

until 350 °C in the first TPD measurement. In contrast, in process ③, the small protrusions between 100 and 200 °C appeared again. This result provides evidence that the small peaks (small protrusions) were caused by adsorption of H_2O immediately after the samples were taken out of the furnace and that the amount of desorbed water (i.e., adsorbed water) saturated during the exposure time. Thus, the small protrusions in the TPD curves can be attributed to physisorbed H_2O and/or chemisorbed Zr-OH bonds at the surface area.

6. Characterization of sol-gel-derived HfO_2 thin films on Si(001) wafers dependent on sol solution: "HCOOH sol" and "HNO_3 sol" HfO_2 thin films

6.1 Crystallization temperature dependent on sol solution of HfO_2 thin films on Si(001) wafers

In sol-gel-derived HfO_2 thin films on Si(001) wafers, the crystallization temperature depends on the composition of the sol solution. Upon preparing the sol solution, (a) a formic acid (HCOOH) or (b) a nitric acid (HNO_3) is used as the catalyst for $Hf(OH)_4$ to form a soluble sol in H_2O, resulting in a hafnia sol solution.

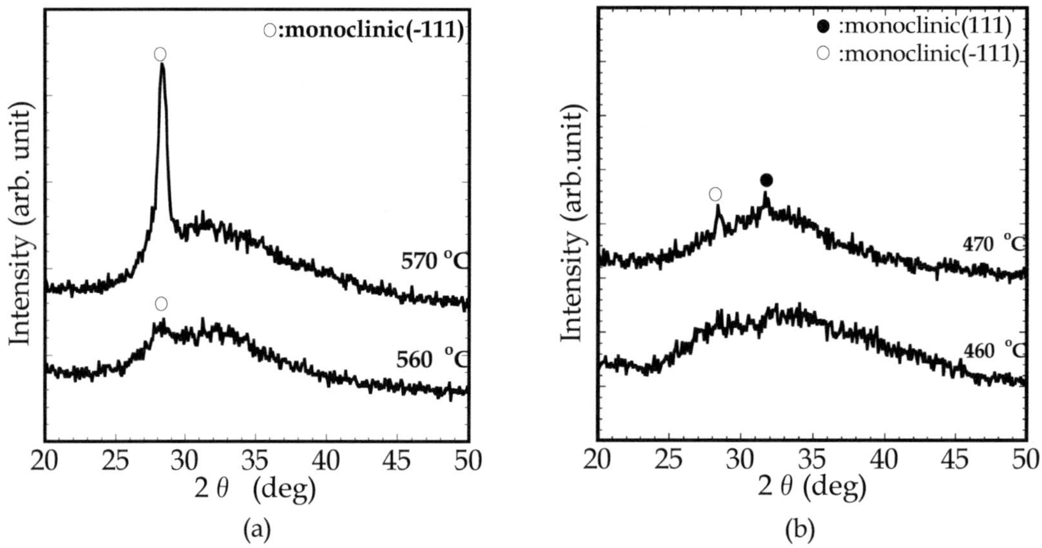

Fig. 12. (a) XRD patterns of "HCOOH sol" HfO_2 films on Si fired at 560 and 570 °C and (b) XRD patterns of "HNO_3 sol" HfO_2 films on Si fired at 460 and 470 °C (Shimizu et al., 2010).

In the X-ray pattern for "HCOOH sol" HfO_2 films fired at temperatures below 560 °C, only halo patterns representing the amorphous state were observed. At 560 °C [Fig. 12(a)], a small diffraction peak was observed at $2\theta = 28°$ in the halo pattern. At 570 °C, the diffraction peak at $2\theta = 28°$ became clearer and higher, indicating that partial crystallization from the amorphous state commenced at 560 °C. The observed diffraction peak was identified to correspond to monoclinic (111) (JCPDS card) and full crystallization was attained at 700 °C.

In contrast, "HNO_3 sol" HfO_2 films fired at 460 °C showed no diffraction peaks (only halo patterns were observed). At 470 °C, two diffraction peaks were identified, corresponding to the monoclinic structures ($\bar{1}11$) and (111) according to the JCPDS card [Fig. 12(b)]. By using a HNO_3 solution as the catalyst, the crystallization temperature was reduced to less than 470 °C compared with 560 °C for the "HCOOH sol" HfO_2 films. The lattice interplanar distances calculated using the Bragg equation were 0.319 and 0.286 nm, in contrast to the reported values of 0.314 and 0.288 nm, respectively (Shimizu et al., 2004). These results probably differed from the crystallization temperature for the monoclinic structure (111) (Nishide et al., 2000) because different sol solutions were employed in each case. The "HCOOH sol" HfO_2 films remained in the amorphous state up to a higher temperature (560 °C) than the "HNO_3 sol" films (crystallized at 470 °C). Based on TPD measurements, HCOOH and HNO_3 desorb at temperatures below 350 °C, indicating that an intrinsic amorphous HfO_2 film without using a catalyst for either film stably exists above 350 °C (Shimizu et al., 2010).

6.2 Thicknesses and refractive indexes dependent on sol solution of HfO_2 thin films

The thickness of the sol-gel-derived HfO_2 films decreased with increasing firing temperature (Figure 13). It is seen that the smallest thickness was 6 nm for the "HCOOH sol" HfO_2 film fired at 700 °C, which is about 1 nm thinner than the thinnest "HNO_3 sol" HfO_2 film. The difference is due to the properties of the catalyst used and this result shows that the "HCOOH sol" HfO_2 film may be suitable for use as the gate insulator of highly integrated CMOS devices. However, its electrical performance should be superior to that of conventional HfO_2 films.

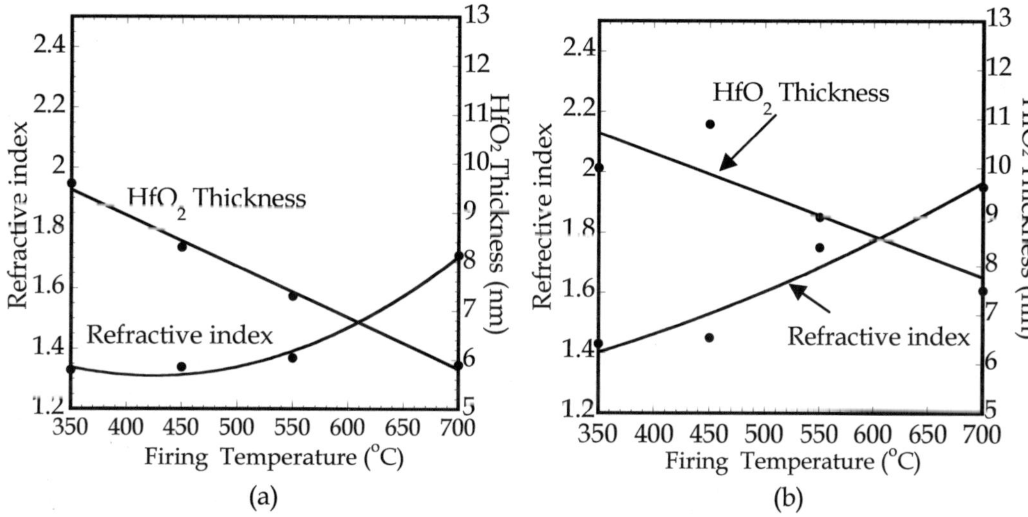

Fig. 13. Thicknesses and refractive indexes of sol-gel-derived HfO_2 films based on both (a) "HCOOH sol" and (b) "HNO_3 sol" fired at 350, 450, 550, and 700 °C for 30 min in air (Shimizu et al., 2010).

The refractive indexes began to increase at approximately 550 °C for the "HCOOH sol" film and at 450 °C for the "HNO_3 sol" film. These temperatures are in good agreement with those at which crystallization occurs, as obtained by XRD analysis [Figures 12(a) and 12(b)]. The maximum refractive indexes obtained were 1.70 for the "HCOOH sol" film and 1.95 for the "HNO_3 sol" film, although the reported value for the HfO_2 crystal (monoclinic) is 2.19. The packing densities of the HfO_2 films were calculated using the Lorentz-Lorentz equation (Nishide et al., 2001),

$$p = \frac{n_f^2 - 1}{n_f^2 + 2} \times \frac{n_m^2 + 2}{n_m^2 - 1}, \tag{2}$$

where p is the packing density, n_f is the refractive index of the film, and n_m is the refractive index of the crystal. The calculated packing densities indicated that nanopores remained in the amorphous state. However, upon the crystallization of the film, the packing densities became greater than those in the amorphous state in both the "HCOOH sol" and "HNO_3 sol" HfO_2 films.

6.3 Surface morphologies of HfO_2 layers for both "HCOOH sol" and "HNO_3 sol" HfO_2 films

Images of surface microstructures were obtained with an atomic force microscope (AFM) for both "HCOOH sol" and "HNO_3 sol" HfO_2 films fired at 350, 450, 550, and 700 °C (Figures 14 and 15). The progress of the microstructure development depended on the firing temperature. The surface of the "HCOOH sol" HfO_2 thin films fired at 350, 450, and 550 °C showed homogeneous glass-like structures. The root mean square (RMS) surface roughness was determined to be 0.13, 0.14, and 0.15 nm at firing temperatures of 350, 450 and 550 °C, whereas it was 0.34 nm at 700 °C, which indicated the presence of grain boundaries caused by crystallization. These values are in good agreement with the XRD and refractive index results. For the "HNO_3 sol" HfO_2 films, the RMS values were 0.14 and 0.15 nm at firing temperatures of 350 and 450 °C, respectively. At firing temperatures of 550 and 700 °C, the RMS values were 0.17 and 0.34 nm, resulting in grain boundaries due to crystallization. In this case, the surface roughness was also due to crystallization.

6.4 TPD spectral analyses of sol-gel-derived HfO_2 thin films based on "HCOOH sol" and "HNO_3 sol"

The desorption of H_2O (m/z = 18) was analyzed by TPD for both "HCOOH sol" and "HNO_3 sol" HfO_2 films on Si(001) wafers fired at 350, 450, 550, and 700 °C for 30 min [Figures 16(a) and 16(b)]. The vertical axis indicates the QMS current and the horizontal axis shows the heating temperature of the samples in the TPD chamber. The film thicknesses ranged approximately between 6 to 10 nm. The overall intensities of the desorption of H_2O in the TPD curves are related to both the "HCOOH sol" and "HNO_3 sol" HfO_2 films. The intensity of both TPD curves decreased with increasing firing temperature. The "HCOOH sol" HfO_2 films fired at 350, 450, and 550 °C, which were amorphous, showed small peaks at approximately 500 °C in the TPD curves. These peaks are presumably associated with crystallization during heating in the TPD chamber, because no corresponding peak was observed in the film fired at 700 °C (crystallization temperature: 560 °C).

TPD curves for the HfO_2 thin films fired at 350 °C using the "HCOOH sol" and "HNO_3 sol" on Si are separated into five Gaussian waveforms shown by dashed lines [Figure 17(a) and 17(b)]. Component (a) is thought to be due to H_2O physically adsorbed (simple adsorption of H_2O) on the surface of the HfO_2. Component (e) can be ascribed to the desorption of H_2O through nanopores in the crystallized HfO_2 film. Component (b) is probably due to the

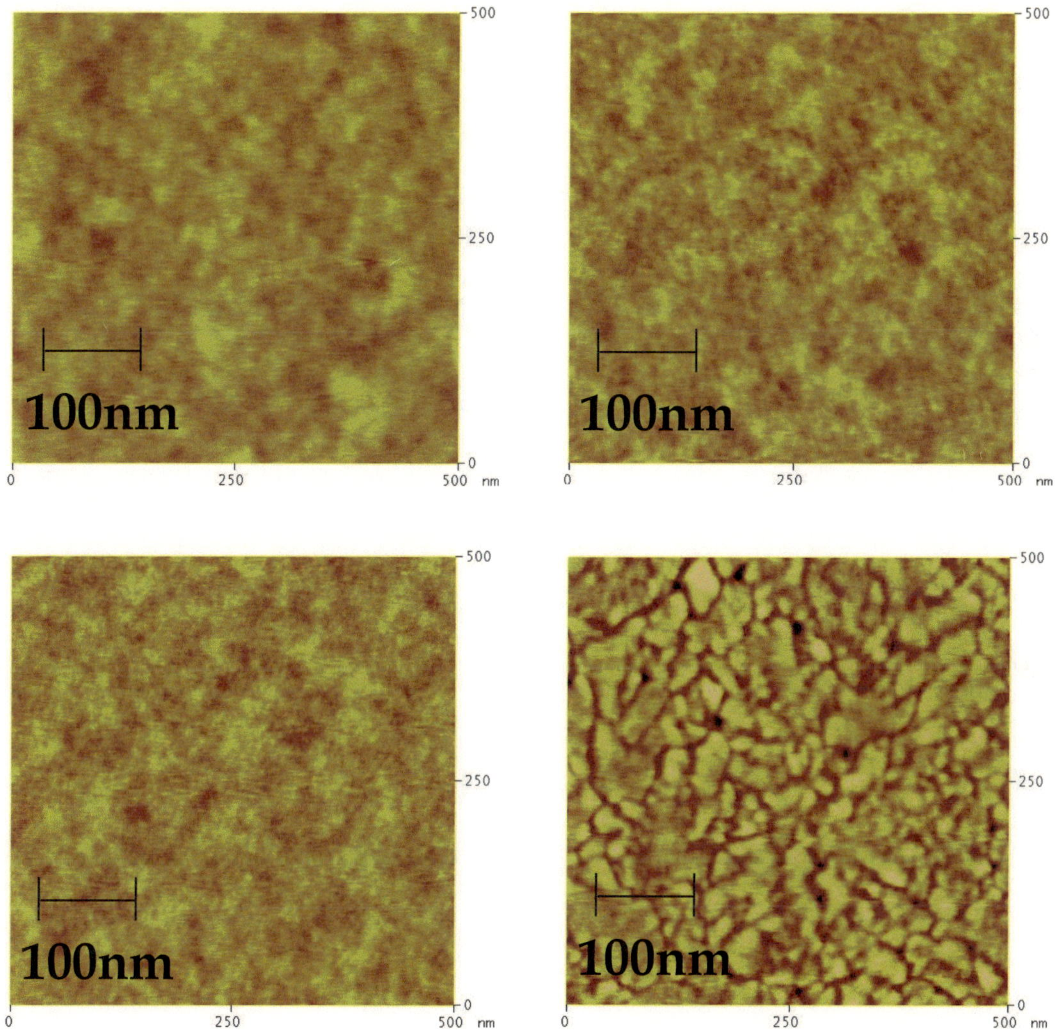

Fig. 14. AFM images showing the surface microstructures of "HCOOH sol" HfO_2 thin films fired at (a) 350, (b) 450, (c) 550 and (d) 700 ^{o}C (Shimizu et al., 2010).

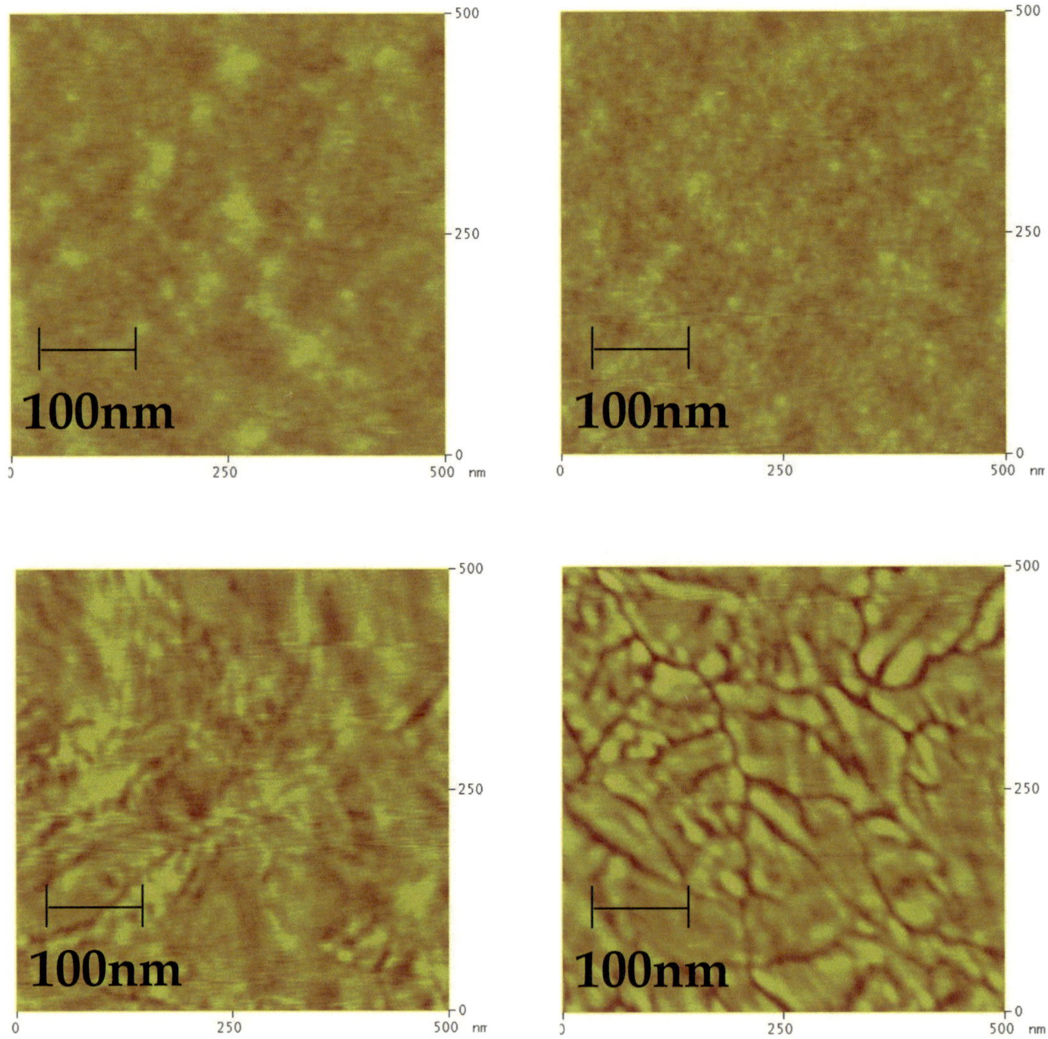

Fig. 15. AFM images showing the surface microstructures of "HNO_3 sol" HfO_2 thin films fired at (a) 350, (b) 450, (c) 550 and (d) 700 ^oC (Shimizu et al., 2010).

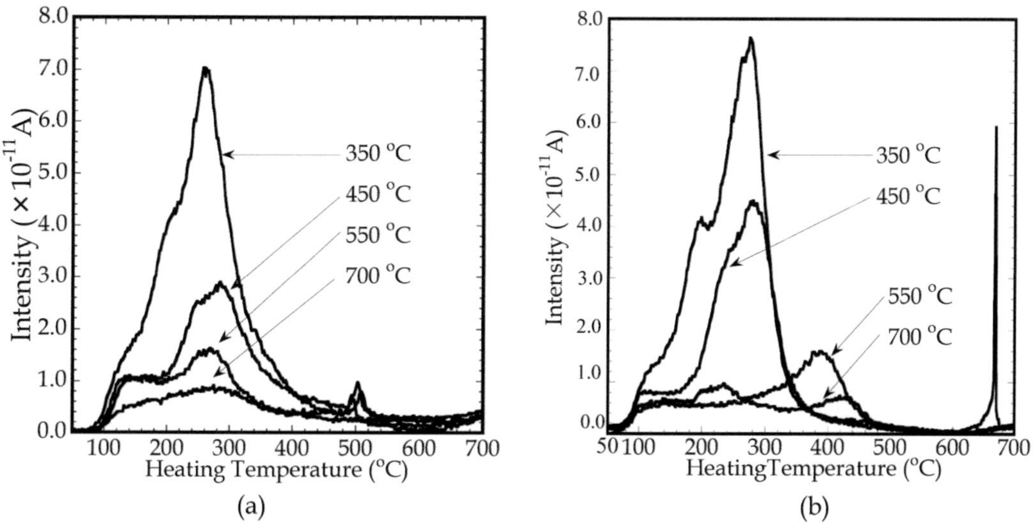

Fig. 16. TPD curves of H_2O (m/z = 18) released from sol-gel-derived HfO_2 thin films fired at 350, 450, 550, and 700 °C for 30 min: (a) HfO_2 films using "HCOOH sol" and (b) HfO_2 films using "HNO_3 sol" on Si. The vertical axis indicates the QMS current and the horizontal axis shows the heating temperature of the samples in the TPD chamber (Shimizu et al., 2010).

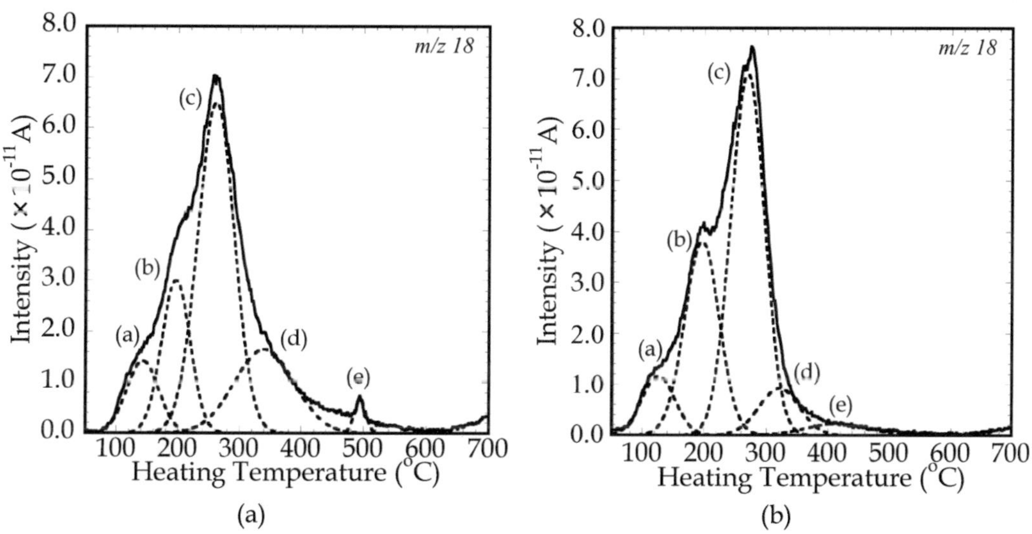

Fig. 17. TPD curves for sol-gel-derived HfO_2 thin films fired at 350 °C using
(a) "HCOOH sol" and (b) "HNO_3 sol" on Si, separated into five Gaussian waveforms shown by the dashed lines (Shimizu et al., 2010).

desorption of H_2O and/or chemically adsorbed Hf-OH bonds on the surface. On the other hand, the desorption of O in the TPD curves has main peaks at ~260 °C and subpeaks at ~350 °C, corresponding to the peak temperatures of components (c) and (d). For components (c) and (d), the desorption of the chemically adsorbed Hf-OH bonds and/or a small amount of O can occur from the nanopores of the HfO_2 film via the reaction OH + OH→H_2O + O. In addition, the H_2O desorption of chemically adsorbed Hf-OH may occur by the abovementioned reaction (≡Hf-OH +HO-Hf≡ → ≡Hf-O-Hf ≡ +H_2O).

HCOOH and/or HNO_3 adsorbs or bonds on the surface of the sol. In the "HCOOH sol", HCOOH ions form a bridge structure coordinated with Hf ions (Nishide et al., 2004). In the TPD measurements performed after the firing process, HCOOH and HNO_3 were found to desorb at less than 350 °C. Thus, "HCOOH sol" and "HNO_3 sol" HfO_2 films in the amorphous state without an acid exist stably on Si wafers above 350 °C. This result may affect the *I-V* characteristics, as will be discussed later.

On the basis of the foregoing results, a speculative schematic diagram of physically adsorbed H_2O and the chemically adsorbed Hf-OH attached to a sol-gel-derived HfO_2 film is shown in Fig. 18. The gaps in the figure correspond to nanopores.

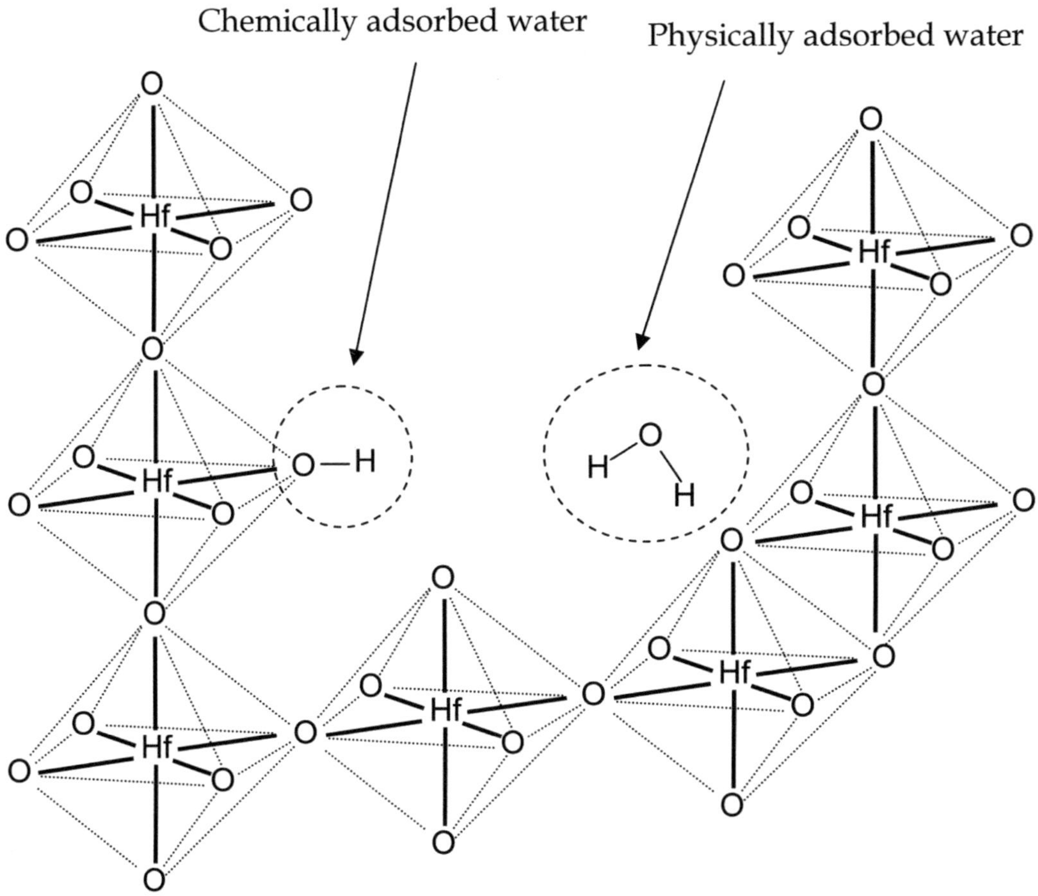

Fig. 18. Speculative schematic diagram of physically adsorbed H_2O and chemically adsorbed Hf-OH attached within the sol-gel-derived HfO_2 film (Shimizu et al., 2010).

6.5 Electrical characterization of both "HCOOH sol" and "HNO_3 sol" HfO_2 thin films on Si(001) wafers

To measure the electrical characteristics (i.e., *I-V* and *C-V* characteristics) of both "HCOOH sol" and "HNO_3 sol" HfO_2 thin films on Si(001) wafers, a 0.4-mm-diameter aluminum (Al) electrode was deposited on the surface of the films. Al/HfO_2/SiO_2/n-Si capacitors were fabricated on the Si wafers using a shadow mask in a vacuum. Using these capacitors, the *I-V* characteristics, i.e., the current vs electric field relationships, were investigated for the "HCOOH sol" and "HNO_3 sol" HfO_2 thin films fired at 350, 450, 550, and 700 °C in air.

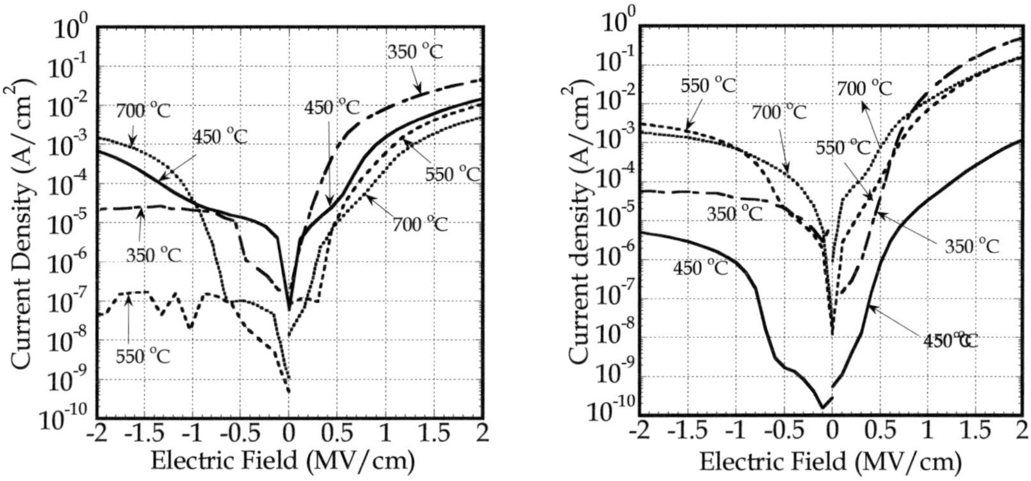

(a) (b)

Fig. 19. *I-V* characteristics of HfO_2 thin films fired at 350, 450, 550, and 700 °C in air using both (a) "HCOOH sol" and (b) "HNO_3 sol (Shimizu et al., 2010)".

The absolute values of the reverse bias are plotted in Figure 19. For the "HCOOH sol" HfO_2 thin films, a small bias dependence on firing temperature was detected for the forward bias condition. In contrast, for the reverse bias condition, the smallest leakage current was observed at a firing temperature of 550 °C (amorphous film) for which the leakage current was ~10^{-7} A/cm^2 in an electric field of -2 MV/cm. These data indicate that the amorphous film is more promising than the crystallized film as a gate insulator. The leakage current was comparable to previously reported results (Suzuki & Kato, 2007, 2009), but was smaller than that of a HfO_2 film deposited using atomic layer deposition (Chiou et al., 2007). At 700 °C, crystallization roughens the layer structure of the film as described in section 4.2 and provides a short leakage path through grain boundaries (Chiou et al., 2007, Zhu et al., 2002). The *I-V* characteristics for both forward and reverse biases in the Al/HfO_2/SiO_2/n-Si structure are unsymmetrical against bias voltages. This is probably because the potential barrier in the band diagram between the Al electrode and the HfO_2 film and that between the HfO_2 film and the SiO_2 film may differ between forward and reverse bias conditions. For the reverse bias condition, the energy slope of each band diagram of the Al/HfO_2/SiO_2/n-Si structure goes upwards and the difference between the work function of Al and the electron affinity of HfO_2 presumably creates a potential barrier against the flow of carriers. Therefore, the flow of electrons or holes may be suppressed by the barrier, resulting in current lower than that in the forward bias condition in which a potential barrier may not exist.

The unsymmetrical *I-V* characteristics are true in the "HNO_3 sol" case. The smallest leakage current in the "HNO_3 sol" HfO_2 thin films was seen for the amorphous films fired at 450 °C, which might be attributable to the smooth surface structure of the film. At 450 °C, the H_2O in the HfO_2 thin film desorbed less compared with that in the amorphous film fired at 350 °C. Therefore, there is some possibility for sol-gel-derived HfO_2 thin films to be used as alternative high-***k*** materials for gate insulators in CMOS devices; however, the amount of H_2O should be reduced to a minimum (Ragnarsson et al., 2009).

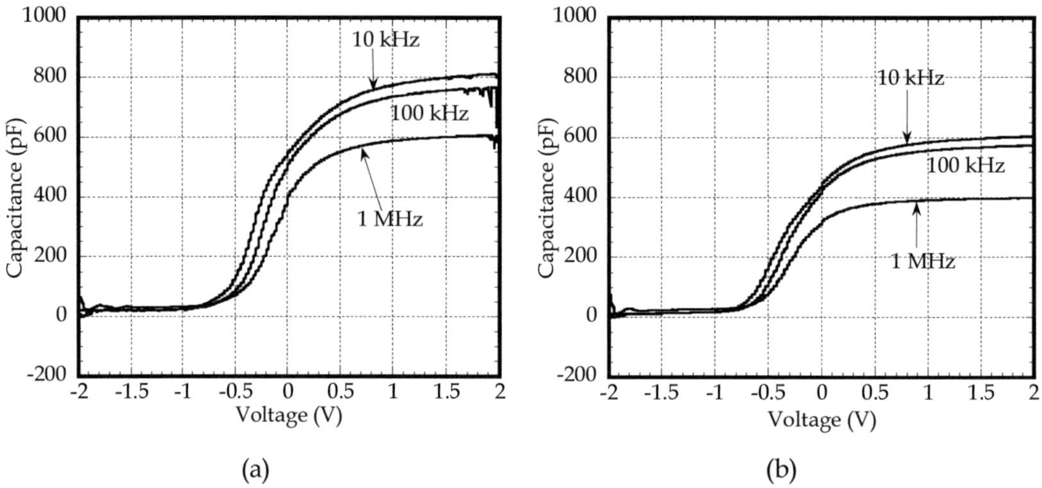

Fig. 20. *C-V* curves for $Al/HfO_2/SiO_2$/n-Si capacitors with HfO_2 films using (a) "HCOOH sol" at a firing temperature of 550 °C and (b) "HNO_3 sol" at 450 °C (Shimizu et al., 2010).

The *C-V* curves for $Al/HfO_2/SiO_2$/n-Si capacitors were examined in relation to the "HCOOH sol" HfO_2 film fired at 550 °C and to the "HNO_3 sol" HfO_2 film fired at 450 °C, respectively. The *C-V* curves are plotted in Figure 20 from – 2 to 2 V, representing the practical range for device operation. The *C-V* curves show a well-defined transition from depletion and inversion to accumulation as the applied voltage was varied from – 2 to 2 V, similar to the *C-V* curves for normal Al/SiO_2/Si capacitors (Nicollian & Brews, 1981). The *C-V* characteristics do not show any dependence on firing temperature, but the capacitance decreases with increasing frequency. On the basis of the well-defined capacitance in the plotting of a *C-V* curve at a frequency of 100 kHz, the relative permittivity ε_{HfO2} of the "HCOOH sol" HfO_2 film was calculated to be 11, with an effective oxide thickness (EOT) of 2.1 nm (HfO_2 film thickness: 7.4 nm). The SiO_2 film thickness was 2 nm, so the relative permittivity ε_{HfO2} was calibrated using that of SiO_2. The relative permittivity was much higher than that of silicon dioxide (SiO_2, 3.9), but is comparable to previously reported results (10~11) (Suzuki & Kato, 2009). The difference in the relative permittivity ε_{HfO2} between the sol-gel HfO_2 film and bulk HfO_2 may be due to the presence of the SiO_2 film and nanopores in the HfO_2 film.

For the "HNO_3 sol" HfO_2 film, the relative permittivity was calculated to be 11 and the EOT was 3.9 nm (HfO_2 film thickness: 10.9 nm). The "HCOOH sol" HfO_2 film is promising, but it requires a relatively higher permittivity and a smaller film thickness to achieve a reasonable EOT for highly integrated CMOS devices. For EOT scaling, the necessity of suppressing the liberation of H_2O from the HfO_2 film at Si oxidation temperatures has been emphasized (Ragnarsson et al., 2009).

The reported *C-V* curves in Fig. 20 show a small reduction with increasing frequency. The relative permittivity decreases with increasing growth temperature of the high-*k* film (ZrO_2) and frequency (Kukli et al., 2001, 2002). In general, the relative permittivity is essentially governed by the polarization of the material, and decreases with increasing frequency. In the present sol-gel-derived HfO_2 films, H_2O, OH groups in the nanopores, and other impurities probably caused electronic and ionic polarizations, thereby giving rise to the possibility of the frequency dependence of the capacitance. One possible way of refining the electrical performance of sol-gel-derived HfO_2 films is to use a firing environment of oxygen, inert gas, or forming gas. Thus, the amount of H_2O, defects, and impurities in sol-gel-derived HfO_2 films must be reduced to make the films applicable as a semiconductor gate insulator material.

7. Characterization of sol-gel-derived and crystalline ZrO_2 thin films on Si(001) wafers

7.1 Crystallinity of sol-gel-derived ZrO_2 thin films on Si(001) wafers

XRD patterns were obtained for sol-gel-derived ZrO_2 films on Si(001) wafers fired at 450, 550, and 700 °C for 30 min (Figure 21). For the ZrO_2 film fired at 450 °C, a halo-like pattern

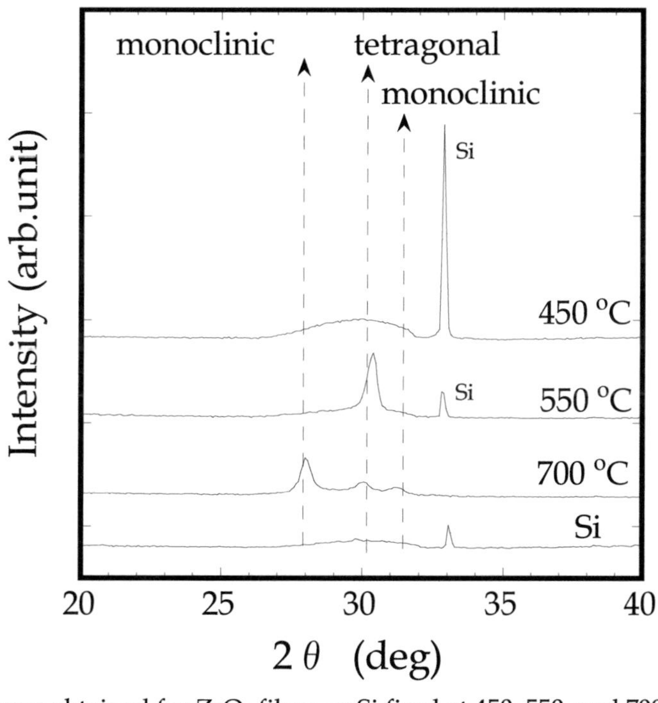

Fig. 21. XRD patterns obtained for ZrO_2 films on Si fired at 450, 550, and 700 °C for 30 min. The XRD pattern for the Si substrate is also shown for reference (Shimizu et al., 2009).

was observed at approximately $2\theta = 30°$, indicating that the film was still amorphous (Liu et al. 2002, Shimizu et al., 2009). The diffraction peak of 33° is ascribed to the Si (001) wafer. At 550 °C, a new peak appeared at $2\theta = 30.3°$, which was determined to be tetragonal (011) (JCPDS card, Liu et al., 2002, Shimizu et al., 2010), and the lattice interplanar distance was calculated to be 0.295 nm. In addition, at 700°C, three peaks at $2\theta = 28$, 30.3, and 31.3° were observed. The two peaks at $2\theta = 28$ and 31.3° were determined to be monoclinic (111) and monoclinic (111), respectively, because the calculated lattice interplanar distances were 0.319 and 0.286 nm, which correspond to the reported values of 0.316 and 0.284 nm. The ZrO_2 thin films fired at 700 °C consisted of a mixed crystal of tetragonal and monoclinic structures. Rapid temperature annealing (RTA) above 700 °C results in a mixture of monoclinic and tetragonal phases (Liu et al., 2002).

7.2 Spectral analyses of sol-gel-derived ZrO_2 thin films by TPD

Figure 22 shows the TPD curves of H_2O ($m/z = 18$) that evolved from the sol-gel-derived ZrO_2 thin films on Si, which were fired at 350, 450, 550, and 700 °C for 30 min. The vertical axis indicates the current value of QMS. The film thicknesses were determined to be 10.2, 9.9, 7.6, and 8.1 nm, respectively. The intensity of the TPD curves decreased as the firing temperature increased, indicating that the amount of H_2O was reduced in the ZrO_2 films on Si(001)wafers. Since the TPD curves were unsymmetrical against the heating temperature,

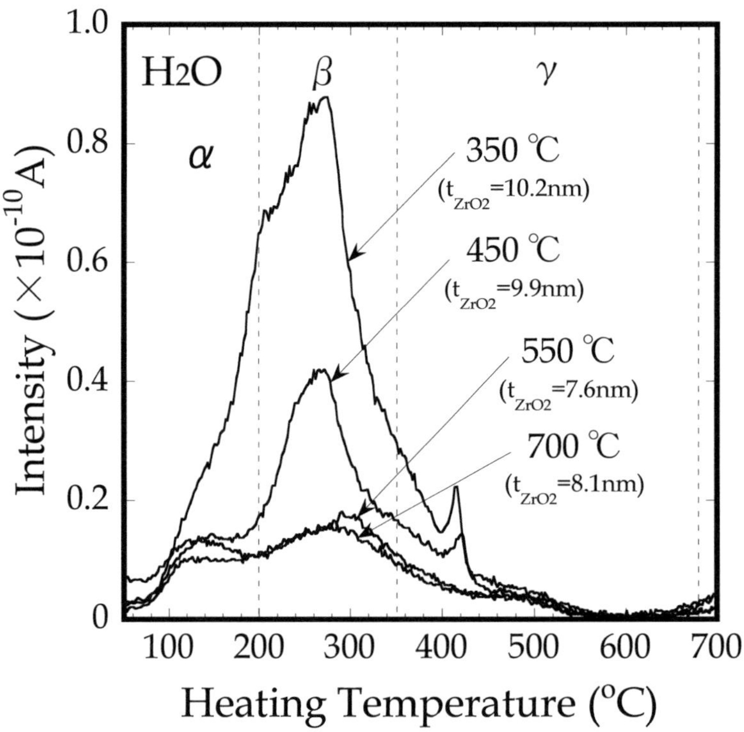

Fig. 22. TPD curves of H_2O ($m/z = 18$) that evolved from sol-gel-derived ZrO_2 thin films on Si fired at 350, 450, 550, and 700 °C for 30 min. The film thicknesses were 10.2, 9.9, 7.6, and 8.1 nm, respectively (Shimizu et al., 2009).

they were classified into three groups on the basis of the TPD results for SiO_2 formed by chemical vapor deposition (Hirashita et al., 1993): α, small peaks (small protrusions) between 100 and 200 °C; β, major peaks between 200 and 350 °C; and γ, small sharp peaks at approximately 410 °C for the samples fired at 350 and 450 °C. The measured TPD curve of H_2O had the main peak at a temperature of 260 °C with an unsymmetrical shape, providing evidence that several desorbed components were present during heating.

In a detailed analysis, the TPD curve for the sample fired at 350 °C was separated into five peak components using a Gaussian-type waveform (Figure 23). Component (a) is presumably due to physisorbed H_2O (mere adsorption of H_2O) on the surface of the ZrO_2 thin films. This was confirmed experimentally as discussed in the next subsection. Component (e) can be attributed to the desorption of H_2O through nanopores of the crystallized ZrO_2 thin film. Component (b) can be ascribed to the desorption of H_2O and/or chemisorbed Zr-OH bonds at the surface area. For components (c) and (d), H_2O desorption may have occurred because of the following reaction (≡ Zr-OH +HO-Zr ≡ → ≡ Zr-O-Zr ≡ +H_2O).

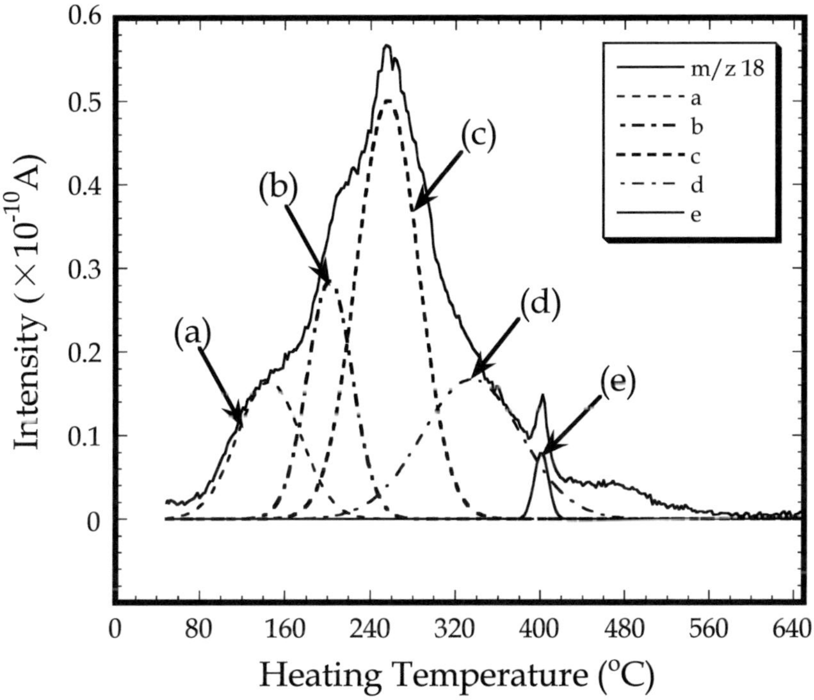

Fig. 23. TPD curve for the sol-gel-derived ZrO_2 thin film fired at 350 °C separated into five peak components using a Gaussian-type waveform as a function of the temperature measured with a thermocouple inside the TPD chamber (Shimizu et al., 2009).

7.3 Refractive indexes and film thicknesses of sol-gel-derived ZrO_2 thin films

The refractive indexes and film thicknesses were determined for sol-gel-derived ZrO_2 films fired at temperatures from 350 to 700 °C (Figure 24). The refractive indexes converged at 2.0, which is in good agreement with deposited ZrO_2 thin films (Moulder, 1995) and monoclinic

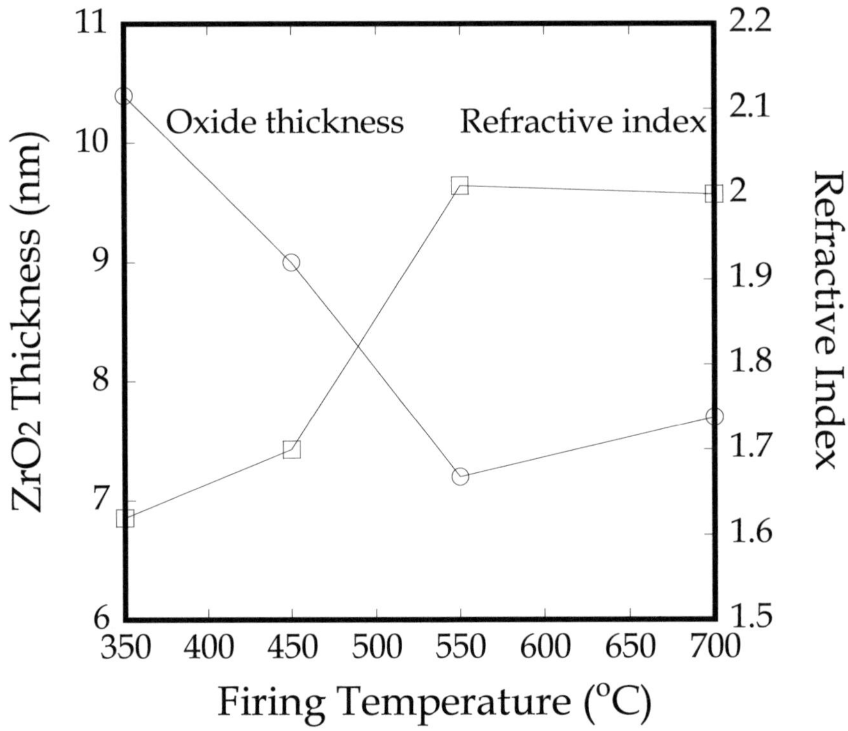

Fig. 24. Refractive indexes and film thicknesses of sol-gel derived ZrO_2 films at firing temperatures from 350 to 700 °C (Shimizu et al., 2009).

ZrO_2 crystals (Niinisto et al., 2004). The packing densities of the ZrO_2 films were calculated using the Lorentz-Lorentz equation (1) (Nishide et al., 2001). The refractive indexes were 1.62 at 350 °C, 1.70 at 450 °C, 2.01 at 550 °C, and 2.00 at 700 °C. The basic refractive index of the ZrO_2 crystal (monoclinic) for calculating the packing density was 2.22 (Yamada et al., 1988). Using this value, the packing densities were estimated to be 0.62 at 350 °C, 0.68 at 450 °C, 0.89 at 550 °C and 0.88 at 700 °C. The packing density of the films increased with increasing firing temperature. This is because more H_2O desorbed at higher firing temperatures and the small gaps of the nanopores were squeezed or evaporated.

7.4 Electrical characteristics of sol-gel-derived ZrO_2 films on Si(001) wafers

The *I-V* characteristics (current density vs electric field) were examined for sol-gel-derived ZrO_2 thin films on Si(001) wafers fired at 350, 450, 550, and 700 °C in air (Figure 25). For the sample fired at 550 °C, the leakage current was smaller than that of the amorphous ZrO_2 thin films fired at 350 and 450 °C. Leakage current deterioration was partially due to the considerable amount of H_2O in the film, but at 700 °C, crystallization was completed, and small surface cracks and surface relief observed with the AFM were responsible for the deterioration. The leakage current (forward bias) for the sample fired at 550 °C was approximately 4×10^{-3} A/cm^2 in an electric field of 1 M/cm, which is one or two orders of magnitude higher than that previously obtained (Chim et al., 2003). This difference is due to the densely compacted ZrO_2 thin film (Chim et al., 2003). For reverse bias, the leakage

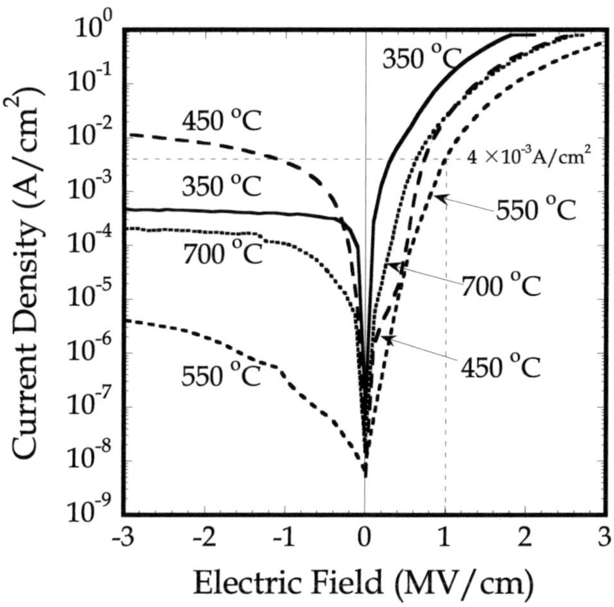

Fig. 25. *I-V* characteristics (i.e., current density vs electric field relationship) for sol-gel-derived ZrO_2 thin films fired at 350, 450, 550, and 700 °C in air. The reverse biases are plotted as absolute values (Shimizu et al., 2009).

current at 550 °C was suppressed more than in the other films measured. Thus, there is some possibility for sol-gel-derived ZrO_2 thin films to be used as an alternative high-***k*** material of gate insulators in densely packed CMOS devices.

To determine the relative permittivity ε_{ZrO2} of the sol-gel-derived ZrO_2 films, the *C-V* curves of the Al/ZrO_2/n-Si capacitors were obtained for the ZrO_2 thin film fired at 550 °C for 30 min. The *C-V* curves are plotted in Figure 26 from – 2 to 2 V, representing the practical range for device operation. The *C-V* curves show a well-defined transition from depletion and inversion to accumulation as the applied voltage was varied from – 2 to 2 V, similar to the *C-V* characteristics of normal Al/SiO_2/Si capacitors (Nicollian & Brews, 1981). The *C-V* characteristics did not show any dependence on firing temperature, but the capacitance decreased with higher frequency. On the basis of the well-defined capacitances in the accumulation region of the *C-V* curves at a frequency of 100 kHz, the relative permittivity ε_{ZrO2} of the sol-gel-derived ZrO_2 film was calculated to be 12 and the EOT was 2.4 nm (ZrO_2 film thickness: 7.4 nm). The relative permittivity was higher than that of silicon dioxide (SiO_2; 3.9) and the EOT was comparable to previously reported results (~2.5 nm) (Chim et al., 2003). The relative permittivity of ZrO_2 formed by atomic layer deposition has been reported to be 23 (Niinisto et al., 2004). The *C-V* curves decline slightly with increasing frequency. The relative permittivity decreases with the growth temperature of ZrO_2 thin films and increasing frequency (Kukli et al., 2001). Relative permittivity is essentially governed by the polarization of the material, so it decreases as the frequency increases. In the sol-gel-derived ZrO_2 film, H_2O, OH groups in nanopores and other impurities probably induced electronic and ionic polarizations, so there is the possibility of the frequency dependence of capacitance. To refine the electrical performance of sol-gel-derived ZrO_2 films, an alternative firing environment such as

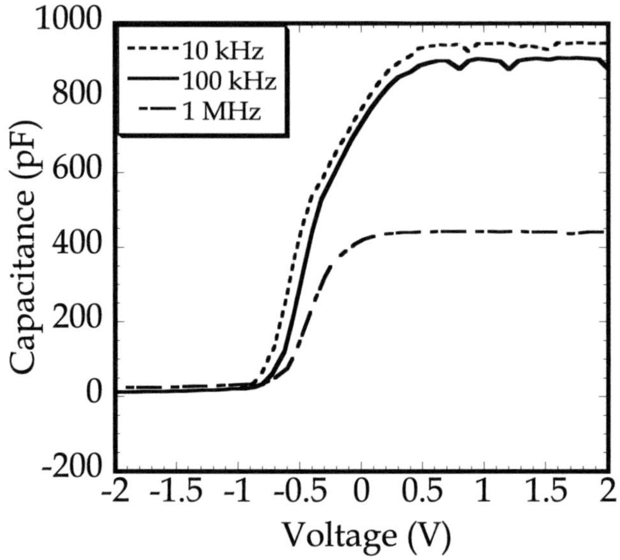

Fig. 26. *C-V* curves for Al/ZrO_2 /n-Si capacitors, showing a well-defined transition from depletion and inversion to accumulation as a function of the applied voltage. The firing temperature of the ZrO_2 film was 550 °C for 30 min (Shimizu et al., 2009).

oxygen, inert gas, or forming gas must be used. Thus, there is some possibility for applying sol-gel-derived ZrO_2 thin films as a semiconductor gate insulator material. To fabricate improved ZrO_2 films, further experiments should be conducted to find an effective way of reducing impurities. Sol-gel-derived Y doped ZrO_2(ZrO_2-Y_2O_3) thin films on Si(001) wafers are also promising.

8. Characterization of sol-gel-derived crystalline ZrO_2 –Y_2O_3 thin films on Si(001) wafers

Sol-gel-derived Y doped ZrO_2 (ZrO_2-Y_2O_3) thin films on Si(001) wafers fired in air between 350 and 700 °C provide electrical characteristics, such as lower leakage current, in MOS capacitors superior to those of sol-gel-derived ZrO_2 thin films (Shimizu & Nishide, 2011). This is attributed to the reduced surface roughness of ZrO_2-Y_2O_3 thin films. The crystallized ZrO_2-Y_2O_3 surface fired at 700 °C clearly shows a crack-free state compared with ZrO_2 thin films. Thus, crystallized ZrO_2-Y_2O_3 thin films can reduce the leakage current, making them a promising material for gate insulators in aggressively scaled CMOS devices.

8.1 Film thicknesses and refractive indexes of sol-gel-derived ZrO_2-Y_2O_3

The film thicknesses and refractive indexes were measured for sol-gel-derived ZrO_2-Y_2O_3 films fired from 350 to 700 °C for 30 min (Figure 27). The film thickness tended to become thinner at temperatures higher than 450 °C (11 and 7 nm thick at 450 and 700 °C, respectively). If this ZrO_2-Y_2O_3 thin film has excellent characteristics with high permittivity, this thickness is suitable for a gate insulator material. Though the refractive indices between 350 and 700 °C increased with increasing firing temperature, the present results were lower than that reported for monoclinic ZrO_2 crystals (Niinisto et al., 2004).

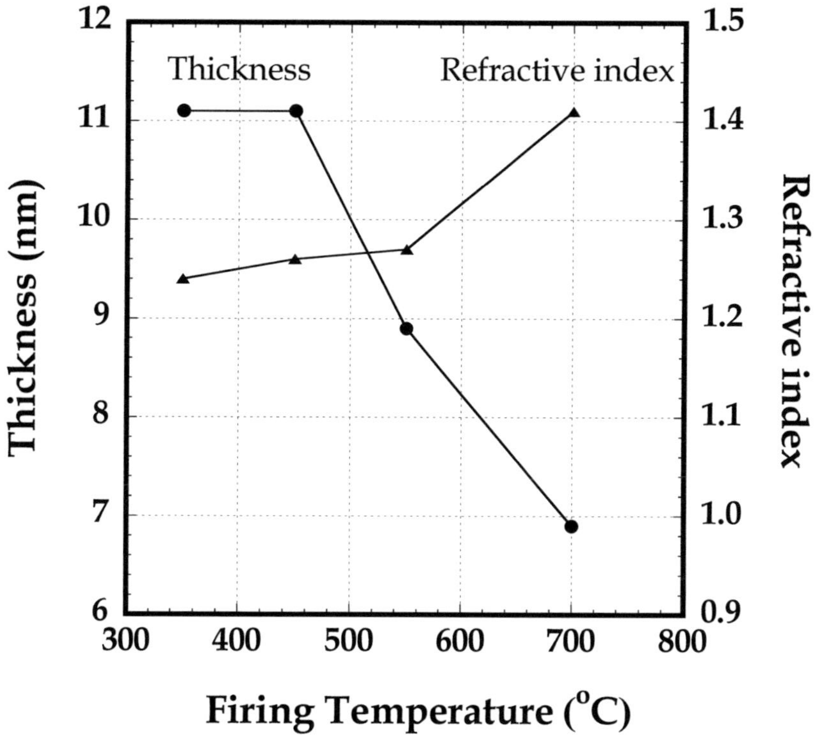

Fig. 27. Film thicknesses and refractive indices of sol-gel-derived ZrO_2-Y_2O_3 films fired at 350, 450, 550 and 700 °C for 30 min (Shimizu & Nishide, 2011).

8.2 AFM-observed surface morphologies of sol-gel-derived ZrO_2-Y_2O_3 thin films

The surface microstructures of ZrO_2-Y_2O_3 thin films fired at 350 and 700 °C were observed with the AFM [Figs. 28(a) and 2(b)]. The morphology depended on the firing temperature. The surface of the ZrO_2-Y_2O_3 thin film fired at 350 °C showed a homogeneous structure [Fig. 28(a)]. The RMS surface roughness was 0.15 nm at 350 °C. The RMS value at 700 °C was 0.24 nm and the surface structure was slightly wavy, but it did not show grain boundaries and/or cracks caused by crystallization [Fig. 28(b)]. Similar results have been reported for crack-free nano- and microcrystalline ZrO_2-Y_2O_3 thin films deposited on sapphire substrates (Peters et al., 2009).

8.3 Electrical characteristics of sol-gel-derived ZrO_2-Y_2O_3 thin films on Si(001) wafers

The *I-V* characteristics (current density vs electric field) were investigated for sol-gel-derived ZrO_2-Y_2O_3 thin films fired at 350 and 700 °C in air, in comparison with those obtained for sol-gel-derived ZrO_2 thin films [Figure 29(a) and 29(b)]. The reverse bias

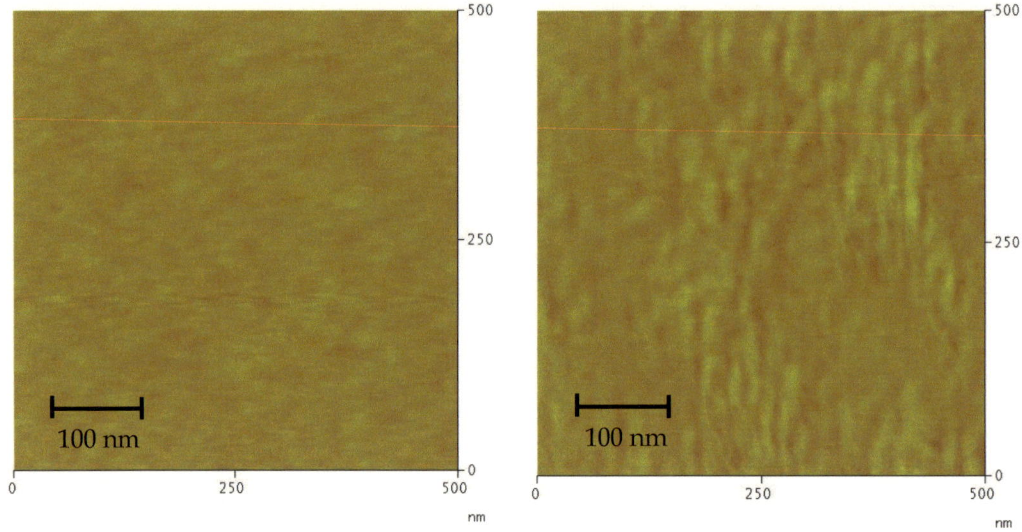

(a) (b)

Fig. 28. AFM images of the surface microstructures of ZrO_2-Y_2O_3 thin films fired at (a) 350 and (b) 700 °C (Shimizu & Nishide, 2011).

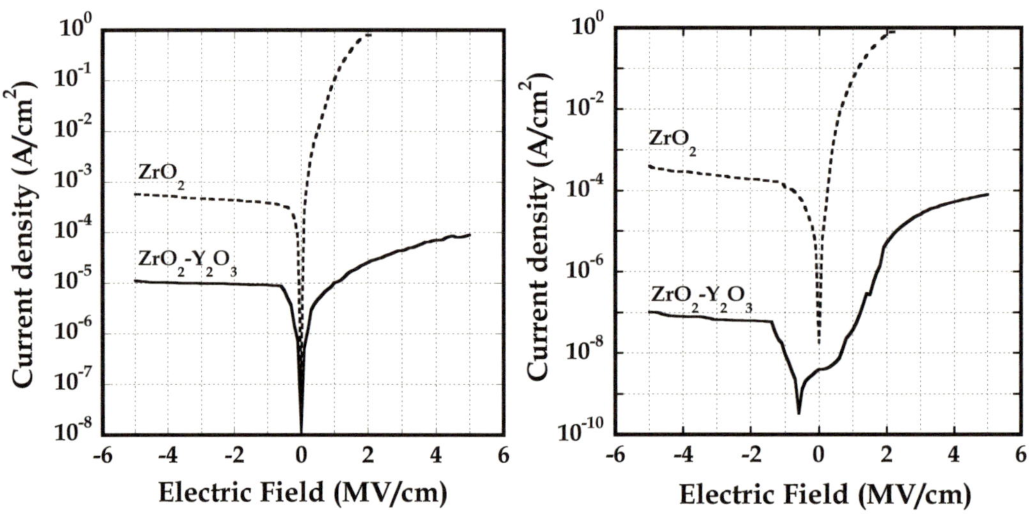

(a) (b)

Fig. 29. *I-V* characteristics (i.e., current density vs electric field relationship) for sol-gel-derived ZrO_2-Y_2O_3 thin films fired at (a) 350 and (b) 700 °C in air, respectively, in comparison with those reported for sol-gel-derived ZrO_2 thin films (Shimizu & Nishide, 2011).

quantities are plotted as absolute values. The leakage current of the Al/ZrO_2-Y_2O_3/Si capacitors was approximately five orders of magnitude lower than that of the ZrO_2 thin films for forward bias at an electric field of 2 MV/cm and three orders of magnitude lower for reverse bias at -2 MV/cm, respectively (Shimizu & Nishide, 2011). This improvement of the leakage current is noteworthy. For the sample fired at 700 °C, a similar reduction was observed for the Al/ZrO_2-Y_2O_3/Si capacitor. This is because the lower surface roughness and crack-free state of the ZrO_2-Y_2O_3 film surface may reduce the leakage current in comparison with the ZrO_2 thin films as described in subsection 5.4. For the ZrO_2-Y_2O_3 thin films fired between 350 and 700 °C, the leakage current of the latter was two orders of magnitude smaller than that of the former [Fig. 29(b)]. This is probably due to the film quality caused by crystallization such as packing density and/or a considerable difference in the amount of H_2O in the film.

The leakage current (forward bias) for the sample fired at 700 °C was approximately 5×10^{-7} A/cm^2 in an electric field of 1 M/cm (Shimizu & Nishide, 2011), which is one or two orders of magnitude lower than previously reported results (Chim et al., 2003). The latter results may be for densely compacted ZrO_2 thin films, because they were fabricated by sputtering in an argon-plus-oxygen gas ambient and annealed at 400 °C in a nitrogen ambient for 5 min. For reverse bias, the leakage current at 700 °C was superior to that of the other measured films. Therefore, there is some possibility for sol-gel-derived ZrO_2-Y_2O_3 thin films to be used as an alternative high-*k* material for gate insulators in miniaturized CMOS devices. However, the film quality must be improved further (Shimizu & Nishide 2011).

8.4 TPD analyses of sol-gel-derived ZrO_2-Y_2O_3 thin films

TPD was used to investigate the desorption of H_2O (m/z = 18) that evolved from sol-gel-derived ZrO_2-Y_2O_3 thin films on Si(001) wafers, which were fired at 350 and 700 °C for 30 min (Figure 30). The vertical axis indicates the current value of QMS. The film thicknesses were 11.1 and 6.9 nm, respectively. The intensity of the TPD curves decreased as the firing temperature increased, indicating that the amount of H_2O was reduced in the ZrO_2-Y_2O_3 thin films on Si(001) wafers. For the ZrO_2-Y_2O_3 thin film fired at 350 °C, the peaks seen at 370 and 400 °C are attributed to equipment noise.

Two TPD curves are close to those of ZrO_2 thin films (Shimizu & Nishide, 2011), except that the sample fired at 350 °C does not show any similar protrusions between 100 and 200 °C like those seen for the ZrO_2 thin film (Figure 22) (Shimizu et al., 2009). The peak was separated into several components using a Gaussian-type waveform (Figure 23), and the waveform indicated by the dashed line is shown as a function of temperature (Figure 30). The desorption temperature of the main peak of the ZrO_2-Y_2O_3 thin film was approximately between 100 and 200 °C. This implies that the TPD peak may be due to physisorbed H_2O (mere adsorption of H_2O). In contrast, at 700 °C, the TPD curves for H_2O desorption are similar in shape to that of the ZrO_2 thin film. The peak from 100 to 200 °C is due to the adsorption of physisorbed H_2O and the main peak at approximately 250 °C is caused by Zr-OH (chemisorbed) (Nishide et al., 2005, Takahashi & Nishide, 2004). The relative permittivity of ZrO_2 formed by atomic layer deposition has been reported to be 23 (Niinisto et al., 2004).

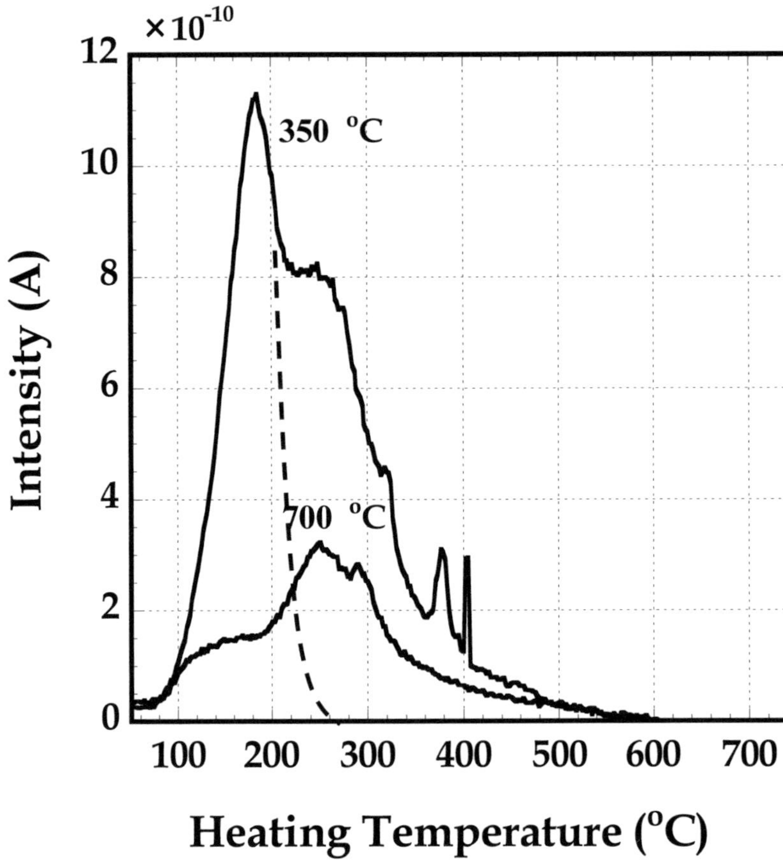

Fig. 30. TPD curves of H_2O (m/z = 18) that evolved from sol-gel-derived ZrO_2-Y_2O_3 thin films on Si(001) wafers, which were fired at (a) 350 and (b) 700 °C for 30 min (Shimizu & Nishide, 2011)

9. Conclusion

Sol-gel-derived HfO_2, ZrO_2 and Y doped ZrO_2(ZrO_2-Y_2O_3) thin films on Si(001) wafers fired in air between 350 and 700 °C were characterized physically, chemically and electrically with the aim of achieving alternative gate insulator materials for advanced CMOS devices. Crystallinity of the sol-gel-derived HfO_2, ZrO_2 films was found to be dependent on the firing temperature and sol solution. The relative permittivity of the films converged to that of bulk HfO_2 and ZrO_2 according to the specific sol solutions and firing temperatures. Residual H_2O and OH groups in the thin films were evaluated in reference to electrical characteristics such as the leakage current of MOS capacitors. The surface of the ZrO_2-Y_2O_3 thin films on Si(001) wafers showed less roughness than the HfO_2 and ZrO_2 thin films, resulting in lower leakage current in MOS capacitors. The leakage current of crystallized ZrO_2-Y_2O_3 thin films was shown to be lower than that of the amorphous state films because of the smooth crystalline surface of the latter in comparison with the ZrO_2 thin films. In conclusion, crystalline sol-gel-derived ZrO_2-Y_2O_3 thin films are postulated to be promising as alternative gate insulator materials of advanced CMOS devices.

10. Acknowledgments

This work was supported by a grant from Nihon University. The authors are also indebted to Dr. Masanori Ikeda for his assistance with the experiments.

11. References

Blanchin, M. G., B. Canut, Y. Lambert, V. S. Teodorescu, A. Barau, & M. Zaharescu (2008) Structure and dielectric properties of HfO_2 films prepared by a sol-gel route, *Journal of Sol-Gel Science Technology*, 47, 165-172.

Chang, S. & R. Doong (2005) ZrO_2 thin films with controllable morphology and thickness by spin-coated sol-gel method, *Thin Solid Films*. 489, 17-22.

Chim, H., W. K., T. H. Ng, B. H. Koh, W. K Chol, J. X Zheng, C. H. Tung, & A. Y. Du (2003) Interfacial and bulk properties of zirconium dioxide as a gate dielectric in metal-insulator-semiconductor structures and current transport mechanism. *Journal of Applied Physics*, 93, 4788-4793.

Chiou, Y-K. C-H. Chang, C-C. Wang, K-y. Lee, T-B. Wu, R. Kwo, & M. Hong (2007) Effect of Al Incorporation in the Thermal Stability of Atomic-Layer-Deposited HfO_2 for Gate Dielectric Applications. *Journal of the Electrical Society*, 154, G99-G102.

Copel, M., E. Gibelyuk & E. Gusev (2000) Structure and stability of ultrathin zirconium oxide layers on Si(001). *Applied Physics Letters*, 76, 436-438.

Feldman, L. C., E. P. Gusev & E. Garfunkel: in *Fundamental Aspects of Ultrathin Dielectrics on Silicon Based Devices*, ed. E. Garfunkel, E. P. Gusev, and A. Vul (Kluwer, Dordrecht, 1998) p. 1.

Dey, S. K., C.-G. Wang, D. Tang, M. J. Kim, R. W. Carpenter, C. Werkhoven & W. Shero (2003) Atomic layer chemical vapor deposition of ZrO_2-based dielectric films: Nanostructure and nanochemistry. *Journal of Applied Physics*, 93, 4144-4157.

Gusev, E. 2005, *Defects in High-k Gate Dielectric Stacks*, Springer.

Feldman, L. C., E. P. Gusev, & E. Garfunkel: in *Fundamental Aspects of Ultrathin Dielectrics on Silicon Based Devices*, ed. E. Garfunkel, E. P. Gusev, and A. Vul (Kluwer, Dordrecht, 1998) p. 1.

Hirashita, N., S. Tokitoh, & H. Uchida (1993) Thermal Desorption and Infrared Studies of Plasm-Enhanced Chemical Vapor Deposited SiO_2 Films with Tetraethyiorthosilicate, *Japanese Journal of Applied Physics*, 32, 1787-1793 .

Huff, H. R. & D. C. Gilmer (2004) High Dielectric Constant Materials-VLSI MOSFET Applications, Springer, Berlin.

Joint Committee for Powder Diffraction Standards , JCPDS card No. 501089, No. 371484 (Joint Committee for Powder Diffraction Standards, Swarthmore, Pennsylvania).

Kozuka, H (2005) Handbook of Sol-Gel Science and Technology Processing, Characterization and Applications, Kluwer Academic Publishers, 1, 247-398.

Kukli, K., K. Forsgren, M. Ritala, M. Leskela, J. Ararik &A. Harsta (2002) Influence of growth temperature on properties of zirconium dioxide films grown by atomic layer deposition. *Journal of Applied Physics*, 95, 1833-1840.

Kukli, K., M. Ritala, J. Ararik, T. Uustare & M. Leskela (2002) Influence of growth temperature on properties of zirconium dioxide films grown by atomic layer deposition. *Journal of Applied Physics*, 95, 1833-1840.

Lee, M. S., C. H. An, J. H. Lim, J.H. Joo, H. J. Lee, & H. Kim (2010) Characteristics of Ce-Doped ZrO_2 Dielectric Films Prepared by a Solution Deposition Process, *Journal of Electrochemical Society,* 157, G142-G146.

Lucovsky, G. (2002) Correlation between electronic structure of transition metal atoms and performance of high-*k* gate dielectrics in advanced Si devices, *Journal of Non-Crystalline Solids* 303, 40-49.

Liu, W. C., D. Wu, A. D. Li, H. Q. Ling, Y. F. Tang, & N. B. Ming (2002) Annealing and doping effects on structure and optical properties of sol-gel-derived ZrO_2 thin films, *Applied Surface Science,* 191, 181-187.

Martinez, E., C. Leroux, N. Benedetto, C. Gaumer, M. Charbonnier, C. Licitra, C. Guedj, F. Fillot & S. Lhostis (2009) Electrical and Chemical Properties of the $HfO_2/SiO_2/Si$ Stack: Impact of HfO_2 Thickness and Thermal Budget. *Journal of Electrochemical Society* 156, G120-G124.

Moulder, J. F., W. F. Stickle, P. E. Sobol & K. D. Bomben (1995) *Handbook of X-ray Photoelectron Spectroscopy,* eds. J. Chastain & R. C. King (Physical Electronics, Minnesota,) 2nd ed., pp.44-45, 168-169.(HfO_2)

Moulder, J. F., W. F. Stickle, P. E. Sobol, & K. D. Bomben (1995) *Handbook of X-ray Photoelectron Spectroscopy,* eds. J. Chastain and R. C. King (Physical Electronics, Minnesota) 2nd ed., p. 44-45, p. 108-109.

Nicollian, E. H. & J. R. Brews: *MOS (Metal Oxide Semiconductor) Physics and Technology* (Wiley, New York, 1981) Chap.4, pp.99.

Niinisto, J., M. Putkonen, L. Niinisto, K. Kukli, M. Ritala, & M. Leskela (2004) Structural and dielectric properties of thin ZrO_2 films on silicon grown by atomic layer deposition from cyclopentadienyl precursor. *Journal of Applied Physics,* 95, 84-92.

Nishide, T., T. Tanaka, & T. Yabe (2004) Temperature programmed desorption analysis of sol-gel-derived titania films, *Journal of thermal analysis and calorimetry.* 90, 373-378.

Nishide, T. & M. Shibata (2001) Orientation and Surface Properties of Sol-Gel Derived Y_2O_3 Films. *Journal of Sol-Gel Science Technology,* 21, 189-193.

Nishide T., S. Honda, M. Matsuura, Y. Ito, & T. Takase (2000) Oriented and Non-Oriented HfO_2-Y_2O_3 Films on Amorphous Quartz Glass Substrates Prepared by Sol-Gel Processes. *Japanese Journal of Applied Physics,* 39, L237-L240.

Nishide, T., S. Honda, M. Matsuura, & M. Ide (2000) Surface, structural and optical properties of sol-gel derived HfO_2 films. *Thin Solid Films,* 371, 61-65.

Nishide, T., T. Yabe, N. Miyabayashi, & M. Sano (2004) Analysis of Firing Processes of Titania Gel Films Fabricated by Sol-Gel Processes. *Thin Solid Films,* 467 43-49.

Nishide, T., T. Meguro, S. Suzuki, & T. Yabe (2005) Ultraviolet Irradiation Hardening of Hafnia Films Prepared by Sol-Gel Processes, *Journal of Ceramic Society of Japan* 113 77-81.

Onili, Y., Y. Iwazaki, M. Hasumi, T. Ueno, & K. Kuroiwa (2009) HfO_2/Si and HfSiO/Si Structures Fabricated by Oxidation of Metal Thin Films. *Japanese Journal of Applied Physics,* 48, 05DA01-1-05DA01-4.

Peters, C., A. Weber, B. Buts, D. Gerthsen, & E. I. Tiffee (2009) Grain-Size Effects in YSZ Thin-Film Electrolytes, *Journal of American Ceramics of Society,* 92, 2017-2024.

Parmentier, B., T. Conard, P. Roussel, S. D. Gendt, & M. M. Heyns (2009) The Importance of Moisture Control for EOT Scaling of Hf-Baced Dielectrics, *Journal of the Electrochemical Society*, 156, H416-H423.

Ragnarsson, L-A., D. P. Brunco, K. Yamamoto, Z. Tokei, G. Pourtois, A. Delabie, B.

Parmentier, T. Conard, P. Roussel, S. De Gendt, & M. M. Heyns (2009) *Journal of the Electrochemical Society*, 156, H416.

Robertson, J. & C. W. Chen (1999) Schottky barrier heights of tantalum oxide, barium strontium titanate, lead titanate, and strontium bismuth tantalite, *Applied Physics Letters*, 74, 1168-1170.

Robertson, J. (2000) Band offsets of wide-band-gap oxides and implications for future electronic devices, *Journal of Vacuum Science Technology*, B 18, 1785-1791.

Shimizu, H., K. Asayama, N. Kawai, & T. Nishide (2004) Material Microcharacterization of Sol-Gel Derived HfO_2 Thin Films on Silicon Wafers, *Japanese Journal Applied Physics*, 43, 6992-6993.

Shimizu, H., T. Sato, S. Konagai, M. Ikeda, T. Takahashi, & T. Nishide (2007) Temperature-Programmed Desorption Analyses of Sol-Gel Deposited and Crystallized HfO_2 Films. *Japanese Journal of Applied Physics*, 46, 4209-4214.

Shimizu, H., S. Konagai, M. Ikeda, & T. Nishide (2009) Characterization of Sol-Gel Derived and Crystallized ZrO_2 Thin Films. *Japanese Journal of Applied Physics*, 48, 101101-1-101101-6.

Shimizu, H., D. Nemoto, M. Ikeda, & T. Nishide (2010) Characterization of Sol-Gel Derived and Crystallized HfO_2 Thin Films Dependent on Sol Solution. *Japanese Journal of Applied Physics*, 49, 121502-1121502-6.

Suzuki, K. & K. Kato (2007) Effects of Modified Precursor Solution on Microstructure of (Y, Yb)MnO_3/HfO_2/Si, *Japan Society of Applied Physics*, 46, 6956-6959.

Suzuki, K. & K. Kato (2009) Characterization of high-*k* HfO_2 films prepared using chemically modified alkoxy-derived solutions, *Japanese Journal of Applied Physics*, 105, 061631-1-061631-3.

Takahashi, T. & T. Nishide (2004) Fabrication and Ultraviolet Irradiation Hardening of Sol-Gel-Derived Hafnia Films. *Journal of Ceramics Society of Japan*, 112 S234-S238.

Tonosaki, A., & T. Nishide (2010) Novel Petal Effect of Hafnia Films Prepared in an Aqueous Solution and Containing Hydroxy Acids. *Applied Physics Express*, 3, 125801-1-125801-3.

Wilk, G. D., R. M. Wallace & J. M. Anthony (2000) Hafnium and zirconium silicates for advanced gate dielectrics, *Journal of Applied Physics*, 87, 484-492.

Wilk, G. D., R. M. Wallace & J. M. Anthony (2001) High-*k* gate dielectrics: Current status and materials properties considerations, *Journal of Applied Physics*, 89, 5243-5275.

Yamada, T., M. Mizuno, T. Ishizuka, & T. Noguchi (1988) Liquidus-curve measurement in the system zirconia-hafnia, *Advanced Ceramics*, 24, 959-964.

Yang, X. M. Xl, X. F. (1996) Sintering behavior of (Y_2O_3-ZrO_2 gels). *Journal of Material Science*, 31, 2697-2703.

Yu, J. J., Q. Fang, J. Y. Zhang, Z. M. Wang & I. W. Boyd (2003) Hafnium oxide layers derived by photo-assisted sol-gel processing, *Applied. Surface Science*, 208-209, 676-681.

Zhan, Z. & H. C. Zeng (1999) A catalyst-free approach for sol-gel synthesis of highly mixed ZrO_2-SiO_2 oxide. *Journal of Non-Crystalline Solids*, 243, 26-38.

Zhu, W. J., T. Tamagawa, M. Gibson, T. Furukawa, & T. P. Ma (2002) Effect of Al Inclusion in HfO_2 on the Physical and. Electrical Properties of the Dielectrics, *IEEE Electron Device Letters*, 23, 649-651.

14

Crystalization in Spinel Ferrite Nanoparticles

Mahmoud Goodarz Naseri[1,2] and Elias B. Saion[1]
[1]Universiti Putra Malaysia,
[2]Malayer University,
[1]Malaysia
[2]Iran

1. Introduction

The enhanced interest of the researchers in nanoobjects is due to the discovery of unusual physical and chemical properties of these objects, which is related to manifestation of so-called 'quantum size effects'. These arise in the case where the size of the system is commensurable with the de-Brogli wavelengths of the electrons, phonons or excitons propagating in them.A key reason for the change in the physical and chemical properties of small particles as their size decreases is the increased fraction of the surface atoms, which occur under conditions (coordination number, symmetry of the local environment, etc.) differing from those of the bulk atoms. From the energy standpoint, a decrease in the particle size results in an increase in the fraction of the surface energy in its chemical potential [1]. Currently, unique physical properties of nanoparticles are under intensive research [2]. A special place belongs to the magnetic properties in which the difference between a massive (bulk) material and a nanomaterial is especially pronounced.The magnetic properties of nanoparticles are determined by many factors, the key of these including the chemical composition, the type and the degree of defectiveness of the crystal lattice, the particle size and shape, the morphology (for structurally inhomogeneous particles), the interaction of the particle with the surrounding matrix and the neighbouring particles. By changing the nanoparticle size, shape, composition and structure, one can control to an extent the magnetic characteristics of the material based on them. However, these factors cannot always be controlled during the synthesis of nanoparticles nearly equal in size and chemical composition; therefore, the properties of nanomaterials of the same type can be markedly different [1].Apart from these factors, the magnetic properties of particles depend on the external conditions: temperature, pressure and, in some cases, the local environment, i.e., the medium in which the particle occurs, in particular, the crystalline or amorphous bulk matrix (for a particle), the local crystal environment (for a single atom) or the substrate (for a film).

2. Ferrites and their structural symmetries

Among the magnetic materials that have found broad practical application in technology, ferrites deserve attention. Ferrite nanoparticles are the most explored magnetic nanoparticles up to date. They are widely used in high-frequency applications, because an AC field does not induce undesirable eddy currents in an insulating material [3,4].To increase the recorded information density, it seems reasonable to obtain nanocrystalline ferrites and to prepare magnetic carriers based on them. Ferrites have three different structural symmetries: garnet, hexagonal and cubic which are determined by the size and charge of the metal ions that balance the charge of the oxygen ions, and their relative amounts [5]. In this review, the focus will be on spinel ferrites nanocrystals because, they are regarded as two of the most important inorganic nanomaterials because of their electronic, optical, electrical, magnetic, and catalytic properties.Moreover, the majority of the important ferrite are spinel ferrite [6].

2.1 Garnets ferrites

The general formula for garnets is $Me_3Fe_5O_{12}$, where Me is one of the rare earth metal ions, including Y, La and Gd. The cubic unit cell contains 8 formula units or 160 atoms, which can be described as a spatial arrangement of 96 O^{2-} with interstitial cations. Yttrium iron garnet $Y_3Fe_5O_{12}$ (YIG) is a well-known garnet. The coordination of the cations is considerably more complex than spinels, with 24 Y^{3+} in dodecahedral sites, 24 Fe^{3+} ions in tetrahedral sites and 16 remaining Fe^{3+} in octahedral sites. Similar to spinels and hexagonal ferrites, a wide range of transition metal cations can substitute Y^{3+} or Fe^{3+}; especially rare earth ions may replace the ions on octahedral and dodecahedral sites. Each type of lattice site will accept other metal ions at dodecahedral sites, octahedral sites and at tetrahedral sites. Thus pentavalent ions such as V^{5+} and As^{5+} can occupy tetrahedral sites, while Ca2+ substitute ions on dodecahedral sites [7].

2.2 Hexagonal ferrites

Hexagonal ferrites are widely used as permanent magnets and are characterized by possesing a high coercivity [4].Their general formula is $MeO \cdot 6Fe_2O_3$ where Me can be Ba, Sr, or Pb. The hexagonal ferrite lattice is similar to the spinel structure, with the oxygen ions closely packed, but some layers include metal ions, which have practically the same ionic radii as the oxygen ions. This lattice has three different sites occupied by metals: tetrahedral, octahedral, and trigonal bi pyramid (surrounded by five oxygen ions).

2.3 Cubic ferrites

Materials which crystallize in the spinel structure, or structures closely related to it, have the general formula AB_2O_4 in which A and B display tetrahedral and octahedral cation sites, respectively, and O indicates the oxygen anion site (Figure1). Spinel ferrites with the general formula MeO Fe_2O_3 or $Me^{II}Fe_2^{III}O_4$ where Me^{II} represents a divalent metal cation such as Mn, Fe, Co, Ni, Cu, Zn, Cd, Mg, or ($0.5Li^{I}$ + $0.5Fe^{III}$), and Fe^{III} is the trivalent iron cation, have the same crystallographic structure as the mineral spinel ($MgAl_2O_4$), which was determined by Bragg [8,9]. The unit cell of spinel ferrite belongs to the cubic structure (space group Oh7-F3dm) and presents itself the cube formed by 8 $MeOFe_2O_3$ molecules and consisting of 32 of O^{2-} anions. The oxygene anions form the close face-centered cube (FCC) packing consisting in 64 tetrahedral (A) and 32 octahedral (B) empty spaces partly populated by Fe^{3+} and Me^{2+} cations [10].

For the interactions the distances Me-O and Me-O-Me play an important role, ten triangular configurations of Me-O-Me are illustrated in Figure2, where the shortest distance is shown as p and other distances are given by q, r, s and t; these distances are also given in Table1, as a function of the variable $\delta = u - 3/8$ that δ is oxygen parameter [11].

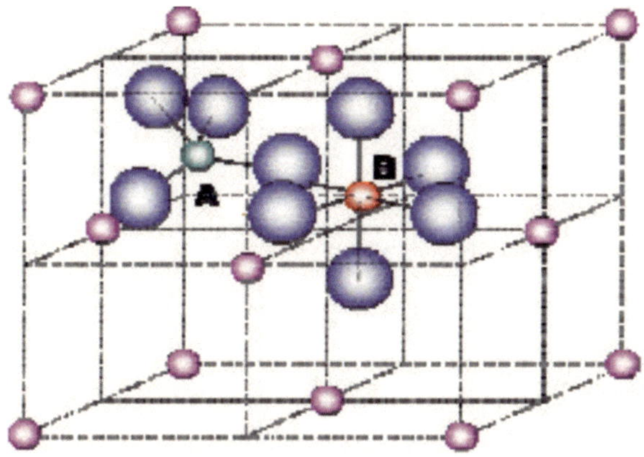

Fig. 1. Schematic of two subcells of a unit cell of the spinel structure, showing octahedral and tetrahedral sites.

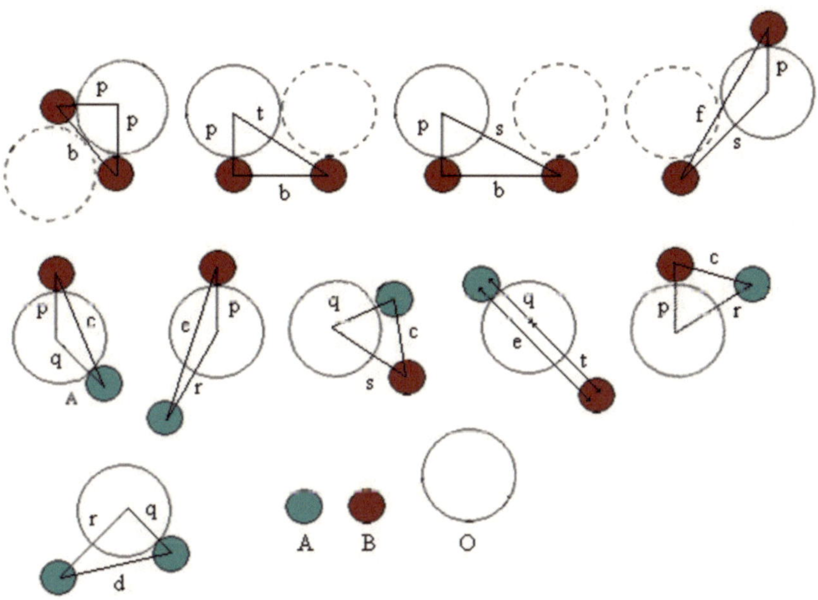

Fig. 2. Interionic distances in the spinel structure for the different types of lattice site interactions

According to the distribution of cations, there are normal, mixed and inversed spinels structures which depended on the fact what kind of ions and in what order A and B take empty spaces. In this review, we will investigate zinc ferrite ($ZnFe_2O_4$), manganese ferrites ($MnFe_2O_4$) and nickel ferrite ($NiFe_2O_4$) from normal, mixed and inverse structures respectively [12].

Me-O distance	Me-Me distance
$p = a\sqrt{\frac{1}{16} - \frac{1}{2}\delta + 3\delta^2}$	$b = \frac{1}{4}a\sqrt{2}$
$q = \left(\frac{1}{8} + \delta\right)\sqrt{3}$	$c = \frac{1}{8}a\sqrt{11}$
$r = a\sqrt{\frac{1}{24} + \frac{1}{4}\delta + 3\delta^2}$	$d = \frac{1}{4}a\sqrt{3}$
$s = a\sqrt{\frac{3}{16} + \frac{1}{2}\delta + 3\delta^2}$	$e = \frac{3}{8}a\sqrt{3}$
$t = a\left(\frac{1}{4} - \delta\right)\sqrt{3}$	$f = \frac{1}{4}a\sqrt{6}$

Table 1. Distances in the spinel lattice as a function of δ=u-3/8

2.3.1 Normal spinel ferrites

Normal spinel structure, where all Me^{2+} ions occupy A sites; structural formula of such ferrites is $Me^{2+}[Fe_2{}^{3+}]\ O_4{}^{2-}$. This type of distribution takes place in zinc ferrites $Zn^{2+}[Fe^{2+}Fe^{3+}]O_4{}^{2-}$. This type spinel ferrite are schematically illustrated in Figure3.

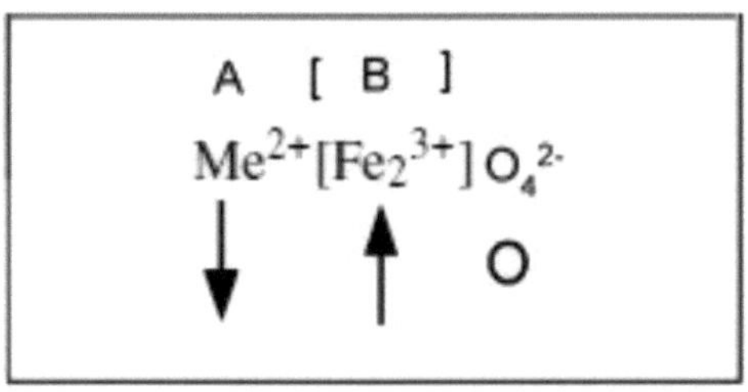

Fig. 3. Cation distribution in normal spinel ferrites.

2.3.2 Mixed spinel ferrites

Mixed spinel structure, when cations Me^{2+} and Fe^{3+} occupy both A and B-positions; structural formula of this ferrite is $Me_{1-\delta}{}^{2+}Fe_{\delta}{}^{3+}\ [Me_{\delta}{}^{2+}Fe_{2-\delta}{}^{3+}]\ O_4{}^{2-}$, where δ is the degree of inversion. $MnFe_2O_4$ represent this type of structure and has an inversion degree of $\delta = 0.2$ and its structural formula therefore is $Mn_{0.8}{}^{2+}Fe_{0.2}{}^{3+}\ [Mn_{0.2}{}^{2+}Fe_{1.8}{}^{3+}]\ O_4{}^{2-}$. [12]. This type spinel ferrite are schematically illustrated in Figure4.

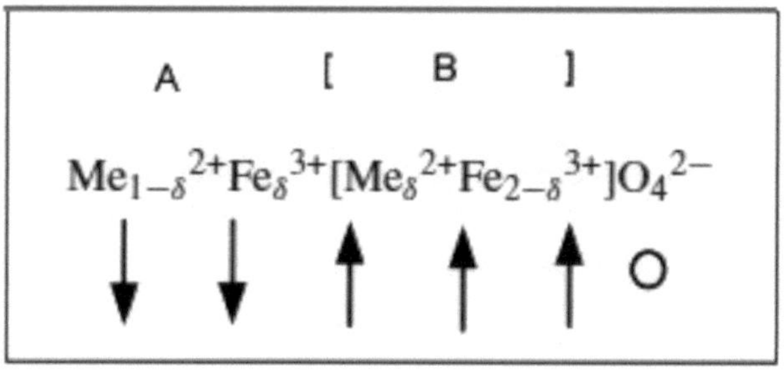

Fig. 4. Cation distribution in mixed spinel ferrites.

2.3.3 Inversred spinel ferrites

Inversed spinel structure, where all Me^{2+} are in B-positions and Fe^{3+} ions are equally distributed betweenAand B-sites: structural formula of these ferrites are $Fe^{3+}[Me^{2+}Fe^{3+}]\ O_4{}^{2-}$. Magnetite Fe_3O_4, ferrites $NiFe_2O_4$ and $CoFe_2O_4$ have inversed spinel structure [12].In the inversed ferrites one half of Fe^{3+} is placed in A-sites and another half in B-sites. Their magnetic moments are mutually compensated and the resulting moment of the ferrite is due to the magnetic moments of bivalent cations Me^{2+} in the B-positions. This type spinel ferrite are schematically illustrated in Figure5.

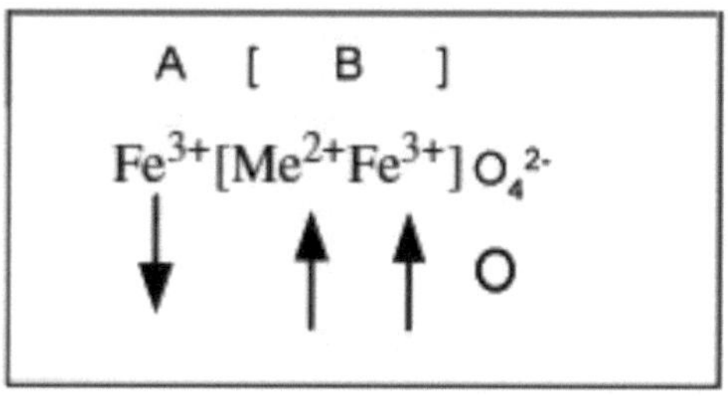

Fig. 5. Cation distribution in inversed spinel ferrites.

2.4 Domains

Domains, which are groups of spins all pointing in the same direction and acting cooperatively are separated by domain walls, which have a characteristic width and energy associated with their formation and existence. The motion of domain walls is a primary means of reversing magnetization. Experimental investigation of the dependence of coercivity on particle size showed a behavior similar to that schematically illustrated in Figure 6 [13].

It was found that the coercivity H_c increases with decreasing grain size D down to values of about 40 nm, independent of the kind of material. The increase of H_c is proportional to 1/D. The reason for this is that in small particles the formation of a closed magnetic flux becomes energetically less favorable so that the magnetic domain size with a uniform magnetization becomes more and more identical with the grain size. This grain size is defined as the first critical size (D_c, which is characteristic of each material) where the multidomain materials change to a monodomain material. This leads to a strong increase of the coercivity (or high remanence) because a change of magnetization in this case cannot happen only by shifting the domain walls which normally requires only a weak magnetic field. As the size of magnetic element scales below20 nm, the transformation from ferromagnetic to superparamagnetic behavior occurs. In the superparamagnetic state of the material, the room temperature thermal energy overcomes the magnetostatic energy well of the domain or the particle, resulting in zero hysteresis. In other words, although the particle itself is a single-domain ferromagnet, the ability of an individual magnetic "dot" to store magnetization orientation information is lost when its dimension is below a threshold. Consequently, the magnetic moments within a particle rotate rapidly in unison, exhibiting the superparamagnetic relation phenomenon.

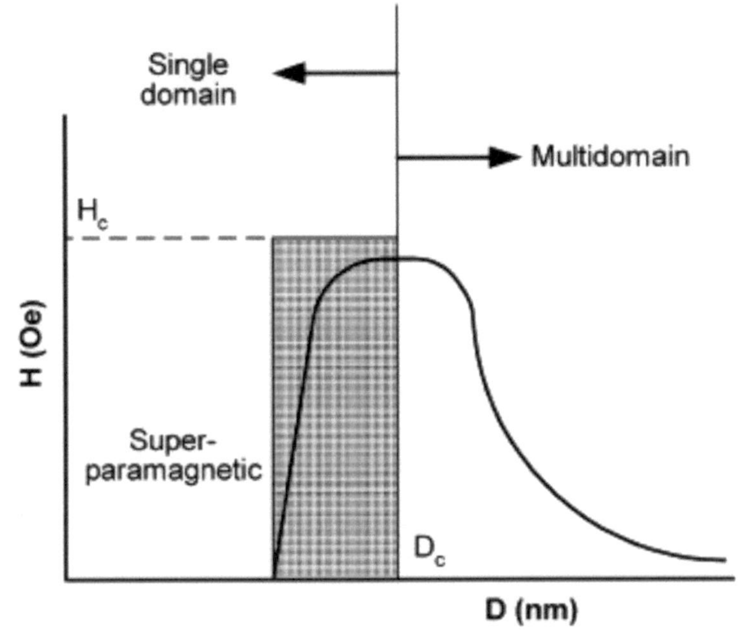

Fig. 6. Qualitative illustration of the behavior of the coercivity in ultrafine systems as the particle size changes, where H is the magnetic field amplitude (Oe) and *D* is the particle diameter (nm).

a. *Magnetostatic or demagnetization energy:* The magnetized material behaves like a magnet, with a surrounding magnetic field. This field acts to magnetize the material in the direction opposite from its own magnetization, causing a magnetostatic energy which depends on the shape of the material. This magnetostatic energy can be reduced by reducing the net external field through the formation of domains inside the material.
b. *Magnetocrystalline anisotropy energy:* In some materials the domain magnetization tends to align in a particular crystal direction (the so-called easy axis). The material is easiest to magnetize to saturation or demagnetize from saturation if the field is applied along an easy axis. The energy difference between aligning the domain in the easy and another direction (hard direction) is called magnetocrystalline anisotropy energy. Anisotropy energy is the energy needed to rotate the moment from the easy direction to

a hard direction. For materials with cubic crystalline structure (such as ferrites), the energy is expressed in terms of anisotropy constants and the direction to which the magnetization rotates.

$$E_k=K_1\sin^2\theta+K_2\sin^4\theta\ldots \text{ (Hexagonal structre)}$$

$$E_k=K_1 (\alpha_1^2 \alpha_2^2+ \alpha_2^2 \alpha_3^2+ \alpha_3^2 \alpha_1^2) + K_2 (\alpha_1^2 \alpha_2^2 \alpha_3^2+\ldots) \text{ (Cubic and Garnet structure)} \quad (1)$$

where K is the anisotropy constant, θ is the angle between the easy axis and the direction of magnetization, and a ; are the direction cosines, which are the ratios of the individual components of the magnetization projected on each axis divided by the magnitude of the magnetization. A crystal is higher in anisotropy energy when the magnetization points in the hard direction rather than along the easy direction. The formation of domains permits the magnetization to point along the easy axis, resulting in a decrease in the net anisotropy energy.

c. *Magnetostrictive energy:* In a magnetic field, the material may change its dimensions on the order of several parts per million. This change in dimension results in what is called magnetostrictive energy, which is lowered by a reduction in the size of the domains, requiring the formation of more domains.

2.5 Magnetization curve and hysteresis loops

The magnetization curve describes the change in magnetization or magnetic flux of the material with the applied field. When a field is applied to a material with randomly oriented magnetic moments, it will be progressively magnetized due to movement of domain boundaries. Initially, when no field is applied, the magnetic dipoles are randomly oriented in domains, thus the net magnetization is zero. When a field is applied, the domains begin to rotate, increasing their size in the case of the domains with direction favorable with respect to the field, and decreasing for the domains with unfavorable direction. As the field increases, the domains continue to grow until the material becomes a single domain, which is oriented in the field direction. At this point, the material has reached saturation (Figure7) [3].As the magnetic field is increased or decreased continuously, the magnetization of the material increases or decreases but in a discontinuous fashion. This phenomenon is called the Barkhausen effect and is attributed to discontinuous domain boundary motion and the discontinuous rotation of the magnetization direction within a domain [14]. The typical magnetization curve can be divided into three regions:

a. *Reversible region:* The material can be reversibly magnetized or demagnetized. Charges in magnetization occur due to rotation of the domains with the field.
b. *Irreversible region:* Domain wall motion is irreversible and the slope increases greatly.
c. *Saturation region:* Irreversible domain rotation. It is characterized by a required large amount of energy to rotate the domains in the direction of the field [5].

If the field is reduced from saturation, with eventual reversal of field direction, the magnetization curve does not retrace its original path, resulting in what is called a hysteresis loop. This effect is due to a decrease of the magnetization at a lower rate. The area inside the hysteresis loop is indicative of the magnetic energy losses during the magnetization process. When the field reaches zero, the material may remain magnetized (i.e., some domains are oriented in the former direction). This residual magnetization is commonly called remanence M_r. To reduces this remanent magnetization to zero; a field with opposite direction must be applied. The magnitute of field required to lower the sample magnetization to zero is called the coercivity H_c (Figure 8). A material can present different hysteresis loops depending on the degree of magnetization. If the maximum magnetization is less than the saturation magnetization, the loop is called a minor loop [3,4].

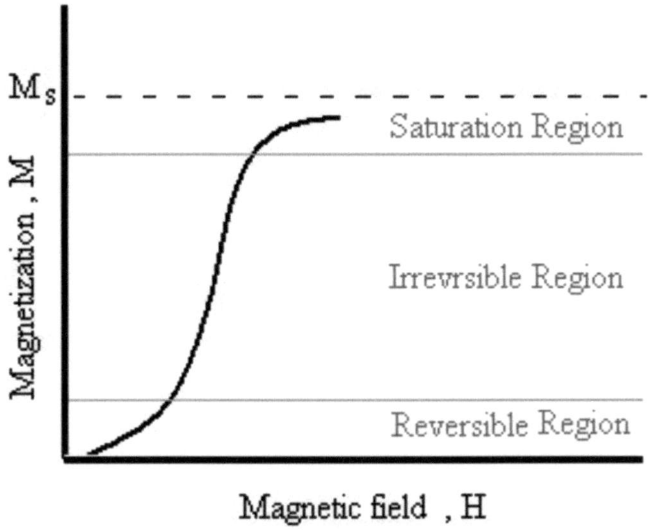

Fig. 7. Magnetization curve with domain configurations at different stages of magnetization.

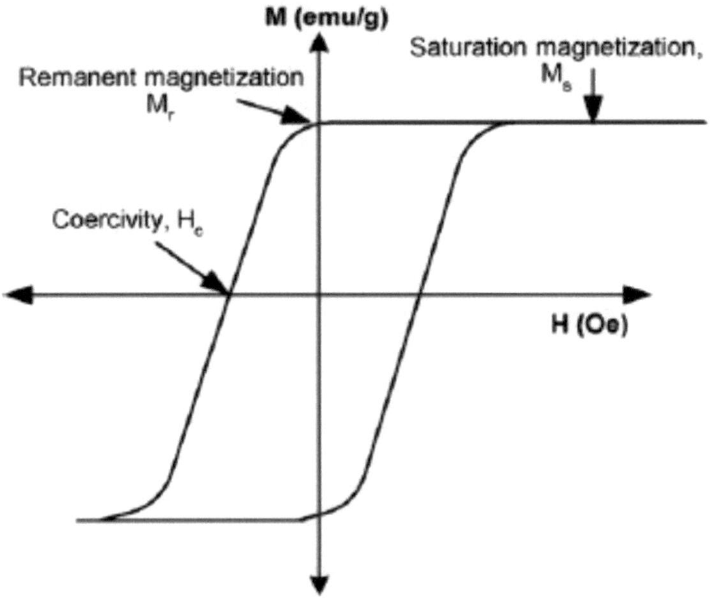

Fig. 8. Hysteresis cycle of a multidomain magnetic material, where*H*is the magnetic field amplitude (Oe) and M is the magnetization of the material (emu/g).

2.6 Magnetic behaviors

Ferrites materials can be classified based on differences between their internal and external flux and the variation of the magnetization M or magnetic induction B when a magnetic field is applied (Figure 9) [3,4]. There are two quantities that relate M and B to H: the susceptibility χ and the permeability μ .

$$\chi = \frac{M}{H}, \tag{2}$$

$$\mu = \frac{B}{H}. \tag{3}$$

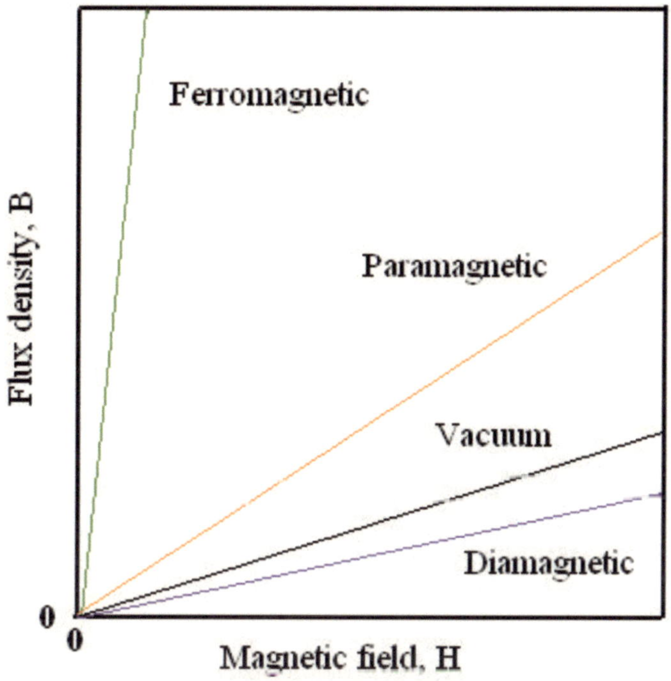

Fig. 9. Representation of the behavior of the flux density with respect to the magnetic field for different classes of magnetic materials.

In SI the permeability μ has units of Henry/m. The susceptibility is a measure of the increase in magnetic moment caused by an applied field, whereas permeability represents the relative increase in flux caused by the presence of the magnetic material [15].

2.6.1 Diamagnetism

Diamagnetism is an inherent result of the orbital motion of the electrons in a magnetic field. It is present when the atom has zero net magnetic moment. In this case the orbital motion generates a field opposite to the applied field (magnetization is directed oppositely to the field, as illustrated in Figure 10), described by a negative susceptibility. These materials tend to move toward regions of weaker field [5, 15].

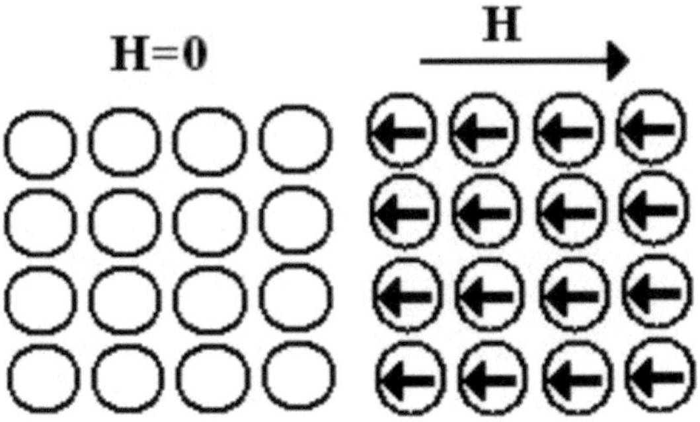

Fig. 10. Atomic dipole configuration for a diamagnetic material.

2.6.2 Paramagnetism

Paramagnetic materials possess a permanent dipole moment due to incomplete cancellation of electron spin and/or orbital magnetic moments. In the absence of an applied magnetic field the dipole moments are randomly oriented; therefore the material has no net macroscopic magnetization. When a field is applied these moments tend to align by rotation towards the direction of the field and the material acquires a net magnetization (Figure 11) [3].

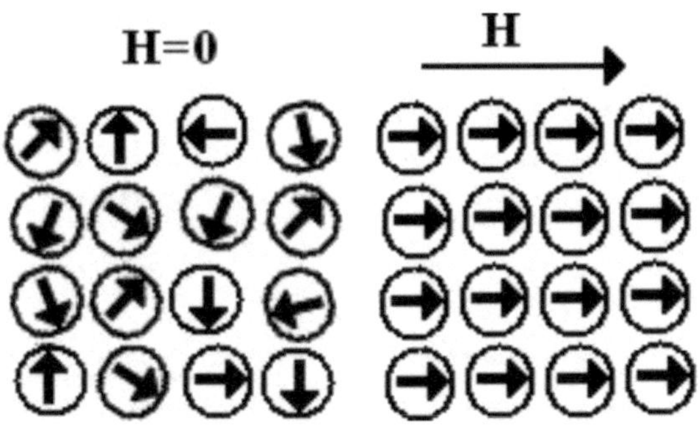

Fig. 11. Schematic of atomic dipoles for a paramagnetic material.

2.6.3 Ferromagnetism and ferrimagnetism

Ferro and ferri-magnetic materials possess a permanent magnetic moment in the absence of an external field and a very large permanent magnetization [3]. In ferromagnetic materials, this permanent magnetic moment is the result of the cooperative interaction of large numbers of atomic spins in what are called domains, regions where all spins are aligned in the same direction (see section2.5).In ferrimagnetic materials, on the other hand, incomplete cancellation of the magnetic dipoles in a domain results in lower permanent magnetization (Figure 12) [5].

The macroscopic magnetization of ferro- and ferri-materials is the sum of the magnetizations of the domains which make up the sample [3]. Ferrimagnets are ionic solids meaning that they are electrically insulating, whereas most ferromagnets are metals (conductors) [4].

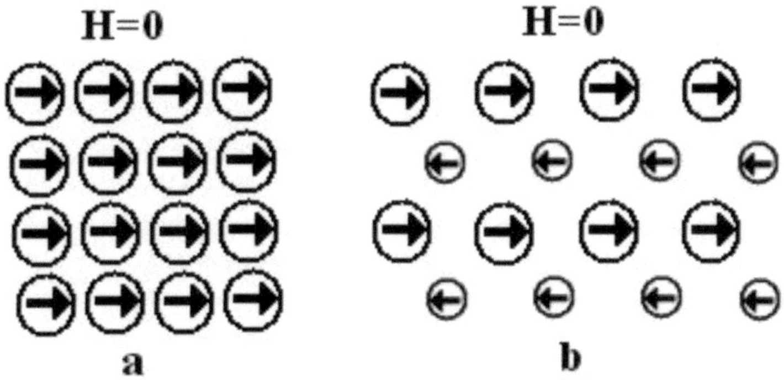

Fig. 12. Ordering of the atomic dipoles in a) ferromagnetic and b) ferrimagnetic material.

2.6.4 Antiferromagnetism

In materials that exhibit antiferromagnetism, the magnetic moments of atoms or molecules, usually related to the spins of electrons, align in a regular pattern with neighboring spins (on different sublattices) pointing in opposite directions. This is, like ferromagnetism and ferrimagnetism, a manifestation of ordered magnetism. Generally, antiferromagnetic order may exist at sufficiently low temperatures, vanishing at and above a certain temperature, the Néel temperature (Neel temperature is the temperature at which an antiferromagnetic material becomes paramagnetic; hence losing its magnetic properties) [16].Above the Néel temperature, the material is typically paramagnetic. Figure13 shows ordering of the atomic dipoles in an antiferromagnetic material

Fig. 13. Ordering of the atomic dipoles in an antiferromagnetic material.

2.6.5 Superparamagnetic

Superparamagnetism is a phenomena by which magnetic materials may exhibit a behavior similar to paramagnetism at temperatures below the Neel or the Curie temperature (The Curie temperature is the temperature at which a ferromagnetic or a ferromagnetic material becomes paramagnetic; hence losing its magnetic properties).Normally, coupling forces in magnetic materials cause the magnetic moments of neighboring atoms to align, resulting in very large internal magnetic fields. At temperatures above the Curie temperature (or the Neel temperature for antiferromagnetic materials), the thermal energy is sufficient to overcome the coupling forces, causing the atomic magnetic moments to fluctuate randomly.

Because there is no longer any magnetic order, the internal magnetic field no longer exists and the material exhibits paramagnetic behavior. Superparamagnetism occurs when the material is composed of very small crystallites (lower than 100 nm). In this case even though the temperature is below the Curie or Neel temperature and the thermal energy is not sufficient to overcome the coupling forces between neighboring atoms, the thermal energy is sufficient to change the direction of magnetization of the entire crystallite. The resulting fluctuations in the direction of magnetization cause the magnetic field to average to zero. The material behaves in a manner similar to paramagnetism, except that instead of each individual atom being independently influenced by an external magnetic field, the magnetic moment of the entire crystallite tends to align with the magnetic field. The energy required to change the direction of magnetization of a crystallite is called the crystalline anisotropy energy (see section 2.5) and depends both on the material properties and the crystallite size. As the crystallite size decreases, so does the crystalline anisotropy energy, resulting in a decrease in the temperature at which the material becomes superparamagnetic [17]. Ordering of the atomic dipoles in a superparamagnetic material is shown in Figure 14.

Fig. 14. Ordering of the atomic dipoles in a superparamagnetic material.

2.7 Classification and applications of ferrites

Ferrites are grouped into two types, soft and hard. This is the classification based on their ability to be magnetized and demagnetized, not their ability to withstand penetration or abrasion. Soft materials are easy to magnetize and demagnetize, so are used for electromagnets, while hard materials are used for permanent magnets. They can also be classified based on their coercive field strength into soft and hard materials [12]. With soft magnetic materials the hysteresis loop is small (low coercive field strength, independent of magnetic field amplitude); with hard magnetic however it is large (high coercive field strength). Table 2 gives a comparative account of both types.

$H_c < 10$ A/Cm: soft magnetic;

$H_c > 300$ A/Cm: hard magnetic (permanent magnets).

Hard ferrite magnets are made in two different magnetic forms: isotropic and oriented. Isotropic magnets are formed to desired shapes, sintered and then magnetized. These exhibit a modest magnetic field and find applications in cycle dynamos and ring magnets. Oriented magnets are formed to shape under a strong magnetic field and then sintered. These exhibit a very strong magnetic field and find applications in loudspeakers, magnets of two wheelers like scooters, etc. [14].

Soft magnetic	Hard magnetic
High saturation magnetization (1-2T)	High saturation magnetization (0.3-1.6T)
Low coercivity (H_c)	High coercivity
High permeability	Not important, but low
Low anisotropy	High anisotropy
Low magnetostriction	Not important
High Curie temperature (T_c)	High T_c
Low losses	High-energy product
High electrical resistivity	Not important

Table 2. The comparative properties of soft and hard magnetic materials

3. Established methods for synthesis of nanocrystalline ferrites

There are two methods for the preparation of magnetic nanoparticles: physical and chemical. The methods of generation of magnetic nanoparticles in the gas or solid phase using high-energy treatment of the material are usually called physical, while the nanoparticle syntheses, which are often carried out in solutions at moderate temperatures are chemical methods. Different routes have become an essential focus of the related research and development activities. Various fabrication methods to prepare spinel ferrites nanocrystals have been reported, e.g., sol-gel methods [18], the ball-milling technique [19], co-precipitation [20], polymeric assisted route [21] the hydrothermal method [22], the reverse micelles process [23], and the micro-emulsion method [24]. Various precipitation agents have been used to produce specific size and shape spinel ferrites nanocrystals, e.g., metal hydroxide in the co-precipitation method, surfactant and ammonia in the reverse micelles process and various micro-emulsion methods, and organic matrices in the sol-gel method. Most of these methods have achieved particles of the required sizes and shapes, but they are difficult to employ on a large scale because of their expensive and complicated procedures, high reaction temperatures, long reaction times, toxic reagents and by-products, and their potential harm to the environment.

3.1 Thermal treatment method

In the present study, spinel ferrites nanocrystals with different structures were prepared from an aqueous solution containing metal nitrates, poly (vinyl pyrrolidon), and deionized water using a relatively low temperature thermal treatment method, followed by grinding and calcinations*. No other chemicals were added to the solution. This method is environmentally friendly in that it neither uses nor produces toxic substances, and it offers the advantages of simplicity, low cost, and low reaction temperatures. The textural and morphological characteristics of the spinel ferrites nanocrystals we prepared were studied with various techniques to determine the influence of calcination temperature on the crystallization, morphology, and particle size distribution of the nanocrystals and to explore other parameters of interest.

* This method was invented by Mahmoud Goodarz Naseri in university Putra Malaysia in 2011

3.1.1 Process of fabrication of metal ferrite nanocrystals

In this study, metal nitrate reagents, poly (vinyl pyrrolidon) (PVP), and deionized water were used as precursors. In addition, a capping agent to control the agglomeration of the particles and a solvent were used. Iron nitrate, $Fe(NO_3)_3{\bullet}9H_2O$, zinc nitrate, $Zn(NO_3)_2{\bullet}6H_2O$, nickel nitrate, $Ni(NO_3)_2{\bullet}6H_2O$, and manganese nitrate $Mn(NO_3)_2{\bullet}6H_2O$, were purchased from Acros Organics with a purity exceeding 99%. PVP (MW = 29000) was purchased from Sigma Aldrich and was used without further purification. An aqueous solution of PVP was prepared by dissolving polymer in deionized water at 363 K, before mixing 0.2 mmol iron nitrate and 0.1 mmol metal nitrate (Fe:M = 2:1) into the polymer solution and constantly stirring for 2 h using a magnetic stirrer until a colorless, transparent solution was obtained. The mixed solution was poured into a glass Petri dish and heated at 353 K in an oven for 24 h to evaporate the water. The dried, orange, solid zinc ferrite that remained was crushed and ground in a mortar to form powder. The calcinations of the powders were conducted at 723, 773, 823, and 873 K for 3 h for the decomposition of organic compounds and the crystallization of the nanocrystals. The processing steps employed separately for the synthesis of each ferrite nanoparticles.

3.2 Characterization

The textural and morphological characteristics of the spinel ferrite nanocrystals we prepared were studied with various techniques to determine the influence of calcination temperature on the crystallization, morphology, and particle size distribution of the nanocrystals and to explore other parameters of interest. The characterization of the prepared spinel ferrites nanoparticles were conducted by using various techniques such as to (TGA), X-ray diffraction (XRD), Fourier transforms infrared spectroscopy (FTIR), transmission electron microscopy (TEM), and vibrating sample magnetometer (VSM) verify the particle size and distribution and to explore other parameters of interest.in this section we introduce these techniques.

3.3 Mechanism of interaction of PVP and metal ions in synthesize of Zn, Mn and Ni ferrite nanoparticle by thermal treatment method

Interactions between the PVP capping agent [25] and metal ions are shown schematically in Figure 15.We have shown the Metal (II) (e.g. Zn, Mn and Ni) and iron (III) ions which are bound by the strong ionic bonds between the metallic ions and the amide group in a polymeric chain. PVP acts as a stabilizer for dissolved metallic salts through steric and electrostatic stabilization of the amide groups of the pyrolidine rings and the methylene groups. Initially, the PVP stabilizer may decompose to as limited extent, thereby producing shorter polymer chains that are capped when they are adsorbed onto the surfaces of metallic ions [26]. The metallic ions, which are well dispersed in the cavities and networks, are created as a result of the shorter polymer chains. These mechanisms continue until they are terminated by the drying step. The influence of PVP is not restricted only to the solution and the drying step; PVP also affects the formation of the nuclei (i.e., nucleation) of the nickel ferrite nanoparticles in the calcination step. In this step, the small nanoparticles with high surface energy levels would become larger via the Ostwald ripening process [27] without the presence of PVP, disrupts steric hindrance, thereby preventing their aggregation. Steric hindrance is a phenomenon that is attributed to large molecular weight (>10,000) and the repulsive forces acting among the polyvinyl groups [28, 29]. These interactions are similar to the stabilization of metallic nanoparticles, i.e., silver and gold [30, 31].

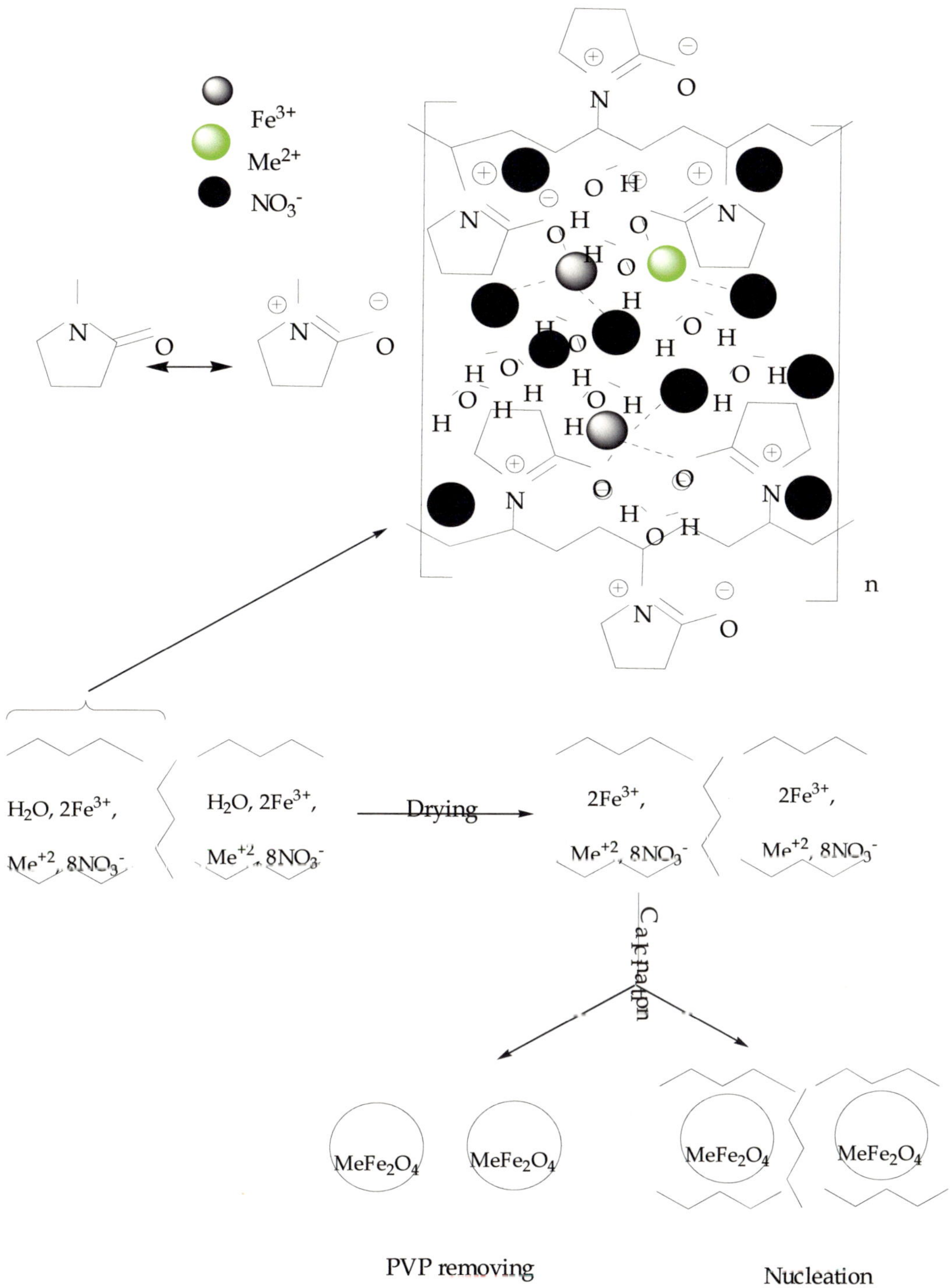

Fig. 15. The proposed mechanism of interactions between PVP and metal ions in the formation of the ferrites nanoparticles.

3.4 Determination of range of calcinations temperature for removing of PVP

Figure 16 shows the simultaneous thermal analyses (TG–DTG) for PVP. It is evident that this polymer exhibited only one mass loss which started at 678 K and its maximum rate decomposition temperature was located at 778 K. This confirms that the majority of the mass loss occurs under 778 K and allows for optimization of the heat treatment program. It is worth noting that several authors reported the thermal degradation of PVP exhibits only one mass loss [32-34].

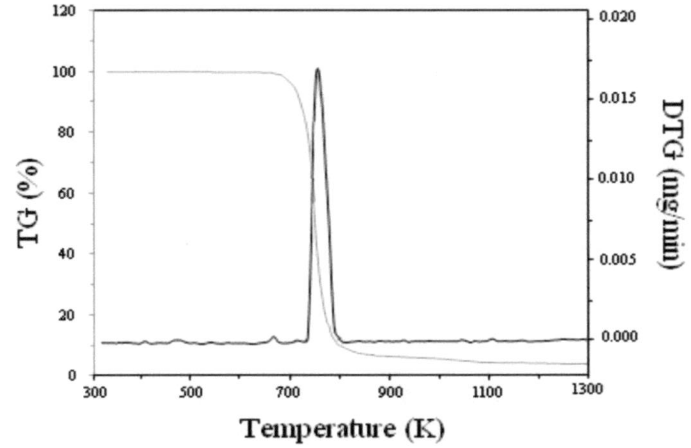

Fig. 16. Thermogravimetric (TG) and thermogravimetric derivative (DTG) curves for: PVP at a heating rate of 10 k/min.

3.5 Determine of optimum parameters of nickel ferrite nanocrystal

To investigate the optimum concentration of PVP in the synthesis of nickel ferrite nanoparticles, we synthesized nickel ferrite nanoparticles with others PVP concentrations of 0, 0.015 and 0.055 gm/ml, and the results are shown in the TEM images in Figure 17 and the FT-IR spectra in Figure 18. Also, optimum temperature for calcinations of nickel ferrite nanoparticle was 723 K because, this temperature was minimum temperature that nanoparticles were pure; furthermore, they have also the lowest particle size with a nearly uniform distribution in shapes.Figure 17a shows that nickel ferrite nanoparticles were formed even in the absence of PVP. However, in this case, it was observed that the nanoparticles did not have a uniform distribution of shapes, and they were aggregated extensively and, in some areas, completely disproportionately distributed. Thus, without the use of PVP in the synthesis of nanoparticles, the small nanoparticles aggregate and produce larger nanoparticles [29] due to high surface energy (as shown earlier in Fig15)

When the concentration of PVP was increased to 0.015 gm/ml, the nickel ferrite nanoparticles that were formed became more regular in shape than in the case without PVP (Figure 17b). But, due to the low concentration of PVP, these nanoparticles also aggregated because there was insufficient PVP to cap them well and prevent their agglomeration.By increasing the PVP concentration to 0.055 gm/ml, the nickel ferrite nanoparticles did not agglomerate, and they were nearly uniform in shape, as shown in Figure 17d. However, in this case, the nickel ferrite nanoparticles ranged in size from 9 to 21 nm, with an estimated average particle size of 12 nm. These results were similar to the results achieved when a PVP concentration of 0.035 gm/ml was used (shown in Figure 17c). But, due to the high concentration of PVP, traces of organic materials were observed at 1254 cm^{-1}, which was attributed to the C-H bending vibration of methylene groups, as shown in Figure 18b while in concentration of 0.035 gm/mlnickel ferrite nanoparticles were pure (Figure 18a).

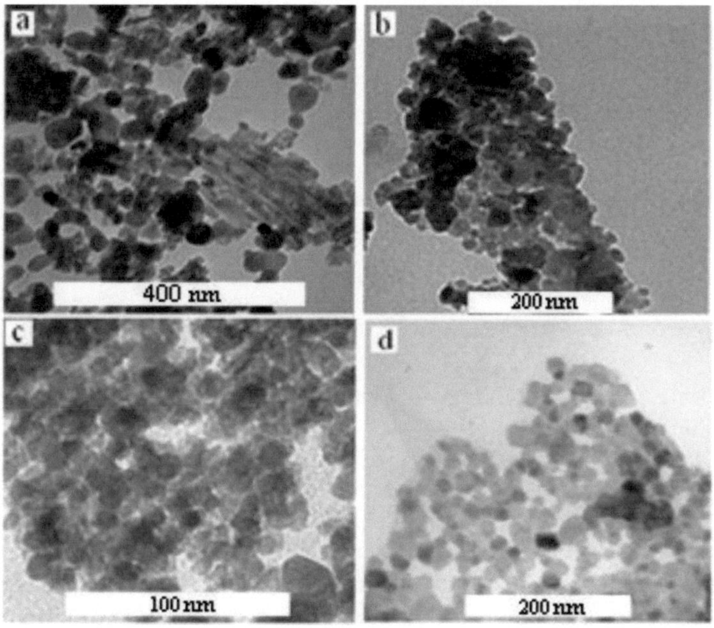

Fig. 17. TEM images of nickel ferrite nanoparticles with PVP concentrations of (a) 0, (b) 0.015, (c) 0.035, and (d) 0.055 gm/ml calcined at 723 K.

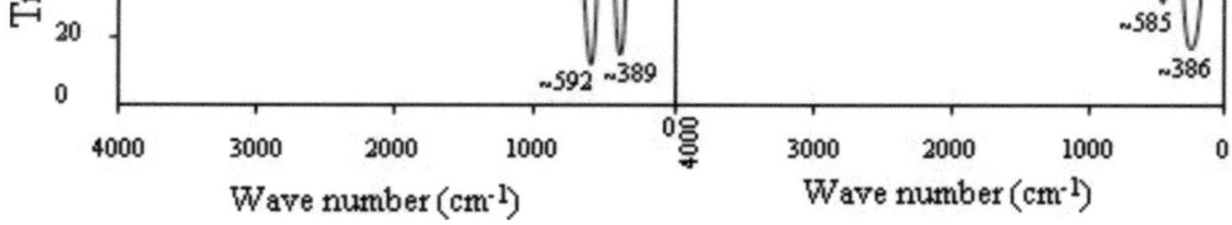

Fig. 18. FTIR spectra of nickel ferrite nanoparticles with PVP concentration of (a) 0.035 and 0.055 gm/ml calcined at 723 K.

So, in fact, it is apparent that the nickel ferrite nanoparticles were contaminated with organic compounds in this case. Therefore, in the thermal treatment method, the optimum concentration of PVP for the synthesis of nickel ferrite nanoparticles is 0.035 gm/ml. That concentration, in combination with the optimum temperature (723 K) provided the conditions required to fabricate pure nickel ferrite nanoparticles that have the smallest particle size. So, as discussed earlier in connection with Figure 15 and as demonstrated in the above discussion of the results obtained, PVP plays three crucial roles in synthesizing nickel ferrite nanoparticles, i.e., (1) the control of the growth of the nanoparticles; (2) the prevention of agglomeration of the nanoparticles; and (3) the production of nanoparticles that have a uniform distribution of shapes [35].

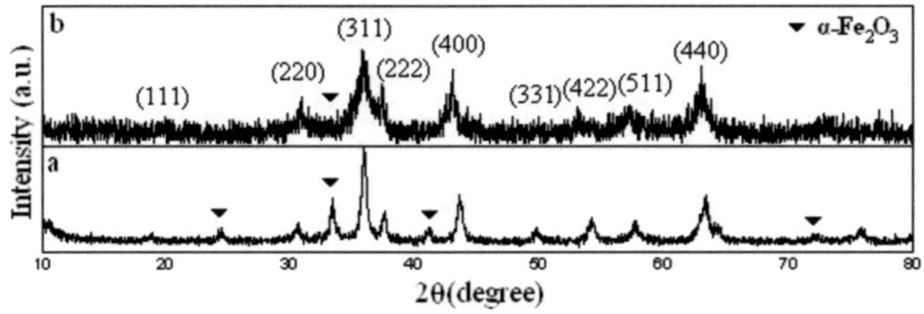

Fig. 19. XRD patterns of nickel ferrite nanoparticles with heating rate of calcination of 10, and 20 K/min calcined at 723.

After our examination, for optimum concentration of PVP (0.035 gm/ml) and optimum temperature for calcinations (723 K) of nickel ferrite nanoparticle, we investigated optimum the time of calcinations and the heating rate of calcinations of nickel ferrite nanoparticles as last optimization.In this investigation, the minimum time that allowed the crystallization to be completed was 3 h as lower than 3 h the crystallization was uncompleted and higher than 3 h particles size increased. The heating rate of calcination was 10 K/min for nickel ferrite nanoparticles calcined at 723K, which was an optimum heating rate. By increasing the heating rate of calcination to 20 K/min, the percent of impure phase of α-Fe_2O_3 increased and the crystallite of nickel ferrite nanoparticles were not as pure as optimum heating rate as shown in Figure19a and Figure 19b.When the heating rate of calcinations was lower than 10 K/min (5 K/min) we wasted long time for calcinations and several neighboring particles fuse together to increase particle sizes by melting their surfaces [36].

Note: We have done exactly these experiments on others ferrites i.e. zinc ferrite and manganese ferrite and have obtained for each of them optimum parameters. But to prevent long or repeated exposures and similar experiments, we reported only the experiments of nickel ferrite nanoparticles and for others ferrites we sufficed only to report of values (Table3).

Pure metal ferrite nanoparticles	Optimum calcinations temperature (K)	Optimum concentration of PVP (gm/ml)	Optimum Heating rate (K/min)	Optimum Calcinations Time (h)
ZnFerrite	873	0.030	10	3
MnFerrite	873	0.030	10	3
NiFerrite	723	0.035	10	3

Table 3. Summary of optimum parameters of metal ferrite nanocrystal prepared by thermal treatment method

3.6 Phase composition and morphology of metal ferrite nanocrystal

XRD and FT-IR were used to characterize the precursors and ferrite nanoparticles calcined at 723, 773, 823 and873 K. The XRD diffraction patterns of the precursor and metal ferrite nanoparticles are shown in Figure20. A broad peak occurred for all samples in the precursor, which does not have sharp diffraction patterns and is still amorphous. The calcined patterns show the reflection planes (111), (220), (311), (222), (400), (331), (422), (511), and (440), which confirm the presence of single-phase in $ZnFe_2O_4$, $MnFe_2O_4$ and $NiFe_2O_4$ with a face-centered cubic structure [37-39].Except for the impure phases of α-Fe_2O_3 (for all the metal ferrite nanoparticles shown in Figure 20a, 20b and 20c) and ZnO (for the zinc ferrite nanopaticles shown in Figure20a), which occur naturally as hematite and zincite, respectively [37, 40], the remaining peaks correspond to the standard pattern of $ZnFe_2O_4$ (cubic, space group: Fd3m, Z = 8; ICDD PDF: 22-1012), $MnFe_2O_4$ (cubic, space group: Fd3m, Z = 8; ICDD PDF: 73-1964) and $NiFe_2O_4$ (cubic, space group: Fd3m, Z = 8; ICDD PDF: 44-1485)[37-39].

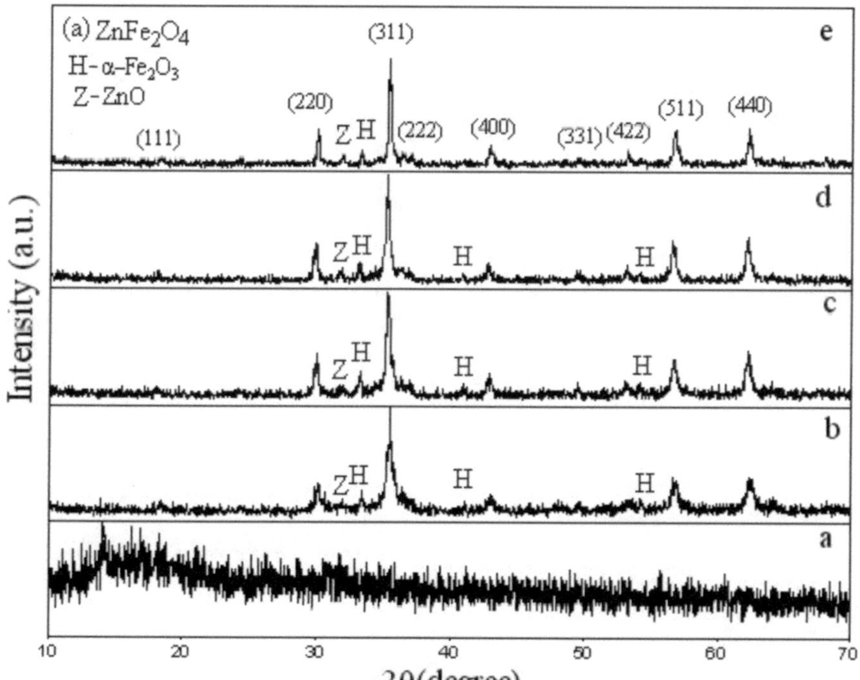

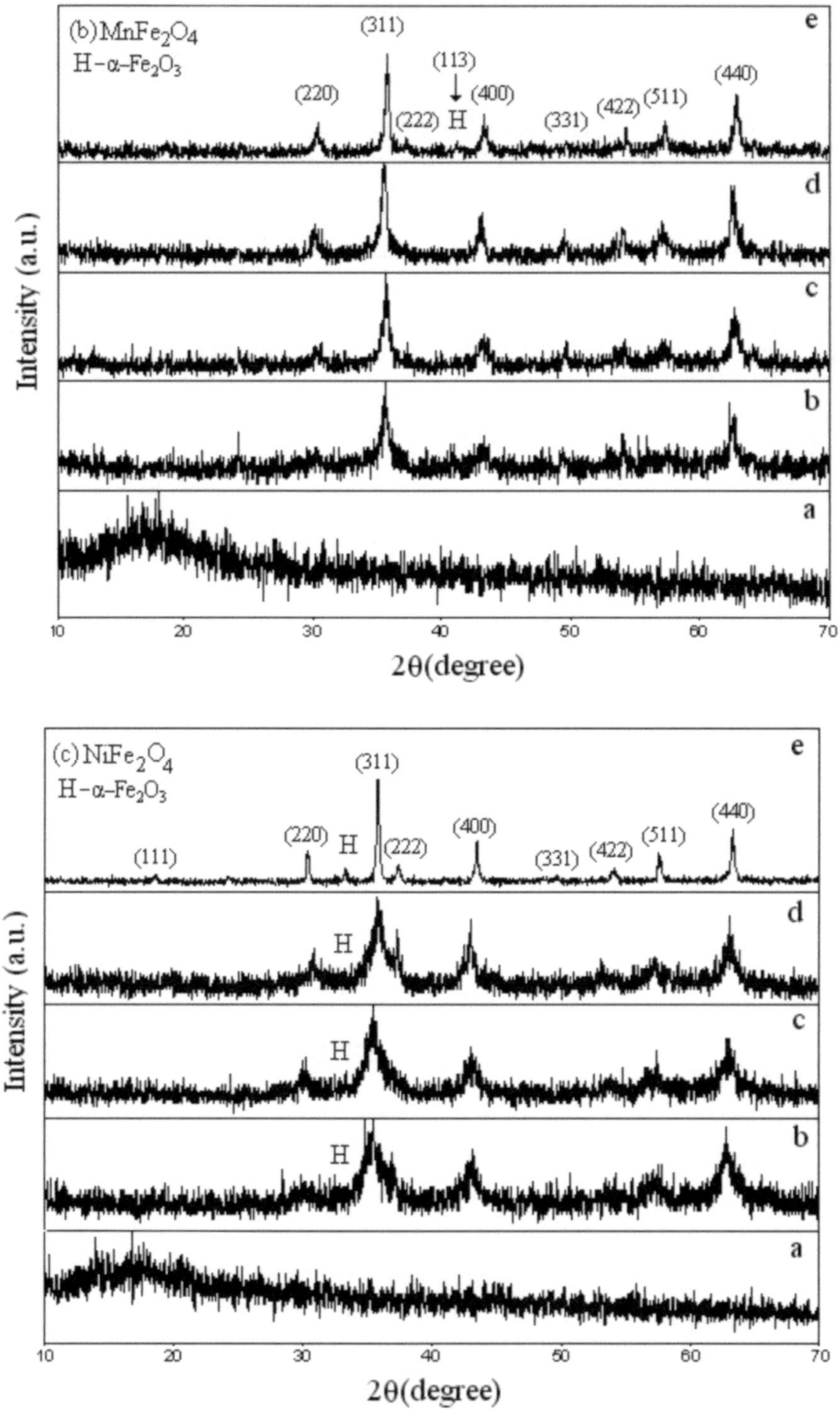

Fig. 20. XRD patterns of precursors and metal ferrite nanoparicles of (a) $ZnFe_2O_4$, (b) $MnFe_2O_4$ and (c) $NiFe_2O_4$ calcined at 723, 773, 823 and 873 K.

The results obtained from XRD were analyzed using the Chekcell program, which calculated lattice parameters of the samples calcined at 723, 773, 823, and 873 K (Table 4). XRD results were analyzed by the Scherer formula:

$$D=0.9\lambda/\ (\beta\cos\theta), \tag{4}$$

where D is the crystallite size (nm), β is the full width of the diffraction line at half the maximum intensity measured in radians, λ is X-ray wavelength, and θ is the Bragg angle [41]. This formula was used to estimate the average particle sizes, which ranged in Table 4.

Specimens $MeFe_2O_4$	Calcination temperature (K)	Average particle size XRD (nm)	Average particle size TEM (nm)	Lattice Parameter (nm)	Saturation magnetization M_s (emu/g)	Coercivity field H_c (Oe)
ZnFerrite 1	723	21	17±7	0.8498	4.49	negligible
ZnFerrite 2	773	24	22±2.5	0.8468	2.66	negligible
ZnFerrite 3	823	31	27±5	0.8471	1.81	negligible
ZnFerrite 4	873	33	31±11	0.8479	0.74	negligible
MnFerrite 1	723	15	12±4	0.8524	3.06	negligible
MnFerrite 2	773	17	15±2	0.8577	6.31	negligible
MnFerrite 3	823	20	17±5	0.8558	7.96	negligible
MnFerrite 4	873	23	22±4	0.8537	15.78	negligible
NiFerrite 1	723	15	12±3	0.8368	21.37	150(-148)
NiFerrite 2	773	27	22±12	0.8373	26.67	107(-112)
NiFerrite 3	823	51	47±11	0.8418	29.05	51(-43)
NiFerrite 4	873	69	67±8	0.8402	34.19	32(-38)

Table 4. Summery of variation of particle sizes, lattice parameters, saturation magnetization and coercivity field with temperature calcinations for metal ferrite nanoparticles calcined at 723, 773, 823 and 873 K.

Figure 21 shows the FT-IR spectrum of the precursor and calcined samples in the wave-number range between 280 and 4000 cm^{-1}.The IR spectra of all calcined samples show the two principle absorption bands in the range of 300-600 cm^{-1} .These two vibration bands Fe↔O and M↔O are corresponded to the intrinsic lattice vibrations of octahedral and tetrahedral coordination compounds in the spinel structure, respectively [42].The bands with peaks around 670 and 850 cm^{-1} were assigned to the formation vibration of C-N=O bending and the C-C ring. The bands in the range of 1200 to 1250 cm^{-1} was associated with C-N stretching vibration, and the appearance of the bands in the range of 1350 to 1450 cm^{-1} was attributed to C-H bending vibration from the methylene groups. Finally, there were bands in the region 1600 to1800 cm^{-1} and around 3400 to 3500 cm^{-1}, which were associated with C=O stretching vibration and N-H or O-H stretching vibration, respectively [43].

The vibrational spectra of the absorption bands of pure $ZnFe_2O_4$ and $MnFe_2O_4$ nanoparticles were observed at 388, and 541 cm^{-1}, and at 404, 502, and 556 $cm^{-1}$for the samples calcined at 873 K (shown in Figure 21a and 21b). In these two ferrite nanoparticles, at the lower temperature of 873 K, however, there was still traces of broadband absorption peaks at 1497,

1761 and 3504 cm-1 due to traces of adsorbed or atmospheric CO_2 and O-H stretching vibration, respectively while in $NiFe_2O_4$ nanoparticles (Figure 21c), at the lower temperature of 723 K, there was still a trace of a broadband absorption peaks due to ester formation as consequence of the scission of the CO_2 and O-H stretching vibration(This is not shown in the Figure21). This suggests that, in thermal treatment method, the calcination temperature of pure nickel ferrite nanoparticles is lower than pure zinc and manganese ferrite nanoparticles [35,44,45].This IR analysis was very useful for establishing the calcination temperature because it removed unwanted ions that may pollute the crystal lattice during preparation.

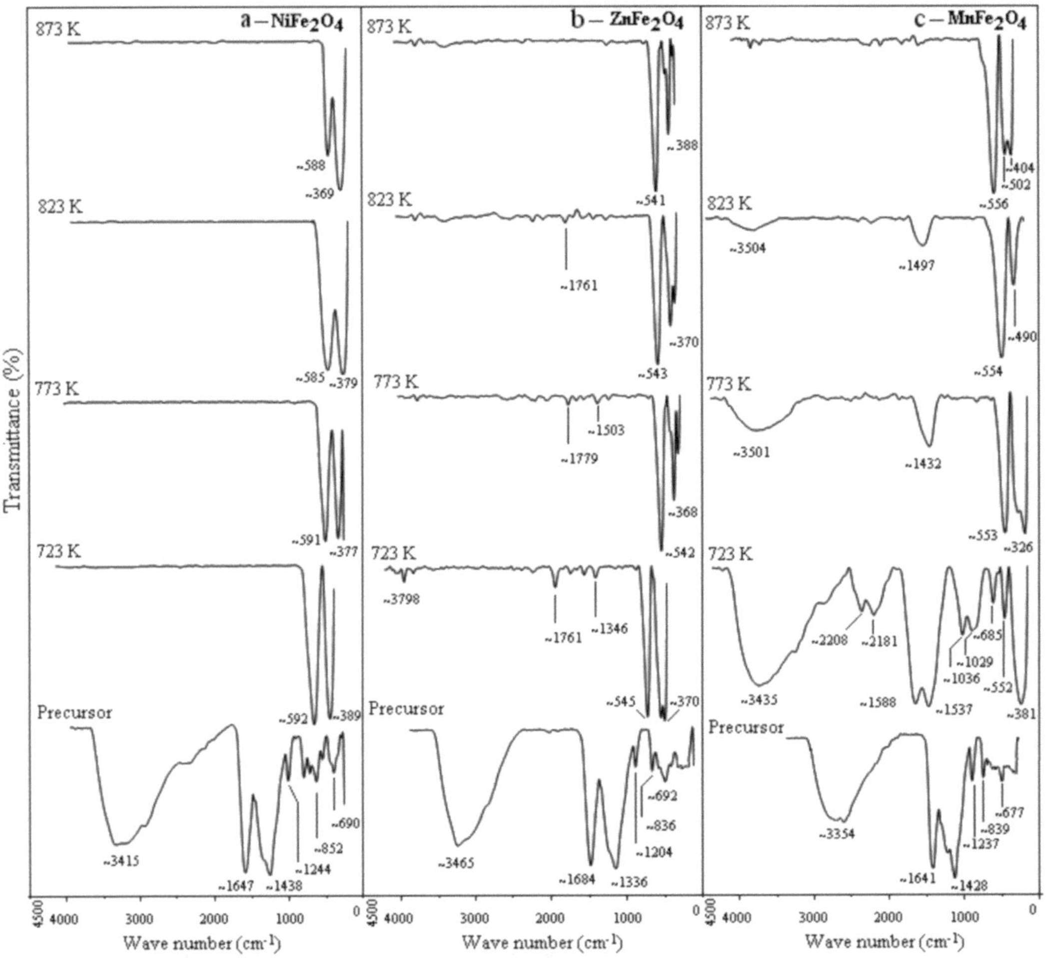

Fig. 21. FT-IR spectra of precursors and metal ferrite nanoparicles of (a) ZnFe2O4, (b) $MnFe_2O_4$ and (c) $NiFe_2O_4$ calcined at 723, 773, 823 and 873 K.

The TEM images (Figures 22-24) show the size, shape, and distribution of $ZnFe_2O_4$, $MnFe_2O_4$ and $NiFe_2O_4$ nanoparticles at different calcination temperatures from 723 to 873 K. The results indicate that the samples prepared by the thermal treatment method were uniform in morphology and particle size distribution. The average particle size of the $ZnFe_2O_4$, $MnFe_2O_4$ and $NiFe_2O_4$ nanoparticles were determined by TEM which increased with the calcinations temperature and they had good agreement with XRD results (Table 4).This suggested that several neighboring particles fused together to increase the particle size by the melting of their surfaces [46]. Particle size enlargement due to grain growth has been observed previously in zinc.manganese and nickel ferrite systems at higher calcination temperatures [35,44,45].

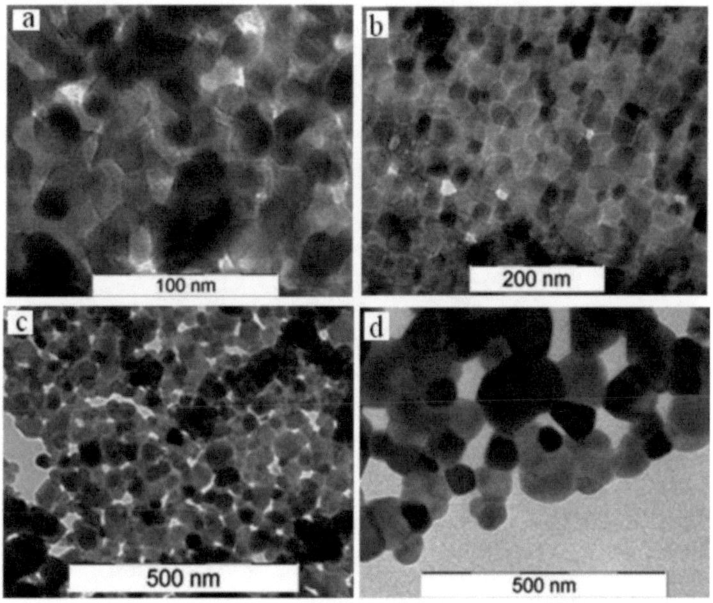

Fig. 22. TEM images of zinc ferrite nanoparticles calcined at (a) 723 (b) 773 (c) 823 (d) 873 K.

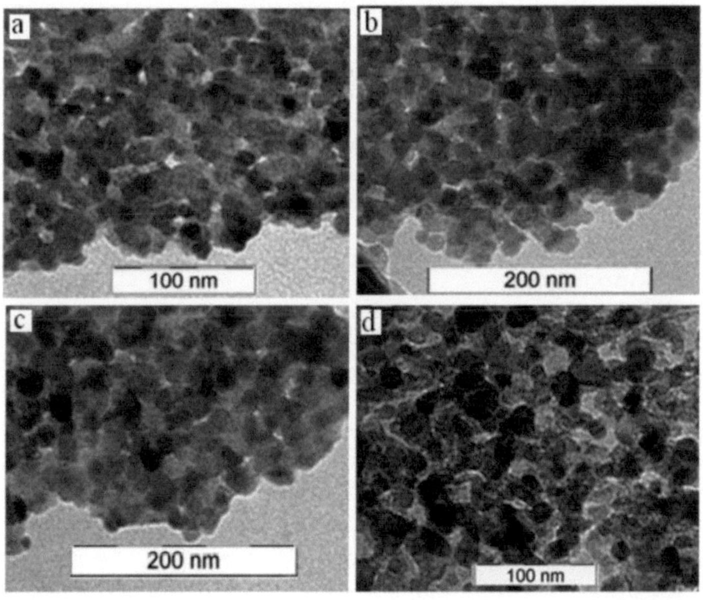

Fig. 23. TEM images of manganese ferrite nanoparticles calcined at (a) 723 (b) 773 (c) 823 (d) 873 K.

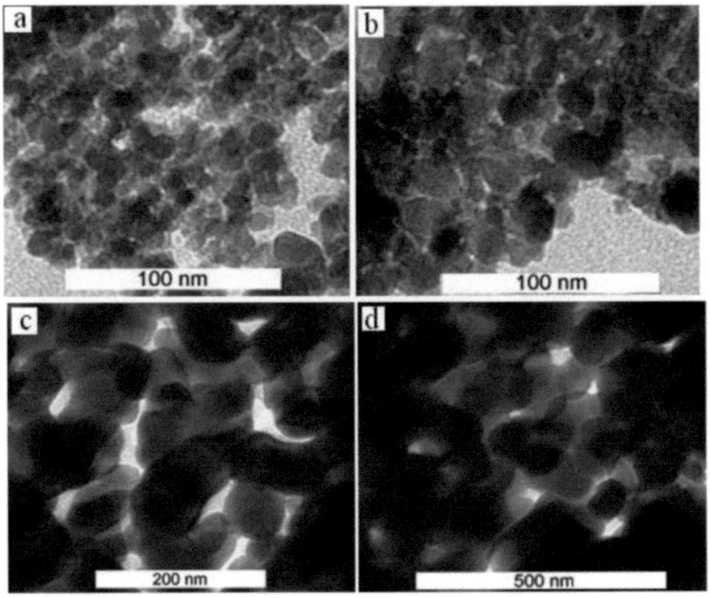

Fig. 24. TEM images of nickel ferrite nanoparticles calcined at (a) 723 (b) 773 (c) 823 (d) 873 K.

3.7 Magnetic properties of MFe_2O_4 nanocrystal obtained VSM

The room temperature (300 K) magnetic properties of the prepared precursors and MFe_2O_4 nanoparticles calcined at different temperatures were investigated by the VSM technique in the range of approximately -15 to +15 kOe. Except for the precursors which were non-magnetic material, the calcined samples exhibited different magnetic behaviors. The room temperature magnetic behaviors of metal ferrite nanoparticles which fabricated by thermal treatment method, can be explained as the results of the four important factors: cationic distribution in spinel structure, the heating rate of calcinations, impurity phase of α-Fe_2O_3, and the surface spin structure of nanoparticles. Although all of these factors can be effective in magnetic behaviors but, their effects on the ferrite nanoparticles with different structures are not similar. We will have a discussion on this matter in next subsections.

3.7.1 Znic ferrite nanoparticles

Figure 25 shows the magnetization curves of precursor and $ZnFe_2O_4$ nanoparticles at (a) 723 (b) 773 (c) 823 and (d) 873 K. Their coercivity fields (H_c) are almost negligible, and all of them exhibit super-paramagnetic behaviours. Table 4 provides the values of saturation magnetization (M_s) of the calcined samples, along with calcinations temperatures and particle sizes. These data make it clear that different parameters were responsible for the saturation magnetization decreasing from 4.49 to 0.74 emu/g when the particle size increased from 17 to 31 nm. Cation inversion is one of the most important parameters that can be effective in the variation of the magnetic properties of zinc ferrite nanoparticles from the properties of the bulk form of the same material. In bulk form, zinc ferrite has an normal spinel structure in which all Zn^{2+} ions are in A sites and Fe^{3+} ions are distributed in B sites [47]. However, in bulk, zinc ferrite only occurs in intra-sub-lattice (B-B) exchange interactions, and it does not have intra-sub-lattice (A-A) exchange interactions or inter-sub-lattice (A-B) super-exchange interactions [48]. Inter-sub-lattice (A-B) super-exchange interactions of the cations are much stronger than the (A-A) and (B-B) interactions [18].Due to the cation inversion, which originates from thermal and mechanical treatment [40], the structure of $ZnFe_2O_4$ transfers from a normal spinel structure to a mixed spinel structure [48]. This cation inversion causes the zinc ferrite nanoparticles to experience inter-sub-lattice (A-B) super-exchange interactions and intra-sub-lattice (A-A) exchange interactions in addition to intra-sub-lattice (B-B) exchange interactions. But, due to the degree of inversion, which is large for smaller size particles, inter-sub-lattice (A-B) super-exchange interactions in smaller size particles occur to a greater extent than in larger size particles. Hence, saturation magnetization increases for smaller size particles [49], using Mossbauer's experiment, showed that the degree of inversion is large in the case of smaller size particles. Also, an impure α-Fe_2O_3 phase was detected by XRD (Figure20a), the heating rate of calcinations and the surface spin structure can be an influence that increases the saturation magnetization in smaller size particles [37].

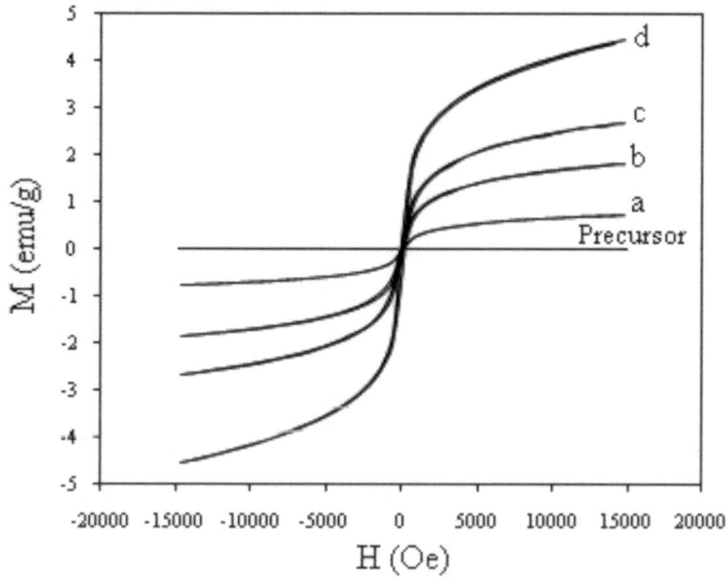

Fig. 25. Magnetization curves at room temperature for precursor and zinc ferrite nanoparticles calcined at (a) 723 (b) 773 (c) 823 (d) 873 K.

3.7.2 Manganese ferrite nanoparticles

Figure 26 shows the curves of magnetization of precursor and $MnFe_2O_4$ nanoparticles which exhibited a typical super-paramagnetic behavior. Table 4 depicts the values of saturation magnetization (M_s) of different samples. When the calcination temperature increased from 723 K to 873 K, the saturation magnetization increased from 3.06 to 15.78 emu/g. This can be attributed to spin canting and surface spin disorder that occurred in these nanoparticles [50].The interactions between the A and B sub-lattices in the spinel lattice system (AB_2O_4) consist of inter-sub-lattice (A-B) super-exchange interactions and intra-sub-lattice (A-A) and (B-B) exchange interactions. Inter-sub-lattice super-exchange interactions of the cations on the

(A-B) are much stronger than the (A-A) and (B-B) intra-sub-lattice exchange interactions [18, 51].As discussed earlier (Figure20b), by increasing the calcination temperature of the $MnFe_2O_4$ nanoparticles, Fe^{3+} ions transferred from B site to A site, so, consequently, the accumulation of Fe^{3+} ions increased in A site; however, the Fe_A^{3+}- Fe_B^{3+} super-exchange interactions increased (Fe_A^{3+}- Fe_B^{3+} interactions were twice as strong as the Mn_A^{2+}- Fe_B^{3+} interactions), and this can lead to an increase in saturation magnetization in $MnFe_2O_4$ nanoparticles [52]. Aslibeiki *et al.* [53] showed that saturation magnetization increases with increasing temperature and particle size in $MnFe_2O_4$ nanoparticles.It has been reported [54] that the spin disorder may occur on the surface of the nanoparticles as well as within the cores of the nanoparticles due to vacant sub-lattice disorder sites (Fe_A^{3+}) and poor crystal structure. The other point that is understood from Table 4 is that the values of saturation magnetization are expressively lower than those reported for the bulk $MnFe_2O_4$ (80 emu/g) [55]. The decrease in saturation magnetization of all the samples compared to that of the bulk is ascribed to the surface effects in these nanoparticles. The existence of an inactive magnetic layer or a disordered layer on the surfaces of the nanoparticles and the heating rate of calcinations can be due to the decrease of saturation magnetization compared to the bulk value [56, 57].

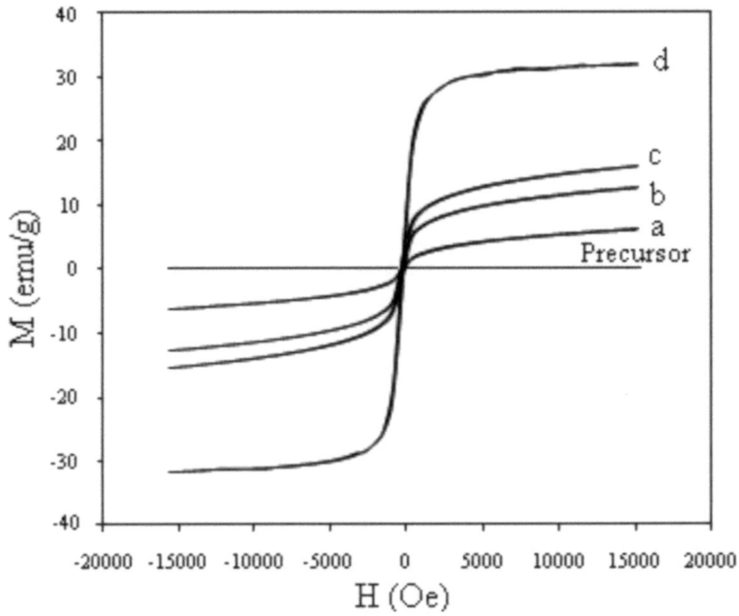

Fig. 26. Magnetization curves at room temperature for precursor and manganese ferrite nanoparticles calcined at (a) 723 (b) 773 (c) 823 (d) 873 K.

3.7.3 Nickel ferrite nanoparticles

Figure 27 (left) exhibits the hysteresis curves of precursor and $NiFe_2O_4$ nanoparticles which exhibited a typical ferromagnetic behavior. It can be seen from this figure that the magnetic properties of nanoparticles depended on the calcinations temperature. Saturation magnetization (M_s) values of 21.37, 26.67, 29.05 and 34.19 emu/g are observed for the nickel ferrite nanoparticles calcined at 723, 773, 823, and 873 K, respectively. There is a clear tendency of M_s increase with the enhancement of crystallinity of the $NiFe_2O_4$ nanoparticles.

As listed in Table 4, the values of M_s for the nickel ferrite nanoparticles were observed to increase with increasing temperature [58, 59]. The largest saturation magnetization was 34.19 emu/g for the sample calcined at 873 K, which is lower than that reported for the multi-domain, bulk nickel ferrite (55emu/g) [60].The decrease in saturation magnetization of these samples, compared to that of bulk material, depends on four factors explained in section 2.15. It seems that in inversed spinel ferrite nanoparticles such as nickel ferrite or cobalt ferrite nanoparticles which fabricated by thermal treatment method, the heating rate of calcination is more important than other parameters that can effectively increase or decrease the saturation magnetization.[35,61]. In our experiments, the heating rate of calcination was 10 K/min for nickel, zinc and manganese ferrite nanoparticles calcined at 723, 773, 823, and 873 K, which was a medium heating rate (Table 3). Therefore, it is possible that calcination at a slower heating rate would allow crystallization to be more complete and the magnetic phase could also increase, resulting in larger saturation magnetization. Sangmanee *et al.* [36] showed that saturation magnetization increases from 9.7 to 56.5 emu/g with decreasing the heating rate of calcination from 20 K/min to 5 K/min in cobalt ferrite nanostructures calcined at 773 K and fabricated by electrospinning. In addition, the appearance of the weakly-magnetic, impure phase of α-Fe_2O_3 (shown in Figure20c) can reduce the saturation magnetization [37].Variations of saturation magnetization with particle size for nickel ferrite nanoparticles are listed in Table 4. The saturation magnetization values of the calcined samples increase with increasing particle size, which may be attributed to the surface effects in these nanoparticles. The surface of the nanoparticles seems to be composed of some distorted or slanted spins that repel the core spins to align the field direction. Consequently, the saturation magnetization decreases for smaller sizes [62-64]. Furthermore, the surface is likely to behave as an inactive and dead layer with inconsiderable magnetization [56, 57]. The variation in the value of the saturation magnetization with particle size also can be resulted from the cation redistribution (interchanging of Ni and Fe ions of the tetrahedral and octahedral sites). This cation redistribution, causes that the structure of $NiFe_2O_4$ transfers from an inverse spinel structure

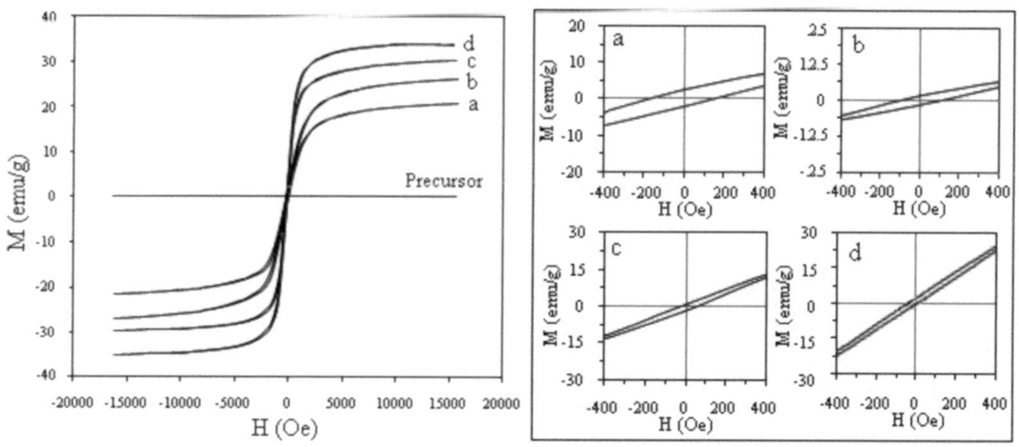

Fig. 27. (left) The magnetization curves of precursor and the nickel ferrite nanoparticles calcined at (a) 723 (b) 773 (c) 823 and (d) 873 K which measured at room temperature in the range of approximately −15 to +15 kOe. Figure27. (right) the expanded field region around the origin for clear visibility of the readers, in the range of approximately −400 to +400 Oe.

to a mixed spinel structure [65].Figure 27 (right) shows the expanded coercivity field (H_c) of region around of origin for clear visibility at room temperature in the range of approximately -400 to +400 Oe. The coercivity field values are listed in Table 4. These variations are not similar with saturation magnetization because, when the particle size increases from 15 to 69 nm, the coercivity field decreases from 150 to 32 Oe at room temperature. Variations of the coercivity field with particle size of nickel ferrite nanoparticles can be elucidated on the basis of domain structure, critical size, and the anisotropy of the crystal [16, 66-67].

Finaly, It is worth noting that the magnetic properties of similar ferrite nanoparticles of the same particle size differ depending on the preparation method used. Table 5 shows some literature values of M_s and H_c that were measured at similar conditions for some spinel ferrite nanoparticles. The data show that the pairs of similar spinel ferrite nanoparticles of the same particle size have different saturation magnetization values and coercivity fields. The results indicate that, in fact, the magnetic properties of ferrites are related primarily to the methods used to prepare them.

Specimens	Preparation method	Average particle size (nm)	Saturation magnetization Ms (emu/g)	Coercivity field Hc (Oe)	References
ZnFerrite	Combustion	20	4	Negligible	[68]
ZnFerrite	Modified sol	~20	1.4	156	[37]
NiFerrite	gel	~9	32.1	59	[69]
NiFerrite	Sol gel	9	11.9	Negligible	[70]
CoFerrite	Coprecipitation	30	77	2000-2700	[71]
CoFerrite	Mechanic alloying Hydrothermal	30	30	Negligible	[72]

Table 5. Magnetic properties of some spinel ferrite nanoparticles reported in the literatures which were measured at room temperature in range of approximately −10 to +10 kOe.

4. Conclusion

We have succeeded in fabricating spinel ferrite nanoparticles such as zinc ferrite ($ZnFe_2O_4$), manganese ferrite ($MnFe_2O_4$) and nickel ferrite ($NiFe_2O_4$) nanocrystals by a thermal treatment method utilizing only metal nitrates and as precursors, deionized water as a solvent and PVP as a capping agent.PVP played three crucial roles in synthesizing spinel ferrite nanoparticles, i.e., (1) the control of the growth of the nanoparticles; (2) the prevention of agglomeration of the nanoparticles; and (3) the production of nanoparticles that have a uniform distribution of shapes. The average particle sizes of metal ferrite nanoparticles were determined by TEM which increased with the calcinations temperature and they had good agreement with XRD results. FT-IR confirmed the presence of metal oxide bands at all temperatures and the absence of organic bands at 873 K for zinc and manganese ferrite nanoparticles and at 723 K for the nickel ferrite nanoparticles.VSM results demonstrated that zinc and manganese ferrite nanoparticles displayed super paramagnetic behaviors while nickel ferrite nanoparticles exhibited ferromagnetic behaviors. The present study also substantiated that, in ferrites, the values of the quantities that were acquired by VSM, such as saturation magnetization and coercivity field, are primarily dependent on the methods of preparation of the ferrites. This simple method, which is cost-effective and environmentally friendly, produces no toxic byproducts and can be used to fabricate pure, crystalline spinel metal ferrite nanocrystals.

Furthermore, this method can be extended to the synthesis of other spinel ferrite nanoparticles of interest.

5. References

[1] Gubin, S.P., Koksharov, Yu. A., Khomutov, G.B., Yurkov, G.Yu. (2005).Magnetic nanoparticles: preparation, structure and properties. Russian Chemical Reviews (74) 489 - 520.
[2] Baraton, M I. (2002).Synthesis, Functionalization, and Surface Treatment of Nanoparticles. University of Limoges, Franch.
[3] Callister, W. (2003). Materials science and engineering an introduction. Sixth ed. New York: JoHn Wiley & Sons, Inc.
[4] Spaldin, N. (2003). Magnetic materials: Fundamentals and device aplications. Cambridge: Cambridge University press.
[5] Goldman, A. (1990).Modern Ferrite Technology. Van Nostrand Reinhold, New York.
[6] Tilley, R.J.D (2004). Understanding solids: the science of materials. John Wiley and Sons, p 376.
[7] Winkler,G.(1971). Crystallography, chemistry and technology of ferrites; in: Magnetic properties of Materials, ed. J Smit, New York, McGraw-Hill.
[8] Bragg, W. H. (1915 a), Nature 95, 561.
[9] Bragg, W. H. (1915 b), Phil. Mag. 30, 305.
[10] Mehdiye, T.R., Gashimov, A.M., and Habibzade, A.A., (2008). Electromagnetic Processes in frequency-dependent resistor sheath.Fizika Cild Xiv №3 p 80-88.
[11] Gorter, E. W. (1954), Philips Res. Rep. 9, 321.
[12] Daliya S. M., and Juang, R. S. (2007). An overview of the structure and magnetism of spinel ferrite nanoparticles and their synthesis in microemulsions.Chemical Engineering Journal, 129 (1-3): 51–65.
[13] Stoner, E.C., Wohlfarth, E. P. (1948). A mechanism of magnetic hysteresis in heterogeneous alloys. IEEE Transactions on Magnetics, 27 (4): 3475 - 3518.
[14] Jiles, D. (1991). Introduction to Magnetism and Magnetic Materials. (First ed) Londond: Chapman & Hall.
[15] Bozorth, R. M. (1993). Ferromagnetism. Editorial Board ,New York.
[16] Cullity, B. D. (1978). Introduction to Magnetic Materials (2nd ed.). Addison-Wesley Publishing.
[17] Cao, X., and Gu, L. (2005). Spindly cobalt ferrite nanocrystals: preparation, characterization and magnetic properties. Nanotechnology, 16 (2): 180-185.
[18] Atif, M., Hasanian, S. K., and Nadeem, M. (2006). Magnetization of sol-gel prepared zinc ferrite nanoparticles: Effects of inversion and particle size. Solid State Communications, 138(8): 416-421.
[19] Jiang, J.Z., Wynn, P., Morup, S., Okada, T., and Berry, F.J. (1999). Nanostructured Materials, 12(5): 737-740.

[20] Shenoy,S.D., Joy, P.A., M.R. Anantharaman (2004). Effect of mechanical milling on the structural, magnetic and dielectric properties of coprecipitated ultrafine zinc ferrite Journal of Magnetism and Magnetic Materials (269) 217-226.
[21] Zhang,D.E. Zhang,X.J. Ni, X.M. Zheng, H.G. and Yang, D.D. (2005) Design and experiment of the self-propelled combine harvester for corn and stalk, J. Magn. Magn. Mater. (292 79-82.
[22] Li, H., Wu, H.Z., Xiao,G.X. (2010). Effects of synthetic conditions on particle size and magnetic properties of NiFe2O4, Powder Technol. (198) 157-166.
[23] Kale, A. Gubbala, S. R. Misra, D.K. (2004). Magnetic behavior of nanocrystalline nickel ferrite synthesized by the reverse micelle technique, J. Magn. Magn. Mater. (277) 350-358.
[24] Hochepied, J.P. Bonville, J.F., and Pileni, M.P.(2000). Non stoichiometric Zinc Ferrite nancrystals: Syntheses and magnetic properties J. Phys. Chem.; (104) 905-912.
[25] Sivakumar, P., Ramesh, R., Ramanand , A., Ponnusamy, S., and Muthamizhchelvan, C. (2011). Synthesis and characterization of NiFe2O4 nanosheet via polymer assisted co-precipitation method. Material. Letters, 65(3): 483-485.
[26] Koebel, M. M., Louis, C., Jones, and Gabor, A.S. (2008). Preparation of size-tunable, highly monodisperse PVP-protected Pt-nanoparticles by seed-mediated growth. Journal of Nanoparticle Research, 10(6): 1063-1069.
[27] Roosen, A.R., Carter. W.C. (1998). Simulations of microstructural evolution: Anisotropic growth and coarsening. Physica A, 261(1): 232-247.
[28] Ghosh, G., Naskar, M. K., Patra, A., and Chatterjee, M. (2006). Synthesis and characterization of PVP-encapsulated ZnS nanoparticles. Optical Materials 28(8-9): 1047-1053.
[29] Shao, H., Y. H., H. Lee, Y. J. Suh , C. O. Kim. (2006). Effect of PVP on the morphology of cobalt nanoparticles prepared by thermal decomposition of cobalt acetate. Current Applied Physics, 6(S1): e195-e197.
[30] Wen-yao, H., Guo.-cai, X. (2010). Characterization of nano-Ag/PVP composites synthesized via ultra-violet irradiation Journal of coal Science and Engineering, 16(2): 188-192.
[31] Tsuji, M., Hashimoto, M., Nishizawa, Y., and Tsuji, T. (2004). Synthesis of gold nanorods and nanowires by microwave-polyol method. Materials. Letters, 58(17-18): 2326-2330.
[32] Kumar, S.V., Musturappa, T.E., Prasannakumar, S., Mahadevan, K.M., Sherigara, B.S. (2007). N-Vinylpyrrolidone and ethoxyethyl methacrylate copolymer: synthesis, characterization and reactivity ratios. Journal of Macromolecular Science, Part A: Pure and Applied Chemistry, 44(11), 1161–1169.
[33] Lau, C., Mi, Y. (2002). A study of blending and complexation of poly (acrylic acid)/poly(vinyl pyrrolidone). Polymer, 43(3): 823–829.
[34] Silva, M.F., Silva,C.A., Fogo,F.C., Pineda,E.A.G., and Hechenleitner A.A.W. (2005). Thermal and FTIR study of polyvinylpyrrolidone/lignin blends. Journal of Therml Analysis and Calorimetry, 79(2): 367–370.
[35] Naseri, M. G. Saion, E.B. Abbastabar Ahangar, H. Hashim, M. and Shaari, A. H.(2011) Simple preparation and characterization of nickel ferrite nanocrystals by a thermal treatment method. Powder Technology 212, 80–88.
[36] Sangmanee, M., and Maensiri, S. (2009). Nanostructures and magnetic properties of cobalt ferrite ($CoFe_2O_4$) fabricated by electrospinning. Applied Physics A, 97(1): 167-177.

[37] Laokul, P., Amornkitbamrung, V., Seraphin ,S., and Maensiri, S. (2011). Characterization and magnetic properties of nanocrystalline $CuFe_2O_4$, $NiFe_2O_4$, $ZnFe_2O_4$ powders prepared by the Aloe vera extract solution. Current Applied Physics, 11(1): 101-108.

[38] L. Zhen, K. He, C.Y. Xu, W.Z. Shao (2008). Synthesis and characterization of single-crystalline $MnFe_2O_4$ nanorods via a surfactant-free hydrothermal route. Journal of Magnetism and Magnetic Materials 320 (2008) 2672– 2675.

[39] S. Bid, P. Sahu, S.K. Pradhan.(2007) Microstructure characterization of mechanosynthesized nanocrystalline $NiFe_2O_4$ by Rietveld's analysis. Physica E (39) 175–184.

[40] Singh, J. P., Srivastava, R. C., Agrawal, H. M., Kumar, R. (2010). Magnetic behaviour of nanosized zinc ferrite under heavy ion irradiation.Nuclear Instruments and Methods in Physics Research Section B: Beam Interactions with Materials and Atoms, 268(9): 1422-1426.

[41] B.D. Cullity, Elements of X-ray Diffraction, (2nd ed.), Addison-Wesley, London,1978, p.102.

[42] R.K. Selvan, C.O. Augustin, L.B. Berchmans, R. Sarawathi, (2003).Combustion synthesis of $CuFe_2O_4$ Mater. Res. Bull. (38) 41-45.

[43] M.I. Lorı´a-Bastarrachea,W. Herrera-Kao, J.V. Cauich-Rodrı´guez, J. M. Cervantes-Uc, H. Va´zquez-Torres, A.A vila-Ortega, J. Therm. Anal. Calorim. DOI 10.1007/s10973-010-1061-9, 2010

[44] Naseri , M. G. Saion,E. B. Ahangar, H. A. Hashim, M. Shaari, A.H. (2011) Synthesis and characterization of manganese ferrite nanoparticles by thermal treatment method. Journal of Magnetism and Magnetic Materials, (323) 1745-1749.

[45] Naseri , M. G. Saion,E. B. Hashim, M. Shaari, A. H. Ahangar, H. A. (2011). Synthesis and characterization of zinc ferrite nanoparticles by a thermal treatment method. Solid State Communications 151 (2011) 1031–1035.

[46] Qu, Y., Yang, Haibin, Yang, N., Fan, Y., Zhu, H., and Zou,G. (2006). The effect of reaction temperature on the particle size, structure and magnetic properties of coprecipitated $CoFe_2O_4$ nanoparticles. Materials. Letters, 60(29-30): 3548-3552.

[47] Mathew, D.S., and Juang, R. S. (2007). An overview of the structure and magnetism of spinel ferrite nanoparticles and their synthesis in microemulsions. Chemical Engineering Journal, 129(1-3): 51-65.

[48] Li, F.S., Wang, L., Wang, J.B., Zhou, Q.G., Zhou, X.Z., Kunkel, H.P., and Williams, G. (2004). Site preference of Fe in nanoparticles of ZnFe2O4. Journal of Magnetism and Magnetic Materials, 268(3): 332-339.

[49] Roy, M.K., Haldar, B., and Verma, H. C. (2006). Characteristic length scales of nanosize zinc ferrite. Nanotechnology, 17(1): 232-237.

[50] Gu, Z., Xiang, X., Fan, G., and Li, Feng. (2008). Facile synthesis and characterization of cobalt ferrite nanocrystals via a simple reduction-oxidation route. Journal of Physical Chemistry C, 112(47): 18459-18466.

[51] Ammar, S., Jouini, N., Fiévet, F., Beji, Z., Smiri, L., Moliné,P., Danot, M., and Grenèche, J.M. (2006). Magnetic properties of zinc ferrite nanoparticles synthesized by hydrolysis in a polyol medium. Journal of Physics: Condensed Matter, 18(39): 9055-9069.

[52] Jianjun, L., Hongming, Y., Guodong, L., Liu, Yanju, and Leng, J. (2010). Cation distribution dependence of magnetic properties of sol-gel prepared $MnFe_2O_4$ spinel ferrite nanoparticles Journal of Magnetism and Magnetic Materials,322(21): 3396-3400.

[53] Aslibeiki, B., H. Salamati, P. K., Eshraghi, M., and Tahmasebi, T. (2010). Superspin glass state in $MnFe_2O_4$ nanoparticles. Journal of Magnetism and Magnetic Materials, 322(19): 2929-2934.
[54] Morales,M.P., Veintemillas-Verdaguer, S., Montero, M.I., and Serna, C.J. (1999). Surface and Internal Spin Canting in γ-Fe_2O_3 Nanoparticles. Chemistry of. Materials, 11(11): 3058.
[55] Brabers, V. A. M. (1995). In Handbook of Magnetic Materials (Vol. 8). New York.
[56] Nogues, J., Sort, J., Langlias, V., Skumryev, V., Suriñach, S., Muñoz, J.S., and Baró, M.D. (2005). Exchange bias in nanostructures. Physics Reports, 422(3): 65-117.
[57] Maaz, K., Mumtaz, A., Hasanain, S.K., and Ceylan, A. (2007). Synthesis and magnetic properties of cobalt ferrite ($CoFe_2O_4$) nanoparticles prepared by wet chemical route. Journal of Magnetism and Magnetic Materials, 308(2): 289-295.
[58] Hanh, N., Quy, O. K., Thuy, N.P., Tung, L.D., and Spinu, L. (2003). Synthesis of cobalt ferrite nanocrystallites by the forced hydrolysis method and investigation of their magnetic properties. Physica B:Condensed Matter, 327(2-4): 382-384.
[59] Maensiri, S., Masingboon, C., Boonchom, B., and Seraphin, S. (2007). A simple route to synthesize nickel ferrite ($NiFe_2O_4$) nanoparticles using egg white. Scripta Materialia, 56(9): 797-800.
[60] Smit, J., and Wijn, H. P. J. (1959). Ferrites-physical properties of ferromagnetic oxides in relation to their technical applications. New York: Wiley.
[61] Naseri, M. G., Saion, E. B., Ahangar, H. A., Shaari, A. H. and Hashim, M.(2010). Simple Synthesis and Characterization of Cobalt Ferrite Nanoparticles by a Thermal TreatmentMethod Journal of Nanomaterials (2010) 1-8.
[62] Berkowitz, A.E., Lahut, J. A., and VanBuren, C.E. (1980). Properties of Magnetic Fluid Particles. IEEE Transactions on Magnetics, MAG-16(2): 184-190.
[63] Berkowitz, A.E., Lahut, J. A., Jacobs, I.S., Levinson, L.M., and Forester, D.W. (1975). Spin pinning at ferrite-organic interfaces. Physical Review Letters, 34(10): 594-597.
[64] Coey, J. M. D. (1971). Noncollinear spin arrangement in ultrafine ferrimagnetic crystallites. Physical. Review. Letters, 27(17): 1140-1142.
[65] Jacob, J., and Khadar., M. A. (2010). Investigation of mixed spinel structure of nanostructured nickel ferrite. Journal of. Applied. Physics 107(11): 114310-114320.
[66] Hadjipanayis, G.C., and Siegel, R.W. (1994). Nanophase Materials-Synthesis-Properties-Applications. Kluwer Academic Publishers, Drodrecht.
[67] Chikazumi, S. (1959). Physics of Magnetism. New York: John Wiley.
[68] Ammar, S., Jouini, N., Fiévet, F., Beji, Z., Smiri, L., Moliné,P., Danot, M., and Grenèche, J.M. (2006). Magnetic properties of zinc ferrite nanoparticles synthesized by hydrolysis in a polyol medium. *Journal of Physics: Condensed Matter, 18*(39): 9055-9069.
[69] George, M., John, A.M., Nair, S.S., Joy, P.A., and Anantharaman, M.R.(2006). Finite size effects on the structural and magnetic properties of sol–gel synthesized $NiFe_2O_4$ powders J. Magn. Magn.Mater. (302) 190–195.
[70] Sepelak, V., Menzel, M., Bergmann, I. Wiebcke, M., Krumeich, F., and Becker, K.D. (2004). Structural and magnetic properties of nanosize mechanosynthesized nickel ferrite. Journal of Magnetism and Magnetic Materials, 272-276(2): 1616-1618.
[71] Zhao, D., Wu, X., Guan, H., and Han, E. (2007). Study on supercritical hydrothermal synthesis of $CoFe_2O_4$ nanoparticles. Journal of Supercritical Fluids, 42(2): 226-233.
[72] Chaoquan, H., Zhenghong, G., and Xiaorui, Y. (2008). One-pot low temperature synthesis of MFe_2O_4 (M = Co, Ni, Zn) superparamagnetic nanocrystals. Journal of Magnetism and Magnetic Materials, 320(8): L70-L73.

Section 4

Bulk Crystallization from Aqueous Solutions

15

Separation of Uranyl Nitrate Hexahydrate Crystal from Dissolver Solution of Irradiated Fast Neutron Reactor Fuel

Masaumi Nakahara
Japan Atomic Energy Agency, Nuclear Fuel Cycle Engineering Laboratories
Japan

1. Introduction

Batch crystallization is widely used for the separation and high purification of organic and inorganic materials in the fine chemical, food, pharmaceutical and biochemical industries. In the atomic power industry, application of crystallization to U purification of the Plutonium Uranium Reduction Extraction (PUREX) first cycle product was attempted in Kernforschungszentrum Karlsruhe (KfK), Germany (Ebert et al., 1989). The feed solution had 240−480 g/dm^3 U concentration and 0.1 g/dm^3 fission products (FPs) concentration in 5−6 mol/dm^3 HNO_3 solution. Reducing conditions were achieved with 2.4 g/dm^3 of U(IV) which was added to change the Pu valence to Pu(IV) which was required for good separation of Pu from U. In a six-stage cascade crystallizer, the feed solution was cooled down in steps from 30 to −30°C in the course of about 30 min. More than 90% of U was recovered in form of uranyl nitrate hexahydrate (UNH) crystals with an average diameter of 0.2 mm, while a much greater proportion of the transuranium (TRU) elements and FPs remained in the mother liquor. The decontamination factors (DFs) of several of the FPs were determined for one crystal step plus several crystal washing operations. The measured DFs of Pu and Cs were 10^2 and 10^3, respectively.

An advanced aqueous reprocessing for a fast neutron reactor fuel cycle named "New Extraction System for TRU Recovery (NEXT)" has been proposed as one fast neutron reactor fuel reprocessing method (Koyama et al., 2009) and is being developed in Japan Atomic Energy Agency (JAEA). On the advanced aqueous reprocessing for fast neutron reactor fuel cycle, it is supposed to recover not only U and Pu but also minor actinides (MAs; Np, Am and Cm) for the efficient utilization of resources. It will be also effective in decreasing the environmental impact because of their long half-life and high radiotoxicity. These elements are loaded in a fast neutron reactor and are burned as core fuel. Figure 1 shows schematic diagram of NEXT process for fast neutron fuel reprocessing. The NEXT consists of highly efficient dissolution of fuel with HNO_3 solution (Katsurai et al., 2009), U crystallization for partial U recovery (Shibata et al., 2009), simplified solvent extraction for U, Pu and Np co-recovery using tri-*n*-butyl phosphate (TBP) as an extractant (Sano et al., 2009), and extraction chromatography for mutual separation of actinide elements and lanthanide elements from a raffinate (Koma et al., 2009). The powdered fuel was dissolved by the highly efficient dissolution process and the dissolver solution was adjusted to high heavy metal concentration. Then, U is recovered as UNH crystals from dissolver solution derived from fast neutron reactor fuel. Since the amount throughput will be reduced in the simplified solvent extraction process, the adoption of the crystallization process is expected to reduce the radioactive waste, equipment, and hot cell volume. In addition, U/Pu ratio in the dissolver solution is adjusted in the crystallization process to be a suitable Pu content for core fuel fabrication. In the NEXT, Np is changed to Np(VI) in the high HNO_3 concentration feed solution and is co-extracted with U and Pu in the simplified solvent extraction system. The FPs in the raffinate obtained from the simplified solvent extraction process is removed using *N,N,N′,N′*-tetraoctyl-3-oxapentane-1,5-diamide (TODGA) absorbent in the extraction chromatography I. The actinide elements such as Am and Cm is recovered from the solution containing actinide and lanthanide elements by chromatography with 2,6-bis-(5,6-dialkyl-1,2,4-triazine-3-yl)pyridine (R-BTP) absorbent in the extraction chromatography II.

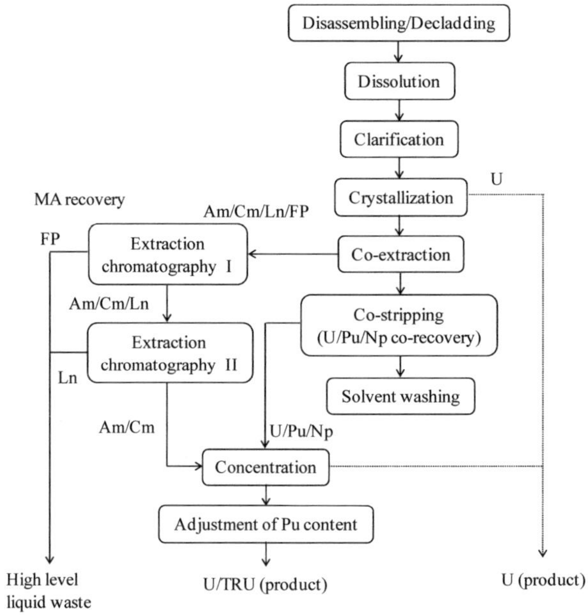

Fig. 1. Schematic diagram of the NEXT process

A dissolver solution of irradiated fast neutron reactor mixed oxide (MOX) fuel in JAEA contains a number of TRU elements and FPs than in KfK. Since U is used as blanket fuel and TRU elements are supposed to recover by other chemical process, it is need to remove TRU elements and FPs from UNH crystals in the U crystallization process. It would be also bring about reduction in the cost for the recovered U storage and the blanket fuel fabrication due to decreased radiation shielding. Therefore, the behavior of TRU elements and FPs in the U crystallization process must be confirmed experimentally.

Since U is recovered as UNH crystal for a blanket fuel fabrication in the U crystallization process, the crystal ratio of U should be evaluated with a dissolver solution of irradiated fast neutron reactor. The crystal ratio of UNH affects HNO_3 concentration in the feed solution. In this study, the feed solution was changed in HNO_3 concentration and the influence on the UNH crystal ratio was examined in the cooling batch crystallization. Two experiments, crystal ratio and the co-existing element behavior, were carried out with a dissolver solution derived from irradiated fast neutron reactor "JOYO" core fuel in a hot cell of the Chemical Processing Facility (CPF), JAEA. Additionally, current status of crystallization apparatus and crystal purification method for the NEXT is described in this paper.

2. Principal of uranium crystallization

In a HNO_3 solution, U ions are crystallized as UNH by the following reaction.

$$UO_2^{2+} + 2NO_3^- + 6H_2O \leftrightarrow UO_2(NO_3)_2 \cdot 6H_2O \quad (1)$$

Figure 2 shows the solubility curves of U in HNO_3 solution (Hart & Morris, 1958). The results represent the mean of two temperatures observed for the first formation and final disappearance of crystal on, respectively, slowly cooling and warming solutions with vigorous agitation. The U ions concentration decreases with decreasing temperature in the solution before reaching the eutectic point, where H_2O and HNO_3 start to crystallize. Thus, U crystallization process should be performed in the right region of the minimum point in this figure. A high HNO_3 solution is desirable for achieving a low U concentration in the solution, therefore yielding more UNH crystals because the eutectic point shifts from right to left as the HNO_3 concentration increases.

In $Pu(NO_3)_4$-HNO_3-H_2O system, the crystallization behavior of plutonyl nitrate hexahydrate (PuNH) was examined. Figure 3 shows the solubility curves of Pu in HNO_3 solution (Yano et al., 2004). The Pu solution was prepared by dissolving PuO_2 powder with 4 mol/dm^3 HNO_3 solution containing 0.05 mol/dm^3 $AgNO_3$ electrochemically. In the experiments, the Pu valence was adjusted as following methods. The valence of Pu was changed to Pu(IV) with a few drops of 100% H_2O_2. On the other hand, the Pu solution was oxidized to Pu(VI) by Ag^{2+} ion and Ag in the solution was separated by ion exchange. The Pu solution was cooled quickly to −20°C and then cooled at −1 °C/min to −55°C. In the Pu(IV) solution appeared to be a green quasi-liquid (crystals in liquid). In all runs, PuNH was not crystallized in the experimental conditions but crystals of H_2O and $HNO_3 \cdot 3H_2O$ were observed. In the NEXT, PuNH would not precipitate solely in the U crystallization process.

The influence of Pu valence in the feed solution was examined in the U crystallization process (Yano et al., 2004). When Pu(IV) existed in the feed solution, the yellow crystal was observed. On the other hand, the appearance of the crystal was orange in the feed solution adjusted so that Pu valence was Pu(VI), this color likely resulting from the mixture of the yellow crystal of UNH and the red crystal of PuNH. Plutonium(VI) in the feed solution was co-crystallized with U(VI) in the course of U crystallization. The crystal yields of Pu were smaller than those of U (Ohyama et al., 2005). The fact that the crystal ratio of Pu is smaller than that of U suggests a mechanism of U-Pu co-crystallization in which U begins to crystallize when the saturation point of U is reached by cooling the feed solution, and then Pu is crystallized on the UNH crystal.

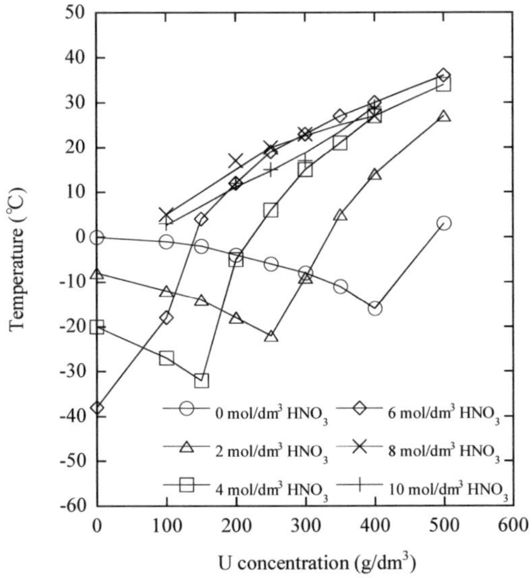

Fig. 2. Solubility of U in HNO_3 solution

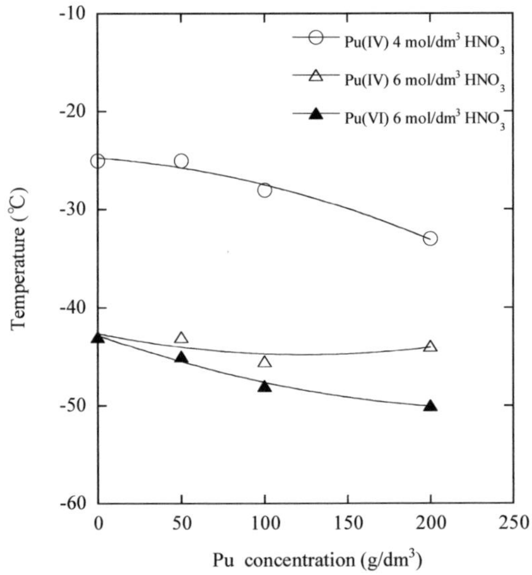

Fig. 3. Solubility of Pu in HNO_3 solution

3. Batch crystallization with dissolver solution of irradiated fast neutron reactor fuel

3.1 Experimental procedure

In the experiments, HNO_3 from Wako Pure Chemical Industries, Ltd., was used without further purification. Irradiated core fuel of the fast neutron reactor "JOYO" Mk-III with an averaged burnup 53 GWd/t and cooling time of 2 y was used for the U crystallization experiments. The sheared pieces of core fuel comprising 166 g of heavy metal were dissolved with 325 cm³ of 8 mol/dm³ HNO_3 solution at 95°C. The valence of Pu in the dissolver solution was changed to Pu(IV) by NO_x gas bubbling. In the crystal ratio experiments, the U and Pu concentrations in the feed solution were approximately 450 and 50 g/dm³, respectively. In co-existing element behavior experiment, the HNO_3 concentration in the feed solution was 4.5 mol/dm³ in the U crystallization process. The $CsNO_3$ solution was prepared by dissolving $CsNO_3$ (Wako Pure Chemical Industries, Ltd.) powder in 2 mol/dm³ HNO_3 solution, and added to the dissolver solution. The Cs concentration was 4.0 g/dm³ in the feed solution.

A schematic diagram of the batch cooling crystallizer is shown in Figure 4. The crystallizer made from Pyrex glass was used for cooling the solution volume capacity was 200 cm³, and it had a cooling jacket for cooling and heating media whose temperature was controlled by a thermostat. The feed solution was placed in the crystal vessel and was initially maintained at about 50°C. The feed solution was cooled from 50 to 4°C while being stirred. The spontaneously nucleated and grown crystalline particles were quickly centrifuged from the mother liquor at 3000 rpm for 20 min. After solid-liquid separation, the UNH crystals were washed using 8 mol/dm³ HNO_3 solution at 4°C and then centrifuged at 3000 rpm for 20 min.

The acidity of the solution was determined by acid-base titration (COM-2500, Hiranuma Sangyo Co., Ltd.) and the Pu valence in the feed solution was confirmed as Pu(IV) by optical spectrometry (V-570DS, JASCO Corporation) of the ultraviolet (UV)-visible region. The U and Pu concentrations were measured by colorimetry. The concentrations of Np, Am and Cm were measured by α-ray spectrometry (CU017-450-100: detector and NS920-8MCA: pulse height analyzer, ORTEC). The FPs concentrations were analyzed by γ-ray spectrometry (GEN10: detector and 92XMCA: pulse height analyzer, ORTEC) and inductively coupled plasma atomic emission spectrometry (ICP-AES; ICPS-7500, Shimadzu Corporation).

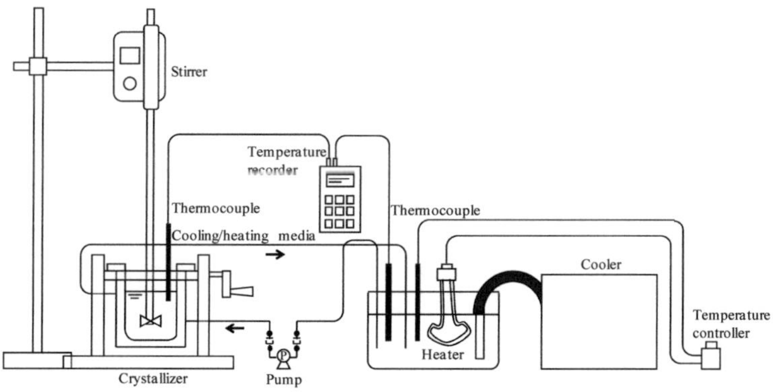

Fig. 4. Schematic diagram of the batch cooling crystallizer

3.2 Crystal ratio of uranyl nitrate hexahydrate

The cooling curve in the U crystallization is shown in Figure 5. The feed solution was placed in the crystallizer and cooled 45.0 to 3.3°C over 150 min. When the temperature of the feed solution reached 23.8°C at 74 min, a small increase in temperature was observed in the feed solution. This indicates the start of crystallization, where heat is released by nucleation.

Figure 6 shows the appearance of UNH crystal recovered the dissolver solution of irradiated fast reactor "JOYO" Mk-III core fuel. After crystal washing, lemon yellow crystals were obtained on a filter.

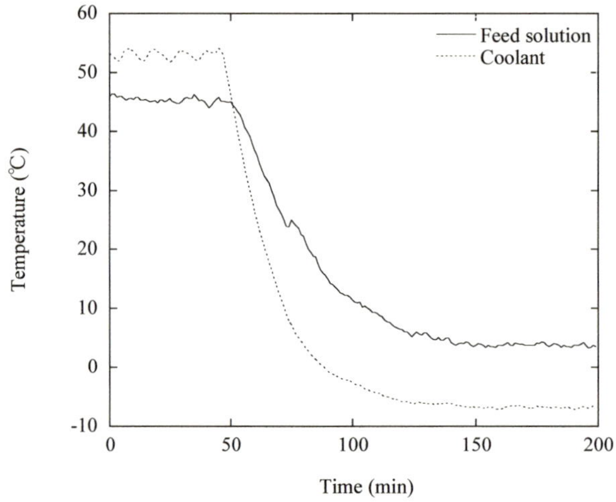

Fig. 5. Cooling curve of feed solution during U crystallization

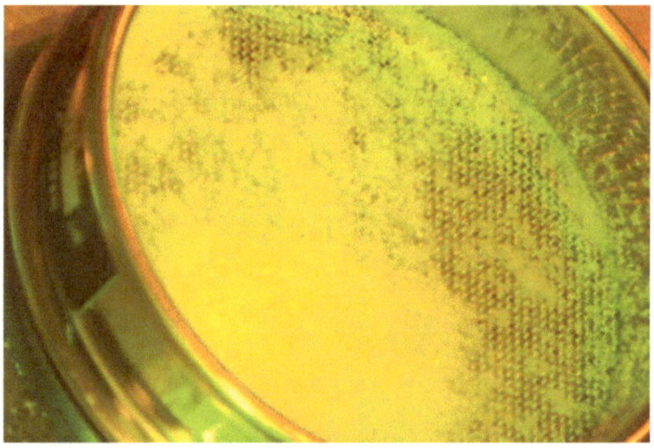

Fig. 6. Appearance of UNH crystal after crystal washing

The crystal ratio of UNH in the dissolver solution of irradiated fast neutron reactor fuel was examined by changing in HNO_3 solution in the feed solution. The crystal ratio, $R_{c,j}$, is calculated by the following equation.

$$R_{c,j} = 1 - \frac{C_{F,H^+} C_{M,j}}{C_{M,H^+} C_{F,j}} \quad (2)$$

where C_{F,H^+} and C_{M,H^+} are H^+ concentration in the feed solution and mother liquor, respectively, and $C_{F,j}$ and $C_{M,j}$ are metal j concentration in the feed solution and mother liquor, respectively. Figure 7 shows the relationship between HNO_3 concentration in the feed solution and UNH crystal ratio in the batch cooling crystallization process. These experimental results show high HNO_3 concentration in the feed solution increased with increasing the UNH crystal ratio in the batch cooling crystallization process. These results were in agreement with the reported experimental data (Hart & Morris, 1958).

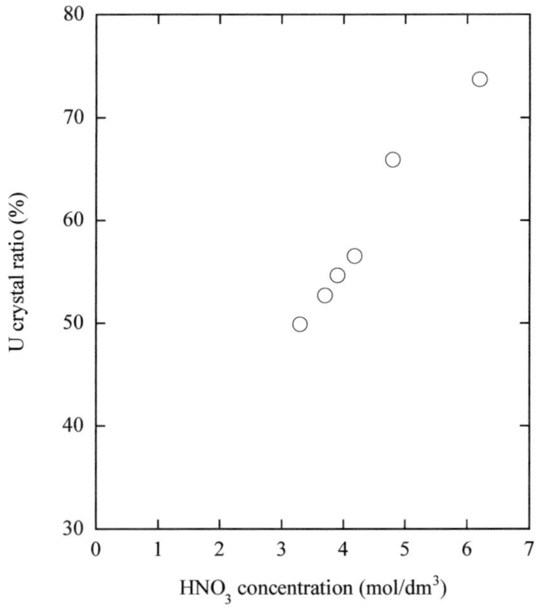

Fig. 7. Relationship between HNO_3 concentration in the feed solution and U crystal ratio

Element	Feed solution (g/dm³)	Decontamination factor (−) Before washing	After washing
U	4.18×10^{2}	–	–
Pu	4.00×10^{1}	8.5	24
Ba	3.70×10^{-2}	3.9	4.7
Nuclide	Feed solution (Bq/cm³)	Decontamination factor (−) Before washing	After washing
^{95}Zr	1.84×10^{7}	7.5	46
^{95}Nb	3.32×10^{7}	8.2	55
^{106}Ru	3.18×10^{8}	13	79
^{125}Sb	6.56×10^{7}	12	129
^{137}Cs	1.04×10^{9}	2.2	3.4
^{144}Ce	2.42×10^{9}	15	164
^{144}Pr	2.42×10^{9}	15	164
^{155}Eu	7.72×10^{7}	14	118
^{237}Np	2.38×10^{4}	3.3	6.7
^{241}Am	9.02×10^{8}	9.8	109
^{242}Cm	2.20×10^{7}	8.2	115

Table 1. Composition of the feed solution and DFs of metals for UNH crystal

3.3 Behavior of transuranium elements and fission products in uranium crystallization process

Table 1 summarizes the composition of feed solution and the DFs of metals in the U crystallization process. The DFs of metals, $\beta_{c,j}$, are calculated by the following equation.

$$\beta_{c,j} = \frac{\frac{C_{F,j}}{C_{F,\mathrm{U}}}}{\frac{C_{P,j}}{C_{P,\mathrm{U}}}} \tag{3}$$

where $C_{F,\mathrm{U}}$ is U concentration in the feed solution, and $C_{P,\mathrm{U}}$ and $C_{P,j}$ are U and metal j concentrations in the UNH crystal, respectively.

Plutonium behavior in the U crystallization process depends on the Pu valence in the feed solution. In this study, the Pu valence in the feed solution was changed to Pu(IV) by NO_x gas bubbling. After the crystallization, almost all the Pu remained in the mother liquor and attached to the surface of the UNH crystal. The mother liquor on the surface of crystal was efficiently removed after the UNH crystal was washed.

The DF of Np was 3.3 and 6.7 before and after washing, respectively. These experimental results implied Np was present in the form of solid impurities in the mother liquor because its DF was not improved by the crystal washing. Generally, Np can exist simultaneously in three stable oxidation state; Np(IV), Np(V) and Np(VI), in a HNO_3 solution. The oxidation states of Np are interconvertible in HNO_3 medium and exhibit different behavior in the reprocessing. Its valence is strongly affected by oxidation and reduction reactions with agents used in the reprocessing and other co-existing ions. All the kinetics of the oxidation and reduction reactions is not elucidated. One of these reactions is the oxidation of Np(V) by HNO_3, which is the principal influential reaction as follows.

$$2\mathrm{NpO_2^+} + 3\mathrm{H^+} + \mathrm{NO_3^-} \leftrightarrow 2\mathrm{NpO_2^{2+}} + \mathrm{HNO_2} + \mathrm{H_2O} \tag{4}$$

In this reaction, HNO_2 plays the important role of oxidation and reduction between Np(V) and Np(VI). This reaction shows that higher HNO_3 and lower HNO_2 concentrations bring about more oxidation of Np(V) to Np(VI). When the U ions crystallize as UNH in HNO_3 solution, it requires a certain amount of H_2O. As a result, the HNO_3 concentration of the mother liquor is higher than that of the feed solution. It brings about more oxidation of Np(V) to Np(VI) in the mother liquor. When the Pu valence is changed to Pu(VI), Pu(VI) is co-crystallize with U(VI). The chemical behavior of Np(VI) is similar to that of Pu(VI), and it is likely to co-crystallize with U(VI) in the course of U crystallization. Since the Np was incorporated into the UNH crystal, it is difficult to remove from the UNH by the crystal washing operation. If the Np valence is adjusted to Np(IV) or Np(V), the behavior of Np would be different from that of Np(VI). The addition of reductant agent, e.g., U(IV), is effective for preventing from the Np oxidation to Np(VI). Thereby, the Np might remain in the mother liquor after cooling the feed solution.

Americium and Cm are supposed to recover by an extraction chromatography process in the NEXT, and are desired to remain in the mother liquor in the U crystallization process.

The experimental results indicated the DFs of Am and Cm for the UNH crystal were 9.8 and 8.2 before washing, respectively. These elements remained in the mother liquor and attached on the surface of the UNH crystal. The adhesion of liquid impurities was washed away with HNO_3 solution. After washing, the DFs of Am and Cm for the UNH crystal increased by a factor of 109 and 115, respectively.

In the experiments, the behavior of Cs was evaluated in the U crystallization process. The DF of Cs showed 2.2 and 3.4 before and after washing. It is reported that alkali metals react with tetravalent actinide elements and form a double salt in a HNO_3 solution (Staritzky & Truitt, 1949). The reaction of Cs and Pu(IV) is expressed by the following equation.

$$2Cs^{+} + Pu(NO_3)_6^{2-} \leftrightarrow Cs_2Pu(NO_3)_6 \tag{5}$$

This reaction indicates that an abundance of $Pu(NO_3)_6^{2-}$ is advantageous for forming of $Cs_2Pu(NO_3)_6$. Figure 8 show the abundance ratio of $Pu(NO_3)_6^{2-}$ in HNO_3 solution (Ryan, 1960). The abundance ratio of $Pu(NO_3)_6^{2-}$ increases with an increase in HNO_3 concentration. In the crystal growth of UNH, a certain amount of H_2O molecules is needed in the feed solution. The mother liquor of HNO_3 concentration is higher than that of feed solution after the U crystallization. Therefore, the formation of $Cs_2Pu(NO_3)_6$ is easy in the course of U crystallization. Anderson reported that double salt of Pu nitrate, $(C_9H_7NH)_2Pu(NO_3)_6$, $Rb_2Pu(NO_3)_6$, $Tl_2Pu(NO_3)_6$, $K_2Pu(NO_3)_6$, $(C_5H_5NH)_2Pu(NO_3)_6$ in addition to $Cs_2Pu(NO_3)_6$ (Anderson, 1949). These materials are less in a dissolver solution of irradiated fast neutron reactor fuel. However, further investigation concerning the double salt of Pu nitrate for the U crystallization process will required experimentally.

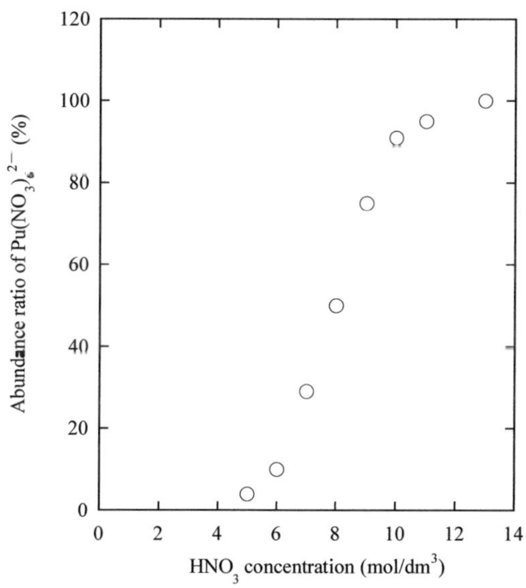

Fig. 8. Abundance ratio of $Pu(NO_3)_6^{2-}$ in HNO_3 solution

Among alkali earth metals, Ba behavior is examined in the cooling batch crystallization. The DFs of Ba was 4.7 after the crystal washing. In the precipitates formation experiments, $Ba_{0.5}Sr_{0.5}(NO_3)_2$ was observed using simulated high level liquid waste solution (Izumida &

Kawamura, 1990). On the other hand, the DF of Sr was decontaminated with the uranyl nitrate solution containing Sr and Ba after the UNH crystal was washed (Kusama et al., 2005). The solubility of Ba is 0.4 g/dm^3 in 400 g/dm^3 of uranyl nitrate solution with 5 mol/dm^3 HNO_3. Therefore, Ba is assumed to precipitate as $Ba(NO_3)_2$ in the U crystallization process. The solid impurities are not removed by the crystal washing with a HNO_3 solution.

Insoluble residues consist of Zr and the elements of platinum group such as Ru in a nuclear fuel reprocessing. Zirconium and Mo precipitate in the form of zirconium molybdate in the dissolution process (Adachi et al., 1990; Lausch et al., 1994; Usami et al., 2010). This compound tends to form at high temperature and with low HNO_3 concentration in the solution. The DF of Zr was high after the crystal washing and Zr remained in the mother liquor. The feed solution was cooled and the acidity in the feed solution increases in the course of U crystallization. Therefore, zirconium molybdate would be difficult to crystallize at low temperature and with high HNO_3 concentration in the mother liquor.

The behavior of Ce, Pr and Eu in the rare earth elements was evaluated in the U crystallization experiments. The DFs of these elements achieved to approximately 10^2 after the crystal washing. Their solubility in the HNO_3 solution was so high that there was no precipitation as solid impurities. Therefore, they remained in the mother liquor during the U crystallization and these elements in the mother liquor that was attached to the surface of the UNH crystal were washed away with the HNO_3 solution.

4. Crystallization apparatus for continuous operation

4.1 Concept of crystallization apparatus

The crystallizer is designed for continuous operation adoption high throughput and equipment scale-up and is developed for the U crystallization in the NEXT (Washiya et al., 2010). Figure 9 shows a schematic diagram of annular type continuous crystallizer. A rotary-driven cylinder has a screw blade to transfer UNH crystal slurry and annular shaped space is formed as crystallization section in between the rotary cylinder and outer cylinder. A dissolver solution of irradiated fast neutron reactor fuel is fed into the annular section from the lower part of the equipment, and the coolant is supplied into the cooling jacket located on the outside cylinder. The dissolver solution is cooled down gradually and is transferred to the outlet in the upper side of the equipment. The UNH slurry is obtained in the annular section and is discharged by the guide blade attached to the rotary cylinder. The mother liquor is separated from the UNH crystal and is discharged from the nozzle located in the more upper side than solution level. The discharged UNH slurry is still accompanied with a little solution. Hence, it needs to be dried by the crystal separator as centrifugal dewatering process.

4.2 Continuous operation with uranyl nitrate solution using annular type continuous crystallizer

The continuous operation experiments were evaluated with a uranyl nitrate solution (Washiya et al., 2010). The feed solution of 450 g/dm^3 U concentration in 5 mol/dm^3 HNO_3 was cooled at 0°C at 20 rph. Afterwards, amount of the crystal stay was increased gradually, and it reached to steady state in 1−2 h. The moisture content in the UNH slurry obtained from the outlet of the slurry was about 40%. The UNH crystal size was approximately 600 μm and is considered to appropriate size for solid-liquid separation. The U concentration in the mother liquor was reached to steady state within 2 h, and crystal ratio of U was about 84%.

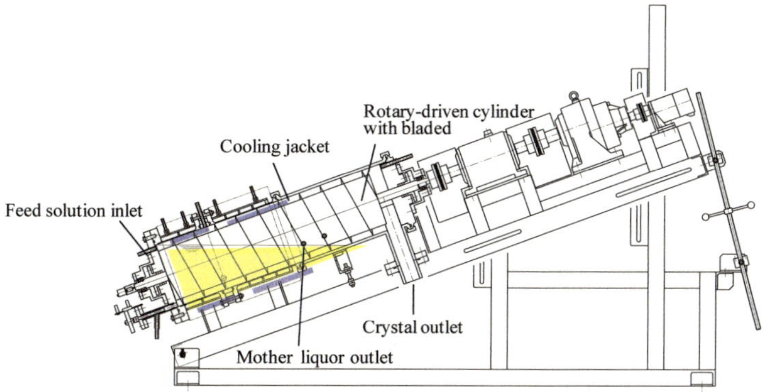

Fig. 9. Schematic diagram of annular type continuous crystallizer

To extract operational failure events comprehensively concerning to the U crystallizer and to clarify their importance, failure mode analysis was carried out by applying Failure Mode and Effects Analysis (FMEA). Significant failure events were identified with failure causes, their effects and probability of these failures were predicted by making use of operation experience. All failure events were evaluated by cause, primary and secondary effects and scored them. As results, crystal accumulation, blockage of mother liquor discharge nozzle, blockage of crystal discharge nozzle were selected as important specific failure events. To investigate how to detect non-steady condition, these three experiments were carried out with screw rotation speed decline, crystal outlet blockage and mother liquor outlet blockage. Also, the resume procedure after non-steady state was examined sequentially to consider countermeasures for each non-steady event (Shibata et al., 2009). The accumulation of UNH crystals can be detected by the torque of the cylinder screw, the liquid level in the annular section and other instruments. These experimental results show that it is possible to recover from non-steady state when the cause of the phenomena such as blockage of crystal outlet is removed by an appropriate operation. The fundamental performance of crystallizer annular type was investigated with uranyl nitrate solution. The experiment will be carried out to confirm the system performance on integrated crystallization system consisting of the engineering-scale crystallizer, crystal separator and related systems.

5. Purification of uranyl nitrate hexahydrate crystal product

5.1 Principal of crystal purification

Generally, crystalline particles produced in crystallizers are often contaminated by the mother liquor which appears on the surface or inside the bodies of the crystals. The UNH crystal recovered from a dissolver solution of irradiated fast neutron reactor fuel is washed away with a HNO_3 solution. Although the TRU elements and FPs on the surface of the UNH crystal are decontaminated by the crystal washing, the inclusions within the crystal and the solid impurities are not removed from the UNH crystal. Therefore, crystal purification method is studied for the purpose of further increasing decontamination performance. One crystal purification method, the grown crystalline particles are purified by heating up to as high as the melting point of the crystal and introducing the mother liquor to the outside of the crystal, which is exhaled along defects and grain boundaries (Zief & Wilcox, 1967; Matsuoka & Sumitani, 1988). This phenomenon is called "sweating" and is applied to organics and metals. The mother liquor and melt in the crystal are discharged by Ostwald ripening and increase in the internal pressure (Matsuoka et al., 1986). The incorporated liquid is expelled from grooves along defects and grain boundaries. It was reported that countless grooves were observed in the organic crystal after sweating (Matsuoka & Sumitani, 1988). The purification of the *p*-dichlorobenzene (*p*-DCB) and *m*-chloronitrobenzene (*m*-CNB) crystalline particles by sweating was experimentally investigated (Matsuoka et al., 1986). The purity of 99.99% was obtained by a single sweating stage at temperatures about 1°C below the melting points of the pure crystals after the duration of 90 or 120 min of sweating. In the batch operation, the UNH crystal purification experiments were carried out with the dissolver solution of MOX fuel containing simulated FPs (Nakahara et al., 2011). Although the DFs of solid impurities such as $Ba(NO_3)_2$ and $Cs_2Pu(NO_3)_6$ did not change in the sweating process, that of Eu increased with increases in temperature and time. In the batch experiments, the DF of Eu increased to approximately 2.4 times after 30 min at 60°C. There results indicated that liquid impurities such as Eu were effectively removed by the sweating method, but solid impurities such as Pu, Cs and Ba were minimally affected in the batch experiments.

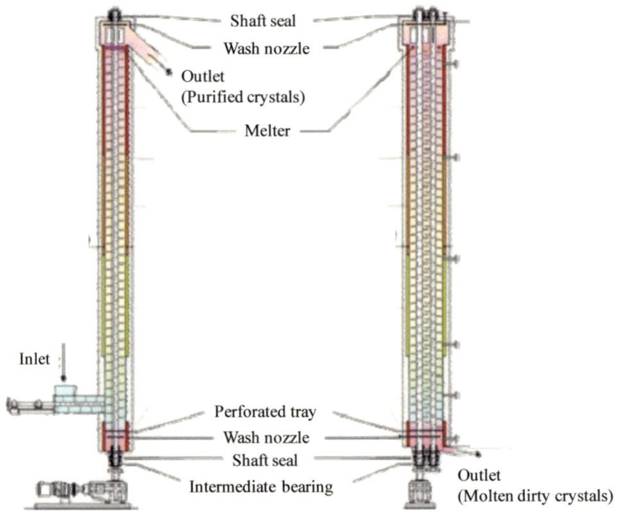

Fig. 10. Schematic diagram of KCP

5.2 Concept of crystal purification apparatus

The crystal purification apparatus, Kureha Crystal Purifier (KCP), has been applied in industrial plants using organic matter (Otawara & Matsuoka, 2002). The schematic diagram of KCP is shown in Figure 10. The apparatus has been developed in the following fashion: feed stock is charged as solids at the bottom of the column, the heating unit is set at the top of the column, and it is possible to contact the melt with crude crystal countercurrently. The KCP features high purity, high yield, energy savings, little maintenance, and a long, stable operation (Otawara & Matsuoka, 2002). The crude crystal is supplied at the bottom of the column and then is carried to the upper side of a column by a double screw conveyor, and then part of the crystal is molten by a heating unit at the top of the column and the melt trickles downward among the crude crystal. The apparatus performs countercurrent contact between the crystal and reflux melts in the course of being conveyed upward, and the crude crystal is washed by a portion of the melt. Therefore, higher DFs of liquid impurities will be obtained by the KCP, because the liquid impurities were washed with melt in addition to sweating effect. The pure crystal product exits from the top of the column.

5.3 Continuous operation with uranyl nitrate solution using kureha crystal purifier

The crystal purification experiments with the KCP were carried out using the UNH crystal recovered from uranyl nitrate solution containing Sr of SUS304L (Yano et al., 2009). Although the DF of liquid impurities, Eu, in the static system was approximately 2.4 at 60°C for 30 min, the DF of Sr was 50 by the KCP. The liquid impurities such as Sr was removed from the UNH crystal not only by the sweating phenomenon but also by washing with U reflux melt, which was produced by the melter at the top of the column of the KCP. On the other hand, the DF of solid impurities, SUS304L, achieved a value of 100 with the KCP. In the static system, the solid impurities remained in the UNH crystal after the sweating operation. In the KCP, the solid impurities were removed from the UNH crystals due to upward movement of the crude crystals from the double screw conveyors; the UNH crystals and solid impurities, which have different densities and particle sizes, separate from each other by gravity and mixing.

6. Conclusion

Experimental studies on the behavior of TRU elements and FPs in the dissolver solution of irradiated fast neutron reactor core fuel were carried out to develop a crystallization method as a part of an advanced aqueous reprocessing. The experimental results show high HNO_3 concentration in the feed solution increased with increasing the UNH crystal ratio in the U crystallization process. Among coexistent elements, Zr, Nb, Ru Sb, Ce, Pr, Eu, Am and Cm remained in the mother liquor at the time of U crystallization. Therefore, portions of these elements in the mother liquor that was attached to the surface of the UNH crystal were washed away with HNO_3 solution. Cesium exhibited different behavior depending on whether Pu was present. Although a high DF of Cs was obtained in the case of uranyl nitrate solution without Pu(IV), Cs was hardly separated at all from the UNH crystal formed from the dissolver solution of irradiated fast neutron reactor core fuel in the case of high Cs concentration in the feed solution. It is likely that a double salt of Pu(IV) and Cs, $Cs_2Pu(NO_3)_6$ precipitated in the course of U crystallization process. Since Ba precipitated as $Ba(NO_3)_2$, its DF was low after the UNH crystal was washed. Neptunium was not removed from the UNH crystal because Np was oxidized to Np(VI) in the feed solution and thus co-crystallized with U(VI). The experimental data on the behavior of TRU elements and FPs will be actually utilized in fast neutron reactor fuel reprocessing. The continuous crystallizer and the KCP were developed, and the apparatus performance was examined with the uranyl nitrate solution containing simulated FPs. In the future, the integrated crystallization system performance will be confirmed for part of U recovery in the NEXT process.

7. Acknowledgment

The author gratefully acknowledges Mr. A. Shibata and Mr. K. Nomura of JAEA for fruitful discussions.

8. References

Adachi, T.; Ohnuki, M.; Yoshida, N.; Sonobe, T.; Kawamura, W.; Takeishi, H.; Gunji, K.; Kimura, T.; Suzuki, T.; Nakahara, Y.; Muromura, T.; Kobayashi, Y.; Okashita, H. & Yamamoto, T. (1990). Dissolution Study of Spend PWR Fuel: Dissolution Behavior and Chemical Properties of Insoluble Residues. *Journal of Nuclear Materials,* Vol. 174, No. 1, (November 1990), pp. 60-71, ISSN 0022-3115

Anderson, H. H. (1949). Alkali Plutonium(IV) Nitrates, In: *The Transuranium Elements, National Nuclear Energy Series IV, Vol. 14B,* G. T. Seaborg, J. J. Katz, W. M. Manning, (Eds.), pp. 964-967, McGraw-Hill Book Co. Inc., New York, USA

Ebert, K.; Henrich, E.; Stahl, R. & Bauder, U. (1989). A Continuous Crystallization Process for Uranium and Plutonium Refinement, *Proceedings of 2nd International Conference on Separation Science & Technology,* pp. 346-352, Paper No. S5b, Hamilton, Ontario, Canada, October 1-4, 1989

Hart, R. G. & Morris, G. O. (1958). Crystallization Temperatures of Uranyl Nitrate-Nitric Acid Solutions, In: *Progress in Nuclear Energy, Series 3, Process Chemistry, Vol. 2,* F. Bruce, (Ed.), pp. 544-545, Pergamon Press, ISSN 0079-6514, New York, USA

Izumida, T. & Kawamura, F. (1990). Precipitates Formation Behavior in Simulated High Level Liquid Waste of Fuel Reprocessing. *Journal of Nuclear Science and Technology,* Vol. 27, No. 3, (March 1990), pp. 267-274, ISSN 0022-3131

Katsurai, K.; Ohyama, K.; Kondo, Y.; Nomura, K.; Takeuchi, M.; Washiya, T. & Myochin, M. (2009). Development of Highly Effective Dissolution Technology for FBR MOX Fuels, *Proceedings of International Conference on The Nuclear Fuel Cycle: Sustainable Options & Industrial Perspectives (GLOBAL 2009),* pp. 108-112, Paper 9219, Paris, France, September 6-11, 2009

Koma, Y.; Sano, Y.; Morita, Y. & Asakura, T. (2009). Adsorbents Development for Extraction Chromatography on Am and Cm Separation, *Proceedings of International Conference on The Nuclear Fuel Cycle: Sustainable Options & Industrial Perspectives (GLOBAL 2009),* pp. 1056-1060, Paper 9325, Paris, France, September 6-11, 2009

Koyama, T.; Washiya, T.; Nakabayashi, H. & Funasaka, H. (2009). Current Status on Reprocessing Technology of Fast Reactor Fuel Cycle Technology Development (FaCT) Project in Japan : Overwiew of Reprocessing Technology Development, *Proceedings of International Conference on The Nuclear Fuel Cycle: Sustainable Options & Industrial Perspectives (GLOBAL 2009),* pp. 46-52, Paper 9100, Paris, France, September 6-11, 2009

Kusama, M.; Chikazawa, T. & Tamaki, Y. (2005). *Estimation Tests for Effecting Factor on Decontamination Property in Crystallization Process,* JNC TJ8400 2005-006, Japan Nuclear Cycle Development Institute, Tokai, Ibaraki, Japan

Lausch, J.; Berg, R.; Koch, L.; Coquerelle, M.; Glatz, J. P.; Walker, C. T. & Mayer, K. (1994). Dissolution Residues of Highly Burnt Nuclear Fuels. *Journal of Nuclear Materials,* Vol. 208, No. 1-2, (January 1994), pp. 73-80, ISSN 0022-3115

Matsuoka, M.; Ohishi, M. & Kasama, S. (1986). Purification of p-Dichlorobenzene and m-Chloronitrobenzene Crystalline Particles by Sweating. *Journal of Chemical Engineering of Japan,* Vol. 19, No. 3, (March 1986), pp. 181-185, ISSN 0021-9592

Matsuoka, M. & Sumitani, A. (1988). Rate of Composition Changes of Organic Soild Solution Crystals in Sweating Operations. *Journal of Chemical Engineering of Japan,* Vol. 21, No. 1, (February 1988), pp. 6-10, ISSN 0021-9592

Nakahara, M.; Nomura, K.; Washiya, T.; Chikazawa, T. & Hirasawa, I. (2011). Removal of Liquid and Solid Impurities from Uranyl Nitrate Hexahydrate Crystalline Particles in Crystal Purification Process. *Journal of Nuclear Science and Technology,* Vol. 48, No. 3, (February 2011), pp. 322-329, ISSN 0022-3131

Ohyama, K.; Yano, K.; Shibata, A.; Miyachi, S.; Koizumi, T.; Koyama, T.; Nakamura, K.; Kikuchi, T. & Homma, S. (2005). Experimental Study on U-Pu Co-crystallization for New Reprocessing Process, *Proceedings of International Conference on Nuclear Energy Systems for Future Generation and Global Sustainability (GLOBAL 2005),* Paper No. 452, Tsukuba, Japan, October 9-13, 2005

Otawara, K. & Matsuoka, T. (2002). Axial Dispersion in a Kureha Crystal Purifier (KCP). *Journal of Crystal Growth,* Vol. 237-239, No. 3, (April 2002), pp. 2246-2250, ISSN 0022-0248

Ryan, J. L. (1960). Species Involved in the Anion-exchange Absorption of Quadrivalent Actinide Nitrates. *The Journal of Physical Chemistry,* Vol. 64, No. 10, (October 1960), pp. 1375-1385, ISSN 0022-3654

Sano, Y.; Ogino, H.; Washiya, T. & Myochin, M. (2009). Development of the Solvent Extraction Technique for U-Pu-Np Co-Recovery in the NEXT Process, *Proceedings of International Conference on The Nuclear Fuel Cycle: Sustainable Options & Industrial Perspectives (GLOBAL 2009),* pp. 158-165, Paper 9222, Paris, France, September 6-11, 2009

Shibata, A.; Kaji, N.; Nakahara, M.; Yano, K.; Tayama, T.; Nakamura, K.; Washiya, T.; Myochin, M.; Chikazawa, T. & Kikuchi, T. (2009). Current Status on Research and Development of Uranium Crystallization System in Advanced Aqueous Reprocessing of FaCT Project, *Proceedings of International Conference on The Nuclear Fuel Cycle: Sustainable Options & Industrial Perspectives (GLOBAL 2009),* pp. 151-157, Paper 9154, Paris, France, September 6-11, 2009

Staritzky, E. & Truitt, A. L. (1949). *Optical and Norphological Crystallography of Plutonium Compounds,* LA-745, Los Almos National Laboratory, Los Alamos, New Mexico, USA

Usami, T.; Tsukada, T.; Inoue, T.; Moriya, N.; Hamada, T.; Purroy, D. S.; Malmbeck, R. & Glatz, J. P. (2010). Formation of Zirconium Molybdate Sludge from an Irradiated Fuel and Its Dissolution into Mixture of Nitric Acid and Hydrogen Peroxide. *Journal of Nuclear Materials,* Vol. 402, No. 2-3, (July 2010), pp. 130-135, ISSN 0022-3115

Washiya, T.; Tayama, T.; Nakamura, K.; Yano, K.; Shibata, A.; Nomura, K.; Chikazawa, T.; Nagata, M. & Kikuchi, T. (2010). Continuous Operation Test at Engineering Scale Uranium Crystallizer. *Journal of Power and Energy Systems,* Vol. 4, No. 1, (March 2010), pp. 191-201, ISSN 1881-3062

Yano, K.; Shibata, A.; Nomura, K.; Koizumi, T.; Koyama, T. & Miyake, C. (2004). Plutonium Behavior Under the Condition of Uranium Crystallization from Dissolver Solution,

Proceedings of International Conference on Advances for Future Nuclear Fuel Cycles (ATALANTE 2004), pp. 1-4, P1-66, Nîmes, France, June 21-25, 2004

Yano, K.; Nakahara, M.; Nakamura, M.; Shibata, A.; Nomura, K.; Nakamura, K.; Tayama, T.; Washiya, T.; Chikazawa, T. & Hirasawa, I. (2009). Reaserch and Development of Crystal Purification for Product of Uranium Crystallization Process, *Proceedings of International Conference on The Nuclear Fuel Cycle: Sustainable Options & Industrial Perspectives (GLOBAL 2009)*, pp. 143-150, Paper 9093, Paris, France, September 6-11, 2009

Zief, M. & Wilcox, W. R. (1967). *Fractional Solidification*, Marcel Dekker Inc., ISBN 978-0824718206, New York, USA

16

Stable and Metastable Phase Equilibria in the Salt-Water Systems

Tianlong Deng
Tianjin Key Laboratory of Marine Resources and Chemistry
College of Marine Science and Engineering
Tianjin University of Science and Technology
TEDA, Tianjin,
CAS Key Laboratory of Salt Lake Resources and Chemistry
Institute of Salt Lakes
Chinese Academy of Sciences
Xining, Qinghai,
People Republic of China

1. Introduction

1.1 Salt lakes and the classification of hydrochemistry

Salt lakes are widely distributed in the world, and some famous salt lake resources are shown in Tables 1 and 2. In China, salt lakes are mainly located in the area of the Qinghai-Xizang (Tibet) Plateau, and the Autonomous Regions of Xinjiang and Inner Mongolia (M.P. Zheng et al., 1989). The composition of salt lake brines can be summarized to the complex salt-water multi-component system (Li - Na – K – Ca - Mg - H - Cl – SO_4 – B_4O_7 – OH- HCO_3 – CO_3 - H_2O).

According to the chemical type of salt lake brines, it can be divided into five types, i.e. chloride type, sulphate type, carbonate type, nitrite type, and borate type among those salt lake resources in the world (Gao et al., 2007).

Chloride type: the component of brines in Death Sea, Mideast and Caerhan Salt Lake in China belongs to the system of chloride type (Na – K – Mg - Cl - H_2O), and the main precipitation of salts are halite (NaCl), sylvite (KCl), carnallite ($KCl{\bullet}MgCl_2{\bullet}7H_2O$), and bischofite ($MgCl_2{\bullet}6H_2O$).

Sulphate type: this kind of salt lake resources is similar with the sea water system (Na – K – Mg – Cl – SO_4 – H_2O), and it can be divided into two kinds of hypotypes i.e. sodium sulphate and magnesium sulphate. As to sodium sulphate hypotype, the Great Salt Lake in America, the gulf of Kara-Bogaz-Golin Urkmenistan, and Da-Xiao Qaidan in China belong to this hypotype with the main deposit of glauberite ($Na_2SO_4{\bullet}CaSO_4$), glauber salt ($Na_2SO_4{\cdot}10H_2O$), halite, galserite ($Na_2SO_4{\bullet}3K_2SO_4$), schonenite ($K_2SO_4{\bullet}MgSO_4{\bullet}6H_2O$), and so on. As to magnesium sulphate hypotype, there are Yunchen Salt Lake in Shanxi Provine and Chaka Salt Lakes in Qinghai Province, China, especially Salt Lakes of the Qaidam Basin in Qinghai Province are a sub-type of magnesium sulphate brines famous for their abundance of lithium, potassium, magnesium and boron resources (Zheng et al., 1989). The main precipitation of salts are halite, glauber salt, blodite ($Na_2SO_4 \cdot MgSO_4 \cdot 7H_2O$), and epsom salt ($MgSO_4 \cdot 7H_2O$).

Carbonate type: this type belongs to the system (Na – K – Cl – CO_3 - SO_4 – H_2O), and Atacama Salt Lake in Chile and Zabuye Salt Lake in Tibet are the famous carbonate type of salt lake. The main precipitated minerals are thermonatrite ($Na_2CO_3 \cdot 10H_2O$), baking soda ($NaHCO_3$), natron ($Na_2CO_3 \cdot 10H_2O$), glauber salt, and halite.

Nitrite type: the brine composition of this type salt lake can be summarized as the system (Na – K – Mg - Cl – NO_3 - SO_4 – H_2O). The type salt lake main locates in the salt lake area in the northern of Chile among the salt lake group of Andes in the South-America, semi and dry salts in Luobubo and Wuzunbulake Lakes in Xinjiang, the northern of China. There are natratime saltier ($NaNO_3$), niter (KNO_3), darapskite ($NaNO_3 \cdot Na_2SO_4 \cdot H_2O$), POTASSIUM-darapskite ($KNO_3 \cdot K_2SO_4 \cdot H_2O$), humberstonite ($NaNO_3 \cdot Na_2SO_4 \cdot 2MgSO_4 \cdot 6H_2O$).

Borate type: it can be divided into carbonate-borate hypotype and sulphate-borate hypotype. Searles Salt Lake in America, Banguo Lake and Zabuye Salt Lakes in Tibet, China belong to the former, and the brines mostly belong to the system (Na - K - Cl - B_4O_7 - CO_3 - HCO_3 - SO_4 - H_2O). In order to prove the industrial development of Searles Lake brines, Teeple (1929) published a monograph after a series of salt-water equilibrium data on Searles lake brine containing carbonate and borate systems. The latter includes Dong-xi-tai Lake, Da-xiao-chaidan Lake and Yiliping Lake in Qinghai Province, Zhachangchaka Lake in Tibet, China. In those lake area, the natural borate minerals of raphite ($NaO \cdot CaO \cdot 3B_2O_3 \cdot 16H_2O$), pinnoite ($MgO \cdot B_2O_3 \cdot 3H_2O$), chloropinnoite ($2MgO \cdot 2B_2O_3 \cdot MgCl_2 \cdot 14H_2O$), inderite ($2MgO \cdot 3B_2O_3 \cdot 15H_2O$), hungchanoite ($MgO \cdot 2B_2O_3 \cdot 9H_2O$), mcallisterite ($MgO \cdot 3B_2O_3 \cdot 7.5H_2O$), kurnakovite ($2MgO \cdot 3B_2O_3 \cdot 15H_2O$) and hydroborate were precipitated (Zheng et al., 1988; Gao et al., 2007). In addition, the concentration of lithium ion exists in the surface brine of salt lakes.

Salt lakes	Death Sea	Great lake, US	Searles lake, US	Atacama, Chile	Caerhan, China	Zabuye, Tibet
Altitude, /m	-400	1280	512	2300	2900	2677
Area, /km^2	1000	3600	1000	1400	5882	120
Dept, m	329	~5	Intragranular brine	Intragranular brine	Intragranular brine	~3
KCl	2×10^9	1×10^8	2.8×10^6	1.1×10^8	3×10^8	6.6×10^7
NaCl	1.2×10^{10}	3.2×10^9	–	–	4.3×10^{10}	2×10^8
$MgCl_2$	2.2×10^{10}	1.2×10^9	–	1.2×10^8	2.7×10^9	5.7×10^8
$MgSO_4$	–	1.7×10^7	–	–	–	–
LiCl	1.7×10^7	3.2×10^6	2.7×10^6	2.8×10^6	–	–
$CaCl_2$	6×10^9	–	–	–	–	–
$CaSO_4$	1×10^8	–	–	–	–	–
$MgBr_2$	1×10^9	–	–	–	3.4×10^5	–
B_2O_3	–	1.9×10^6	3×10^7	1.6×10^7	5.5×10^6	1.8×10^6
WO_3	–	–	7.5×10^4	–	–	–

Table 1. Basic data of salt lakes and their salt reserves in the world. unit, /t (Song, 2000)

Type of lithium resources	Country and section	Lithium storage capacity, (Li_2O)
Salt lakes	Uyuni, Bolivia	More than 19 million tons
	Silver and Searles, US	More than 10 million tons
	Caerhan and Caida, China	10 million tons
	Kata Baca, Argentina	Sever million tons
Type of crystalline rocks	America	6.34 million tons
	Chile	4.3 million tons
	Canada	6.6 million tons
	Greenbusbse, Australian	6 million tons

Table 2. Statistical distribution of the lithium reserves in the world (Song, 2000; Zhao, 2003)

1.2 Phase equilibria of salt-water systems

It is essential to study the stable and metastable phase equilibria in multi-component systems at different temperatures for its application in the fields of chemical, chemical engineering such as dissolution, crystallization, distillation, extraction and separation.

1.2.1 The stable phase equilibria of salt-water systems

The research method for the stable phase equilibria of salt-water system is isothermal dissolution method. It is worthy of pointing out that the status of the stable phase equilibrium of salt-water system is the in a sealed condition under stirring sufficiently, and the speed of dissolution and crystallization of equilibrium solid phase is completely equal with the marker of no change for the liquid phase composition. As to the thermodynamic stable equilibrium studies aiming at sea water system (Na - K - Mg - Cl - SO_4 - H_2O), J.H. Vant'hollf (1912) was in the earliest to report the stable phase diagram at 293.15 K with isothermal dissolution method.

In order to accelerate the exploiting of Qaidam Basin, China, a number of the stable phase equilibria of salt-water systems were published at recent decades (Li et al., 2006; Song, 1998, 2000; Song & Du, 1986; Song & Yao, 2001, 2003).

1.2.2 The metastable phase equilibria of salt-water systems

However, the phenomena of super-saturation of brines containing magnesium sulfate, borate is often found both in natural salt lakes and solar ponds around the world. Especially for salt lake brine and seawater systems, the natural evaporation is in a autogenetic process with the exchange of energy and substances in the open-ended system , and it is controlled by the radiant supply of solar energy with temperature difference, relative humidity, and air current, etc. In other word, it is impossible to reach the thermodynamic stable equilibrium, and it is in the status of thermodynamic non-equilibrium.

For the thermodynamic non-equilibrium phase diagram of the sea water system as called "solar phase diagram" in the first, N.S. Kurnakov (1938) was in the first to report the experimental diagrams based on the natural brine evaporation, and further called "metastable phase diagram" for the same system (Na - K - Mg - Cl - SO_4 - H_2O) at (288.15, 298.15, and 308.15) K was reported on the basis of isothermal evaporation method (Jin, et al., 1980, 2001, 2002; Sun, 1992). Therefore, the metastable phase equilibria research is essential to predict the crystallized path of evaporation of the salt lake brine.

The isothermal evaporation phase diagrams of the sea water system at different temperature show a large difference with Vant'hoff stable phase diagram. The crystallization fields of leonite ($MgSO_4.K_2SO_4.4H_2O$), and kainite ($KCl.MgSO_4.3H_2O$) are all disappear whereas the crystallization field of picromerite ($MgSO_4.K_2SO_4.6H_2O$) increases by 20-fold, which is of great importance for producing potassium sulfate or potassium-magnesium fertilizer.

Therefore, in order to separate and utilize the mixture salts effectively by salt-field engineering or solar ponds in Qaidam Basin, studies on the phase equilibria of salt-water systems are focused on the metastable phase equilibria and phase diagrams at present years (Deng et al., 2011; Deng et al., 2008a-g; Deng, et al., 2009a-c; Wang & Deng, 2008, 2010; Li & Deng, 2009; Li et al., 2010; Liu et al., 2011; Meng & Deng, 2011; Guo et al., 2010; Gao & Deng, 2011a-b; Wang et al., 2011a-b).

1.2.3 Solubility prediction for the phase equilibria of salt-water systems

Pitzer and co-workers have developed an ion interaction model and published a series of papers (Pitzer, 1973a-b, 1974a-b, 1975, 1977, 1995, 2000; Pabalan & Pitzer, 1987) which gave a set of expressions for osmotic coefficients of the solution and mean activity coefficient of electrolytes in the solution. Expressions of the chemical equilibrium model for conventional single ion activity coefficients derived are more convenient to use in solubility calculations (Harvie & Weare, 1980; Harvie et al.1984; Felmy & Weare, 1986; Donad & Kean, 1985).

In this chapter, as an example, the stable and metastable phase equilibria in the salt-water system (NaCl - KCl – $Na_2B_4O_7$ - $K_2B_4O_7$ - H_2O), which is of great importance to describe the metastable behavior in order to separate and purify the mixture salts of borax and halo-sylvite were introduced in detail. The stable phase diagrams of the sub-ternary systems (NaCl - $Na_2B_4O_7$ - H_2O), (KCl –$K_2B_4O_7$ - H_2O), ($Na_2B_4O_7$ - $K_2B_4O_7$ - H_2O) at 298.15 K and the metastable phase diagrams of the sub-ternary systems (NaCl - $Na_2B_4O_7$ - H_2O) at 308.15 K for the mentioned reciprocal quaternary system were systematically studied on our previous researches under several scientific funding supports. The theoretical prediction for the stable solubility of this reciprocal quaternary system was also briefly introduced based on the ion-interaction model.

2. Apparatus

2.1 Apparatus for the stable phase equilibria in the salt-water system

Stable phase equilibria are the thermodynamic equilibria. In order to reach the isothermal dissolve equilibrium, the apparatus mainly contains two parts i.e. constant temperature installing and equilibrator. Therefore, experimental apparatus depends on the target of temperature. Generally, thermostatic water-circulator bath is used under normal atmospheric temperature, and thermostatic oil-circulator bath is chosen at higher level temperature. Under low temperature, the refrigerator or freezing saline bath is commonly used. Figure 1 shows the common used equalizer pipe with a stirrer. The artificial synthesis complex put in the pipe to gradually reach equilibria under vigorous stirring. In order to avoid the evaporation of water, the fluid seal installing is needed, and the sampling branch pipe is also needed to seal. Usually, for aqueous quaternary system study, a series of artificial synthesis complex, normally no less than 30, was needed to be done one by one the experimental time consume is equivalence large. At present, a thermostatic shaker whose temperature could controlled with temperature precision of ± 0.1 K can be used for the measurement of stable phase equilibrium (Deng et al., 2002; Deng, 2004). The advantage is that a series artificial synthesis complexes which is loading in each sealed bottle can be put in and vigorous shaking together.

In this study, the stable phase equilibria system ($NaCl$ - KCl – $Na_2B_4O_7$ - $K_2B_4O_7$ - H_2O) at 298.15 K, a thermostatic shaker (model HZQ-C) whose temperature was controlled within 0.1 K was used for the measurement of phase equilibrium.

2.2 Apparatus for the metastable phase equilibria in the salt-water system

The isothermal evaporation method was commonly used, and Figure 2 is our designed isothermal evaporation device in our laboratory (Guo et al., 2010). The isothermal evaporation chamber was consisted of evaporating container, precise thermometer to keep the evaporating temperature as a constant and electric fan to simulate the wind in situ, and the solar energy simulating system with electrical contact thermograph, electric relay and heating lamp. The temperature controlling apparatus is made up of an electric relay, an electrical contact thermograph and heating lamps.

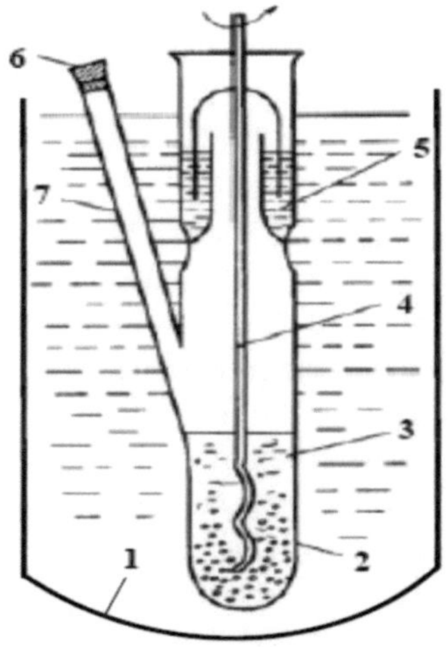

Fig. 1. Apparatus of equalizer pipe. 1, thermostatic water-circulator bath; 2, pipe body; 3, assay; 4, stirrer; 5, fluid seal; 6, rubber seal lock; 7, sampling branch pipe.

In this example of the metastable phase equilibria system ($NaCl$ - KCl – $Na_2B_4O_7$ - $K_2B_4O_7$ - H_2O) at 308.15 K, the isothermal evaporation box was used. In an air-conditioned laboratory, a thermal insulation material box (70 cm long, 65 cm wide, 60 cm high) with an apparatus to control the temperature was installed. When the solution temperature in the container was under (308.15 ± 0.2) K, the apparatus for controlling the temperature formed a circuit and the heating lamp began to heat. Conversely, the circuit was broken and the heating lamp stopped working when the temperature exceeded 308.15 K. Therefore, the temperature in the box could always be kept to (308.15 ± 0.2) K. An electric fan installed on the box always worked to accelerate the evaporation of water from the solutions.

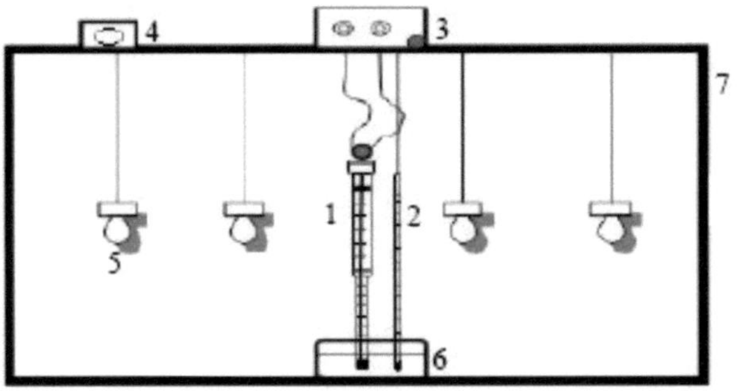

Fig. 2. The schematic diagram of the isothermal evaporation chamber. 1, electrical contact thermograph; 2, precise thermometer; 3, electric relay; 4, electric fan; 5, heating lamp; 6, evaporating container; 7, isothermal container.

Of course, the experimental conditions of an air flow velocity, a relative humidity, and an evaporation rate were controlled as similar as to those of the climate of reaching area in a simulative device.

3. Experimental methods

3.1 Reagents

For phase equilibrium study, reagents used should be high-purity grade otherwise the re-crystallized step was needed. For the stable and metastable phase equilibria in the salt-water system ($NaCl$ - KCl – $Na_2B_4O_7$ - $K_2B_4O_7$ - H_2O), the chemicals used were of analytical grade, except borax which was a guaranteed reagent (GR), and were obtained from either the Tianjin Kermel Chemical Reagent Ltd. or the Shanghai Guoyao Chemical Reagent Co. Ltd: sodium chloride ($NaCl$, ≥0.995 in mass fraction), potassium chloride (KCl, ≥ 0.995 in mass fraction), borax ($Na_2B_4O_7 \cdot 10H_2O$, ≥ 0.995 in mass fraction), potassium borate tetrahydrate ($K_2B_4O_7 \cdot 4H_2O$, ≥ 0.995 in mass fraction), and were re-crystallized before use. Doubly deionized water (DDW) with conductivity less than 1.2×10^{-4} $S \cdot m^{-1}$ and pH 6.60 at 298.15 K was used to prepare the series of the artificial synthesized brines and chemical analysis.

3.2 Analytical methods

3.2.1 The chemical analysis of the components in the liquids

For phase equilibrium study in this phase equilibrium system ($NaCl$ - KCl – $Na_2B_4O_7$ - $K_2B_4O_7$ - H_2O), the composition of the potassium ion in liquids and their corresponding wet solid phases was analyzed by gravimetric methods of sodium tetraphenyl borate with an uncertainty of ≤ ±0.0005 in mass fraction; Both with an uncertainty of ≤ ± 0.003 in mass fraction, the concentrations of chloride and borate were determined by titration with mercury nitrate standard solution in the presence of mixed indicator of diphenylcarbazone and bromphenol blue, and by basic titration in the presence of mannitol, respectively (Analytical Laboratory of Institute of Salt Lakes at CAS, 1982). The concentration of sodium ion was calculated by subtraction via charge balance.

3.2.2 The measurements of the physicochemical properties

For the physicochemical properties determinations, a PHS-3C precision pH meter supplied by the Shanghai Precision & Scientific Instrument Co. Ltd was used to measure the pH of the equilibrium aqueous solutions (uncertainty of ± 0.01). The pH meter was calibrated with standard buffer solutions of a mixed phosphate of potassium dihydrogen phosphate and sodium dihydrogen phosphate (pH 6.84) as well as borax (pH 9.18); the densities (ρ) were measured with a density bottle method with an uncertainty of ± 0.2 $mg.cm^{-3}$. The viscosities (η) were determined using an Ubbelohde capillary viscometer, which was placed in a thermostat at (308.15 ± 0.1) K. No fewer than five flow times for each equilibrium liquid phase were measured with a stopwatch with an uncertainty of 0.1 s to record the flowing time, and the results calculated were the average. An Abbe refractometer (model WZS-1) was used for measuring the refractive index (n_D) with an uncertainty of ± 0.0001. The physicochemical parameters of density, refractive index and pH were also all placed in a thermostat that electronically controlled the set temperature at (308.15 ± 0.1) K.

3.3 Experimental methods of phase equilibria

3.3.1 Stable phase equilibria

For the stable equilibrium study, the isothermal dissolution method was used in this study. The series of complexes of the quaternary system were loaded into clean polyethylene bottles and capped tightly. The bottles were placed in the thermostatic rotary shaker, whose temperature was controlled to (298.15 ± 0.1) K, and rotated at 120 rpm to accelerate the equilibrium of those complexes. A 5.0 cm^3 sample of the clarified solution was taken from the liquid phase of each polyethylene bottle with a pipet at regular intervals and diluted to 50.0 cm^3 final volumes in a volumetric flask filled with DDW. If the compositions of the liquid phase in the bottle became constant, then equilibrium was achieved. Generally, it takes about 50 days to come to equilibrium.

3.3.2 Metastable phase equilibria

The isothermal evaporation method was used in metastable phase equilibria study. According to phase equilibrium composition, the appropriate quantity of salts and DDW calculated were mixed together as a series of artificial synthesized brines and loaded into clean polyethylene containers (15 cm in diameter, 6 cm high), then the containers were put into the box for the isothermal evaporation at (308.15 ± 0.2) K. The experimental conditions with air flowing velocity of 3.5-4.0 m/s, relative humidity of 20-30%, and evaporation rate of 4-6 mm/d are presented, just like the climate of the Qaidam Basin. For metastable evaporation, the solutions were not stirred, and the crystal behavior of solid phase was observed periodically. When enough new solid phase appeared, the wet residue mixtures were taken from the solution. The solids were then approximately evaluated by the combined chemical analysis, of XP-300D Digital Polarizing Microscopy (Shanghai Caikon Optical Instrument Co,. Ltd., China) using an oil immersion, and further identification with X-ray diffraction (X'pert PRO, Spectris. Pte. Ltd., The Netherlands). Meanwhile, a 5.0 cm^3 sample of the clarified solution was taken from the liquid phase of each polyethylene container through a filter pipette, and then diluted to a 250.0 cm^3 final volume in a volumetric flask filled with DDW for the quantitative analysis of the compositions of the liquid phase. Some other filtrates were used to measure the relative physicochemical properties individually according to the analytical method. The remainder of the solution continued to be evaporated and reached a new metastable equilibrium.

4. Experimental results

4.1 Mineral identification for the solid phase

For mineral identification when enough new solid phase appeared either in the stable equilibrium system or in the metastable equilibrium system, the wet residue mixtures were taken from the solution according to the experimental method. Firstly, as to the minerals of $Na_2B_4O_7 \cdot 10H_2O$ and $K_2B_4O_7 \cdot 4H_2O$, the former belongs to monoclinic system, and the dual optical negative crystal i.e. 2v(-) whereas the later belongs to trimetric system, and the dual optical positive crystal i.e. 2v(+). Secondly, to the minerals NaCl and KCl, they can be identified through the property of refractive index. The refractive index of NaCl is higher than that of KCl. Observed with a XP-300D Digital Polarizing Microscopy using an oil immersion method, the crystal photos of the single and orthogonal polarized light on representative solid phases in the invariant points (NaCl + KCl + $Na_2B_4O_7 \cdot 10H_2O$) and ($Na_2B_4O_7 \cdot 10H_2O$ + $K_2B_4O_7 \cdot 4H_2O$ + KCl) are presented in Figure 3.

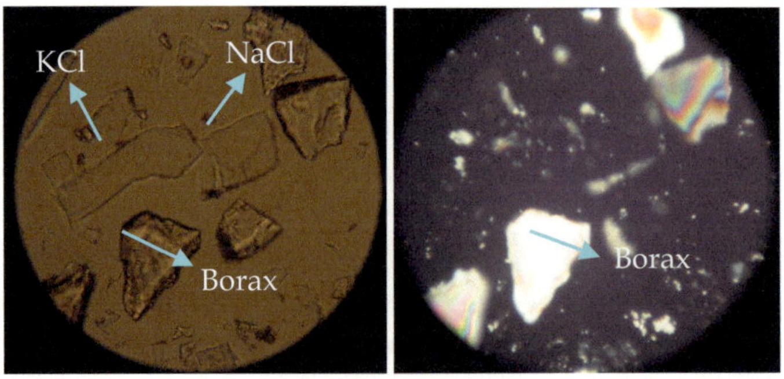

single polarized light (10×10) orthogonal polarized light (10×10)

(a) Invariant point (NaCl + KCl + Borax)

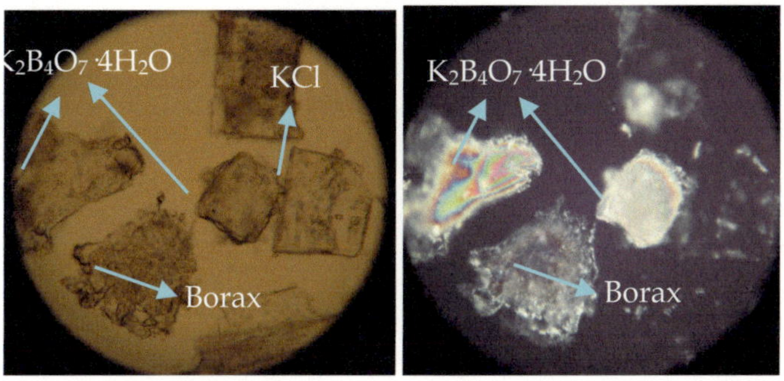

single polarized light (10×10) orthogonal polarized light (10×10)

(b) Invariant point (KCl + Borax + $K_2B_4O_7 \cdot 4H_2O$)

Fig. 3. Identification of the invariant points for the solid phase in the reciprocal system ($NaCl$ - KCl – $Na_2B_4O_7$ - $K_2B_4O_7$ - H_2O) with a polarized microscopy using an oil-immersion method. (a), the invariant point ($NaCl$ + KCl + $Na_2B_4O_7 \cdot 10H_2O$); (b), the invariant point (KCl + $Na_2B_4O_7 \cdot 10H_2O$ + $K_2B_4O_7 \cdot 4H_2O$).

The metastable equilibria solid phases in the two invariant points are further confirmed with X-ray diffraction analysis, and listed in Figure 4, except in the invariant points (NaCl + KCl + Borax) in Figure 4a which shows that the minerals KCl, NaCl, $Na_2B_4O_7 \cdot 10H_2O$ and a minor $Na_2B_4O_7 \cdot 5H_2O$ are existed. The minor of $Na_2B_4O_7 \cdot 5H_2O$ maybe is formed due to the dehydration of $Na_2B_4O_7 \cdot 10H_2O$ in the processes of transfer operation and/or grinding.

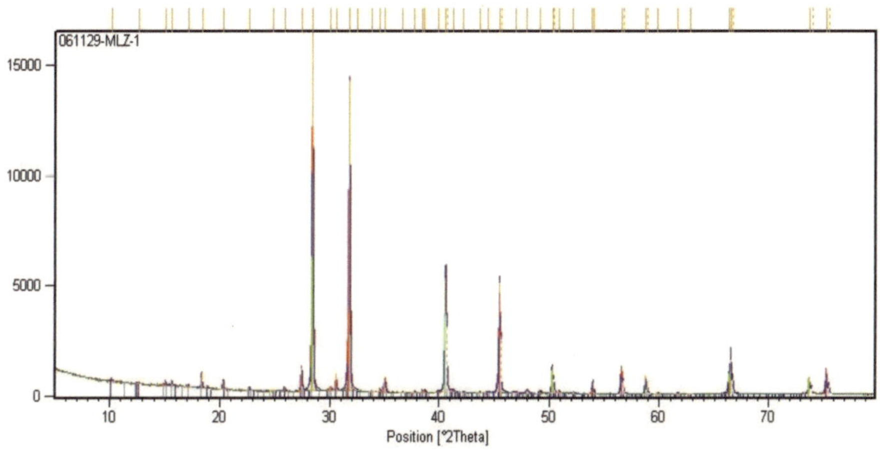

No.	Visible	Ref. Code	Chemical Formula	Score	Scale Factor	Semi-Quant/%
1	True	01-075-0296	KCl	49	0.369	22
2	True	01-075-0296	NaCl	45	0.732	57
3	True	01-075-0296	$B_4O_5(OH)_4(Na_2(H_2O)_8$	39	0.030	20
4	True	01-075-0296	$Na_2B4O_7(H_2O)_5$	13	0.007	1

(a), the X-ray diffraction photograph and the analytical data for the invariant point (NaCl + KCl + Borax)

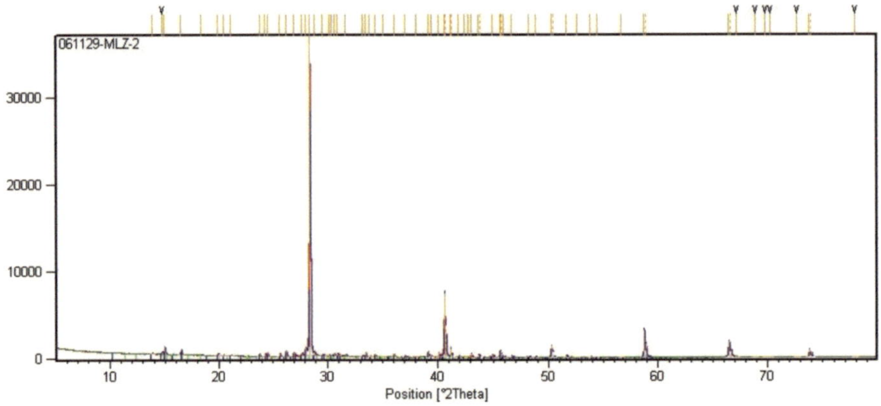

No.	Visible	Ref. Code	Chemical Formula	Score	Scale Factor	Semi-Quant/%
1	True	01-072-1540	KCl	39	0.650	72
2	True	01-076-0753	$K_2(B_4O_5(OH)_4)(H_2O)_2$	37	0.011	22
3	True	01-074-0339	$B_4O_5(OH)_4(Na_2(H_2O)_8$	10	0.004	5

(b), the X-ray diffraction photograph and the analytical data for the invariant point (KCl + Borax + $K_2B_4O_7{\cdot}4H_2O$)

Fig. 4. The X-ray diffraction data of the invariant points. (a), the invariant point (NaCl + KCl + $Na_2B_4O_7{\cdot}10H_2O$); (b), the invariant point ($Na_2B_4O_7{\cdot}10H_2O$ + $K_2B_4O_7{\cdot}4H_2O$ + KCl).

4.2 Stable phase equilibrium of the quaternary system (NaCl - KCl – $Na_2B_4O_7$ - $K_2B_4O_7$ - H_2O) at 298.15 K

The stable phase equilibrium experimental results of solubilities of the quaternary system (NaCl - KCl – $Na_2B_4O_7$ - $K_2B_4O_7$ - H_2O) at 298.15 K were determined, and are listed in Table 3, respectively. On the basis of the Jänecke index (J_B, J_B/[mol/100 mol($2Na^+ + 2K^+$)]) in Table 3, the stable equilibrium phase diagram of the system at 298.15 K was plotted and shown in Figure 5.

4.3 Metastable phase equilibrium of the quaternary system (NaCl - KCl – $Na_2B_4O_7$ - $K_2B_4O_7$ - H_2O) at 308.15 K

The experimental results of the metastable solubilities and the physicochemical properties of the quaternary system (NaCl - KCl – $Na_2B_4O_7$ - $K_2B_4O_7$ - H_2O) at 308.15 K were determined, and are listed in Tables 4 and 5, respectively. On the basis of the Jänecke index (J_B, J_B/[mol/100 mol($2Na^+ + 2K^+$)]) in Table 4, the metastable equilibrium phase diagram of the system at 308.15 K was plotted (Figure 6).

No.	Composition of the solution 100 w_B*				Jänecke index J_B, /[mol/100mol(2Na$^+$+2K$^+$)]			Equilirium solid phase**
	Na^+	K^+	Cl^-	$B_4O_7^{2-}$	$J(2Cl^-)$	$J(2K^+)$	$J(H_2O)$	
1	10.25	0.00	15.80	0.00	100.00	0.00	1807.5	NaCl
2	0.00	13.87	12.58	0.00	100.00	100.00	2310.8	KCl
3	0.72	0.00	0.00	2.41	0.00	0.00	34623.3	N10
4	0.00	4.74	0.00	9.42	0.00	100.00	7864.1	K4
5, E_1	8.02	5.85	17.68	0.00	100.00	30.03	1526.2	NaCl+KCl
6, E_2	9.60	0.00	14.33	1.04	96.79	0.00	1997.5	NaCl+N10
7, E_3	0.00	15.87	12.35	4.48	85.85	100.00	1807.8	KCl +K4
8, E_4	1.15	4.36	0.00	12.54	0.00	69.24	5649.7	N10 + K4
9	1.38	6.11	0.92	14.77	11.98	72.33	3947.5	N10 + K4
10	1.57	11.74	12.35	1.59	24.64	72.41	3496.9	N10 + K4
11	1.59	7.55	3.83	11.97	41.25	73.74	3181.6	N10 + K4
12	3.50	11.39	12.24	7.62	96.67	45.60	1427.4	N10 + K4
13, E	4.72	9.62	14.55	3.17	90.05	54.55	1672.8	KCl + N10 + K4
14	1.13	13.44	12.65	2.78	90.59	88.46	2123.7	KCl+K4
15	1.57	11.74	12.35	1.59	94.45	81.45	2191.0	KCl+K4
16	4.18	10.56	14.96	2.33	93.36	59.79	1670.9	KCl+K4
17	4.15	9.65	13.96	2.61	92.13	57.77	181.0	KCl+K4
18	5.27	9.43	15.50	2.58	93.83	51.84	1602.8	KCl + N10
19	6.11	9.83	17.52	1.79	95.55	48.48	1386.6	KCl + N10
20	6.40	9.10	17.52	1.32	96.67	45.60	1427.4	NaCl + KCl
21	7.50	7.13	17.73	0.67	98.30	35.89	1462.7	NaCl + KCl
22, F	6.16	9.64	17.55	1.52	96.19	47.94	1406.5	NaCl+KCl+N10
23	9.73	1.05	15.35	1.35	96.05	6.12	1830.4	NaCl+N10

* w_B is in mass fraction; ** K4, $K_2B_4O_7 \cdot 4H_2O$; N10, $Na_2B_4O_7 \cdot 10H_2O$.

Table 3. Stable solubilities of the system ($NaCl - KCl - Na_2B_4O_7 - K_2B_4O_7 - H_2O$) at 298.15 K

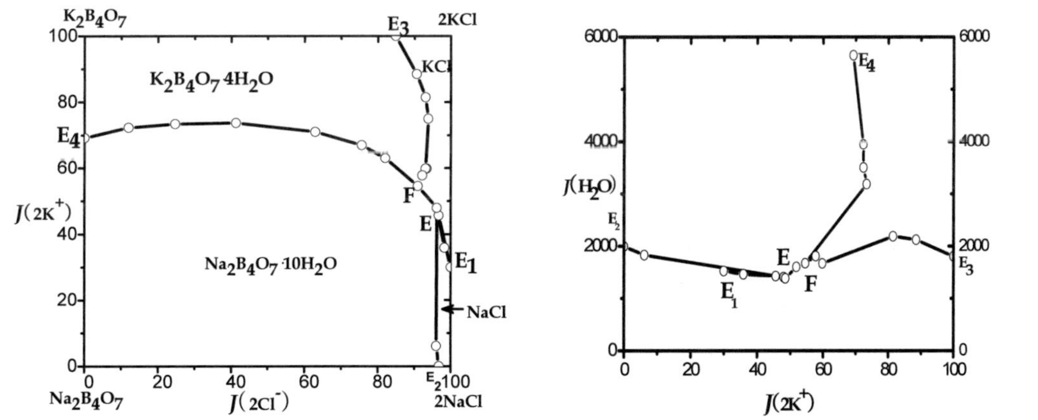

(a) dry-salt phase diagram (b) water-phase diagram

Fig. 5. Stable phase diagram of the quaternary system ($NaCl - KCl - Na_2B_4O_7 - K_2B_4O_7 - H_2O$) at 298.15 K. (a), dry-salt phase diagram; (b), water-phase diagram.

No.	Composition of the solution $100w_B$				Jänecke index J_B, /[mol/100mol($2Na^+ + 2K^+$)]			Equilibrium solid phase*
	Na^+	K^+	Cl^-	$B_4O_7^{2-}$	$J(2Cl^-)$	$J(2K^+)$	$J(H_2O)$	
A	1.30	0.00	0.00	4.40	0.00	0.00	2090.17	N10
B	10.47	0.00	16.15	0.00	100.00	0.00	1788.31	NaCl
C	0.00	14.87	13.49	0.00	100	100	18468.05	KCl
D	0.00	3.87	0.00	19.26	0.00	100	4311.59	K4
1, E'_1	7.85	6.56	18.06	0.00	100	32.94	1471.69	NaCl+KCl
2	7.84	6.54	17.76	0.58	98.54	32.89	1469.10	NaCl+KCl
3	7.85	6.60	17.66	0.95	97.6	33.09	1456.10	NaCl+KCl
4, E′	7.87	6.56	17.17	1.99	94.98	32.89	1445.77	NaCl+KCl+N10
5, E'_2	10.62	0.00	15.80	1.28	96.47	0.00	1736.76	NaCl +N10
6	9.88	1.51	15.89	1.59	95.63	8.25	1684.84	NaCl +N10
7	9.36	2.87	16.24	1.74	95.33	15.27	1612.49	NaCl +N10
8	9.10	3.62	16.50	1.81	95.22	18.97	1566.69	NaCl +N10
9	8.15	5.38	16.57	1.94	94.92	27.96	1532.30	NaCl+N10
10, E'_3	0.00	15.22	11.83	4.34	85.66	100	1955.76	KCl+K4
11	0.49	14.75	12.16	4.34	85.99	94.61	1899.70	KCl+K4
12	0.54	14.73	12.29	4.15	86.66	94.16	1894.74	KCl+K4
13	1.22	13.68	12.51	3.88	87.60	86.86	1893.49	KCl+K4
14	1.86	12.88	12.45	4.60	85.56	80.30	1844.98	KCl+K4
15	2.65	12.41	12.51	6.19	81.57	73.39	1699.85	KCl+K4
16, F′	2.84	12.36	12.06	7.74	77.34	71.90	1641.11	KCl+K4+N10
17, E'_4	2.01	8.65	0.00	24.00	0.00	71.66	2346.01	N10+K4
18	1.89	9.86	1.78	22.09	15.03	75.41	2134.79	N10+K4
19	1.74	9.55	4.49	15.01	39.58	76.36	2000.89	N10+K4
20	1.66	10.78	6.11	13.66	49.49	79.22	2160.56	N10+K4
21	1.79	12.05	8.55	11.27	62.43	79.81	1906.21	N10+K4
22	1.87	12.73	9.48	10.86	65.66	79.97	1773.28	N10+K4
23	2.28	12.65	10.89	8.97	72.68	76.57	1712.53	N10+K4
24	2.78	12.61	8.30	11.95	75.94	72.72	1838.60	N10+K4
25	3.11	11.86	12.46	6.78	80.09	69.15	1664.48	N10+KCl
26	3.26	11.41	13.13	4.90	85.45	67.31	1723.59	N10+KCl
27	3.64	10.91	13.59	4.20	87.64	63.8	1717.14	N10+KCl
28	4.06	10.43	14.15	3.45	89.98	60.17	1699.65	N10+KCl
29	4.45	10.17	14.57	3.30	90.62	57.36	1652.70	N10+KCl
30	5.27	9.25	15.16	2.97	91.80	50.79	1604.93	N10+KCl
31	6.17	8.23	15.94	2.27	93.91	43.95	1562.35	N10+KCl

* K4, $K_2B_4O_7 \cdot 4H_2O$; N10, $Na_2B_4O_7 \cdot 10H_2O$; w_B, in mass fraction.

Table 4. Metastable solubilities of the quaternary system ($NaCl - KCl - Na_2B_4O_7 - K_2B_4O_7 - H_2O$) at 308.15 K

No.*	Density ρ, /(g·cm^{-3})	pH	Refractive index	Viscosity η/(mPa·s)
A	1.0441	–**	1.3405	–
B	1.1935	–	1.3800	–
C	1.1857	–	1.3742	–
D	1.2003	–	1.3678	–
1, E'_1	1.2300	5.63	1.3869	1.1241
2	1.2414	7.29	1.3872	1.1452
3	1.2433	7.72	1.3880	1.1930
4, 'E	1.2524	7.81	1.3890	1.2620
5, E'_2	1.2060	–	1.3802	–
6	1.2172	8.22	1.3836	1.3023
7	1.2274	8.66	1.3860	1.2865
8	1.2313	9.21	1.3862	1.2829
9	1.2398	8.51	1.3879	1.2729
10, E'_3	1.2270	9.55	1.3782	0.8499
11	1.2347	9.43	1.3798	0.8443
12	–	–	–	–
13	–	–	1.3792	0.8844
14	1.2433	9.02	1.3814	1.0625
15	1.2536	9.10	1.3840	1.0876
16, F'	1.2700	9.30	1.3863	1.1963
17, E'_4	1.3040	10.36	1.3868	2.8279
18	1.3206	10.12	1.3868	2.8032
19	1.2533	9.94	1.3803	1.5661
20	1.2600	9.62	1.3823	1.4548
21	1.2636	9.58	1.3840	–
22	1.2764	9.53	–	1.3900
23	1.2784	9.47	1.3856	–
24	–	–	–	–
25	1.2555	9.08	1.3857	1.1682
26	1.2484	9.51	1.3842	1.0812
27	1.2409	9.48	1.3828	1.0480
28	1.2348	9.02	1.3827	1.0252
29	1.2366	9.04	1.3835	1.0421
30	1.2404	8.65	1.3841	1.0809
31	1.2381	8.38	1.3849	1.0130

* Corresponding to the no. column in Table 4; ** not determined.

Table 5. Physicochemical properties of the metastable reciprocal quaternary system ($NaCl$ - KCl – $Na_2B_4O_7$ - $K_2B_4O_7$ - H_2O) at 308.15 K

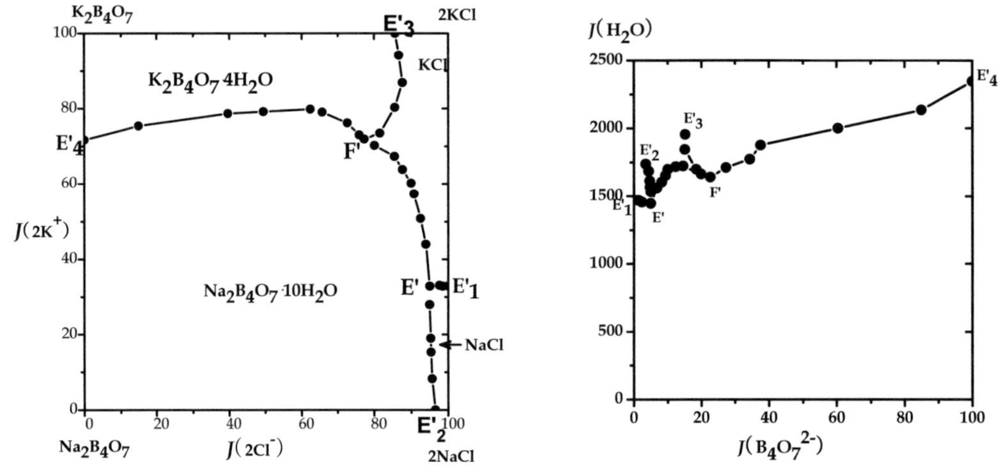

(a) dry-salt phase diagram (b) water-phase diagram

Fig. 6. Metastable phase diagram of the quaternary system (NaCl - KCl – $Na_2B_4O_7$ - $K_2B_4O_7$ - H_2O) at 308.15 K. (a), dry-salt phase diagram; (b), water-phase diagram.

On the basis of physicochemical property data of the metastable system (NaCl - KCl – $Na_2B_4O_7$ - $K_2B_4O_7$ - H_2O) at 308.15 K in Table 5, the diagram of physicochemical properties versus composition was drawn and shows in Figure 7. The physicochemical properties of the metastable equilibrium solution vary regularly with the composition of borate mass fraction. The singular point on every curve of the composition versus property diagram corresponds to the same invariant point and on the metastable solubility.

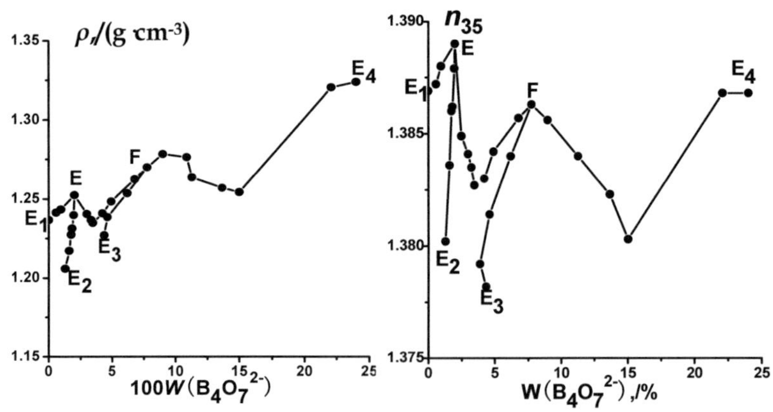

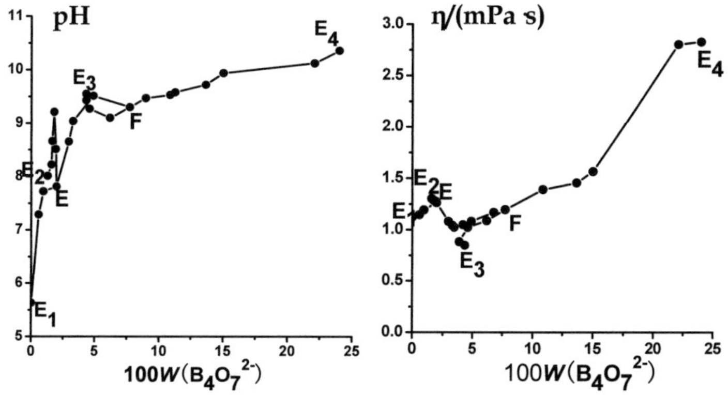

Fig. 7. Diagram of physicochemical properties versus composition for the metastable quaternary system (NaCl - KCl – $Na_2B_4O_7$ - $K_2B_4O_7$ - H_2O) at 308.15 K

4.3 Comparison of the stable and metastable phase diagram of the quaternary system (NaCl - KCl – $Na_2B_4O_7$ - $K_2B_4O_7$ - H_2O)

A comparison of the dry-salt diagrams of the metastable phase equilibrium at 308.15 K and the stable phase equilibrium at 298.15 K for the same system is shown in Figure 8. The metastable crystallization regions of borax and potassium chloride are both enlarged while the crystallized area of other minerals existed is decreased. When compared with the stable system, the solubility of borax in water in the metastable system is increased from 3.13 % to 5.70 %. The metastable phenomenon of borax is obvious in this reciprocal quaternary system.

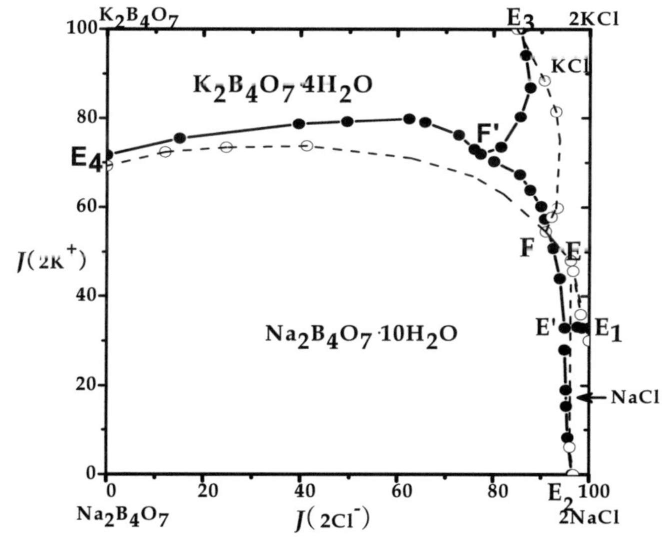

Fig. 8. Comparison of the metastable phase diagram at 308.15 K in solid line and the stable phase diagram at 298.15 K in dashed line for the quaternary system (NaCl - KCl – $Na_2B_4O_7$ - $K_2B_4O_7$ - H_2O). -●-, metastable experimental points; -○-, stable experimental points.

5. Solubility theoretic prediction of salt-water system

5.1 Ion-interaction model

As to any electrolyte, its thermodynamic prosperity varied from weak solution to high concentration could be calculated through 3 or 4 Pitzer parameters. Pitzer ion-interaction model and its extended HW model of aqueous electrolyte solution can be briefly introduced in the following (Pitzer, 1975, 1977, 2000; Harvie & Wear, 1980; Harvie et al., 1984; Kim & Frederich, 1988a-b).

As to the ion-interaction model, it is a semiempirical statistical thermodynamics model. In this model, the Pitzer approach begins with a virial expansion of the excess free energy of the form to consider the three kinds of existed potential energies on the ion-interaction potential energy in solution.

$$G^{ex} / (n_w RT) = f(I) + \sum_i \sum_j \lambda_{ij}(I) m_i m_j + \sum_i \sum_j \sum_k \mu_{ijk} m_i m_j m_k + \ldots \quad (1)$$

Where n_w is kilograms of solvent (usually in water), and m_i is the molality of species i (species may be chosen to be ions); i, j, and k express the solute ions of all cations or anions; I is ion strength and given by

$I = \frac{1}{2}\sum m_i z_i^2$, here z_i is the number of charges on the i-th solute.

The first term on the right in equation (1) is the first virial coefficients. The first virial coefficients i.e. the Debye-Hückel limiting law, $f(I)$, is a function only of ionic strength to express the long-range ion-interaction potential energy of one pair of ions in solution and not on individual ionic molalities or other solute properties.

Short-range potential effects are accounted for by the parameterization and functionality of the second virial coefficients, λ_{ij}, and the third virial coefficients, μ_{ijk}. The quantity λ_{ij} represents the short-range interaction in the presence of the solvent between solute particles i and j. This binary interaction parameter of the second virial coefficient does not itself have any composition dependence for neutral species, but for ions it is dependent it is ionic strength.

The quantity μ_{ijk} represents short-range interaction of ion triplets and are important only at high concentration. The parameters μ_{ijk} are assumed to by independent of ionic strength and are taken to be zero when the ions i, j and k are all cations or all anions.

Taking the derivatives of equation ? with respect to the number of moles of each components yields expressions for the osmotic and activity coefficients.

5.1.1 For pure electrolytes

For the pure single-electrolyte MX, the osmotic coefficient defined by Pitzer (2000):

$$\phi - 1 = |z_M z_X| f^{\phi} + m \frac{2 v_M v_X}{v} B_{MX}^{\phi} + m^2 \frac{2(v_M v_X)^{3/2}}{v} C_{MX}^{\phi} \quad (2)$$

φ is the osmotic coefficient; Z_M and Z_X are the charges of anions and cautions in the solution. m is the molality of solute; υ_M, υ_X, and υ ($\upsilon = \upsilon_M + \upsilon_X$) represent the stiochiometric coefficients of the anion, cation, and the total ions on the electrolyte MX.

In equation (1), f^{φ}, B^{φ}_{MX} and C^{φ}_{MX} are defined as following equations. In equation (1a), here b is a universal empirical constant to be equal 1.2 $kg^{1/2}\cdot mol^{-1/2}$.

$$f^{\phi} = -\frac{A^{\phi} I^{1/2}}{1 + bI^{1/2}} \tag{2a}$$

For non 2-2 type of electrolytes, such as several 1-1-,2-1-, and 1-2-type pure salts, the best form of B^{φ}_{MX} is following (Pitzer, 1973):

$$B^{\phi}_{MX} = \beta^{(0)}_{MX} + \beta^{(1)}_{MX} e^{-\alpha\sqrt{I}} \tag{2b}$$

For 2-2 type of electrolytes, such as several 3-1- and even 4-1-type pure salts, an additional term is added (Pizter, 1977):

$$B^{\phi}_{MX} = \beta^{(0)}_{MX} + \beta^{(1)}_{MX} e^{-\alpha_1\sqrt{I}} + \beta^{(2)}_{MX} e^{-\alpha_2\sqrt{I}} \tag{2c}$$

$$A^{\phi} = \frac{1}{3}\left(\frac{2\pi N_0 \rho_W}{1000}\right)^{1/2}\left(\frac{e^2}{DkT}\right)^{3/2} \tag{2d}$$

A^{φ} is the Debye-Hückel coefficient for the osmotic coefficient and equal to 0.3915 at 298.15 K. Where, N_0 is Avogadro's number, d_w and D are the density and static dielectric constant of the solvent (water in this case) at temperature and e is the electronic charge. k is Boltzmann's constant. In equation (1b), $\beta^{(0)}_{MX}$, $\beta^{(1)}_{MX}$, C^{φ}_{MX} are specific to the salt MX, and are the single-electrolyte parameters of MX. The universal parameters $\alpha = 2.0$ $kg^{1/2}\cdot mol^{-1/2}$ and omit $\beta^{(2)}_{MX}$ for several 1-1-,2-1-, and 1-2-type salts at 298.15 K. As salts of other valence types, the values $\alpha_1 = 1.4$ $kg^{1/2}\cdot mol^{-1/2}$, and $\alpha_2 = 12$ $kg^{1/2}\cdot mol^{-1/2}$ were satisfactory for all 2-2 or higher valence pairs electrolytes at 298.15 K. The parameter $\beta^{(2)}_{MX}$ is negative and is related to the association equilibrium constant.

The mean activity coefficient $\gamma_{\pm}$ is defined as:

$$ln\,\gamma_{\pm} = |z_M z_X| f^{\gamma} + m\frac{2v_M v_X}{v} B^{\gamma}_{MX} + m^2\frac{2(v_M v_X)^{3/2}}{v} C^{\gamma}_{MX} \tag{3}$$

$$f^{\gamma} = -A^{\phi}[I^{1/2}/(1 + bI^{1/2}) + (2/b)ln(1 + bI^{1/2})] \tag{3a}$$

$$B^{\gamma}_{MX} = B_{MX} + B^{\phi}_{MX} \tag{3b}$$

$$B_{MX} = \beta^{(0)}_{MX} + \beta^{(1)}_{MX} g(\alpha_1 I^{1/2}) + \beta^{(2)}_{MX} g(\alpha_2 I^{1/2})] \tag{3c}$$

$$g(x) = 2[1 - (1 + x)exp(-x)/x^2] \tag{3d}$$

$$C_{MX}^{\gamma} = 3C_{MX}^{\phi} / 2 \tag{3e}$$

5.1.2 For mixture electrolytes

In order to treat mixed electrolytes, the following sets of equations are identical with the form used by Harvie & Weare (1984) for modeling the osmotic coefficient and the activity coefficient of a neutral electrolyte based on Pitzer Equations.

$$\begin{aligned}\sum_i m_i(\phi-1) = {} & 2(-A^{\phi}I^{3/2} / (1+1.2I^{1/2}) + \sum_{c=1}^{N_c}\sum_{a=1}^{N_a} m_c m_a (B_{ca}^{\phi} + ZC_{ca}) \\ & + \sum_{c=1}^{N_c-1}\sum_{c'=c+1}^{N_c} m_c m_{c'} (\Phi_{cc'}^{\phi} + \sum_{a=1}^{N_a} m_a \psi_{cc'a}) + \sum_{a=1}^{N_a-1}\sum_{a'=a+1}^{N_a} m_a m_{a'} (\Phi_{aa'}^{\phi} \\ & + \sum_{c=1}^{N_c} m_c \psi_{aa'c}) + \sum_{n=1}^{N_n}\sum_{c=1}^{N_c} m_n m_c \lambda_{nc}\end{aligned} \tag{4}$$

$$\begin{aligned}ln\,\gamma_M = {} & z_M^2 F + \sum_{a=1}^{N_a} m_a (2B_{Ma} + ZC_{Ma}) + \sum_{c=1}^{N_c} m_c (2\Phi_{Mc} + \sum_{a=1}^{N_a} m_a \psi_{Mca}) + \\ & \sum_{a=1}^{N_a-1}\sum_{a'=a+1}^{N_a} m_a m_{a'} \psi_{aa'M} + |z_M| \sum_{c=1}^{N_c}\sum_{a=1}^{N_a} m_c m_a C_{ca} + \sum_{n=1}^{N_n} m_n (2\lambda_{nM})\end{aligned} \tag{5}$$

$$\begin{aligned}ln\,\gamma_X = {} & z_X^2 F + \sum_{c=1}^{N_c} m_c (2B_{cX} + ZC_{cX}) + \sum_{a=1}^{N_a} m_a (2\Phi_{Xa} + \sum_{c=1}^{N_c} m_c \psi_{Xac}) + \\ & \sum_{c=1}^{N_c-1}\sum_{c'=c+1}^{N_c} m_c m_{c'} \psi_{cc'X} + |z_X| \sum_{c=1}^{N_c}\sum_{a=1}^{N_a} m_c m_a C_{ca} + \sum_{n=1}^{N_n} m_n (2\lambda_{nX})\end{aligned} \tag{6}$$

$$ln\,\gamma_N = \sum_{c=1}^{N_c} m_c (2\lambda_{nc}) + \sum_{a=1}^{N_a} m_a (2\lambda_{na}) \tag{7}$$

In equations (3), (4), (5) and (6), the subscripts M, c, and c' present cations different cations; X, a, and a' express anions in mixture solution. N_c, N_a and N_n express the numbers of cations, anions, and neutral molecules; r_M, Z_M, m_C and r_X, Z_X, m_a, Φ present the ion activity coefficient, ion valence number, ion morality, and the permeability coefficient; γ_n, m_n, λ_{nc}, and λ_{na} express activity coefficient of neutral molecule, morality of neutral molecule the interaction coefficient between neutral molecules with cations c and anion a.

In equations from (3) to (6), the function symbols of $F, C, Z, A^{\phi}, \psi, \Phi, B^{\phi}, B$ are as following, respectively:

1. The term of F in equations (4) to (5) depends only on ionic strength and temperature. The defining equation of F is given by equation (7).

$$\begin{aligned}F = {} & -A^{\phi}[I^{1/2} / (1+1.2I^{1/2}) + 2/1.2 ln(1+1.2I^{1/2})] + \sum_{c=1}^{N_c}\sum_{a=1}^{N_a} m_c m_a B'_{ca} \\ & + \sum_{c=1}^{N_c-1}\sum_{c'=c+1}^{N_c} m_c m_{c'} \Phi'_{cc'} + \sum_{a=1}^{N_a-1}\sum_{a'=a+1}^{N_a} m_a m_{a'} \Phi'_{aa'}\end{aligned} \tag{8}$$

2. The single-electrolyte third virial coefficient, C_{MX}, account for short-range interaction of ion triplets and are important only at high concentration. These terms are independent of ionic strength. The parameters C_{MX} and C^{ϕ}_{MX}, the corresponding coefficients for calculating the osmotic coefficient, are related by the equation (1-6) (Pitzer & Mayorga, 1973):

$$C_{MX} = C^{\phi}_{MX} / (2|Z_M Z_X|^{1/2}) \tag{9}$$

3. The function Z in the equation (8) is defined by:

$$Z = \sum_i |z_i| m_i \tag{10}$$

Where, m is the molality of species i, and z is its charge.

4. A^{φ} is the Debye-Hückel coefficient for the osmotic coefficient and equal to 0.3915 at 298.15 K, and it is decided by solvent and temperature as equation (1d).
5. The third virial coefficients, $\psi_{i,j,k}$ in equations (3) to (5) are mixed electrolyte parameters for each cation-cation-anion and anion-anion-cation triplet in mixed electrolyte solutions.
6. The parameters $B^{\phi}_{CA}, B, B'_{CA}$ which describe the interaction of pair of oppositely charged ions represent measurable combinations of the second virial coefficients. They are defined as explicit functions of ionic strength by the following equations (Kim & Frederick, 1988).

$$B^{\phi}_{CA} = \beta^{(0)}_{CA} + \beta^{(1)}_{CA} exp(-\alpha_1 I^{1/2}) + \beta^{(2)}_{CA} exp(-\alpha_2 I^{1/2}) \tag{11}$$

$$B_{CA} = \beta^{(0)}_{CA} + \beta^{(1)}_{CA} g(\alpha_1 I^{1/2}) + \beta^{(2)}_{CA} g(\alpha_2 I^{1/2}) \tag{12}$$

$$B'_{CA} = [\beta^{(1)}_{CA} g'(\alpha_1 I^{1/2}) + \beta^{(2)}_{CA} g'(\alpha_2 I^{1/2})] / I \tag{13}$$

Where the functions g and g′ in equations (10), (11) and (12) are defined by

$$g(x) = 2[1-(1+x)exp(-x)] / x^2 \tag{14}$$

$$g'(x) = -2[1-(1+x+x^2/2)exp(-x)] / x^2 \tag{15}$$

In equations (13) and (14), $x = \alpha_1 I^{1/2}$ or $= \alpha_2 I^{1/2}$.

In Pitzer's model expression in Eqns. (10) to (12), α is a function of electrolyte type and does not vary with concentration or temperature. Following Harvie et al. (1984), when either cation or anion for an electrolyte is univalent, the first two terms in equations (10) to (12) are considered, $\beta^{(2)}_{CA}$ can be neglect and α_1 =2.0 $kg^{1/2} \cdot mol^{-1/2}$, α_2 = 0 at 298.15 K. For higher valence type, such as 2-2 electrolytes for these higher valence species accounts for their increased tendency to associate in solution, the full equations from (10) to (12) are used, and α_1=1.4 $kg^{1/2} \cdot mol^{-1/2}$ and α_2=12 $kg^{1/2} \cdot mol^{-1/2}$ at 298.15 K.

7. $\Phi^{\phi}_{ij}, \Phi_{ij}, \Phi'_{ij}$ which depend upon ionic strength, are the second virial coefficients, and are given the following form (Pitzer, 1973).

$$\Phi^{\phi}_{ij} = \theta_{ij} + {}^{E}\theta_{ij} + I\,{}^{E}\theta'_{ij} \tag{16}$$

$$\Phi_{ij} = \theta_{ij} + {}^{E}\theta_{ij} \tag{17}$$

$$\Phi'_{ij} = {}^{E}\theta'_{ij} \tag{18}$$

In equations (15), (16) and (17), $\theta_{i,j}$ is an adjustable parameter for each pair of anions or cations for each cation-cation and anion-anion pair, called triplet-ion-interaction parameter. The functions, ${}^{E}\theta_{ij}$ and ${}^{E}\theta'_{ij}$ are functions only of ionic strength and the electrolyte pair type. Pitzer (1975) derived equations for calculating these effects, and Harvie and Weare (1981) summarized Pitzer's equations in a convenient form as following:

$${}^{E}\theta_{ij} = (Z_iZ_j / 4I)[J(x_{ij}) - J(x_{ii})/2 - J(x_{jj})/2] \tag{19}$$

$${}^{E}\theta'_{ij} = -({}^{E}\theta_{ij} / I) + (Z_iZ_j / 8I^2)[x_{ij}J'(x_{ij}) - x_{ii}J'(x_{ii})/2 - x_{jj}J'(x_{jj})/2] \tag{20}$$

$$x_{ij} = 6Z_iZ_jA^{\phi}I^{1/2} \tag{21}$$

In equations (18) and (19), $J(x)$ is the group integral of the short-range interaction potential energy. $J'(x)$ is the single-order differential quotient of $J(x)$, and both are independent of ionic strength and ion charges. In order to give the accuracy in computation, $J(x)$ can be fitted as the following function:

$$J(x) = 1/4x - 1 + 1/x\int_0^{\infty}[1 - exp(-x/y e C_1x^{-C_2} \cdot exp(-C_3x^{C_4})]^{-1} \tag{22}$$

$$\begin{aligned} J'(x) = {} & [4 + C_1x^{-C2} \cdot exp(-C_3x^{C_4})]^{-1} \\ & + [4 + C_1x^{-C_2}\, exp(-C_3x^{C_4})]^{-2}[C_1x\, exp(-C_3x^{C_4})(C_2x^{-C_2-1} + C_3C_4x^{C_4-1}x^{-C_2})] \end{aligned} \tag{23}$$

In equations (21) and (22), $C_1 = 4.581$, $C_2 = 0.7237$, $C_3 = 0.0120$, $C_4 = 0.528$.

Firstly, x_{ij} can be calculated according to equation (20), and $J(x)$ and $J'(x)$ were obtained from equations (21) and (22), and then to obtained ${}^{E}\theta_{ij}$ and ${}^{E}\theta'_{ij}$ from equations (18) and (19); finally, $\Phi^{\phi}_{ij}, \Phi_{ij}, \Phi'_{ij}$ can be got through equations from (15) to (17). Using the values of $\Phi^{\phi}_{ij}, \Phi_{ij}, \Phi'_{ij}$, the osmotic and activity coefficients of electrolytes can be calculated via equations from (3) to (6).

Using the osmotic coefficient, activity coefficient and the solubility products of the equilibrium solid phases allowed us to identify the coexisting solid phases and their compositions at equilibrium.

On Pitzer ion-interaction model and its extended HW model, a numbers of papers were successfully utilized to predict the solubility behaviors of natural water systems, salt-water systems, and even geological fluids (Felmy & Weare, 1986; Kim & Frederich, 1988a, 1988b; Fang et al., 1993; Song, 1998; Song & Yao, 2001, 2003; Yang, 1988, 1989, 1992, 2005).

By the way, additional work has centered on developing variable-temperature models, which will increase the applicability to a number of diverse geochemical systems. The primary focus has been to broaden the models by generating parameters at higher or lower temperatures (Pabalan & Pitzer, 1987; Spencer et al., 1990; Greenberg & Moller, 1989).

5.2 Model parameterization and solubility predictions

As to the borate solution, the crystallized behavior of borate salts is very complex. The coexisted polyanion species of borate in the liquid phase is difference with the differences of boron concentration, pH value, solvent, and the positively charged ions. The ion of $B_4O_7^{2-}$ is the general statistical express for various possible existed borates. Therefore, the structural formulas of $Na_2B_4O_7 \cdot 10H_2O$ and $K_2B_4O_7 \cdot 4H_2O$ in the solid phases of the quaternary system ($NaCl$ - KCl – $Na_2B_4O_7$ - $K_2B_4O_7$ - H_2O) are $Na_2[B_4O_5(OH)_4] \cdot 8H_2O$ and $K_2[B_4O_5(OH)_4] \cdot 2H_2O$, respectively. Borate in the liquid phase corresponding to the equilibrium solid phase maybe coexists as $B_4O_5(OH)_4^{2-}$, $B_3O_3(OH)_4^-$, $B(OH)_4^-$, and son on due to the reactions of polymerization or depolymerization of boron anion.

Therefore, in this part of predictive solubility of the quaternary system ($NaCl$ - KCl – $Na_2B_4O_7$ - $K_2B_4O_7$ - H_2O), the predictive solubilities of this system were calculated on the basis of two assumptions: Model I: borate in the liquid phase exists all in statistical form of $B_4O_7^{2-}$ i.e. $B_4O_5(OH)_4^{2-}$; Model II: borate in the liquid phase exists as various boron species of $B_4O_5(OH)_4^{2-}$, $B_3O_3(OH)_4^-$, $B(OH)_4^-$.

The necessary model parameters for the activity coefficients of electrolytes in the system at 298.15 K were fit from obtained osmotic coefficients and the sub-ternary subsystems by the multiple and unary linear regression methods.

5.2.1 Model I for the solubility prediction

Model I: Suppose that borate in solution exists as in the statistical expression form of $B_4O_7^{2-}$ i.e. $B_4O_5(OH)_4^{2-}$, and the dissolved equilibria in the system could be following:

$$Na_2B_4O_7 \cdot 10H_2O = 2Na^+ + B_4O_5(OH)_4^{2-} + 8H_2O$$

$$K_2B_4O_7 \cdot 4H_2O = 2K^+ + B_4O_5(OH)_4^{2-} + 2H_2O$$

$$NaCl = Na^+ + Cl^-$$

$$KCl = K^+ + Cl^-$$

So, the dissolved equilibrium constants can be expressed as:

$$K_{N10} = (m_{Na^+} \cdot \gamma_{Na^+})^2 \cdot (m_{B4} \cdot \gamma_{B4}) \cdot a_w^{\ 8} \qquad (24)$$

$$K_{K4} = (m_{K^+} \cdot \gamma_{K^+})^2 \cdot (m_{B4} \cdot \gamma_{B4}) \cdot a_w^{\ 2} \qquad (25)$$

$$K_{NaCl} = (m_{Na^+} \cdot \gamma_{Na^+}) \cdot (m_{Cl^-} \cdot \gamma_{Cl^-}) \qquad (26)$$

$$K_{KCl} = (m_{K^+} \cdot \gamma_{K^+}) \cdot (m_{Cl^-} \cdot \gamma_{Cl^-}) \quad (27)$$

And the electric charge balance exists as:

$$m_{Na^+} + m_{K^+} = m_{Cl^-} + 2m_{B_4O_5(OH)_4^{2-}} \quad (28)$$

Where, K, r, m, and a_w express equilibrium constant, activity coefficient, and water activity, and N10, K4 instead of the minerals of $Na_2B_4O_7 \cdot 10H_2O$, $K_2B_4O_7 \cdot 4H_2O$ (the same in the following), respectively. Then, the equilibria constants K are calculated with μ^0/RT and shown in Table 6.

The single salt parameters $\beta^{(0)}$, $\beta^{(1)}$, $C^{(\varphi)}$ of NaCl, KCl, $Na_2[B_4O_5(OH)_4]$, and $K_2[B_4O_5(OH)_4]$, two-ion interaction Pitzer parameters of $\theta_{Na,\ K}$, $\theta_{Cl,\ B4O5(OH)4}$ and the triplicate-ion Pitzer parameters of $\Psi_{Cl,\ B4O5(OH)4,\ Na}$, $\Psi_{Cl,\ B4O5(OH)4,\ K}$, $\Psi_{Na,\ K,\ Cl}$, $\Psi_{Na,\ K,\ B4O5(OH)4}$ in the reciprocal quaternary system at 298.15 K were chosen from Harvie et al. (1984), Felmy & Weare (1986), Kim & Frederick (1988), and Deng (2001) and summarized in Tables 7 and 8.

According to the equilibria constants and the Pitzer ion-interaction parameters, the solubilities of the quaternary system at 298.15 K have been calculated though the Newton's Iteration Method to solve the non-linearity simultaneous equations system, and shown in Table 9.

5.2.2 Model II for the solubility prediction

Model II: Suppose that borate in solution exists as in various boron species of $B_4O_5(OH)_4^{2-}$, $B_3O_3(OH)_4^-$, $B(OH)_4^-$ to further describe the behaviors of the polymerization and depolymerization of borate anion in solution, and the dissolved equilibria in the system could be following:

$$Na_2B_4O_7 \cdot 10H_2O = 2Na^+ + B_4O_5(OH)_4^{2-} + 8H_2O$$

$$K_2B_4O_7 \cdot 4H_2O = 2K^+ + B_4O_5(OH)_4^{2-} + 2H_2O$$

$$B_4O_5(OH)_4^{2-} + 2H_2O = B_3O_3(OH)_4^- + B(OH)_4^-$$

$$NaCl = Na^+ + Cl^-$$

$$KCl = K^+ + Cl^-$$

So, the dissolved equilibrium constants can be expressed as:

$$K_{N10} = (m_{Na^+} \cdot \gamma_{Na^+})^2 \cdot (m_{B4} \cdot \gamma_{B4}) \cdot a_w^{\ 8} \quad (29)$$

$$K_{K4} = (m_{K^+} \cdot \gamma_{K^+})^2 \cdot (m_{B4} \cdot \gamma_{B4}) \cdot a_w^{\ 2} \quad (30)$$

$$K_{B4B3B} = \frac{(m_{B3} \cdot \gamma_{B3}) \cdot (m_B \cdot \gamma_B)}{(m_{B4} \cdot \gamma_{B4}) \cdot a_w^{\ 2}} \quad (31)$$

$$K_{NaCl} = (m_{Na^+} \cdot \gamma_{Na^+}) \cdot (m_{Cl^-} \cdot \gamma_{Cl^-}) \quad (32)$$

$$K_{KCl} = (m_{K^+} \cdot \gamma_{K^+}) \cdot (m_{Cl^-} \cdot \gamma_{Cl^-}) \quad (33)$$

Where, B4, B3 and B to instead of $B_4O_5(OH)_4^{2-}$, $B_3O_3(OH)_4^-$, and $B(OH)_4^-$ for short; K_{B4B3B} expresses the equilibrium constant of the polymerized species reaction of $B_4O_5(OH)_4^{2-}$, $B_3O_3(OH)_4^-$, $B(OH)_4^-$.

And the electric charge balance exists as:

$$m_{Na^+} + m_{K^+} = m_{Cl^-} + 2m_{B_4O_5(OH)_4^{2-}} + m_{B_3O_3(OH)_4^-} + m_{B(OH)_4^-} \quad (34)$$

From this reaction of B4, B3 and B, i.e. $B_4O_5(OH)_4^{2-} + 2H_2O = B_3O_3(OH)_4^- + B(OH)_4^-$, the molalities of B3 and B are in equal. In the meantime, we suppose that two-ion and triplicate-ion interaction of different boron species would be weak, and the mixture ions parameters of different boron species should be ignored.

Similar as in model I, then, the equilibrium constant *K* existed solid phase is calculated with μ^0/RT, and also shown in Table 6, where another four possible borate salts of $NaB_3O_3(OH)_4$, $NaB(OH)_4$, $KB_3O_3(OH)_4$, $KB(OH)_4$ were also listed. The single salt parameters, binary ion interaction parameters, triplet mixture parameters and more parameters of $\theta_{Cl,B3O3(OH)4}$, $\theta_{Cl,B(OH)4}$, $\Psi_{Cl,B3O3(OH)4,Na}$ and $\Psi_{Cl,B(OH)4,Na}$ were considered, and shown in Table 8. According to the equilibria constants and the Pitzer ion-interaction parameters, the solubilities of the quaternary system at 298.15 K have been calculated though the Newton's Iteration Method to solve the non-linearity simultaneous equations system, and shown in Table 10. In fact, this theoretic calculation for the reciprocal quaternary system is equivalence of the calculated solubilities for the six-component system (Na - K - Cl - $B_4O_5(OH)_4$ - $B_3O_3(OH)_4$ - $B(OH)_4$ – H_2O). It is worthy saying that although the concentrations of Na^+, K^+, Cl^-, $B_4O_5(OH)_4^{2-}$, $B_3O_3(OH)_4^-$, $B(OH)_4^-$ in molalities could be got (Table 10), the concentrations including $B_4O_5(OH)_4^{2-}$, $B_3O_3(OH)_4^-$, $B(OH)_4^-$ should be all inverted into the concentration of $B_4O_7^{2-}$ when the Jänecke index of $B_4O_7^{2-}$ calculation.

Species	μ^0/RT	Refs	Species	μ^0/RT	Refs
H_2O	-95.6635		$B(OH)_4^-$	-465.20	
Na^+	-105.651	Harvie et al., 1984	$Na_2B_4O_5(OH)_4 \cdot 8H_2O$	-2224.16	Felmy & Weare, 1986
K^+	-113.957		$K_2B_4O_5(OH)_4 \cdot 2H_2O$	-1663.47	
Cl^-	-52.955				
$B_4O_5(OH)_4^{2-}$	-1239.10	Felmy & Weare, 1986	NaCl	-154.99	Harvie et al., 1984
$B_3O_3(OH)_4^-$	-963.77		KCl	-164.84	

Table 6. μ^0/RT of species in the system ($NaCl$ - KCl – $Na_2B_4O_7$ - $K_2B_4O_7$ - H_2O) at 298.15 K

On the basis of the calculated solubilities, a comparison diagram among model I, model II, experimental values for the reciprocal quaternary system at 298.15 K are shown in Figure 9.

Cation	Anion	$\beta_{MX}^{(0)}$	C_{MX}^{ϕ} $\beta_{MX}^{(1)}$	$C_{MX}^{(\phi)}$	Refs
Na^+	Cl^-	0.07722	0.25183	0.00106	Kim & Frederick, 1988
Na^+	$B_4O_5(OH)_4^{2-}$	-0.11	-0.40	0.0	
Na^+	$B_3O_3(OH)_4^-$	-0.056	-0.91	0.0	Felmy & Weare, 1986
Na^+	$B(OH)_4^-$	-0.0427	0.089	0.0114	
K^+	Cl^-	0.04835	0.2122	-0.00084	Harvie et al., 1984
K^+	$B_4O_5(OH)_4^{2-}$	-0.022	0.0	0.0	
K^+	$B_3O_3(OH)_4^-$	-0.13	0.0	0.0	Felmy & Weare, 1986
K^+	$B(OH)_4^-$	0.035	0.14	0.0	

Table 7. Single-salt Pitzer parameters in the system ($NaCl$ - KCl – $Na_2B_4O_7$ - $K_2B_4O_7$ - H_2O) at 298.15 K

Parameters	Values	Refs
$\theta_{Na+, K+}$	-0.012	Harvie et al., 1984
$\theta_{Cl^-, B4O5(OH)42-}$	0.074	
$\theta_{Cl^-, B3O3(OH)4^-}$	0.12	Felmy & Weare, 1986
$\theta_{Cl^-, B(OH)4^-}$	-0.065	
$\theta_{B4O5(OH)42-, B3O3(OH)4^-}$	—	—
$\theta_{B4O5(OH)42-, B(OH)4^-}$	—	—
$\theta_{B3O3(OH)4^-, B(OH)4^-}$	—	—
$\Psi_{Cl^-, B4O5(OH)42-, Na+}$	0.025	
$\Psi_{Cl^-, B3O3(OH)4^-, Na+}$	-0.024	Felmy & Weare, 1986
$\Psi_{Cl^-, B(OH)4^-, Na+}$	-0.0073	
$\Psi_{B4O5(OH)42-, B3O3(OH)4^-, Na+}$	—	—
$\Psi_{B4O5(OH)42-, B(OH)4^-, Na+}$	—	—
$\Psi_{B3O3(OH)4^-, B(OH)4^-, Na+}$	—	—
$\Psi_{Cl^-, B4O5(OH)42-, K+}$	0.0185245	Deng, 2004
$\Psi_{Cl^-, B3O3(OH)4^-, K+}$	—	—
$\Psi_{Cl^-, B(OH)4^-, K+}$	—	—
$\Psi_{B4O5(OH)42-, B3O3(OH)4^-, K+}$	—	—
$\Psi_{B4O5(OH)42-, B(OH)4^-, K+}$	—	—
$\Psi_{B3O3(OH)4^-, B(OH)4^-, K+}$	—	—
$\Psi_{Na+, K+, Cl^-}$	-0.0018	Harvie et al., 1984
$\Psi_{Na+, K+, B4O5(OH)42-}$	0.289823	Deng, 2004
$\Psi_{Na+, K+, B3O3(OH)4^-}$	—	—
$\Psi_{Na+, K+, B(OH)4^-}$	—	—

Table 8. Mixing ion-interaction Pitzer parameters in the system ($NaCl$ - KCl – $Na_2B_4O_7$ - $K_2B_4O_7$ - H_2O) at 298.15 K

Though the theoretical calculation on the basis of model II, it was found that the boron species are mainly existed $B_3O_3(OH)_4^-$ and $B(OH)_4^-$ while the concentration of $B_4O_5(OH)_4^{2-}$ is very low when the total concentration of boron is low in weak solution. This result demonstrated that the polymerization or depolymerization behaviors of borate are complex.

No.	Composition liquid phase molality, /(mol/kgH$_2$O)				Jänecke index, J/(mol/100mol dry salts)		Equilibrium solid phases*
	Na^+	K^+	Cl^-	$B_4O_7^{2-}$	$J(2Na^+)$	$J(B_4O_7^{2-})$	
1	0.2748	1.2843	0.00	0.7796	17.62	100.00	N10+K4
2	0.2768	1.3032	0.2000	0.6900	17.52	87.34	N10+K4
3	0.2804	1.3279	0.3500	0.6292	17.43	78.24	N10+K4
4	0.2863	1.3619	0.5000	0.5741	17.37	69.66	N10+K4
5	0.3014	1.4496	0.7800	0.4855	17.21	55.45	N10+K4
6	0.3177	1.5386	1.0000	0.4282	17.11	46.13	N10+K4
7	0.3603	1.7573	1.4400	0.3388	17.02	32.00	N10+K4
8	0.4297	2.0854	2.0000	0.2575	17.08	20.48	N10+K4
9	0.5028	2.4030	2.5000	0.2029	17.30	13.97	N10+K4
10	0.5858	2.7319	3.0000	0.1589	17.66	9.58	N10+K4
11	0.6774	3.0673	3.5000	0.1224	18.09	6.54	N10+K4
12	0.8903	3.7442	4.5000	0.06725	19.21	2.90	N10+K4
13,A1	1.0298	4.1687	5.1107	0.04392	19.81	1.69	N10+K4+KCl
14	0.00	4.8834	4.7149	0.08423	0.00	3.45	K4+KCl
15	0.1500	4.7688	4.7699	0.07444	3.05	3.03	K4+KCl
16	0.3000	4.6592	4.8262	0.06649	6.05	2.68	K4+KCl
17	0.4500	4.5534	4.8835	0.05996	8.99	2.40	K4+KCl
18	0.6000	4.4509	4.9418	0.05457	11.88	2.16	K4+KCl
19	0.7500	4.3511	5.0009	0.05009	14.70	1.96	K4+KCl
20	0.9000	4.2536	5.0609	0.04636	17.46	1.80	K4+KCl
21	4.8000	2.2523	7.0507	8.13E-4	68.06	0.023	N10+KCl
22	4.3000	2.4641	6.7621	9.67E-4	63.57	0.029	N10+KCl
23	3.6000	2.7821	6.3794	0.00139	56.41	0.044	N10+KCl
24	3.2000	2.9750	6.1713	0.00185	51.82	0.060	N10+KCl
25	2.8000	3.1758	5.9705	0.00263	46.86	0.088	N10+KCl
26	2.4000	3.3843	5.7762	0.00405	41.49	0.14	N10+KCl
27	2.0000	3.6007	5.5869	0.00686	35.71	0.25	N10+KCl
28	1.6000	3.8255	5.3992	0.01316	29.49	0.49	N10+KCl
29	5.2183	1.9000	7.1163	9.75E-4	73.31	0.027	N10+NaCl
30	5.4046	1.5000	6.9014	0.00159	78.28	0.046	N10+NaCl
31	5.6432	1.0000	6.6369	0.00313	84.95	0.094	N10+NaCl
32	5.8894	0.5000	6.3762	0.0066	92.17	0.21	N10+NaCl
33	6.1479	0.00	6.1178	0.01504	100.00	0.49	N10+NaCl
34,B1	5.1148	2.1256	7.2389	7.53E-4	70.64	0.021	N10+NaCl+KCl
35	5.1147	2.1259	7.2394	6.00E-4	70.64	0.017	NaCl+KCl
36	5.1145	2.1264	7.2401	4.00E-4	70.63	0.011	NaCl+KCl
37	5.1143	2.1269	7.2408	2.00E-4	70.63	0.0055	NaCl+KCl
38	5.1142	2.1273	7.2415	0.00	70.62	0.00	NaCl+KCl

Table 9. Calculated solubility data of the system ($NaCl$ - KCl – $Na_2B_4O_7$ - $K_2B_4O_7$ - H_2O) at 298.15 K on the basis of Model I. * N10, $Na_2B_4O_7 \cdot 10H_2O$; K4, $K_2B_4O_7 \cdot 4H_2O$.

No.	Composition liquid phase molality, /(mol/kgH$_2$O)*						Jänecke index, **J /(mol/100mol dry-salt)**		Equilibrium solid phases
	Na$^+$	K$^+$	Cl$^-$	B4	B3	B	J(2Na$^+$)	J(B$_4$O$_7^{2-}$)	
1	0.3362	1.4923	0.00	0.7084	0.2058	0.2058	18.38	100.00	N10+K4
2	0.3394	1.5158	0.2000	0.6362	0.1914	0.1914	18.28	89.22	N10+K4
3	0.3431	1.5428	0.3500	0.5868	0.1812	0.1812	18.19	81.44	N10+K4
4	0.3485	1.5771	0.5000	0.5413	0.1715	0.1715	18.10	74.03	N10+K4
5	0.3628	1.6608	0.7800	0.4667	0.1551	0.1551	17.93	61.45	N10+K4
6	0.3777	1.7431	1.0000	0.4165	0.1439	0.1439	17.81	52.85	N10+K4
7	0.4168	1.9432	1.4400	0.3347	0.1253	0.1253	17.66	38.98	N10+K4
8	0.4818	2.2464	2.0000	0.2562	0.1079	0.1079	17.66	26.69	N10+K4
9	0.5519	2.5451	2.5000	0.2018	0.0967	0.0967	17.82	19.28	N10+K4
10	0.6326	2.8591	3.0000	0.1574	0.0884	0.0884	18.12	14.08	N10+K4
11	0.7234	3.1822	3.5000	0.1207	0.0821	0.0821	18.52	10.38	N10+K4
12	0.9355	3.8426	4.5000	0.0654	0.0736	0.0736	19.58	5.82	N10+K4
13,A2	1.0770	4.2295	5.0794	0.04305	0.0705	0.0705	20.30	4.28	N10+K4+KCl
14	0.00	4.9588	4.6844	0.08694	0.0502	0.0502	0.00	5.53	K4+KCl
15	0.1500	4.8452	4.7369	0.07661	0.0525	0.0525	3.00	5.17	K4+KCl
16	0.3000	4.7368	4.7905	0.06822	0.0549	0.0549	5.96	4.89	K4+KCl
17	0.4500	4.6326	4.8449	0.06133	0.0576	0.0576	8.85	4.68	K4+KCl
18	0.6000	4.5320	4.8999	0.05563	0.0604	0.0604	11.69	4.52	K4+KCl
19	0.7500	4.4342	4.9558	0.05089	0.0633	0.0633	14.47	4.41	K4+KCl
20	0.9000	4.3390	5.0122	0.04692	0.0665	0.0665	17.18	4.33	K4+KCl
21	4.8000	2.2665	7.0005	8.237E-4	0.0322	0.0322	67.93	0.93	N10+KCl
22	4.3000	2.4815	6.7150	9.779E-4	0.0323	0.0323	63.41	0.98	N10+KCl
23	3.6000	2.8052	6.3356	0.00141	0.0334	0.0334	56.20	1.09	N10+KCl
24	3.2000	3.0021	6.1289	0.00188	0.0347	0.0347	51.60	1.18	N10+KCl
25	2.8000	3.2078	5.9289	0.00268	0.0368	0.0368	46.61	1.31	N10+KCl
26	2.4000	3.4226	5.7346	0.00415	0.0398	0.0398	41.22	1.51	N10+KCl
27	2.0000	3.6474	5.5442	0.0071	0.0445	0.0445	35.41	1.83	N10+KCl
28	1.6000	3.8845	5.3532	0.0138	0.0519	0.0519	29.17	2.40	N10+KCl
29	5.2316	1.9000	7.0663	9.963E-4	0.0316	0.0316	73.36	0.91	N10+NaCl
30	5.4169	1.5000	6.8530	0.00163	0.0303	0.0303	78.31	0.92	N10+NaCl
31	5.6544	1.0000	6.5902	0.00321	0.0289	0.0289	84.97	0.96	N10+NaCl
32	5.8998	0.5000	6.3310	0.00679	0.0276	0.0276	92.19	1.07	N10+NaCl
33	6.1580	0.00	6.0739	0.0155	0.0266	0.0266	100.00	1.37	N10+NaCl
34,B2	5.1668	2.1309	7.2304	7.463E-4	0.0329	0.0329	70.80	0.92	N10+KCl+NaCl
35	5.1612	2.1307	7.2318	6.00E-4	0.0294	0.0294	70.78	0.82	NaCl+KCl
36	5.1523	2.1303	7.2339	4.00E-4	0.0239	0.0239	70.75	0.67	NaCl+KCl
37	5.1410	2.1296	7.2364	2.00E-4	0.0168	0.0168	70.71	0.47	NaCl+KCl
38	5.1142	2.1273	7.2415	0.00	0.00	0.00	70.62	0.00	NaCl+KCl

Table 10. Calculated solubility data of the system ($NaCl$ - KCl – $Na_2B_4O_7$ - $K_2B_4O_7$ - H_2O) at 298.15 K on the basis of Model II. * B4, B3, B express for $B_4O_5(OH)_4^{2-}$, $B_3O_3(OH)_4^-$, $B(OH)_4^-$; N10, $Na_2B_4O_7 \cdot 10H_2O$; K4, $K_2B_4O_7 \cdot 4H_2O$.

In Figure 9, compared with Models I and II, the calculated values in the boundary points and the cosaturated point of ($Na_2B_4O_7 \cdot 10H_2O$ + KCl + NaCl) based on model II were in good agreement with the experimental data. However, in the cosaturated point of ($Na_2B_4O_7 \cdot 10H_2O$ + $K_2B_4O_7 \cdot 4H_2O$ + KCl), a large difference on the solubility curve still existed. Reversely, the predictive result based on model II closed to the experimental curve. There were two possible reasons: one is that the structure of borate in solution is very complex, an the Pitzer's parameters of borate salts is scarce; the other one is the high saturation degree of borate, the difference between the experimental equilibrium constant and the theoretic calculated equilibrium constant was large enough.

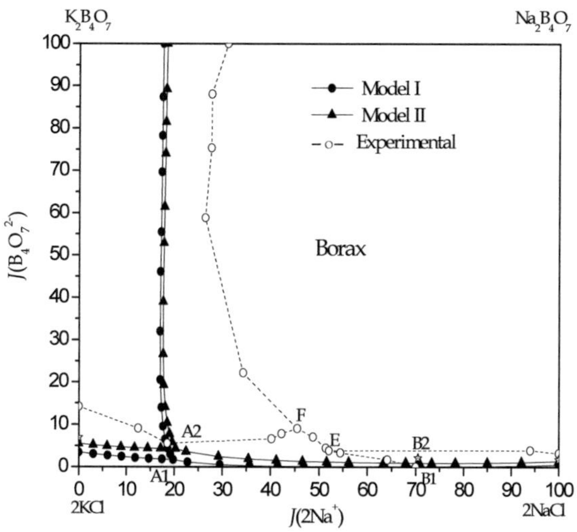

Fig. 9. Comparison of the experimental and calculated phase diagram of the quaternary system (NaCl - KCl – $Na_2B_4O_7$ - $K_2B_4O_7$ - H_2O) at 298.15 K. -●-, Calculated based on Model I; -▲-, Calculated based on Model II; -○-, Experimental.

6. Acknowledgments

Financial support from the State Key Program of NNSFC (Grant.20836009), the NNSFC (Grant. 40773045), the "A Hundred Talents Program" of the Chinese Academy of Sciences (Grant. 0560051057), the Specialized Research Fund for the Doctoral Program of Chinese Higher Education (Grant 20101208110003), The Key Pillar Program in the Tianjin Municipal Science and Technology (11ZCKFGX2800) and Senior Professor Program of Tianjin for TUST (20100405) is acknowledged. Author also hopes to thank all members in my research group and my Ph.D. students Y.H. Liu, S.Q. Wang, Y.F. Guo, X.P. YU, J. Gao, L.Z. Meng, DC Li, Y. Wu, and D.M. Lai for their active contributions on our scientific projects.

7. References

Analytical Laboratory of Institute of Salt Lakes at CAS. (1988). *The analytical methods of brines and salts*, 2nd ed., Chin. Sci. Press, ISBN 7-03-000637-2, pp. 35-41 & 64-66, Beijing

Deng, T.L. (2004). Phase equilibrium on the aqueous five-component system of lithium, sodium, potassium, chloride, borate at 298.15 K, *J. Chem. Eng. Data,* Vol. 47, No. 9, pp. 1295-1299, ISSN 0021-9568

Deng, T.L.; Li, D. (2008a). Solid-liquid metastable equilibria in the quaternary system ($NaCl + LiCl + CaCl_2 + H_2O$) at 288.15 K, *J. Chem. Eng. Data,* Vol. 53, No. 11, pp. 2488–2492, ISSN 0021-9568

Deng, T.L.; Li, D.C. (2008b). Solid-liquid metastable equilibria in the quaternary system ($NaCl + KCl + CaCl_2 + H_2O$) at 288.15 K, *Fluid Phase Equilibria,* Vol. 269, No. 1-2, pp. 98-103, ISSN 0378-3812

Deng, T.L.; Li, D.C.; Wang, S.Q. (2008c). Metastable phase equilibrium on the aqueous quaternary system ($KCl - CaCl_2 - H_2O$) at (288.15 and 308.15) K, *J. Chem. Eng. Data,* Vol. 53, No. 4, pp. 1007-1011, ISSN 0021-9568

Deng, T.L.; Meng, L.Z.; Sun, B. (2008d). Metastable phase equilibria of the reciprocal quaternary system containing sodium, potassium, chloride and borate ions at 308.15 K, *J. Chem. Eng. Data,* Vol. 53, No. 3, pp. 704-709, ISSN 0021-9568

Deng, T.L.; Wang, S.Q. (2008e). Metastable phase equilibrium in the reciprocal quaternary system ($NaCl + MgCl_2 + Na_2SO_4 + MgSO_4 + H_2O$) at 273.15 K, *J. Chem. Eng. Data,* Vol. 53, No. 12, pp. 2723-2727, ISSN 0021-9568

Deng, T.L.; Wang, S.Q.; Sun, B. (2008f). Metastable phase equilibrium on the aqueous quaternary system ($KCl + K_2SO_4 + K_2B_4O_7 + H_2O$) at 308.15 K, *J. Chem. Eng. Data,* Vol. 53, No. 2, pp. 411-414, ISSN 0021-9568

Deng, T.L.; Yin, H.A.; Tang, M.L. (2002). Experimental and predictive phase equilibrium of the quary System at 298K, *J. Chem. Eng. Data,* Vol. 47, pp. 26-29, ISSN 0021-9568

Deng, T.L.; Yin, H.J.; Guo, Y.F. (2011). Metastable phase equilibrium in the aqueous ternary system ($Li_2SO_4 + MgSO_4 + H_2O$) at 323.15 K, *J. Chem. Eng. Data,* Vol. 56, No. 9, pp. 3585-3588, ISSN 0021-9568

Deng, T.L.; Yin, H.J.; Li, D.C. (2009a). Metastable phase equilibrium on the aqueous ternary system ($K_2SO_4 - MgSO_4 - H_2O$) at 348.15 K, *J. Chem. Eng. Data,* Vol. 54, No. 2, pp. 498-501, ISSN 0021-9568

Deng, T.L.; Yu, X.; Li, D.C. (2009b). Metastable phase equilibrium on the aqueous ternary system ($K_2SO_4 - MgSO_4 - H_2O$) at (288.15 and 308.15) K, *J. Solution Chem,* Vol. 38, No. 1, pp. 27-34, ISSN 0095-9782

Deng, T.L.; Yu, X.; Sun, B. (2008g). Metastable phase equilibrium in the aqueous quaternary system ($Li_2SO_4 - K_2SO_4 - MgSO_4 - H_2O$) at 288.15 K, *J. Chem. Eng. Data,* Vol. 53, No. 11, pp. 2496–2500, ISSN 0021-9568

Deng, T.L.; Zhang, B.J.; Li, D.C.; Guo, Y.F. (2009c). Simulation studies on the metastable phase equilibria in the aqueous ternary systems ($NaCl - MgCl_2 - H_2O$) and ($KCl - MgCl_2 - H_2O$) at 308.15 K, *Front. J. Chem. Eng.,* Vol. 3, No. 2 , pp. 172-175, ISSN 2095-0179

Donad, S.P.; Kean, H.K. (1985). The application of the Pitzer equations to 1-1 electrolytes in mixed solvents, *J. Solution Chem,* Vol. 14, No. 4, pp. 635-651, ISSN 0095-9782

Fang, C.H.; Song, P.S.; Chen, J.Q. (1993). Theoretical calculation of the metastable phase diagram of the quinary system($Na + K + Cl + SO_4 + CO_3 + H_2O$) at 25°C, *J. Salt Lakes Res.,* Vol. 1, No. 2, pp. 16-22, ISSN1008-858X

Felmy, A.R.; Weare, J.H. (1986). The prediction of borate mineral equilibria in nature waters: Application to Searles Lake, California, *Geochim. Cosmochim. Acta,* Vol. 50, No. 10, pp. 2771-2783, ISSN 0016-7037

Gao, J.; Deng, T.L. (2011a). Metastable phase equilibrium in the aqueous quaternary system ($LiCl + MgCl_2 + Li_2SO_4 + MgSO_4 + H_2O$) at 308.15 K, *J. Chem. Eng. Data,* Vol. 56, No. 4, pp. 1452-1458, ISSN 0021-9568

Gao, J.; Deng, T.L. (2011b). Metastable phase equilibrium in the aqueous quaternary system ($MgCl_2 + MgSO_4 + H_2O$) at 308.15 K, *J. Chem. Eng. Data,* Vol. 56, No. 5, pp. 1847-1851, ISSN 0021-9568

Gao, S.Y.; Song, P.S.; Xia, S.P.; Zheng, M.P. (2007). *Salt Lake Chemistry – A new type of lithium and boron salt lake,* Chinese Science Press, ISBN 978-7-03-016972-3, Beijing

Greenberg, J.P.; Moller, N. (1989). The prediction of mineral solubilities in natural waters: A chemical equilibrium model for the Na-K-Ca-Cl- SO_4-H_2O system to high concentration from 0°C to 250°C, *Geochim. Cosmochim. Acta,* Vol. 53, No. 9, pp. 2503-2518, ISSN 0016-7037

Guo, Y.F.; Yin, H.J.; Wu, X.H.; Deng, T.L. (2010). Metastable phase equilibrium in the aqueous quaternary system ($NaCl + MgCl_2 + Na_2SO_4 + MgSO_4 + H_2O$) at 323.15 K, *J. Chem. Eng. Data,* Vol. 55, No. 10, pp. 4215-4220, ISSN 0021-9568

Harvie, C.E.; Moller, N.; Weare, J.H. (1984). The prediction of mineral solubilities in natural waters: The Na-K-Mg-Ca-H-C1-SO_4-OH-HCO_3-CO_3-CO_2-H_2O system to high ionic strength at 25°C, *Geochim.Cosmochim. Acta,* Vol. 48, No. 4, pp. 723-751, ISSN 0016-7037

Harvie, C.E.; Weare, J.H. (1980). The prediction of mineral solubilities in natural waters: the Na-K-Mg-Ca-Cl-SO_4-H_2O system from zero to high concentration at 25°C, *Geochim. Cosmochim. Acta,* Vol. 44, No. 7, pp. 981-997, ISSN 0016-7037

Jin, Z. M.; Xiao, X. Z.; Liang, S. M. (1980). Study of the metastable equilibrium for pentanary system of (Na^+, K^+, Mg^{2+}),(Cl^-, SO_4^{2-}), H_2O. *Acta Chim. Sinica,* Vol. 38, No. 2, pp. 313-321, ISSN 0567-7351

Jin, Z. M.; Zhou, H. N.; Wang, L. S. (2001). Study on the metastable phase equilibrium of Na^+, K^+, Mg^{2+}//Cl^-, SO_4^{2-} -H_2O quinary system at 35℃. *Chem. J. Chin. Univ.* Vol. 22, No. 3, pp. 634-638, ISSN 0251-0790

Jin, Z. M.; Zhou, H. N.; Wang, L. S. (2002). Studies on the metastable phase equilibrium of Na^+, K^+, Mg^{2+}//Cl^-, SO_4^{2-} -H_2O quinary system at 15℃. *Chem. J. Chin. Univ.* Vol. 23, No. 3, pp. 690-694, ISSN 0251-0790

Kim, H-T.; Frederick, W.J. (1988a). Evaluation of Pitzer ion interaction parameters of aqueous electrolytes at 25 ℃.1.Single salt parameters, *J. Chem. Eng. Data,* Vol. 33, No. 2, pp. 177-184, ISSN 0021-9568

Kim, H-T.; Frederick, W.J. (1988b). Evaluation of Pitzer ion interaction parameters of aqueous electrolytes at 25℃.2.Ternary mixing parameters, *J. Chem. Eng. Data,* Vol. 33, No. 3, pp. 278-283, ISSN 0021-9568

Kurnakov, N. S. (1938). Solar phase diagram in the regions of potassium salts for the five-component system of ($NaCl - MgSO_4 - MgCl_2 - KCl - H_2O$ at (288.15, 298.15, and 308.15) K, *Eehv. Fieh. Khim. Analieha,* Vol.10, No. 5, pp. 333-366.

Li, D.C.; Deng, T.L. (2009). Solid-liquid metastable equilibria in the quaternary system ($NaCl + KCl + CaCl_2 + H_2O$) at 308.15 K, *J. Therm. Anal. Calorim,* Vol. 95, No. 2, pp. 361-367, ISSN 0021-9568

Li, Y.H.; Song, P.S.; Xia, S.P. (2006). Solubility prediction for the HCl–$MgCl_2$ –H_2O system at 40℃ and solubility equilibrium constant calculation for HCl·$MgCl_2$·$7H_2O$ at 40℃, *CALPHAD,* Vol. 30, No. 1, pp. 61-64, ISSN 0364-5916

Li, Z.Y.; Deng, T.L.; Liao, M.X. (2010). Solid-liquid metastable equilibria in the quaternary system Li_2SO_4 + $MgSO_4$ + Na_2SO_4 + H_2O at T = 263.15 K, *Fluid Phase Equilibria,* Vol. 293, No.1, pp. 42-46, ISSN 0378-3812

Liu, Y.H.; Deng, T.L.; Song, P.S. (2011).Metastable phase equilibrium of the reciprocal quaternary system LiCl + KCl + Li_2SO4 + K_2SO_4 + H_2O at 308.15 K, *J. Chem. Eng. Data,* Vol. 56, No. 4, pp. 1139-1147, ISSN 0021-9568

Meng, L.Z.; Deng, T.L. (2011). Solubility prediction for the system of ($MgCl_2$–$MgSO_4$–MgB_4O_7–H_2O) at 298.15 K using the ion-interaction model, *Russ. J. Inorg.* Chem, Vol. 56, No. 8, pp. 1-4, ISSN 0036-0236

Pabalan, R.T.; Pitzer, K.S. (1987). Thermodynamics of concentrated electrolyte mixtures and the prediction of mineral solubilities to high temperatures for mixtures in the system Na-K-Mg-Cl-SO_4-OH-H_2O, *Geochim. Cosmochim. Acta,* Vol. 51, No. 9, pp. 2429-2443, ISSN 0016-7037

Pitzer, K.S. (1973a). Thermodynamics of electrolytes I: theoretical basis and general equation. *J. Phys. Chem,* Vol. 77, No. 2, pp. 268-277, ISSN 1932-7447

Pitzer, K.S. (1973b). Thermodynamics of electrolytes. II. Activity and osmotic coefficients for strong electrolytes with one or both ions univalent, *J. Phys. Chem,* Vol. 77, No. 19, pp. 2300-2308, ISSN 1932-7447

Pitzer, K.S. (1974a). Thermodynamics of electrolytes. III. Activity and osmotic coefficients for 2-2 electrolytes, *J. Solution Chem,* Vol. 3, No. 7, pp. 539-546, ISSN 0095-9782

Pitzer, K.S. (1974b). Thermodynamics of electrolytes. IV. Activity and osmotic coefficients for mixed electrolytes, *J. Am. Chem. Soc,* Vol. 96, No. 18, pp. 5701-5707, ISSN 0002-7863

Pitzer, K.S. (1975). Thermodynamics of electrolytes. V. Effects of higher-order electrostatic terms, *J. Solution Chem,* Vol. 4, No. 3, pp. 249-265, ISSN 0095-9782

Pitzer, K.S. (1977). Electrolytes theory-improvements since Debye-Hückel, *Account Chem. Res,* Vol. 10, No. 10, pp. 371-377, ISSN 0013-4686

Pitzer, K.S. (1995). *Semiempirical equations for pure and mixed electrolytes. Thermodynamics,* 3rd ed.; McGraw-Hill Press, ISBN 0-07-050221-8, New York, Sydney, Tokyo, Toronto

Pitzer, K.S. (2000). *Activity coefficients in electrolyte solutions,* 2rd ed.; CRC Press, ISBN 0-8493-5415-3, Boca Raton, Ann Arbor, Boston, London

Song, P.S. (1998). Calculation of the metastable phase diagram for sea water system, *J. Salt Lakes Res.,* Vol. 6, No. 2-3, pp. 17-26, ISSN1008-858X

Song, P.S. (2000). Salt lakes and their exploiting progresses on the relatively resources. *J. Salt Lakes Res..* Vol. 8, No. 1, pp. 1-16, ISSN1008-858X

Song, P.S.; Du, X.H. (1986). Phase equilibrium and properties of the saturated solution in the quaternary system $Li_2B_4O_7$-Li_2SO_4-LiCl-H_2O at 25°C, *Chinese Science Bulletin,* Vol. 31, No. 19, pp. 1338-1343, ISSN 0567-7351

Song, P.S.; Yao, Y. (2001). Thermodynamics and phase diagram of the salt lake brine system at 25°C I. Li^+, K^+, Mg^{2+}//Cl^-, SO_4^{2-}–H_2O system, *CALPHAD,* Vol. 25, No. 3, pp. 329-341, ISSN 0364-5916

Song, P.S.; Yao, Y. (2003). Thermodynamic model for the salt lake brine system and its applications–I. Applications in physical chemistry for the system Li^+, Na^+, K^+, Mg^{2+}/Cl^-, SO_4^{2-} – H_2O, *J. Salt Lakes Res.*, Vol. 11, No. 3, pp. 1-8, ISSN1008-858X

Spencer, R.J.; Moller, N.; Weare, J. (1990). The prediction of mineral solubilities in natural waters: A chemical equilibrium model for the Na-K-Ca-Mg-Cl-SO_4-H_2O system at temperatures blow 25°C, *Geochim. Cosmochim. Acta,* Vol. 54, No. 4, pp. 575-590, ISSN 0016-7037

Su, Y. G.; Li, J.; Jiang, C. F. (1992). Metastable phase equilibrium of K^+, Na^+, Mg^{2+}//Cl^-, SO_4^{2-} - H_2O quinary system at 15℃. *J. Chem. Ind. Eng.* Vol. 43, No. 3, pp. 549-555, ISSN 0438-1157

Teeple, J.E. (1929). *The development of Searles Lake brines: with equilibrium data,* American Chemical Society Monograph Series, Chemical Catalog Company Press, ISBN 0-07-571934-6, New York.

Vant'hoff, J.H. (1912). Solubilities of the five-component system of (NaCl - $MgSO_4$ – MgCl2 – KCl - H_2O at 298.15 K, in: Howard L. Silcock. (1979). Solubilities of inorganic and organic compounds, Vol.3 Part 2: Ternary and multi-component systems of inorganic substances, Pergamon Press, ISBN 0-08-023570-0, Oxford, New York, Toronto, Sydney, Paris, Frankfurt.

Wang, S.Q.; Deng, T.L. (2008). Solid-liquid isothermal evaporation phase equilibria in the aqueous ternary system (Li_2SO_4+ $MgSO_4$+ H_2O) at T = 308.15 K, *Journal of Chemistry Thermodynamics,* Vol. 40, No. 6, pp. 1007-1011, ISSN 0021-9614

Wang, S.Q.; Deng, T.L. (2010). Metastable phase equilibria of the reciprocal quaternary system containing lithium, sodium, chloride, and sulfate ions at 273.15 K, *J. Chem. Eng. Data,* Vol. 55, No. 10, pp. 4211-4215, ISSN 0021-9568

Wang, S.Q.; Guo, Y.F.; Deng, T.L. (2011a). Solid-liquid metastable equilibria in the aqueous ternary system containing lithium, magnesium, and sulfate ions at 273.15 K, *Proceedings of CECNet,* Vol. 6, pp. 5243-5246, XianNing, China, ISBN 978-1-61284-470-1, April 16-18, 2011

Wang, S.Q.; Guo, Y.F.; Deng, T.L. (2011b). Solubility predictions for the reciprocal quaternary system (NaCl + MgCl2 + Na2SO4 + MgSO4 + H2O) at 283.15 K using Pitzer ion-interaction model, *Proceedings of CECNet,* Vol. 2, pp. 1661-1664, XianNing, China, ISBN 978-1-61284-470-1, April 16-18, 2011

Yang, J.Z. (1988). Thermodynamics of electrolyte mixtures. Activity and osmotic coefficient with higher-order limiting law for symmetrical mixing, *J. Solution Chem,* Vol., No. 7, pp. 909-924, ISSN 0095-9782

Yang, J.Z. (1989). The application of the ion-interaction model to multi-component 1-1 type electrolytes in mixed solvents, *J. Solution Chem,* Vol. 3, No. 2, pp. 201-210, ISSN 0095-9782

Yang, J.Z. (1992). Thermodynamics of amino acid dissociation in mixed solvents. 3: Glycine in aqueous glucose solutions from 5 to 45°C, *J. Solution Chem,* Vol. 11, No. 9, pp. 1131-1143, ISSN 0095-9782

Yang, J.Z. (2005). Medium effect of an organic solvent on the activity coefficients of HCl consistent with Pitzer's electrolyte solution theory, *J. Solution Chem,* Vol. 1, No. 1, pp. 71-76, ISSN 0095-9782

Zhao, Y.Y. (2003). Chinese salt lake resources and the exploiting progresses. *Mineral Deposits,* Vol. 22, No. 1, pp. 99-106, ISSN 0258-7106

Zheng, M.P.; Xiang, J.; Wei, X.J.; Zheng, Y. (1989). *Saline lakes on the Qinghai-Xizang (Tibet) Plateau*, Beijing Science and Technology Press, ISBN 7-5304-0519-5, Beijing
Zheng, X. Y.; Tang, Y.; Xu, C.; Li, B.X.; Zhang, B.Z.; Yu, S.S. (1988). *Tibet saline lake*; Chinese Science Press, ISBN 7-03-000333-0, Beijing

17

"Salt Weathering" Distress on Concrete by Sulfates?

Zanqun Liu[1,2,3], Geert De Schutter[2], Dehua Deng[1,3] and Zhiwu Yu[1,3]
[1]*School of Civil Engineering, Central South University, Changsha, Hunan,*
[2]*Magnel Laboratory for Concrete Research, Department of Structural Engineering, Ghent University, Ghent,*
[3]*National Engineering Laboratory for High Speed Railway Construction, Changsha, Hunan,*
[1,3]*P.R China*
[2]*Belgium*

1. Introduction

Salt weathering, also called salt crystallization or physical salt attack, is defined as the basic degradation mechanism that a porous material, such as stone and masonry, undergoes at and near the Earth's surface [1]. The parts of porous materials in contact with relatively dry air near the Earth's surface will be severely deteriorated but the parts buried in salts environment look sound.

Generally, the idea of sulfate attack on concrete means that a complex physiochemical process including several harmful productions formation through chemical reaction, such as ettringite and gypsum, following the crystal growth of these productions in cracks or pores resulting in concrete damage. However, another concept was given more and more attention that "salt weathering/physical salt attack" on concrete partially exposed to environment specially containing Na_2SO_4 or $MgSO_4$. ACI (American Concrete Institute) created a new subcommittee, ACI 201-E (Salt Weathering/Physical Salt Attack) in 2009. In 2011, an ballot was performed to discuss if it is necessary to separate the "physical salt attack" from chapter 6 "sulfate attack" as chapter 8 for ACI 201.2R. There were also more and more reports discussing this topic [2-9]. It seems that this topic will be high interest and relevance for the concrete community.

Certainly, concrete is also a kind of porous material. When partially exposed to an environment containing salts (especially sodium sulfate), such as in the case of a foundation, dam, column, flatwork and tunnel, a large amount of efflorescence will appear on the surface of the concrete accompanied with a similar scaling manner as salt weathering distress on masonry, showing a freezing-and-thawing-like deterioration on the surface of concrete [2] (Fig. 1). Therefore, concrete technologists logically and involuntarily define this phenomenon as salt weathering distress on concrete or physical attack on concrete.

Apparently, it seems reasonable to attribute salt weathering to the decay of concrete partially exposed to sulfate environment. Concrete technologists subjectively accepted that salt weathering or salt crystallization cannot be avoided in concrete, because concrete is also a kind of porous material similar to stone. However, in effect, some field and indoor research results of "salt weathering" distress on concrete have shown a number of appearances opposite to the basic principles of salt weathering on porous materials. Therefore, it is necessary and imperative to present this problem to avoid further confusion.

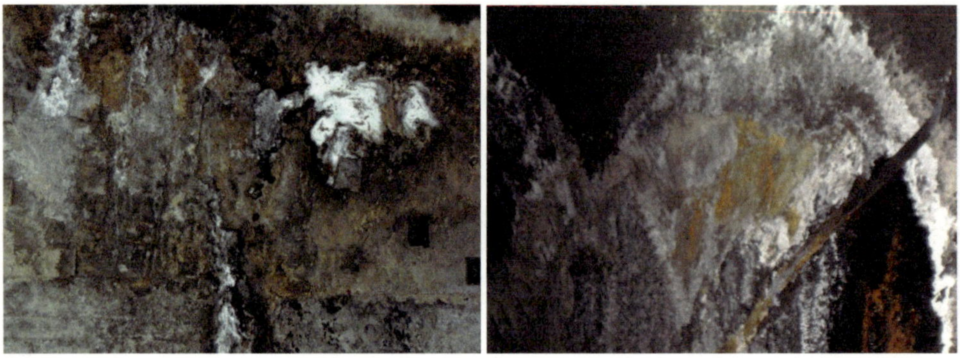

Fig. 1. Deterioration of railway tunnel (Southwestern Region, China)

This review paper includes three parts. First, the basic principles of salt weathering on porous materials are reviewed. Second, some field and indoor tests of "salt weathering" on concrete by sulfates are presented. Some appearances, which were generated by "salt weathering" on concrete but were opposite to the basic principles of salt weathering on porous materials, are analyzed in detail. Several points that need further study are presented in the third part.

2. Salt weathering distress on porous materials

2.1 Salt crystallization in pore

The work of Carl W. Correns on crystallization pressure is undoubtedly a milestone in the field of durability of porous materials [10], and the equation (Eq. (1)) exhibited in his paper written in 1949 for crystallization is broadly used and quoted.

$$P = \frac{RT}{v}\ln\left(\frac{C}{Cs}\right) \tag{1}$$

Where R is the ideal gas constant, T is the absolute temperature, v is the molar volume, C is the concentration of solution, and Cs is the concentration of saturated solution. C/Cs is the supersaturation.

The above equation indicates that supersaturation is the key factor for crystallization. The supersaturation should be maintained during the process of salt crystallization. The crystal will grow until the supersaturation is consumed. He also pointed out that a thin layer/film of aqueous solution always remains between the crystal and the internal solid walls of the porous network. The thin layer allows the solute to diffuse from the pore solution to the crystal surface that is growing against the pore wall. If this thin layer did not exist, the crystal would go into contact with the pore wall, the growth would stop and no crystallization pressure would form [11]. Diffusion through this thin layer will equalize the concentration at the tip of the crystal and in the gap between the side of the crystal and the pore wall [12] [13]. The concentration and mobility of ions within this gap have a profound impact on the crystallization stress [14].

On the other hand, for a crystal, when the equilibrium is established between solution and crystal, the solubility product will satisfy:

$$\gamma_{cl}\kappa_{cl} = \frac{RT}{v}\ln(\frac{C}{Cs}) \tag{2}$$

Where, γ_{cl} is the crystal /liquid interfacial energy; κ_{cl} is the surface curvature of crystal. Eq. (2) means two facts: a smaller spherical crystal is in equilibrium with a higher concentration than a larger flat crystal (equilibrium growth). The larger crystal (a relatively flat crystal) will grow and consume the supersaturation. Consequently, the smaller crystal will dissolve and the liberated solution will diffuse to the larger crystal (non-equilibrium growth) [14].

For equilibrium growth, a confined crystal can only exert stress if it is in contact with a pore solution that is supersaturated with respect to the unloaded face of the crystal [15]. The stress can be obtained by Eq. (3) [11]:

$$\sigma_W = \gamma_{cl}(\kappa_{cl}^C - \kappa_{cl}^E) \tag{3}$$

Where, κ_{cl}^E is the curvature of the pore entrance (labeled point E), and κ_{cl}^C is the curvature of other internal points (labeled point C) (Fig.2).

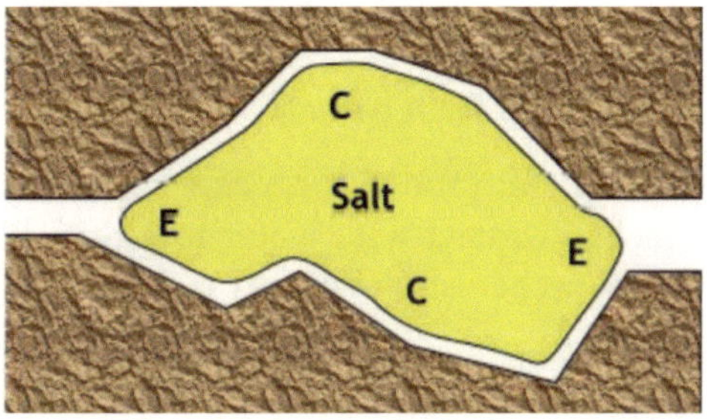

Fig. 2. Schematic of crystal of salt growing in a pore [14]

Because κ_{cl}^E is less convex (positive) than κ_{cl}^C, the compressive strength is negative, but it creates a tensile stress in the hoop direction around the pore. This tensile stress is the destructive "crystallization pressure" A high equilibrium crystallization pressure requires a confined crystal in a pore of any geometry with a very small pore entrance [16]. Therefore, the stones with a bimodal pore size distribution are extremely susceptible to salt attack [17-19].

For non-equilibrium growth, all of the crystals in internal pores of a matrix with a distribution of pore sizes are unstable with respect to macroscopic crystals that nucleate in large voids. During the drying (evaporation) or in the presence of a sharp temperature gradient, the smaller crystals will dissolve and feed the growth of the larger one, reaching another equilibrium. During this equilibrium, a high transient stress can be produced (Eq. (4)) [14].

$$\sigma_W = \frac{\gamma_{CL}}{r_s} - \frac{R_g T}{V_m}\ln(\frac{C}{Cs}) \tag{4}$$

Where, r_s is the radius of the small pore entrance.

The duration and intensity of the transient crystallization pressure depend on three factors [14]: (1) the rate of supply of solute; (2) the rate of growth of crystal; (3) the rate of diffusion of solute to macro-pores. High evaporation can result in high supersaturation, and increase the growth of crystal and result in a high transient stress [14] [20], leading to severe damage by salt crystallization.

The supersaturation can be produced by cooling, evaporation and drying and wetting cycle. If the temperature dependence of the solubility of a salt is high, a drop of temperature can result in supersaturation. Supersaturation caused by evaporation always occurs when one face of the porous material is in contact with the solution and the other face is exposed to relatively dry conditions, i.e., the salt weathering process.

As to the relationship between strength and durability of porous materials, it always shows positive correlation [21-23]: porous materials with higher strength can suffer stronger salt crystallization distress.

2.2 Characteristics of salt weathering distress

In the process of salt weathering, efflorescence and sub-efflorescence will occur. Efflorescence always occurs on the surface of the material, and shows little or no damage. On the contrary, sub-efflorescence forms under the material surface and results in significant damage [24-26]. Some interesting studies showed that addition of ferrocianides ($[Fe(CN)_6]^{4-}$) can promote NaCl efflorescence growth as opposed to sub-efflorescence growth in porous stones, and minimize salt damage [27, 28].

Wick action is the transport of water (and any species it may contain) through a concrete (porous material) element face in contact with water to a drying face with less than 100% relative humidity of air [29]. The mechanism involves capillary sorption and evaporation.

During the process of wick action, if there is no evaporation, the solution level can increase through capillary rise in the concrete according to Eq. 5: [30]

$$h = \frac{2\gamma_{LV} \cdot \cos\theta}{rg\rho} \tag{5}$$

where h is the height of capillary rise, γ_{LV} is the liquid/vapor interfacial energy, θ is the contact angle, r is the pore radius, g is the gravitational acceleration, and ρ is the density of the solution. In the case of water in concrete $\cos\theta \approx 1$, γ_{LV} is ~ 400 mJ m^{-2}, and r is the typically 10~100nm. Therefore, h is about 1-10m. However, the pores will easily lose water due to evaporation. The pores of 10 um will start to empty when the relative humidity is lower than 95%. So, when the relative humidity is lower, h will decrease. After some time, a state of equilibrium (wet-drying interface) may be reached. Then the rate of water entering the concrete by capillary sorption matches the rate of water leaving the opposite face of the concrete element by water vapour diffusion.

If the water is containing salts, these salts cannot be carried by the vapour and therefore build up at this position. This concentration effect causes back-diffusion of salt away from the wet-dry interface. If the salts concentration near the wet-dry interface ever exceeds the solubility of the salt compounds present, precipitation is likely to occur [31-34]. The absorption-diffusion relationship can be described by the definition of the Peclet number [34]:

$$Pe \equiv \frac{hL}{\theta_m D_c} \tag{6}$$

Where, h (m^3 m^{-2} s^{-1}) is the drying rate, L (m) the length of the sample, and θ_m (m^3 m^{-3}) the maximum fluid content by capillary saturation. D_c (m^2 s^{-1}) is the diffusion coefficient of the ions in the moisture in the porous medium. For Pe<<1 diffusion dominates and the ion-profiles will be uniform, whereas for Pe>>1 absorption dominates and ions will be accumulated at the drying surface.

Y. T. Puyate et al discussed the chloride transport due to wick action in the concrete in detail [31-34]. One vital conclusion is that it was the vapour pressure of the solution and the relative humidity of air which control the position of the wet-dry interface [33]. The position of dry-wet interface locates in the inner of the concrete faced to a low relative humidity situation (0%) (Pe>>1), and a sharp peak of chloride concentration exceeding the saturation value occurs [34]. In contrast, in a high relative humidity condition (78%) (Pe<<1), the location of the interface is close to the concrete surface [31, 32]. Therefore, high evaporation can induce severe crystallization distress.

Nuclear magnetic resonance (NMR) is used to study the crystallization of sodium chloride due to wick action. Measuring the moisture and ion profile in a fired-clay brick cylinder (Ø 20×45mm), an efflorescence pathway diagram is plotted [35, 36] as shown in Fig. 3.

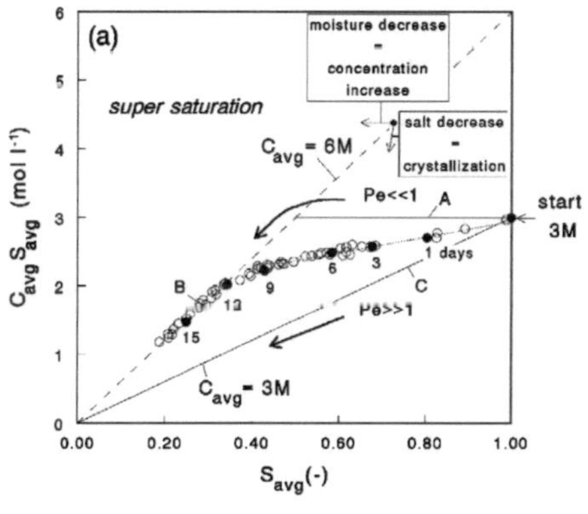

Fig. 3. Efflorescence pathway diagram [35]

According to this diagram (Fig.3), when Pe<<1, i.e. very slow drying or high relative humidity, the ion profile remains homogeneous and for some time no crystallization will occur. The average NaCl concentration slowly increases until the complete sample has reached saturation, forming a high concentration pore solution zone. When Pe>>1, i.e. very fast drying or low relative humidity, ions are directly advected with the moisture to the top of the sample and a saturation peak will build up with a very small width. If the rate of crystallization is high enough, i.e. if there are enough nucleation sites at the top to form crystals, the average NaCl concentration of the solution in the sample itself will remain constant at nearly the initial concentration.

The mechanism of efflorescence is the crystals growing at a free surface: the crystals in the pores cannot be stable and will dissolve and diffuse towards the atmosphere (an infinite pore) (Eq.2). Because the crystals are in contact with the solution only in their bases, they cannot grow laterally but form long needles like whisker [14]. This is the reason why efflorescence is un-harmful for the porous materials. Sub-efflorescence precipitates when the evaporative flux is greater than the capillary flux in the porous materials where the solution is supplied by the capillary suction and evaporation [37].

In summary, the efflorescence and sub-efflorescence of salt weathering distress on the porous material can be schematically shown in Fig. 4.

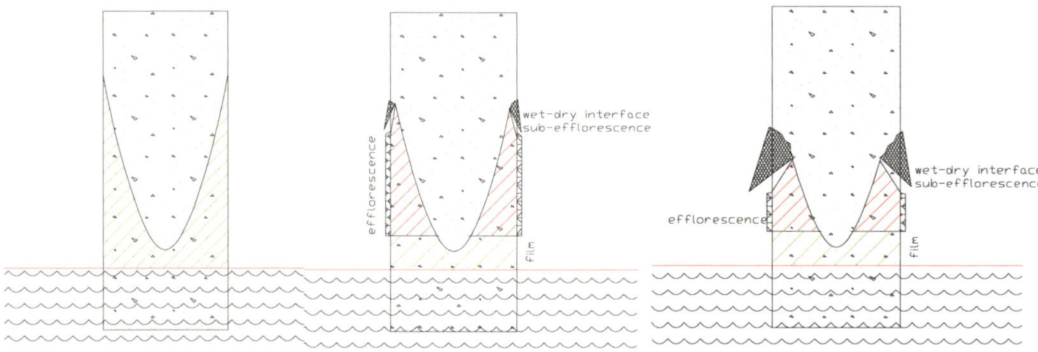

(a) (b) (c)

Fig. 4. Schematic of salt weathering distress on porous material.
(a) no evaporation condition, (b) low evaporation condition, (c) high evaporation condition

As we know, the capillary absorption just occurs in the interconnected pores between air and water. When a porous element is partially subjected to the sulfate solution under no evaporation condition, a pore solution zone will be formed as shown in Fig. 4(a). The solution cannot rise from the solution surface to the top of the element by capillary absorption due to few or no interconnected pores from the bottom to the top in a relatively long distance. The interconnected pores can form from the solution surface to the side surface of the element in a relatively short distance, resulting in the generation of capillary absorption.

Under a low evaporation condition, the wet-dry interface can occur in the tip of the pore solution zone, where the rate of evaporation is fast compared with the rate of solution rise, because solution rises into the bulk at a rate that decreases with height. At the same time the sulfate concentration of pore solution will slowly increase until the complete sample has reached saturation, forming a high concentration pore solution zone, where the efflorescence occurs. Near the solution a liquid film occurs on the surface of the element, where the rate of rise if fast compared with the evaporation and the sulfate concentration is close to the exposure solution [20]. In this case the deterioration due to salt crystallization is minor (as shown in Fig. 4 (b)).

Under a low relative humidity condition, due to the high evaporation rate, the position of the wet-dry interface will move to the inner part of the element and closer to the bottom of the element, where a saturation peak will build up with a very small width, forming supersaturation, and resulting in sub-efflorescence and more severe deterioration. The breadth of efflorescence zone decreases and the average concentration of the pore solution will remain constant at nearly the initial concentration of exposure solution (as shown in Fig. 4 (c)).

2.3 Crystallization of Na_2SO_4 and $MgSO_4$

Sodium sulfate is known to be a salt that causes the worst crystallization decay on porous materials and has become widely used in accelerated durability testing [38]. However, the sodium sulfate system is complicated, because under different conditions (temperature and relative humidity), it will form two stable phases (thenardite, Na_2SO_4 and mirabilite, $Na_2SO_4 \cdot 10H_2O$) or one metastable phase (heptahydrate, $Na_2SO_4 \cdot 7H_2O$) [12, 13] [39]. The metastable phase ($Na_2SO_4 \cdot 7H_2O$) is formed during the rehydration of the anhydrous sodium sulfate phase (Na_2SO_4) to the nucleation of mirabilite. Prior to mirabilite [12] [39], the crystallization pressure exerted by heptahydrate does not cause damage under the condition of the cooling experiments [36], and it can not be observed in building stone [13].

Regarding the damage caused by the crystallization of sodium sulfate, there are two views. One school thinks that the crystallization of thenardite is more destructive [40], because the crystallization of thenardite can generate higher pressure than mirabilite at the same supersaturation [41]. However, more and more experimental results support another school that the dissolution of thenardite producing a solution highly supersaturated with respect to mirabilite will cause the precipitation of mirabilite and result in the damage of porous materials [13] [38]. I.e. the transformation between thenardite and mirabilite can generate severe large crystallization pressure, resulting in porous materials damage.

The only naturally occurring members of the $MgSO_4 \cdot nH_2O$ series on Earth are epsomite ($MgSO_4 \cdot 7H_2O$, 51 wt% water), hexahydrite ($MgSO_4 \cdot 6H_2O$, 47 wt% water) and kieserite ($MgSO_4 \cdot H_2O$, 13 wt% water). In aqueous systems, epsomite is stable at T below 48.4°C, hexahydrite is stable in the T range 48.4–68 °C, and kieserite is stable at T > 68 °C [42]. Thus, at the normal temperature, the crystallization of epsomite ($MgSO_4 \cdot 7H_2O$,) is the distress reason.

2.4 Summary

In summary, according to above review, the following basic principles of salt weathering on porous materials can be concluded:

1. Supersaturation is the key factor for salt crystallization. During the process of salt weathering the supersaturation must be maintained at a high level.
2. High evaporation results in the formation of strong sub-efflorescence, causing severe deterioration. Low evaporation results in weak crystallization distress but causes the formation of a pore solution zone with high concentration in the part of porous materials in contact with air.

3. Experimental studies of "salt weathering" on concrete

3.1 Long term field tests

Since 1940, a long-term study of "salt weathering" on concrete was carried out by the Portland Cement Association (PCA) [43-45]. Thousands of concrete beams (152×152×762mm) were laid horizontally to a depth of 75 mm in sulfate rich soils (about 5.6% sulfate ion by weight of soil) basins in Sacramento, California. About 10 to 12 wetting and drying cycles are carried out every year. As the experiment progressed, commercial salts were added into the soils to replenish losses through leakage, overflow, and possibly other undetermined causes. Water was added to the basins just before the soils began to show drying to maintain the soils saturated.

Three reports were published [43-45]: the first [43] provided the experimental results of initial set of beams resistance to sulfate attack between 1940 and 1949. The second [44] described the performance development of concrete beams for 5 years field exposure to soils containing sodium sulfate. The third [45] introduced the experimental results for 16 years exposure.

Some other five years field tests were carried out by Irassar and Di Maio [46]. Concrete cylinders with the size of Ø150 × 300mm were buried at half height in a soil containing approximately 1% sodium sulfate. There are several important common experimental observations of the above field experiences:

1. The parts of the beams above ground, regardless of their cement content, cement composition, mineral additions, surface treatments and type of coarse aggregates, were deteriorated severely. The parts of the beams under ground, however, show little or no deterioration;
2. Pozzolanic additions, such as fly ash, furnace slag or silica fume, play a negative role in the performance of concrete exposed to these conditions;
3. The water-to-cementitious material ratio (W/CM) is the primary factor affecting the durability and performance of concrete in contact with sulfate soils: applying a low W/CM ratio results in a higher resistance to sulfate attack;
4. According to XRD, optical microscopy and SEM analysis, a large amount of chemical sulfate attack products, such as ettringite, gypsum and thaumasite, were identified in the upper part of concrete in contact with air. However, the samples for these tests were drilled with water [45, 46].

Concerning the deterioration mechanism, researchers attributed the failure of concrete to physical attack or salt crystallization. However, in effect the experimental results cannot be explained by salt weathering. Fig. 5 is the evolution of visual rating of the upper part of concrete in contact with air, obtained by Irassar and Di Maio [46].

In the test, the mix proportions of different mixes (the ratio of water : binder : sand : aggregate) were almost the same with different dosages of fly ash, slag and natural pozzolan. From Fig.5, we can deduce some interesting observations:

1. Comparing concrete H1 and H2, the difference between them was the air content, namely 1.3% resp. 4.4%. Thus, if the damage mechanism of the upper part of concrete was caused due to salt weathering, there would be a big difference in visual rating of these two concretes due to the different pore structures. However, after 5 years exposure the visual ratings were almost the same.

2. As to the role of mineral additions, with the increase of dosage of fly ash and natural pozzolan, the concrete cylinders showed worse visual observation. This may be explained by the fact that salt crystallization in smaller pores can form higher crystallization stress due to the refinement of mineral additions. On the other hand, if we compare concrete H3 and H5, H4 and H6, we can find that with the same dosage the natural pozzolan (H5 and H6) played a more negative role in concrete damage than fly ash (H3 and H4). However the compressive strength of reference cylinders of H5 and H6 were higher than H3 and H4 respectively. The tests of PCA also showed similar results. This appearance cannot be explained by the salt weathering.
3. Comparing the normal concrete and blended concrete, Irassar et al [46] attributed an increase in capillary suction height caused by the pore size refinement of mineral addition to the more severe deterioration of blended concrete. However, this is in conflict with the following observations. Following the explanation based on the height of capillary sorption, concrete with a low W/C ratio should be more susceptible to salt weathering than with high W/C ratio. In the paper by Nehdi and Hayek [47], we can find that the sorption height of mortar with W/C of 0.45 is higher than with W/C of 0.6. If the above explanation is right, the mortar with low W/C (0.45) should be more susceptible to damage than the mortar with high W/C (0.6). Obviously, this conclusion is opposite to the result of field tests.

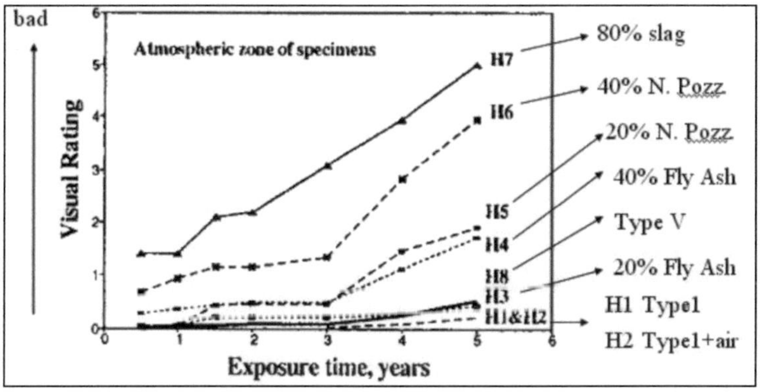

Fig. 5. Evolution of visual rating at the atmosphere part of concrete [46]

Besides, there are also some field cases, in which a wide variety of efflorescence salts (sodium sulfate, sodium chloride and magnesium sulfate) were routinely observed on the evaporating surface of the foundation concrete in South California [48, 49] However, the damage was not caused by salt crystallization. Much of the cement paste had lost its integrity, mainly as a result of the removal of portlandite, de-calcification of the calcium silicate phases and the ultimate replacement of calcium with magnesium in many of the cementitious compounds. Many reaction products, in particular magnesium silicate hydrate, brucite, Friedel's salt ($3CaO \cdot Al_2O_3 \cdot CaCl_2 \cdot 10H_2O$), sodium carbonate and thaumasite were found at some depth within the concrete.

Another interesting case [50] concerns the slabs of Yongan Dam, which is in Keshi, Xinjiang, P.R. China, where the land is arid and rich in various kinds of salts, especially sulfate salts. The slabs were constructed in August - October 2003. The air temperature fluctuates significantly: the highest temperature is up to 37 °C and at night it is only 15-20 °C. Within this temperature range, the transformation between thenardite and mirabilite can occur. However, it was found that the slabs became gray and mushy throughout the thickness where they were in contact with groundwater due to thaumasite sulfate attack.

In summary, according to the above analysis the appearances of long term field tests did not show convincing evidences to support "salt weathering" causing the deterioration of concrete partially exposed to sodium sulfate environment.

3.2 Indoor tests

Rodriguez-Navarro and Doehne [26] studied the effect of evaporation on salt weathering distress on stone. After 30 days of exposure to a saturated sodium sulfate solution at constant 20 °C, larger amounts of efflorescence and lower weight losses were observed when the crystallization took place at a relative humidity of 60% instead of at 30% RH. H. Haynes and his coworkers [9] carried out some tests partially exposing the same concrete cylinders to 5% NaCl and Na_2CO_3 solutions. Severe damage was observed for concrete cylinders, which were placed in constant environment at 20 °C and 54% relative humidity from day 28 to day 530, and then at 20 °C and 32% RH from day 530 to day 1132. The specimens kept in a constant environment at 20 °C and 82% relative humidity from day 28 to day 1132 looked sound. These observations can be explained by the fact that a low relative humidity results in more evaporation, leading to sub-efflorescence [20] that forms deep in the material and results in significant damage [24,25] as explained in the section 2.2.

However, contradictory observations can be found with respect to concrete exposed to sodium sulfate solution [51]. Two concrete specimens with the same mixture proportions were partially immersed in 10% Na_2SO_4, solution. One specimen was placed at 80%±5% RH and the other was placed at 30% ±5%. After 75 days of exposure to a constant temperature of 25 °C, the specimens at 80% RH showed signs of deterioration first over a very large area, starting from above the solution level. On the contrary, at 30% RH, the zone of deterioration was narrower and was situated at a certain distance above the solution level. In this case, the first sign of deterioration was a crack, not spalling (shown in Fig. 6).

Fig. 6. Concrete cylinders exposed to sodium sulfate solution for 75 days [51]

Similar results were also observed in the tests performed by H. Haynes and his coworkers [2]. Narrower spalling zone was found in case of concrete specimens exposed to constant environment at 20 °C and 54% relative humidity from day 28 to day 530, and then at 20 °C and 32% RH from day 530 to day 1132. Extensive spalling zone was found under constant environment at 20 °C and 82% relative humidity from day 28 to day 1132. These tests will be discussed in detail as follow.

3.2.1 H. Haynes tests [2, 9]

H. Haynes and his coworkers performed very important and systemical tests about the salt weathering distress on concrete. In the two papers [2, 9], different ambient conditions were created within storage cabinets whose temperature and relative humidity were controlled. The concrete cylinders (Ø76 × 145mm) were partially exposed to 5% Na_2SO_4, $NaCO_3$ and NaCl solutions. A partial submergence condition was achieved by wetting the specimen to a height of 25 mm. At the height of 50mm, a plastic cover to the container functioned as a quasi-vapor retarder to minimize evaporation. The plastic cover did not touch the cylinder. Hence, within the region of 25 to 50mm the cylinder was exposed to a moist environment. Above 50mm (2 in.), the concrete was exposed to ambient environmental conditions. In the test program the author said that "the sulfate solution and tap water were replaced on a monthly basis; however, replacements for evaporation loss were provided at 2-week intervals. Much of the solution evaporated in the 40 °C and 31% relative humidity environment where, in general, at the end of 2 weeks, minor amounts of solution remained; and at times, no solution remained".

The tests were divided into two Phases for 3.1 years. The performance of concrete cylinders under five storage conditions was studied in detail, the exposures were:

Condition 1: steady at 20°C and 54% relative humidity from 28 to 530 days (Phase I), and then 20°C and 32% relative humidity from 530 to 1132 days (Phase II),

Condition 2: steady at 20°C and 82% relative humidity from 28 to 530 days (Phase I), and then from 530 to 1132 days (Phase II),

Condition 3: 40°C and 74% relative humidity from 28 to 406 days (Phase I), and then 40°C and 31% relative humidity from 406 to 1132 days (Phase II),

Condition 4: 2-week cycles between 20°C and 54% relative humidity and 20°C and 82% relative humidity from 28 to 530 days (Phase I), and then 2-week cycles between 20°C and 31% relative humidity and 20°C and 82% relative humidity from 530 to 1132 days(Phase II),

Condition 5: exposed to 2-week cycles between 20°C and 82% relative humidity and 40°C and 74% relative humidity from 28 to 406 days(Phase I), and then 2-week cycles between 20°C and 82% relative humidity and 40°C and 31% relative humidity from 406 to 560 (847) days (Phase II).

The effects of Na_2SO_4, Na_2CO_3 and NaCl were compared. The visual observation was photographed, the average mass of scaling materials was collected, the species of concrete were identified by petrographic analysis, and chemical analysis was employed to study the ions distribution. According to the experimental results, they concluded that salt weathering plays the predominant role in concrete damage. However, there are a few questionable points about the relationship between the evidences and the conclusion based on the basic principles.

3.2.1.1 Visual observation of concrete cylinders

According to the photographs of visual observation at the end of exposure, we can find that:

1. When concrete cylinders were exposed to Na_2SO_4 solution, as abovementioned the concrete cylinders exposed to high relative humidity condition were deteriorated more severely than low relative humidity condition (as shown in Fig. 7).

Condition 1 Condition 2

Fig. 7. Visual observation of concrete deterioration. Condition 1: steady at 20°C and 54% relative humidity from 28 to 530 days (Phase I), and then 20°C and 32% relative humidity from 530 to 1132 days; Condition 2: steady at 20°C and 82% relative humidity from 28 to 530 days (Phase I), and then from 530 to 1132 days [2]

Correspondingly, Fig. 8 shows the visual observation of concrete exposed to Na_2CO_3 and NaCl solutions under Condition 1 and Condition 2.

(Na_2CO_3)

Condition 1 Condition 2

(NaCl)

Condition 1 Condition 2

Fig. 8. Visual observation of concrete deterioration exposed to Na_2CO_3 and NaCl solutions [9]

2. Under Condition 3 (40°C /74%RH, 40°C /31%), concrete showed the most significant scaling in the case of NaCl, on the contrary, the concrete showed least scaling in the cases of Na_2CO_3 and Na_2SO_4 solutions. These appearances are contradictory because the concretes should show similar scaling manners due to the salt weathering in case of Na_2CO_3 and NaCl solutions. If the mechanism of concrete damage is also salt weathering in case of Na_2SO_4 solution, the concrete should also show similar scaling manners in NaCl solutions. If the mechanism of concrete damage is the chemical sulfate attack, the scaling manners of concrete cylinders by Na_2CO_3 and Na_2SO_4 solutions should show big difference. The possible reason of the above contradictory appearances may be the fast evaporation rate of Na_2CO_3 and Na_2SO_4 solutions. At high ambient temperature, Na_2CO_3 and Na_2SO_4 solutions would dry up soon, but some NaCl solution would remain. The tests under Condition 3 should be further studied to avoid the effect of evaporation of solution.

3.2.1.2 Mass of scaling materials

During Phase I from 28 days to 406 or 530 days: the worst damage occurred under Condition 5 (cycle 20°C/82% RH and 40°C/70% RH) in the case of Na_2SO_4 solution (the mass of scaled material was about 16g). However, in the case of Na_2CO_3 solution the worst damage appeared under Condition 1 (20°C/54% RH) (the mass of scaled material was just about 2.8g). This means that high RH can accelerate concrete damage by Na_2SO_4.

During Phase II from 406 or 530 days to the end of tests:

1. In the case of Na_2SO_4 solution the mass of scaled material under Condition 1(20°C/32% RH) was less than Condition 2(20°C/82% RH). The opposite appearance was observed in case of Na_2CO_3: the mass of scaled material was about 22g under Condition 1(20°C/32% RH) and about 1g under Condition 2 (20°C/82% RH). Similar appearance also observed in the case of NaCl solution.
2. The mass of scaled material of concrete under Condition 1 (20°C/32% RH) was less than Condition 4(cycle 20°C/82% RH and 20°C/31% RH) in case of Na_2SO_4. However, the mass of scaled material of concrete under Condition 1 (20°C/32% RH) was almost the same as Condition 4(cycle 20°C/82% RH and 20°C/31% RH) in the case of Na_2CO_3.

Corresponding to Phase I, this also means that high RH is in favor of the deterioration effect of Na_2SO_4 on concrete.

3.2.1.3 Petrographic analysis

According to petrographic analysis, abundant gypsum deposits were detected in large and small cracks, microcracks and voids near the surface of concrete. However, the authors presented two points to show that gypsum cannot result in concrete damage:

1. "Although trace amounts of gypsum were found near the outer surfaces, gypsum formation is a one-time occurrence, whereas crystallization of mirabilite and thenardite occurred repeatedly due to the biweekly cyclic changes in environmental conditions. Hence, the cycles of mirabilite and thenardite crystallization appear to be responsible for any significant expansion force". It is not clear why the authors thought that "gypsum formation is a one-time occurrence". According to above review, the solution can be drawn into the concrete continuously during wick action. In the presence of sulfate, gypsum crystals can continuously grow. Moreover, the authors pointed out that the pH value of the pore solution in the concrete should have been reduced due to carbonation, whereas for gypsum formation, the pH value of solution should be less than 11.9 [52].
2. "Despite the extensive alteration of the microstructure and the formation of gypsum, the concrete below the solution line was mostly intact with no mass loss, whereas there was substantial mass loss at the surface of the cylinder above the solution line. If gypsum did not cause scaling below the solution line, there is little reason to suspect that gypsum would cause scaling above the solution line. This indicated that salt crystallization alone, or in conjunction with gypsum, caused the scaling above the solution line. As salt crystallization by itself is known to damage rocks, the presence of gypsum is not necessary". This point seems reasonable, however, the authors did not give the quantitative analysis of gypsum. Because according to abovementioned wick action, a much higher concentration pore solution will be formed in the cylinder above the solution than under the solution, resulting in more severe sulfate attack and forming more gypsum.

3.2.1.4 Chemical analysis

Several cylinders were cut vertically to obtain a 25 mm thick midsection slice. This slice was then cut vertically into two 17 mm exterior sections and one 34 mm interior section. Starting at the bottom, the vertical sections were cut horizontally into six pieces 25 mm each. These pieces were crushed and pulverized to minus No. 50 mesh. The SO_3 contents and Na_2O contents were determined. The distributions of SO_3 contents and Na_2O contents are schematically shown in Fig. 9.

The salt distribution in the concrete cylinder provides a powerful evidence supporting that salt weathering is not the major mechanism causing concrete damage.

Based on the above review on the basic principles of salt weathering, supersaturation is the key factor for the salt crystallization. The salt crystals will deposit from the solution during the process of salt crystallization, however, the salt concentration of pore solution must be maintained high for the formation of supersaturation during the whole process. I.e. the salt contents should be highest where salt crystallization distress occurs in the concrete. This is the reason why the position where Na_2O content is highest corresponds with most severe deterioration of concrete exposed to Na_2CO_3 and NaCl solutions. However the situation is opposite in the case of Na_2SO_4, the position, where SO_3 content is highest, locates on the top portion of cylinder that shows little or no deterioration (as shown the black line).

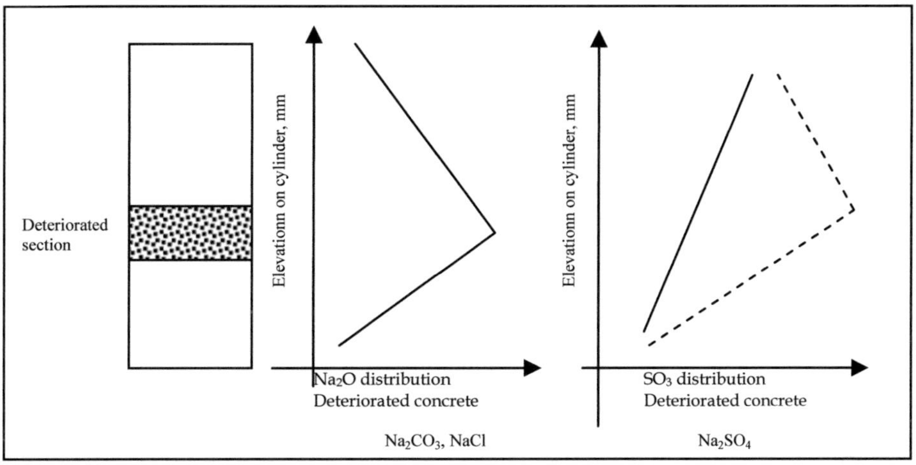

Fig. 9. Schematic of salt distribution in the concrete cylinders

In summary, based on the above analysis of indoor tests, two conclusions can be deduced:

1. Concrete partially exposed to Na_2SO_4 is susceptive to being more severely deteriorated under high RH environment than low RH environment. This appearance is in conflict with the basic principles of salt weathering.
2. The most severe deterioration does not occur in the portion of concrete containing the highest sulfates content. This is also in conflict with the basic principles of salt weathering.

3.3 Our tests [53, 54]

The starting point of our tests is to find a trace of salt crystals in the concrete as a direct evidence by means of XRD, SEM and EDS [53, 54]. Sulfate crystals can be easily identified in stone [23, 55]. However, in case of concrete elements, it is hard to identify them. Concrete technologists always attribute this to the coring and sawing operations when preparing samples for experimental analysis, as lapping water can readily dissolve salts from original and treated surfaces [2, 3, 4]. However this is not the main cause for the problem. Samples also can be taken in a dry manner to avoid the influence of water. Furthermore, In our study, the tests were designed to avoid the influence of water within the detection process of sulfate crystals.

Cement paste and cement - fly ash paste specimens and normal concrete specimens were partially exposed to Na_2SO_4 and $MgSO_4$ solution under constant and fluctuating storage conditions respectively. After a period of exposure, the specimens were moved out from the solution and did not touch solution or water any more. The surface of the specimens was cleared by a thin blade and a soft brush. The samples for XRD and SEM were dried in a vacuum container with silica gel.

3.3.1 Cement paste partially exposed to Na_2SO_4 and $MgSO_4$ solution under a constant storage condition

The test of cement paste specimens (20×20×150mm) partially exposed to Na_2SO_4 and $MgSO_4$ solution under a constant storage condition (20°C and 60%RH) is based on the study performed by Ruiz-Agudo [42] in which limestone specimens were partially submerged in a 19.4 g/100ml sodium sulfate solution and a 33.5 g/100ml magnesium sulfate solution respectively and located in a controlled environment (20°C±2°C, and 45%±5% RH). Results showed that the limestone specimens were severely damaged in both cases. While salt weathering by sodium sulfate consisted of a detachment of successive stone layers, magnesium sulfate induced the formation and propagation of cracks within the bulk stone. Thenardite (Na_2SO_4) and epsomite ($MgSO_4 \cdot 7H_2O$) crystals were identified by ESEM in the pores of limestone.

Before immersion minor shrinkage cracks were observed in the cement paste specimens. These cracks were focused upon in detail because narrow micro-fissures appear to be important in the decay process due to the effectiveness of crystallization pressure generated by salt growth [37]. So, if crystallization is the mechanism of decay of cement paste, salt crystallization should first occur in the shrinkage cracks and sodium sulfate or magnesium sulfate crystals should be identified in these cracks.

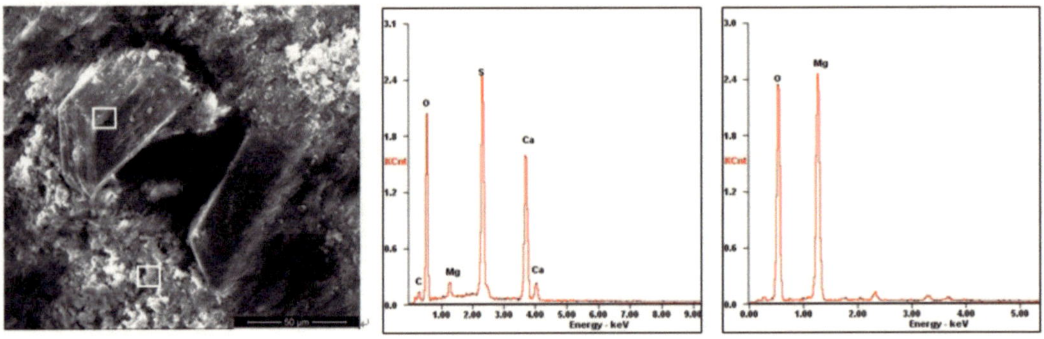

(a) (b) (c)

Fig. 10. ESEM and EDS analysis of white substance on the surface of shrinkage crack [54]
(a) ESEM image of white substance (b) EDS analysis of prismatic crystal
(c) EDS analysis of flocculent crystals

However, based on the micro-analysis results the sulfate crystals were not detected in the atmospheric part of the paste partially exposed to Na_2SO_4 solution. On the contrary large amounts of ettringite crystals, the main chemical sulfate attack product, were identified as the reason for paste spalling. Another important observation is that a layer of white substance was formed on the surface of shrinkage cracks in the atmospheric part of the paste partially exposed to $MgSO_4$ solution. Two distinct crystals can be distinguished: prismatic crystals surrounded by flocculent crystals. According to the EDS analysis, the prismatic crystal is gypsum and the flocculent crystal is brucite (shown in Fig.10). The products of this white substance in the shrinkage cracks are the same as the products in the interfacial zone of concrete fully immersed in magnesium sulfate solution [56].

3.3.2 Cement – fly ash paste partially exposed to Na_2SO_4 solution under a constant storage condition

Cement - fly ash paste specimens (20×20×150mm) were partially exposed to Na_2SO_4 solution. After 5 months exposure under the constant storage condition (20°C and 60%RH), some cracks were found near the upper edge above solution level of the cement-fly ash paste specimen. Along the crack, small pieces were carefully broken off using a thin blade. The ESEM image of a small piece is shown in Fig. 11.

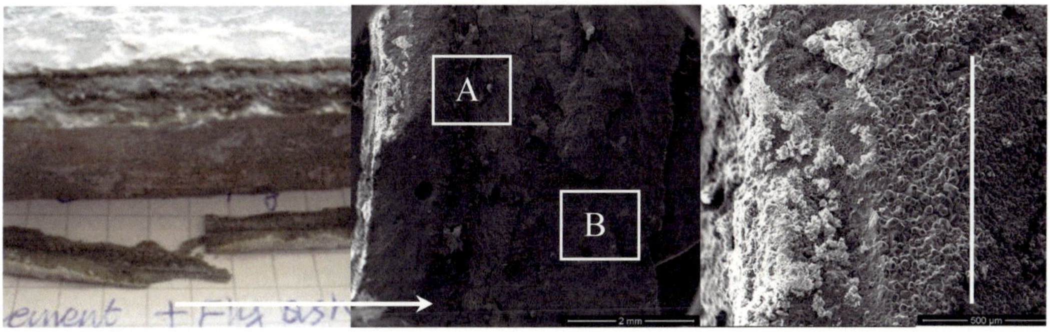

Fig. 11. Cracks in cement - fly ash paste [54]

The middle image is the zoomed surface of a small piece with magnification of 25 times. The left side is the outer surface of paste in contact with air, and zone A is the surface of a crack. Zone B is a small point in the bulk of paste.

Two distinct parts can be observed at the right and left hand side of the white line in Zone A. At the right side a large amount of dense granular crystals cover the surface (Fig. 12), while at the left side porous crystals can be found accompanied with white substance (Fig. 13). It can be found that the crystals at left and right sides are both calcite. However, some calcite crystals at left side are peeled off and crystal caves are left. Some crystals are honeycombed with small pores. According to the EDS analysis, Na and S are also present. Obviously, the crystallization of sodium sulfate results in damage of the calcite crystals. At the right side, the calcite crystals show no damage.

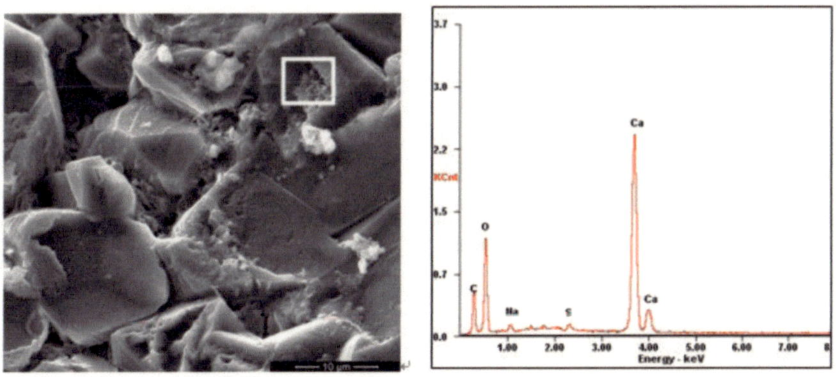

Fig. 12. ESEM image and EDS analysis of f the granular crystal at left side [54]

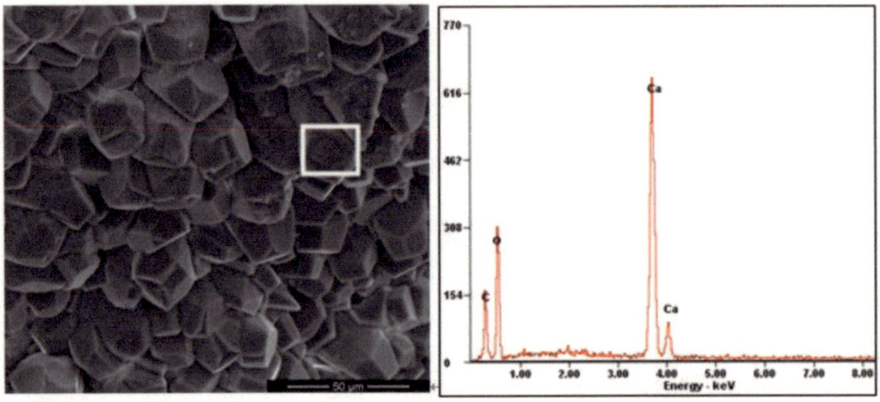

Fig. 13. ESEM image and EDS analysis off the granular crystal at right side [54]

According to the above observation, two conclusions can be drawn:

1. If salt crystallization is causing crack formation, the salt crystals should be identified at the right side in Fig. 11 to form sub-efflorescence instead of in the area in contact with air.
2. Salt crystallization can occur in the calcite crystals.

As we know, the crack formation is attributed to some expansive products present in the paste. When a small piece was broken off along the crack, the inner zone B on the surface of piece was a weak part in the paste and the source of crack initiation. The analysis of the products in this zone can disclose the real reason for the crack formation. Fig. 14 shows the ESEM image of the surface of Zone B.

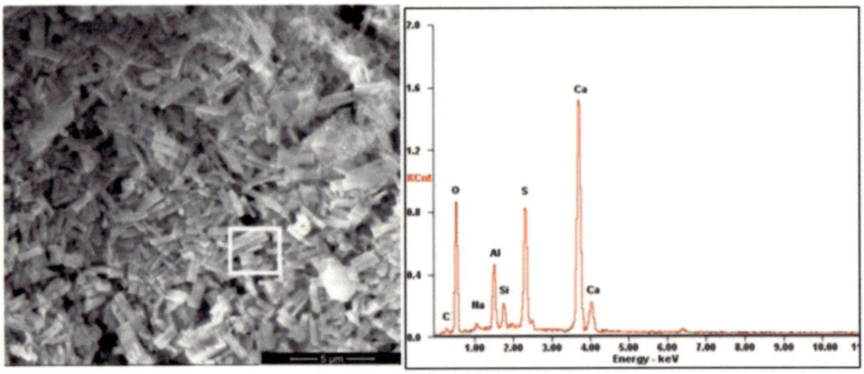

Fig. 14. ESEM and EDS analysis of white square [54]

In Fig. 14, a large amount of short needle crystals are found in this zone. According to the EDS analysis, there are Ca, Al, Si, S, Na, and O elements. Combining the XRD analysis, thenardite was not identified and the crystals were ettringite.

3.3.3 Cement – fly ash paste partially exposed to Na_2SO_4 solution under a fluctuating condition

Specimens (10×40×150 mm) were placed in a fluctuating condition (40±2°C and 35±5% RH for 24 hours, 10±1°C and 85±5% RH for 24 hours). After 3 cycles they were broken into several small pieces along some cracks. We checked the products on the surface of a crack. Fig. 15 shows the SEM image and EDS analysis.

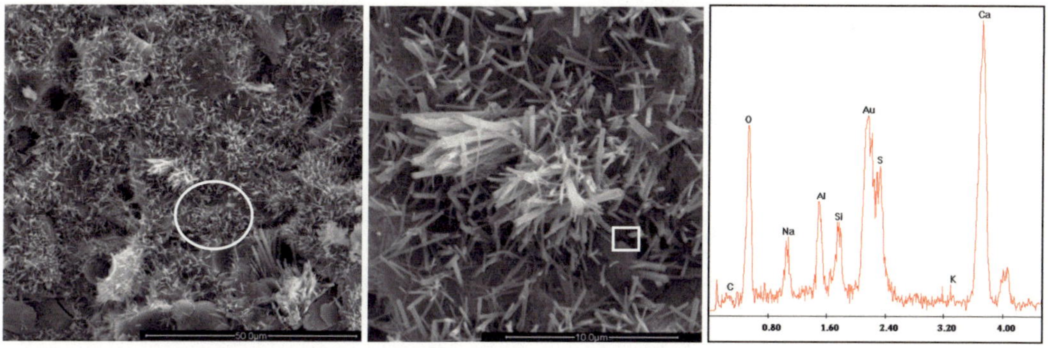

Fig. 15. SEM image and EDS analysis of the surface of a crack [54]

It can be found that a large amount of needle-like crystals grow on the surface like a hedgehog. Some pores are filled with a cluster of the needle-like crystals. Based on the EDS analysis and combining the XRD analysis, the needles are ettringite.

3.3.4 Normal concrete partially exposed to Na_2SO_4 and $MgSO_4$ solution under a constant storage condition

As we know, in concrete the weak interfacial transition zone (ITZ) plays a particularly important and even determining role in the main characteristic of concrete. A number of full immersion tests already showed that concrete deterioration occurred first in the ITZ by sulfate attack [56-60]. In this test, the concrete was made with just cement and aggregate to emphasize the role of ITZ in concrete deterioration. The concrete specimens (10×40×150 mm) were partially exposed to Na_2SO_4 and $MgSO_4$ for 8 months. The results showed that: (1) the harmful effect of $MgSO_4$ is much weaker than Na_2SO_4. This appearance also cannot be explained by the mechanism of salt weathering. This will be discussed in detail in section 4; (2) in the case of Na_2SO_4, damage also initiated in the ITZ. A large amount of gypsum crystals were formed on the surface of cement paste of ITZ in the upper part of concrete above solution (shown in Fig. 16).

Besides, the effect of carbonation on the salt weathering on concrete was studied. Before exposure a group of concrete cylinders were placed in an accelerated carbonation chamber with 10% CO_2 concentration at 20 ± 2 °C and 60% ± 5% RH for 14 days. Then, these cylinders were partially exposed to Na_2SO_4 solution. After 8 months exposure, the carbonated cylinders were deteriorated more severely than normal concrete (shown in Fig. 17).

During the process of cleaning the surface of cylinders, a lot of small mortar pieces could be easily brushed off. According to the XRD analysis (Fig.18) Na_2SO_4 crystals and $CaCO_3$ crystals were present in the mortar. This appearance also means that salt crystallization can occur in the $CaCO_3$ crystals.

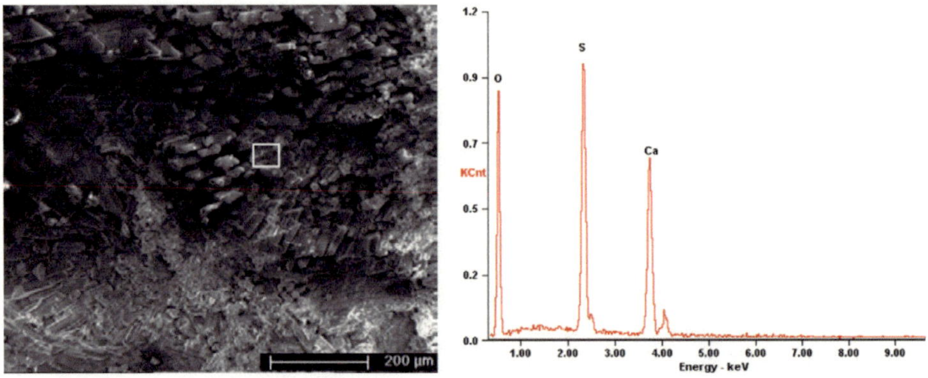

Fig. 16. ESEM image and EDS analysis of surface of cement paste [53]

Fig. 17. Visual observation of normal and carbonated concrete specimens exposed to sodium sulfate solution [53]

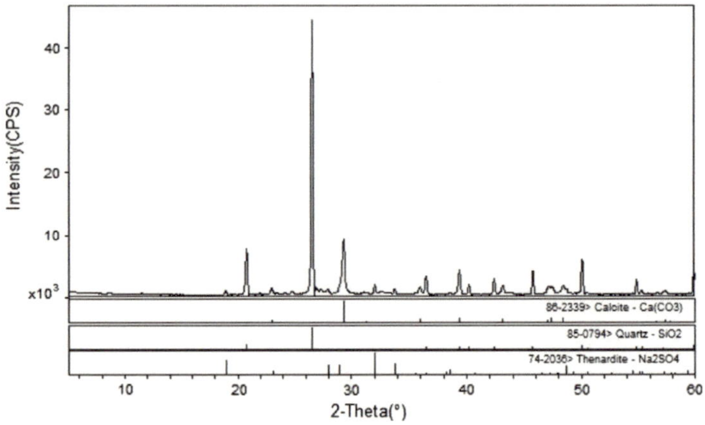

Fig. 18. XRD pattern of mortar [53]

In summary, according to the above test results, two main conclusions can be deduced:

1. Sulfate crystals cannot be identified in the cement paste or concrete partially exposed to Na_2SO_4 and $MgSO_4$ solutions. The chemical reaction products, ettringite, gypsum and brucite, were the determining factors for material damage.

2. Salt crystallization can occur in the calcite crystals, the carbonated products of concrete.

3.4 Summary

The "salt weathering" on concrete was just received a lot of attention in the recent years. Based on the above analysis of a limited number of research reports, the experimental results already showed convincing appearances that were completely opposite to the basic principles of salt weathering distress on porous materials. On the contrary, the experimental results of long term field tests and indoor tests rather tended to indicate that chemical sulfate attack is the mechanism for the concrete damage.

4. Further research

According to the limited literature review, the conclusion can be made that the so-called "salt weathering" on concrete in effect is rather chemical sulfate attack. In order to systematically disclose the principles of this appearance, some issues should be further studied.

4.1 Study of the reason why salt crystallization cannot occur in concrete

The reason why salt cannot occur in concrete may be explained as follows:

Sulfates likely do not crystallize in a cement paste because in the highly alkaline pore solution other less soluble salts, e.g. ettringite, or gypsum, are preferably precipitated according to chemical equilibria theory. Salt crystallization in porous materials is difficult because salt crystallization occurrence has to reach and even exceed a threshold-supersaturation. However, the chemical reactions in pore solution can occur regardless of the sulfate concentration and decrease the possibility of physical attack due to consuming sulfates and decreasing the sulfate concentration of pore solution, moreover, high concentration solution will increase the rate of chemical reaction. This will make it is very difficult that the pore solution reaches supersaturation.

In Fig. 9, the SO_3 distribution showed some powerful evidence that the sulfates were consumed, resulting in the severest concrete damage. I.e. if there was no chemical reaction and if it were salt weathering causing concrete damage, the ion distribution curves of Na_2SO_4 should show similar features to Na_2O distribution of Na_2CO_3 and NaCl. Certainly, this explanation is not convincing enough to disclose the mechanism. Further studies may be performed through thermodynamic calculation to check the negative effect of chemical reaction on the supersaturation formation.

4.2 Study of the concentration of solution on the formation of pore solution zone in concrete

In the previous tests, an opposite appearance to salt weathering was that the concrete was susceptive to be damaged under a higher relative humidity condition. Combining the role of relative humidity in wick action and chemical sulfate attack, it can be explained that a wider sulfate pore solution can be formed in the upper part of concrete in contact with moist air and chemical sulfate attack occurring in the pore solution zone resulted in concrete damage.

In our previous study [61], the pore solution expression test method was used to squeeze the pore solution in the cement paste. Cement paste samples were partially exposed to 10% Na_2SO_4 solution under the constant storage condition (20°C and 60% RH). The sulfate concentrations of pore solution in the lower part under solution (labeled L), film zone (labeled M) and efflorescence zone (labeled U) (shown in Fig. 4) were measured respectively. Fig. 19 gives the results.

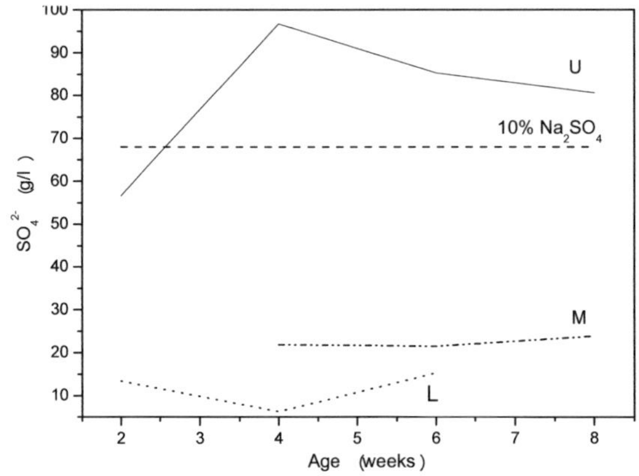

Fig. 19. SO_4^{2-} concentration of different parts of the cement paste partially exposed to the 10% Na_2SO_4 solution [61]

The results confirm that a pore solution zone can be formed in the efflorescence zone in the concrete, and the sulfate concentration was much higher than the lower part under solution and even the exposure solution (10% by mass). The strong chemical reactions occurring in this high concentration pore solution cause severe concrete decay. This also confirms the wick action theory.

Certainly, the ambient temperature and relative humidity of environment are always fluctuating. The boundary, the sulfate concentration and the formation time of pore solution zone were controlled by the evaporation rate due to the interactive effect of temperature and relative humidity. To study the pore solution zone formation needs further study.

4.3 Study of chemical sulfate attack mechanism

As we know, the main hydrated phases of cement paste are calcium silicate hydrate (C-S-H), calcium hydroxide (CH), calcium aluminate hydrate (C-A-H) ettringite (AFt) and mono-sulfoaliuminate (AFm). However, these three hydrated phases are not stable in the external environment containing sulfates. The following reactions can occur [62]:

$$Ca(OH)_2 + C\text{-}S\text{-}H + SO_4^{2-} + H_2O \rightarrow CaSO_4 \cdot 2H_2O \tag{7}$$

$$Ca(OH)_2 + C\text{-}S\text{-}H + MgSO_4 + H_2O \rightarrow CaSO_4 \cdot 2H_2O + Mg(OH)_2 + SiO_2 \cdot xH_2O \tag{8}$$

$$3CaO \cdot Al_2O_3 \cdot Ca(OH)_2 \cdot (12\text{-}18)H_2O + SO_4^{2-} \cdot 2H_2O + H_2O \rightarrow 3CaO \cdot Al_2O_3 \cdot 3CaSO_4 \cdot 32H_2O \tag{9}$$

$$Ca(OH)_2 + C\text{-}S\text{-}H + SO_4^{2-} + CO_3^{2-} + H_2O \rightarrow CaSiO_3 \cdot CaCO_3 \cdot CaSO_4 \cdot H_2O \tag{10}$$

The main reaction products are gypsum, ettringite, thaumasite, brucite and silica gel. Gypsum and ettringite are the common products of sulfate attack. Brucite and silica gel are found in case of magnesium sulfate. Thaumasite is formed when CO_3^{2-} is presented.

However, the product formation depends on the exposure conditions, such as sulfate content and pH value of sulfate environment, temperature and relative humidity.

Concerning gypsum, Bellmann et al have discussed the influence of sulfate concentration and pH value of solution on the gypsum formation in detail [52]. They indicate that portlandite will react to gypsum at a minimal sulfate concentration of approximately 1400 mg/l (pH=12.45). With rising pH, higher concentrations of sulfate ions are needed for the reaction to proceed. Between pH values of 12.45 and 12.7, the sulfate concentration slowly increases, whereas it rises dramatically from that level on. In solutions in which sodium ions are the counterpart of the hydroxide ions, the precipitation of gypsum can take place until pH values of approximately **12.9.** Beyond that mark, a further increase of the sulfate concentration is unable to lead to the formation of gypsum [52].

Concerning ettringite, ettringite is not stable in an environment with pH value below 11.5-12.0. At this low pH range, ettringite decomposes and forms gypsum [63, 64].

Concerning thaumasite, a number of experimental studies show that a high pH value (above 10.5) is in favor of the thaumasite formation [65-67]. If the pH value drops below 10.5 and even further towards 7, thaumasite is unstable, calcite and another calcium-bearing phase will be generated in the field cases [68, 69]. Thaumasite formation needs a relatively cold condition (below 15°C) [70].

In summary, the sulfate concentration, pH value and temperature control the reaction products.

Normally, in the full immersion test, 5% sulfate solutions stored at 23.0 ± 2.0°C are used in laboratories [71]. Compared to ground water in the field, a 5% sulfate solution used in the tests is much more concentrated [72]. Thus, the concrete immersed in the 5% sulfate solution can be regarded as an accelerated test. However, a high sulfate contents pore solution (higher than 5% and 10%) can be formed in the concrete in contact with air (Fig. 19) [61]. Due to concrete carbonation the pH value of pore solution in the concrete will decrease. The ambient temperature during the process of salt weathering in the field is always fluctuating. The exposure conditions of "salt weathering" on concrete are different from the full immersion tests. According to the XRD analysis, the results of long term field tests and indoor tests indicated that gypsum likely was the main reaction products and responsible for the concrete damage [2, 45, 46, 53]. Certainly, the mechanism of chemical sulfate attack should be further and systematically studied.

4.4 Study of the role of mineral additions in "Salt weathering" on concrete

An important result of long term field tests is the negative role of mineral additions in the concrete sulfate resistance. Normally, the indoor tests [74-80] always showed that the mineral additions can improve the sulfate resistance of cementitious materials based on the full immersion in 5% sulfate solutions stored at 23.0 ± 2.0°C. However, the long term field tests showed that the mineral additions accelerated the concrete decay.

As pointed out by Mehta [73], when concrete is fully immersed in the sulfate solution, for the prevention of sulfate attack "control of permeability is more important than control of the chemistry of cement". The pore size refinement due to mineral additions will prevent the sulfates to penetrate into concrete and lighten the negative effect of sulfate attack on concrete. Therefore, a number of previous researches all supported the idea that the mineral additions, such as fly ash, slag powder, silica fume and metakaolin, play a positive role in making sulfate-resisting concrete [74-80], not depending on the exposure conditions (sodium sulfate, magnesium sulfate, or ammonium sulfate).

However, in the case of partial immersion the pore size refinement due to mineral addtions will contribute to an increase in the capillary sorption height following Eq. 5. The pore solution expression tests showed that this process can draw more sulfates into fly ash concrete than into normal concrete, resulting in a pore solution with a higher sulfate concentration as shown in Fig. 20 [61].

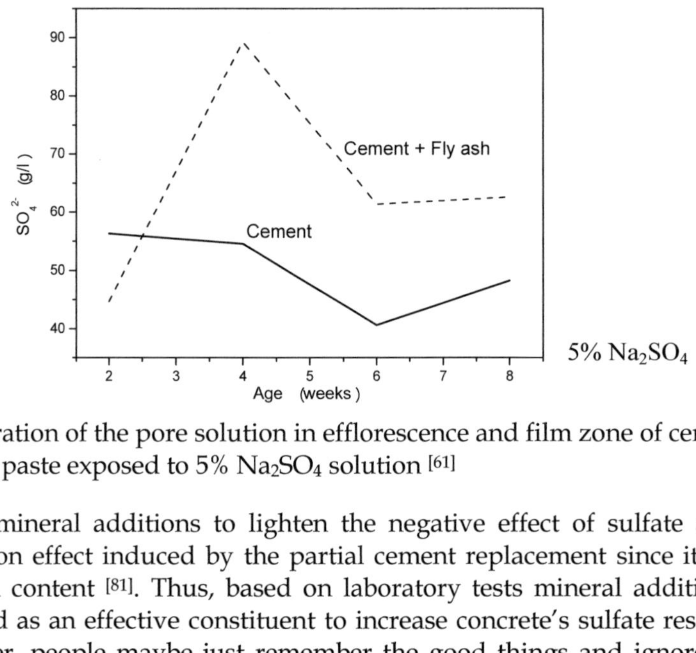

Fig. 20. SO_4^{2-} concentration of the pore solution in efflorescence and film zone of cement paste and cement-FA paste exposed to 5% Na_2SO_4 solution [61]

Another reason for mineral additions to lighten the negative effect of sulfate attack on concrete is the dilution effect induced by the partial cement replacement since it entails a reduction in the C_3A content [81]. Thus, based on laboratory tests mineral additions have always been regarded as an effective constituent to increase concrete's sulfate resistance in the field [82]. However, people maybe just remember the good things and ignore the bad ones. In the 1960s and 1970s extensive studies at the U.S. Bureau of Reclamation had reminded that [83, 84] concretes containing 30 percent low-calcium fly ashes showed greatly improved sulfate resistance to a standard sodium sulfate solution. However, the use of high-calcium fly ashes generally reduced the sulfate resistance. The high-calcium fly ashes containing highly reactive alumina in the form of C_3A or $C_4A_3\hat{S}$ are therefore less suitable than low-calcium fly ashes for improving the sulfate resistance of concrete. Taylor also pointed out that if slag has low alumina content, it improves the sulfate resistance, but with a high content of alumina, the reverse is the case [85,86]. M. Nehdi[47] also pointed out that it should not be overlooked that fly ash contains a large amount of reactive aluminum and that binders with an increased Al_2O_3 content can be more susceptive to the formation of ettringite. P. Nobst and J. Stark [87] carried out a very interesting test. Hardened cement pastes modified by different mineral additions were ground to a fineness of <200 µm and were mixed with stoichiometric parts of high quality gypsum powder ($CaSO_4 \cdot 2H_2O$) and chemically produced calcite ($CaCO_3$) as well as with an excess of 20% of distilled water to investigate the thaumasite formation without the physical obstacle. The products identification showed unexpected results: (1) concerning cement-FA paste, the amount of ettringite increased with an increasing Al_2O_3 content at 20 °C while at 6 °C fly ash promoted a little more thaumasite formation; (2) slag cements which are generally classified as high sulfate resisting cements showed the most intensive thaumasite formation; (3) the use of micro-silica strongly accelerated the thaumasite formation. These findings indicate that mineral additions have a potentially negative effect in the concrete's resistance to sulfate attack depending on the exposure conditions. In the paper [87], the negative effect emerged due to no physical obstacle.

In the fly ash the aluminum phase existing as solid glass spheres is stable, but can be activated in a thermal, mechanical or chemical way [88]. It should be noted that Na_2SO_4 is an effective activator which is often used to activate the pozzolanic fly ash reaction in cement-fly ash pastes [89, 90]. What is worse, the ambient temperature may rise and also play a positive role in activating the aluminum phase, promoting the ettringite formation. According to the thermal analysis results [61], the cement and cement - fly ash (25%) pastes were immersed in the 5% Na_2SO_4 solution at 30 °C for 6 months. The amount of ettringite in the cement and cement-FA pastes amounted to 0.173 mg /mg and 0.217 mg/mg respectively. On the other hand, more gypsum was also detected in the cement - fly ash paste than cement paste. Fig. 21 is the thermal analysis of pastes exposed to 15% Na_2SO_4 solution at 30°C for 6 months. More ettringite and gypsum were generated in the cement - fly ash paste than in cement paste. Moreover, according to the wick action, 15% Na_2SO_4 can be formed in the upper portion of concrete in contact with air during the process of "salt weathering" on field concrete.

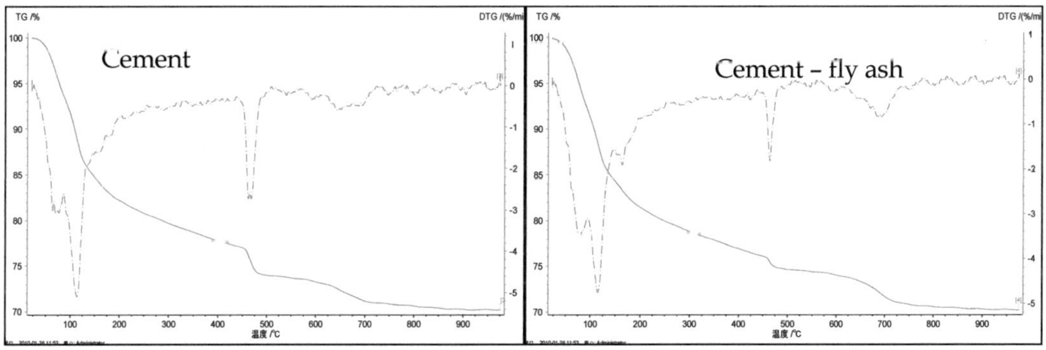

Fig. 21. Thermal analysis of pastes partially immersed in the 15% Na_2SO_4 solution at 30°C for 6 months [61]

In the paper [87], the negative effect of mineral additions on concrete sulfate resistance opposite to the normal results was attributed to no physical obstacle. In the process of "salt weathering" on concrete, a similar no physical obstacle appearance also can be defined. As abovementioned, solution goes into concrete by capillary suction. For porous materials, the capillary suction is a kind of active process, i.e. the solution is invited into the concrete. This can also be regarded as a no physical obstacle process. The sulfates can homogeneously distribute in the cement paste similar to the alkali activated cement in which Na_2SO_4 and powders are mixed before adding water, resulting in Na_2SO_4 homogeneously distributing in the cement paste.

In summary, concerning the role of mineral additions in the sulfate attack on partially exposed concrete, the exposure conditions and the solution transport mechanism are different from the full immersion cases. It needs further research.

4.5 Study of the effect of different kinds of sulfates in "salt weathering" on concrete

As abovementioned, Ruiz-Agudo [42] studied salt weathering distress of limestone specimens submerged in sodium sulfate and magnesium sulfate solutions respectively. The results showed that these two sulfates both severely damaged stone.

In the full immersion attack, because of the simultaneous significant decomposition of the C-S-H gel that accompanies the formation gypsum and ettringite, admittedly, people think the overall corrosive action of magnesium sulfate is greater than that of sodium sulfate [81, 91]

However, during the process of "salt weathering" on concrete, the test results showed the opposite appearance. Nehdi and Hayek [47] observed the appearances of the cement mortar partially exposed to 10% sodium sulfate and magnesium sulfate solution in a RH cycling between 32±3% and >95% condition respectively. The results showed that a large amount of efflorescence covers the surface when mortar is exposed to sodium sulfate solution. On the contrary, the surface of mortar subjected to magnesium sulfate solution is clean. It seems that sodium sulfate performs more corrosive effect on mortar than magnesium sulfate. The tests [53] also showed the same results. The aggregates and cement paste were completely separated in the upper part of concrete after 8 months exposure. However, the samples exposed to magnesium sulfate solution showed little damage. As shown in Fig. 22.

Fig. 22. Visual observation of concrete specimens partially exposed to Na_2SO_4 and $MgSO_4$ solutions [53]

This appearance may indicate that the concrete damage cannot be explained by salt weathering. First, $MgSO_4$ showed a harmful effect on stone due to salt weathering. As a porous material concrete should also show a similar scaling manner. Secondly, Fig. 23 shows the surface tensions of NaCl, Na_2SO_4 and $MgSO_4$ [92]. According to Eq. 5 the equilibrium heights of capillary rise of sodium sulfate and magnesium sulfate should be almost the same, showing similar efflorescence zone due to salt weathering. The reason for the opposite appearance of $MgSO_4$ in concrete may be the insoluble brucite due to chemical reaction that blocks the capillary. The role of sulfates in the "salt weathering" on concrete also needs further research.

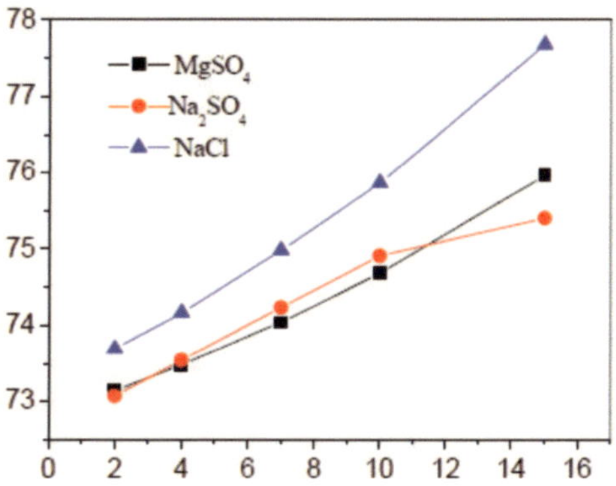

Fig. 23. Surface tensions of NaCl, Na_2SO_4 and $MgSO_4$ [92]

4.6 Study of the role of concrete carbonation in "salt weathering" on concrete

The negative effect of carbonation on corrosion of reinforcing steel in concrete is well known. As to the sulfate attack on concrete, as a result of carbonation, the total porosity would be reduced and the permeability of concrete could be improved [93-94]. So, Gao [95] pointed out that the carbonation layer could mitigate diffusion of sulfate ions to some extent in the full immersion situation.

However, when the concretes are partially exposed to sulfate solutions, the situation may be different. V.T. Ngala [94] studied the effect of carbonation on the ratio of capillary to total porosity of cement paste. The results showed that the capillary pore fraction greatly was improved after carbonation (shown in Fig. 24). This will promote the capillary suction of concrete, forming a more severe sulfate pore solution in the concrete and resulting more severer concrete damage.

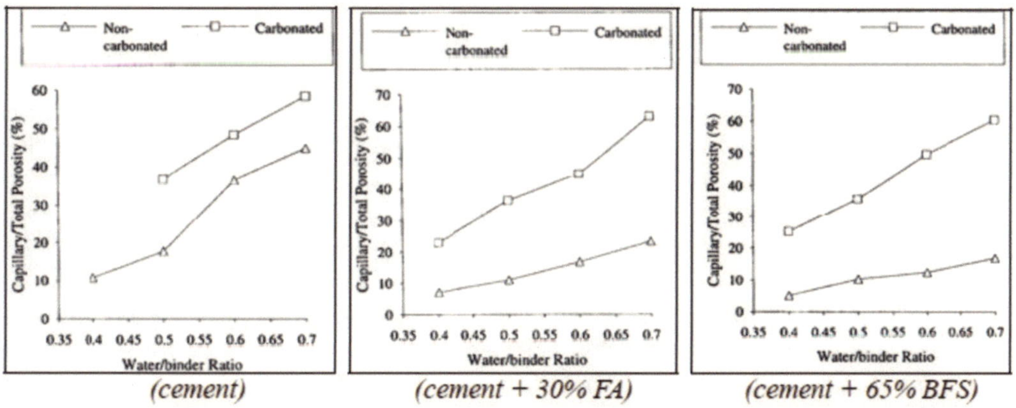

Fig. 24. Ratio of capillary to total porosity of non-carbonated and carbonated paste [94]

From Fig. 24, compared to cement paste, the ratios of capillary pores fraction of cement + 30% FA and cement + 65% BFS were higher than cement paste after carbonation. Some research showed that blended concrete has high carbonation rate [96], high degree of carbonation [97] or large carbonation depth [98] compared to the ordinary cement concrete. The carbonation susceptibility of blended concrete may be another reason for the negative effect of mineral addition on sulfate resistance of partially exposed concrete.

Besides, according to the review of indoor tests of "salt weathering" on concrete, two experimental results were observed showing that sulfate crystallization can be detected in the calcite crystals, the carbonation products of concrete (Fig. 12) and that carbonation could accelerate the concrete damage (Fig. 17). It might be that the efflorescence also occurs after concrete carbonation.

In summary, the effect of carbonation on sulfate resistance of partially exposed concrete is not clear. Further research will contribute to disclose the mechanism of "salt weathering" on concrete.

5. Conclusions

"Salt weathering" on concrete by sulfates is a deceptive and misleading phenomenon. In this paper, according to the comparison between the basic principles of salt weathering on porous materials and the abnormal appearances of "salt weathering" on concrete, the conclusion can be drawn that the salt weathering distress is not the major reason causing concrete damage when partially exposed to the sulfate environment. Chemical sulfate attack occurring in a high concentration pore solution is more likely the degradation mechanism for concrete deterioration similar to the full immersion cases of sulfate attack on concrete.

6. Acknowledgements

This work was financially supported by the National Science Foundation of P. R. China under contract #50378092, the scholarship from CSC (China Scholarship Council) and the co-funding from Ghent University of Belgium. The research was performed under a Bilateral Cooperation Agreement between Ghent University of Belgium and Central South University of P.R. China.

7. References

[1] D. Benavente, M.A. García Del Cura, A. Bernabéu, S. Ordóñez, Quantification of salt weathering in porous stones using an experimental continuous partial immersion method, Eng. Geol 2001; 59(3-4):313-325.

[2] Harvey Haynes, Robert O'Neill, Michael Neff, and P. Kumar Mehta, Salt weathering distress on concrete exposed to sodium sulfate environment, ACI Mater. J. 2008, 105(1):35-43.

[3] K. J.Folliard, P. Sandberg, Mechanisms of Concrete Deterioration by Sodium Sulfate Crystallization, Durability of Concrete, SP-145, American Concrete Institute, Farmington Hills, MI, 1994: 933-945.

[4] Haynes, H.; O'Neill, R.; and Mehta, P. K. Concrete Deterioration from Physical Attack by Salts, Concr. Int. 1996, 18 (1):63-68.
[5] W. G. Hime, R. A. Martinek, L. A. Backus, S. L. Marusin, Salt Hydration Distress, Concr. Int. 2001.23(10): 43-50.
[6] Thaulow, Niels, Sahu, Sadananda. Mechanism of concrete deterioration due to salt crystallization, Mater. Charact.2004, 53(2-4):123-127.
[7] YANG Quanbing，YANG Qianrong. Effects of salt-crystallization of sodium sulfate on deterioration of concrete, J. Chin. Chem. Soc 2007, 35(7): 877-880+885 (Chinese).
[8] MA Kunlin，XIE Youjun，LONG Guangcheng，LIU Yunhua. Deterioration characteristics of cement mortar by physical attack of sodium sulfate, J. Chin. Chem. Soc 2007, 35(10):1376-1381.
[9] Harvey Haynes, Robert O'Neill, Michael Neff, and P. Kumar Mehta, Salt Weathering of Concrete by Sodium Carbonate and Sodium Chloride, ACI Mater. J. 2010, 107(3):256-266
[10] Robert J. Flatt, Michael Steiger, George W. Scherer, A commented translation of the paper by C.W. Correns and W. Steinborn on crystallization pressure, Environ. Geol. 2007, 52(2): 221-237.
[11] George W. Scherter, crystallization in pores, Cem. Concr. Res. 1999, 29(8): 1347-1358.
[12] Rosa M. Espinosa Marzal, George W. Scherer, Crystallization of sodium sulfate salts in limestone, Environ. Geol. 2008, 56(3-4): 605-621.;
[13] Robert J. Flatt, Salt damage in porous materials: How high supersaturations are generated, J. Cryst. Growth 2002, 242(3-4): 435-454.
[14] George W. Scherer, Factors affecting crystallization pressure, International RILEM TC 186-ISA workshop and internal sulfate attack and delayed ettringite formation, 2002, Villars, Switzerland.
[15] Michael Steiger, Crystal growth in porous materials - I: The crystallization pressure of large crystals, J. Cryst. Growth 2005, 282(3-4):455-469.
[16] Michael Steiger, Crystal growth in porous materials - II: Influence of crystal size on the crystallization pressure, J. Cryst. Growth 2005, 282(3-4):470-481.
[17] Joerg Ruedrich, Siegfried Siegesmund, Salt and ice crystallization in porous sandstones, Environ. Geol. 2007, 52(2): 343-367.
[18] R.M. Espinosa, L. Franke, G.. Deckelmann, Model for the mechanical stress due to the salt crystallization in porous materials, Constr. Build. Mater. 2008, 22(7):1350-1367.
[19] G. Cultrone, L.G.Russo, C. Calabrò, M. Uroševic, A. Pezzino, Influence of pore system characteristics on limestone vulnerability: A laboratory study, Environ. Geol. 2008, 54(6):1271-1281.
[20] George W. Scherer, Stress from crystallization of salt, Cem. Concr. Res. 2004, 29(9): 1613-1624.
[21] D.Benavente, N. Cueto, J. Martínez-Martínez, M.A. García Del Cura, J.C. Cañaveras, The influence of petrophysical properties on the salt weathering of porous building rocks, Environ. Geol. 2007, 52(2):197-206.
[22] D. Benavente, M.A. Garcia del Cura, R. Fort, S. Ordónez, Durability estimation of porous building stones from pore structure and strength, Environ. Geol. 2004, 74(1-2): 113-127.

[23] D. Benavente, J. Martínez-Martínez, N.Cueto, M.A. García-del-Cura, Salt weathering in dual-porosity building dolostones, Environ. Geol. 2007, 94(3-4): 215-226.
[24] V. Lopez-Acevedo, C. Viedma, V. Gonzalez, A. La Iglesia, Salt crystallization in porous construction materials. II. Mass transport and crystallization processes, J. Cryst. Growth 1997, 182(1-2): 103-110.
[25] Miguel Gomez-Heras, Rafael Fort, Patterns of halite (NaCl) crystallisation in building stone conditioned by laboratory heating regimes, Environ. Geol. 2007, 52(2): 239-247.
[26] C. Rodriguez-Navarro, E. Doehne, Salt weathering: influence of evaporation rate, supersaturation and crystallization pattern. Earth Surf Processes and Landforms 1999, 24(2-3): 91-209.
[27] C. Rodriguez-Navarro, L. Linares-Fernandez, E. Doehne, E. Sebastian, Effects of ferrocyanide ions on NaCl crystallization in porous stone, J. Cryst. Growth 2002, 243(3-4): 503-516.
[28] S. Charles, E. Doehne, The evaluation of crystallization modifiers for controlling salt damage to limestone, J. J Cult. Herit. 2002 3 (3) 205–216.
[29] N.R. Buenfeld, M-T. Shurafa - Daoudi, I. M. Mcloughin, Chloride transport due to wick action in concrete RILEM International Workshop on Chloride Penetration into Concrete 1995:315-324.
[30] J. Francis Yong, Sidney Mindess, Robert J. Gray, Arnon Bentur, The science and technology of civil Engineering materials, Chinese Architecture & Building Press, 2006.
[31] Y.T. Puyate, C.J. Lawrence, Steady state solutions for chloride distribution due to wick action in concrete, Chem. Eng. Sci. 2000, 55(16): 3329-3334.
[32] Y.T. Puyate, C.J. Lawrence, N.R. Buenfeld, I.M. McLoughlin, Chloride transport models for wick action in concrete at large Peclet number, Phys. Fluids. 1998, 10(3): 566-575.
[33] Y.T. Puyate, C.J. Lawrence, Wick action at moderate Peclet number, Phys. Fluids. 1998, 10(8): 2114-2116.
[34] Y.T. Puyate, C.J. Lawrence, Effect of solute parameters on wick action in concrete, Chem. Eng. Sci., 1999, 54(19):4257-4265.
[35] L. Pel, H. Huinink, K. Kopinga, Ion transport and crystallization in inorganic building materials as studied by nuclear magnetic resonance, Appl. Phys. Lett. 2002, 81(15): 2893-2895.
[36] L.Pel, H. Huinink, K. Kopinga, R.P.J.Van Hees, O.C.G. Adan, Efflorescence pathway diagram: Understanding salt weathering, Constr. Build. Mater. 2004, 18(5): 309-313.
[37] C. Cardell, D.Benavente, J. Rodríguez-Gordillo, Weathering of limestone building material by mixed sulfate solutions. Characterization of stone microstructure, reaction products and decay forms, Mater. Charact. 2008, 59 10): p 1371-1385.
[38] Nicholas Tsui, Robert J. Flatt, George W. Scherer. Crystallization damage by sodium sulfate, J. J Cult. Herit. 2003 4 (2): 109–115.
[39] Genkinger, Selma, Putnis, Andrew, Crystallisation of sodium sulfate: Supersaturation and metastable phases, Environ. Geol. 2007, 52(2): 295-303.

[40] C. Rodriguez-Navarro, E. Doehne, E. Sebastian, How does sodium sulfate crystallize? Implications for the decay and testing of building materials, Cem. Concr. Res. 2000, 30(10): 1527-1534.

[41] E.M. Winkler, P.C. Singer, Crystallization pressure of salt in stone and concrete, Geol. Soc Am. 1972, 83(11): 3509-351.

[42] E. Ruiz-Agudo, F. Mees, P. Jacobs, C. Rodriguez-Navarro, The role of saline solution properties on porous limestone salt weathering by magnesium and sodium sulfates, Environ. Geol. 2007, 52(2):305-317.

[43] F.R. Mcmillan, T.E. Stantion, I.L. Tyler, W. C. Hansen. Long-Time Study of Cement Performance in Concrete, chapter 5. Concrete exposed of sulfate solis, Portland Cement Association 1949.

[44] D. Stark. Durability of concrete in sulfate-rich soils, Research and Development Bulletin, vol. RD O97, Portland Cement Association, 1989.

[45] D. Stark Performance of Concrete in Sulfate Environments, RD129, Portland Cement Association 2002.

[46] E.F. Irassar, A. Di Maio, O.R. Batic , Sulfate attack on concrete with mineral admixtures, Cem. Concr. Res. 1996, 26(1):113-123.

[47] M. Nehdi, M. Hayek, Behavior of blended cement mortars exposed to sulfate solutions cycling in relative humidity, Cem. Concr. Res. 2005, 35(4): 731-742.

[48] Norah Crammond, The occurrence of thaumasite in modern construction - a review, Cem. Concr. Compo. 2002, 24(3-4): 393-402.

[49] P.W. Brown, April Doerr, Chemical changes in concrete due to the ingress of aggressive species, Cem. Concr. Res. 2000, 30(3): 411-418.

[50] Mingyu, Hu, Fumei, Long, Mingshu, Tang, The thaumasite form of sulfate attack in concrete of Yongan Dam, Cem. Concr. Res. 2006, 36,(10): 2006-2008.

[51] Chiara F. Ferraris, Paul E. Stutzman, Kenneth A. Snyder, Sulfate Resistance of Concrete: A New Approach, Research and Development Information PCA R&D, Serial No. 2486, 2006.

[52] Bellmann Frank, Möser Bernd, Stark Jochen, Influence of sulfate solution concentration on the formation of gypsum in sulfate resistance test specimen, Cem. Concr. Res. 2006, 36(2): 358-363.

[53] Zanqun Liu, Geert De Schutter, Dehua Deng, Zhiwu Yu, Micro-analysis of the role of interfacial transition zone in "salt weathering" on concrete, Constr. Build. Mater. 2010, 24(11): 2052–2059.

[54] Zanqun Liu, Dehua Deng, Geert De Schutter, Zhiwu Yu, Micro-analysis of "salt weathering" on cement paste, accepted by Cem. Concr. Compo. for publish.

[55] D. Benavente, García del Cura, M.A.,García-Guinea, J, Sánchez-Moral, S, Ordóñez, S, Role of pore structure in salt crystallisation in unsaturated porous stone, J. Cryst. Growth 2004, 260 (3-4): 532-544.

[56] LIU Zanqun, XIAO Jia, HUANG Hai, YUAN Qiang, DENG Dehua. Physicochemical Study on the Interface

[57] Zone of Concrete Exposed to Different Sulfate Solutions, J.Wuhan Univ.Technol. (Materials Science Edition) 2006, 21(z1):167-175.

[58] Faran J. Introduction: the transition zone - discovery and development. ITZ in concrete RILEM report 11. London: E&FN Spon; 1996.
[59] Shenyang, Zhongzi Xu, Mingshu Tang, The process of sulfate attack on cement mortars. Adv Cem Based Mater 1996;4(1):1–5.
[60] Bonen David. Micro-structural study of the effect produced by magnesium sulfate on plain and silica fume-bearing Portland cement mortars. Cem. Concr. Res. 1993;23(3):541–55.
[61] Santhanam Manu, Cohen Menashi D, Olek Jan. Mechanism of sulfate attack: afresh look - part I: summary of experimental results. Cem. Concr. Res. 2002;32(6):915–21.
[62] Zanqun Liu, Study of the basic mechanisms of sulfate attack on cementitious materials, 2010, Central and South University China and Ghent University Belgium.
[63] M. Collepardi, A state-of-the-art review on delayed ettringite attack on concrete, Cem. Concr. Compo. 2003 25(4-5): 401-407.
[64] [63]Manu Santhanam, Menashi D, Jan OLek. Sulfate attack research — whither now? Cem. Concr. Res. 2004, 31(8): 1275-1296.
[65] P. Kumar Mehta. sulfate attack on concrete: separating myths from reality, Concr. Int. 2000, 28 (8): 57- 61.
[66] Q. Zhou, J. Hill, E.A. Byars, et al, The role of pH in thaumasite sulfate attack, Cem. Concr. Res. 2006, 36(1): 160-170.
[67] Jallad, Karim N., Santhanam, Manu, Cohen, Menashi D. Stability and reactivity of thaumasite at different pH levels, Cem. Concr. Res. 2003, 33(3): 433-437.
[68] N.J. Crammond, The thaumasite form of sulfate attack in the UK, Cem. Concr. Res. 2003, 25(7): 809-818.
[69] P. Hagelia, R.G. Sibbick, N.J. Crammond, C.K. Larsen, Thaumasite and secondary calcite in some Norwegian concretes, Cem. Concr. Compo. 2003, 25(8):1131-1140.
[70] P. Hagelia, R.G.. Sibbick, Thaumasite Sulfate Attack, Popcorn Calcite Deposition and acid attack in concrete stored at the Blindtarmen test site Oslo, from 1952 to 1982, Mater. Charac. 2009, 60(7):686-699.
[71] Deng De-Hua, Xiao Jia, Yuan Qiang, On thaumasite in cementitious materials, Jianzhu Cailiao Xuebao/J. Build. Mater. 2005, 8(4): 400-409 (Chinese).
[72] AETM C 1012 - 04, Standard Test Method for Length Change of Hydraulic-Cement Mortars Exposed to a Sulfate Solution.
[73] H. Haynes, Sulfate Attack on Concrete: Laboratory versus Field Experience, Concr. Int. 2002, 24(7): 64-70.
[74] P.K. Mehta, Sulfate attack on concrete: a critical review, Materials Science of Concrete, vol.III, Amer. Ceramic Society 1993:105- 130.
[75] H.T. Cao, L. Bucea, A. Ray, S. Yozghatlian, The effect of cement composition and pH of environment on sulfate resistance of Portland cements and blended cements, Cem. Concr. Compo. 1997, 19(2): 161-171.
[76] S. Miletic, M.Ilic, S. Otovic, R.Folic, Y. Ivanov, Phase composition changes due to ammonium-sulphate: Attack on Portland and Portland fly ash cements, Constr. Build. Mater. 1999, 13(3): 117-127.
[77] Rodriguez-Camacho, Redz E., Uribe-Afif, R., Importance of using the natural pozzolans on concrete durability, Cem. Concr. Res. 2003, 32(12): 1851-1858.

[78] El Sokkary, T.M., Assal, H.H., Kandeel, A.M., Effect of silica fume or granulated slag on sulphate attack of ordinary portland and alumina cement blend, Ceram. Int. 2004, 30(2): 133-138.
[79] P. Chindaprasirt, S. Homwuttiwong, V. Sirivivatnanon, Influence of fly ash fineness on strength, drying shrinkage and sulfate resistance of blended cement mortar, Cem. Concr. Res. 2004, 34(7): 1087-1092.
[80] Hanifi Binici, Orhan Aksogan, Sulfate resistance of plain and blended cement, Cem. Concr. Compo. 2006, 28(1): 39-46.
[81] Nabil M. Al-Akhras, Durability of metakaolin concrete to sulfate attack, Cem. Concr. Res. 2005, 36(9): 1727-1734.
[82] Omar S Al-Amoudi Baghabra, Attack on plain and blended cements exposed to aggressive sulfate environments, Cem. Concr. Compo. 2002, 24(3-4): 305-316.
[83] GB/T 50476-2008, Code for durability design of concrete structures (Chinese standard).
[84] P. Kumar Mehta, Concrete: structure, properties, and materials, Second Edition, Prentice Hall College Div; 2nd Revised edition edition (November 1992);
[85] Monteiro, Paulo J.M., Kurtis, Kimberly E., Time to failure for concrete exposed to severe sulfate attack, Cem. Concr. Res. 2003, 33(7): 987-993,.
[86] R.S. Collop, H. F. W.Taylor, Microstructural and microanalytical studies of sulfate attack III: Sulfate-resisting cement: reactions with sodium and magnesium sulfate solution. Cem. Concr. Res. 1995, 25(7): 1581-1590.
[87] R.S. Collop, H. F. W.Taylor, Microstructural and microanalytical studies of sulfate attack. V. Comparison of different slag blends, Cem. Concr. Res. 1996, 26(7) 1029-1044.
[88] P. Nobst, J. Stark, Investigations on the influence of cement type on thaumasite formation, Cem. Concr. Compos.2003, 25(8): 899–906.
[89] Wu, Zichao,Naik, Tarun R., Chemically activated blended cement, ACI Mater. J. 2003, 100(5): 434-440.
[90] Shi, Caijun, and Day, Robert L. Pozzolanic reaction in the presence of chemical activators. Part I. Reaction kinetics, Cem Concr Res. 2000, 30(1):51-58.
[91] Shi, Caijun, and Day, Robert L., Pozzolanic reaction in the presence of chemical activators: Part II. Reaction products and mechanism, Cem Concr Res. 2000, 30(4): 607-613.
[92] Omar S. Baghabra Al-Amoudi, Mohammed Maslehuddin, Effect of magnesium sulfate and sodium sulfate on the durability performance of plain and blended cement, ACI Mater. J. 1995, 92(1): 15-24.
[93] Ma Kunlin, Mechanism and Evaluation Method of Salt Crystallization Attack on Concrete, Central South University PhD thesis, 2009;
[94] V.T. Ngala, C.L. Page, Effect of carbonation on pore structure and diffusional properties of hydrated cement paste, Cem Concr Res. 1997, 27(7): 995-1007.
[95] Ha-Won Song, Seung-Jun Kwon, Permeability characteristics of carbonated concrete considering capillary pore structure, Cem Concr Res. 2007, 37 (6): 909–915;
[96] GAO Rundong · ZHAO Shunbo , LI Qingbin, Deterioration Mechanisms of Sulfate Attack on Concrete under the Action of Compound Factors, Jianzhu Cailiao Xuebao/J. Build. Mater. 2009, 12(1): 41-46 (Chinese).

[97] P. Sulapha, S.F. Wong, T.H. Wee, S. Swaddiwudhipong, Carbonation of concrete containing mineral admixtures, J. Mater. Civ. Eng. 2003, 15(2): 134-143.
[98] Monkman, Sean,, Shao, Yixin, Assessing the carbonation behavior of cementitious materials, J. Mater. Civ. Eng. 2006, 18(6): 768-776.
[99] Marlova P. Kulakowski, Fernanda M. Pereira, Denise C.C. Dal Molin Carbonation-induced reinforcement corrosion in silica fume concrete, Constr. Build. Mater. 2009, 23(3): 1189-1195.

18

Crystallization, Alternation and Recrystallization of Sulphates

Joanna Jaworska
Adam Mickiewicz University
Poland

1. Introduction

Sulphates as well as silicates and carbonates are one of the most common minerals on the Earth's surface. They cover about 25% of continents surface (Blatt et al. 1980; Ford &Williams, 1989). Their recent sedimentary environments are the terrains of: the southern Mediterranean coast - coastal salt lakes of Marocco, Libya, Tunisia and Egypt, Gulf of Kara Bogaz (Caspian Sea), Persian Gulf - coastal sabkhas of UAE (special Abu Dhabi Emirate) and Qatar, Texas and California (Death Valley), salt lakes of South and Central Australia and salt lakes, salinas and salares of South America.

Annual total world production of gypsum in 2010 exceeded 146 million metric tones (http://minerals.usgs.gov/minerals/pubs/commodity/).

First of all, the sulphates are represented by two kinds of calcium sulphate - gypsum ($CaSO_4 \cdot 2H_2O$) and anhydrite ($CaSO_4$); mainly the first one creates deposits that are of economical value; it is used in the construction industry as bond material and to control the bonding speed, in casting and modelling and also in medicine (surgery and stomatology), during the production of paper. Its properties influence the parameters and quality of materials which it consist in. In construction/building industry the semi-hydrated gypsum is used as a result of frying in temperatures about 160°C (150-190°C) with sufficient amount of added water, the material bonds and hardens - the reaction is exothermic and the gypsum's volume increases of about 1%. Bassanite ($CaSO_4$ ½ H_2O), calcium sulphate semi-hydrate, is also known.

Rarely we can find the sulphates of: strontium (celestine), barium (barite), potassium (e.g. polyhalite), sodium (e.g. mirabilite, glauberite), magnesium (e.g. epsomite, kieserite), copper (e.g. brochantite, chalcanthite) and others. Most of gypsum and anhydrite on Earth are of evaporate origin, they are formed in specific order as a result of precipitation of the calcium sulphate inside the gradually drying sea basin (deep or shallow), lake, by the coastal lagoons, bays or sabkhas (indications of hot and arid climate). They are also the products of volcanic exhalations or low temperature hydrothermal processes, as well as of oxidation of sulphide deposits. The sulphates are also found above the salt mirror of diapirs, where they form the secondary deposit as the harder soluble residuum after the salt leaching - they constitute the main component of so-called gypsum or anhydrite-gypsum cap-rock.

The average precipitation rate of sulphates (gypsum and anhydrite) in the evaporite basin is ca. 0.5-1.2 mm/year and requires the evaporation of few to few tens cm high (2 m) column of water.

Probably, the oldest documented sulphate pseudomorphs are 3. 45 billion years old and come from West Australia (Pilbara), cm-size growth and interpreted to replace gypsum (Barley et al., 1979; Buick & Dunlop, 1990); only slightly younger are pseudomorphs after swallowtail gypsum - 3.4 billion years old - from S. Africa, Kaapvall Craton (Wilson & Versfeld, 1994).

1.1 Gypsum: $CaSO_4 \cdot 2H_2O$

Crystal system: monoclinic, hardness: 2, density: 2.3-2.4 g/cm^2
soluble: in water, in HCl and in concentrated solution of H_2SO_4
contains impurities: Ba, Sr, deposit grains where it crystallizes, bituminous substances
habit: platy, columnar, fibrous, needle-like, lenticular; forms massive aggregates and twins - swallowtail (figs. 1.,2.,3.,7. and 10.), usually colourless, might be coloured by Fe compounds
particular varieties:

- alabaster - fine-grained, sugar-like variety used in sculpture (fig. 4.),
- selenite - large well-crystallized varieties with dimensions reaching few m (fig. 5.); usually colourless
- spar - fibrous variety with semi-gloss, filling fissures and fractures (fig. 7.)
- desert rose - flower-like form of rounded gypsum aggregates (fig. 8.), occurring in the deserts as a result of ascent of the underground water rich in sulphates; it contains embedded sand grains built-in during the fast crystal growth.

Fig. 1. Platy gypsum (Petunia Bukta, Spitsbergen) phot. J. Jaworska

Fig. 2. Fibrous gypsum (Germany) phot. J. Jaworska

Fig. 3. Columnar to needle-like gypsum (Polkowice, Poland) phot. J. Jaworska

Fig. 4. Alabaser (Ukraine) phot. J. Jaworska

Primarily, gypsum that crystallizes in the evaporite basins forms usually medium or coarse grains; sometimes the lamination occurs, reflecting the changes in the basin (water composition, water level). Among the gypsum laminas, biolaminae appear; they are formed in the neritic zones and can be either deformed by periodical droughts (mudcraks) or ruptured by crystallizing sulphates (teepee-like structures, see fig. 9). In deeper zones of the basin, sabre-like gypsum (fig. 14.) can crystallize; these are elongated gypsum crystals, 20-30 cm long, distorted in one direction due to the demersal current activity (they constitute the perfect indicators of paleocurrents). Selenite gypsum is an exceptional feature; it forms under stable conditions at the depth of few to several m (figs. 16. and 17.) and reaches the dimensions of 3.5-4 m usually, but even up to 10 m. In deeper zones, laminated gypsum forms; sometimes with the ripplemark remains or even turbidites and slump structure with fragments of older, more lithified gypsum.

Fig. 5. Selenite (Busko-Zdrój, Poland) phot. J. Jaworska

Fig. 6. Gypsum twins - swallowtail (Dymaczewo Stare, Poland) phot. J. Jaworska

1.2 Anhydrite: $CaSO_4$

crystal system: orthorhombic; hardness: 3.5, density: 2.98 g/cm^2
hardly soluble in: HCl and concentrated H_2SO_4
contains impurities: Ba, Sr

habit: platy, columnar, fibrous; the crystal size rarely exceeds 0.5-1 mm (fig. 11.); sometimes crystals grown in caverns and fractures appear; massive aggregates (fig. 13.), rare radiant aggregates exceptionally reach the length of few cm

usually colourless crystals
particular varieties:
- enterolithic anhydrite
- bluish, fibrous variety, resembling twisted viscera (regional mining name, see fig. 12.).

Recently, the gypsum precipitates from among calcium sulphates; whereas anhydrite crystallizes very rarely - the only locations of its recent crystallization are: the Persian Gulf coast, lakes: Elton and Inger, Death Valley and Clayton Playa (Nevada).

Fig. 7. Fibrous -spar gypsum in clay-slate (Niwnice, Poland) phot. J. Jaworska

Fig. 8. Desert rose; phot. J. Jaworska

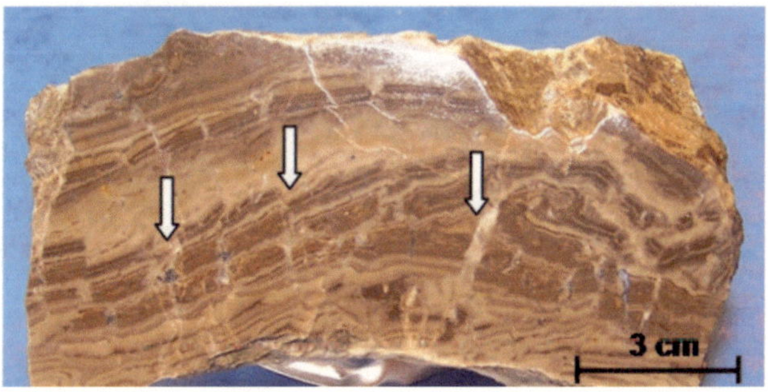

Fig. 9. Biolaminas deformed by crystallizing sulphates (near Ostrówka quarry, Poland) phot. J. Jaworska

Fig. 10. Lenticular gypsum (Wapno cap-rock, Poland) phot. J. Jaworska

Fig. 11. Anhydrite crystals (from Dębina salt dome, Poland), phot. A. Kyc

12. Enterolithic anhydrite (Wieliczka mine, Poland) phot. J. Jaworska

2. Crystallization and alternation: Hydratation and dehydratation (gypsification, anhydritization)

2.1 Crystallization

In most of the cases during evaporation processes, the gypsum crystallizes first, than the anhydrite (higher concentration of solution, 5-6 times higher than the normal sea water

Fig. 13. Anhydrite rock - massie agreggates (Niwnice, Poland) phot. J. Jaworska

Fig. 14. Sabre-like gypsum (Nida region, Poland) phot. J. Jaworska

Fig. 15. Laminated gypsum (Niemeyer quarry, Germany) phot. J. Jaworska

salinity and in temperature about 40°C). Not until the concentration of solution reaches values close to NaCl concentration, the only phase of calcium sulphate which crystallizes and accompanies the rock salts is anhydrite; even if the temperature does not exceed 18°C. The thick rock salt deposits seldom form salt pillows together with salt swells and diapirs; their roof surfaces are located close to the Earth's surface (at the boundary of the salt mirror) and easily undergo leaching, leaving less soluble residue of - among the others - anhydrite grains and next the anhydrite sandstone (fig. 20.), forming so-called cap-rock that forms the natural cover of the salt deposit. The anhydrite sandstone can - depending on the conditions - undergo further transformation typical for this very mineral (fig. 21.).

Fig. 16. and 17. Outcrop of 2.5-3 m senlenite gypsums, regional named szklica (Nida region, Poland) phot. J. Jaworska

Fig. 18. Second native sulphur in gypsum rock (Niemeyer quarry, Germany) phot. J. Jaworska

In the recent evaporation basins mainly the gypsum precipitates; anhydrite crystallizing under more extreme conditions occurs more rarely. Whereas among the sediments - particularly at the depths of few hundreds to few thousand meters - the anhydrite dominates. In many cases the anhydrite occurs as a product of the dehydration of gypsum; usually it is easily recognized pseudomorph of gypsum (e.g. selenite gypsum). The primary anhydrite, as well as the secondary one (dehydrate), as a result of tectonic processes, intense weathering of the overburden, climate changes etc., can be placed within the range of the underground or subsurface water (ground, meteoric) - where the hydration processes occur resulting in substitution of anhydrite by gypsum.

Fig. 19. Gypsum-karst (Nida region, Poland). phot. J. Jaworska

Fig. 20. Anhydrite sandstone; phot. J. Jaworska

2.2 Alternation

The sulphates - mainly the products of the hypergenic processes - very easily undergo the diagenetic processes, in which the dominant role is played by: hydration (gypsification) of anhydrite and dehydration (anhydritization) of gypsum; both processes are reversible and the reaction takes place as follows:

$$CaSO_4 \bullet 2H_2O \leftrightarrows CaSO_4 + 2H_2O$$

gypsum ⇆ anhydrite + water

There are many factors affecting the start and course of this reaction:

1. temperature and environmental pressure - depending on:
- climate (for sulphates on the surface or close below it)
- depth of the deposits - thickness of the overburden,
- geothermal gradient of the area where the deposits occur - geotectonic environment and lithology of the overburden (thermal conductivity of the overburden),
2. chemical composition and concentration of solution, pore fluid pressure and the activity of water,
3. presence of micro-organisms and organisms (changes in Eh),
4. presence of cracks and pores in the sulphates as well as in the surrounding rocks.

Fig. 21. Lenticular gypsum with anhydrite inclusions; phot. J. Jaworska

Fig. 22. Fine-crystalline gypsum; phot. J. Jaworska

2.2.1 Conditions

Anhydrite under surface conditions or close to the surface can be formed as a result of intense heating (over 50°C) of primary gypsum by the sun under hot and arid conditions.

When the gypsum deposits are buried, their transformation into anhydrite can theoretically start at the depth of about 450-500 m (Murray, 1964; Hardie, 1967; Jowett et al. 1993); those are the depths where temperature reaches 20°C, so the dehydration should not appear, however it is compensated by high overburden pressure (10 MPa; Kubica, 1972); on the other hand, according to Sonnenfeld (1984), gypsum can be found at the depth of 1200 m; and according to Ford and Williams (2007) even at 3000 m. The depth of the gypsum dehydration among others is modified by the geotectonic environment and the lithology of the overburden. The weakly heat conducting overburden, e.g. schists and gneisses, upon the areas seismically active, volcanic, causes the increase of the hydration speed - anhydrite can substitute the gypsum already at the depth of about 400 m; whereas well conducting overburden, e.g. rock salt of the cratonic areas, causes the process of transformation of the gypsum into anhydrite to occur hypothetically at the depth of even 4 km (Jowett et al., 1993). But the anhydrite gypsification process during the exhumation occurs usually at the depth of about 100-150 m (Murrey, 1964; Klimchouk &Andrejchuk, 1996). It starts either when the anhydrite appears in the area of influence of the ground water, or when it is exposed to rain water.

Fig. 23. Lenticular gypsum; phot. J. Jaworska

The crystallization process of calcium sulphates, as well as their gypsification or anhydritization are affected by the solutions (and their pressure). The NaCl solution occurring in the pore fluids plays special role; it modifies the temperature of the gypsum-anhydrite phase transformation. If the composition of pore fluids corresponds to the composition of sea water, the water activity (αH_2O) is 0.93 and the transformation of gypsum into anhydrite occurs at the temperature of 52°C; however if the pore fluids are NaCl saturated, then the water activity reaches 0.75 and the transformation occurs at 18°C (Jowett et al., 1993). The temperature of gypsum-anhydrite transformation is increased by: the presence of alkaline metal ions (Conley and Bundy, 1958) up to 98°C and the solution of $CaSO_4$ up to 95°C, but with lack of the anhydrite nuclei (Posnjak, 1940). Additionally it is necessary to take into account the regime of pore fluids pressure; if it is hydrostatic, then the temperature of the gypsum transformation decreases along with depth from 52°C under surface conditions to about 40°C at the depth of 3 km, and in the case of the lithostatic regime - rises to about 58°C at 2 km (Jowett et al., 1993).

Fig. 24. Grain boundary migration between two gypsum crystals; phot. J. Jaworska

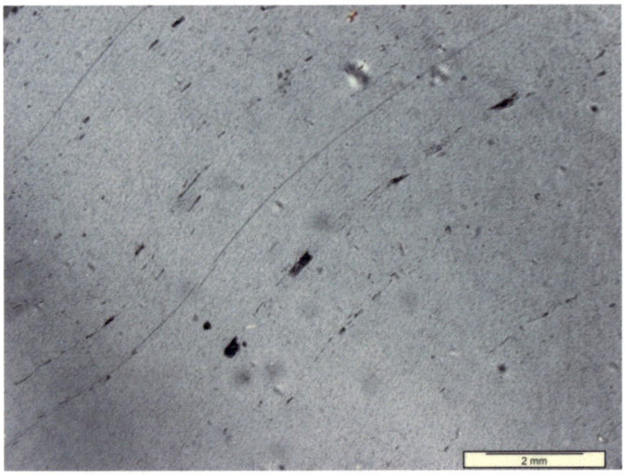

Fig. 25. Large gypsum with kink folds; phot. J. Jaworska

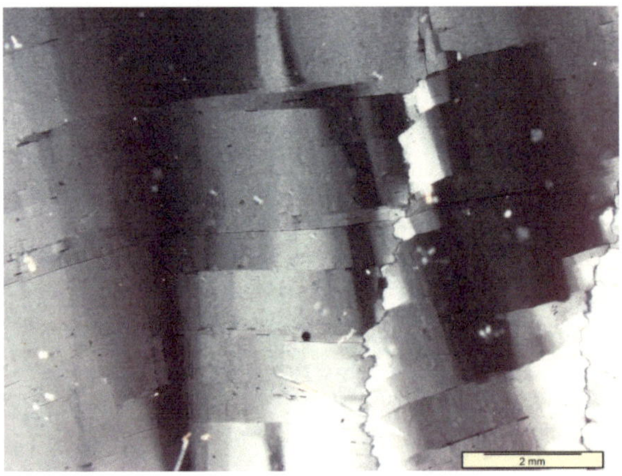

Fig. 26. 'Kink bands' and result of subgrain rotation in gypsum; phot. J. Jaworska

Fig. 27. 'Kink bands' and result of subgrain rotation in gypsum; phot. J. Jaworska

Shahid et al. (2007) comparing the crystallization and transformation conditions of sulphates in salt lakes and sabkhas in north Africa (Libia) and those from the Persian Gulf (Abu Dhabi) noted that while the climate is comparable, in the first case the anhydrite occurs very rarely, unlike in the area of the Arabian Peninsula. The main causes of this difference are the geochemical environment conditions: in the African sabkhas and salt lakes, the environment is more reducing and there is an occurrence of the organic material, the hydrogen sulphide releases and the sediment is dark; while sabkhas from the Persian gulf are more oxidised with lack of hydrogen sulphide - the sediment is light. The presence of fractures and joints in sediments/rocks surrounding the sulphates, as well as the microfractures and pores in the sulphates themselves strongly affect the start of the gypsification and anhydritization. Those free spaces allows the water to migrate and solutions to start and catalyse the course of processes.

2.2.2 Time

The anhydritization and gypsification (dehydration and hydration) under natural conditions can occur very quickly: within few years (Farnsworth, 1925) or even within one year (Moiola & Glover, 1965); and experiments showed that even within several/several dozen of days (i.e. Sievert et al., 2005), what depends on physical and chemical conditions under which the process occurs. We can see for ourselves the speed of these processes, when inside a brick (ceramic material) we note the anhydrite grains, which with infiltrating water are being gypsificated and expand destroying the material – the damage of walls occurs even within several years.

2.2.3 Volume

The volumetric change comes along with hydration and dehydration processes of the sulphates – the increase of volume of anhydrite by its gypsification is about 30-50% according to Petijohn (1957), and according to Azam (2007) - close to 63%. Whereas the gypsum anhydritization decreases its volume of about 39% (Azam, 2007); sometimes it occurs together with many alterations, especially of the primary rock structure. The different situation takes place in case of sulphate deposits which already contain water; according to Farnsworth (1924), 1000g of gypsum fills 431 cm^3, while the sum of anhydrite and water needed to form the same amount of gypsum fills 473 cm^3, 9% more - then under natural conditions, when the anhydrite deposit is porous/fractured and water supersaturated, the gypsification process can result not in increase but decrease of volume of the newly formed rock.

2.2.4 Models of gypsification and anhydritisation

According to Hardie (1967) there are three models describing transformation of gypsum into anhydrite (or backwards - anhydrite into gypsum):

1. dissolution of gypsum, and furthermore precipitation of anhydrite (during anhydritisation) or dissolution of anhydrite and later precipitation of gypsum (during gypsification);
2. direct dehydration of gypsum, that is loosing of the crystallization water (during anhydritisation) or adding the water - hydration of anhydrite (during gypsification). This mechanism results in change of the rock volume;
3. dehydration or hydration with mid stage, with participation of bassanite (mineral rarely occurring in nature). During the hydration, the reaction (occurring very slowly) is as follows:

$$4CaSO_4{\cdot}\tfrac{1}{2}H_2O \rightarrow CaSO_4{\cdot}2H_2O + 3CaSO_4$$

bassanite gypsum anhydrite

2.2.4.1 Anhydritization

Petrichenko (1989) stated that the process of anhydritisation of gypsum began with its dissolution. This process is accompanied by the appearance of the nuclei of the new mineral phase - bassanite. During the second stage, bassanite transforms into anhydrite. The structural rearrangement of this mineral occurs, resulting in increase of thickness at the cost of length. Sheets (plates) of anhydrite crystals form with corroded edges. However in case of the presence of anhydrite "nuclei", the bassanite does not form, but anhydrite continues its crystallization at the cost of the calcium sulphate from dissolved gypsum. On the basis of examination of the inclusions in minerals, Petrichenko (1989) determined the conditions of the origin of anhydrite: this process takes place in the presence of concentrated brine solutions and under the conditions of high pressure and temperature, but not above 40-50°C.

Depending on time and speed of the sulphates transformation there are three kinds of the process: syndepositional, early- and late-diagenetic. The syndepositional anhydritisation occurring during the deposit formation, in shallow basin, sabkhas, in the subsurface environment, causes the substitution of gypsum to take place so fast that the anhydrite remain in its primary form. Anhydritisation during the later stages, according to the solutions of lower salinity, causes the primary sedimentary structures to disappear and the nodular structures to form - gypsum is substituted by incohesive mass of fine anhydritic strips and water, whereas the anhydritisation under the influence of highly concentrated brines can lead to the preservation of the primary gypsum pseudomorphs (Peryt, 1996; Warren, 1999), especially apparent in the coarse-crystalline gypsum forming "the grass-like selenite".

2.2.4.2 Gypsification

The process of hydration was described in detail by Sievert et al. (2005):

1. during the first quick phase, there is an initial partial dissolution of $CaSO_4$ and adsorption of hydratem Ca^{2+} and SO_4^{2-} ions at the surface of anhydrite;
2. during the second - the slower one, there is an increase of thickness of adsorbed layer
3. during the third phase, there is a crack formation in the absorber layer and counter migration of H_2O and Ca^{2+}, SO_4^{2-} ions;
4. during the fourth phase - the formation of gypsum nuclei at the surface of anhydrite occurs and in the end gypsum crystals are formed.

This process takes place in the presence of water (in the active phreatic zone), in temperatures below 40°C (process takes place faster in lower temperatures), and its speed depends on the presence of chemical activators, for example K_2SO_4, $MgSO_4 \cdot 7H_2O$ or H_2SO_4 (Sievert et al., 2005) and CO_2, which speeds and eases the hydration. At first, it covers the most fractured parts of the rock, taking place along the cracks and grain boundaries. As a result of hydration, the anhydrite rock transforms into gypsum rock with fine-grained (alabaster), fibrous, porphyroblastic texture (Warren, 1999), coarse/lenticular-crystalline gypsum (sometimes with preserved relic of the anhydritic precursor) - they result from the dissolution of primary sulphates (fine-crystalline anhydrites); see fig. 22. and 23. The secondary gypsum can also be formed as a pseudomorph of the primary anhydrite (e.g. the floor of the cap-rock) or the coarse-crystalline gypsum (selenitic gypsum), which underwent anhydritisation and furthermore gypsification - in this case, despite the multi-stage characteristics of the diagenetic processes, the primary rock structure is preserved. There is an example of the Zechstein (Permian) sulphates, which were uplifted close to the surface as a result of diapirism, and further incorporated into a cap-rock, while being anhydritised and later gypsificated (Jaworska & Ratajczak, 2008).

2.3 Inclusions

Inclusions or remains of the primary precursor minerals (e.g. the remains of anhydrite in gypsum) can appear in the primary as well as in the secondary sulphates. Particularly valuable are the inclusions in the primary minerals which can be liquid, solid, gaseous, or even organic. They reach diameters between few and several hundred of µm. Sometimes they are arranged zonally, rhythmically - as the crystal grew. Among the inclusions:

a. solid - most often occur: clay minerals, quartz, chalcedony, barite, halite, carbonates - calcite, dolomite, magnesite
b. liquid - mainly the chlorine-sulphate solutions of various mineralization,

Part of the solutions can be saturated with gases (CO_2, N_2, CH_4, H_2 and H_2S), e.g. originating from the organic decomposition (Petrichenko et al., 1995). For example, in the badenian gypsum of Carpathian Foredeep, the presence of: fragments of characean algas, filamentous algas, and colony of unicellular cyanobacterium, insects, coccoids, and multicellular organisms - most probably fungi, has been confirmed. The good state of preservation of these microorganism tissues indicates anaerobic conditions during gypsum precipitation (Petrichenko et al., 1995). The detailed inclusion analyses led to a series of conclusions on the environment, chemical (basin type: open sea or inland ?; brine type: e.g. Na- (Ca)-SO_4-Cl or Mg-Na-(Ca)-SO_4-Cl or Na-CO_3-SO_4-Cl ?) and biochemical conditions during the sulphate sedimentation; the variations of the solution chemical composition (e.g. indication of the fresh sea water inflow direction). In addition, the analyses of one-phase liquid inclusions provide information on the water temperature in the crystallization basin.

2.4 Calcitization

Sulphates, as well as gypsum and anhydrite can undergo calcification by:

a. bacterial reduction in deposits rich in organic substances – the most effective process,
- sulphates are altered by S-reducing bacteria to form H_2S, pyrite and other sulphides, native S and calcite (Holster, 1992)

b. infiltration of meteoric water rich in carbonate ions – occurs during the sulphates exposition onto water activity (Warren, 1999),

c. thermal reduction of sulphates – late diagenetic process, occurs in temperatures over 100°C, under atoxic conditions and with presence of the hydrocarbons (Machel, 1987).

Dissolution of sulphates in presence of hydrocarbons leads to biogenic SO_4 reductions and calcite precipitation according to reactions:

$$CH_4 + CaSO_4 \rightarrow CaCO_3 + H_2S + H_2O$$

Calcitization of the sulphates can be a multi-stage process (Scholle et al., 1992), which begins with (1) dissolution (or at least corrosion) of anhydrite, (2) hydration of anhydrite and gypsum formation, (3) dissolution of gypsum (this process can be accompanied by the formation of collapse breccia), and afterwards (4) precipitation of calcite inside free spaces and pores arisen after leached sulphates. Sometimes sulphur is the secondary product of calcitization of sulphates (see fig. 18.).

Generally, the gypsum – more easily than the anhydrite – can be substituted by calcite. In case where this process occurs in bigger scale, the post-gypsum limestones form. They can occur in the highest parts of the cap-rock, covering the upper parts of some diapirs - upon the area of Costal Gulf the shallowest subsurface cap-rock levels are usually formed as the calcitic deposits and therefore named as calcitic cap-rocks. However, the microscopic analyses of the cap-rock deposits demonstrated that among the secondary coarse-grained gypsum with the anhydrite remains, the calcification process starts exactly with these anhydrite inclusions, not with the gypsum.

2.5 Polyhalitization

The sulphate rocks can also undergo the polyhalitization process. It proceeds during the early stages of the diagenesis of evaporites as a result of infiltration of hot brines into the sulphate deposits (in the peripheral zones of the evaporite basins): halite saturated, with high contents of Mg^{2+} and K^+ (originating from the dissolution of the potassium salts in the local salt pans), sulphate-rich (Peryt, 1995 and 1996). This process starts from the edges of the grain/crystal and proceeds with deep embayments into the core – the anhydrite/gypsum grain disintegrates into smaller parts that undergo polyhalitization more easily (Stańczyk, 1970).

2.6 Dissolution and Karst

Sulphates – gypsum in particular – are common ingredients of the lithosphere and often occur close to the Earth's surface. Additionally, the gypsum easily undergo physical weathering (is soft and has ductile rheology), as well as chemical (dissolves in water). Gypsum dissolution rates reach 29 mm/year and have been measured in Ukraine (Klimchouk & Aksem, 2005). Therefore upon the areas of gypsum deposits karst processes and forms occur (fig. 19.). Gypsum-karst features commonly develop along bedding planes, joint or fractures; sometimes up to 30 m below the Earth's surface. The evidence is the presence of: caves, sinkholes, karren, disappearing streams and springs, collapse structures (Johnson, 2008). One of the longest reported gypsum caves is D.C. Jaster Cave (SW Oklahoma, USA) where main passage is 2,413 m long but total length of all the passages reaches 10,065 m (Johnson, 2008). Speleothems in gypsum caves may provide information about paleoclimate and climate changes in the past, because in arid or semi-arid climates, the speleothems in gypsum cave are mainly composed of gypsum, whereas in contrast, in humid or tropical climate – of carbonate (calcite). The dating of speleothems could provide the paleoclimatic data relating to:

a. dry periods, when gypsum speleothems were deposited,
b. wet periods in arid zone, when calcite speleothems were deposited (Calaforra et al., 2008).

Gypsum-karst area could be dangerous and should be monitored due to the risk of danger. Some sinkholes and collapse structures, commonly being few hundreds m wide and tens of m deep, may cause the loss of human lives and damages, e.g. in Spain in Oviedo and Calatayud situated on cavernous gypsum area, direct economic losses by collapse events were estimated to be 18 mln euro in 1998 and 4.8 mln euro in 2003 (Gutiérrez et al., 2004 and 2008).

The process of the sulphates dissolution is visible not only in developement of karst features; it reveals itself in the smaller scale for example in development of stylolites as a result of pressure solution. The development of the stylolitization process has been usually described among the carbonate rocks - mainly limestones; in the evaporites the stylolites are exceptional. Bäuerle et al. (2000) took under consideration the problem of stylolites genesis in the main anhydrite deposits located in the salts of the Gorleben diapir (Germany). Detailed studies of these forms led to estimation of the amount of dissolved material thanks to the measurements of the maximum amplitudes of the stylolitic sutures visible inside the core. The calculations showed that over 26% rock mass were dissolved. Moreover the microscopic observations indicated the gaps in the sutures – the sutures were 'cut' by the anhydrite crystals formed as pseudomorphs after gypsum. This fact proves that the stylolitization had developed before the gypsum underwent anhydritization. In the article summary, the authors plotted the conditions of the stylolites formation in sulphates, especially in gypsum as the primary deposit where such forms appear. The process requires:

a. the presence of interbeds different than the sulphate rocks; the lithological heterogeneity,
b. the presence of overburden in which the increase of thickness and its chemical characteristics favour the conditions where the lower gypsum is under conditions balancing between pressure solution and the gypsum-anhydrite transformation.

3. Geochemystry of sulphates

The analysis of chemical (including isotopes) contents of the sulphate rocks leads to the conclusions regarding their genesis and diagenetic transformations; e.g. strontium and boron.

3.1 Strontium (Sr)

Sr can substitute Ca ions in minerals (mainly in carbonates and sulphates) or create their own minerals (celestine or strontianite), which very often occur dispersed in the marine sediments. High level of Sr characterizes the rocks formed during the final stage of carbonates' sedimentation and during the first stage of calcium sulphates crystallization; generally in sulphates the Sr content increases in direct proportion to the brine concentration (Rosell et al., 1988). The primary gypsum precipitated from evaporated seawaters is expected to have a Sr content of 0.1-0.2% (Ichikuni & Setsuko Musha, 1978) and the one from K-Mg brines - the content of 0.97% Sr (Usdowsky, 1973). Butler (1973) thinks that gypsum precipitated from the celestine saturated solution should consist about 0.09% Sr, and anhydrite – about 0.24% Sr, but primary selenitic gypsum from the Eastern Betics basin contents strontium only between 493-625 ppm Sr (Warren, 2006) and primary Zechstein (Permian) anhydrites content 0.61% Sr (Polański & Smulikowski 1969).

Multimodal distribution of Sr compound in the primary sulphates (gypsum in particular) profiles indicates various sources of this element and multi-stage process of its concentration. The Miocene selenite gypsum from the southern border of the Holy Cross Mts. shows high Sr content (averagely 1300-1500 up to max. 2575 ppm); and scarce variations of the content indicate only episodic salinity fluctuations of the basin, probably connected with the inflow of fresh sea or meteoric water; the gypsum was formed in the sub-aqueous environment characterized by high salinity, whereas the laminated stromatolitic gypsum is characterized by high variations of Sr content (from max. 3695 to 179 ppm), simultaneously indicating high salinity fluctuations (Kasprzyk, 1993).

Strontium can also originate from diagenetic processes: bacterial sulphates reduction, dissolution and recrystallization – they may favour the liberation of strontium ions from the sulphate and could locally form higher concentrations within the other sulphate rocks (Kasprzyk, 1994).

Apparent decrease of Sr concentration occurs during rock transformation in the open system with unbounded circulation of the solution in free pore spaces, whereas the residual products of these transformations are often enriched in strontium. During the hydration of anhydrite, gypsum shows limited ability of Sr ions incorporation into its crystal lattice and is not able to incorporate them completely. Dissolution and recrystallization purify gypsum and anhydrite from impurities, and activate strontium lowering its content in newly created mineral comparing to the primary mineral, i.e. some secondary gypsums from Wapno Salt Dome consist only 159 ppm Sr (Jaworska & Ratajczak, 2008), primary anhydrite from which it has been created consist 1700 ppm Sr.

3.2 Boron (B)

B likewise Sr is a sensitive indicator of changing conditions in the evporite sedimentary environment, as its concentration in the sediment depends on the salinity.

Systematic increase of B content in the profile of sulphate sediments indicates progressive increase of basin salinity during the crystallization of successive generations of sulphates - evaporites containing the highest amounts of B originate from the most concentrated solutions. Any decrease/variation/fluctuation of this element concentration indicates fresh (sea or meteoric) water supply to the evaporite basin.

Sea water contains 4,45 ppm of boron, mainly in the form of undissociated ortho-boric acid. Solutions of this element deriving from the terrigenic sediments, submarine exhalations and decomposing clay minerals (especially illite) constitute the source of borate ions in the sedimentary basins. Ions of BO_3^{2-} can isomorphically replace SO_4^{2-} and form their own minerals (borates, e.g. boracite).

The highest B concentrations are noted during the latest stages of evaporation - when the K-Mg salts precipitate accompanied (under favour conditions) by borates' crystallization. The B content in sulphate rocks (gypsum, as well as anhydrite) can fluctuate between 2 and 5500 ppm; in the Zechstein anhydrite the content ranges from 16 to 500 ppm, and in polyhalite reaches 800 ppm (Pasieczna, 1987) - generally, there are high B contents noted in polyhalite.

Sulphates can be analysed from the point of view of Mn and Fe contents; increased concentrations of both elements usually indicate the terrigenic deposit (siliciclastic sediments, clay minerals) supply into the sedimentary basin.

3.3 Isotopes

Another indication of the genesis and diagenesis of sulphates are the isotopic analyses of $^{87}Sr/^{86}Sr$ ratio, S ($\delta^{34}S$) and O ($\delta^{18}O$) in SO_4, and in the case of gypsum, also O ($\delta^{18}O$) of the crystallization water. $\delta^{34}S$ and $\delta^{18}O$ in SO_4 does not change despite of many transformations, the sulphate molecule maintain its primary isotopic composition, what allows to determine the primary sedimentary conditions, but dynamic and multiple transformation can affect the $\delta^{18}O$ of crystallization water, so in gypsum we have to indicate two $\delta^{18}O$ - in SO_4 and H_2O.

3.3.1 Sulphur (S)

The present-day $^{34}S/^{32}S$ ($\delta^{34}S$) ratio of sulphates in oceanic water is constant and reaches +20±0.5‰ with respect to V-CDT (Pierre, 1988) and the fractionation between dissolved sulphates in oceanic water and crystallized sulphates is negligible (Thode & Monster, 1965; Raab & Spiro, 1991). $\delta^{34}S$ was changing in the geological past and its general trends are known as the sulphur-isotope age curve (Claypool et al., 1980). This curve allows to define the time of evaporate crystallization.

3.3.2 Oxygen (O)

The present-day $^{18}O/^{16}O$ ($\delta^{18}O$) ratio of sulphates in oceanic water reaches 9.5±0.5‰ with respect to V-SMOW (Longinelli & Craig, 1967) but during crystallization of the oceanic sulphates, the $\delta^{18}O$ is raised up to 3.5‰ (Lloyd, 1968; Pierre, 1988) and $\delta^{18}O$ value of this sulphates reaches 13.0±0.5‰.

Primary gypsum and its crystallization water are formed in isotopic equilibrium with the mother brine (Sofer, 1978), but gypsum can easy loose its original crystallization water during further dehydration and hydration. During hydration sulphates interact with meteoric-, ground-, residual or sea water and gypsum absorbs this new, fresh or sometimes mixed primary water. In the areas of several-, several dozen of m long profiles consisting gypsum rocks, basing on the determination of $\delta^{18}O$ of their crystallization water, it is possible to indicate the type and range of individual water types which affected the sulphates. E.g. in profiles of the cap-rock of the Wapno and Mogilno salt diapirs (Jaworska, 2010) there is gypsum, which shows $\delta^{18}O$ of crystallization water indicating the influence of: cold period - post-glacial water – $\delta^{18}O$ reaches values from -11 up to -13‰ in the lowest part of the profile (Wapno and Mogilno), recent (or similar to) meteoritic water - $\delta^{18}O$ reaches values of -9 to -10‰ (Wapno), cap-rock water - $\delta^{18}O$ reaches -4.3 to -6.6‰ (Mogilno), „mixing" water or warmer period water - $\delta^{18}O$ is -5.6‰ (Wapno) and from -6.9 to -8.7‰ (Mogilno).

The presence of water described as recent or originated from the colder periods inside the lowest and the middle parts of the cap-rock is very important for further management plan of such salt structure. The influence of present day water or the water from colder periods in the lowest part of the cap-rock indicates free flow of surface water into the area of so called salt mirror; the presence of this water in the middle part of the cap-rock indicates the occurrence of cracks, fractures and karst forms in cap-rock body. In consequence it means, that such cap-rock is not a hermetic cover and does not fulfil the requirements for a seal which protects the rock salt and salt mirror against inflow of freshwater. This information is of great importance for salt structures which are prepared for underground disposal of radioactive waste or for the storage of hydrocarbons, as well as salt mine.

3.3.3 Strontium (Sr)

The $^{87}Sr/^{86}Sr$ ratio of modern oceanic water is uniform and reaches 0.70901 (Burke et al., 1982) but has been changing in time. Main reasons of these irregular changes were contribution of Sr with high $^{87}Sr/^{86}Sr$ ratios from continents and input of Sr with low $^{87}Sr/^{86}Sr$ ratios from active mid –oceanic ridges (Veizer, 1989; Chaudhuri & Clauer, 1992). The general trends and variations of the marine Sr isotopes during the Phanerozoic carbonates are known (Burke et al., 1982) and this curve (the same as S-curve) allows us to study the age of evaporates precipitation. In evaporites the $^{87}Sr/^{86}Sr$ ratios reflect the isotopic composition of the brines or diagenetic fluids. Strontium does not fractionate (Holster, 1992).

Present-day strontium isotope ratio equilibrated between ^{87}Sr-depleted young oceanic basalts and hydrothermal activity along mid-oceanic ridges (ca. 0.7035) and ^{87}Sr-enriched continental sediments (from old continental granites) transported into the basin by wind and rivers (ca. 0.7119 and more; Chaudhuri & Clauer, 1992; Dickin, 2005). It is the same reason why primary Sr isotopic ratio of evaporites could not be the same as that of contemporaneous sea water - e.g. sediments may have deposited in closed basin with inflow of continental water and continental Sr - the Sr ratio of such sulphates is higher than the one of contemporaneous ocean water, so any variation of Sr isotopic composition may relate to the paleohydrology of the basin. Additionally, variations of Sr isotopic ratio may be explain by contamination with more radiogenic Sr or by diagenesis (Hess et al., 1986; Saunders et al., 1988; Chaudhuri & Clauer, 1992).

4. Recrystallization

In the classic approach recrystallization means the transformation of fine-crystalline minerals/rocks into coarse-crystalline ones and makes sometimes the continuation of the recovery process, when the mineral/rock or the whole material tries to loose the excess of the internal energy generated during the deformation/strain, when the crystal lattice defects occur. During those processes the shape and size of grains change and the crystallographic axes rotate; they are also accompanied by progressive loss and disappearance of the primary rock texture/structure.

In the case of recrystallization of cap-rock gypsum, a reverse process can be generally observed (looking upwards) - the size reduction of the mineral grains (dominant or subordinate components).

The boundaries between adjacent fine gypsum grains are usually blurred and irregular, what results from transformation of the larger grains into smaller ones, which successively become individual.

The recrystallization of gypsum can occur via: grain boundary migration or subgrain rotation. The grain boundary migration is characteristic for the mineral grains with large variety of lattice defects density, whereas the subgrain rotation occurs in grains with uniformly dispersed defects (Passchier & Trouw, 1998).

4.1 Grain boundary migration

If the adjacent grains differs in defects density, the defect-poor one bulges into the defect-rich one; see fig. 24. It results in the removal of grains with many dislocations. It also enables the spontaneous crystallization and the growth of new grains - "nuclei" (either defectless or with few dislocations) inside the defect-rich grain; these fine new grains are called 'subgrains' as well.

4.2 Subgrain rotation

The deformation bands formed during the recovery tighten progressively, creating a grid determined by subgrain walls that developed successively within the grain. The subgrains are fragments of larger grain with fine boundaries. As a result of rotation, the crystalline axis of the subgrain becomes slightly misoriented relating to the axes of the adjacent subgrains or the main grain/crystal; the misorientation angle usually reaches max. 5° (FitzGerald et al., 1983; White & Mawer, 1988 fide Passchier & Trouw, 1998). During the rotation recrystallization the mylonitic and porphyroblastic/porphyroclastic rocks are formed.

Another (however not so common) mechanism of subgrain development can be observed in the rocks of the gypsum cap-rock - the process is called kinking and leads to formation of 'kink bands' (Means & Ree, 1988 fide Passchier & Trouw, 1998), which are represented by narrow accumulation of kink folds; see fig. 25. They are formed in brittle-ductile system and correspond to the initial shearing along the planes oblique to the dense anisotropic planes (sedimentary, metamorphic, lattice anisotropy) under the influence of parallel (to those planes) or close to parallel compression at rather high surrounding pressure (Dadlez & Jaroszewski, 1994). This process has been observed in few mm to few cm lenticular, cigar-shaped gypsum crystals; see fig. 26. and 27.

5. Summary

Sulphates are common minerals; they are easy crystallized, alternated and recrystallized.

Distinct variation of isotope ratios of sulphur, oxygen and strontium in the sea water sulphates in time enables their use to determine:

- the age of evaporite deposits;
- the sulphates' origin (marine or non-marine?, and primary or secondary minerals?) and
- in the case of gypsum (oxygen analysis of crystallization water), the determination of paleoclimatic conditions (also the time) when the gypsification occurred due to water particles accretion or isotopic composition exchange of water in gypsum.

The liquid inclusions analysis in the primary evaporites enables determination of chemical composition of primary solutions/brines from which the sulphates crystallized, as well as the temperature of water.

The analysis of the primary minerals remains constituting the impurities in the secondary crystals enables determination of the diagenetic processes taking place in the evaporite deposits (including the mineral precursor for the secondary crystal), and the direction and cause of diagenetic transformations (e.g. anhydrite gypsification: primary mineral - anhydrite, cause - presence of fresh or low-mineralized water in the deposit, e.g. as a result of tectonic uplift and exposition to the activity of shallow underground water).

The crystal shape, form and texture of gypsum and anhydrite sediments indicate the environmental conditions of their formation such as: basin bathymetry (shallow or deep zones of the basin), water oxygenation, either stability or dynamics of the environment (e.g. turbidity currents, sea-level fluctuations - in case of high variability and low thickness of separate sulphate lithotypes in the profile).

Trace elements analysis in sulphates:

1. Sr, B contents: constant increase of their contents in the profile indicate stable evaporation conditions; their variations episodes connected with the fresh water inflows to the evaporite basin and its dilution;

2. Mn and Fe contents: elevated concentrations of both elements indicate the supply of terrigenic sediments to the basin.

6. References

Azam S. 2007 - Study on the geological and engineering aspects of anhydrite/gypsum transition in the Arabian Gulf coastal deposits. Bull. Eng. Geol. Env., 66: 177–185.

Barley M.E., Dunlop J.S.R., Glover J.E. & Groves D.I. 1979 – Sedimentary evidence for an Archaean shallow-water volcanic-sedimentary facies, eastern Pilbara Block, Western Australia. Earth and Planetary Science Letters, 43: 74-84.

Blatt H., Middleton G. & Murray R. 1980 - *Origin of Sedimentary Rocks* (2nd ed.). Prentice-Hall, Inc. Englewood Cliffs, 782 pp.

Buick R. & Dunlop J.S.R. 1990 – Evaporitic sediments of early Archaean age from the Warrawoona Group, North Pole, Western Australia. Sedimentology, 37: 247-277.

Burke W.H., Denison R.E., Hetherington E.A., Koepnick R.E., Nelson H.F. & Otto J.B. 1982 – Variations of sea water $^{87}Sr/^{86}Sr$ throughout Phanerozoic time. Geology, 10: 516-519.

Butler G.P. 1973 – Strontium geochemistry of modern and ancient calcium sulphate minerals [in:] Purser B.H. (ed.) The Persian Gulf. Springer, New York, Heidelberg, Berlin, 423-470.

Calaforra J.M., Forti P. & Fernandez-Cortes A. 2008 – Speleothems in gypsum caves and their paleoclimatological significance. Environ. Geol., 53: 1099-1105.

Chaudhuri S. & Clauer N. 1992 – History of marines evaporites: constraints from radiogenic isotopes. Lecture Notes in Earth Sciences, 43: 177-198.

Claypool G.E., Holster W.T., Kaplan I.R., Sakai H. an& Zak I. 1980 – The age curves of sulfur and oxygen isotopes in marine sulfate and their mutual interpretation. Chem. Geol., 28: 199-260.

Conley R.F. & Bundy W.M. 1958 - Mechanism of gypsification. Geochim. Cosmochim. Acta, 15: 57-72.

Dadlez R. & Jaroszewski W. 1994 - *Tektonika*. PWN, Warszawa, 743 pp.

Dickin A.P. 2005 – *Radiogenic isotope geology* (2 ed.). Cambridge University Press, 67 pp.

Farnsworth M. 1924 - Effects of temperature and pressure on gypsum and anhydrite. U. S. Bur. Mines Rept. Inv. 2654

Farnsworth M. 1925 – The hydration of Anhydrite. Industrial and Engineering Chemistry, 17, 9: 967-970.

FitzGerald J.D., Etheridge M.A. & Vernon R.H. 1983 - Dynamic recrystallization in a naturally deformed albite. Text Microstruct, 5: 219-237.

Ford D.C. & Williams P.W. 1989 – *Karst geomorphology and geology*. London, Chapman and Hall., 601 pp.

Ford D.C. & Williams P.W. 2007 - *Karst hydrogeology and geology* (2sec ed.). Wiley, 562 pp.

Gutiérrez F., LuchaP. & Guerrero 2004 – La dolina de colapso de la casa azul de Calatayud (noviembre de 2003). Origen, efectos y pronóstico [in:] Benito G., Díez-Herrero A. (eds.) Riesgos naturales y antrópicos en Deomorfología, VII Reunión Nacional de Geomorfología, Toledo, pp. 477-488.

Gutiérrez F., Cooper A.H. & Johnson K.S. 2008 - Identification, prediction, and mitigation of sinkhole hazards in evaporite karst areas. Environ. Geol., 53: 1007-1022.

Hardie L.A. 1967 - The gypsum-anhydrite equilibrium at one atmosphere pressure. Am. Mineral., 52: 171-200.

Ichikuni M. & Setsuko Musha 1978 - Partition of strontium between gypsum and solution. Chemical Geology, 21 (3-4): 359-363.

Hess J., Bender M.L. & Schilling J.G. 1986 - Evolution of the ratio strontium-87 to strontium-86 in seawater from Cretaceous to present. Science, 231: 979-984.

Holster W.T. 1992 - Stable isotope geochemistry of sulfate and chloride rocks. Lecture Notes in Earth Sciences, 43: 153-176.

Jaworska J. 2010 - An oxygen and sulfur isotopic study of gypsum from the Wapno Salt Dome cap-rock (Poland). Geological Quarterly, 54, 1: 25-32.

Jaworska J. & Ratajczak R. 2008 - Geological structure of the Wapno Salt Dome in Wielkopolska (western Poland). Prace Państwowego Instytutu Geologicznego, Warszawa, 190, 69 pp. (in Polish, with English summary).

Johnson K.S. 2008 - Evaporite-karst problems and studies in the USA. Environ. Geol., 53: 937-943.

Jowett E.C., Cathles III L.M. & Davis B.W. 1993 - Predicting depths of gypsum dehydration in evaporitic sedimentary basin. AAPG Bull., 77, 3: 402-413

Kasprzyk A. 1993 - Prawidłowość występowania strontu w gipsach mioceńskich południowego obrzeżenia Gór Świętokrzyskich. Przegląd Geologiczny, 41, 6: 416-421. (in Polish with English summary).

Kasprzyk A. 1994 - Distribution of strontium in the Badenian (Middle Miocene) gypsum deposits of the Nida area, southern Poland. Geological Quarterly, 38, 3: 497-512.

Kasprzyk A., 2003 - Sedimentological and diagenetic patterns of anhydrite deposits in the Badenian evaporite basin of the Carpathian Foredeep, southern Poland. Sedimentary Geology, 158 (3-4): 167-194.

Klimchouk A.B. & Aksem S.D. 2005 - Hydrochemistry and solution rates in gypsum karst: case study from the Western Ukraine. Environ. Geology, 48: 307-319.

Klimchouk A. & Andrejchuk V. 1996 - Sulphate rocks as an arena for karst development. International Journal of Speleology, 25 (3-4): 9-20.

Kubica B. 1972 - O procesie dehydratacji gipsów w zapadlisku przedkarpackim. Przegląd Geologiczny, 20, 4: 184-188.

Lloyd R.M. 1968 - Oxygen isotope behavior in the sulfate-water system. J. Geophys. Res., 73: 6099-6110.

Longinelli A. & Craig H. 1967 - Oxygen-18 variations in sulfate ions and sea water and saline lakes. Science, 156: 56-59.

Machel H.G. 1987 - Sadle dolomite as a by-product of chemical compaction and thernochemical sulfate reduction. Geology, 15: 936-940.

Means W.D. & Ree J.H. 1988 - Seven types of subgrain boundaries in octachloropropane. J. Struct. Geol., 10: 765-770.

Moiola R.J. & Glover E.D. 1965 - Recent anhydrite from Clayton Playa, Nevada. Am. Mineralogists, 50: 2063-2069.

Murray R.C. 1964 - Origin and diagenesis of gypsum and anhydrite. J. Sed. Petrol., 34, 3: 512-523.

Pasieczna A. 1987 – Mineralogical and geochemical analysis of the Zrchstein sulphate deposits of the Puck Bay region. Archiwu Mineralogiczne, 43, 1: 19-40. (in Polish with English summary).

Passchier C.W. and Trouw R.A.J. 1998 – *Microtectonics* (2ed ed.). Springer – Verlag, Berlin, Heidelberg.

Peryt T.M. 1995 – Geneza złóż polihalitu w cechsztynie rejonu Zatoki Puckiej w świetle badań sedymentologicznych i geochemicznych. Przegląd Geologiczny, 43, 12: 1041-1044.

Peryt T.M. 1996 - Diageneza ewaporatów. Przegląd Geologiczny, 44, 6: 608-611.

Petrichenko O.I. 1989 - *Epigenez evaporitov*. Naukova Dumka, Kiev, 62 pp.

Petrichenko O.I., Peryt T.M., Poberezski A.W. & Kasprzyk A. 1995 – Inkluzje mikroorganizmów w kryształach badeńskich gipsów Przedkarpacia. Przegląd Geologiczny, 43, 10: 859-862.

Pettijohn F.J. 1957 - *Sedimentary rocks* (2ed ed.). Harper & Bros., New York, 718 pp.

Pierre C. 1988 - Applications of stable isotope geochemistry to study of evaporites. In: Schreiber BC (ed). Evaporites and Hydrocarbons. Columbia University Press, New York, pp 300-344.

Polański A. & Smulikowski K. 1969 – *Geochemia*. Wydawnictwa Geologiczne, Warszawa.

Posnjak E. 1940 - Deposition of calcium sulfate from sea water. Am. J. Sci., 238: 559-568.

Raab M. & Spiro B. 1991 - Sulfur isotopic variation during seawater evaporation with fractional crystallization. Chemical Geology, 86: 323-333.

Rosell L., Ortí F., Kasprzyk A., Playà E. & Peryt T.M. 1998 - Strontium geochemistry of Miocene primary gypsum: Messinian of southeastern Spain and Sicily and Badenian of Poland. Journal of Sedimentary Research, 68: 63–79.

Saunders J.A. 1988 – Pb-Zn-Sr mineralization in limestone caprock, Tatum salt dome, Mississippi. Trans. Gulf Coast Assoc. Geol. Soc., 38: 569-576.

Scholle P.A., Ulmer D.S. & Melim L.A. 1992 – Late-stage calcites in the Permian Capitan Formation and its equivalents, Delaware Basin margin, west Texas and New Mexico: evidence for replacement of precursor evaporates. Sedimentology, 39: 207-234.

Shahid S.A., Abdelfattah M.A. & Wilson A. 2007 - A Unique Anhydrite Soil in the Coasta Sabkha of Abu Dhabi Emirate. Soil Surv. Horiz., 48: 75-79.

Sievert T., Wolter A. & Singh N.B. 2005 - Hydratation of anhydrite of gypsum ($CaSO_4$.II) in ball mill. Cement and Concerete Research, 35: 623-630.

Sonnenfeld P. 1984 - *Brines and evaporates*. Academic Press, London, 613 pp.

Stańczyk I. 1970 – Polihalit w kopalniach soli regionu kujawskiego. Acta Geologica Polonica, 10, 4: 305-820.

Stewart F.H., 1968 – Geochemistry of marine evaporate deposits. Geological Society America Special Paper, 88: 539-540.

Thode H.G. & Monster J. 1965 - Sulfur isotope geochemistry of petroleum evaporites in ancient seas. AAPG Mem., 4: 367-377.

Usdowsky E. 1973 - Das geochemische Verhalten des Strontiums bei der Genese und Diagenese von Ca-Karbonat-und Ca-Sulfat-Mineralen. Contrib. Minerl. Petrol., 38: 177-195.

Veizer J. 1989 - Strontium isotopes in seawater through time. Ann. Rev. Earth Plan. Sci., 17: 141-167.

Warren 1999 - *Evaporites – their Evolution and Economy*. Blackwell Science, 438 pp.

Warren 2006 - *Evaporites: Sediments, Resources and Hydrocarbons*. Springer, 1035 pp.

White J.C. & Mawer C.K. 1988 - Dynamic recrystallisation and associated exsolution in perthites: evidence of deep crystal thrusting. J. Geophys. Res., 93: 325-337.

Wilson A.H. & Versfeld J.A. 1994 - The Elary Archaean Nondweni greenstone belt, southern Kaapvaal Craton, South Africa; Part I, Stratigraphy, sedimentology, mineralization and depositional environment: Precambrian Research, 67: 243-276.

http://minerals.usgs.gov/minerals/pubs/commodity/

Section 5

General Issues in Crystallization

Synthetic Methods for Perovskite Materials – Structure and Morphology

Ana Ecija, Karmele Vidal, Aitor Larrañaga,
Luis Ortega-San-Martín and María Isabel Arriortua
[1]*Universidad del País Vasco/Euskal Herriko Unibertsitatea (UPV/EHU), Facultad de Ciencia y Tecnología, Dpto. Mineralogía y Petrología, Leioa,*
[2]*Pontificia Universidad Católica del Perú (PUCP), Dpto. Ciencias, Sección Químicas, Lima,*
[1]*Spain*
[2]*Perú*

1. Introduction

Solid state chemistry thrives on a rich variety of solids that can be synthesized using a wide range of techniques. It is well known that the preparative route plays a critical role on the physical and chemical properties of the reaction products, controlling the structure, morphology, grain size and surface area of the obtained materials (Cheetham & Day, 1987; Rao & Gopalakrishnan, 1997). This is particularly important in the area of ABO_3 perovskite compounds given that they have for long been at the heart of important applications. From the first uses of perovskites as a white pigments, $PbTiO_3$ in the 1930's (Robertson, 1936) to the MLC capacitors (mostly based in substituted $PbTi_{1-x}Zr_xO_3$ or $BaTiO_3$ materials) in which today's computers rely on to operate, synthetic methods have been a key factor in the optimization of their final properties (Pithan et al., 2005).

Traditionally, most of these ceramic materials have been prepared from the mixture of their constituent oxides in the so called solid state reaction, "shake and bake" or ceramic method, a preparative route for which high temperature is a must in order to accelerate the slow solid–solid diffusion (Fukuoka et al., 1997; Inaguma et al., 1993; Safari et al., 1996). Despite its extended use in practically all fields in which perovskite-structured materials are needed, not all applications are better off with this method since the low kinetics and high temperature also yield samples with low homogeneity, with the presence of secondary phases and with uncontrolled (and typically large) particle size of low surface area which are undesired for some applications such as in gas sensors or in catalysis where small particles and high surface area are needed (Bell et al., 2000). This conventional route, however, is widely employed due to its simplicity and low manufacturing cost. Nevertheless, with appropriate optimisation, as when soft-mechanochemical processing is used prior to calcinations at high temperature (Senna, 2005), the method results in high quality single phase perovskites that can be used in electroceramic applications.

Alternative routes to the solid-state reaction method are wet chemical synthetic methods such as co-precipitation (with oxalates, carbonates, cyanides or any other salt precursors), combustion (including all variants from low to high temperature), sol–gel (and all of its modifications with different chelating ligands), spray-pyrolysis, metathesis reactions, etcetera (Patil et al., 2002; Qi et al., 2003; Royer et al., 2005; Segal, 1989; Sfeir et al., 2005). In all cases the idea is to accelerate the pure phase formation, a goal that is achieved due to the liquid media which permits the mixing of the elements at the atomic level resulting in lower firing temperatures. Other advantages of these methods are the possibility of having controlled particle size, morphology and improvement in surface area.

In most cases, the final microstructure of the sample is the key issue in choosing the synthetic method, but phase purity is also a must, and is sometimes overlooked when authors praise their particular synthetic method (Kakihana & Yoshimura, 1999). This was pointed out by Polini et al. (Polini et al., 2004) in the case of the preparation of substituted $LaGaO_3$ phases for SOFC cathodes: methods that supposedly have been developed to improve the scalability and uniformity of the samples, such as the Pechini method (a particular case of the sol-gel method), do not always result in a single phase of the crystalline sample required. Similar cases are common in the literature, as in the case of $La_{1-x}Sr_xMnO_{3-\delta}$ phases (Conceição et al., 2009). This clearly indicates that there is not such a thing as the ideal synthetic method: every method has its advantages and disadvantages.

Ideally, as many as possible synthetic methods should be tried and optimised for each compound of interest in order to obtain better crystals with the proper microstructure. But this is obviously time consuming and very costly. Consequently, researchers usually choose to follow the general trends that have been observed to work in a particular area of interest. As a result, each field has its preferences. For example, the ceramic method, widely used at the beginning of the first years of the high-Tc superconductors was soon replaced because it almost always resulted in non-stoichiometric products with some undesired phases that complicated the interpretation of the superconducting properties. These materials got so much attention in the late 1980's and early 1990's that completely new synthesis methods were introduced including many modifications of sol-gel methods with the ample use of alcoxides as precursors (Petrikin and Kakyhana, 2001). In this case, the synthetic route consisted on the preparation of mixed coordination compounds with alcoxy ligands in aqueous media which ensured good distribution of all metals involved and yielded purer superconducting oxides at relatively lower temperatures than before.

On the other hand, combustion methods (glycine-nitrate, urea based, and other modifications) have been proposed as one of the most promising methods to prepare perovskite oxide powders to be used as cathode materials in Solid Oxide Fuel Cell technology. (Bansal & Zhong, 2006; Berger et al., 2007; Dutta et al., 2009; Liu & Zhang, 2008). This method consist on a highly exothermal self-combustion reaction between the fuel (usually glycine, urea or alanine) and the oxidant (metal nitrates), that produces enough heat to obtain the ceramic powders. Compared to the ceramic method this synthetic route has much faster reaction times and lower calcination temperatures leading to powders with large compositional homogeneity and nanometric particle sizes, which are desired characteristics for this type of application.

For some applications, such as in multiferroic materials based devices, it is the crystal symmetry of the multiferroic what matters. In these materials the presence or not of a centre of symmetry is crucial for the observation of ferroelectricity. With this regard, there are cases, as in some $AMnO_3$ perovskites (A=Y or Dy), in which the synthetic route determines whether an orthorhombic compound with a centre of symmetry (i.e. non ferroelectric) or a hexagonal phase without the centre (i.e. ferroelectric) is formed (Carp et al., 2003; Dho et al., 2004). Consequently, preparative conditions have to be carefully selected in order to obtain crystal phases with the adequate structure. The use of more than one synthesis method is thus worth trying in all cases.

In this work three different groups of perovskite compounds have been prepared and their crystal structure and microstructure have been studied using X-ray diffraction (XRD) and scanning electron microscopy (SEM). Each group of samples had its own structural characteristics so, prior to choosing one synthetic approach, trials were carried out using different methods. In all cases, the final method chosen was the one that maximised phase purity and resulted in better properties. Here we demonstrate that phase pure samples susceptible to be compared depending on the desired characteristics can be obtained using different synthetic methods.

2. Experimental

Up to four different synthetic methods (solid state reaction, glycine-nitrate route, sol-gel and freeze-drying) have been used to synthesize a group of 14 perovskite compounds (Figure 1), which have the potential for their use in different applications.

Compound	Synthetic method	Variable		Label
$La_{0.50}Pr_{0.30}Sr_{0.20}FeO_{3-\delta}$	Solid state reaction	x in A	0.2	LPS20
$La_{0.40}Nd_{0.30}Sr_{0.23}Ca_{0.07}FeO_{3-\delta}$	Solid state reaction		0.3	LNSC30
$La_{0.20}Pr_{0.40}Sr_{0.26}Ca_{0.14}FeO_{3-\delta}$	Solid state reaction		0.4	LPSC40
$La_{0.19}Pr_{0.31}Sr_{0.26}Ca_{0.24}FeO_{3-\delta}$	Solid state reaction		0.5	LPSC50
$La_{0.19}Pr_{0.21}Sr_{0.26}Ca_{0.34}FeO_{3-\delta}$	Solid state reaction		0.6	LPSC60
$La_{0.18}Pr_{0.12}Sr_{0.26}Ca_{0.44}FeO_{3-\delta}$	Solid state reaction		0.7	LPSC70
$La_{0.20}Sr_{0.25}Ca_{0.55}FeO_{3-\delta}$	Solid state reaction		0.8	LSC80
$La_{0.50}Ba_{0.50}FeO_{3-\delta}$	Glycine-nitrate route	$\langle r_A \rangle$ (Å)	1.34	LB134
$La_{0.34}Nd_{0.16}Sr_{0.12}Ba_{0.38}FeO_{3-\delta}$	Glycine-nitrate route		1.31	LNSB131
$La_{0.04}Nd_{0.46}Sr_{0.24}Ba_{0.26}FeO_{3-\delta}$	Glycine-nitrate route		1.28	LNSB128
$La_{0.05}Sm_{0.45}Sr_{0.32}Ba_{0.18}FeO_{3-\delta}$	Solid state reaction		1.25	LSSB125-ss
	Glycine-nitrate route		1.25	LSSB125-gn
$Nd_{0.8}Sr_{0.2}Mn_{0.9}Co_{0.1}O_3$	Sol-gel	x in B	0.1	NSMC10-sg
	Freeze-drying		0.1	NSMC10-fd
$Nd_{0.8}Sr_{0.2}Mn_{0.8}Co_{0.2}O_3$	Sol-gel		0.2	NSMC20-sg
	Freeze-drying		0.2	NSMC20-fd
$Nd_{0.8}Sr_{0.2}Mn_{0.7}Co_{0.3}O_3$	Sol-gel		0.3	NSMC30-sg
	Freeze-drying		0.3	NSMC30-fd

x: doping level in $Ln_{1-x}A_xFeO_{3-\delta}$ and $Nd_{0.8}Sr_{0.2}(Mn_{1-x}Co_x)O_3$ series; $\langle r_A \rangle$: average ionic radius of A-cation in the $Ln_{0.5}A_{0.5}FeO_3$.

Table 1. Nominal compositions, synthetic methods, and labels of the studied perovskites.

These compounds have been divided into three groups and their compositional details are summarised in Table 1. Preparation procedures are detailed below.

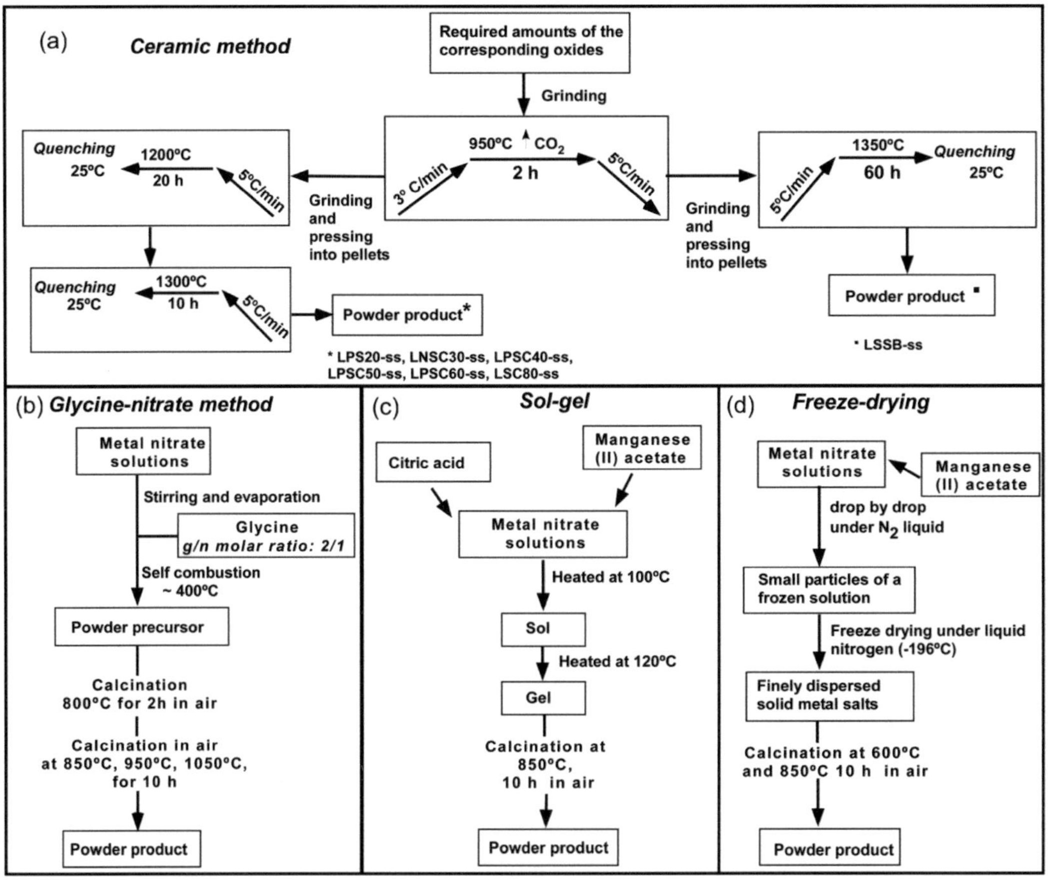

Fig. 1. Flowcharts for the: (a) ceramic, (b) glycine-nitrate, (c) sol-gel and (d) freeze-drying methods used to obtain the perovskite compounds shown in the present work.

The selection of each method and composition for each series of perovskites was based on the desired applications that are described later.

2.1 Ceramic solid state reaction

The compounds prepared via the ceramic route were obtained from mixing stoichiometric amounts of the raw oxides with 2-propanol in an agate mortar. Starting materials were always oxides of high purity such as La_2O_3 (99.99%), Sm_2O_3 (99.999%), Pr_2O_3 (99.9%), Gd_2O_3 (99.99%), BaO (99.99%), SrO (99.9%), CaO (99.9%) and Fe_2O_3 (99.98%). Afterwards, these mixtures were shaped into pellets and were fired in air at 950ºC for 2h. The products obtained were ground, pelletized again and fired at higher temperatures. The flowchart shown in Figure 1a details the heat treatments required in each case until pure samples were obtained.

2.2 Glycine-nitrate route

The preparations of the perovskite compounds with the general composition $Ln_{0.5}A_{0.5}FeO_{3-\delta}$ by glycine-nitrate route involved the use of some nitrates instead of the oxides as starting materials: $Ba(NO_3)_2$ (99.99%), $Sr(NO_3)_2$ (99.9%) and $Fe(NO_3)_3$ (99.98%). The oxides used were La_2O_3 (99.99%), Sm_2O_3 (99.999%), Gd_2O_3 (99.99%). The oxides were dissolved in diluted nitric acid to obtain the corresponding nitrates and the metal nitrates dissolved in distilled water. The solutions were mixed in a 1 litre glass beaker, under constant stirring and placed on a hot plate to evaporate the water excess. After a significant reduction of the solution volume the glycine was added. The amount of glycine used was calculated in order to obtain a glycine/nitrate molar ratio of 2:1. This amino acid acts as complexing agent for metal cations and as the fuel for the combustion reaction. The resulting viscous liquid was auto-ignited by putting the glass beaker directly in a preheated plate (at ~400°C). The obtained powders were pelletized and fired at 800°C for 2 hours to remove the carbon residues. The heatings at temperatures above 800°C were repeated until pure phases were obtained. A flowchart with more details and heat treatments involved in this synthesis is shown in Figure 1b.

2.3 Sol–gel

The sol-gel method was used for the oxides of general composition $Nd_{0.8}Sr_{0.2}(Mn_{1-x}Co_x)O_3$ (x = 0.1, 0.2 and 0.3). Initially, the oxide Nd_2O_3 (99.9%) was dissolved in aqueous nitric acid followed by the addition of $Sr(NO_3)_2$ (99%), $Co(NO_3)_2.6H_2O$ (99%) and $Mn(C_2H_3O_2)_2.4H_2O$ (99%). Citric acid was then used as the quelating agent and ethylene glycol as the sol forming product. The solution was slowly evaporated in a sand bath for 24 h and the gel obtained was subjected to successive heat treatments at the temperature of 850°C (with intermediate grindings). Each firing was of 10 h and was carried out under nitrogen atmosphere. The flowchart for this synthesis is shown in Figure 1c.

2.4 Freeze-drying technique

In the freeze drying method, standardized nitrate solutions were mixed according to the stoichiometry of the final products: $Nd_{0.8}Sr_{0.2}(Mn_{1-x}Co_x)O_3$ (x = 0.1, 0.2 and 0.3). The starting materials were Nd_2O_3 (99.9%) which had to be dissolved in diluted nitric acid before the addition of the other compound; $Sr(NO_3)_2$ (99%); $Co(NO_3)_2.6H_2O$ (99%) and $Mn(C_2H_3O_2)_2.4H_2O$ (99 %).

The mixture for freeze-drying method was frozen drop-by-drop under liquid nitrogen and subjected to freeze drying at P = 5.10^{-2} mbar. Thermal decomposition was achieved by slow heating in air up to 600°C. The pure phases were obtained after repeated heatings at 850°C (with intermediate grindings), each of 10 h, under nitrogen atmosphere. A flow chart for this method is shown in Figure 1d.

2.5 Characterization

Room temperature X-ray powder diffraction data were collected in the $18 \leq 2\theta \leq 110°$ range with an integration time of 10 s/0.02° step. A Bruker D8 Advance diffractometer equipped with a Cu tube, a Ge (111) incident beam monochromator (λ = 1.5406 Å) and a Sol-X energy dispersive detector were used for the samples obtained by glycine-nitrate route and LSSB-ss. A Philips X'Pert-MPD X-ray diffractometer with secondary beam graphite monochromated Cu-Kα radiation was used for the samples obtained by ceramic solid-state, sol-gel and freeze-drying techniques.

All samples were single phase without detectable impurities. The crystal structure was refined by the Rietveld method (Rietveld, 1959) from X-ray powder diffraction data using GSAS software package (Larson & Von Dreele, 1994).

Microstructural analysis was carried out in a JEOL JSM 6400 scanning electron microscope (SEM) using a secondary electron detector at 30 kV and 1.10^{-10} A for the LPS20, LNSC30, LPSC40, LPSC50, LPSC60, LPSC70 samples and a JEOL JSM-7000F at 3 kV and 11.10^{-12} A for the rest of samples.

3. Results

3.1 Characterization of $Ln_{1-x}A_xFeO_{3-\delta}$ (Ln=La, Nd, Pr; A=Sr, Ca) perovskites with 0.2≤x≤0.8

The $(La_{1-x}Sr_x)FeO_{3-\delta}$ (LSF) perovskite system exhibits high electronic and oxide ion conductivities at high temperatures, which make it attractive for solid oxide fuel cell (SOFC) cathodes. Several works (Ecija et al., 2011; Rodríguez-Martínez & Attfield, 1996; Vidal et al., 2007 and references therein) have shown that the physical properties of these perovskite materials are very sensitive to changes in the doping level (x) the average size of the A cations ($<r_A>$) and the effects of A cation size disorder ($\sigma^2(r_A)$).

The synthesis of these compounds allows us to study the effect of the variation of the doping level x on the properties of the perovskites with general formula $Ln_{1-x}A_xFeO_{3-\delta}$ (Ln=La, Nd, Pr; A=Sr, Ca) applied as SOFC cathodes. This has been achieved by keeping constant both the average size ($<r_A>$) and the size mismatch ($\sigma^2(r_A)$) to 1.22 Å and 0.003 Å^2, respectively.

For the preparation of this series the solid state reaction route has been chosen due to its simplicity to obtain perovskite phases in the same synthetic conditions.

3.1.1 Structural study

The room temperature X-ray powder diffraction patterns of these compounds are shown in Fig. 2a. A structural transition from orthorhombic symmetry (space group *Pnma*) for samples with x≤0.4 to rhombohedral symmetry (space group *R-3c)* for 0.5≤x≤ 0.7 compounds, and finally, to a mixture of rhombohedral *R-3c* and cubic *Pm-3m* perovskite phases for the x = 0.8 composition was observed. Representative Rietveld fits to the X-ray diffraction data for the samples LPS20, LNSC50 and LNSC80 are shown in Fig. 2b, 2c and 2d, respectively.

The dependence with the doping level x of the cell parameters and cell volume per formula unit and atomic distances and bond angles for all samples are represented in Figure 3a and b, respectively. There is a systematic decrease in volume, cell parameters and <Fe-O> distances with increasing doping level across the series. Given that in the system studied the A-site mean ionic radius $<r_A>$ has been kept constant as the doping level increases, the observed effect can be solely associated to a reduction of the Fe-site mean ionic radius as it oxidises from Fe^{3+} to Fe^{4+}, with smaller radius ($\langle r_{Fe} \rangle$, $r_{Fe^{3+}}$=0.645 Å and $r_{Fe^{4+}}$ =0.585 Å) (Shannon, 1976). Details of these effects are given elsewhere (Vidal et al., 2007).

The increase of the <A-O> distances and <Fe-O-Fe> bond angles with increasing doping level (x) can be explained due to the structural transition produced with x: when passing from orthorhombic (*Pnma,* LPS20) to a mixture of rhombohedral and cubic (*R-3c* + *Pm-3m,* LSC80) the octahedra that compose the perovskite structure reduce its cooperative tilting and the structure "expands".

These results are in nice agreement with other structural studies of related perovskites in which similar structural transitions with doping level were observed (Blasco et al., 2008; Dann et al., 1994; Tai et al., 1995).

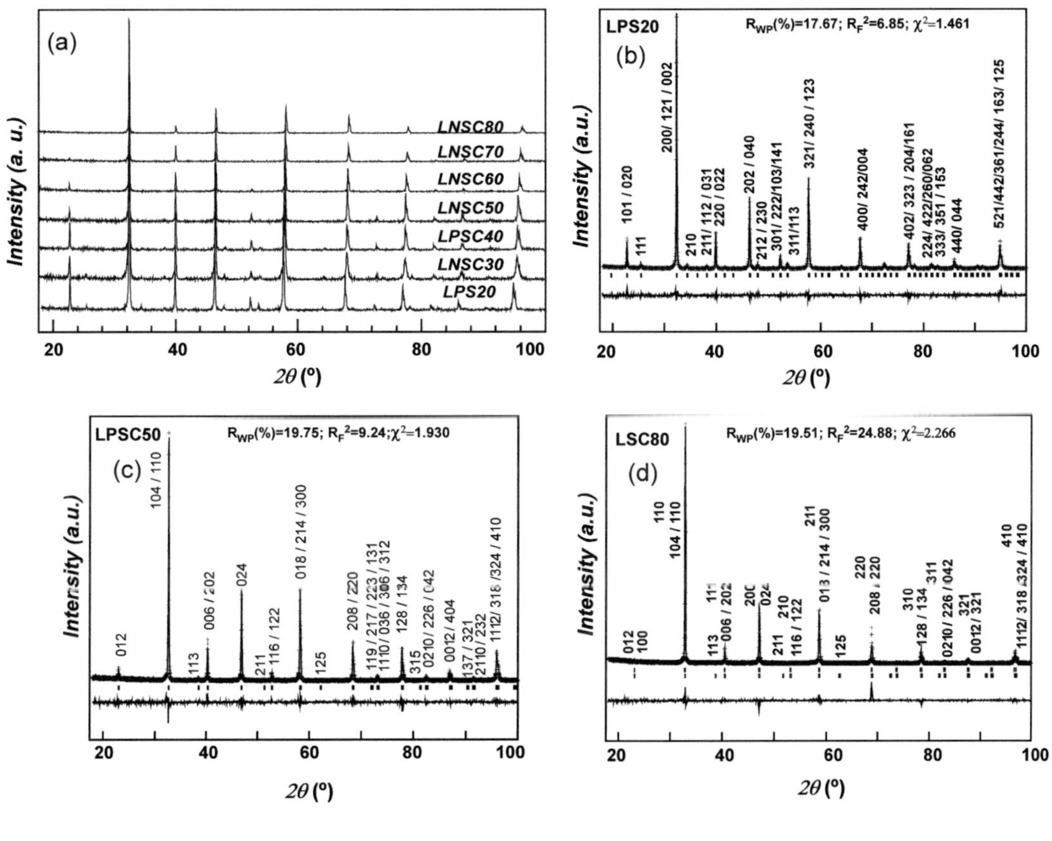

Fig. 2. (a) X-ray diffraction patterns for the series $Ln_{1-x}A_xFeO_{3-\delta}$ with x=0.2 to x=0.8, all obtained by the ceramic route. Rietveld fits to the X-ray diffraction data for samples LPS20 (b), LPSC50 (c) and LSC80 (d).

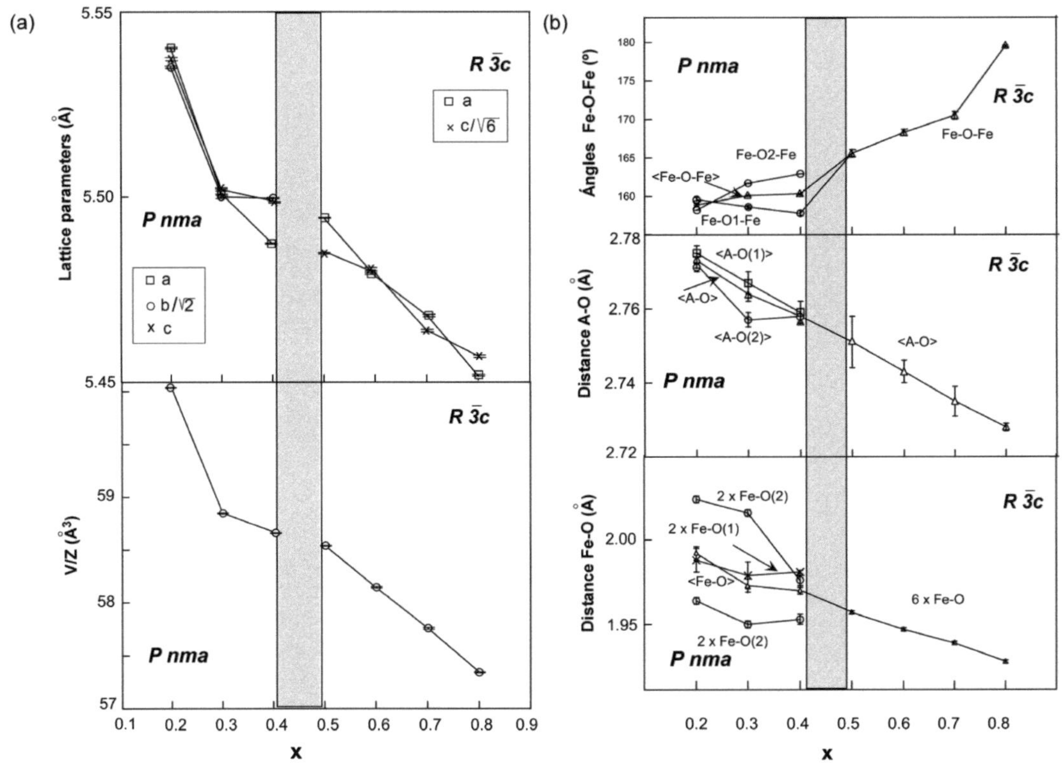

Fig. 3. (a) Variation of the unit cell parameters and volume per formula unit with doping, x. (b) Dependence of the mean atomic distances and bond angles with doping, x. Shaded area indicates the x range where the structural transition takes place.

3.1.2 Morphological study

Microstructure of bulk samples was study by scanning electron microscopy (SEM). Images of the sintered bars at 1300ºC are shown in Fig. 4.

These micrographs show different particle size distributions with grain sizes ranging between 0.33 and 2.83 µm for sample LPS20, to 5 and 37 µm for sample LSC70.

The dispersion in particle sizes is larger for values of x≥0.4. As observed previously (Liou, 2004a, 2004b) this increase in particle size with the doping level seems to be a result of a change in the melting point of the samples that decreases increasing alkaline-earth cation content. According to Kharton et al. (Kharton et al., 2002) this effect results in a liquid-phase assisted by sintering and an enhanced grain growth.

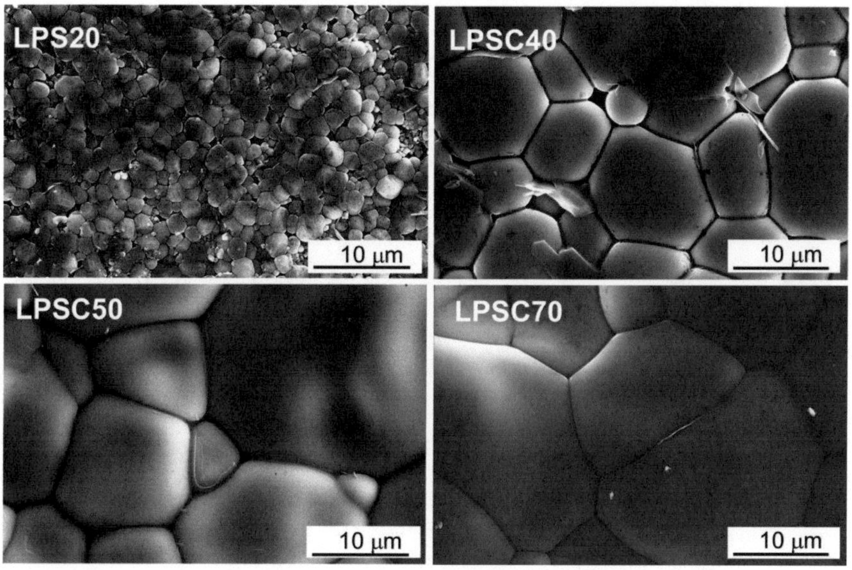

Fig. 4. (a) Scanning electron microscopy (SEM) images obtained at the same magnification for all $Ln_{1-x}A_xFeO_{3-\delta}$ compositions as a function of doping level x.

3.2 $Ln_{0.5}A_{0.5}FeO_{3-\delta}$ (Ln= La, Nd, Sm; A= Ba, Sr) 1.25≤ $\langle r_A \rangle$ ≤1.34 Å perovskites

As in the case of the previous family of iron perovskites, this new series of compounds are of interest for their use as mixed ionic electronic conducting materials, mainly from the point of view of cathodes for Solid Oxide Fuel Cells (SOFC), although they could also be used as ceramic membranes for oxygen separation. In the present case, the degree of lanthanide substitution was fixed to x=0.5 given that previous studies have shown that it is precisely at this degree of substitution when electronic and ionic conductivity are maximised (Hansen, 2010; Vidal et al., 2007).

In the present series of compounds the effect of the $\langle r_A \rangle$ variation on the properties of four different phases, $Ln_{0.5}A_{0.5}FeO_{3-\delta}$ (Ln=La, Nd, Sm; A=Ba, Sr) was evaluated. For this series (Table 1), $\langle r_A \rangle$ has been varied between 1.34 and 1.25 Å keeping x and $\sigma^2(r_A)$ constant, with values of 0.5 and 0.0161 Å^2, respectively.

Prior to choosing a synthesis method for all samples, two different methods were tried for one of the samples, $La_{0.05}Sm_{0.45}Sr_{0.18}Ba_{0.32}FeO_{3-\delta}$, for which the structural parameters and morphology were evaluated. The ceramic and glycine-nitrate routes were used in the present case.

3.2.1 Influence of the synthetic method on the structure and morphology of the $La_{0.05}Sm_{0.45}Sr_{0.18}Ba_{0.32}FeO_{3-\delta}$

Laboratory X-ray diffraction data at room temperature for $La_{0.05}Sm_{0.45}Ba_{0.5}FeO_{3-\delta}$ obtained by ceramic and glycine-nitrate routes were extremely similar, both samples being pure to the detection limits of the technique. XRD patterns were fitted by the Rietveld method (Figure 5) considering a rhombohedral symmetry (space group *R*-3*c*) in both cases.

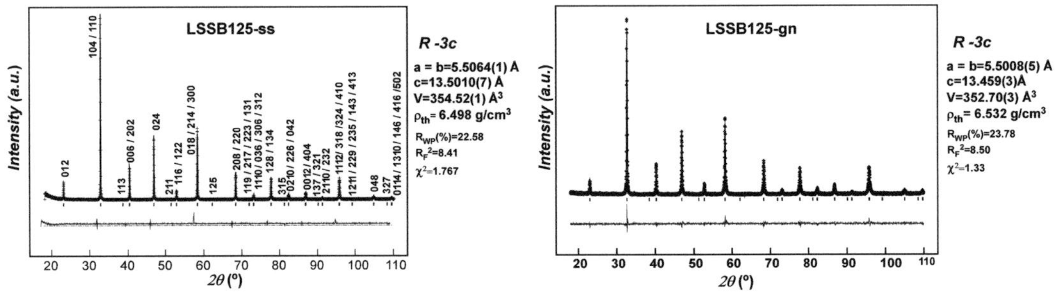

Fig. 5. Rietveld fits to room XRD patterns for LSSB-ss and LSSB-gn samples. In each case, lattice parameters (a, b, c), unit cell volume (V) and theoretical density (ρ) are included.

From the fits to the XRD data it was observed that both phases are nearly identical: they crystallise in the same space group and does not show significant difference among lattice parameters, cell volume or density values.

The morphological study, however, shows a different picture. As shown in Figure 6, where SEM micrographs taken at the same amplification for both samples are presented, their microstructure is quite different. The sample prepared following the ceramic route presents a microstructure with heterogeneous grain sizes and shapes, in which particles range from ~ 2 to approximately 8 μm in diameter. On the other hand, the average grain size of the sample obtained by glycine-nitrate route is about 200 nm, almost an order of magnitude smaller. The higher calcination temperature and longer reaction time required to obtain the samples in the ceramic process can explain the bigger grain size showed for these samples (Melo Jorge et al., 2001).

Fig. 6. SEM micrographs taken on the surface of the $La_{0.05}Sm_{0.45}Sr_{0.18}Ba_{0.32}FeO_{3-\delta}$ phases obtained by the glycine-nitrate (gn) and solid state reaction (ss) methods.

For the application as SOFC cathodes, samples with small and homogeneous particle sizes are usually preferred. As a consequence, the glycine-nitrate process was considered a more appropriate technique for preparing the perovskite samples of this series: $Ln_{0.5}A_{0.5}FeO_{3-\delta}$ (Ln= La, Nd, Sm; A= Ba, Sr) with $1.25 \leq \langle r_A \rangle \leq 1.34$ Å.

3.2.2 Structural study

Room temperature X-ray diffraction patterns of the LB134, LNSB131, LNSB128, LSSB125-gn samples show that all the samples are single phase compounds (Figure 7a). A shift of the diffraction maxima to lower diffraction angles (2θ) with decreasing $<r_A>$ anticipates an increase of the lattice parameters. X-ray powder diffraction patterns were indexed using a cubic symmetry (*Pm-3m* space group) for the LB134 and LNSB131 samples and a rhombohedral symmetry (*R-3c* space group) in the case of LNSB128 and LSSB125-gn compounds. Figure 7b shows the Rietveld fits to the XRD patterns for two samples.

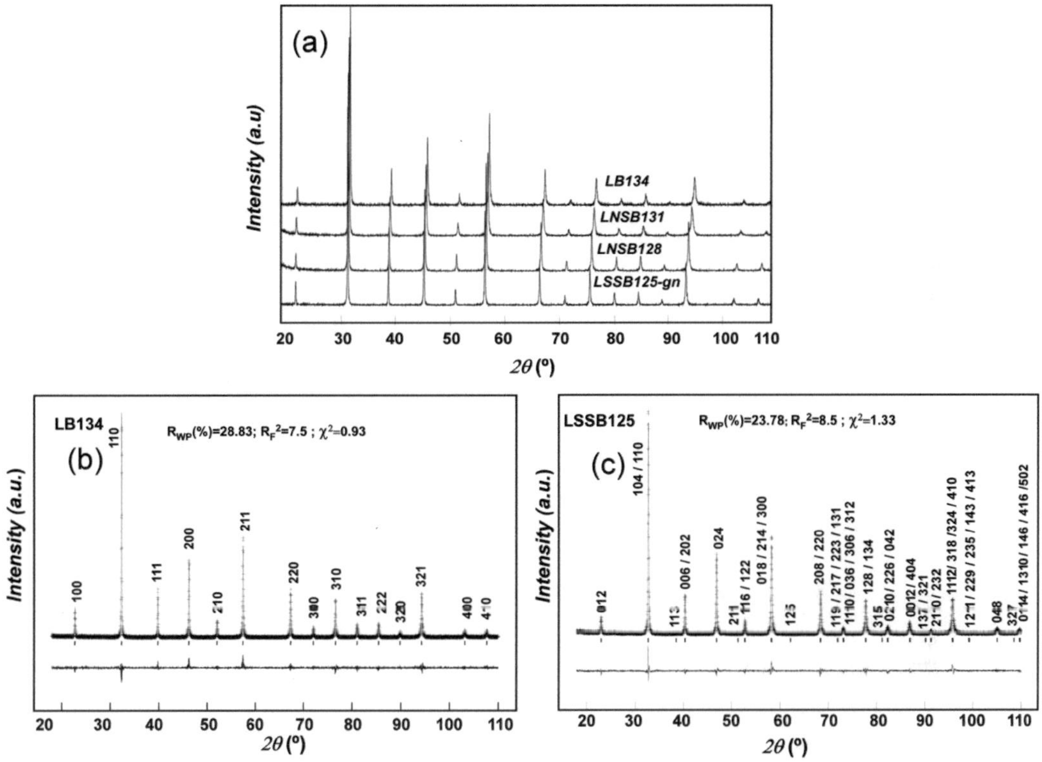

Fig. 7. (a) X-ray powder diffraction patterns at room temperature for all samples. (b) and (c) show the Rietveld fits to the X-ray powder diffraction patterns at room temperature for LB134-ss and LSS125-ss samples.

The $<r_A>$ dependence of the lattice parameters, unit cell volume, main bond distances and Fe-O-Fe bond angles are shown in Figure 8. As it can be observed, lattice parameters and unit cell volume decrease with decreasing the average A-site ionic radius. Given that the doping level has been fixed (x=0.5), changes in the ratio Fe^{3+}/Fe^{4+} and, therefore, in the Fe-site ionic radius ($<r_{Fe}>$), are not expected. Consequently, the decrease in lattice parameters is ascribed to the variation of $<r_A>$. A more detailed analysis is given elsewhere (Ecija et. al, 2011). This also explains the decrease of the <A-O> and <A-Fe> mean distances with decreasing $<r_A>$ (Figure 8b). Although $<r_{Fe}>$ has been kept constant, there is a slight reduction of the <Fe-O> distances as $<r_A>$ decreases, which is a consequence of the tilting in the FeO_6 octahedra due to the rhombohedral distortion. The decrease of the <Fe-O-Fe> bond angles is the result of the same effect (Woodward, 1998).

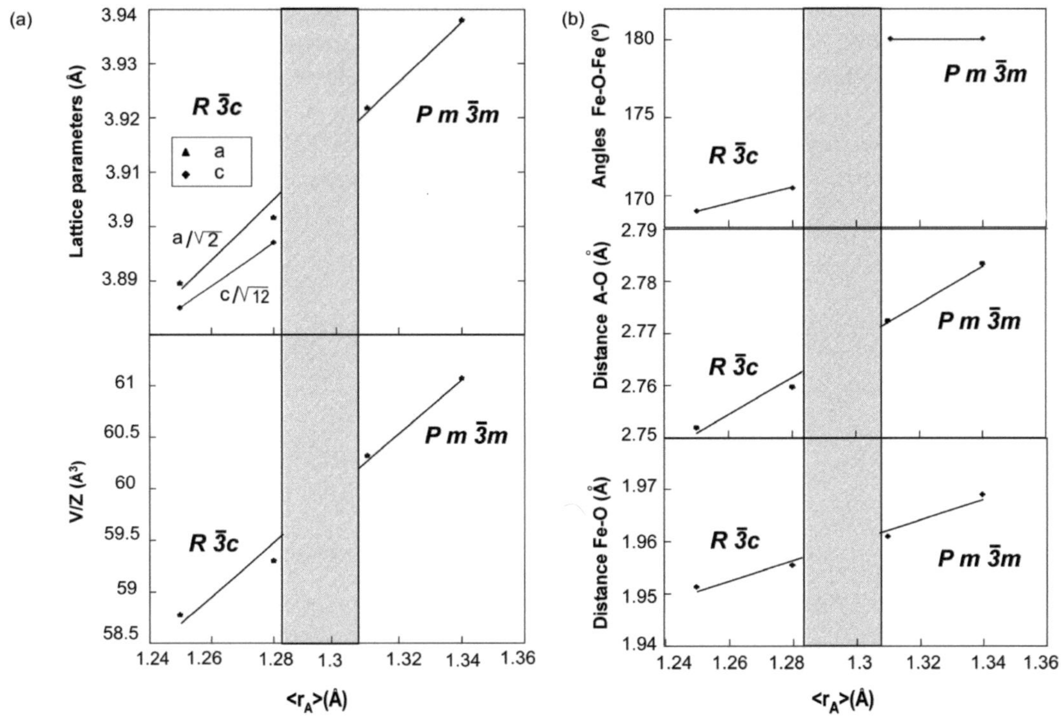

Fig. 8. <r_A> dependence of (a) the unit cell parameters and volume per formula unit; and (b) the main average bond lengths (<A-Fe>, <A-O>, <Fe-O>) and Fe-O-Fe bond angle. Shaded area indicates the x range where the structural transition takes place.

3.2.3 Morphological study

Figure 9 shows the SEM micrographs of the LB134, LNSB131, LNSB128, LNSB125-gn bulk samples after the last heating at 1050°C.

In this series of samples no significant differences can be found in the morphology and average particle size. All samples present some agglomeration and fine grain size. Image analysis of the micrographs has allowed us to determine that the average grain size of the samples is in the range of 150-250 nm.

3.3 Synthesis and characterization of $Nd_{0.8}Sr_{0.2}(Mn_{1-x}Co_x)O_3$ perovskites with x = 0.1, 0.2, 0.3

The hole-doped perovskite manganese oxides with general formula $Ln_{1-x}A_xMnO_3$ (Ln= La, Pr, Nd; A= Ca, Sr, Ba, Pb; x<0.5) draw considerable attention in the late 1990´s due to their colossal magneto-resistance (CMR) effect at low temperatures (Rao, 1998). In the search for new CMR materials it was observed that doping on Mn site by other transition metal elements, such as Cr, Fe, Co and Ni, was an effective way to obtain new materials, which also helped to understand the new phenomenon (Tai, 2000; Takeuchi, 2002; Ulyanov, 2007). Phase morphology, highly dependent on preparative conditions, was also observed to play an important role in the effect: low temperature CMR was improved as the grain size was reduced (Das, 2002). In order to help in this area, it was decided to study the effects of the synthesis method in a series of perovskite compounds with the general formula $Nd_{0.8}Sr_{0.2}(Mn_{1-x}Co_x)O_3$ ($0.1 \leq x \leq 0.3$). The sol-gel and freeze-drying techniques were used in order to compare their structural, morphological and magnetic properties. Details of the later (magnetic properties) can be found elsewhere (Vidal et al., 2005).

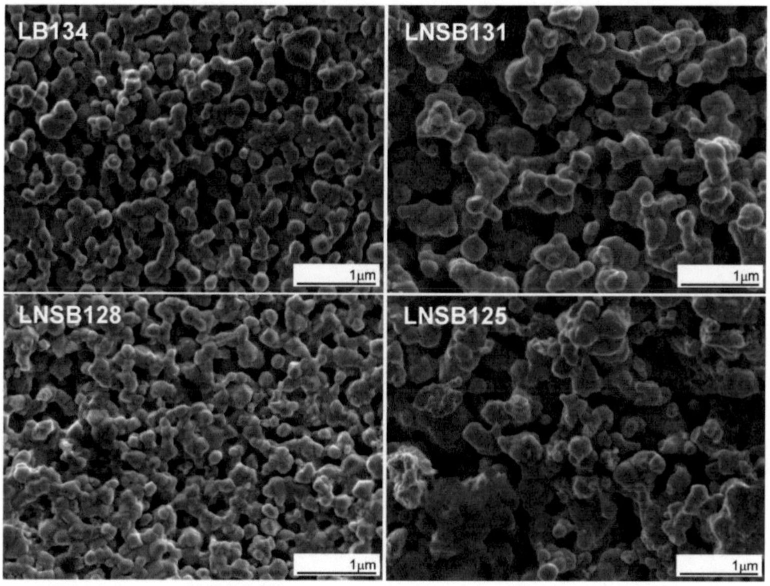

Fig. 9. SEM micrographs of the surface of the LB134, LNSB131, LNSB128, LSSB125-gn bulk samples.

3.3.1 Structural study

The X-ray powder diffraction patterns (XRPD) of all the compounds studied in this section were indexed in the orthorhombic space group *Pmna* irrespective of the synthesis method used. Fig. 10 shows the XRD and Rietveld refinement profiles for all phases with the formula $Nd_{0.8}Sr_{0.2}(Mn_{1-x}Co_x)O_3$.

When the lattice parameters and cell volume are compared a slight decrease with increasing cobalt content is observed in both cases. This effect is related to the changes in sizes of the B site atoms upon doping (Meera et al., 2001; Pollert et al., 2003). The oxidation states of Mn and Co in the $AMn_{1-x}Co_xO_3$ systems has been for long debated (Goodenough et al., 1961; 1997; Toulemonde et al., 1998; Troyanchuk et al., 2000), and different mixtures of Mn^{4+}-Mn^{3+} and Co^{3+}-Co^{2+} have been proposed. The most likely combination, based in spectroscopic and magnetic results, seems to indicate that the cobalt is introduced as Co^{2+} (and not as Co^{3+}) thus causing a mixed state (4+ and 3+) in manganese. This mixture of oxidation states has also been proved useful for Pollert et al. to explain magnetic and electrical properties of the series $Nd_{0.8}Na_{0.2}Mn_{1-x}Co_xO_3$ ($x \leq 0.1$), $Pr_{0.8}Na_{0.2}Mn_{1-x}Co_xO_3$ ($x \leq 0.2$) (Pollert et al., 2003a, 2003b) and $La_{0.8}Na_{0.2}Mn_{1-x}Co_xO_3$ ($x \leq 0.2$) (Pollert et al., 2004). On the other hand, as the Co content increases, the LSMC materials would shift from the oxygen excessive region to the oxygen deficient region, which would also contribute to the decrease in the unit volume. Wandekar et al. (Wandekar et al., 2009) carried out a detailed study on the crystal structure and conductivity of Co substituted LSM system and showed that the change in the ion radius of B-site element plays a predominant role at low Co content and the increase in oxygen vacancy becomes dominant at high Co content.

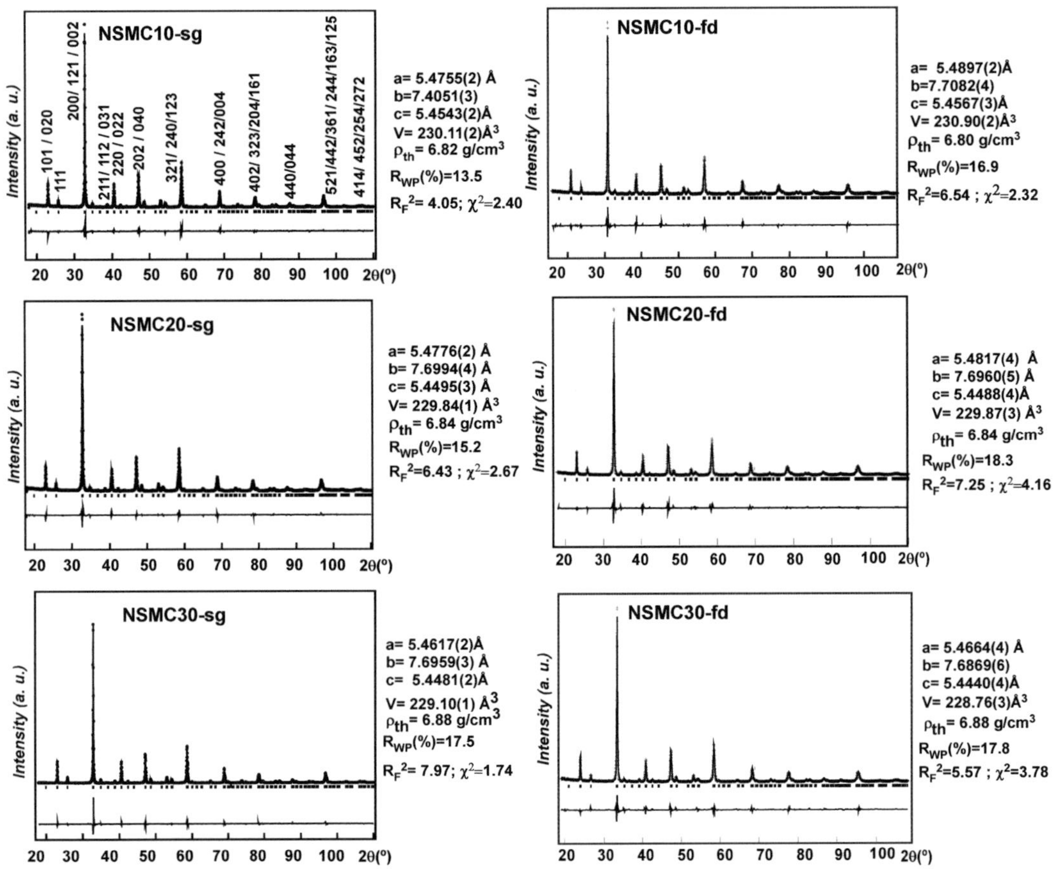

Fig. 10. Rietveld fits to the X-ray diffraction data in the orthorhombic *Pnma* space group for the $Nd_{0.8}$ $Sr_{0.2}$ $Mn_{1-x}Co_xO_3$ compounds. The labels "sg" and "fd" indicates that samples were prepared using the sol gel and freeze-drying techniques, respectively. In each case, lattice parameters (a, b, c), unit cell volume (V) and theoretical density (ρ) data are inset.

In consequence, considering only the high spin states of these elements and assuming nearly oxygen stoichiometric phases at room temperature, the observed reduction of the lattice volume in the present samples is consistent with the gradual appearance of Mn^{4+} (of smaller size than Mn^{3+} and Co^{2+} (Shannon, 1976) and so with the reduction of the mean B-site ionic radii. This explanation is also satisfactory for the same effect in the compounds $La_{0.7}Na_{0.3}Mn_{1-x}Co_xO_3$ where the authors had assumed the presence of only Mn^{3+}–Co^{3+} (Meera et al., 2001).

Table 2 shows details of the averge interatomic <B-O>, <A-O> and <A-B> distances together with <B-O-B> bond angles. As it can be observed <B-O> and <A-B> distances decrease as the cobalt concentration increases. These tendencies can be explained following the same reasoning indicated before, which basically concludes that cobalt enters in the structure in an oxidation state with smaller radius than manganese. No significant changes are observed in the <A-O> distances or in the <B-O-B> angles. This indicates that, despite the observed changes in the B size, the cobalt doping does not cause a noticeable distortion in the perovskite structure.

Variables		NSMC10		NSMC20		NSMC30	
		sg	fd	sg	fd	sg	fd
distance (Å)	<A-B>	3.346	3.350	3.345	3.345	3.341	3.339(1)
	<A-O1>	2.749(1)	2.756(1)	2.748(1)	2.751(2)	2.748(1)	2.747(2)
	<A-O2>	2.588(1)	2.592(1)	2.588(2)	2.593(2)	2.587(1)	2.591(2)
	<A-O>	2.668	2.674	2.668	2.672	2.667	2.671
	<B-O1>	1.954(4)	1.959(4)	1.950(4)	1.958(6)	1.958(4)	1.957(4)
	<B-O2>	1.968(1)	1.970(2)	1.967(2)	1.965(2)	1.963(2)	1.960(2)
	<B-O>	1.961	1.964	1.958	1.961	1.957	1.958
bond angle (°)	B-O1-B	160.8(4)	159.3(2)	161.5(4)	158.6(2)	158.5(3)	158.3(2)
	B-O2-B	157.9(4)	158.3(4)	158.2(4)	159.0(1)	158.5(3)	159.4(1)
	<B-O-B>	159.3	158.8	159.8	158.8	158.5	158.8

Table 2. Mean atomic distances and bond angles for the series of ABO_3 (A = $Nd_{0.8}$ $Sr_{0.2}$; B = $Mn_{1-x}Co_x$; x = 0.1, 0.2 and 0.3) samples prepared by the two methods described in the text.

In the same way, no significant changes are observed between the compounds synthesised by the sol-gel or the freeze-drying techniques.

3.3.2 Morphological study

The SEM micrographs of the present series of compounds are shown in Fig. 11. All pictures were taken on sintered bars after they were prepared by the sol-gel and freeze-drying methods.

As observed, the $Nd_{0.8}Sr_{0.2}(Mn_{1-x}Co_x)O_3$ samples prepared by the sol-gel route show a homogeneous particle size morphology distributed in small agglomerates. The grain size of the nearly spherical particles decreases with Co content from ~ 200 nm for the sample with x = 0.1 to ~ 100 nm for x = 0.3. In the case of the compounds obtained by the freeze-drying technique the morphology of the particles is also spherical and homogeneous but the grain size is quite stable in all phases and slightly higher (~250 nm) than in the previous case.

According to some studies (Kuharuangrong et al., 2004), Co doping of 20% does not usually influence the grain size of LSM, but 40 % mol Co significantly reduces the grain size (about 5-10 μm) of the $La_{0.84}Sr_{0.16}Mn_{1-x}Co_xO_3$ (x= 0, 0.2, 0.4) samples prepared by conventional oxide mixing process. However, some reduction in the particle size was observed in the case of $La_{0.67}Pb_{0.33}Mn_{1-x}Co_xO_3$ (x= 0 and 0.3) compounds, which were also prepared by the ceramic route (Dhahri et al., 2010). The present observation of a reduction in the particle size of the $Nd_{0.8}Sr_{0.2}(Mn_{1-x}Co_x)O_3$ phases in one of the methods but not in the other would point towards the synthesis method as the responsible for this effect, rather than the amount of cobalt. In this case, however, the observed decrease of the grain size is less significant than in previous studies.

Fig. 11. Secondary electron SEM micrographs obtained for all $Nd_{0.8}Sr_{0.2}(Mn_{1-x}Co_x)O_3$ compositions prepared by (a) sol-gel (sg) and (b) freeze-drying (fd) methods.

4. Conclusion

Different synthetic methods were used to prepare several perovskite type compounds. Phase-pure perovskites were obtained in all cases, irrespective of the method of synthesis and even when more than one method was used. It is to note that no differences in the overall crystal structure were observed in any of the cases when the method was changed. Moreover, changes in the crystal structure of the compounds within each series could be perfectly explained considering only compositional (and resulting structural) variables such as the mean A-site ionic radius or the A-site disorder, none of them being influenced by the synthesis route. This is true for the $Ln_{1-x}A_xFeO_{3-\delta}$ series where changes from orthorhombic to cubic are consistent with the change in the doping level x and also for the $Ln_{0.5}A_{0.5}FeO_3$ where the change was observed from rhombohedral to cubic symmetry when the mean A-site radius, $<r_A>$, was varied.

At least in the present group of compounds this seems to indicate that, properly used, and with no other factors present (such as in the case of the growth of samples on crystallographycally oriented substrates), the structural characteristics can be studied and properly addressed without having into account the synthesis method.

It is no news that, compared with the solid state reaction, the glycine-nitrate route, sol-gel and freeze-drying techniques require lower calcination temperatures to yield pure crystals of perovskite phases, with the resulting energy saving (very important nowadays). The disadvantage is, however, the fact that they are more time consuming and require more controlled synthesis conditions. In the same way, it is clearly observed that the ceramic method always yields phases with higher particle sizes but this is not a disadvantage if the final compounds are not to be used in an application that requires high specific area. It is to note that in the case of the $Nd_{0.8}Sr_{0.2}(Mn_{1-x}Co_x)O_3$, grain size was observed to change depending on the synthesis route. Although no significant differences were observed in the crystal structure, sol-gel method resulted in smaller grain sizes than the freeze drying method. In consequence, this seems to be a consequence of the preparative method which reinforces the idea that, if the microstructure of the sample is the key issue, the use of several synthesis methods is always worth trying.

5. Acknowledgments

This work has been financially supported by the Ministerio de Ciencia e Innovación PSE-12000-2009-7 (MICINN) and MAT 2010-15375; Consejería de Industria, Innovación, Comercio y Turismo (SAIOTEK 2011) and by the Consejería de Educación, Universidades e Investigación (IT-177-07) of the Basque Goverment of Gobierno Vasco/Eusko Jaurlaritza. Technical and human support provided by SGIker (UPV/EHU, MICINN, GV, EJ, ESF) and Alternative Generation Systems Group of Technological Research Centre is gratefully acknowledged.

6. References

Bansal, N.P. & Zhong, Z. (2006). Combustion Synthesis of $Sm_{0.5}Sr_{0.5}CoO_{3-x}$ and $La_{0.6}Sr_{0.4}CoO_{3-x}$ Nanopowders for Solid Oxide Fuel Cell Cathodes. *Journal of Power Sources*, Vol. 158, No. 1, (July 2006), pp. 148-153, ISSN: 0378-7753

Bell, R.J.; Millar, G.J. & Drennan. (2000). J. Influence of Synthesis Route on the Catalytic Properties of $La_{1-x}Sr_xMnO_3$. *Solid State Ionics,* (June 2000), Vol. 131, No. 3, 4, pp. 211-220, ISSN: 0167-2738

Berger, D.; Matei, C.; Papa, F.; Macovei, D.; Fruth, V. & Deloume, J.P. (2007). Pure and Doped Lanthanum Manganites Obtained by Combustion Method. *Journal of the European Ceramic Society*, Vol. 27, No. 13-15, (March 2007), pp. 4395-4398, ISSN: 0955-2219

Blasco, J.; Aznar, B.; García, J.; Subías, G.; Herrero-Martín, J. & Stankiewicz, J. (2008). Charge Disproportionation in $La_{1-X}Sr_xFeo_3$ Probed by Diffraction and Spectroscopic Experiments. *Physical Review B*, Vol. 77, N° 5, (February 2008), pp. 054107-1-054107-10, ISSN: 1098-0121

Carp, O.; Patron, L.; Ianculescu, A.; Pasuk, J. & Olar, R. (2003). New Synthesis Routes for Obtaining Dysprosium Manganese Perovskites. *Journal of Alloys and Compounds*, Vol. 351, No. 1-2, (March 2003), pp. 314-318, ISSN: 0925-8388

Cheetham, A.K. & Day, P. (1987). *Solid State Chemistry: Techniques*, Oxford University Press, ISBN: 0198551657, Oxford

da Conceicao, L.; Silva, C.R.B.; Ribeiro, N.F.P. & Souza, M.M.V.M. (2009). Influence of the Synthesis Method on the Porosity, Microstructure and Electrical Properties of $La_{0.7}Sr_{0.3}MnO_3$ Cathode Materials. *Materials Characterization*, Vol. 60, No. 12, (December 2009), pp. 1417-1423, ISSN: 1044-5803

Dann, S.E; Currie, D.B.; Weller, M.T.; Thomas, M.F. & Al-Rawwas, A.D. (1994). The Effect of Oxygen Stoichiometry on Phase Relations and Structure in the System $La_{1-x}Sr_xFeO_{3-\delta}$ ($0 \le x \le 1, 0 \le \delta \le 0.5$). *Journal of Solid State Chemistry*, Vol. 109, No. 1, (March 1994), pp. 134-144, ISSN: 0022-4596

Das, S.; Chowdhury, P.; Gundu Rao, T. K.; Das, D. & Bahadur, D. (2002). Influence of Grain Size and Grain Boundaries on the Properties of $La_{0.7}Sr_{0.3}Co_xMn_{1-x}O_3$. *Solid State Communications*, Vol. 121, No. 12, (March 2002), pp. 691-695, ISSN: 0038-1098

Dhahri, N.; Dhahri, A.; Cherif, K.; Dhahri, J.; Taibi, K. & Dhahri, E. Structural, Magnetic and Electrical properties of $La_{0.67}Pb_{0.33}Mn_{1-x}Co_xO_3$ (0□x□0.3). *Journal of Alloys and Compounds*, Vol. 496, No. 1-2, (October 2010), pp. 69-74, ISSN: 0925-8388

Dho, J.; Leung, C.W.; MacManus-Driscoll, J.L. & Blamire, M.G. (2004). Epitaxial and Oriented $YMnO_3$ Film Growth by Pulsed Laser Deposition. *Journal of Crystal Growth*, Vol. 267, No. 3-4, (July 2004), pp. 548-553, ISSN: 0022-0248

Dutta, A.; Mukhopadhyay, J. & Basu, R.N. (2009). Combustion Synthesis and Characterization of LSCF-based Materials as Cathode of Intermediate Temperature Solid Oxide Fuel Cells. *Journal of the European Ceramic Society*, Vol. 29, No.10, (July 2009), pp. 2003-2011, ISSN: 0955-2219

Ecija, A.; Vidal, K.; Larrañaga, A.; Martínez-Amesti, A.; Ortega-San-Martín, L. & Arriortua, M.I. (2011). Characterization of $Ln_{0.5}M_{0.5}FeO_{3-\delta}$ (Ln=La, Nd, Sm; M=Ba, Sr) Perovskites as SOFC Cathodes. *Solid State Ionics*, Vol. 201, No. 1, (October 2011), pp. 35-41, ISSN: 0167-2738

Fukuoka, H.; Isami, T. & Yamanaka, S. (1997). Superconductivity of Alkali Metal Intercalated Niobate with a Layered Perovskite Structure. *Chemistry Letters*, Vol. 8, (April 1997), pp. 703-704, ISSN: 0366-7022

Goodenough, J.B.; Wold, A.; Wold, R.J.; Arnott, R.J. & Menyuk, N. (1961). Relationship Between Crystal Symmetry and Magnetic Properties of Ionic Compounds Containing Mn^{3+}. *Physical Review*, Vol. 124, (October 1961), pp. 373-384, ISSN: 0031-899X

Hansen, K.K. (2010). Electrochemical Reduction of Nitrous Oxide on $La_{1-x}Sr_xFeO_3$ Perovskites. *Materials Research Bulletin*, Vol. 45, No. 9, (May 2010), pp. 1334-1337, ISSN: 0025-5408

Inaguma, Y.; Liquan, C.; Itoh, M.; Nakamura, T.; Uchida, T.; Ikuta, H. & Wakihara, M. (1993) High Ionic Conductivity in Lithium Lanthanum Titanate. *Solid State Communications*, Vol. 86, No. 10, (June 1993), pp. 689-693, ISSN: 0038-1098

Kakihana, M. & Yoshimura, M. (1999). Synthesis and Characteristics of Complex Multicomponent Oxides Prepared by Polymer Complex Method. *Bulletin of the Chemical Society of Japan*, Vol. 72, No. 7, (1999), pp. 1427-1443, ISSN: 0009-2673

Kharton, V.V.; Shaulo, A.L.; Viskup, A.P.; Avdeev, M.; Yaremchenko, A.A.; Patrakeev, M.V.; Kurbakov, A.I.; Naumovich, E.N. & Marques, F.M.B. (2002). Perovskite-like System $(Sr,La)(Fe,Ga)O_{3-\delta}$: Structure and Ionic Transport under Oxidizing Conditions. *Solid State Ionics*, Vol. 150, No. 3-4, (October 2002), pp. 229-243, ISSN: 0167-2738

Kuharuangrong, S.; Dechakupt, T. & Aungkavattana, P. (2004). Effects of Co and Fe Addition on the Properties of Lanthanum Strontium Manganite. *Materials Letters*, Vol. 58, No. 12-13, (May 2004), pp. 1964-1970, ISSN: 0167-577X

Larson, A.C. & Von Dreele R.B. (1994). *GSAS: General Structure Analysis System, LAUR*, pp. 86–748

Liang, J.J. & Weng, H.S. (1993). Catalytic Properties of Lanthanum Strontium Transition Metal Oxides ($La_{1-x}Sr_xBO_3$; B = Manganese, Iron, Cobalt, Nickel) for Toluene Oxidation. *Industrial & Engineering Chemistry Research*, Vol. 32, No. 11, (November 1993), pp. 2563-2572, ISSN: 0888-5885

Liou, Y.C. (2004). Effect of Strontium Content on Microstructure in $(La_xSr_{1-x})FeO_3$ Ceramics. *Ceramics International*, Vol. 30, No. 5, (March 2004), pp. 667-669, ISSN: 0272-8842

Liou, Y.C. (2004). Microstructure Development in $(La_xSr_{1-x})MnO_3$ Ceramics. *Materials Science & Engineering, B: Solid-State Materials Advance Technology*, Vol. 108, No. 3, (April 2004), pp. 278-280, ISSN: 0921-5107

Liu, B. & Zhang, Y. (2008). $Ba_{0.5}Sr_{0.5}Co_{0.8}Fe_{0.2}O_3$ Nanopowders Prepared by Glycine-nitrate Process for Solid Oxide Fuel Cell Cathode. *Journal of Alloys and Compounds*, Vol. 453, No. (1-2), (April 2008), pp. 418-422, ISSN: 0925-8388

Meera, K.V.K.; Ravindranayh, V. & Rao M.S.R. (2001). Magnetotransport Studies in $La_{0.7}Ca_{0.3}Mn_{1-x}M_xO_3$ (M = Co, Ga). *Journal of Alloys and Compounds*, Vol. 326, No. 1-2, (August 2001), pp. 98-100, ISSN: 0925-8388

Melo Jorge, M.E.; Correia dos Santos, A. & Nunes, M.R. (2001). Effects of Synthesis Method on Stoichiometry, Structure and Electrical Conductivity of $CaMnO_{3-\delta}$. *International. Journal of Inorganic Materials*, Vol. 3, No. 7, (November, 2001), pp. 915-921, ISSN: 14666049

Narlikar, A. (2001). Essential Chemistry of High-Tc Cuprate Synthesis through the Solution Precursor Methods, In: *Studies of High Temperature Superconductors*, Nova Science Publishers, New York, ISBN: 1-59033-026-9

Patil, K. C.; Aruna, S. T. & Mimani, T. (2002). Combustion Synthesis: an Update. *Current Opinion in Solid State & Materials Science*, Vol. 6, No. 6, (December 2002), pp. 507-512, ISSN: 1359-0286

Pithan, C., Hennings, D. & Waser, R. (2005). Progress in the Synthesis of Nanocrystalline $BaTiO_3$ Powders for MLCC, *International Journal of Applied Ceramic Technology*, Vol. 2, No. 1, (January 2005), pp. 1–14, ISSN: 1546-542X

Polini, R.; Pamio A. & Traversa. E. (2004). Effect of Synthetic Route on Sintering Behavior, Phase Purity and Conductivity of Sr- and Mg-doped $LaGaO_3$ Perovskites. *Journal of the European Ceramic Society*, Vol. 24, No. 6, (June 2004), pp. 1365-1370. ISSN: 0955-2219

Pollert, E.; Hejtmánek, J. ; Knížek, K.; Maryško M.; Doumerc, J.P.; Grenier J.C. & Etourneau, J. (2003). Insulator-metal Transition in $Nd_{0.8}Na_{0.2}Mn_{(1-x)}Co_xO_3$ Perovskites. *Journal of Solid State Chemistry*, Vol. 170, No. 2, (February 2002), pp. 368-373, ISSN: 0022-4596

Pollert, E.; Hejtmánek, J.; Jirák, Z.; Knízek, K. & Marysko, M. (2003). Influence of Co Doping on Properties of $Pr_{0.8}Na_{0.2}Mn_{(1-y)}Co_yO_3$ Perovskites. *Journal of Solid State Chemistry*, Vol. 174. No. 2, (September 2003) pp. 466-470, ISSN: 0022-4596

Pollert, E.; Hejtmánek, J.; Jirák, Z.; Knížek, K.; Maryško, M.; Doumerc, J.P.; Grenier, J.C; & Etourneau, J. (2004). Influence of the Structure on Electric and Magnetic Properties of $La_{0.8}Na_{0.2}Mn_{1-x}Co_xO_3$ Perovskites. *Journal of Solid State Chemistry*, Vol. 177, No. 12, (December 2004), pp. 4564-4568, ISSN: 0022-4596

Qi, X; Zhou, J.; Yue, Z.X.; Gui, Z.L. & Li, L.T. (2003). A Simple Way to Prepare Nanosized $LaFeO_3$ Powders at Room Temperature. *Ceramics International*, Vol. 29, No. 3, (June 2003), pp. 347-349, ISSN: 0272-8842

Rao, C.N.R. & Gopalakrishnan, J. (1997). *New Directions in Solid State Chemistry*, 2nd ed., Cambridge University Press, ISBN: 0521495598, Cambridge

Rao, C.N.R. & Raveau, B. (1998). *Colossal Magnetoresistance, Charge Ordering and Related Properties of Manganese Oxides*, World Scientific, ISBN: 9810232764, Singapure.

Rietveld, H.M. (1969). A Profile Refinement Method for Nuclear and Magnetic Structures. *Journal of Applied Crystallography*, Vol. 2, (June 1969), pp. 65-71, ISSN: 0021-8898

Rodríguez-Martínez, L.M. & Attfield, J.P. (1996). Cation Disorder and Size Effects in Magnetoresistive Manganese Oxide Perovskites. *Physical Review B - Condensed Matter and Materials Physics*, Vol. 54, No. 22, (December 1996), pp. R15622–R15625, ISSN: 0163-1829

Royer, S.; Berube, F.S. & Kaliaguine, S. (2005). Effect Of The Synthesis Conditions on the Redox And Catalytic Properties in Oxidation Reactions of $LaCo_{1-x}Fe_xO_3$, *Applied Catalysis A: General*, Vol. 282, No. 1-2, (March 2005), pp. 273-284, ISSN: 0926-860X

Safari, A.; Panda, R. K. & Janas, V. F. (1996). Ferroelectricity. Materials, Characteristics, and Applications. *Key Engineering Materials*, Vol. 122-124, (1996), pp. 35-69, ISSN: 1013-9826

Segal, D. (1989). *Chemical Synthesis of Advanced Ceramic Materials*, Cambridge University Press, ISBN: 9780521354363, Cambridge

Senna, M. (2005). A Straight Way toward Phase Pure Complex Oxides. *Journal of the European Ceramic Society*, Vol. 25, No. 12, (March 2005), pp. 1977-1984, ISSN: 0955-2219

Sfeir, J.; Vaucher, S.; Holtappels, P.; Vogt, U.; Schindler, H.-J.; Van Herle, J.; Suvorova, E.; Buffat, P.; Perret, D.; Xanthopoulos, N. & Bucheli, O. (2005). Characterization of Perovskite Powders for Cathode and Oxygen Membranes Made by Different Synthesis Routes. *Journal of the European Ceramic Society*, Vol. 25, No.12, (March 2005), pp. 1991-1995, ISSN: 0955-2219

Shannon, R.D. (1976). Revised Effective Ionic Radii and Systematic Studies of Interatomic Distances in Halides and Chalcogenides, *Acta Crystallographica*, Vol. A32, No. 5, (September 1976), pp. 751-767, ISSN: 0567-7394

Skinner, S.J. (2001). Recent Advances in Perovskite-type Materials for Solid Oxide Fuel Cell Cathodes. *International Journal of Inorganic Materials*, Vol. 3, (March 2001), pp. 113-121, ISSN: 1466-6049

Tai, L.-W; Nasrallah, M.M.; Anderson H.U.; Sparlin, D.M & Sehlin, S.R. (1995). Structure and electrical properties of $La_{1-x}Sr_xCo_{1-y}Fe_yO_3$. Part 1. The system $La_{0.8}Sr_{0.2}Co_{1-y}Fe_yO_3$, *Solid State Ionics*, Vol. 76, No. 3-4, (March 1995), pp. 259–271, ISSN: 0167-2738

Tai, M.F.; Lee, F.Y. & Shi, J.B. (2000). Co Doping Effect on the Crystal Structure, Magnetoresistance and Magnetic Properties of an $(La_{0.7}Ba_{0.3})(Mn_{1-x}Co_x)O_3$ System with x=0-1.0., *Journal of Magnetism and Magnetic Materials,* Vol. 209, No. 1-3, (February 2000), pp. 148-150, ISSN: 0304-8853

Takeuchi, J.; Hirahara, S.; Dhakal, T.P; Miyoshi, K. & Fujiwara, K. (2002). Cr-doping Effect on the Perovskite (Nd,Sr)MnO_3 Single Crystals. *Physica B: Condensed Matter,* Vol. 312-313, (March 2002), pp. 754-756, ISSN: 0921-4526

Toulemonde, O.; Studer, F.; Barnabé, A.; Maignan, A. Martin, C. & Raveau, B. (1998). Charge States of Transition Metal in "Cr, Co and Ni" Doped $Ln_{0.5}Ca_{0.5}MnO_3$ CMR Manganites. *European Physical Journal B: Condensed Matter Physics,* Vol. 4, No. 2, (April 1998) pp. 159-167, ISSN: 1434-6028

Troyanchuk, I.O.; Lobanovsky, L.S.; Khalyavin, D.D.; Pastushonok, S.N. & Szymczak, H. (2000). Magnetic and Magnetotransport properties of Co-doped Manganites with Perovskite Structure. *Journal of Magnetism and Magnetic Materials,* Vol. 210, No. 1-2, (February 2000), pp. 63-72, ISSN: 0304-8853

Ulyanov, A.N.; Kim, J.S.; Shin, G.M.; Song, K.J.; Kang, Y.M. & Yoo, S.I. (2007). $La_{0.7}Ca_{0.3}Mn_{0.95}M_{0.05}O_3$ Manganites (M ¼ Al, Ga, Fe, Mn, and In): Local Structure and Electron Configuration Effect on Curie. Temperature and Magnetization. *Physica B: Condensed Matter,* Vol. 388, No. 1-2, (January 2007), pp. 16-19, ISSN: 0921-4526

Vidal, K.; Lezama, L.; Arriortua, M.I.; Rojo, T.; Gutiérrez, J. & Barandiarán. J.M. (2005). Magnetic characterization of $Nd_{0.8}Sr_{0.2}(Mn_{1-x}Co_x)O_3$ perovskites. *Journal of Magnetism and Magnetic Materials,* Vol. 290-291, No. 2, (April 2005), pp. 914-916, ISSN: 0304-8853

Vidal, K: Rodríguez-Martínez, L.M.; Ortega-San-Martín, L.; Díez-Linaza, E.; Nó, M.L.; Rojo, T.; Laresgoiti, A. & Arriortua, M.I. (2007). Isolating the Effect of Doping in the Structure and Conductivity of $(Ln_{1-x}M_x)FeO_{3-\delta}$ Perovskites. *Solid State Ionics,* Vol. 178, No. 21-22, (July 2007), pp. 1310-1316, ISSN: 0167-2738

Vidal, K.; Rodríguez-Martínez, L.M.; Ortega-San-Martin, L.; Martínez-Amesti, A.; Nó, M.L.; Rojo, T.; Laresgoiti, A. & Arriortua, M.I. (2009). The effect of doping in the electrochemical performance of $(Ln_{1-x}M_x)FeO_{3-\delta}$ SOFC cathodes. *Journal of Power Sources,* Vol. 192, No. 1, (July 2009), pp. 175-179, ISSN: 0378-7753

Vidal, K.; Rodríguez-Martínez, L.M.; Ortega-San-Martin, L.; Nó, M.L.; Rojo, T.; Laresgoiti, A. & Arriortua, M.I. (2010). $Ln_{0.5}M_{0.5}FeO_{3-\delta}$ Perovskites as Cathode for Solid Oxide Fuel Cells: Effect of Mean Radius of the A-Site Cations. *Journal of the Electrochemical Society,* Vol. 157, No. 8, (June 2010), pp. A919-A-924, ISSN: 0013-4651

Vidal, K.; Rodríguez-Martínez, L.M.; Ortega-San-Martin, L.; Nó, M.L.; Rojo, T.; Laresgoiti, A. & Arriortua, M.I. (2011). Effect of the A Cation Size Disorder on the Properties of an Iron Perovskite Series for their Use as Cathodes for SOFCs. *Fuel Cells,* Vol. 11, No. 1, (February 2011), pp. 51-58, ISSN: 1615-6846

Wandekar, R.V.; Wani, B.N. & Bharadwaj, S.R. (2009). Crystal Structure, Electrical Conductivity, Thermal Expansion and Compatibility Studies of Co-substituted Lanthanum Strontium Manganite System. *Solid State Science,* Vol. 11, No. 1, (January 2009), pp. 240-250, ISSN: 1293-2558

Woodward, P.M.; Vogt, T.; Cox, D.E.; Arulraj, A.; Rao, C.N.R.; Karen, P. & Cheetham, A. K. (1998). Influence of Cation Size on the Structural Features of $Ln_{1/2}A_{1/2}MnO_3$ Perovskites at Room Temperature. *Chemmistry of Materials*, Vol. 10, No. 11, (October 1998), pp. 3652-3665, ISSN: 0897-4756

20

Phase Behavior and Crystal Structure of Binary Polycyclic Aromatic Compound Mixtures

Jinxia Fu[1,*], James W. Rice[2] and Eric M. Suuberg[2]
[1]*Brown University Department of Chemistry, Providence, RI*
[2]*Brown University School of Engineering, Providence, RI*
USA

1. Introduction

Polycyclic aromatic hydrocarbons (PAHs) are a class of compounds that consist of multiple fused aromatic rings. Concerns have been raised regarding PAHs due to their known health effects(Luthy et al., 1994; Sun et al., 2003). In addition PAHs, chlorinated and brominated polycyclic aromatic hydrocarbons (ClPAHs and BrPAHs) are of interest commercially and of concern for their environmental effects (Shiraishi et al., 1985; Haglund et al., 1987; Nilsson and Ostman, 1993; Koistinen et al., 1994a, b; Ishaq et al., 2003; Kitazawa et al., 2006; Horii et al., 2008; Horii et al., 2009; Ohura et al., 2009; Ni et al., 2010; Ohura et al., 2010). The thermodynamic properties of pure PAHs have been widely studied for more than 50 years (Szczepanik et al., 1963; Wakayama and Inokuchi, 1967; Murray and Pottie, 1974; De Kruif, 1980; Mackay et al., 1982; Bender et al., 1983; Sonnefeld et al., 1983; Hansen and Eckert, 1986; Sato et al., 1986; Hinckley et al., 1990; Nass et al., 1995; Oja and Suuberg, 1997; Ruzicka et al., 1998; Chickos and Acree, 1999; Shiu and Ma, 2000; Burks and Harmon, 2001; Lei et al., 2002; Mackay et al., 2006; Odabasi et al., 2006; Goldfarb and Suuberg, 2008b, a, c; Ma et al., 2010). However, PAHs and halogenated polycyclic aromatic hydrocarbons (HPAHs) often exist as solid and/or liquid mixtures. Therefore it is also important to understand the phase behavior and crystal structures of these PAH and HPAH mixtures.

Phase behavior involving solid-liquid equilibrium is the basis for crystallization in chemical and materials engineering. Binary mixture systems can have up to three degrees of freedom according to the Gibbs phase rule,

$$F=C-P+2 \tag{1}$$

where F is the degrees of freedom, C is the number of components, and P is the number of phases. Therefore, the equilibrium of binary systems is determined by three variables such as temperature, pressure, and composition, and this is of course increased by one compositional variable for each additional component.

More than half of the true binary organic mixture systems in the literature exhibit simple eutectic behavior (Matsuoka, 1991) (see Figure 1(A)), while about 10% of binary solid systems form solid solutions (Matsuoka, 1991) (see Figure 1(B)), in which the atoms or molecules of one of the components occupy sites in the crystal lattice of the other component without modifying its crystal structure. Additionally, about a quarter of these systems form intermolecular compounds (Matsuoka, 1991), such as monotectics (see Figure 1(C)). However, only limited research has been done on binary organic mixture systems, especially PAH binary mixture systems. Moreover, crystal morphology, i.e., polymorphs, racemates, and structural isomers, also affect the phase diagram and may induce non-ideal solid-liquid equilibrium.

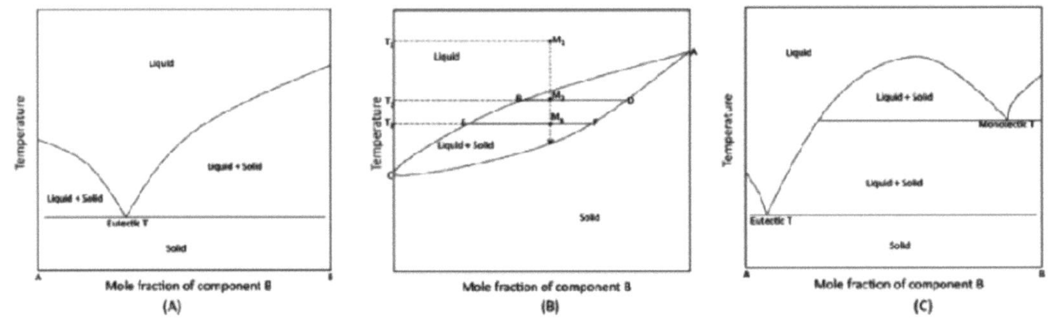

Fig. 1. Phase diagram of eutectic (A), solid solution (B), and monotectic (C) systems.

2. Eutectic systems

Figure 1(A) shows a phase diagram of a typical eutectic mixture system, which has a minimum melting temperature, i.e. a eutectic point. The eutectic point of a binary condensed mixture is defined as the temperature at which a solid mixture phase is in equilibrium with the liquid phase and a eutectic is generally considered to be a simple mechanical mixture of the solid and liquid (Rastogi and Bassi, 1964).

As in many other binary alloy mixtures, most PAH binary mixture systems exhibit eutectic behavior. Table 1 lists the eutectic point and eutectic concentration for about 50 binary PAH-containing mixture systems, in which at least one of the components is a PAH. The shape of the phase diagram for most of these binary mixture systems is similar to the phase diagram of anthrancene + pyrene mixture system (see Figure 2), except for a few systems, whose eutectic concentration is quite close to a pure component, such as in the naphthalene + chrysene system and phenanthrene + chrysene system.

For the studies preformed by this group on the anthracene + pyrene system (Rice et al., 2010), mixtures were prepared at various compositions by using a melt and quench-cool technique (Fu et al., 2010). Generally, the melting points and enthalpies of fusion of these PAH binary mixtures were found to often actually be independent of mixture preparation techniques. The liquidus and thaw points were determined according to the method proposed by Pounder an Masson (Pounder and Masson, 1934). The thaw temperature is the temperature at which the first droplet of liquid appears in a mixture-containing capillary. The liquidus temperature is the maximum temperature at which both solid crystals and liquid are observed to coexist. Above this temperature, there is only liquid phase present.

System	T_{fus1}/K	T_{fus2}/K	x_1	T_E/K
Naphthalene(1) + α-Naphthylamine(2) (Rastogi and Rama Varma, 1956)	353.5	323.2	0.360	301.3
Naphthalene(1) + α-Naphthol(2) (Rastogi and Rama Varma, 1956)	353.5	368.2	0.487	327.7
Naphthalene(1) + Phenanthrene(2) (Rastogi and Rama Varma, 1956; Rastogi and Bassi, 1964)	353.5	373.2	0.558	321.3
Naphthalene(1) + 2-methylnaphthalene(2) (Szczepanik et al., 1963)	353.5	307.6	0.362	298.7
Naphthalene(1) + Thionaphthene(2) (Szczepanik et al., 1963; Szczepanik and Ryszard, 1963)	353.5	305.2	0.063	302.4
Naphthalene(1) + Biphenyl(2) (Szczepanik et al., 1963; Szczepanik and Ryszard, 1963)	353.5	343.7	0.442	312.4
Naphthalene(1) + 2,6-dimethylnaphthalene(2)(Szczepanik and Ryszard, 1963)	353.5	383.2	0.665	333.7
Naphthalene(1) + 2,3-dimethylnaphthalene(2)(Szczepanik and Ryszard, 1963)	353.5	377.2	0.666	327.4
Naphthalene(1) + Acenaphthene(2) (Szczepanik et al., 1963; Szczepanik and Ryszard, 1963)	353.5	368.5	0.564	324.6
Naphthalene(1) + Fluorene(2) (Szczepanik et al., 1963; Szczepanik and Ryszard, 1963)	353.5	388.2	0.613	330.2
Naphthalene(1) + Phenanthrene (2) (Szczepanik et al., 1963; Szczepanik and Ryszard, 1963)	353.5	373.2	0.552	323.2
Naphthalene(1) + Fluoranthene (2) (Szczepanik et al., 1963; Szczepanik and Ryszard, 1963)	353.5	383.2	0.612	331
Naphthalene(1) + Pyrene (2) (Szczepanik et al., 1963; Szczepanik and Ryszard, 1963)	353.5	423.2	0.746	339.2
Naphthalene(1) + Chrysene(2) (Szczepanik et al., 1963; Szczepanik and Ryszard, 1963)	353.5	528.2	0.971	351.4
Biphenyl(1) + Fluorene(2) (Szczepanik et al., 1963; Szczepanik and Ryszard, 1963)	343.7	388.2	0.909	340.8
Biphenyl(1) + Acenaphthene(2) (Szczepanik et al., 1963; Szczepanik and Ryszard, 1963)	343.7	368.5	0.641	319.3

Diphenylene oxide(1) + Acenaphthene(2) (Szczepanik et al., 1963; Szczepanik and Ryszard, 1963)	359.2	368.5	0.578	326
Fluorene(1) + Acenaphthene(2) (Szczepanik et al., 1963; Szczepanik and Ryszard, 1963)	388.2	368.5	0.431	338.6
Fluorene(1) + 2,3,6-trimethylnaphthalene(2) (Szczepanik et al., 1963; Szczepanik and Ryszard, 1963)	388.2	375.2	0.658	361.6
Phenanthrene(1) + Biphenyl(2) (Szczepanik et al., 1963; Szczepanik and Ryszard, 1963)	373.2	343.7	0.691	324.8
Phenanthrene(1) + Acenaphthene(2) (Szczepanik et al., 1963; Szczepanik and Ryszard, 1963)	373.2	368.5	0.495	327.5
Phenanthrene(1) + 2,3,6-trimethylnaphthalene(2) (Szczepanik et al., 1963; Szczepanik and Ryszard, 1963)	373.2	375.2	0.704	331.5
Phenanthrene(1) + Fluorene(2) (Szczepanik et al., 1963; Szczepanik and Ryszard, 1963)	373.2	388.2	0.637	368.7
Phenanthrene(1) + Pyrene(2) (Szczepanik et al., 1963; Szczepanik and Ryszard, 1963)	373.2	423.2	0.747	354.7
Phenanthrene(1) + 3-methylphenanthrene(2) (Szczepanik et al., 1963; Szczepanik and Ryszard, 1963)	373.2	332	0.318	309.7
Phenanthrene(1) + 4,5-dimethylphenanthrene(2) (Szczepanik et al., 1963; Szczepanik and Ryszard, 1963)	373.2	388.2	0.621	342.6
Phenanthrene(1) + Fluoranthene (2) (Szczepanik et al., 1963; Szczepanik and Ryszard, 1963)	373.2	383.2	0.532	347.7
Phenanthrene(1) + Chrysene(2) (Szczepanik et al., 1963; Szczepanik and Ryszard, 1963)	373.2	528.2	0.957	369.2
Anthracene(1) + Carbazole(2) (Szczepanik et al., 1963; Szczepanik and Ryszard, 1963)	489.8	518	0.943	488.4
Anthracene(1) + 2-methylanthracene(2) (Szczepanik et al., 1963; Szczepanik and Ryszard, 1963)	489.8	472.2	0.108	471.5
Anthracene(1) + Chrysene(2) (Szczepanik et al., 1963; Szczepanik and Ryszard, 1963)	489.8	528.2	0.662	464.6

Anthracene(1) + Pyrene(2) (Szczepanik et al., 1963; Szczepanik and Ryszard, 1963)	489.8	423.2	0.221	404.6
Carbazole(1) + Fluoranthene(2) (Szczepanik et al., 1963; Szczepanik and Ryszard, 1963)	518	383.2	0.119	377.3
Carbazole(1) + Pyrene(2) (Szczepanik et al., 1963; Szczepanik and Ryszard, 1963)	518	423.2	0.154	409.1
Carbazole(1) + Chrysene(2) (Szczepanik et al., 1963; Szczepanik and Ryszard, 1963)	518	528.2	0.578	480.6
Fluoranthene(1) + Acenaphthene(2) (Szczepanik et al., 1963; Szczepanik and Ryszard, 1963)	383.2	368.5	0.433	336.9
Fluoranthene(1) + Fluorene(2) (Szczepanik et al., 1963; Szczepanik and Ryszard, 1963)	383.2	388.2	0.516	342.7
Fluoranthene(1) + 2-methylanthracene (2) (Szczepanik et al., 1963; Szczepanik and Ryszard, 1963)	383.2	472.2	0.794	368.7
Fluoranthene(1) + Pyrene(2) (Szczepanik et al., 1963; Szczepanik and Ryszard, 1963)	383.2	423.2	0.800	368.3
Fluoranthene(1) + Chrysene(2) (Szczepanik et al., 1963; Szczepanik and Ryszard, 1963)	383.2	528.2	0.952	379.6
Pyrene(1) + Chrysene(2) (Szczepanik et al., 1963; Szczepanik and Ryszard, 1963)	423.2	528.2	0.855	405.7
Acenaphthene (1) + 1,2-dimethylbenzene(2) (Szczepanik et al., 1963; Szczepanik and Ryszard, 1963)	368.5	298.7	0.055	295
Acenaphthene (1) + 1,2,4,5-tetramethylbenzene(2) (Szczepanik et al., 1963; Szczepanik and Ryszard, 1963)	368.5	352.3	0.423	323.6
Acenaphthene (1) + 2-methylnaphthalene(2) (Szczepanik et al., 1963; Szczepanik and Ryszard, 1963)	368.5	307.5	0.212	290.9
Acenaphthene (1) + 2,6-dimethylnaphthalene (2) (Szczepanik et al., 1963; Szczepanik and Ryszard, 1963)	368.5	383.2	0.598	339.7
Acenaphthene (1) + 2,7-dimethylnaphthalene (2) (Szczepanik et al., 1963; Szczepanik and Ryszard, 1963)	368.5	370.2	0.531	333.9

Acenaphthene (1) + Naphthalene (2) (Szczepanik et al., 1963; Szczepanik and Ryszard, 1963)	368.5	353.2	0.417	323.2
Acenaphthene (1) + Phenanthrene(2) (Szczepanik et al., 1963; Szczepanik and Ryszard, 1963)	368.5	372.5	0.492	329
Acenaphthene (1) + Fluorene(2) (Szczepanik et al., 1963; Szczepanik and Ryszard, 1963)	368.5	387.2	0.582	337.7
Acenaphthene (1) + Anthracene(2) (Szczepanik et al., 1963; Szczepanik and Ryszard, 1963)	368.5	489.7	0.914	361.2

Table 1. Melting temperatures of previously reported binary PAH eutectic systems

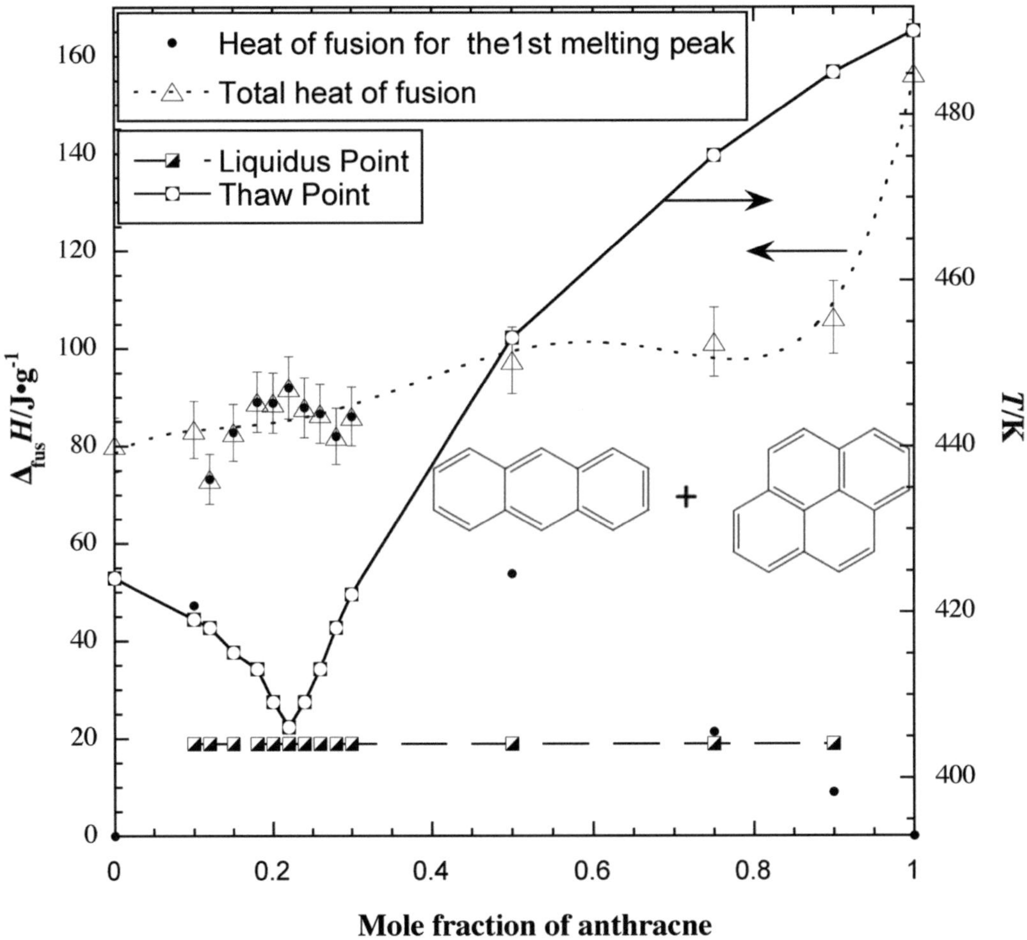

Fig. 2. Phase diagram and enthalpy of fusion of the anthracene (1) + pyrene (2) system (Rice et al., 2010).

The eutectic point for the anthracene (1) + pyrene (2) system occurs at 404 K at $x_1 = 0.22$ (see Figure 2). Only solid state exists below the thaw curve, i.e. eutectic temperature, and only liquid state exists above the liquidus curve. The areas between these two curves exhibit the coexistence of both solid and liquid phases.

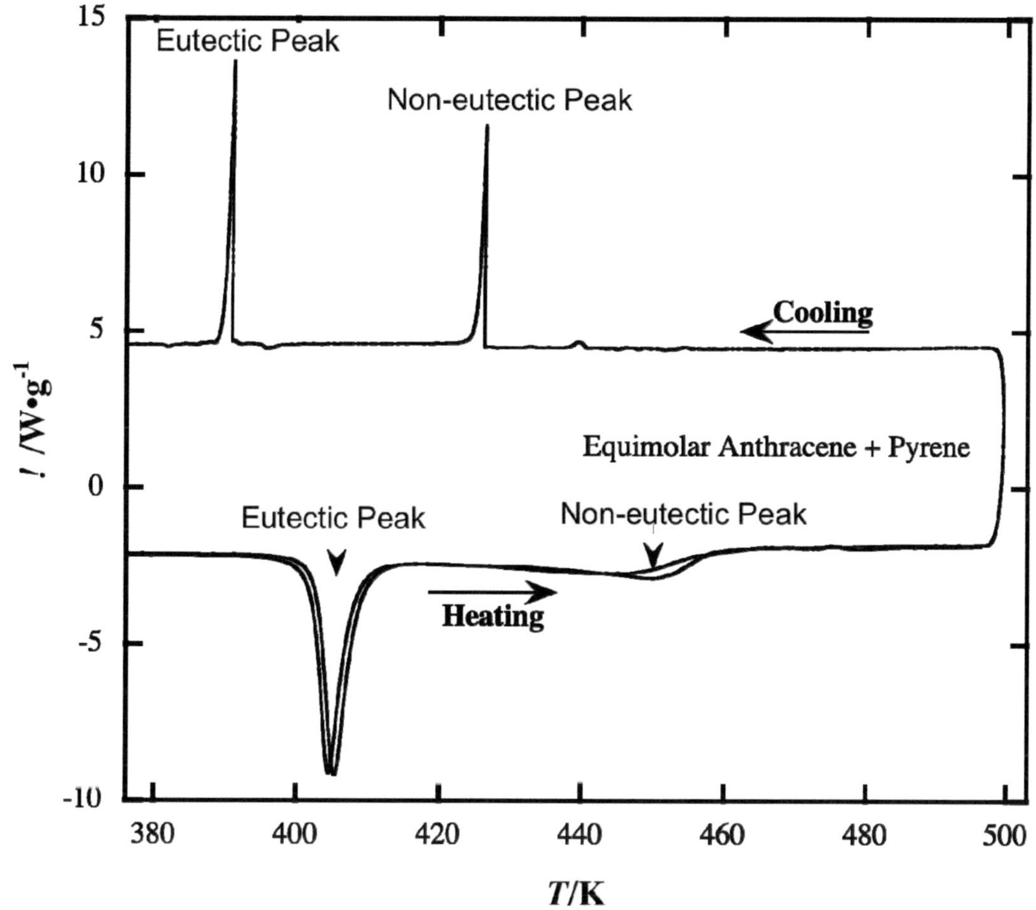

Fig. 3. Full DSC scan of an equimolar anthracene (1) + pyrene (2) mixture (Rice et al., 2010).

Figure 2 also displays the correlation between phase behavior and enthalpy of fusion, $\Delta_{fus}H$ for the system. The $\Delta_{fus}H$ observed for a DSC peak near the eutectic temperature of 404 K indicates the heat input for the initial melting of a eutectic solid phase to occur. The total $\Delta_{fus}H$ shown in Figure 2 is a summation of both endothermic phase transition peaks observed in the DSC scan, i.e. the eutectic phase melting and the non-eutectic phase melting (see Figure 3). It is worth noting that the total $\Delta_{fus}H$ is very similar to that of pure pyrene over a wide range of compositions and thus the $\Delta_{fus}H$ for both pure pyrene and the eutectic mixture are very similar. This means that when the mixture contains only a modest amount of anthracene, energetically it behaves quite similarly to pure pyrene, and this persists until the mixture is nearly pure anthracene (see Figure 2). There is a slight increase in fusion enthalpy when the mixtures are enriched in anthracene beyond the eutectic composition, but the shift is only modest as compared with the increase of fusion enthalpy to that of pure anthracene (see Figure 2). This indicates that the ability of anthracene to reach a lower energy crystalline configuration is significantly impeded by the presence of relatively small amounts of pyrene.

Additionally, Powder X-ray diffraction patterns for the same anthracene (1) + pyrene (2) system were also obtained. Figure 4 shows that the crystal structure of the eutectic mixture is similar to that of pyrene because peaks at 10.6, 11.6, 14.9, 16.3, 18.2, 23.3, 24.7 and 28.0 degree are all retained in the mixture diffraction pattern. This is consistent with the DSC result that implies that the $\Delta_{fus}H$ of the eutectic is very close to that of pure pyrene, and indicates that the crystal structures of the eutectic mixture and pure pyrene are similar. Likewise, Figure 4 shows that the crystal structure of a mixture at $x_1 = 0.90$ is comparable to that of pure anthracene.

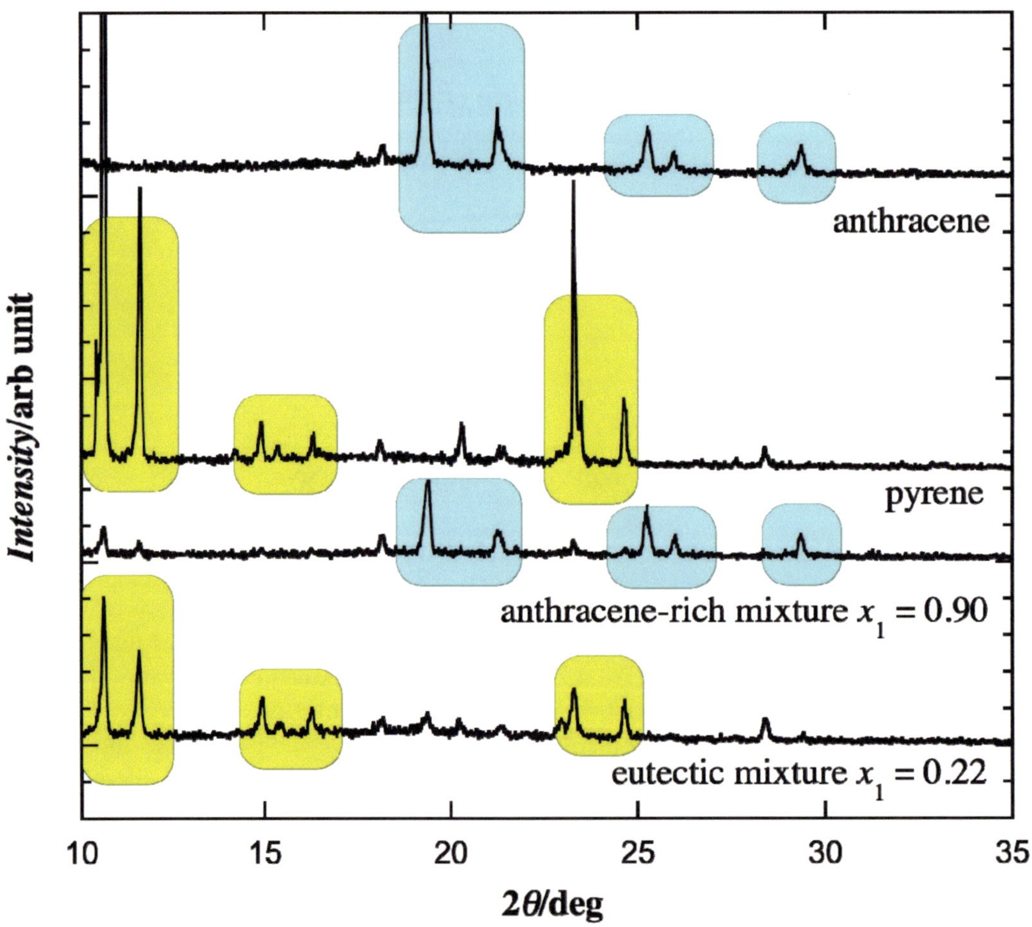

Fig. 4. X-ray diffraction patters of pure components and mixtures of anthracene (1) + pyrene (2) (Rice et al., 2010).

3. Monotectic systems

In contrast to eutectic systems, in which both components solidify below eutectic temperature, a monotectic reaction is characterized by the breakdown of a liquid into one solid and one liquid phase (Singh et al., 1985), i.e. one liquid phase decomposes into a solid phase and a liquid phase when the temperature is below the monotectic temperature. Figure 1(C) shows the phase diagram of a typical monotectic system. The monotectic composition is determined by the intersection of a liquidus line and a liquid miscibility gap (Singh et al., 1985). Generally, monotectic systems are less studied than eutectic systems.

Binary organic mixtures with PAHs can form monotectic systems. Table 2 lists the monotectic and eutectic point of a few monotectic forming PAH systems. Monotectic systems are characterized by monotectic, eutectic and upper consolute temperatures, though the upper consolute temperature is often not reported. The monotectic temperature, t_M, is the temperature at monotectic composition and the upper consolute temperature is the highest melting temperature of the mixture system, i.e. the critical point where the two liquid phases having identical composition become indistinguishable.

System	T_{fus1}/K	T_{fus2}/K	x_M	T_M/K	x_E	T_E/K
2,4-Dinitrophenol(1) + Naphthalene(2) (Singh et al., 2001; Singh et al., 2007)	378.2	353.2	0.316	357.7	0.838	344.2
Succinonitrile (1) + Pyrene(2) (Rai and Pandey, 2002)	330.2	423.2	0.025	416.5	0.744	328.5
Succinonitrile (1) + Phenanthrene (2) (Singh et al., 1985)	330.2	373.2	0.225	363.2[a]	0.975	~328.2
p-benzoquinone(1) + Pyrene(2) (Gupta and Singh, 2004)	388.2	423.2	0.324	392.2	0.792	376.2
m-dinitrobenzene(1)+ Pyrene(2) (Gupta and Singh, 2004)	362.2	423.2	0.301	363.2	0.702	361.2
m-nitrobenzoic acid(1) +Pyrene(2) (Gupta and Singh, 2004)	413.2	423.2	0.902	413.2	0.299	403.2

Table 2. Melting temperatures of previously reported binary PAH monotectic systems

Rai and Pandey studied the phase behavior of succinonitrile (1) + pyrene (2) mixture system (Rai and Pandey, 2002), which is a typical monotectic system (Figure 5). The enthalpy of fusion of pyrene, 17.65 kJ mole^{-1} (Chickos and Acree, 1999), is much higher than that of succinonitrile, 3.7 kJ mole^{-1} (Rai and Pandey, 2002). The monotectic point is 416.5 K (143.3°C) at x_1=0.025. The eutectic temperature is 328.5 K (55.4°C) at x_1=0.744 and the upper consolute temperature, t_C (465.2 K, 192.0°C), is 48.7 K above the monotectic point. When x_1 is between monotectic and eutectic composition, the two liquids, L_1 (rich in pyrene) and L_2 (rich in succinonitrile) are mutually immiscible. However, if the temperature is above the consolute temperature, there is complete miscibility in liquid state, i.e. only one liquid phase exists.

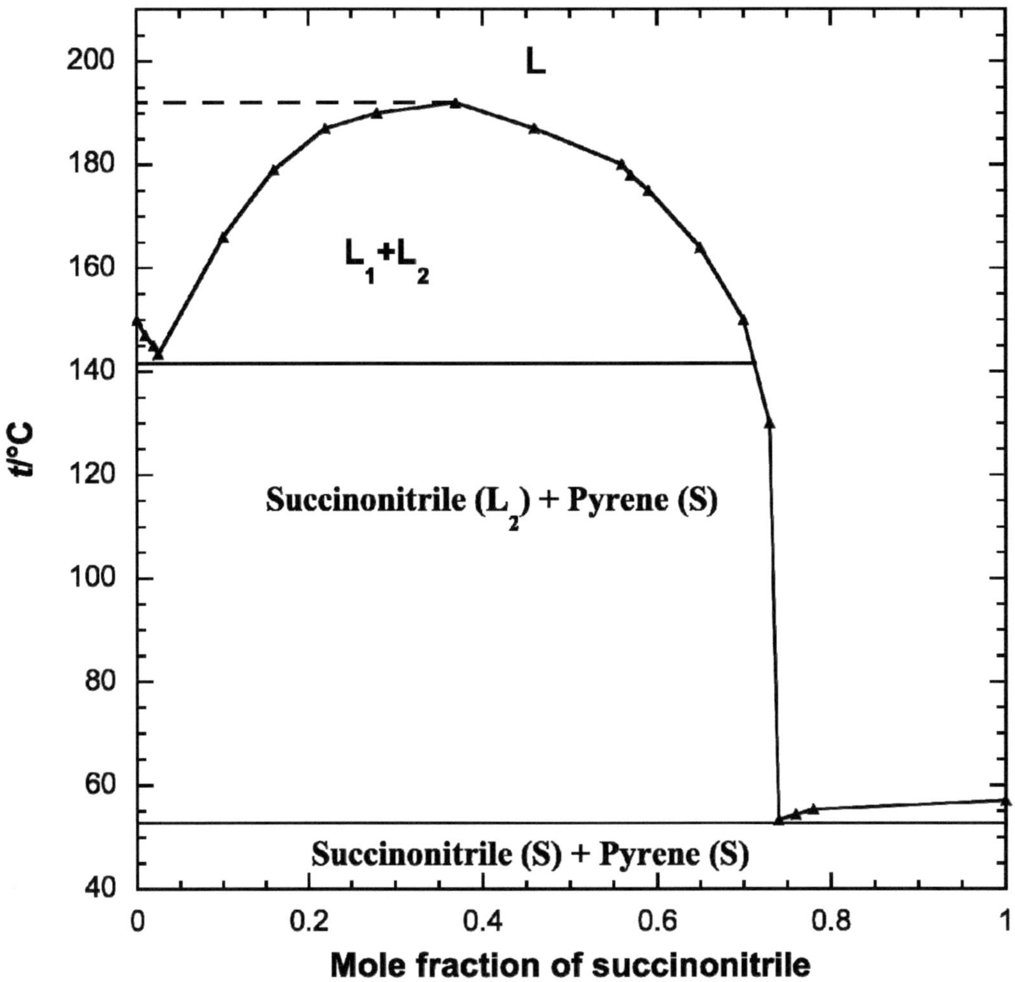

Fig. 5. Phase diagram of succinonitrile (1) + pyrene (2) mixture system (Rai and Pandey, 2002).

4. Solid solution

A solid solution is a solid mixture in which one or more atoms and/or molecules of one of the components occupies sites in the crystal lattice of the other component without significantly changing its crystal structure, even though the lattice parameter may vary. So this kind of system has a homogenous crystalline structure and is also called isomorphic system, because the components are completely miscible in both the liquid and solid phases. Figure 1(B) shows the phase behavior of a binary mixture system that forms a solid solution. In the diagram, the curve ABC and ADC are the liquidus and solidus curves, respectively. The area above ABC curve represents the region of homogeneous liquid solutions and the area below ADC curve represents the region of homogeneous solid solution. The area enclosed by ABCD is the region of liquid + solid solution. For instance, a mixture M_1 at temperature $T_1$is cooled to temperature T_2, the mixture M_2 becomes a mixture of liquid B and solid D. If M_2 is further cooled to temperature T_3, the liquid composition changes continuously from B to E along the liquidus curve, while the solid composition changes from D to F along the solidus curve. Additionally, the Hume-Rothery rules, named after William Hume-Rothery, are used to describe the conditions under which an element can dissolve in a metal and form a solid solution.

Szczepanik and Skalmowski (Szczepanik et al., 1963; Szczepanik and Ryszard, 1963) studied the phase behavior of over 60 PAH binary mixture systems, and demonstrate that PAH mixture systems also form solid solution, as shown by naphthalene + 1-methynaphthanlene, naphthalene + anthracene, phenanthrene + anthracene, phenanthrene + carbazole, anthracene + acridine, anthracene + fluoranthene, and chrysene + 1,2-benzanthracene systems. It is not known whether the the Hume-Rothery rules still work for PAH mixtures. However, it is worth noting that the number of such systems is small, compared with the number of eutectic-forming systems.

5. Systems with complex phase behavior

Due to the large molecular mass and complexity of the crystal structure of PAHs, i.e. polymorphism and racemate, the phase behavior of some of the PAH binary mixtures may be different from the above described three phase behaviors. Three complicated PAH binary mixture systems, i.e. anthracene + benzo[a]pyrene system (Rice and Suuberg, 2010), pyrene + 9,10-dibromoanthracene system (Fu et al., 2010), and anthracene + 2-bromoanthracene are introduced here.

5.1 Anthracene + benzo[a]pyrene system

Benzo[a]pyrene has a much larger molecular mass compared to pyrene, which leads to phase behavior in the anthracene (1) + benzo[a]pyrene (2) system (Rice and Suuberg, 2010) that is different from that of the anthracene + pyrene system. The phase diagram of anthracene (1) + benzo[a]pyrene (2) system (see Figure 6) indicates an eutectic-like mixture behavior. A eutectic-like phase is formed near x_1 = 0.26 between 414 and 420 K. There is however always a gap between the thaw curve and the lowest liquidus temperature, which is distinct from true eutectic behavior such as in Figure 1(A) or Figure 2. Therefore, mixtures of anthracene and benzo[a]pyrene form a single, amorphous, solid eutectic-like phase at x_1 = 0.26 that lacks any organized crystal structure and which melts throughout the 414 to 420 K temperature range. This region of phase transition, represented by the shaded region of Figure 6, is not rate dependant and is observed in both the DSC and melting temperature analysis for all combinations of anthracene + benzo[a]pyrene, providing evidence that this region represents the melting temperature range of a single, amorphous, solid phase. This conclusion is also supported by the X-ray diffraction results.

Powder X-ray diffraction studies were conducted to study the crystal structures of the anthracene (1) + benzo[*a*]pyrene (2) system (see Figure 7). The eutectic-like mixture lacks any organized crystal structure because the few peaks that exist in the X-ray pattern are not well defined and do not rise much above the baseline. Additionally, there is no real similarity between the eutectic mixture scan and those of the pure components. This result is consistent with the melting point studies that imply that the mixtures form a single, amorphous solid phase at the eutectic composition.

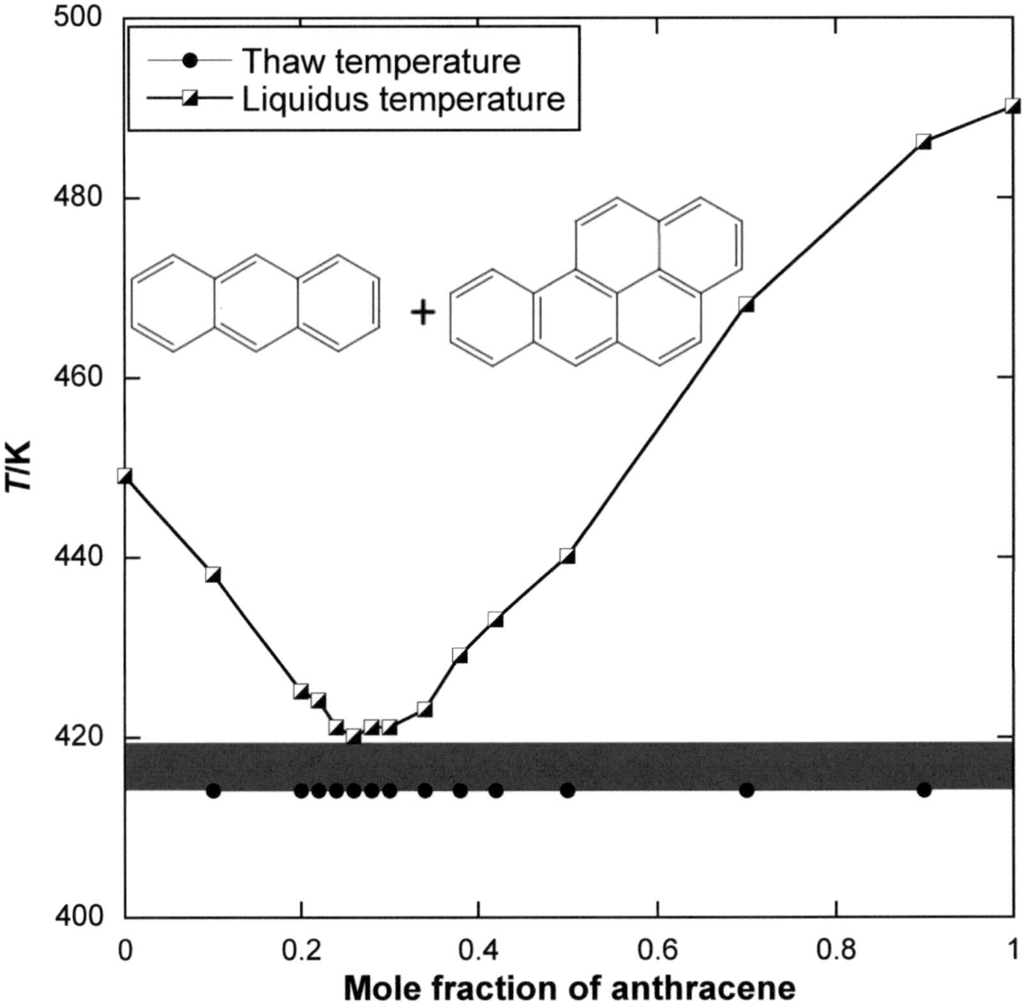

Fig. 6. Phase diagram of anthracene (1) + benzo[a]pyrene (2) system (Rice and Suuberg, 2010).

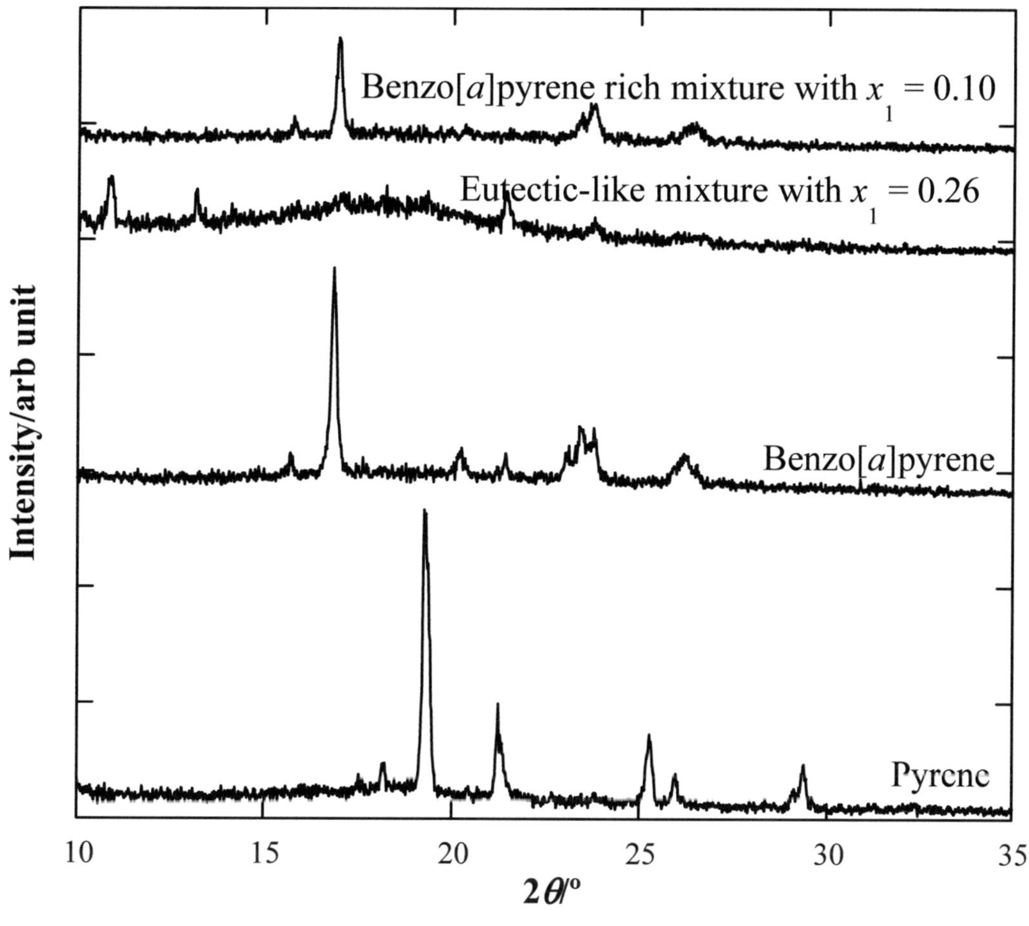

Fig. 7. X-ray diffraction patters of pure components and mixtures of anthracene (1) + benzo[*a*]pyrene (2) system (Rice and Suuberg, 2010).

5.2 Pyrene + 9,10-dibromoanthracene system

The influence of halogen substitution on the interaction energy between PAH molecules has also been investigated. Unlike the anthracene + pyrene mixture system, bromine substitution on anthracene induces a different kind of interaction in the pyrene (1) + 9,10-dibromoanthracene (2) mixture system, which also results in non-idealities in solid-liquid equilibrium (see Figure 8). The surface area and volume of the 9,10-dibromoanthracene molecule is much larger than that of pyrene.

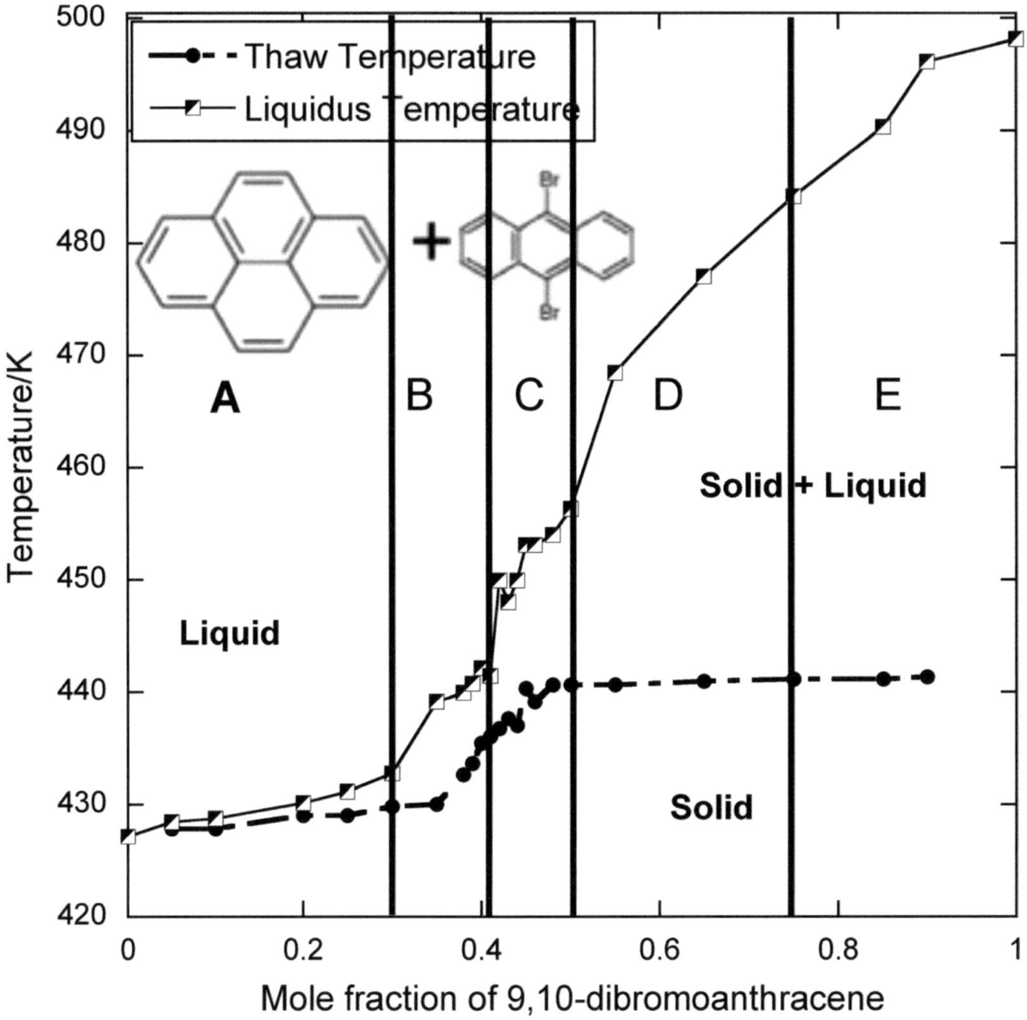

Fig. 8. Phase diagram of pyrene (1) + 9,10-dibromoanthracene (2) mixture system (Fu et al., 2010).

The phase diagram of this system can be crudely divided into 5 regions. The mixtures with relatively low mole fraction of 9,10-dibromoanthracene (< 0.30), in region A, form a pyrene like phase. When the mole fraction of 9,10-dibromoanthracene is between 0.30-0.41, in region B, the mixtures transition from a pyrene-like phase to two phases that both have low melting temperatures. The divergence of the liquidus and thaw curve is 2-9 K. In region C, mixtures containing about x_2 = 0.41-0.50 also show two-phase character and start to transition to 9,10-dibromoanthracene behavior. Mixtures with x_2 = 0.50-0.75, in region D, also have two phases with 9,10-dibromoanthracene like behavior and high melting temperature. Only one of the phases evolves while the other gives a constant low melting temperature (corresponding to the thaw point). In region E, a 9,10-dibromoanthracene like phase is defined based upon the thermal behavior, shown below.

The full heating, cooling and reheating scan of a pyrene + 9,10-dibromoanthracene mixture at x_2 = 0.48 (in region C) is shown in Figure 9, where Φ is heat flow in the DSC. During the heating scan, two peaks appear at 428 K and 440 K, which indicates the two-phase character of the mixture. Two peaks are also observed in the cooling scan, in which the 9,10-dibromoanthracene like phase crystallizes first at 418 K, and then the pyrene like phase crystallizes at 410 K. The cooling scan also suggested two-phase behavior of the mixture just as did the melting behavior. When reheated, the phase transition enthalpies and associated temperatures matched those of the initial heating scan.

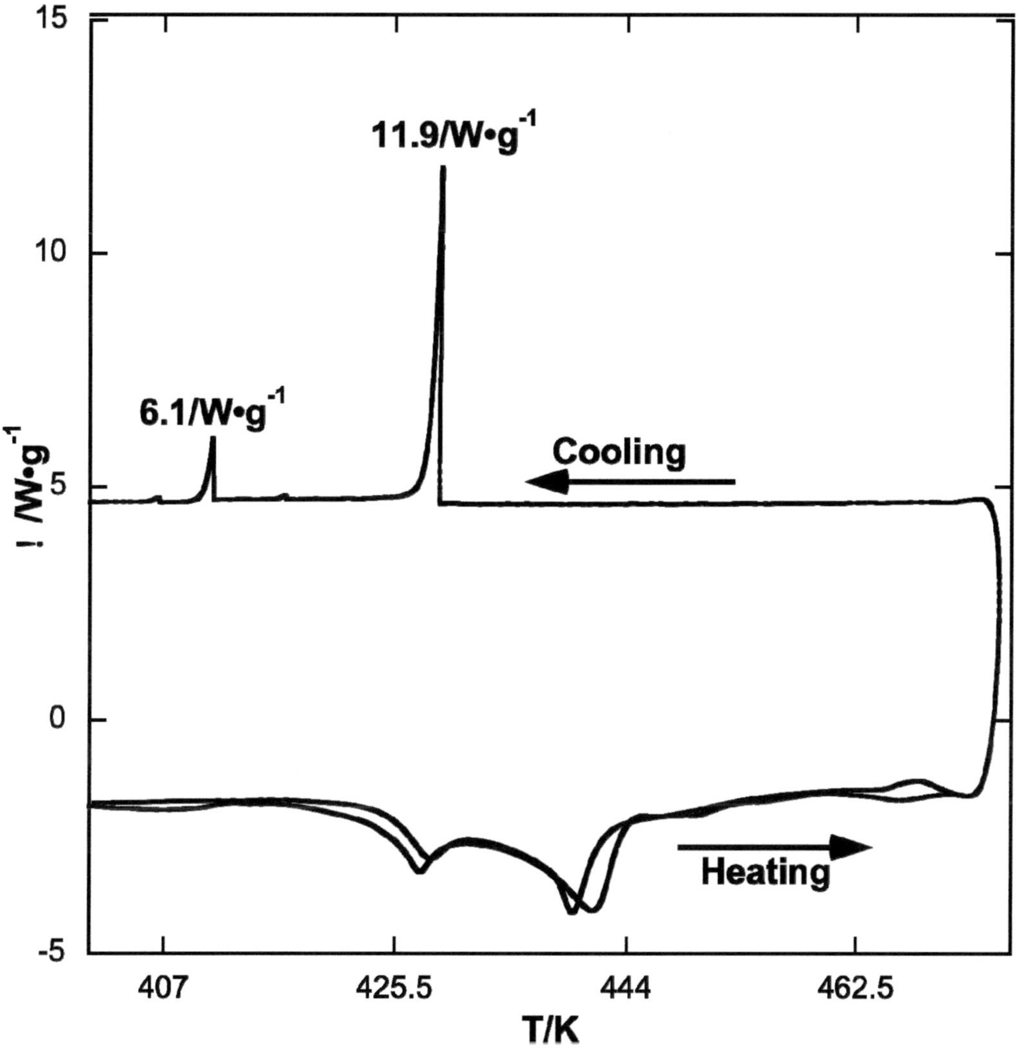

Fig. 9. Full DSC scan of a pyrene (1) + 9,10-dibromoanthracene (2) mixture at x_2 = 0.48 (Fu et al., 2010).

The temperature and enthalpy of crystallization (subcooled), shown in Figure 10, correspond to the results obtained from the phase diagram. Mixtures with a mole fraction of 9,10-dibromoanthraene 0.30-0.75, in regions B, C and D, have two-phase character, which is observed as two distinct phase-transition peaks during the cooling procedure. Region E showed two-phase melting behavior, but in the DSC experiments of Figure 10, the low temperature crystallization peak was absent. Likewise, region B showed two distinct melting peaks, whereas in the DSC experiment only a single peak was observed.

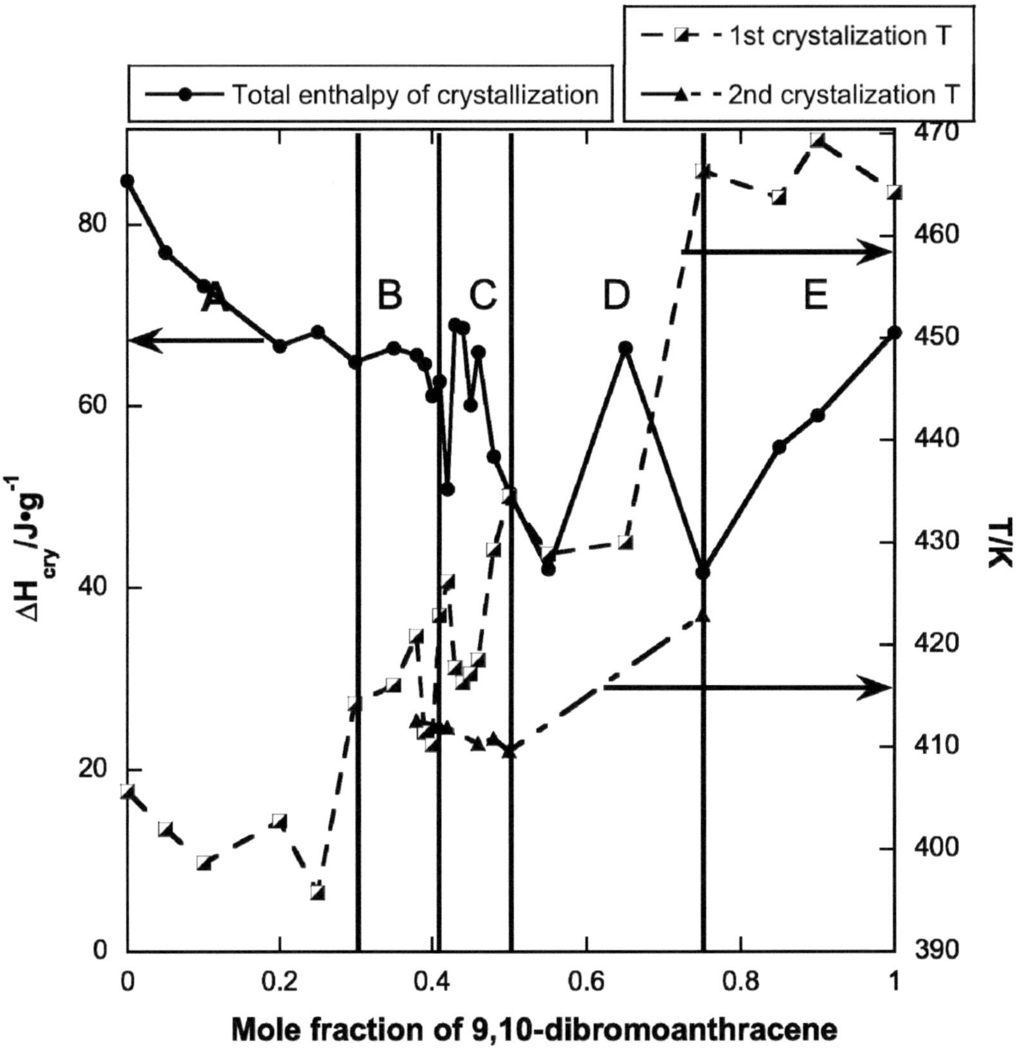

Fig. 10. Crystallization temperature and total enthalpy of crystallization of pyrene (1) + 9,10-dibromoanthracene (2) mixtures. 1st crystallization temperature is the higher temperature peak in the DSC cooling scan, and 2nd crystallization temperature is the lower temperature peak in the DSC cooling scan (Fu et al., 2010).

Since the enthalpies of crystallization of the mixtures with 9,10-dibromoanthracene mole fractions of 0.55 and 0.75 are significantly lower than that of other mixtures, these are at a higher energy state and are less stable than other mixtures with nearby compositions. Moreover, the mixture with 0.65 mole fraction of 9,10-dibromoanthracene is in a more stable state than those mixtures with 0.55 and 0.75 mole fraction of 9,10-dibromoanthracene meaning that around a 2:1 molar ratio of 9,10-dibromoanthracene to pyrene, there exists a particular lower energy configuration.

The powder X-ray diffraction method was used to study the crystal structures of the pyrene and 9,10-dibromoanthracene mixtures (see Figure 11). The results are qualitative. For the 9,10-dibromoanthracene rich mixture at the region D-E boundary, at $x_2 = 0.75$ (curve E), the XRD data show a 9,10-dibromoanthracene like microstructure though there are distinct differences from 9,10-dibromoanthracene. The pyrene rich mixture in region A at $x_2 = 0.25$ (curve A) has the pyrene like microstructure. However, the mixture at $x_2 = 0.65$ (curve D) reflects neither pyrene nor 9,10-dibromoanthracene like microstructure, and in fact is amorphous.

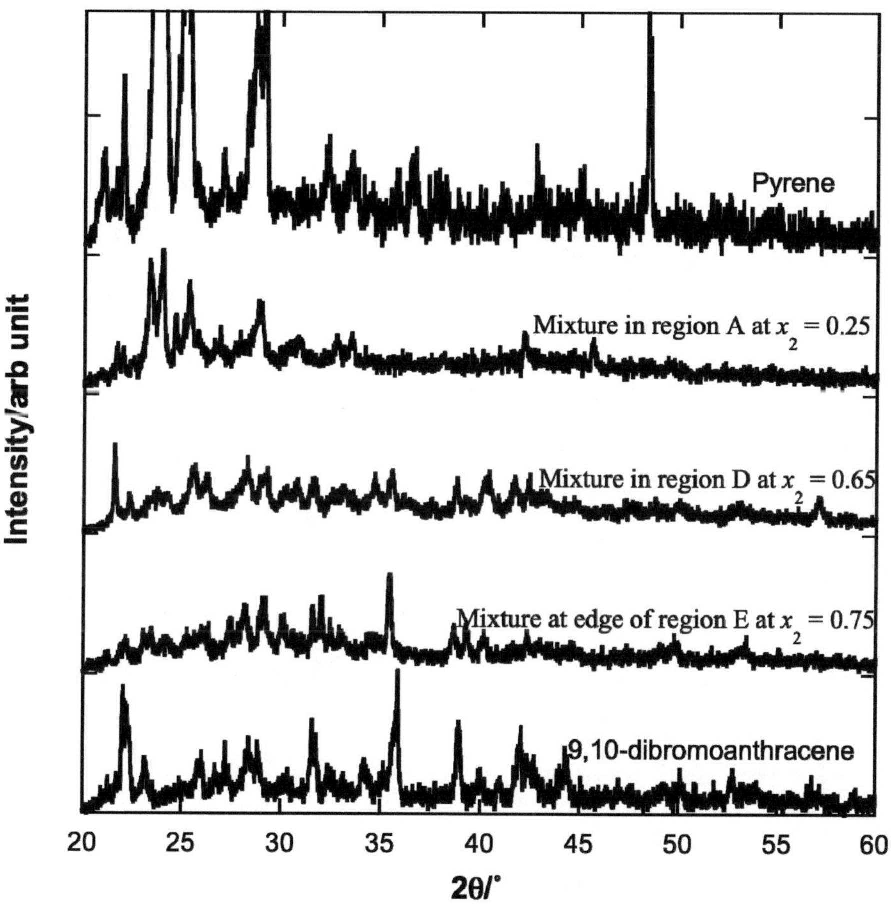

Fig. 11. X-ray diffraction patters of pure components and mixtures of pyrene (1) + 9,10-dibromoanthracene(2) (Fu et al., 2010).

5.3 Anthracene + 2-bromoanthracene system

The influence of bromine substitution on thermochemical properties of PAH mixture systems was further investigated by studying the anthracene (1) + 2-bromoanthracene (2) system. The crystal structure is changed by addition one bromine atom on the aromatic ring. Moreover, the surface area and volume of 2-bromoanthracene is about 10% bigger than those of anthracene.

The solid-liquid equilibrium diagram of anthracene (1) + 2-bromoanthracene (2) system is shown in Figure 12. The diagram suggests the non-ideality of the anthracene + 2-bromoanthracene system. The melting temperature range (thaw to completion) of these mixtures at any given composition is observed to be 1.1 - 2.6 K. The reported solid-liquid equilibrium melting temperature is here taken as the thaw temperature, in Figure 12. The lowest solid-liquid equilibrium temperature for the system is 477.65 K at x_1 = 0.74, and the melting temperature range of this mixture is 1.8 K.

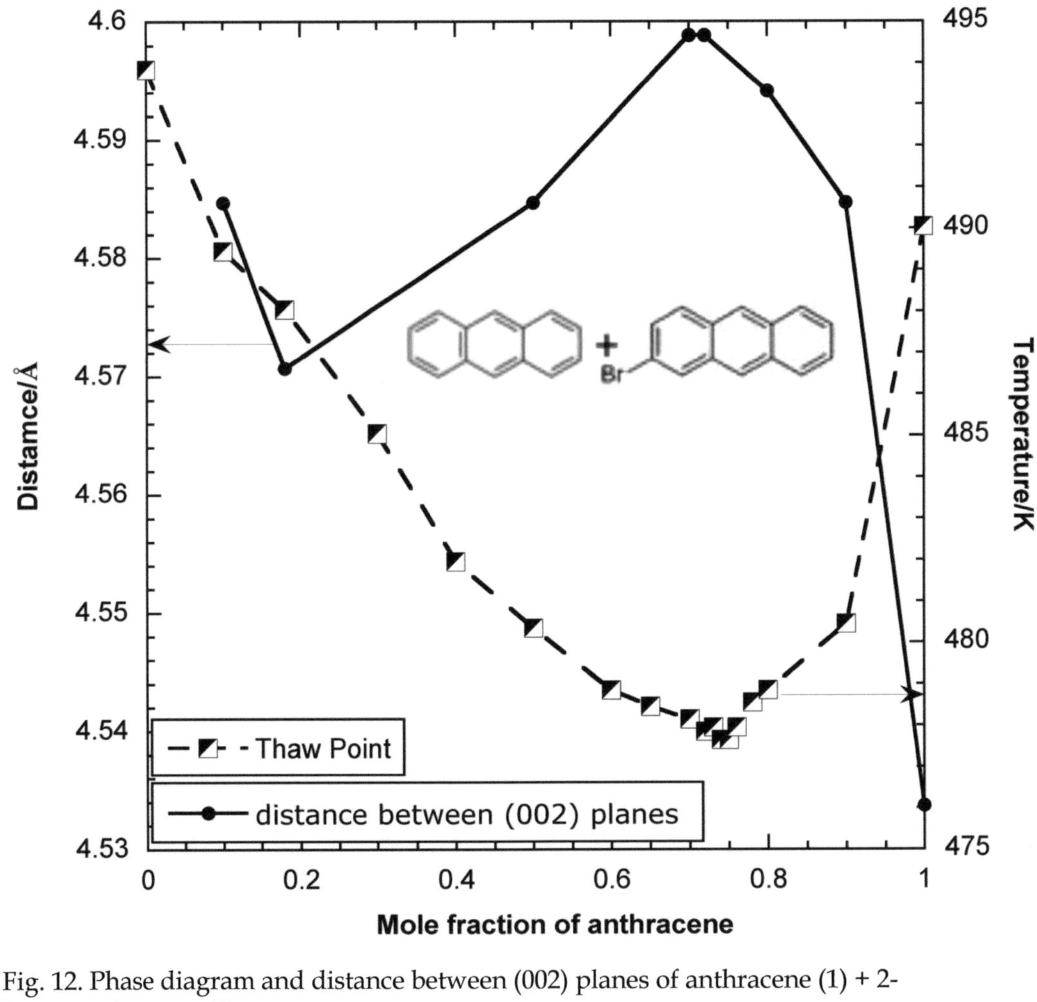

Fig. 12. Phase diagram and distance between (002) planes of anthracene (1) + 2-bromoanthracene (2) system.

The powder X-ray diffraction method was also used to study the crystal structures of pure anthracene, 2-bromoanthracene and their mixtures (see Figure 13). The lattice structure of anthracene crystals is monoclinic with a = 8.44 Å, b = 5.99 Å, c = 11.11 Å, β = 125.4° (Jo et al., 2006). The strong diffraction peak at 19.58° in pure anthracene corresponds to the (002) plane, and the spacing between the 002 planes is 4.53 Å. With the increase of the mole fraction of 2-bromoanthracene, x_2, in the mixture, the (002) plane spacing starts to shift to lower values. Moreover, a new diffraction peak occurs near 2θ = 17° with increasing x_2 in the mixture. This indicates that new mixture crystals are formed. The new peak appears at 2θ = 16.38° when x_1 = 0.70 roughly corresponding to the lowest solid-liquid equilibrium melting point. With increase of x_1, the peak position increases from 16.38° to 17.06° and disappears in pure anthracene. The diffraction data for mixtures with x_1 = 0.50 and 0.10 indicate relatively amorphous structures.

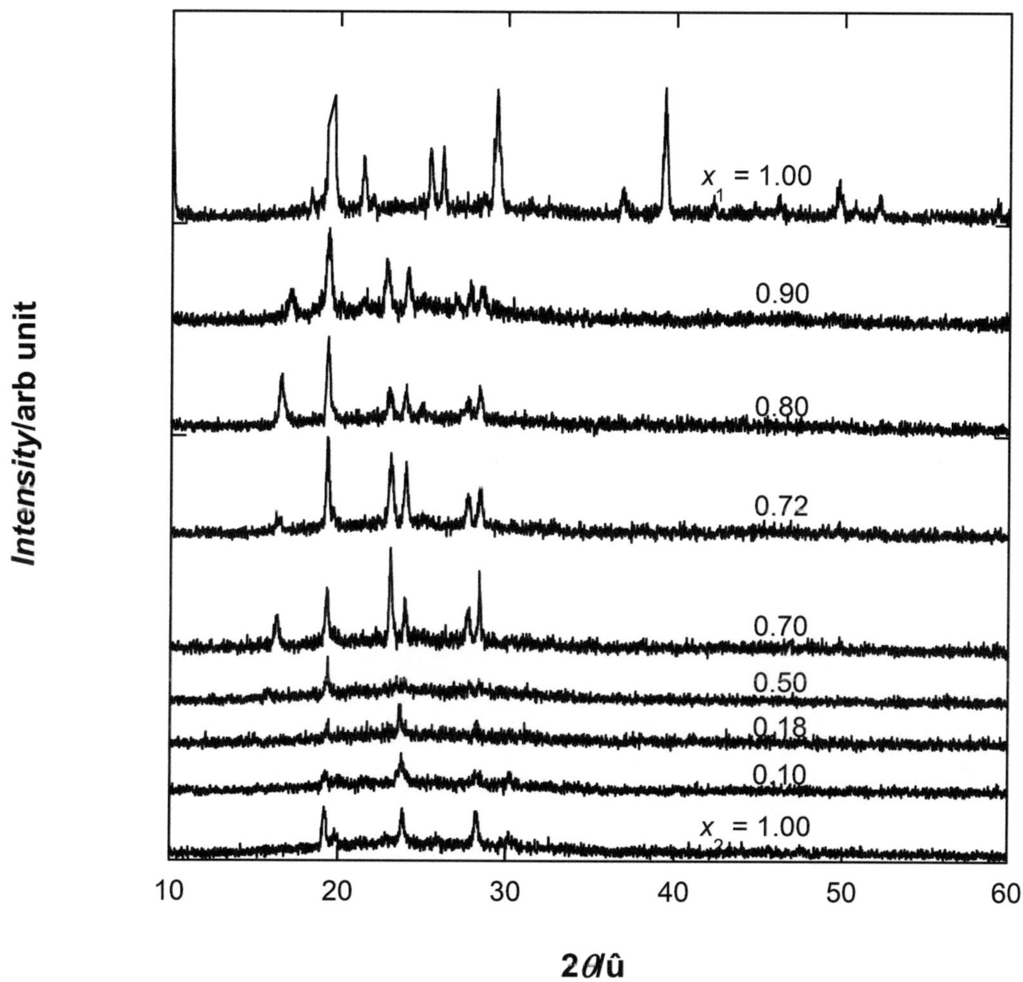

Fig. 13. X-ray diffraction patters of pure components and mixtures of anthracene (1) + 2-bromoanthracene(2).

The distance between 002 planes in the pure anthracene, pure 2-bromoanthracene and mixtures can be calculated by Bragg's law

$$n\lambda = 2d\sin\theta \qquad (2)$$

where n is an integer, λ is the wavelength of the incident wave, d is the spacing between the planes in the atomic lattice, and θ is the angle between the incident ray and the scattering planes.

Figure 12 also shows changes of the distance between 002 planes in this system, which demonstrates that the spacings between 002 planes are stretched by adding 2-bromoanthracene into anthracene. The distance between 002 planes reaches a maximum when the mixture is near the lowest melting solid-liquid equilibrium point, which is in good agreement with the thermodynamic data in Figure 12, indicating the formation of the least stable solid state near the lowest solid-liquid equilibrium point. Interestingly, the mixture at $x_1 = 0.18$ gives a local minimum in the (002) plane spacing.

6. Conclusions

The phase behaviors of binary PAH-containing mixtures are complicated. Most of these mixture systems are eutectic systems, which have a behavior like the anthracene + pyrene system. Fewer binary PAH-containing mixtures can form monotectic and solid solution systems, such as succinonitrile + pyrene system and phenanthrene + anthracene system.

The phase behaviors of binary PAC mixtures are complicated and non-ideal. Mixtures with large PAHs, such as benzo[*a*]pyrene, can exhibit a gap between the thaw curve and liquidus curve. Halogen substitution (bromine substitution) also has significant effect on the thermochemical behaviors of binary PAC mixtures. Bromine substitution on anthracene results in non-ideal phase behavior in pyrene + 9,10-dibromoanthracene and anthracene + 2-bromoanthracene systems.

7. Acknowledgement

This project was supported by Grant Number P42 ES013660 from the National Institute of Environmental Health Sciences (NIEHS)/NIH, and the contents are solely the responsibility of the authors and do not necessarily represent the official views of the NIEHS/NIH.

8. References

Bender, R., Bieling, V., Maurer, G., 1983. The vapour pressures of solids: anthracene, hydroquinone, and resorcinol. The Journal of Chemical Thermodynamics 15, 585-594.

Burks, G.A., Harmon, T.C., 2001. Volatilization of solid-phase polycyclic aromatic hydrocarbons from model mixtures and lampblack-contaminated soils. J Chem Eng Data 46, 944-949.

Chickos, J.S., Acree, W.E., 1999. Estimating solid-liquid phase change enthalpies and entropies. J Phys Chem Ref Data 28, 1535-1673.

De Kruif, C., 1980. Enthalpies of sublimation and vapor-pressure of 11 polycyclic-hydrocarbons. J Chem Thermodyn 12, 243-248.

Fu, J., Rice, J.W., Suuberg, E.M., 2010. Phase behavior and vapor pressures of the pyrene + 9,10-dibromoanthracene system. Fluid Phase Equilibr 298, 219-224.

Goldfarb, J.L., Suuberg, E.M., 2008a. The effect of halogen hetero-atoms on the vapor pressures and thermodynamics of polycyclic aromatic compounds measured via the Knudsen effusion technique. J Chem Thermodyn 40, 460-466.

Goldfarb, J.L., Suuberg, E.M., 2008b. Vapor pressures and enthalpies of sublimation of ten polycyclic aromatic hydrocarbons determined via the Knudsen effusion method. J Chem Eng Data 53, 670-676.

Goldfarb, J.L., Suuberg, E.M., 2008c. Vapor pressures and thermodynamics of oxygen-containing polycyclic aromatic hydrocarbons measured using Knudsen effusion. Environ Toxicol Chem 27, 1244-1249.

Gupta, R.K., Singh, R.A., 2004. Thermochemical and microstructural studies on binary organic eutectics and complexes. Journal of Crystal Growth 267, 340-347.

Haglund, P., Alsberg, T., Bergman, A., Jansson, B., 1987. Analysis of Halogenated Polycyclic Aromatic-Hydrocarbons in Urban Air, Snow and Automobile Exhaust. Chemosphere 16, 2441-2450.

Hansen, P.C., Eckert, C.A., 1986. An Improved Transpiration Method for the Measurement of Very Low Vapor-Pressures. J Chem Eng Data 31, 1-3.

Hinckley, D.A., Bidleman, T.F., Foreman, W.T., Tuschall, J.R., 1990. Determination of Vapor-Pressures for Nonpolar and Semipolar Organic-Compounds from Gas-Chromatographic Retention Data. J Chem Eng Data 35, 232-237.

Horii, Y., Khim, J.S., Higley, E.B., Giesy, J.P., Ohura, T., Kannan, K., 2009. Relative Potencies of Individual Chlorinated and Brominated Polycyclic Aromatic Hydrocarbons for Induction of Aryl Hydrocarbon Receptor-Mediated Responses. Environ Sci Technol 43, 2159-2165.

Horii, Y., Ok, G., Ohura, T., Kannan, K., 2008. Occurrence and profiles of chlorinated and brominated polycyclic aromatic hydrocarbons in waste incinerators. Environ Sci Technol 42, 1904-1909.

Ishaq, R., Naf, C., Zebuhr, Y., Broman, D., Jarnberg, U., 2003. PCBs, PCNs, PCDD/Fs, PAHs and Cl-PAHs in air and water particulate samples - patterns and variations. Chemosphere 50, 1131-1150.

Jo, S., Yoshikawa, H., Fujii, A., Takenaga, M., 2006. Surface morphologies of anthracene single crystals grown from vapor phase. Appl Surf Sci 252, 3514-3519.

Kitazawa, A., Amagai, T., Ohura, T., 2006. Temporal trends and relationships of particulate chlorinated polycyclic aromatic hydrocarbons and their parent compounds in urban air. Environ Sci Technol 40, 4592-4598.

Koistinen, J., Paasivirta, J., Nevalainen, T., Lahtipera, M., 1994a. Chlorinated Fluorenes and Alkylfluorenes in Bleached Kraft Pulp and Pulp-Mill Discharges. Chemosphere 28, 2139-2150.

Koistinen, J., Paasivirta, J., Nevalainen, T., Lahtipera, M., 1994b. Chlorophenanthrenes, Alkylchlorophenanthrenes and Alkylchloronaphthalenes in Kraft Pulp-Mill Products and Discharges. Chemosphere 28, 1261-1277.

Lei, Y.D., Chankalal, R., Chan, A., Wania, F., 2002. Supercooled liquid vapor pressures of the polycyclic aromatic hydrocarbons. J Chem Eng Data 47, 801-806.

Luthy, R.G., Dzombak, D.A., Peters, C.A., Roy, S.B., Ramaswami, A., Nakles, D.V., Nott, B.R., 1994. Remediating Tar-Contaminated Soils at Manufactured-Gas Plant Sites. Environ Sci Technol 28, A266-A276.

Ma, Y.G., Lei, Y.D., Xiao, H., Wania, F., Wang, W.H., 2010. Critical Review and Recommended Values for the Physical-Chemical Property Data of 15 Polycyclic Aromatic Hydrocarbons at 25 degrees C. J Chem Eng Data 55, 819-825.

Mackay, D., Bobra, A., Chan, D.W., Shiu, W.Y., 1982. Vapor-Pressure Correlations for Low-Volatility Environmental Chemicals. Environ Sci Technol 16, 645-649.

Mackay, D., Shiu, W.Y., Ma, K.-C., Lee, S.C., 2006. Physical-Chemical Properties and Environmental Fate for Organic Chemicals. CRC Press.

Matsuoka, M.G., J. (ed); Davey, R.J (ed); Jones A. (ed), 1991. Advances in Industrial Crystallization. Butterworth-Heinemann, Oxford.

Murray, J.J., Pottie, R.F., 1974. The Vapor Pressures and Enthalpies of Sublimation of Five Polycyclic Aromatic Hydrocarbons. Canadian Journal of Chemistry 52, 557–563.

Nass, K., Lenoir, D., Kettrup, A., 1995. Calculation of the Thermodynamic Properties of Polycyclic Aromatic-Hydrocarbons by an Incremental Procedure. Angew Chem Int Edit 34, 1735-1736.

Ni, H.G., Zeng, H., Tao, S., Zeng, E.Y., 2010. Environmental and Human Exposure to Persistent Halogenated Compounds Derived from E-Waste in China. Environ Toxicol Chem 29, 1237-1247.

Nilsson, U.L., Ostman, C.E., 1993. Chlorinated Polycyclic Aromatic-Hydrocarbons - Method of Analysis and Their Occurrence in Urban Air. Environ Sci Technol 27, 1826-1831.

Odabasi, M., Cetin, E., Sofuoglu, A., 2006. Determination of octanol-air partition coefficients and supercooled liquid vapor pressures of PAHs as a function of temperature: Application to gas-particle partitioning in an urban atmosphere. Atmos Environ 40, 6615-6625.

Ohura, T., Morita, M., Kuruto-Niwa, R., Amagai, T., Sakakibara, H., Shimoi, K., 2010. Differential Action of Chlorinated Polycyclic Aromatic Hydrocarbons on Aryl Hydrocarbon Receptor-Mediated Signaling in Breast Cancer Cells. Environ Toxicol 25, 180-187.

Ohura, T., Sawada, K.I., Amagai, T., Shinomiya, M., 2009. Discovery of Novel Halogenated Polycyclic Aromatic Hydrocarbons in Urban Particulate Matters: Occurrence, Photostability, and AhR Activity. Environ Sci Technol 43, 2269-2275.

Oja, V., Suuberg, E.M., 1997. Development of a nonisothermal Knudsen effusion method and application to PAH and cellulose tar vapor pressure measurement. Anal Chem 69, 4619-4626.

Pounder, F.E., Masson, I., 1934. Thermal analysis, and its application to dinitrobenzenes. J. Chem. Soc., 1357-1360.

Rai, U.S., Pandey, P., 2002. Crystallization behaviour of metal-nonmetal monotectic alloys; Succinonitrile-pyrene system. Progress in Crystal Growth and Characterization of Materials 45, 59-64.

Rastogi, R.P., Bassi, P.S., 1964. Mechanism of Eutectic Crystallization. The Journal of Physical Chemistry 68, 2398-2406.

Rastogi, R.P., Rama Varma, K.T., 1956. Solid–liquid equilibria in solutions of non-electrolytes. Journal of the Chemical Society, 2097-2101.

Rice, J.W., Fu, J., Suuberg, E.M., 2010. Anthracene + Pyrene Solid Mixtures: Eutectic and Azeotropic Character. Journal of Chemical & Engineering Data, doi: 10.1021/je100208e.

Rice, J.W., Suuberg, E.M., 2010. Thermodynamic study of (anthracene + benzo[a]pyrene) solid mixtures. The Journal of Chemical Thermodynamics 42, 1356-1360.

Ruzicka, K., Mokbel, I., Majer, V., Ruzicka, V., Jose, J., Zabransky, M., 1998. Description of vapour-liquid and vapour-solid equilibria for a group of polycondensed compounds of petroleum interest. Fluid Phase Equilibria 148, 107-137.

Sato, N., Inomata, H., Arai, K., Saito, S., 1986. Measurement of Vapor-Pressures for Coal-Related Aromatic-Compounds by Gas Saturation Method. J Chem Eng Jpn 19, 145-147.

Shiraishi, H., Pilkington, N.H., Otsuki, A., Fuwa, K., 1985. Occurrence of Chlorinated Polynuclear Aromatic-Hydrocarbons in Tap Water. Environ Sci Technol 19, 585-590.

Shiu, W.Y., Ma, K.C., 2000. Temperature dependence of physical-chemical properties of selected chemicals of environmental interest. I. Mononuclear and polynuclear aromatic hydrocarbons. J Phys Chem Ref Data 29, 41-130.

Singh, N.B., Rai, U.S., Singh, O.P., 1985. Chemistry of eutectic and monotectic; phenanthrene-succinonitrile system. Journal of Crystal Growth 71, 353-360.

Singh, N.B., Srivastava, A., Singh, N.P., Gupta, A., 2007. Molecular interaction between naphthalene and 2,4-dinitrophenol in solid state. Molecular Crystals and Liquid Crystals 474, 43-+.

Singh, N.B., Srivastava, M.A., Singh, N.P., 2001. Solid-liquid equilibrium for 2,4-dinitrophenol plus naphthalene. J Chem Eng Data 46, 47-50.

Sonnefeld, W.J., Zoller, W.H., May, W.E., 1983. Dynamic Coupled-Column Liquid-Chromatographic Determination of Ambient-Temperature Vapor-Pressures of Polynuclear Aromatic-Hydrocarbons. Anal Chem 55, 275-280.

Sun, C.G., Snape, C.E., McRae, C., Fallick, A.E., 2003. Resolving coal and petroleum-derived polycyclic aromatic hydrocarbons (PAHs) in some contaminated land samples using compound-specific stable carbon isotope ratio measurements in conjunction with molecular fingerprints. Fuel 82, 2017-2023.

Szczepanik, Richard, Skalmowski, Wlodzimierz, 1963. Effects of crystal growth and volatilization of tar components on the aging of prepared tar. III. Solid-liquid phase relationship of the components of raw coal tar and prepared road tar. Bitumen, Teere, Asphalte, Peche 14, 506,508-512,514.

Szczepanik, Ryszard, 1963. Two- and multicomponent, solid-liquid systems formed by aromatic hydrocarbons, anthraquinone, and coal-tar fractions. Chem. Stosowana Ser. A 7, 621-660.

Wakayama, N., Inokuchi, H., 1967. Heats of sublimation of polycyclic aromatic hydrocarbons and their molecular packings. 40, 2267-2271.

21

Structure of Pure Aluminum After Endogenous and Exogenous Inoculation

Tomasz Wróbel
Silesian University of Technology, Foundry Department
Poland

1. Introduction

The phenomenon of crystallization following after pouring molten metal into the mould, determines the shape of the primary casting (ingot) structure, which significantly affects on its usable properties.

The crystallization of metal in the mould may result in three major structural zones (Fig.1) (Barrett, 1952; Chalmers, 1963; Fraś, 2003; Ohno, 1976):

- zone of chilled crystals (grains) formed by equiaxed grains with random crystallographic orientation, which are in the contact area between the metal and the mould,
- zone of columnar crystals (grains) formed by elongated crystals, which are parallel to heat flow and are a result of directional solidification, which proceeds when thermal gradient on solidification front has a positive value,
- zone of equiaxed crystals (grains) formed by equiaxed grains with random crystallographic orientation in the central part of the casting. The equiaxed crystals have larger size than chilled crystals and are result of volumetric solidification, which proceeds when thermal gradient has a negative value in liquid phase.

Depending on the chemical composition, the intensity of convection of solidifying metal, the cooling rate i.e. geometry of casting, mould material and pouring temperature (Fig.2), in the casting may be three, two or only one structural zone.

Due to the small width of chilled crystals zone, the usable properties of casting depend mainly on the width and length of the columnar crystals, the size of equiaxed crystals and content of theirs zone on section of ingot, as well as on interdendritic or interphase distance in grains such as eutectic or monotectic. For example, you can refer here to a well-known the Hall-Peth law describing the influence of grain size on yield strength (Fig.3) (Adamczyk, 2004):

$$\sigma_y = \sigma_0 + k \bullet d^{-1/2} \tag{1}$$

where:
σ_y – yield strength, MPa,

σ_o - approximate yield strength of monocrystal, for Al amount to 11,1 MPa,
k - material constant characterizing the resistance of grain boundaries for the movement of dislocations in the initial stage of plastic deformation (strength of grain boundaries), for Al amount to 0,05 $MN \cdot m^{-3/2}$,
d - grain size, mm.

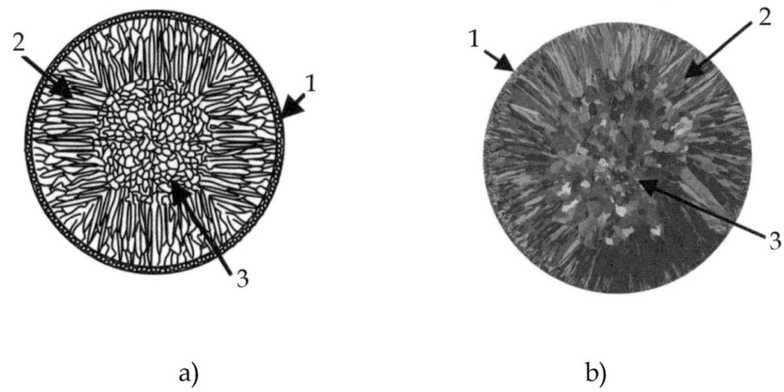

Fig. 1. The primary structure of ingot: a - scheme, b - real macrostructure;
1 - chilled crystals zone, 2 - columnar crystals zone, 3 - equiaxed crystals zone

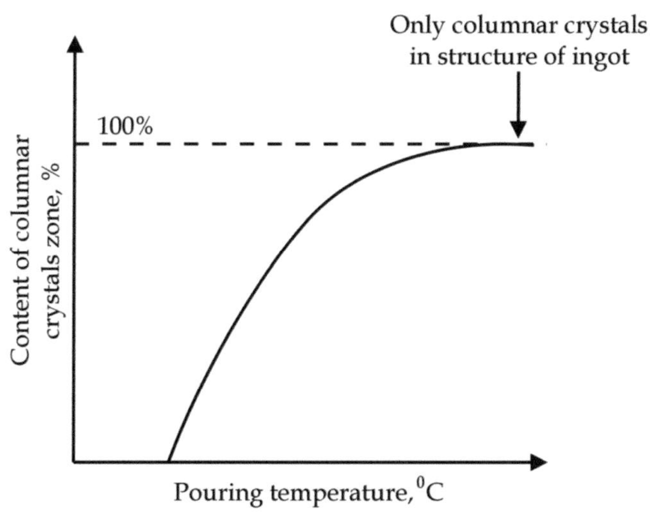

Fig. 2. The influence of pouring temperature on primary structure of ingot (Fraś, 2003)

Because the Hall-Peth law concerns only metals and alloys with the structure of solid solutions, therefore the solidification of alloys with eutectic transformation for example from Al-Si group in describing the influence of refinement of structure on the value of yield strength should take into account the value of interphase distance in eutectic (Paul et al., 1982; Tensi & Hörgerl, 1994; Treitler, 2005):

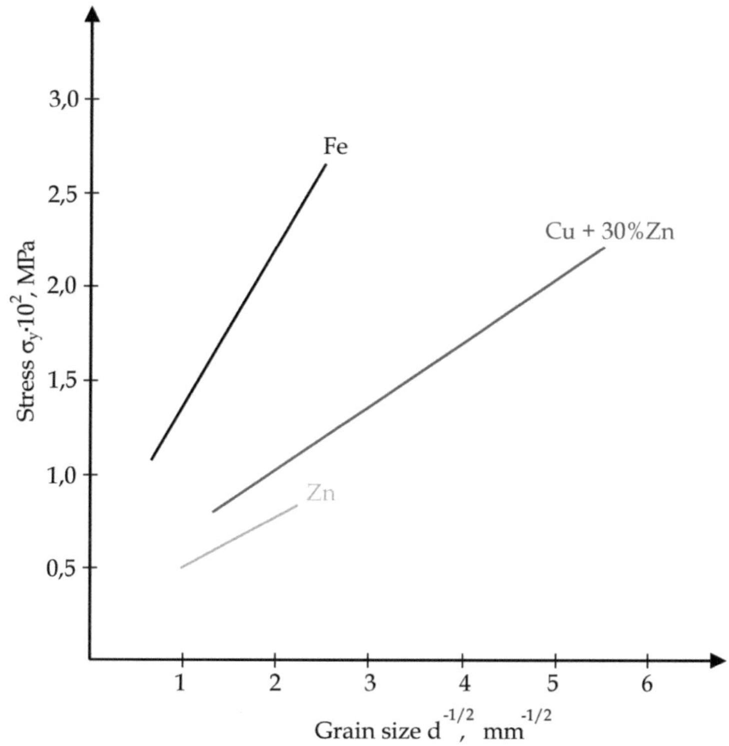

Fig. 3. The influence of grain size on yield strength of Fe, Zn and α brass (Adamczyk, 2004)

$$\sigma_y = \sigma_0 + k_1 \bullet d^{-1/2} + k_2 \bullet \lambda^{-1/2} \tag{2}$$

where:
σ_y - yield strength, MPa,
σ_o - approximately yield strength of monocrystal, MPa
k_1 and k_2 - material constants, $MN \cdot m^{-3/2}$,
d - grain size, mm,
λ - interphase distance in eutectic, mm.

The primary structure of pure metals independently from the crystal lattice type creates practically only columnar crystals (Fig.4) (Fraś, 2003). According to presented data, this type of structure gives low mechanical properties of castings and mainly is unfavourable for the plastic forming of continuous and semi-continuous ingots, because causing forces extrusion rate reduction and during the ingot rolling delamination of external layers can occur (Szajnar & Wróbel, 2008a, 2008b).

This structure can be eliminated by controlling the heat removal rate from the casting, realizing inoculation, which consists in the introduction of additives to liquid metal and/or influence of external factors for example infra- and ultrasonic vibrations or electromagnetic field.

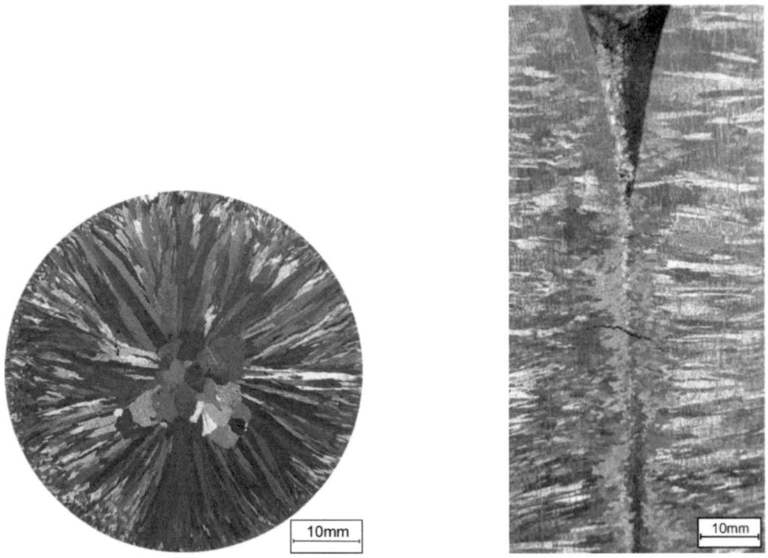

a) b)

Fig. 4. Macrostructure of ingot of Al with a purity of 99,7%: a – transverse section, b – longitudinal section

2. Endogenous inoculation of pure aluminum structure

In aim to obtain an equiaxed and fine-grained structure, which gives high mechanical properties of castings, can use inoculation, which occurs in introducing into metal bath of specified substances, called inoculants (Fraś, 2003). Inoculants increase grains density as result of creation of new particles in consequence of braking of grains growth velocity, decrease of surface tension on interphase boundary of liquid – nucleus, decrease of angle of contact between the nucleus and the base and increase of density of bases to heterogeneous nucleation (Fraś, 2003; Jura, 1968). The effectiveness of this type of inoculation depends significantly on crystallographic match between the base and the nucleus of inoculated metal. This crystallographic match is described by type of crystal lattice or additionally by index (Fraś, 2003):

$$\xi = \left((1 - \frac{x_b - x_n}{x_n}) \right) \bullet 100\% \qquad (3)$$

where:

ξ - match index,
x_b, x_n – parameter of crystal lattice in specified direction, suitable for base and nucleus.

When the value of index (ξ) is closer to 100%, it the more effective is the base to heterogeneous nucleation of inoculated metal.

Therefore active bases to heterogeneous nucleation for aluminum are particles which have high melting point i.e. TiC, TiN, TiB, TiB_2, AlB_2 i Al_3Ti (Tab.1) (Easton & StJohn, 1999a,

1999b; Fjellstedt et al., 2001; Fraś, 2003; Guzowski et al., 1987; Hu & H. Li, 1998; Jura, 1968; Kashyap & Chandrashekar, 2001; H. Li et al., 1997; P. Li et al., 2005; McCartney, 1988; Murty et al., 2002; Naglič et al., 2008; Pietrowski, 2001; Sritharan & H. Li, 1996; Szajnar & Wróbel, 2008a, 2008b; Whitehead, 2000; Wróbel, 2010; Zamkotowicz et al., 2003).

Phase	Melting point (circa) [°C]	Type of crystal lattice	Parameters of crystal lattice [nm]
Al	660	Cubical A1	a = 0,404
TiC	3200	Cubical B1	a = 0,431
TiN	3255	Cubical B1	a = 0,424
TiB	3000	Cubical B1	a = 0,421
TiB_2	2900	Hexagonal C32	a = 0,302 c = 0,321
AlB_2	2700	Hexagonal C32	a = 0,300 c = 0,325
Al_3Ti	1400	Tetragonal $D0_{22}$	a = 0,383 c =0,857

Table 1. Characteristic of bases to heterogeneous nucleation of aluminum (Donnay & Ondik, 1973)

Moreover the effectiveness of inoculants influence can be assessed on the basis of the hypothesis presented in the paper (Jura, 1968). This hypothesis was developed at the assumption that the fundamental physical factors affecting on the crystallization process are the amount of give up heat in the crystallization process on the interphase boundary of liquid - solid and the rate of give up heat of crystallization. After analyzing the results of own researches, the author proposed to determine the index (α), which characterizes the type of inoculant.

$$\alpha = \frac{(\Delta E_k/\nu)_s}{(\Delta E_k/\nu)_p} \bullet W \tag{4}$$

where:

ΔE_k – heat of crystallization of 1 mol of inoculant or inoculated metal, J/mol,
ν - characteristic frequency of atomic vibration calculated by the Lindemman formula, 1/s,
s - symbol of inoculant, p - symbol of inoculated metal.
W - parameter dependent on the atomic mass of inoculant and inoculated metal.
On the basis of equation (4) additives can be divided into three groups:
At $\alpha > 1$ - additives which inhibit crystals growth by the deformation of the crystallization front, thus are effective inoculants.
At $\alpha = 1$ - additives do not affect on structure refinement.
At $\alpha < 1$ - additives which accelerate crystals growth, favoring consolidation of the primary structure of the metal, thus are deinoculants.

In case of inoculation of Al the index α = 2.35 for inoculant in form of Ti and 1.76 for inoculant in form of B.

In case of aluminum casting inoculants are introduced in form of master alloy AlTi5B1. This inoculant has Ti:B ratio equals 5:1. This Ti:B atomic ratio, which corresponds to the mass content of about 0.125% Ti to about 0.005% B, assures the greatest degree of structure refinement (Fig.5). For this titanium and boron ratio bases of type TiB_2 and Al_3Ti are created (Fig.6) (Easton & StJohn, 1999a, 1999b; Fjellstedt et al., 2001; Guzowski et al., 1987; Hu & H. Li, 1998; Kashyap & Chandrashekar, 2001; H. Li et al., 1997; P. Li et al., 2005; Murty et al., 2002; Naglič et al., 2008; Pietrowski, 2001; Sritharan & H. Li, 1996; Szajnar & Wróbel, 2008a, 2008b; Whitehead, 2000; Wróbel, 2010). Type and amount of bases to heterogeneous nucleation of aluminum depend on Ti:B ratio. For example given in paper (Zamkotowicz et al., 2003) the possibility of application of master alloy AlTi1.7B1.4, which has Ti:B ratio equals 1.2:1 is presented. This ratio allows to increase in amount of fine phases TiB_2 and AlB_2 along with the Al_3Ti phase decrease.

Moreover minimum quantities of carbon and nitrogen, which come from metallurgical process of aluminum, create with inoculant the bases in form of titanium carbide TiC and titanium nitride TiN (Fig.7) (Pietrowski, 2001; Szajnar & Wróbel, 2008a, 2008b).

Additionally, because there is a possibility of creation the bases to heterogeneous nucleation of aluminum in form of TiC phase without presence of bases in form of borides, in the practice of casting the inoculation with master alloy AlTi3C0.15 is used (Naglič et al., 2008; Whitehead, 2000). However, on the basis of results of own researches was affirmed that assuming of introducing to Al with a purity of 99,5% the same quantity of Ti i.e. 25ppm, the result of structure refinement caused by master alloy AlTi3C0.15 is weaker than caused by master alloy AlTi5B1 (Fig.8).

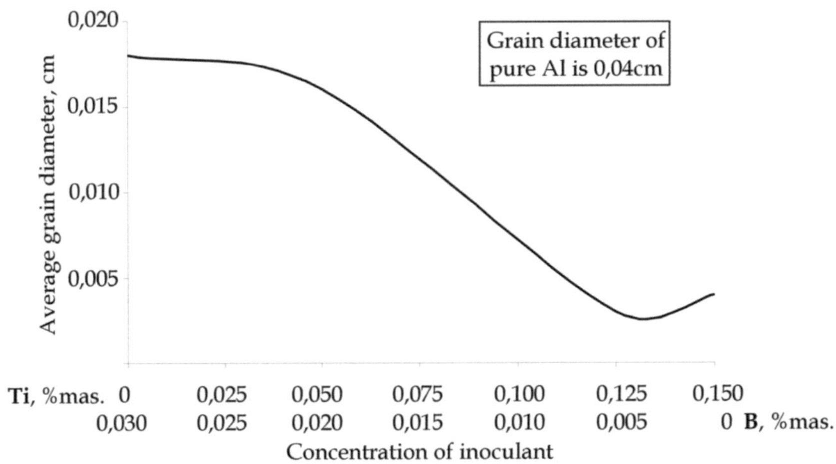

Fig. 5. Influence of Ti and B contents on the average size of Al ingots (H. Li et al., 1997)

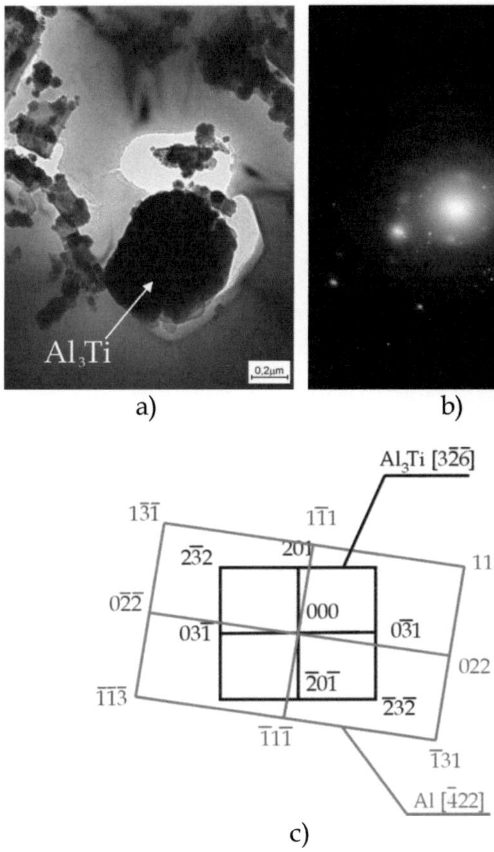

Fig. 6. Structure of thin foil from pure Al after inoculation with (Ti+B), a) TEM bright field mag. 30000x , b) diffraction pattern from the area as in Fig. a, c) analysis of the diffraction pattern from Fig. b

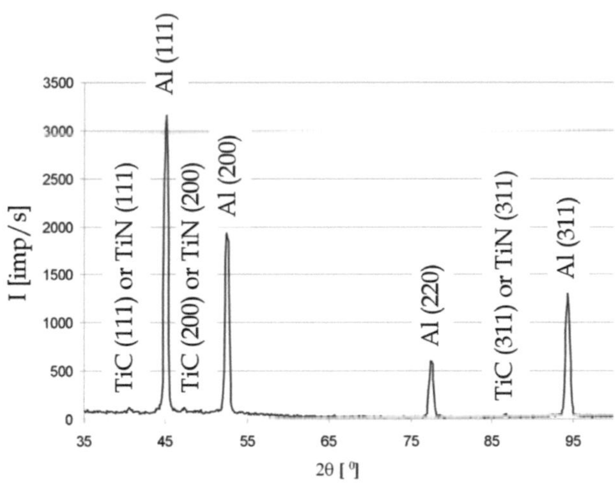

Fig. 7. Result of X-ray diffraction of Al with a purity of 99,5% after inoculation with Ti

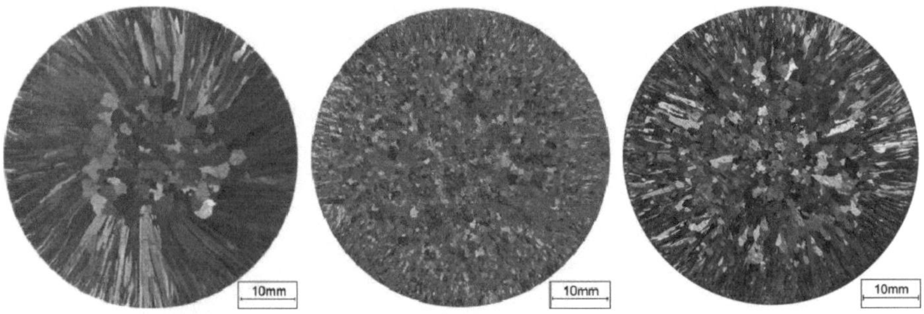

a) b) c)

Fig. 8. Macrostructure of ingot of Al with a purity of 99,5%: a - in as-cast condition, b - after inoculation with (Ti+B), c - after inoculation with (Ti+C)

However, this undoubtedly effective method of inoculation of primary structure of ingot is limited for pure metals, because inoculants decrease the degree of purity specified in European Standards, and Ti with B introduced as modifying additives are then classify as impurities. Moreover, inoculants, mainly Ti which segregates on grain boundary of Al (Fig.9) influence negatively on physical properties i.e. electrical conductivity of pure aluminum (Fig.10) (Wróbel, 2010).

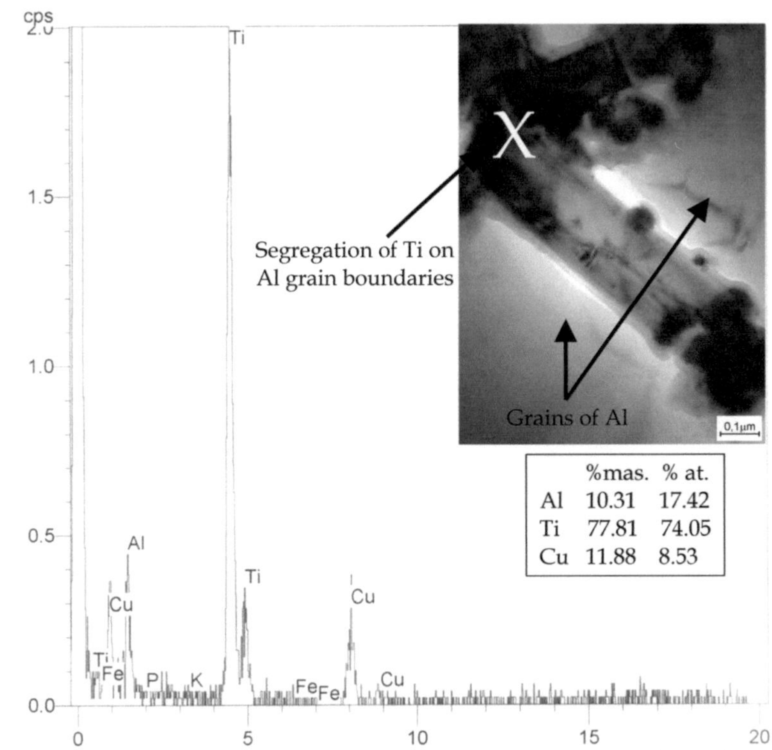

Fig. 9. Segregation of Ti on grain boundaries of Al

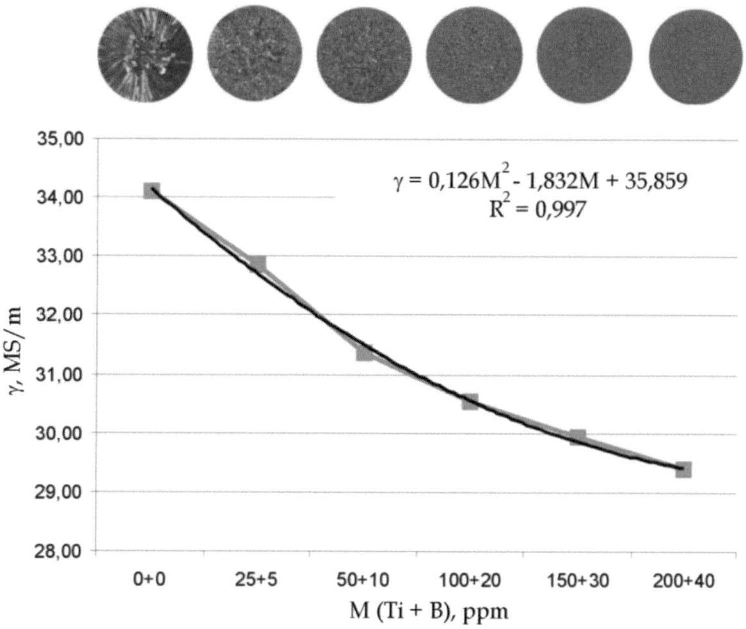

Fig. 10. The influence of quantity of inoculants in form of Ti and B on electrical conductivity γ of Al with a purity of 99,5%

Moreover the presence of the bases to heterogeneous nucleation in form of hard deformable phases for example titanium borides in structure in aluminum, generate possibility of point cracks formation (Fig. 11) and in result of this delamination of sheet (foil) during rolling (Keles & Dundar, 2007).

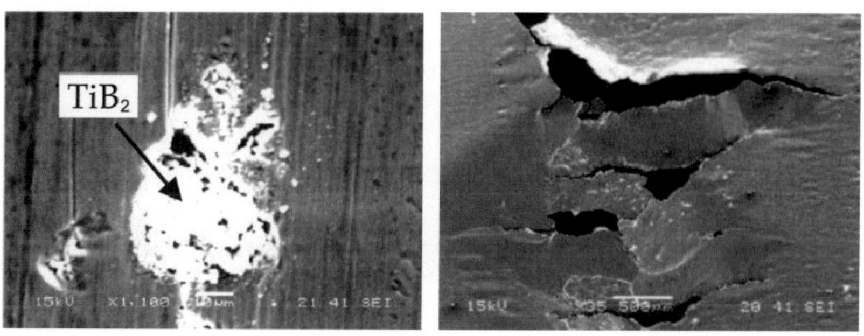

a) b)

Fig. 11. Phase TiB_2 in structure of pure Al (Fig. a) and produced in result from its present crack in sheet (foil) during rolling (Fig. b) (Keles & Dundar, 2007)

Therefore important is the other method of inoculation, which consists of influence of electromagnetic field (Asai, 2000; Campanella et al., 2004; Doherty et al., 1984; Gillon, 2000; Griffiths & McCartney, 1997; Harada, 1998; Szajnar, 2004, 2009; Szajnar & Wróbel, 2008a, 2008b, 2009; Vives & Ricou, 1985; Wróbel, 2010) or mechanical vibrations (Abu-Dheir et al., 2005; Szajnar, 2009) on liquid metal in time of its solidification in mould.

3. Exogenous inoculation of pure aluminum structure

First research works on the application of stirring of liquid metal at the time of its solidification in order to improve the castings quality were carried out by Russ Electroofen in 1939 and concerned the casting of non-ferrous metals and their alloys (Wróbel, 2010). In order to obtain the movement of the liquid metal in the crystallizer in the researches carried out at this period of time and also in the future, a physical factor in the form of a electromagnetic field defined as a system of two fields i.e. an electric and magnetic field was introduced. The mutual relationship between these fields are described by the Maxwell equations (Sikora, 1998).

Generated by the induction coil powered by electric current intensity (I_0) electromagnetic field affects the solidifying metal induces a local electromotive force (E_m), whose value depends on the local velocity of the liquid metal (V) and magnetic induction (B) (Gillon, 2000).:

$$E_m = \overline{V} \times \overline{B} \tag{5}$$

This is a consequence of the intersection of the magnetic field lines with the current guide in form of liquid metal. It also leads to inducing an eddy current of intensity (I) in liquid metal (Gillon, 2000; Vives & Ricou, 1985):

$$\overline{I} = \sigma\left(\overline{V} \times \overline{B}\right) \tag{6}$$

where:

σ - electrical conductivity proper to the liquid metal.

The influence of the induced current on the magnetic field results in establishing of the Lorenz (magnetohydrodynamic) force (F) (Gillon, 2000; Vives & Ricou, 1985):

$$\overline{F} = \overline{I} \times \overline{B} \tag{7}$$

that puts liquid metal in motion e.g. rotary motion in the direction consistent with the direction of rotation of the magnetic field. Strength (F) has a maximum value when the vector (V) and (B) are perpendicular (Fig.12).

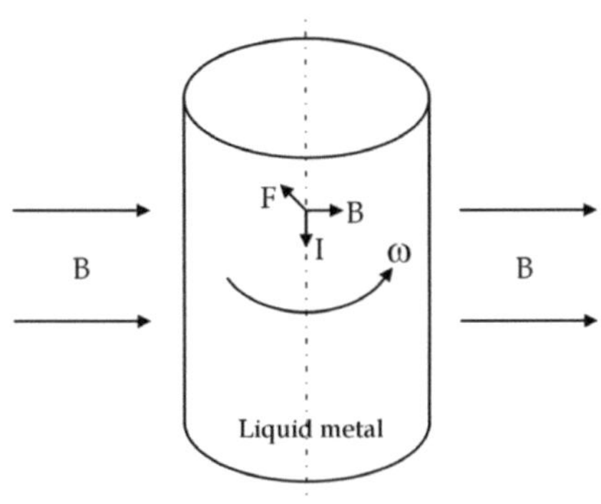

Fig. 12. Scheme of electromagnetic field influence on the liquid metal

In addition, as presented in the paper (Szajnar, 2009) the rotating velocity of the liquid metal (V) is inversely proportional to the density of the metal (ρ), because with some approximation we can say that (Fig.13):

$$\overline{V} \approx \frac{\overline{F}}{\rho} or \frac{\overline{B}}{\rho} \quad (8)$$

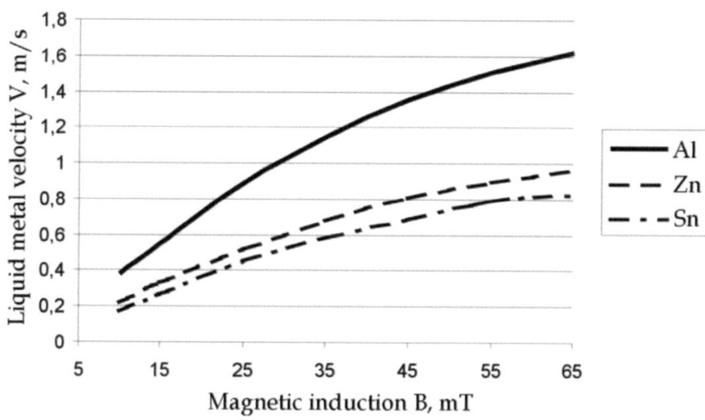

Fig. 13. Dependence of peripheral velocity of liquid metal (V) in a cylindrical mould of inside diameter 45mm on magnetic induction (B) for example metals (Szajnar, 2009)

Forced liquid metal movement influences by diversified way on changes in structure of casting i.e. by changes of thermal and concentration conditions on crystallization front, which decrease or completely stops the velocity of columnar crystals growth (Szajnar, 2004, 2009) and by (Campanella et al., 2004; Doherty et al., 1984; Fraś, 2003; Ohno, 1976; Szajnar, 2004, 2009; Szajnar & Wróbel, 2008a, 2008b, 2009; Wróbel, 2010):

- tear off of crystals from mould wall, which are transferred into metal bath, where they can convert in equiaxed crystals,
- fragmentation of dendrites by coagulation and melting as result of influences of temperature fluctuation and breaking as a result of energy of liquid metal movement,
- crystals transport from the free surface to inside the liquid metal,
- crystals from over-cooled outside layer of the bath are transported into liquid metal.

One of the hypotheses regarding the mechanism of dendrites fragmentation caused by the energy of the movement of liquid metal is presented in work (Doherty et al., 1984). It is based on the assumption of high plasticity of growing dendrites in the liquid metal, which in an initial state are a single crystal with specified crystallographic orientation (Fig.14a). The result of liquid metal movement is deformation (bending) of plastic dendrite (Fig.14b), which caused creation of crystallographic misorientation angle Θ (Fig.14c). Created high-angle grain boundary ($\Theta > 20°$) has the energy γ_{GZ} much greater than double interfacial energy of solid phase - liquid phase $\gamma_{S\text{-}L}$. In result of unbalancing and satisfying the dependence $\gamma_{GZ} > 2\gamma_{S\text{-}L}$ the grain boundary is replaced by a thin layer of liquid metal. This leads to dendrite shear by liquid metal along the former grain boundary (Fig.14d). Dendrite fragments of suitable size after moving into the metal bath can transform into equiaxed crystals.

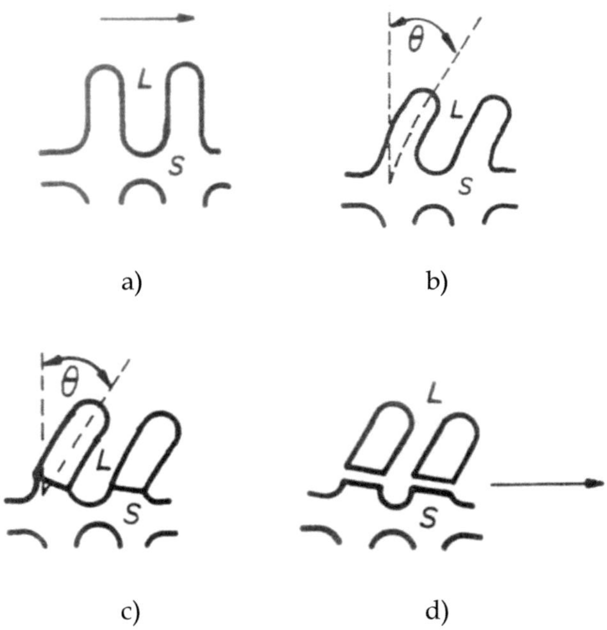

Fig. 14. Schematic model of the grain boundary fragmentation mechanism: a - an undeformed dendrite, b - after bending, c - the reorganization of the lattice bending to give grain boundaries, d - for $\gamma_{GZ} > 2\ \gamma_{S\text{-}L}$ the grain boundaries have been "wetted" by the liquid phase (Doherty et al., 1984).

The influence of electromagnetic field on liquid metal in aim of structure refinement (Fig.15), axial and zonal porosity elimination and obtaining larger homogeneity of structure, was applied in permanent mould casting (Griffiths & McCartney, 1997; Szajnar & Wróbel, 2008a, 2008b, 2009; Wróbel, 2010) and mainly in technologies of continuous (Adamczyk, 2004; Gillon, 2000; Harada, 1998; Miyazawa, 2001; Szajnar et al., 2010; Vives & Ricou, 1985) and semi-continuous casting (Guo et al., 2005).

Fig. 15. Macrostructure of ingot of Al with a purity of 99,5% after cast with influence of rotating electromagnetic field

In case of continuous ingots of square and circular transverse section, rotating electromagnetic field induction coils are used. Rotating electromagnetic field forces rotational movement of liquid metal in perpendicular planes to ingot axis (Fig.16a). Whereas, mainly for flat ingots, longitudinal electromagnetic field induction coils are used, which forced oscillatory movement of liquid metal in parallel planes to ingot axis (Fig.16b) (Adamczyk, 2004; Miyazawa, 2001).

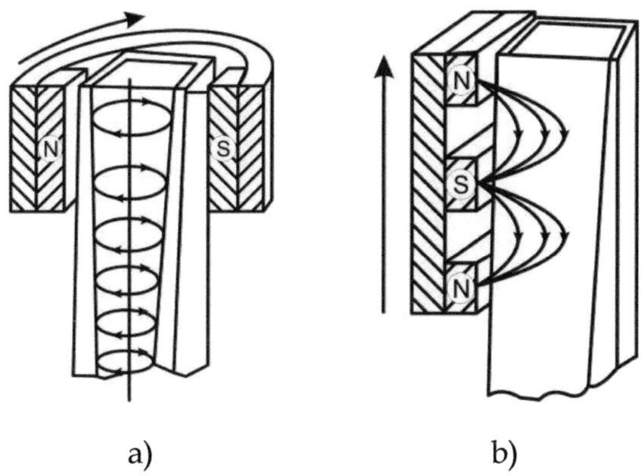

Fig. 16. Scheme of an electromagnetic stirrer (induction coil) forced rotational (a) and oscillatory movement of liquid metal (Adamczyk, 2004)

Whereas the authors of paper (Szajnar & Wróbel, 2008a, 2008b) suggests the use of reversion in the direction of electromagnetic field rotation during permanent mould casting. The advantage of casting in rotating electromagnetic field with reversion compared to casting in rotating field, mainly based on the fact that the liquid metal located in the permanent mould and put in rotary-reversible motion practically does not create a concave meniscus, and thus is not poured out off the mould under the influence of centrifugal forces. Moreover, the influence of this type of field combines impact of high amplitude and low frequency vibration with action of rotating electromagnetic field. Also important is double-sided bending of growing crystals, causing the creation in the columnar crystals zone of ingot characteristic crystals so-called corrugated (Fig.17).

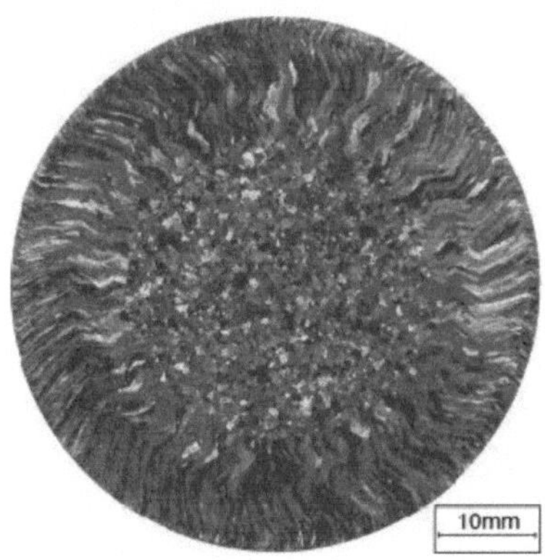

Fig. 17. Macrostructure of ingot of Al with a purity of 99,5% after cast with influence of rotating electromagnetic field with reversion

However in papers (Szajnar, 2004, 2009; Szajnar & Wróbel, 2008a, 2008b, 2009) was shown that influence of forced movement of liquid metal by use of electromagnetic field to changes in structure of pure metals, which solidify with flat crystallization front is insufficient. The effective influence of this forced convection requires a suitable, minimal concentration of additives i.e. alloy additions, inoculants or impurities in casting. Suitable increase of additives concentration causes at specified thermal conditions of solidification, occurs in change of morphology of crystallization front according to the scheme shown in Figure 18.

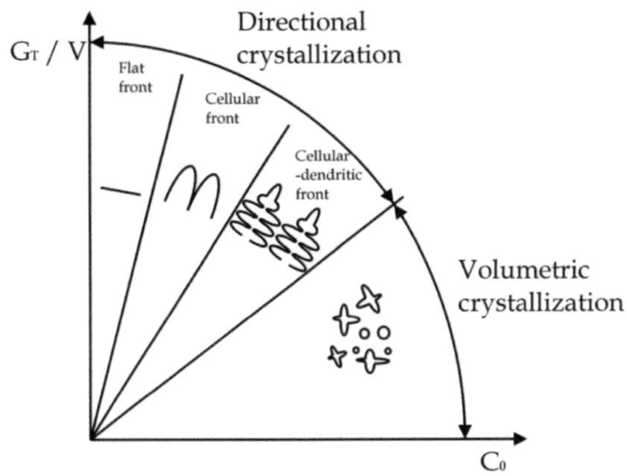

Fig. 18. Scheme of relationship between thermal and concentration conditions and type of crystallization; C_0 - concentration of additives, G_T - thermal gradient on crystallization front, V - velocity of crystallization (Fraś, 2003)

However it should be noted that, based on the latest results of author researches was affirmed that in some cases it is possible to obtain a sufficient refinement degree of pure aluminum structure in result of inoculation carried out only with the use of an electromagnetic field. Because it shows a possibility of increasing the force, which creates movement of liquid metal and in result of this the velocity of its rotation in mould, not only by increasing the value of magnetic induction according to the dependences (7) and (8), but also by increasing the frequency of the current supplied to the induction coil (Fig.19).

The effect of refinement of structure of Al with a purity of 99,5% caused by the rotating electromagnetic field produced by the induction coil supplied by current with frequency different from the network i.e. 50Hz is presented in Table 2. On the basis of macroscopic metallographic researches, which lead to the calculation of the equiaxed crystals zone content on transverse section of ingot (SKR) and average area of macro-grain in this zone (PKR) was affirmed, that application of frequency of supply current $f \leq 50$Hz does not guarantee favourable transformation of pure aluminum structure (Fig.20 and 21). Whereas induction coil supplied with frequency of current larger than power network, mainly 100Hz generates rotating electromagnetic field, which guarantees favourable refinement of structure, also in comparison to obtained after inoculation with small, acceptable by European Standards amount of Ti and B i.e. 25 and 5ppm (Tab.2).

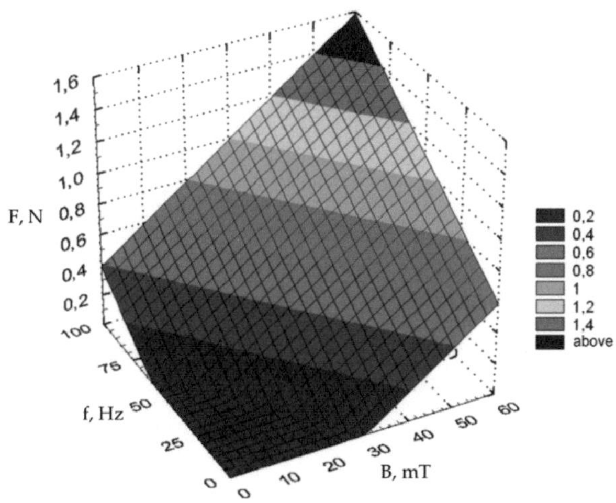

Fig. 19. The influence of magnetic induction (B) and frequency (f) of the current supplied to the induction coil on force value (F), which creates movement of liquid metal

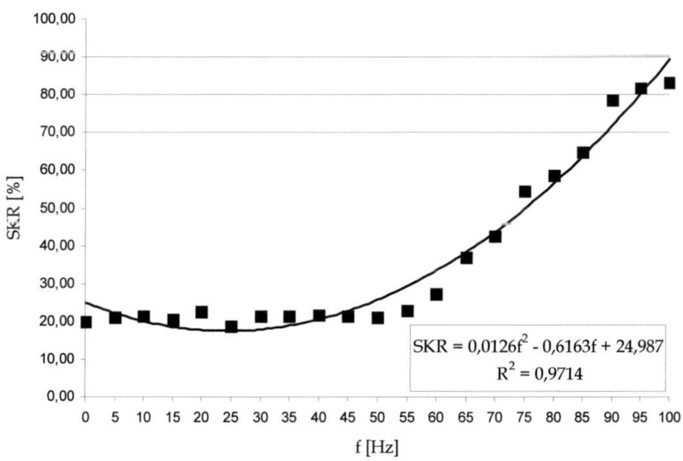

Fig. 20. The influence of current frequency (f) supplied to the induction coil on equiaxed crystals zone content (SKR) on transverse section of pure Al ingot

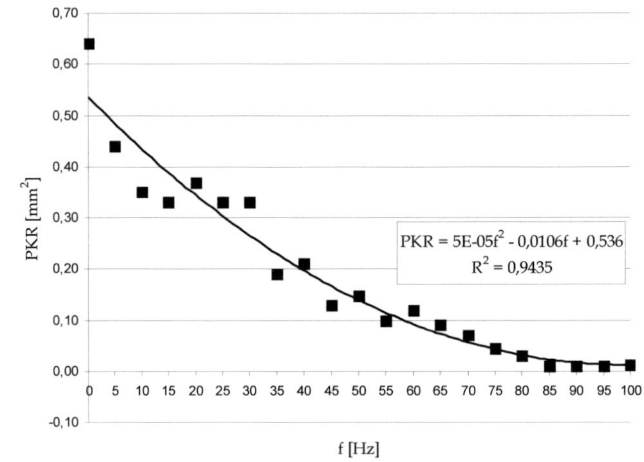

Fig. 21. The influence of current frequency (f) supplied to the induction coil on average area of equiaxed crystal (PKR) of pure Al ingot

No.	Cast parameters			Refinement parameters		Macrostructure of ingot
	B [mT]	f [Hz]	(Ti+B) [ppm]	SKR [%]	PKR [mm²]	
-1-	-2-	-3-	-4-	-5-	-6-	-7-
1	-	-	-	19,94	0,64	
2	-	-	25+5	80,30	0,42	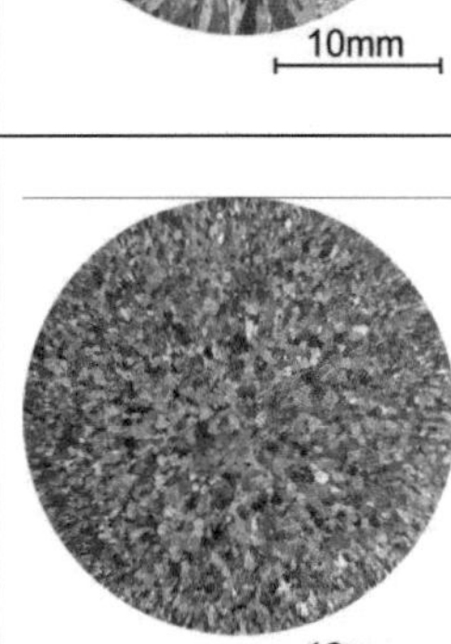

No.	Cast parameters			Refinement parameters		Macrostructure of ingot
	B [mT]	f [Hz]	(Ti+B) [ppm]	SKR [%]	PKR [mm²]	
-1-	-2-	-3-	-4-	-5-	-6-	-7-
3		5		21,01	0,44	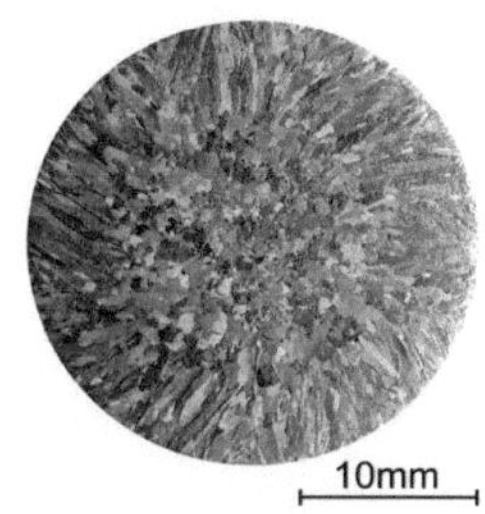
4	60	10	-	21,36	0,35	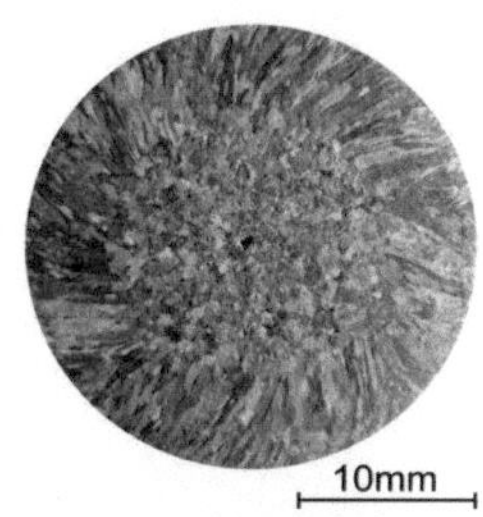
5		15		20,66	0,33	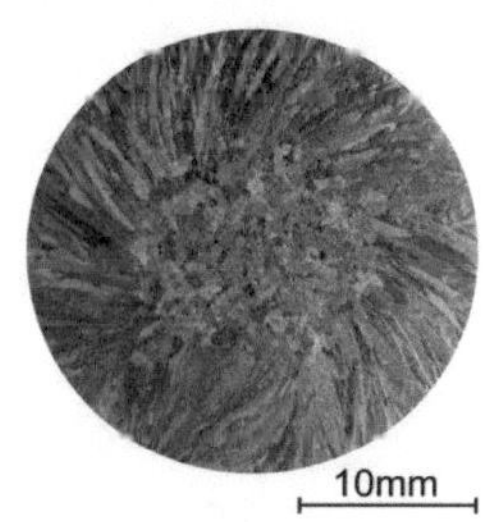
6	60	20	-	22,63	0,37	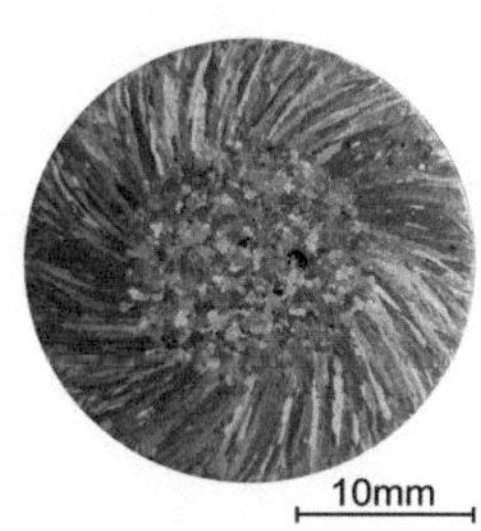

No.	Cast parameters			Refinement parameters		Macrostructure of ingot
	B [mT]	f [Hz]	(Ti+B) [ppm]	SKR [%]	PKR [mm²]	
-1-	-2-	-3-	-4-	-5-	-6-	-7-
7		25		18,90	0,33	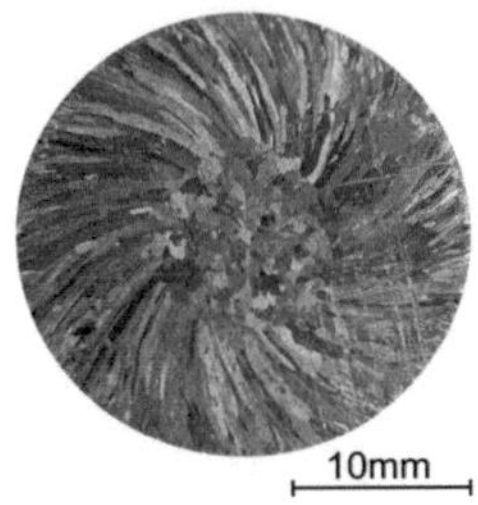
8		30		21,42	0,33	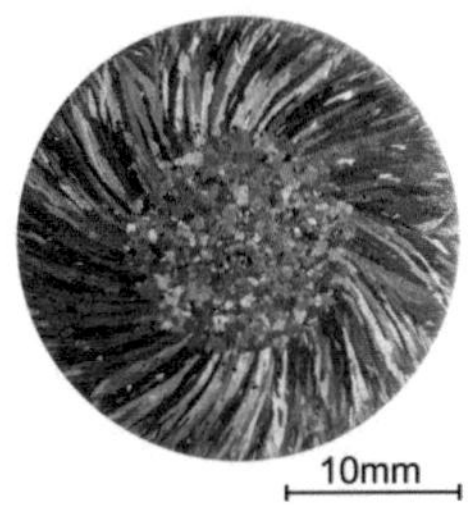
9		35		21,44	0,19	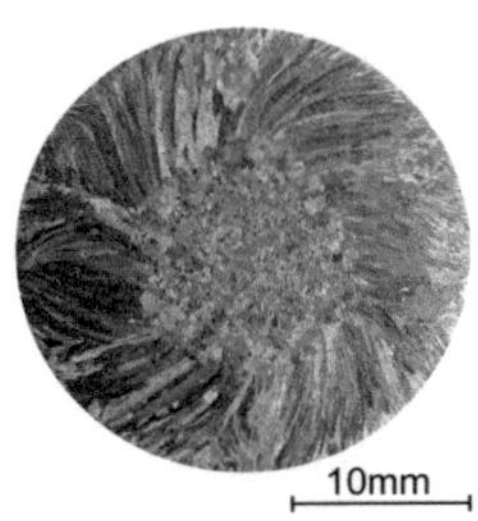
10		40		21,68	0,21	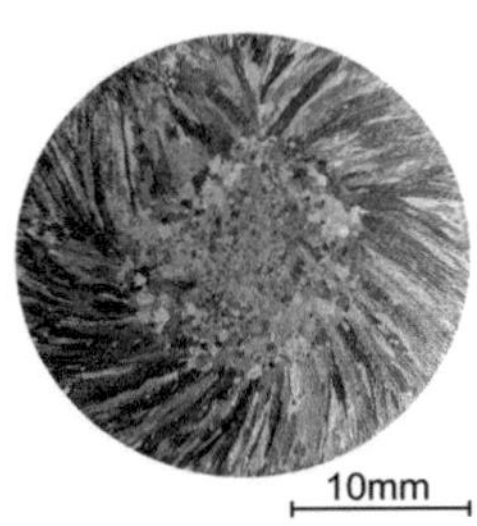

<table>
<tr><th rowspan="2">No.</th><th colspan="3">Cast parameters</th><th colspan="2">Refinement parameters</th><th rowspan="2">Macrostructure of ingot</th></tr>
<tr><th>B [mT]</th><th>f [Hz]</th><th>(Ti+B) [ppm]</th><th>SKR [%]</th><th>PKR [mm²]</th></tr>
<tr><td>-1-</td><td>-2-</td><td>-3-</td><td>-4-</td><td>-5-</td><td>-6-</td><td>-7-</td></tr>
<tr><td>11</td><td rowspan="4">60</td><td>45</td><td rowspan="4">-</td><td>21,46</td><td>0,13</td><td>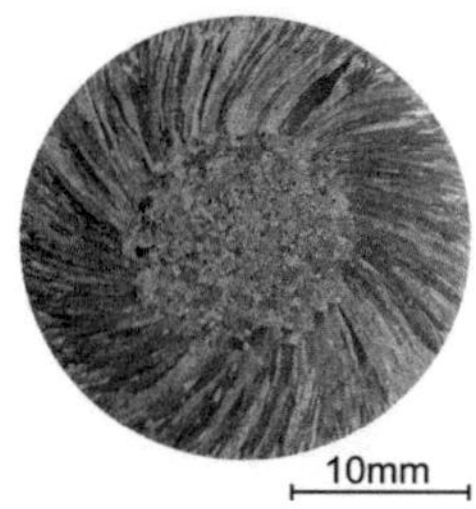</td></tr>
<tr><td>12</td><td>50</td><td>21,21</td><td>0,15</td><td>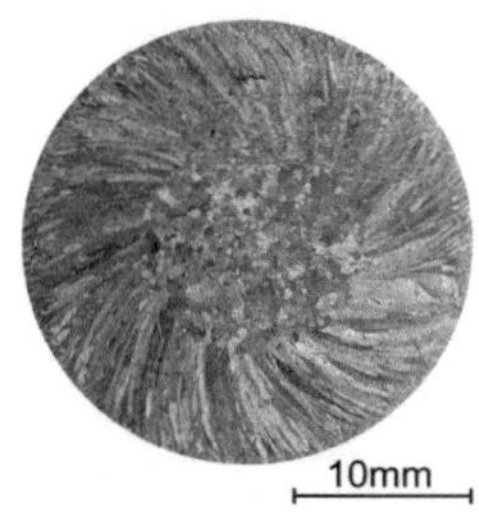</td></tr>
<tr><td>13</td><td>55</td><td>22,87</td><td>0,10</td><td>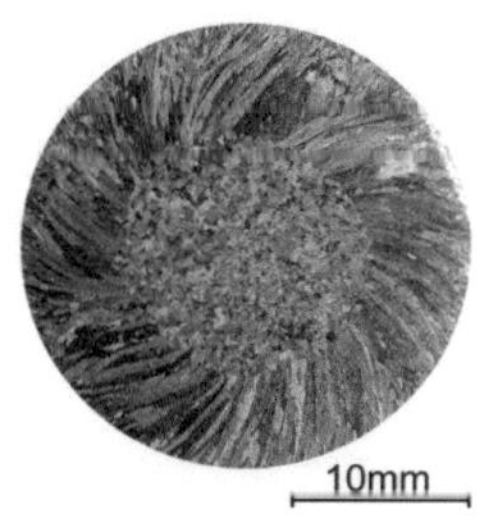</td></tr>
<tr><td>14</td><td>60</td><td>27,22</td><td>0,12</td><td>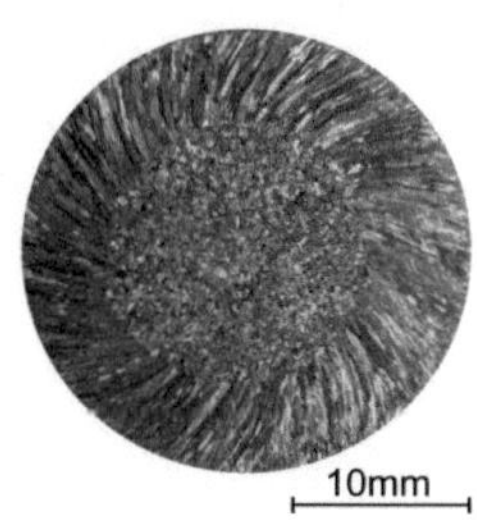</td></tr>
</table>

No.	Cast parameters			Refinement parameters		Macrostructure of ingot
	B [mT]	f [Hz]	(Ti+B) [ppm]	SKR [%]	PKR [mm²]	
-1-	-2-	-3-	-4-	-5-	-6-	-7-
15		65		37,05	0,09	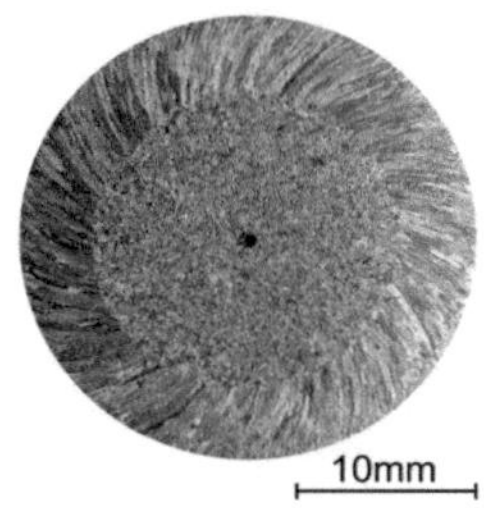
16		70		42,53	0,07	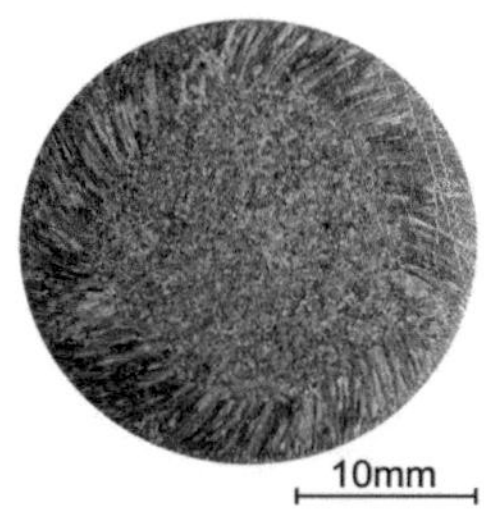
17	60	75	-	54,63	0,04	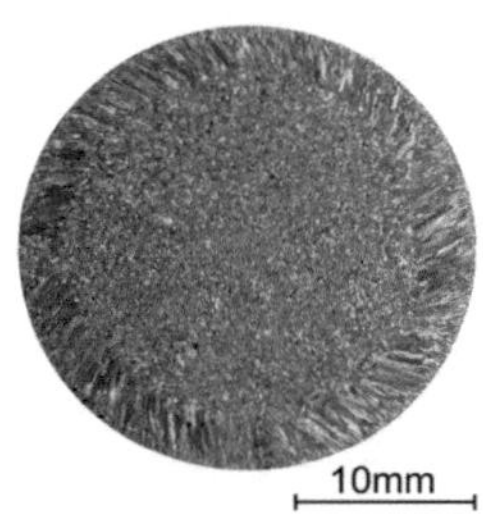
18		80		58,56	0,03	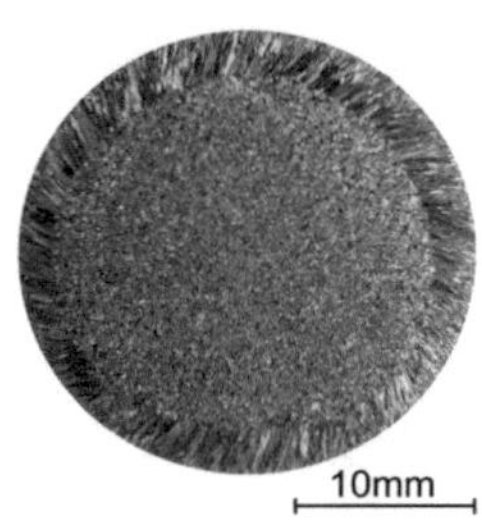

No.	Cast parameters			Refinement parameters		Macrostructure of ingot
	B [mT]	f [Hz]	(Ti+B) [ppm]	SKR [%]	PKR [mm²]	
-1-	-2-	-3-	-4-	-5-	-6-	-7-
19	60	85	-	64,70	0,01	
20		90		78,67	0,01	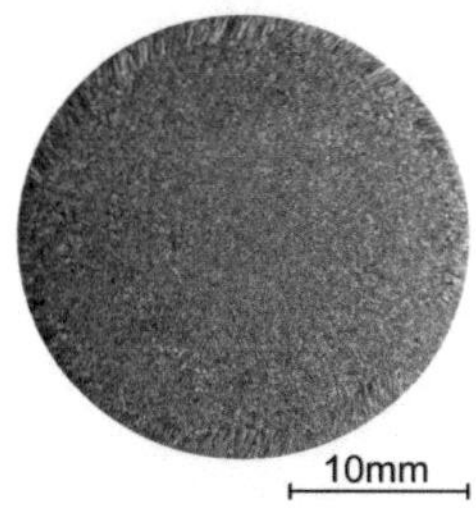
21		95		81,78	0,01	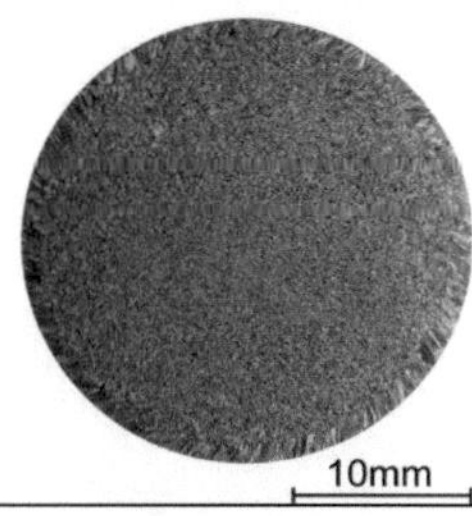
22		100		83,36	0,01	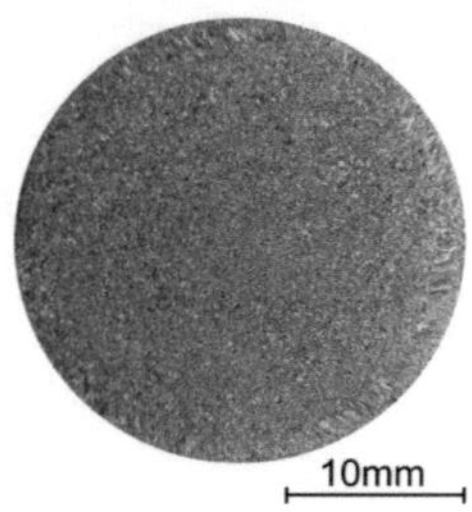

Table 2. The influence of rotating electromagnetic field on structure of Al with a purity of 99,5%

4. The influence of exogenous inoculation on the result of endogenous inoculation in pure aluminum

In the practice of casting is also a problem of connection of endogenous inoculation i.e. realized by use of additives, for example Ti and B with exogenous inoculation i.e. realized by use of electromagnetic field. However as presented in papers (Szajnar & Wróbel, 2008a, 2008b, Wróbel, 2010) occurs that the phenomenon of convection transport (rejection) of impurities for example Cu and inoculants for example Ti from crystallization front into metal bath volume in result of intensive, forced by electromagnetic field the movement of liquid metal. This leads to an increase in density of bases to heterogeneous nucleation of aluminum and in consequence to increase in density of grains in the central area of ingot. Results of determination of Cu and Ti concentration in near-surface and central areas of investigated ingots with use of emission optical spectrometry is a proof of such reasoning. On their basis was affirmed, that in ingot of Al with a purity of 99,5% which was cast under the influence of electromagnetic field and with (Ti + B) inoculation, the Cu and Ti concentration in central area increase was observed (Fig.22a). Whereas in Al, which was cast only with (Ti + B) inoculation, the Cu and Ti concentrations in the near-surface and central areas of ingot are similar (Fig.22b).

The second proof of convection transport (rejection) of Cu and Ti from crystallization front into liquid metal volume is the analysis of macrostructure of investigated ingots and counting of all macro-grains in equiaxed crystals zone. Macrostructure of ingot of Al with a purity of 99,5%, which was cast with the combined effect of the electromagnetic field and with (Ti+B) inoculation has smaller equiaxed crystals zone than the ingot which was cast only with the influence of endogenous inoculation, but the first ingot has a smaller size of macro-grain in its equiaxed crystals zone than the ingot which was cast only with (Ti+B) inoculation (Fig.23).

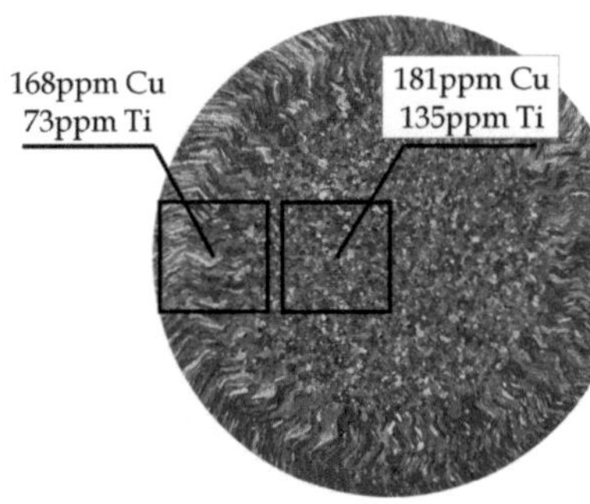

a)

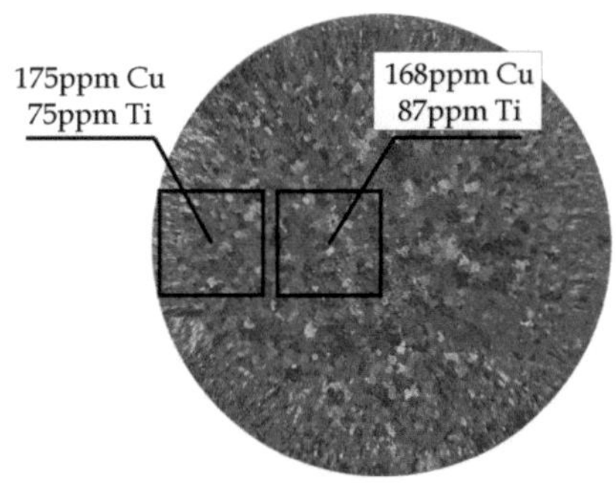

b)

Fig. 22. Concentration of Cu and Ti in near-surface and central areas of Al with a purity of 99,5%ingots: a – after common exogenous (electromagnetic field) and endogenous (Ti + B) inoculation, b – only after endogenous (Ti + B) inoculation

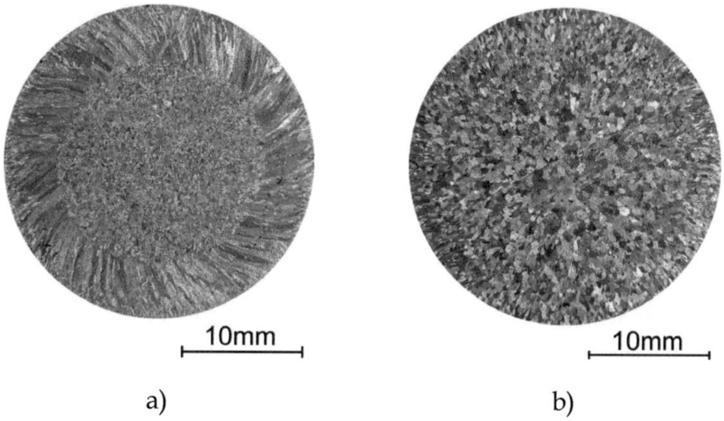

Fig. 23. Macrostructure of ingot of Al with a purity of 99,5%: a – after common exogenous (electromagnetic field) and endogenous (Ti + B) inoculation, b – only after endogenous (Ti + B) inoculation

Based on conducted calculations of number of macro-grains in equiaxed crystals zone following formula was formulated:

$$n_{ex+en} > n_{ex} + n_{en} \tag{9}$$

where:

n_{ex+en} – number of macro-grains in equiaxed crystals zone of ingot which was cast with common influence of exogenous (electromagnetic field) and endogenous (Ti + B) inoculation,

n_{ex} – number of macro-grains in equiaxed crystals zone of ingot which was cast only with influence of exogenous (electromagnetic field) inoculation,

n_{en} – number of macro-grains in equiaxed crystals zone of ingot which was cast only with influence of endogenous (Ti + B) inoculation.

Summarize, was affirmed that application of common exogenous (electromagnetic field) and endogenous (Ti + B) inoculation strengthens effect of structure refinement in comparison with application of one type of inoculation, only if is used skinning of ingot surface i.e. machining in aim of columnar crystals zone elimination.

5. Conclusions

In conclusion can say, that even endogenous inoculation with small amount of (Ti + B) strongly increase on refinement in pure aluminum structure. It results from reactions, which proceed between inoculating elements and inoculated metal or charge impurities. These reactions lead to formation of active bases to heterogeneous nucleation of aluminum as high melting small particles of type TiB, TiB_2, AlB_2, Al_3Ti and TiC or TiN, which have analogy in crystal lattice with Al.

However on the basis of conducted analysis of the literature and results of authors researches it was affirmed, that the rotating electromagnetic field generated by induction coil supplied by current with frequency larger than power network, influences liquid metal in time of its solidification in mould, guarantees refinement of structure of pure Al without necessity of application of inoculants sort Ti and B.

This method of exogenous inoculation is important, because Ti and B decrease the degree of purity and electrical conductivity of pure aluminum. Moreover Ti and B are reason of point cracks formation during rolling of ingots.

Presented method of inoculation by use of electromagnetic field is possible to apply in conditions of continuous casting because it allows producing of ingots from aluminum of approx. 99,5% purity with structure without columnar crystals, which are unfavourable from point of view of usable properties.

6. Acknowledgment

Project financed from means of National Science Centre.

7. References

Abu-Dheir, N.; Khraisheh, M.; Saito, K. & Male, A. (2005). Silicon morphology modyfication in the eutectic Al-Si alloy using mechanical mold vibration. *Materials Science and Engineering:A*, Vol.393, No.1-2, (September 2004), pp. 109-117, ISSN 0921-5093

Adamczyk, J. (2004). *Engineering of metallic materials*, Publishers of Silesian University of Technology, ISBN 83-7335-223-6, Gliwice, Poland

Asai, S. (2000). Recent development and prospect of electromagnetic processing of materials. *Science and Technology of Advanced Materials*, Vol.1, No.4, (September 2000), pp. 191-200, ISSN 1468-6996

Barrett, C. (1952). *Structure of metals, Metallurgy and Matallurgical Engineering Series*, McGraw-Hill Book Co. Inc., New York, USA.

Campanella, T.; Charbon, C. & Rappaz, M. (2004). Grain refinement induced by electromagnetic stirring: a dendrite fragmentation criterion. *Metallurgical and Materials Transactions A*, Vol.35, No.10, (December 2003), pp. 3201-3210, ISSN 1073-5623

Chalmers, B. (1963). The structure of ingot. *Journal of the Australian Institute of Metals*, Vol.8, No.6, pp. 255-263, ISSN 0004-9352

Doherty, R.; Lee, H. & Feest, E. (1984). Microstructure of stir-cast metals. *Materials Science and Engineering*, Vol.65, (January 1984), pp. 181-189, ISSN 0025-5416

Donnay, J. & Ondik, H. (1973). *Crystal date – Determinative Tables*, NSRDS – Library of Congress, Washington, USA

Easton, M. & StJohn, D. (1999). Grain refinement of aluminum alloys: Part I. The nucleant and solute paradigms – a review of the literature. *Metallurgical and Materials Transactions A*, Vol.30, No.6, (February 1998), pp. 1613-1623, ISSN 1073-5623

Easton, M. & StJohn, D. (1999). Grain refinement of aluminum alloys: Part II. Confirmation of, and a mechanism for, the solute paradigm. *Metallurgical and Materials Transactions A*, Vol.30, No.6, (February 1998), pp. 1625-1633, ISSN 1073-5623

Fjellstedt, J.; Jarfors, A. & El-Benawy, T. (2001). Experimental investigation and thermodynamic assessment of the Al-rich side of the Al-B system. *Materials & Design*, Vol.22, No.6, (February 2000), pp. 443-449, ISSN 0261-3069

Fraś, E. (2003). *Crystallization of metals*, WNT, ISBN 83-204-2787-8, Warsaw, Poland

Gillon, P. (2000). Uses of intense d.c. magnetic fields in materials processing. *Materials Science and Engineering:A*, Vol.287, No.2, (August 2000), pp.146-152, ISSN 0921-5093

Griffiths, W. & McCartney, D. (1997). The effect of electromagnetic stirring on macrostructure and macrosegregation in the aluminium alloy 7150. *Materials Science and Engineering:A*, Vol.222, No.2, (May 1996), pp.140-148, ISSN 0921-5093

Guo, S.; Cui, J.; Le, Q. & Zhao, Z. (2005) The effect of alternating magnetic field on the process of semi-continuous casting for AZ91 billets. *Materials Letters*, Vol.59, No.14-15, (November 2004), pp. 1841-1844, ISSN 0167-577X

Guzowski, M.; Sigworth, G. & Sentner, D. (1987). The role of boron in the grain refinement of aluminum with titanium. *Metallurgical and Materials Transactions A*, Vol.18, No.5, (May 1985), pp. 603-619, ISSN 1073-5623

Harada, H.; Takeuchi, E.; Zeze, M. & Tanaka, H. (1998). MHD analysis in hydromagnetic casting process of clad steel slabs. *Applied Mathematical Modeling*, Vol.22, No.11, (July 1997), pp. 873-882, ISSN 0307-904X

Hu, B. & Li, H. (1998) Grain refinement of DIN226S alloy at lower titanium and boron addition levels, *Journal of Materials Processing Technology*, No.74, (October 1996), pp. 56-60, ISSN 0924-0136

Jura, S. (1968). *Modeling research of inoculation process in metals*, Publishers of Silesian University of Technology, Gliwice, Poland

Kashyap, K. & Chandrashekar, T. (2001). Effects and mechanism of grain refinement in aluminium alloys. *Bulletin of Materials Science*, Vol.24, No.4, (April 2001), pp. 345-353, ISSN 0250-4707

Keles, O. & Dundar, M. (2007). Aluminum foil: its typical quality problems and their causes. *Journal of Materials Processing Technology*, Vol. 186, No.1-3, (December 2006), pp. 125-137, ISSN 0924-0136

Li, H.; Sritharan, T.; Lam, Y. & Leng, N. (1997). Effects of processing parameters on the performance of Al grain refinement master alloy Al-Ti and Al-B in small ingots. *Journal of Materials Processing Technology*, Vol.66, No.1-3, (October 1995), pp. 253-257, ISSN 0924-0136

Li, P.; Kandalova, E. & Nikitin, V. (2005). Grain refining performance of Al-Ti master alloy with different microstructures. *Materials Letters*, Vol.59, No.6, (December 2004), pp. 723-727, ISSN 0167-577X

McCartney, D. (1988). Discussion of "The role of boron in the grain refinement of aluminum with titanium". *Metallurgical and Materials Transactions A*, Vol.19, No.2, (July 1987), pp. 385-387, ISSN 1073-5623

Miyazawa, K. (2001). Continuous casting of steels in Japan. *Science and Technology of Advanced Materials*, Vol.2, No.1, (June 1999), pp.59-65, ISSN 1468-6996

Murty, B.; Kori, S. & Chakraborty, M. (2002). Grain refinement of aluminium and its alloys by heterogeneous nucleation and alloying. *International Materials Reviews*, Vol.47, No.1, pp. 3-29, ISSN 1743-2804

Naglič, I.; Smolej, A. & Doberšek, M. (2008). Remelting of aluminium with the addition of AlTi5B1 and AlTi3C0.15 grain refiners. *Metalurgija*, Vol.47, No.2, (January 2007), pp. 115-118, ISSN 0543-5846

Ohno, A. (1976). *The solidification of metals*, Chijin Shokan Co. Ltd, Tokyo, Japan

Paul, J.; Exner, H. & Müller-Schwelling, D. (1982). Microstructure and mechanical properties of cast and heat-treated eutectic Al-Si alloys, *Metallkunde*, Vol.1, No.43, pp. 50-55, ISSN 0044-3093

Pietrowski, S. (2001). *High-silicon aluminum alloys*, Publishers of Technical University of Lodz, ISBN 83-7283-029-0, Łódź, Poland

Sikora, R. (1998). *Theory of electromagnetic field*, WNT, ISBN 83-204-2226-4, Warsaw, Poland

Sritharan, T. & Li, H. (1996). Optimizing the composition of master alloys for grain refining aluminium, *Scripta Materialia*, Vol.35, No.9, (February 1996), pp. 1053-1056, ISSN 1359-6462

Szajnar, J. (2004). The columnar crystals shape and castings structure cast in magnetic field. *Journal of Materials Processing Technology*, Vol.157-158, (December 2004), pp. 761-764, ISSN 0924-0136

Szajnar, J. (2009). *The influence of selected physical factors on the crystallization process and casting structure*, Archives of Foundry Engineering, ISBN 1897-3310, Katowice-Gliwice, Poland

Szajnar, J.; Stawarz, M.; Wróbel, T.; Sebzda, T.; Grzesik, B. & Stępień, M. (2010). Influence of continuous casting conditions on grey cast iron structure, *Archives of Materials Science and Engineering*, Vol.42, No.1, (January 2010), pp. 45-52, ISSN 1897-2764

Szajnar, J. & Wróbel, T. (2008). Inoculation of pure aluminium aided by electromagnetic field. *Archives of Foundry Engineering*, Vol.8, No.1, (July 2007), pp. 123-132, ISSN 1897-3310

Szajnar, J. & Wróbel, T. (2008). Inoculation of pure aluminum with an electromagnetic field, *Journal of Manufacturing Processes*, Vol.10, No.2, (September 2008), pp. 74-81, ISSN 1526-6125

Szajnar, J. & Wróbel, T. (2009). The use of electromagnetic field in the process of crystallization of castings, In: *Progress in the theory and practice of foundry*, Szajnar, J., pp. 399-418, Archives of Foundry Engineering, ISBN 978-83-929266-0-3, Katowice-Gliwice, Poland

Tensi, H. & Hörgerl, J. (1994) Metallographic studies to quality assessment of alloys Al-Si. *Metallkunde*, Vol.10, No.73, pp. 776-781, ISSN 0044-3093

Treitler, R. (2005). *Calculating the strenght of casting and extruded alloys aluminum - magnesium*, Universitätsverlag Karlsruhe, ISBN 3-937300-94-5, Karlsruhe, Germany

Vives, C. & Ricou, R. (1985). Experimental study of continuous electromagnetic casting of aluminum alloy. *Metallurgical and Materials Transactions B*, Vol.16, No.2, (July 1983), pp. 377-384, ISSN 1073-5615

Whitehead, A. (2000). Grain refiners (modifiers) of the Al-Ti-C type – their advantages and application. *Foundry Review*, Vol.50, No.5, pp. 179-182, ISSN 0033-2275

Wróbel, T. (2010). Inoculation of pure aluminum structure with use of electromagnetic field, In: *The tendency of optimization of production system in foundries*, Pietrowski, S., pp. 253-262, Archives of Foundry Engineering, ISBN 978-83-929266-1-0, Katowice-Gliwice, Poland

Zamkotowicz, Z.; Stuczński, T.; Augustyn, B.; Lech-Grega, M. & Wężyk, W. (2003). Sedimentation of intermetallic compounds in liquid aluminum alloys of type AlSiCu(Ti), In: *Nonferrous Metals – Science and Technology*, Bonderek Z., CCNS, pp. 77-82

22

Phosphoramidates: Molecular Packing and Hydrogen Bond Strength in Compounds Having a $P(O)(N)_n(O)_{3-n}$ (n = 1, 2, 3) Skeleton

Mehrdad Pourayoubi[1], Fahimeh Sabbaghi[2], Vladimir Divjakovic[3] and Atekeh Tarahhomi[1]
[1]*Department of Chemistry, Ferdowsi University of Mashhad, Mashhad,*
[2]*Department of Chemistry, Zanjan Branch, Islamic Azad University, Zanjan,*
[3]*Department of Physics, Faculty of Sciences, University of Novi Sad, Novi Sad,*
[1,2]*Iran*
[3]*Serbia*

1. Introduction

Compounds containing the $P(O)(N)_n(O)_{3-n}$ (n = 1, 2, 3), $P(O)(N)_m(O)_{2-m}X$ (m = 1, 2, X = C, Cl, F, S etc.) and $P(O)(O)_3$ moieties are among the well-studied inorganic compounds [an interested reader may find many examples of compounds with the mentioned skeletons through a CSD search, [1]]. *N,N,N',N',N",N"*-hexamethyl phosphoric triamide (HMPA, Scheme 1) is an important polar aprotic solvent with a high-dielectric constant [2] and an excellent ligand for interaction with hard metal-cations [3].

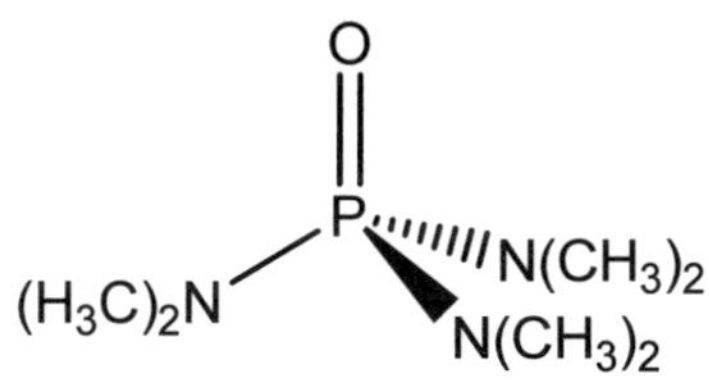

Scheme 1. *N,N,N',N',N",N"*-hexamethyl phosphoric triamide

Tabun, $NCP(O)[N(CH_3)_2][OCH_2CH_3]$ (Scheme 2), Sarin, $FP(O)(CH_3)[OCH(CH_3)_2]$ and Soman, $CH_3P(O)(F)[OCH(CH_3)(C(CH_3)_3)]$ are among the well-known "nerve agents" that act as acetylcholinesterase enzyme (AChE) inhibitors in human body and mammals [4].

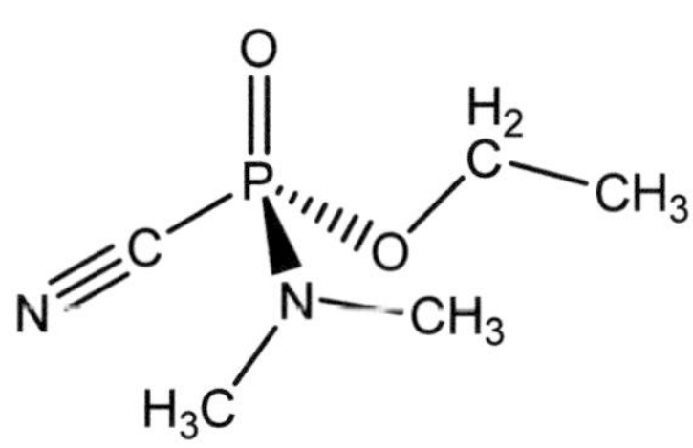

Scheme 2. Tabun, a nerve agent

Some researchers focus on decontamination of such compounds under UV-irradiation or in the presence of nano-oxides or nano-photocatalysts under sun-light [5]. The flame retardancy of some phosphoric esters was studied [6] and some phosphoramidates have therapeutic applications in the treatments of HIV and cancer [7]. Some pure chemists have interested to the NMR consideration [8], chemical calculation [9] and crystallography [10] of such compounds. A few bio-inorganic chemists have worked on the prediction of the biological properties of compounds based on their structures, with the related software programs such as PASS [11], and the evaluation of some relationships between structural features and biological activities [12]. In our laboratory, we centralize on the synthesis of new phosphorus-nitrogen and phosphorus-oxygen compounds and on obtaining their suitable single crystals for the X-ray crystallography experiments [13-64].

A schematic classification for the compounds having a $P(O)(N)_n(O)_{3-n}$ (n = 1, 2, 3) skeleton is shown in Scheme 3.

The numbers of the reported crystal structures in each family are presented in Scheme 3. The central box (blue) indicates the overall number of phosphoramidates having a $P(O)(N)_n(O)_{3-n}$ (n = 1, 2, 3) skeleton; the more well-studied categories of phosphoramidates are shown as green boxes in the top and bottom of the central box namely: a) phosphoric triamides (having a P(O)(N)(N)(N) or P(O)(NHC(O))(N)(N) fragment), and b) amidophosphoric acid esters (containing a P(O)(O)(N)(N) or P(O)(O)(O)(N) skeleton).

As a nitrogen bonded H atom is very important in the H-bond pattern consideration, in the sub-categories, the presence or the absence of this H atom is clarified. In the applied notation, for example, the $P(O)(NH)_3$ and $P(O)(N)_3$ denote to the presence of secondary and tertiary nitrogen atoms, respectively. The phosphoramidates containing a $P(O)NH_2$ moiety are distinguished in the left side box directly related to the central box.

The less-studied (so far) related compounds i.e. c) the proton-transfer and phosphate salts and the acids, and d) the anhydride compounds with a P(O)(O)P(O) skeleton are shown in the right and the left of the central box.

In this flowchart, the skeletons of 643 compounds -which their crystal structures were deposited- have been collected. In this classification, the phosphoramidates containing the phosphorus-carbon and the phosphorus-halogen bonds have not been considered.

2. Synthesis and purification of phosphoramidates and phosphoric acid esters

The reaction of phosphorus(V)-halogen compounds of the type $P(O)X_{3-n}Y_n$ (X = halide, Y = another group such as amide, alkoxide and so on, and n = 0, 1 and 2) with primary or secondary amines leads to the formation of phosphorus(V)-nitrogen compounds. The promotion of this reaction needs to the presence of an excess amount of amine as an HX scavenger or the presence of another acid scavenger such as tertiary amines [61] or pyridine [8] (Scheme 4). In this strategy, removing of the hydrohalide salt of the organic base is a challenging task in the purification process.

The purification may be performed by stirring the crude product in water to remove the amine hydrohalide or pyridinium halide and/or may be done by selecting the solvent

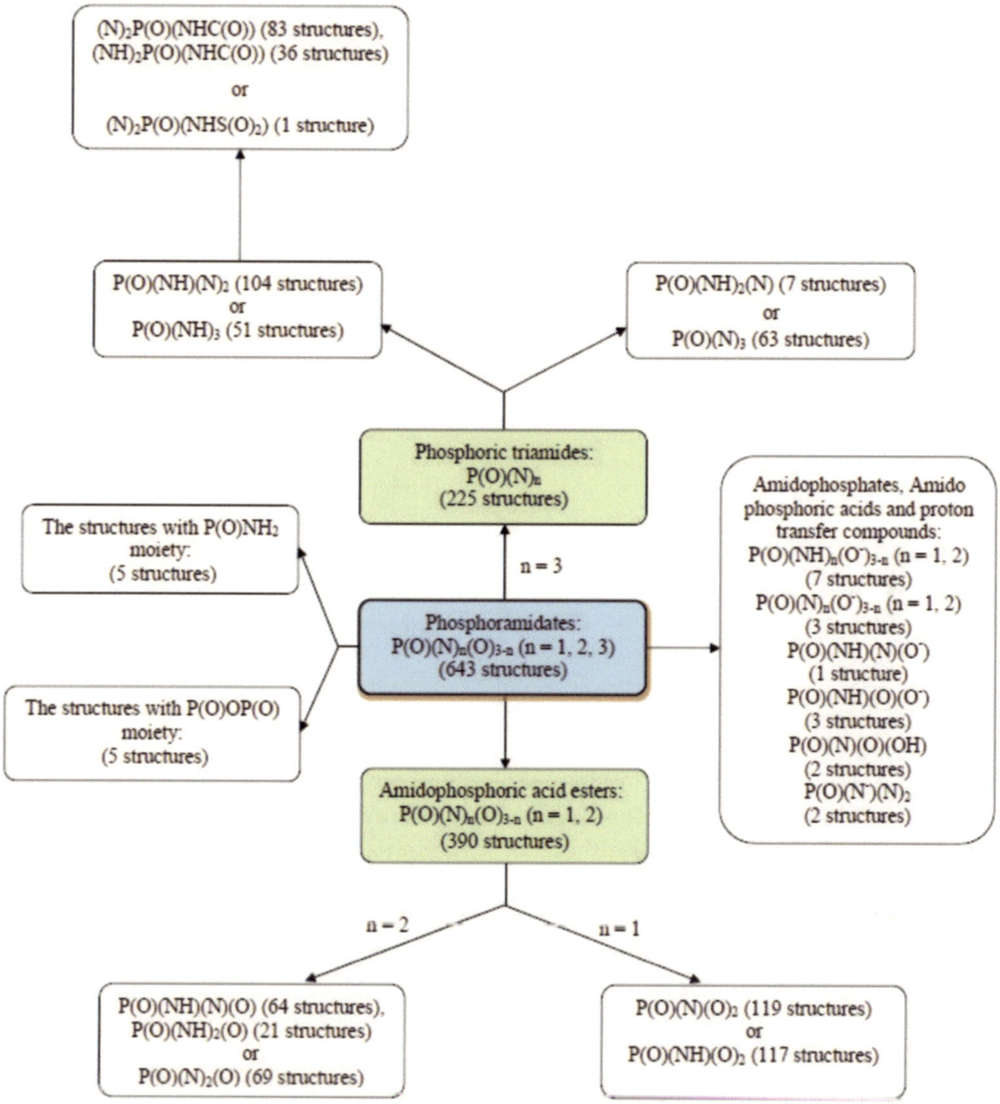

Scheme 3. The classification of compounds having a $P(O)(N)_n(O)_{3-n}$ (n = 1, 2, 3) skeleton

which the salt is as precipitate (and the product is soluble) and then the filtering off the salt. Moreover, if more than twice mole ratio of amine relative to each P-X bond is used, removing the un-reacted amine should be done in the purification process, too, which may be performed by stirring the crude product in a diluted hydrochloric acid [65].

Setzer and co-workers reported the synthesis of 1,3,2-oxazaphospholane from the reaction between (lR,2R)-(-)-pseudoephedrine, phenyl dichlorophosphate and triethylamine in ethyl acetate. Triethylamine hydrochloride was filtered off and the solvent removed from the filtrate under reduced pressure [66].

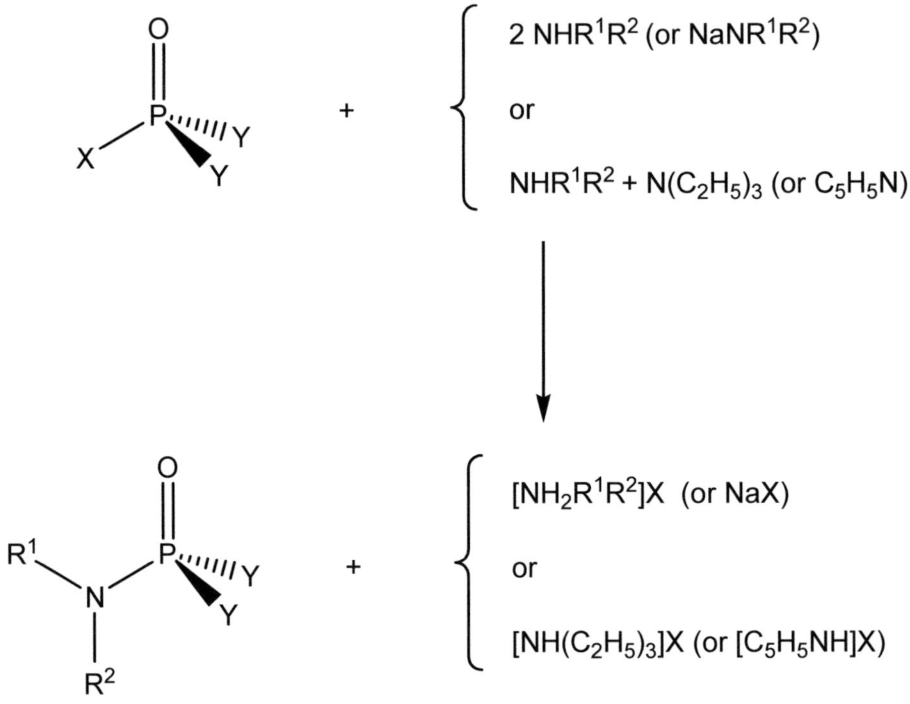

Scheme 4. A common route for the synthesis of phosphoramidates

This method may be developed to the reaction between phenols and phosphorus-chlorine compounds. Selecting of a suitable solvent, which triethylamine hydrochloride or the other salt is low-soluble, develops the synthesis of some initial phosphorus-chlorine compounds such as $[(CH_3)(C_6H_{11})N]P(O)Cl_2$ [67], $[4\text{-}CH_3C_6H_4NH]P(O)Cl_2$ [68], $[C_6H_5O][4\text{-}CH_3C_6H_4NH]P(O)Cl$ [69] and so on. For example, *para*-toluidine hydrochloride is relatively insoluble in CH_3CN; so, the reaction of $P(O)Cl_3$ or $[C_6H_5O]P(O)Cl_2$ with $4\text{-}CH_3\text{-}C_6H_4NH_2$ (1:2 mole ratio) respectively leads to the formation of $[4\text{-}CH_3C_6H_4NH]P(O)Cl_2$ [68] and $[C_6H_5O][4\text{-}CH_3C_6H_4NH]P(O)Cl$ [69] which are soluble in acetonitrile, whereas *para*-toluidine hydrochloride is simply filtered off.

Selection of a suitable solvent for such reactions leads to avoid from the time tedious purification methods such as column chromatography. Recently, we are developing this simple strategy for the synthesis of new phosphorus-chlorine compounds such as $[C_6H_5O]P(O)[NHC_6H_{11}]Cl$, $CF_3C(O)NHP(O)[NHC_6H_4(4\text{-}CH_3)]Cl$, $[C_6H_{11}NH]P(O)Cl_2$ and $[(C_6H_5CH_2)_2N]P(O)Cl_2$ [70].

With starting from $P(O)Cl_3$ or PCl_5 as initial phosphorus-chlorine compounds to reaction with an amine, surely a dry solvent is needed. A fully de-watered solvent is obtained by refluxing of a relatively dry solvent in the presence of a very efficient drying agent such as P_2O_5 (for CCl_4 and $CHCl_3$) or sodium (for CH_3OH, C_2H_5OH, C_6H_6 and $C_6H_5CH_3$) and distilling the totally dried solvent. However, it seems that the sensitivity of a $YP(O)Cl_2$ starting material (Y = amide, alkoxy, phenoxy and so on) is very reduced to the moisture and the solvent which was dried with a moderate desiccant (such as $CaCl_2$) is good for the synthesis. For a bulky amine such as *iso*-propylbenzyl amine or di-cyclohexyl amine as a nucleophile, it seems that a totally-dried solvent is better; of course, it needs to approve with further experiments.

In the case of *iso*-propylbenzyl amine as nucleophile, the reactions with $[C_6H_5O]P(O)Cl_2$, $[C_6H_5O]_2P(O)Cl$ or 4-F-$C_6H_4C(O)NHP(O)Cl_2$ were not successful to prepare the pure $[C_6H_5O]P(O)[N(CH(CH_3)_2)(CH_2C_6H_5)]_2$, $[C_6H_5O]_2P(O)[N(CH(CH_3)_2)(CH_2C_6H_5)]$ and 4-F-$C_6H_4C(O)NHP(O)[N(CH(CH_3)_2)(CH_2C_6H_5)]_2$; however, the crystal structures of two polymorphs of $[NH_2(CH(CH_3)_2)(CH_2C_6H_5)]Cl$ were obtained [71,72]. With using this amine, the compounds [4-NO_2-$C_6H_4C(O)NH]P(O)[N(CH(CH_3)_2)(CH_2C_6H_5)]_2$ [51], $[NH_2(CH(CH_3)_2)(CH_2C_6H_5)][CCl_3C(O)NHP(O)(O)[OCH_3]]$ [40] and $[NH_2(CH(CH_3)_2)(CH_2C_6H_5)][CF_3C(O)NHP(O)(O)(N(CH(CH_3)_2)(CH_2C_6H_5))]$ [73] were prepared which structurally studied, too.

In the case of $[NH_2(CH(CH_3)_2)(CH_2C_6H_5)][CCl_3C(O)NHP(O)(O)[OCH_3]]$ salt, for example, it seems that the presence of a few amount of H_2O in solvent (or environment) leads to the formation of $CCl_3C(O)NHP(O)(OH)Cl$ which the proton-transfer reaction with the amine produces $[NH_2(CH(CH_3)_2)(CH_2C_6H_5)][CCl_3C(O)NHP(O)(O)Cl]$ and then crystallization in methanol replaces the Cl with OCH_3. Moreover, from the reaction of $P(O)(OC_6H_5)Cl_2$ and $NH(C_6H_{11})_2$, the related pure amido phosphoric acid ester was not achieved; however, the crystal structure of $[(C_6H_{11})_2NH_2]^+Cl^-$ was obtained [74].

We are going to try to synthesize neutral phosphoramidate compounds with this and the other bulky amines. A similar feature was observed for the reaction of $POCl_3$ with *tert*-butyl cyclohexyl amine in $CHCl_3$ under reflux condition, where the salt $[NH_2($*tert*-$C_4H_9)(C_6H_{11})][PO_2Cl_2]$ was obtained [75].

The moisture led to the formation of some undesirable but interesting products such as $X_2P(O)OP(O)X_2$ (X = $(CH_3)_3CNH$ [76], $C_6H_4(2\text{-}CH_3)NH$ [48] and $C_6H_4(4\text{-}CH_3)NH$ [77]) from the reaction of $P(O)Cl_3$ and corresponding amine (1 to 6 or more mole ratio) , and also formation of [3-F-$C_6H_4C(O)NH][(CH_3)_3CNH]P(O)(O)P(O)[NHC(CH_3)_3][NHC(O)C_6H_4(3\text{-}F)]$ [78] from the reaction of 3-F-$C_6H_4C(O)NHP(O)Cl_2$ and *tert*-butyl amine. Another salt, [*tert*-$C_4H_9NH_2][CF_3C(O)NHP(O)(O)NH($*tert*-$C_4H_9)].0.333CH_3CN.0.333H_2O$, was also obtained [79].

N-methyl cyclohexyl amine showed an interesting feature in some examples which may be accidental needing to further considerations. In the reaction of 4-$CH_3C_6H_4S(O)_2NHP(O)Cl_2$ with an excess amount of $NH(CH_3)(C_6H_{11})$ (1:5 mole ratio), the product is a proton-transfer compound, $[NH_2(CH_3)(C_6H_{11})][4\text{-}CH_3\text{-}C_6H_4S(O)_2NP(O)[N(CH_3)(C_6H_{11})]_2]$ [23]; furthermore, the crystal structures of $[NH_2(CH_3)(C_6H_{11})][CF_3C(O)NP(O)[N(CH_3)(C_6H_{11})]_2]$ [80] and $[NH_2(CH_3)(C_6H_{11})][CCl_3C(O)NP(O)[N(CH_3)(C_6H_{11})]_2]$ [81] were obtained, but in an effort to preparation of their alkaline complexes.

Synthesis of such proton-transfer compounds through stirring a mixture of a few examined amines (NHR^1R^2) and a synthesized phosphoric triamide ($CF_3C(O)NHP(O)[NR^1R^2]_2$) were not successful; however, this also needs to some further experiments.

Another strategy for preparation of phosphoramidate compounds is the application of sodium amide salts which produces sodium halide as a by-product [82] (Scheme 4).

The P-N bond formation between an amide, of the type $RC(O)NH_2$, and a phosphorus(V) site may be performed *via* a two-stages reaction, which is shown for PCl_5 in Scheme 5, showing the reaction of PCl_5 with an amide and then the treatment of HCOOH. Moreover, a few efforts have been devoted to the synthesis of $RS(O)_2NHP(O)Cl_2$ by a similar procedure [23].

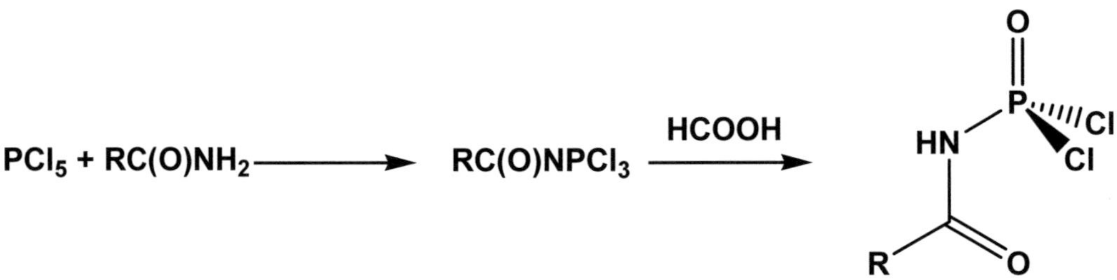

Scheme 5. Synthesis of $RC(O)NHP(O)Cl_2$

The simple mentioned methods for the preparation of phosphoramidates from the reaction of phosphorus-chlorine compounds and amines may be extended to the diamines or amino alcohols to produce cyclic [83] or bridged compounds [61].

The preparation of some compounds containing the P-Cl bonds, such as 4-$CH_3C_6H_4OP(O)Cl_2$ and $[(CH_3)_2N]P(O)Cl_2$, were performed through the reaction between corresponding phenol derivatives or amine hydrochloride salts [for example *para*-cresol or dimethylamine hydrochloride salt for the mentioned phosphorus-chlorine compounds] with an excess amount of $POCl_3$ and then the removal of the remaining $POCl_3$ in a reduced pressure [32,47].

Some compounds were synthesized by the reaction of P-H compounds (such as dimethylphosphine oxide, $(CH_3)_2P(O)H$ and *N,N*-disubstituted derivatives of 5,6-benzo-2H-2-oxo-1,3,2λ^4-diazaphosphorinan-4-one, Scheme 6) with ketones [84].

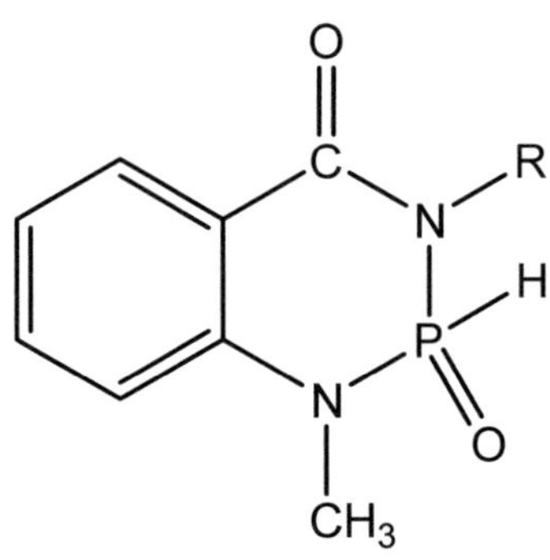

Scheme 6. 5,6-benzo-2H-2-oxo-1,3,2λ^4-diazaphosphorinan-4-one

Wan and Modro developed the synthesis of a bicyclic phosphoric triamide (Scheme 7) *via* the base-promoted cyclization of the corresponding 3-(2-chloroethyl)-2-oxo-1-aryl-2-arylamino-1,3,2-diazaphospholidine [85].

Mbianda and co-workers reported the solvolysis of 1-oxo-2,8-diphenyl-2,5,8-triaza-1 λ^5-phosphabicyclo[3.3.0]octane under base-promoted alcoholysis and acid-catalyzed alcoholysis. Scheme 8 shows the two different products of such solvolysis reactions [86].

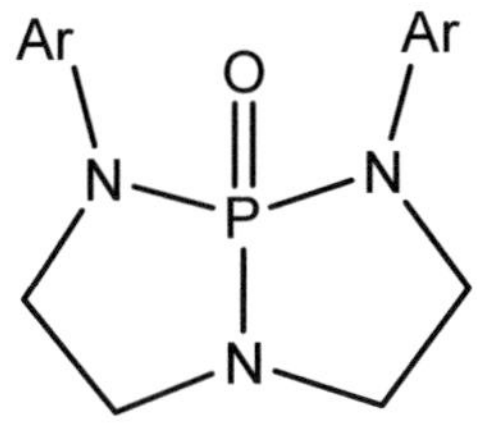

Scheme 7. A bicyclic phosphoric triamide

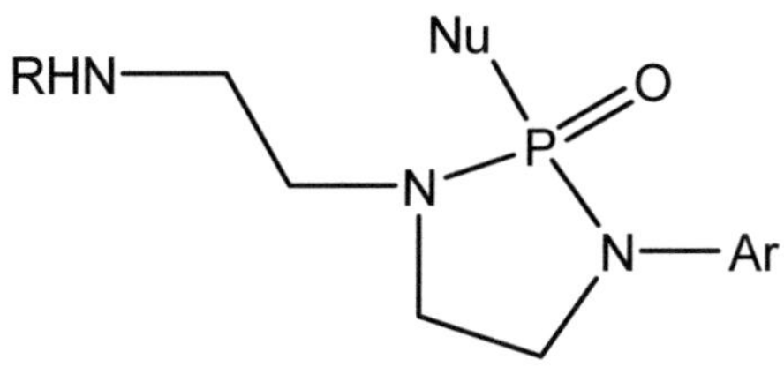

Base-promoted alcoholysis

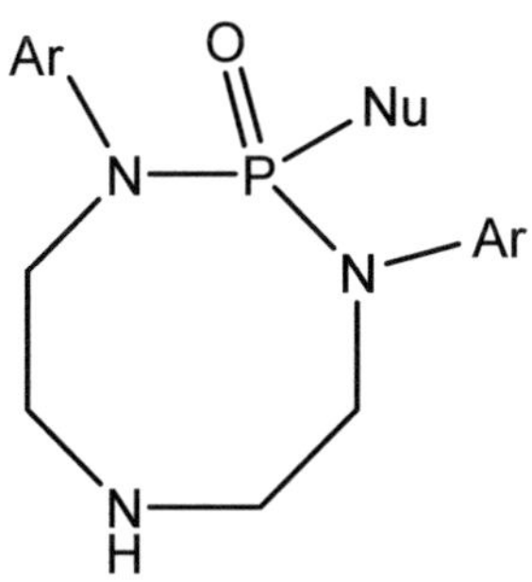

Acid-catalyzed alcoholysis

Scheme 8. The obtained products from the solvolysis of 1-oxo-2,8-diphenyl-2,5,8-triaza-1 λ^5-phosphabicyclo[3.3.0]octane under base-promoted alcoholysis and acid-catalyzed alcoholysis

3. Crystallization of phosphoramidates

The convenient solvents for obtaining suitable single crystals for the studied compounds may be $CH_3C(O)CH_3$, $CHCl_3$, $CHCl_3/n\text{-}C_7H_{16}$, CH_2Cl_2, CH_3CN, CH_3CN/CH_3OH, $CH_3CN/CHCl_3$, CH_3OH, CH_3OH/H_2O, $C_2H_5OH/n\text{-}C_6H_{14}$, $(CH_3)_2CHOH/n\text{-}C_6H_{14}$, $(CH_3)_2NC(O)H/CHCl_3$, $(CH_3)_2NC(O)H/CH_3OH$ and $n\text{-}C_6H_{14}$. The crystal may be obtained at room temperature after slow evaporation of the solvent.

4. General features of phosphoramidate compounds

Compounds with formula $RC(O)NHP(O)[NR^1R^2]_2$ *and* $RC(O)NHP(O)[NHR^1]_2$

The four different groups linked to the P atom result in a distorted tetrahedral configuration; as one instance, the bond angles around the P atom of $P(O)[NHC(O)CF_3][NHCH_2C_6H_4(2\text{-}Cl)]_2$ range from 102.67(12)° to 117.60(12)° [87]. In the

C(O)NHP(O) moiety, however, the carbonyl and phosphoryl groups are separated from each other with one N-H unit, but the terms *syn, gauche* and *anti* were used for the description of the C(O) orientation versus P(O) in the literatures [10,26]. Up to now, both *gauche* and *anti* orientations were found for phosphoric triamide compounds having a C(O)NHP(O) skeleton, respectively in 14 and 98 structures (from the 119 structurally reported compounds, 7 cifs are not available). Among them, for the acyclic compounds of the type $RC(O)NHP(O)[NHR^1]_2$, merely, the *anti* orientation was reported, so far.

Fig. 1. indicates a general view of compounds with formula $CF_3C(O)NHP(O)[NHR]_2$.

In the C(O)NHP(O) moiety, the P—N bond is longer and the O—P—N angle is contracted compared with the respective values in the $[P(O)NHR]_2$ section. For the phosphoramidate compounds, each N atom bonded to phosphorus has a sp^2 character which is reflected in the C—N—P angles of the C(O)NHP(O) or C—NH—P moiety or sum of the surrounding angles around the tertiary nitrogen atom (C—N—C + C—N—P + P—N—C). The deviation of this summation from 360° (to a lower value) has been used to show the deviation of nitrogen atom environment from planarity. This may be also illustrated with the distance between the position of N atom from the plane crossing from the directly attached atoms to nitrogen, i.e. C, C and P. In $[C_6H_5O]_2P(O)[NC_4H_8N]P(O)[OC_6H_5]_2$ (Fig. 2) which belongs to the amidophosphoric acid ester family, the N atom shows some deviation from planarity and it is 0.25(1) Å above (or below) the CCP plane [61]. For the phosphoramidate compounds, the P-N bonds are shorter than the P-N single bond and the P=O bond are longer than the normal P=O bond [88].

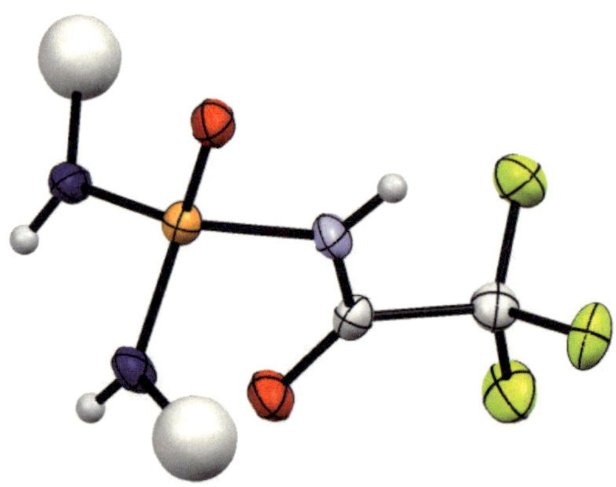

Fig. 1. A typical view for a compound with formula $CF_3C(O)NHP(O)[NHR]_2$ [Color key: O atoms are red, the N atom of C(O)NHP(O) is light blue, the other amido N atoms are dark blue, F atoms are yellowish green., C and H atoms are light grey and P atom is orange]; the R substituents are shown as big balls.

In 1,3-diazaphosphorinane compounds, the P=O bond is placed in an equatorial position and the aliphatic six-membered rings adopt conformation between chair and envelope (Fig. 3).

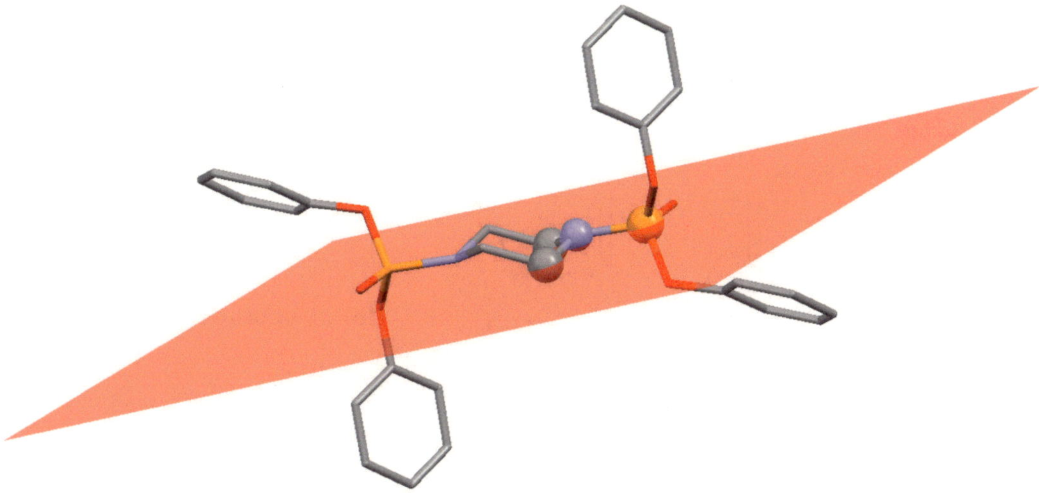

Fig. 2. A view of the PCC mean plane which is crossed from the phosphorus and carbon atoms shown as balls in the right-side $[C_6H_5O]_2P(O)(NC_2H_4)$ moiety of $[C_6H_5O]_2P(O)[NC_4H_8N]P(O)[OC_6H_5]_2$ (the molecule is organized around an inversion center located at the centre of the piperazine ring), the N atom environment shows some deviation from planarity. The balls representation denote to the P (orange), N (blue) and C (grey) atoms.

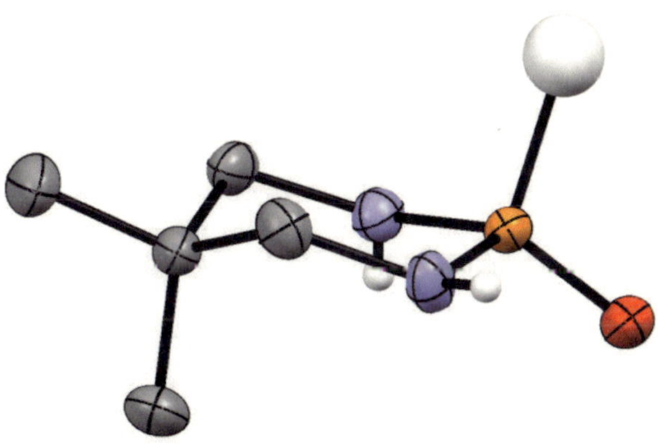

Fig. 3. A general view of a 1,3-diazaphosphorinane, a six-membered ring heterocyclic phosphorus compound (the carbon-bonded H atoms were omitted for clarity). The grey big ball in the figure may be RC(O)NH, RNH or the other moieties.

The hydrogen bond pattern of compounds having the $C(O)NHP(O)(N)_2$ and $C(O)NHP(O)(NH)_2$ skeletons may be predictable with considering the following "empirical rules":

1. In the reported compounds, the nitrogen atoms bonded to P don't involve in hydrogen bonding interaction as an acceptor (due to their low Lewis base characteristic). Scheme 9 illustrates the possible H-donor sites and H-acceptor centers in the structure of compounds having the $C(O)NHP(O)(N)_2$ and $C(O)NHP(O)(NH)_2$ skeletons.

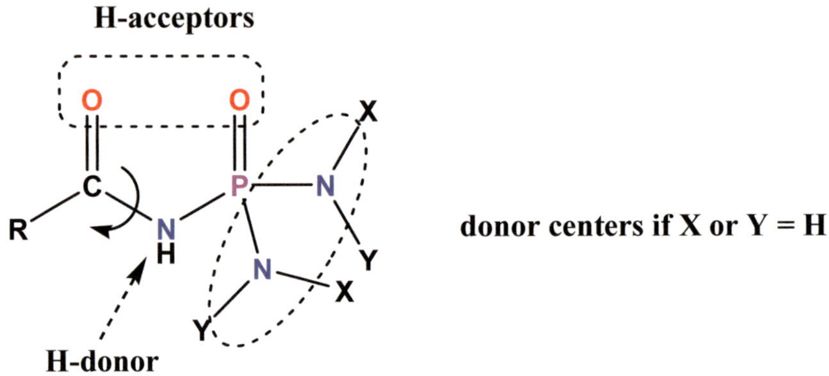

Usually the N atoms don't involve in hydrogen bonding as an acceptor

Scheme 9. The possible H-donor sites and H-acceptor centers in the $C(O)NHP(O)(N)_2$ and $C(O)NHP(O)(NH)_2$ skeletons (the R, X or Y groups may also be contained the additional H-donor or H-acceptor sites in their structures which may be involved in the H-bond pattern, the curved arrow shows that the orientation of C=O versus P=O may change)

2. The P=O is a better H-acceptor than the C=O counterpart.
3. In compounds having a $C(O)NHP(O)(N)_2$ skeleton, i.e. with formula $RC(O)NHP(O)[NR'R'']_2$, both *gauche* and *anti* orientations of P=O *versus* C=O have been found, so, two kinds of packing are expectable which are seen: a) a 1-D chain for a *gauche* orientation, and b) a dimeric aggregate (as an $R_2^2(8)$ loop; for H-bond motifs of phosphoric triamide, see: ref. [10]) with C_i or C_1 symmetry for *anti*. The unique NH proton interacts with the oxygen atom of PO, whereas the CO does not cooperate in HB. Such H-bond patterns may also be expectable for the other phosphoramidate compounds having a P(O)NH group, Scheme 10; however, the other H-bond patterns have also been observed in the other sub-categories with a P(O)NH moiety which will be noted, later.
4. In compounds having a $C(O)NHP(O)(NH)_2$ skeleton, only an *anti* situation has been found in acyclic molecules; however, in diazaphosphorinane molecules both conformations were found.
5. In the crystal packing of acyclic compounds having a $C(O)NHP(O)(NH)_2$ skeleton, adjacent molecules are often linked *via* $N_{C(O)NHP(O)}$−H…O=P and N−H…O=C (or (N−H)$_2$…O=C) hydrogen bonds, building $R_2^2(8)$ and $R_2^2(12)$ rings (or $R_2^2(8)$ and $R_2^2(12)/R_2^1(6)$) in a linear arrangement (Scheme 11). However, the existence of a PO…HNR interaction has been observed for a few compounds as tri-centered PO[…$H_{C(O)NHP(O)}$N][…HNR] and PO[…HNR][…HNR] hydrogen bonding, where the oxygen atom of the phosphoryl group acts as a double H-acceptor (for a definition of a double-H bond acceptor, see: ref. [89]).

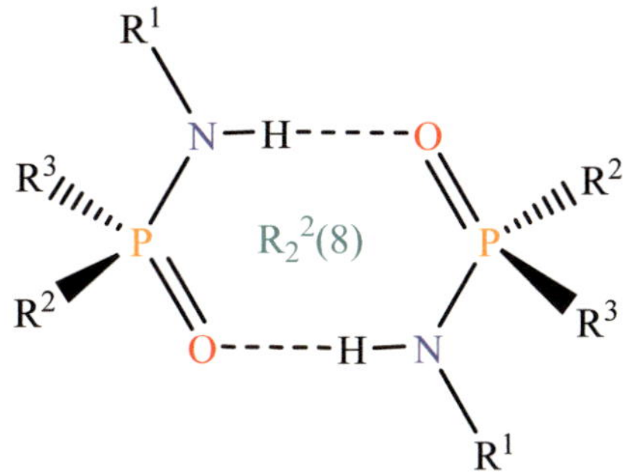

R^1 = alkyl, aryl or C(O)R'
R^2 and R^3 = OR, NR'R"

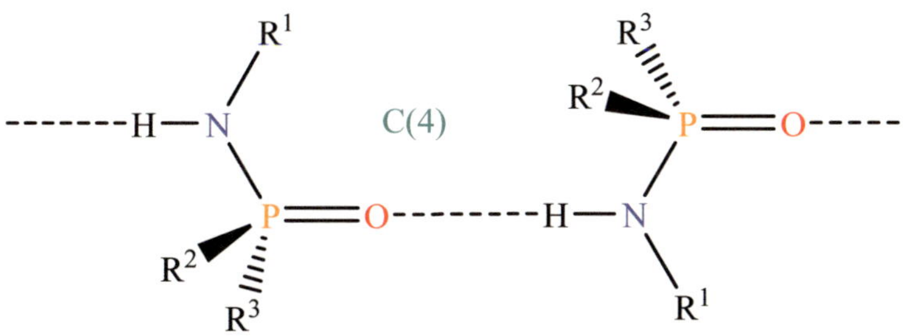

R^1 = alkyl, aryl or C(O)R, R^2 = OR or NR'R", R^3 = OR, NHR' or NR"R'"

Scheme 10. The observed hydrogen bond patterns in compounds having a P(O)NH skeleton (such as $C(O)NHP(O)[N]_2$)

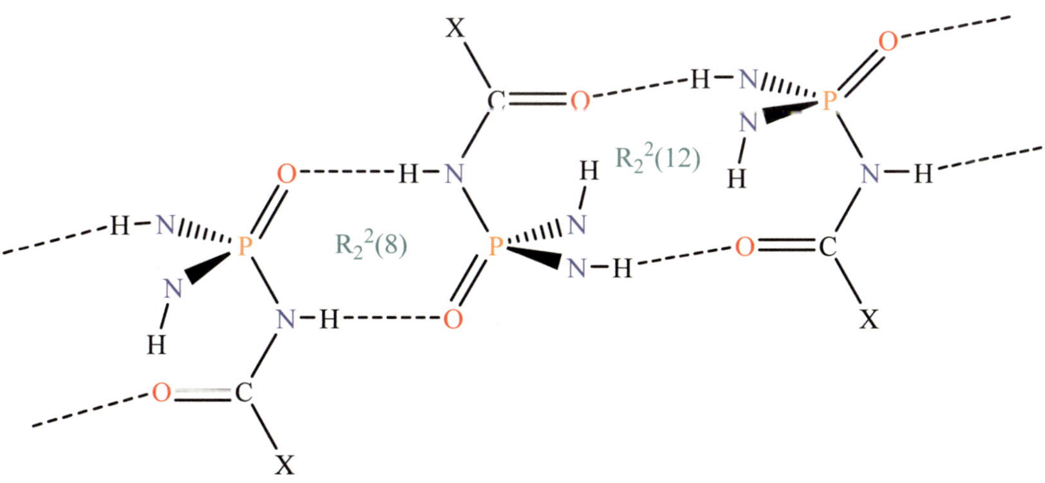

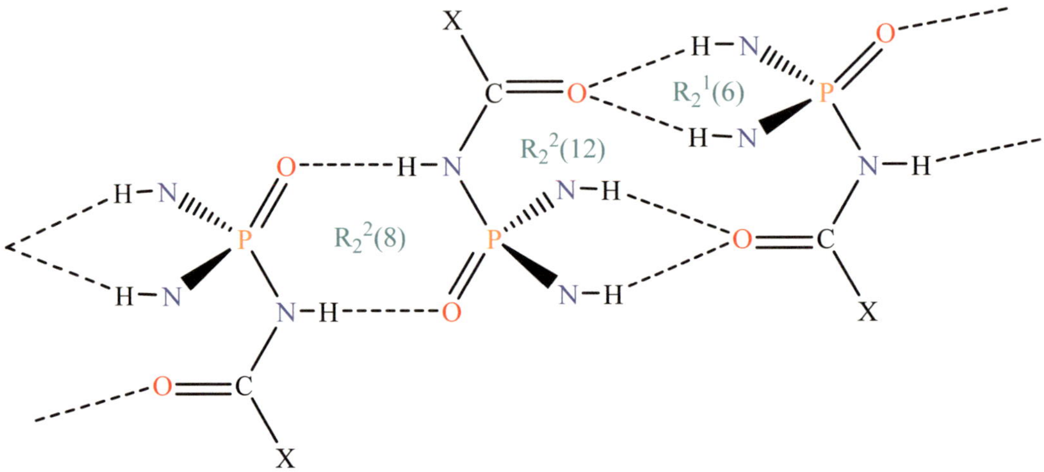

Scheme 11. A sequence of $R_2^2(8)$ and $R_2^2(12)$ (top), a sequence of $R_2^2(8)$ and $R_2^2(12)/R_2^1(6)$ (bottom) rings in compounds having a C(O)NHP(O)$(NH)_2$ skeleton: in these H-bond patterns, the P=O...H-$N_{C(O)NHP(O)}$ and C=O...H-N or C=O...$(H\text{-}N)_2$ exist

In most cases of compounds having a C(O)NHP(O)$(NH)_2$ skeleton (containing two H-acceptors–three H-donors), the HBs lead to a 1-D chain. Different 1-D ladder arrangements with tetramer motifs and a linear arrangement with two different kinds of motifs (dimer and tetramer) were also observed. Therefore, two H-donor sites ($HN_{C(O)NHP(O)}$ and one of the HNR) participate with two O atoms in the intermolecular HBs, the other HNR may act as the three following manners: (a) in an intramolecular HB with C(O), (b) in a weaker HB with P(O) as the above mentioned tri-centered HB and (c) without cooperation in any HB.

6. A sequence of $R_2^2(10)$ (or $R_2^2(10)/R_2^1(6)$ or $R_2^2(10)/S_6$) may be expected when the C=O is hydrogen-bonded to the $N_{C(O)NHP(O)}$-H unit and the P=O interacts with the N_{amide}-H unit. If the remaining N_{amide}-H unit is involved in an intramolecular hydrogen bond with the oxygen of carbonyl, the $R_2^2(10)/S_6$- graph-set is formed, in this case the oxygen atom of carbonyl acts as a double-H acceptor. In the case of involving this N-H unit in the H-bonding interaction with the oxygen of phosphoryl, the $R_2^2(10)/R_2^1(6)$ is formed (Scheme 12).
7. In a solvated molecule [90], the hydrogen-bond pattern may not be predictable, but some previously mentioned rules may be beneficial, Scheme 13.
8. The investigation for the phosphoric triamide containing a C(O)NHP(O)$(NH)_2$ skeleton shows that in eleven structures the carbonyl oxygen atom acts as a double-H acceptor *via* C(O)...(H−N)(H−N) grouping (in the $R_2^2(12)/R_2^1(6)$ motifs) and in ten structures the phosphoryl oxygen atom acts as a double-H acceptor *via* P(O)...(H−N)(H−N) or P(O)...(H−N)(H−$N_{C(O)NHP(O)}$) groups in the $R_2^2(12)/R_2^1(6)$ rings or $R_2^2(10)/R_2^1(6)$ ring or in the 1-D ladder arrangement). In the other such phosphoramidate compounds the remaining N−H unit doesn't cooperate in H-bonding interaction. The unique found structure of phosphoramidate with a C(O)NHP(O)Cl_2 skeleton is $CF_3C(O)NHP(O)Cl_2$ [91].

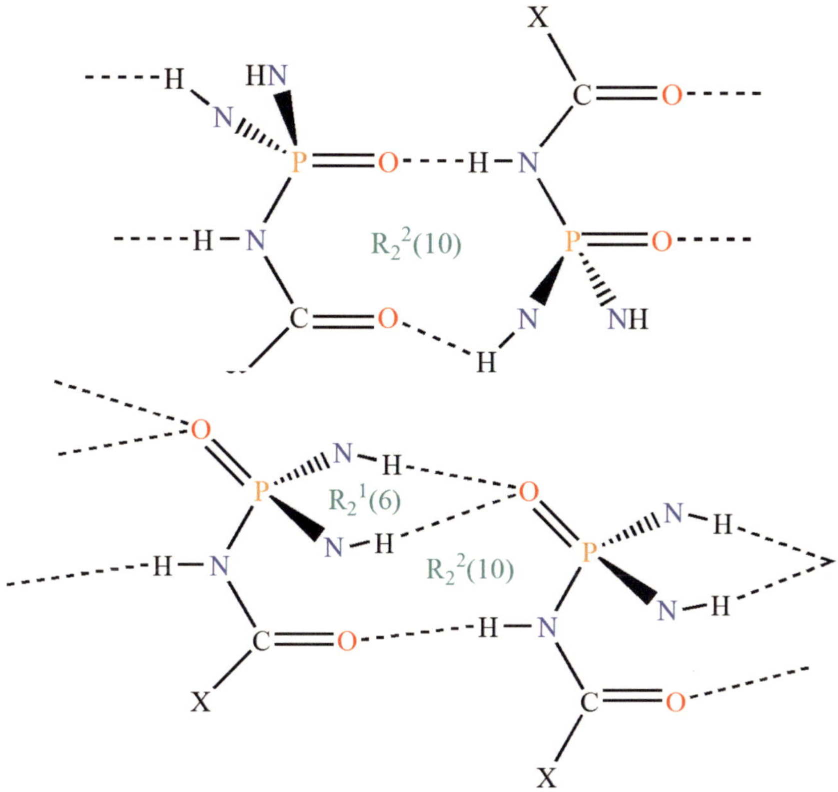

Scheme 12. A sequence of $R_2^2(10)$ (top) and $R_2^2(10)/R_2^1(6)$ rings (bottom) in compounds having a C(O)NHP(O)$(NH)_2$ skeleton; in the existing examples of $R_2^2(10)/S_6$, the N-H…O angle is less than 110°

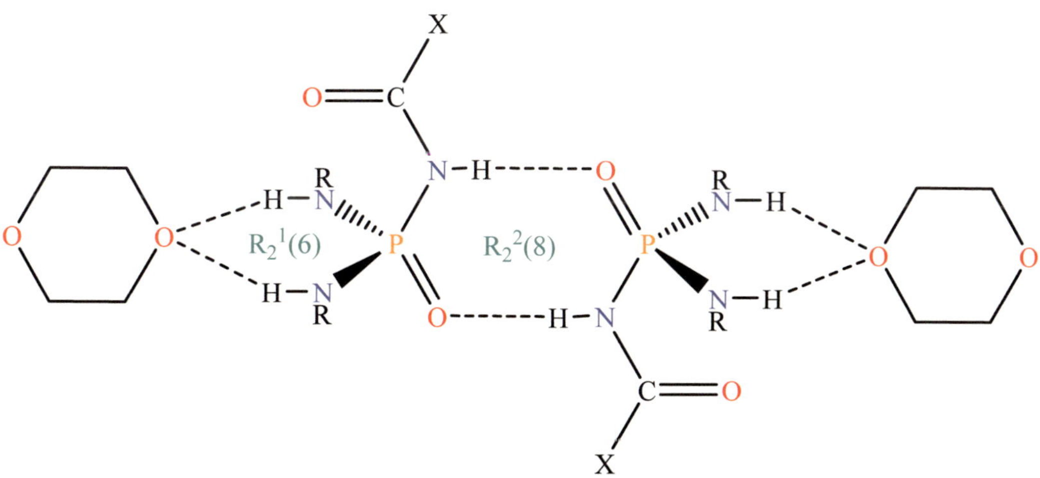

Scheme 13. A four-component cluster in the solvated molecule $CCl_3C(O)NHP(O)[NHC_6H_{11}]_2.C_4H_8O_2$(X = CCl_3, R = C_6H_{11})

Proton transfer compounds

In the proton transfer compounds containing a $\{[C(O)NP(O)][N]_2\}^-$ skeleton (Scheme 14), the P-N bond in the [C(O)NP(O)] fragment is shorter than the two other P-N bonds, one example for such compounds is $[C_6H_{11}NH_2CH_3][CF_3C(O)NP(O)[N(CH_3)(C_6H_{11})]_2]$ in which the phosphoryl and carbonyl groups are staggered [O-P-N-C = 64.8(3)°] [80].

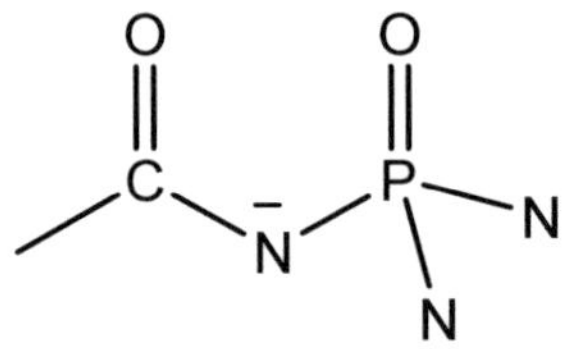

Scheme 14. The $\{[C(O)NP(O)][N]_2\}^-$ skeleton

The hydrogen bonds (Scheme 15) in such compounds, of the type charge-assisted and also polarization-assisted HBs [92], are strong, reflecting in the distances between the donor and acceptor atoms. Scheme 16 shows a polarization-assisted hydrogen bond in a neutral phosphoramidate, for comparison.

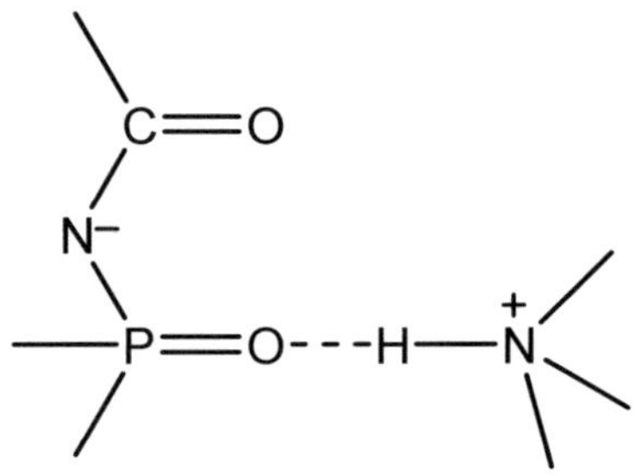

Scheme 15. Contribution of two factors in strengthening of hydrogen bonds (charge and polarization) in the proton-transfer compounds: charge and polarization-assisted hydrogen bonds

Two reported crystal structures of this category, $[C_6H_{11}NH_2CH_3][CF_3C(O)NP(O)[N(CH_3)(C_6H_{11})]_2]$ [80] and $[C_6H_{11}NH_2CH_3][4\text{-}CH_3\text{-}C_6H_4S(O)_2NP(O)[N(CH_3)(C_6H_{11})]_2]$ [20], have a similar HB pattern as a centrosymmetric four-component cluster involving two anions and two cations which interact through N—H...O hydrogen bonds (Fig. 4).

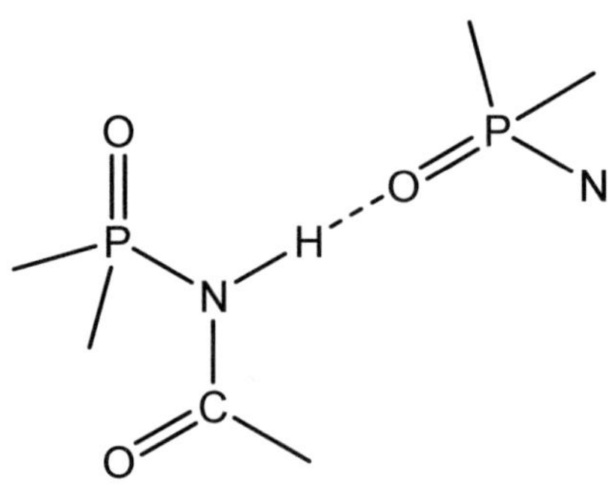

Scheme 16. A polarization-assisted hydrogen bond in a neutral phosphorus compound

Compounds with formula $[R^1R^2N][R^3R^4N][R^5R^6N]P(O)$, $[RNH][R^1R^2N][R^3R^4N]P(O)$, $[R^1NH][R^2NH][R^3R^4N]P(O)$, $[R^1NH][R^2NH][R^3NH]P(O)$ or more complicated phosphoric triamide compounds

Tris-alkyl (aryl) amido phosphates of the formula $[R^1R^2N]_3P(O)$, with three equal amido substituents linked to the P atom, are easily prepared from a one-pot reaction between phosphoryl chloride and corresponding amine. The single crystal X-ray determinations were performed for $[R^1R^2N]$ = $NHCH_3$ (KABVAL) [93], $N(CH_3)_2$ (POTJAJ) [94], $NHC(CH_3)_3$ (KABVEP) [93], NHC_6H_5 (KEQLUO) [95], $NHCH_2C_6H_5$ (TOKXIB) [96] and $NHC_6H_4(4\text{-}OCH_3)$ (WAWNIS) [97] and also for the substituents shown in Scheme 17 [98-105]. Moreover, a few other phosphoramide compounds which each contains a triamido moiety (like for example compounds with refcodes EDEVAK [106] and NUVSEC [107] (Schemes 18 and 19) and some co-crystal compounds (for example BARHMP [108] and VAFRIE [109], see Schemes 20 and 21) were reported.

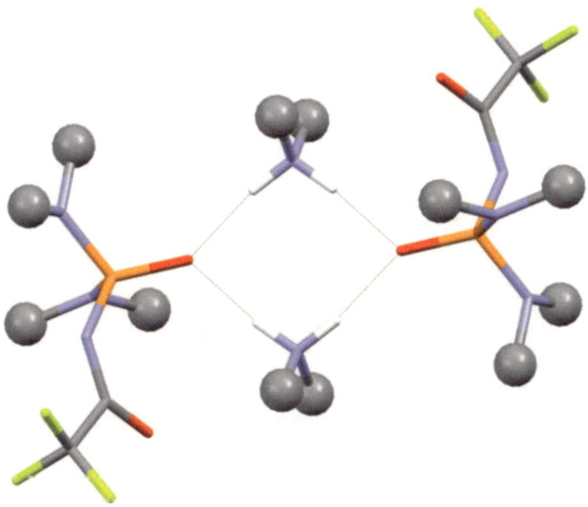

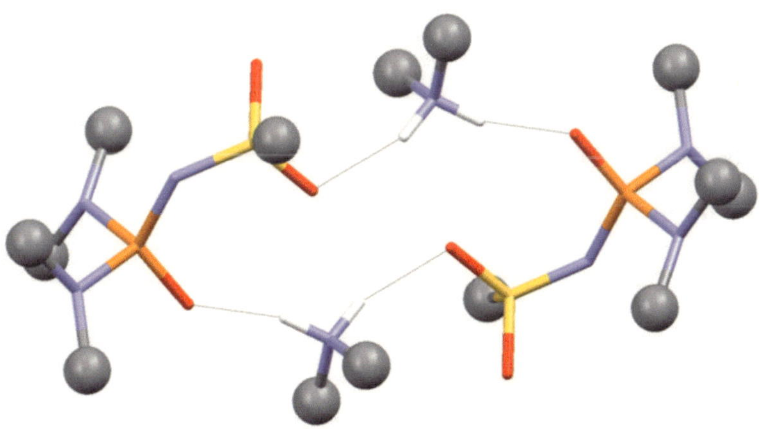

Fig. 4. A view of the H-bonded centrosymmetric four-component cluster in the crystal packing of $\{C_6H_{11}NH_2CH_3\}^+\{CF_3C(O)NP(O)[N(CH_3)(C_6H_{11})]_2\}^-$ (top) and $\{C_6H_{11}NH_2CH_3\}^+\{4\text{-}CH_3C_6H_4S(O)_2NP(O)[N(CH_3)(C_6H_{11})]_2\}^-$ (bottom); the N−H···O hydrogen bonds are shown

as dotted lines. The H atoms not involved in hydrogen bonding have been omitted for the sake of clarity and the 4-CH_3-C_6H_4 (bottom) and C_6H_{11} and CH_3 substituents are shown as balls (the N…O distances are 2.771(3) & 2.804(3) Å and 2.648(4) & 2.864(4) Å, respectively).

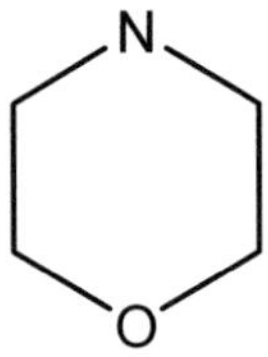	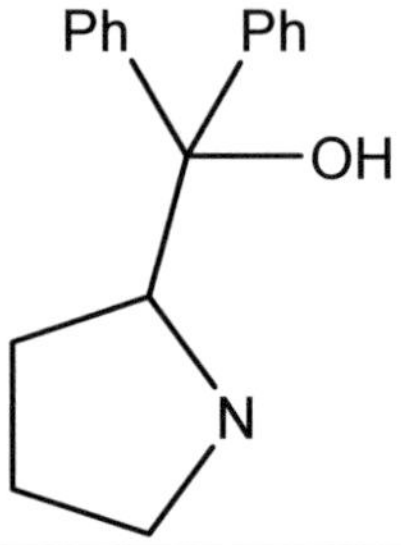	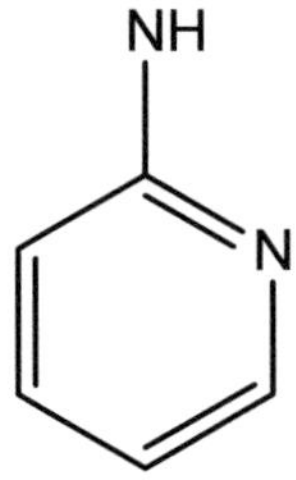	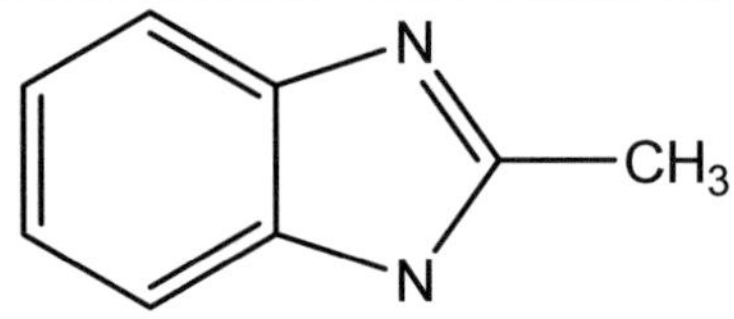
BIVYAG [98]	JEKLER [99]	LAFNAI [100]	QONBEB [101]
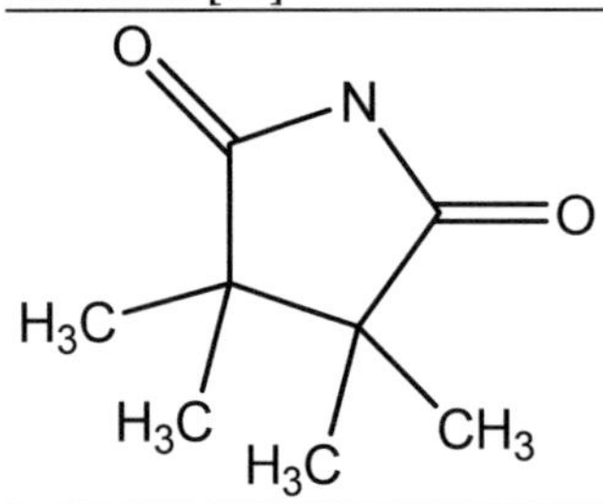	$NHC_6H_4(4\text{-}CH_3)$	$NHC_6H_4(3\text{-}CH_3)$	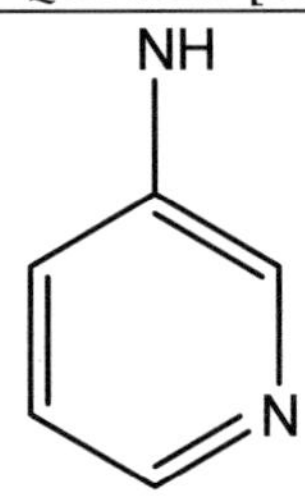
[1]QUFJIL [102]	[2]TAZPOX [103]	[3]TOLZPO [104]	[4]OLOCEY [105]

[1]There is the hydrogen-bonded amide molecule in the structure, i.e. the formula is $C_{24}H_{36}N_3O_7P_1,C_8H_{13}N_1O_2$. [2]There is the solvent C_2H_5OH molecule in the structure, i.e. the formula is $C_{21}H_{24}N_3O_1P_1,C_2H_6O_1$. [3]The formula is $C_{21}H_{24}N_3O_1P_1,2(C_7H_{10}N_1^{1+}),2(Cl_1^{1-})$. [4]The formula is $C_{15}H_{15}N_6O_1P_1,H_2O_1$.

Scheme 17. The structurally investigated compounds of the formula $[R^1R^2N]_3P(O)$ or $[R^1R^2N]_3P(O).B$, where B is a hydrogen-bonded species to phosphoric triamide (the related amido moieties and the CSD refcode are presented)

In the crystal packing of molecules having a $P(O)(NH)_3$ skeleton, hydrogen bonded 1-D chain, 1-D ladder, 2-D layer and 3-D arrangements were found. Three different types of 1-D arrangement are formed respectively through a P=O…$(H\text{-}N)_3$ or P=O…$(H\text{-}N)_2$ groups or *via* the P=O…H-N hydrogen bond. In the two latter cases, respectively one and two N-H units don't cooperate in the hydrogen bond interaction. One example of a linear arrangement, involving the N−H···O and N−H···N HBs, is also found in the structure of $P(O)(NH-C_5H_4N)_3$ (LAFNAI) [100] in which the pyridine nitrogen atom is involving in the HB pattern as an acceptor, too. As, the phosphoryl oxygen atom may cooperate in H-bonding interaction as a double- or a triple- acceptor, some examples of 2-D and 3-D arrangements have also been found in this class of compounds.

Some phosphoric triamides $[R^1R^2N][R^3R^4N][R^5R^6N]P(O)$ (where merely tertiary nitrogen atoms exist in the structure of molecule) have been reported (for example see: Scheme 22 [110]). Structures with a $P(O)(N)_3$ skeleton, where "N" is a tertiary nitrogen atom, do not show any classical (normal) hydrogen bonding in their crystal packing if the substituent involving the N atom doesn't contain the hydrogen linked to an electronegative atom.

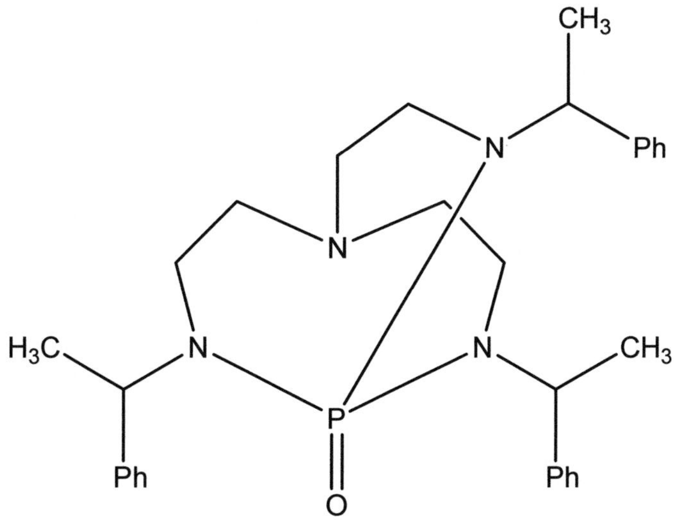

Scheme 18. 4,6,9-Tris(1-phenylethyl)-1,4,6,9-tetraaza-5-phosphabicyclo(3.3.3) undecane P-oxide (refcode: EDEVAK [106])

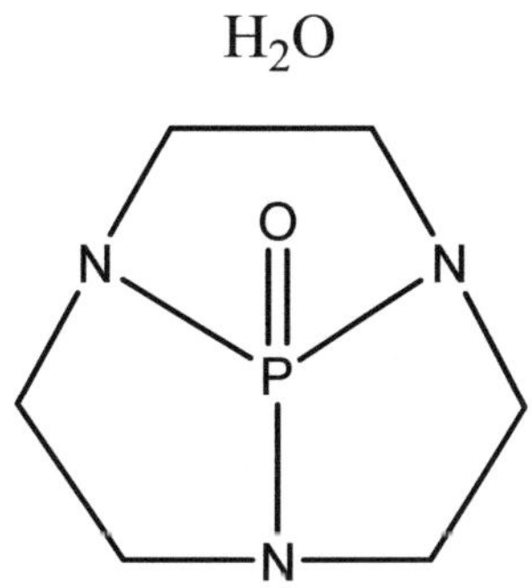

Scheme 19. 10-Oxo-10-phospha-1,4,7-triazatricyclo(5.2.1.0^{4,10})decane monohydrate (NUVSEC [107])

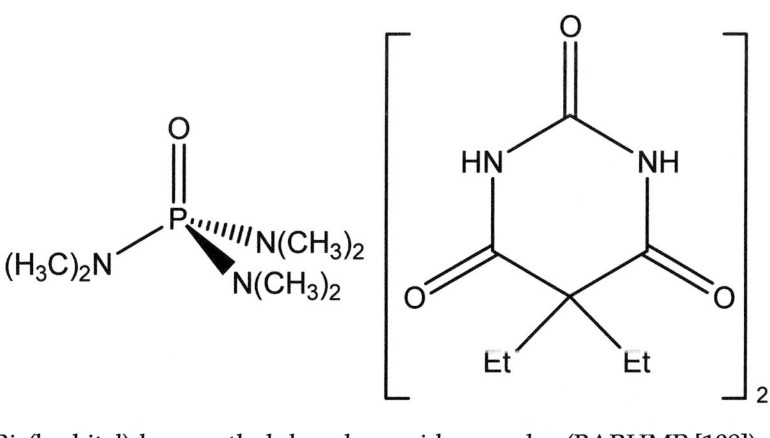

Scheme 20. Bis(barbital)-hexamethylphosphoramide complex (BARHMP [108])

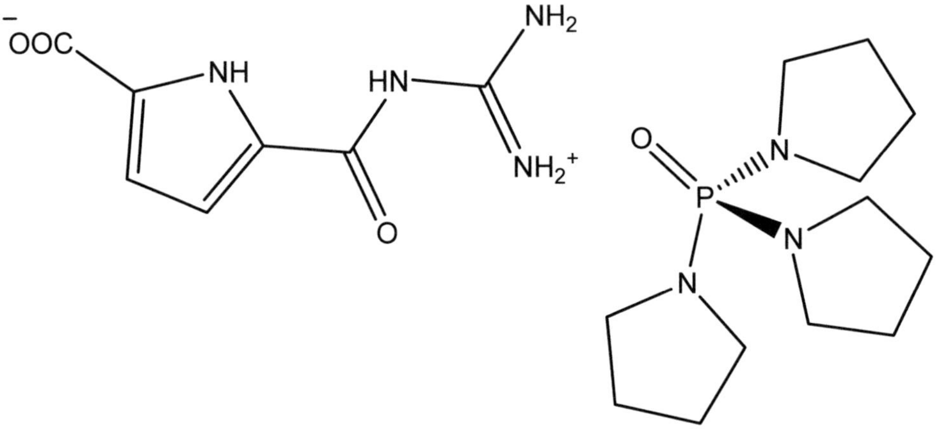

Scheme 21. 5-(Guanidiniocarbonyl)pyrrole-2-carboxylate tris(pyrrolidino)phosphine oxide solvate (VAFRIE [109])

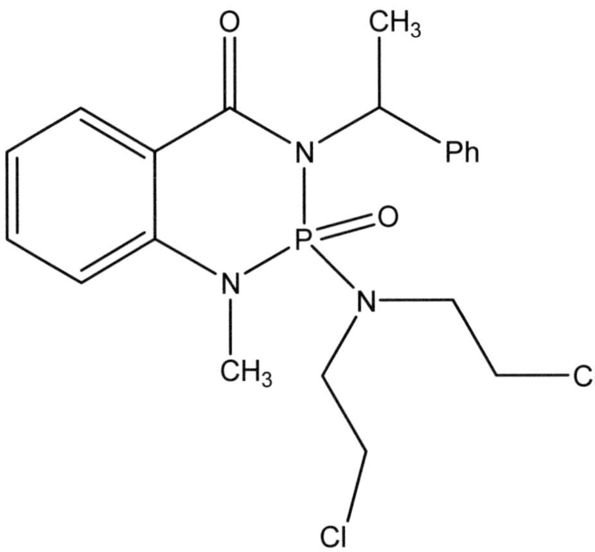

Scheme 22. Refcode BEJNEJ [110]

A search on the CSD shows that the nitrogen atoms bound to phosphorus in phosphoramidate compounds aren't involved in normal H-bonding interaction as an acceptor due to the deviation of each N atom environment from pyramidality after binding to P and decreasing its Lewis base character with respect to the initial amine; so that, merely one example (refcode: HESCEO [111], Scheme 23), belonging to the diazaphosphorinane family, is observed so far with the donor…acceptor (N…N) distance of 3.258(8) Å in which it may be considered as a weak N−H…N−P hydrogen bond.

In compounds having a P(O)(NH)(N)$_2$ skeleton, only the P(O)NH unit cooperates in a HB interaction; so, two expectable HB patterns are the H-bonded dimer (with C_i symmetry; a dimer with C_1 symmetry has not reported, so far) and the 1-D chain (Scheme 10). Usually, an H-bonded dimer forms when the P(O) group and the N−H unit have a *syn* orientation with respect to one another. However, in one structure (refcode: DIYMED [112], Scheme 24) with the *syn* orientation of P(O) *versus* N−H, the molecules are aggregated as a one dimensional H-bonded chain.

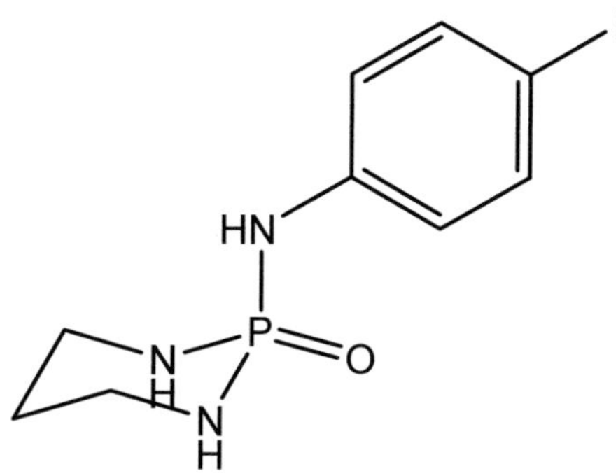

Scheme 23. One N atom of the diazaphosphorinane ring cooperates in hydrogen bonding interaction as an H-acceptor [111]

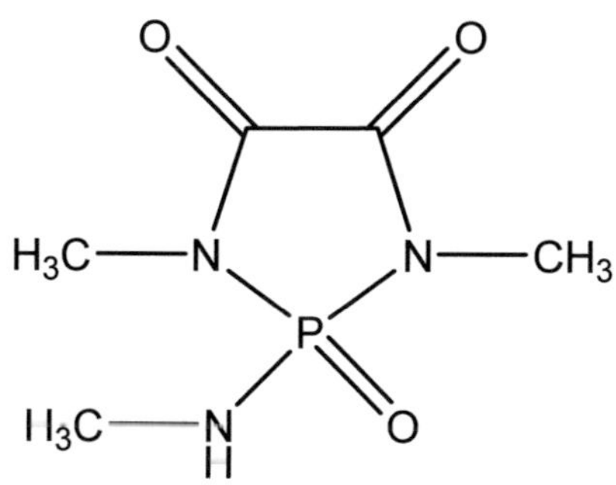

Scheme 24. Refcode DIYMED [112]

If the P(O) adopts an *anti* orientation *versus* NH, an extended 1-D chain arrangement is expectable through the intermolecular PO···HN hydrogen bond which is found for the most of reported compounds. One example without any N−H···O HB (BIFDUP [113], Scheme 25) and one example as H-bonded tetramer (XAVXEY, Scheme 26) were also found.

In compounds having an $(N)P(O)(NH)_2$ skeleton, three different linear arrangements were observed: a) through P(O)···H−N hydrogen bonds in which one N−H unit doesn't cooperate in H-bonding (NUVROL [107], Scheme 27 and HIVLII [67], Scheme 28), b) through $R_2^1(6)$ (in $[(CH_3)_2N]P(O)[NHC_6H_5]_2$ [52] and the compound with refcode MIFYIJ [114], Scheme 29 (top)) and c) through $R_2^2(8)$ rings ($[(CH_3)_2N]P(O)[NHC_5H_9]_2$ [62] and the compound with refcode IKASAP [47], Scheme 29 (bottom)), two latter cases *via* P(O)···(H−N)(H−N) grouping in which the phosphoryl oxygen atom acts as a double-H acceptor.

In this series, some other H-bond motifs were observed in compounds having an NH_2 moiety instead of NHR moiety (BIXFOE [115], GOMDOB [116]), Scheme 30.

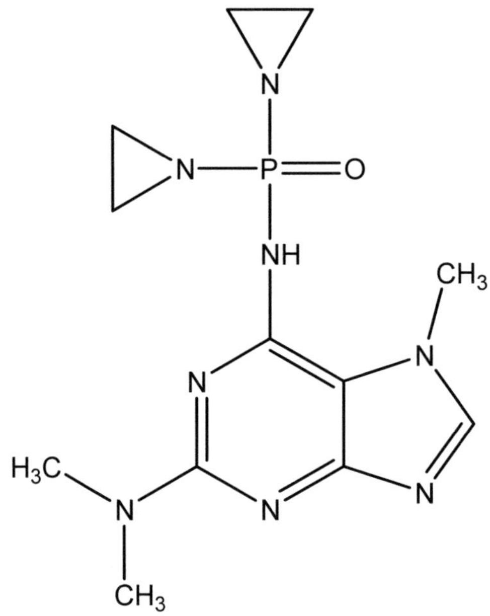

Scheme 25. Refcode BIFDUP [113]

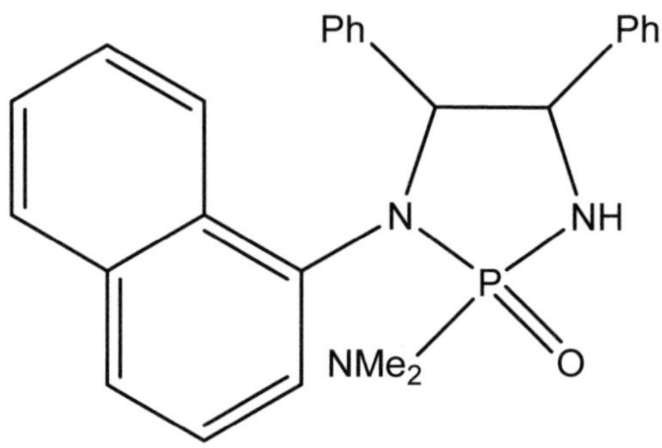

Scheme 26. Refcode XAVXEY (any reference to a journal, book and so on was not found for this structure)

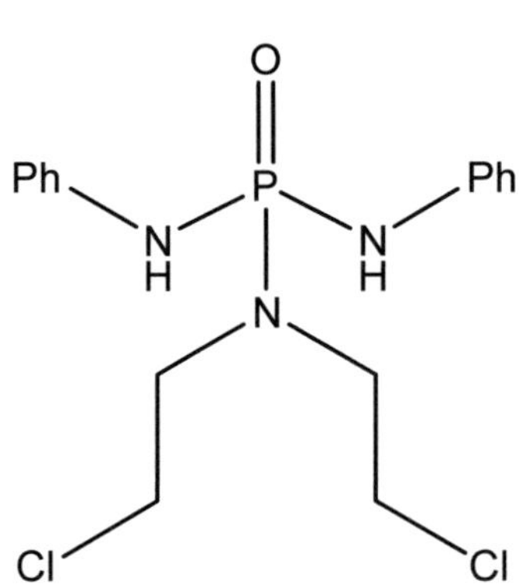

Scheme 27. Refcode NUVROL [107]

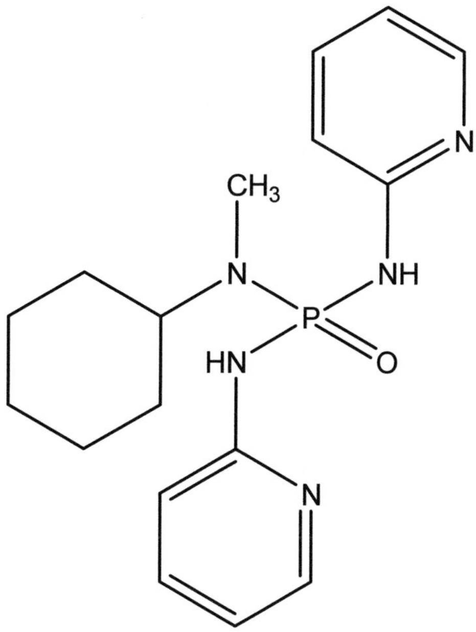

Scheme 28. Refcode HIVLII [67]

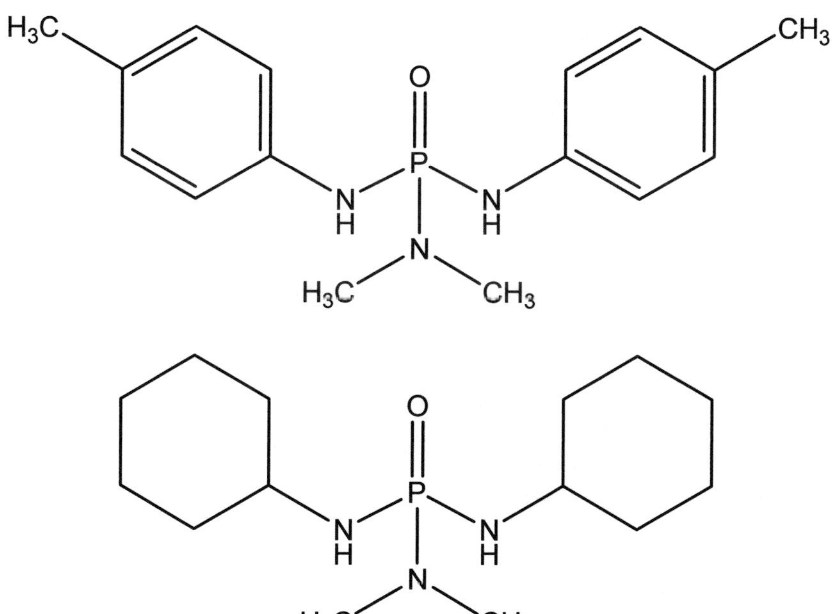

Scheme 29. Refcodes MIFYIJ (top) [114] and IKASAP (bottom) [47]

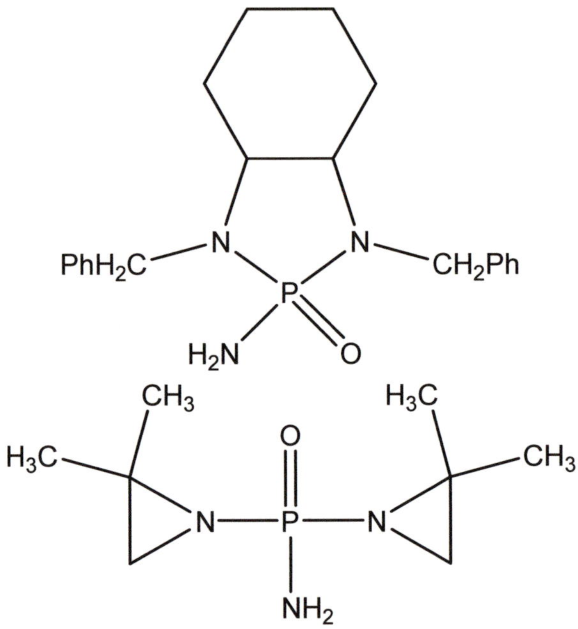

Scheme 30. Refcodes BIXFOE (top) [115] and GOMDOB (bottom) [116]

Compounds with formula $(R^1O)(R^2O)(R^3R^4N)P(O)$, $(R^1O)(R^2O)(R^3NH)P(O)$, $(R^1O)(R^2R^3N)(R^4R^5N)P(O)$, $(R^1O)(R^2NH)(R^3R^4N)P(O)$ and $(R^1O)(R^2NH)(R^3NH)P(O)$

The tetrahedral configuration of phosphorus atom is significantly distorted as it has been noted for the other phosphoramides and their chalco-derivatives [117]. For example, the bond angles around the P atom of $(4\text{-}CH_3\text{-}C_6H_4O)(C_6H_{11}NH)_2P(O)$ [116] vary in the range from 101.48(10)° [for $O_{phenoxy}-P-N1$ angle] to 118.58(9)° [for $O_{phosphoryl}-P-N2$ angle]. The $C-O-P$ angle is 123.52(15)°. A general view of a compound with formula $(4\text{-}CH_3\text{-}C_6H_4O)(RNH)_2P(O)$ is shown in Fig. 5.

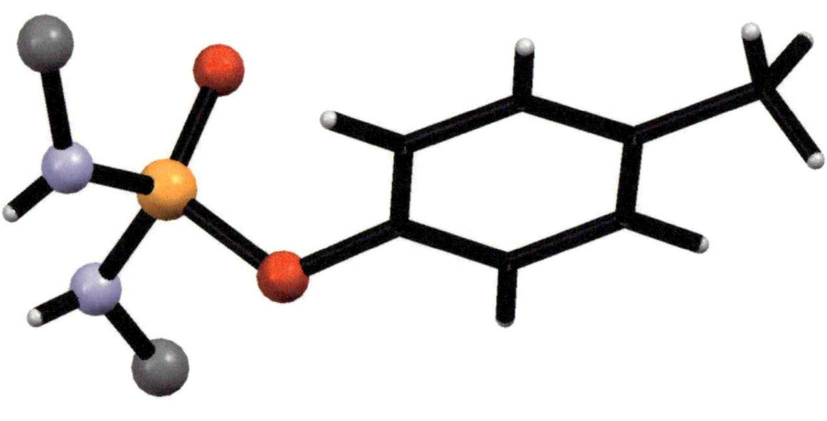

Fig. 5. A general view of a compound with formula $(4\text{-}CH_3\text{-}C_6H_4O)(RNH)_2P(O)$, the R moieties are shown as grey balls.

Similar to the other phosphoramidates, each N atom bonded to phosphorus doesn't involve in any HB as an acceptor, showing its low Lewis-base character. In most cases, it has a nearly planar environment [26]. Of course, the nitrogen atom' environment of some substituents, such as aziridinyl, like for example in the compound with refcode GOMDOB [118] (Scheme 30 (bottom)) shows some deviation from planarity, but such N atom doesn't cooperate in hydrogen bonding interaction, too. Moreover, the oxygen atom of the phenoxy or alkoxy groups in the 47 structures with an (O)P(=O)(NH)(N) skeleton (like for example in $4\text{-}CH_3\text{-}C_6H_4OP(O)[N(CH_3)_2][NHC(CH_3)_3]$, GUDGIW: [119]) doesn't cooperate in the HB interaction, as it can not compete with the phosphoryl oxygen atom for H-accepting from the unique H-donor site in the molecule. Furthermore, among the 106 deposited structures with an (O)(O)P(=O)(NH) skeleton, only the structure of $[CH_3O]_2P(O)[NHCH(CH(CH_3)(OC(O)CH_3))(C(O)(C(NN)(COOC_2H_5)))]$ (IJUMAB: [120]) shows N-H...O(CH_3) not N-H...O(P) hydrogen bond (in this consideration, some structures with unavailable cifs were not enumerated). So, such compounds [if the substituents linked to the N or/and O atoms don't contain any H-acceptor or H-donor centers] may be almost always considered as compounds with "one H-acceptor (the oxygen of phosphoryl) and one H-donor sites", both in the P(O)NH group.

The oxygen atom of OR moiety in some examples of compounds with a higher H-donor sites, such as compounds containing an (O)P(=O)(NH)(NH) skeleton, however, it has a lower H-acceptability than the phosphoryl oxygen atom, is enforced to involve in the HB interaction ($[4\text{-}CH_3\text{-}C_6H_4O]P(O)[NHC_6H_4(4\text{-}CH_3)]_2$: MUBPIJ, [36]). The better H-acceptability of the phosphoryl O atom than that of the RO moiety, in some cases for example in $[C_6H_5O]P(O)[NHC_6H_{11}][NHC_6H_4(4\text{-}CH_3)]$ (ERUFIH: [69]), leads to act it as a double-H acceptor. In the molecular packing of $[C_6H_5O]P(O)[NHC_6H_{11}]_2.CH_3OH$ (HIVLOO, [67]), a linear arrangement is formed through a P(O)[...H−O][...H−N] grouping, where, the P(O) group acts as a double H-acceptor, the OH unit belongs to the solvent methanol.

In the 2-D H-bonded arrangement for diazaphosphorinane $4\text{-}CH_3C_6H_4OP(O)X$ [X = $NHCH_2CH_2CH_2NH$ (KIVXIX)] and $C_6H_5OP(O)Y$ [Y = $NHCH_2C(CH_3)_2CH_2NH$ (KIVXOD)], the P(O) functions as a double-H acceptor [121]. In the other cases, both O atoms are involved in the HB interactions with two N−H units (or the other H-donor site(s) in the molecule or in the crystal), in which the P(O) forms a stronger HB. Typically, in the crystal packing of $[4\text{-}CH_3\text{-}C_6H_4O]P(O)[NHC_6H_4(4\text{-}CH_3)]_2$ (MUBPIJ, [36]), the N...O(P) = 2.805(2) Å & N...O(C_6H_4-4-CH_3) = 3.068(2) Å and of $C_6H_5OP(O)[NHC_6H_4(4\text{-}CH_3)][NHCH_2C_6H_5]$, in a recently published paper by Pourayoubi *et al.*, 2011 [43], these distances are 2.761(3) Å & 3.127(3) Å, respectively.

Scheme 31 illustrates the contribution of P(O) as a double-H atom acceptor (top) and cooperation of both oxygen atoms (bottom) in hydrogen bond pattern in compounds having a P(O)(O)(NH)(NH) moiety.

In the crystal packing of compounds with the general formula $(R^1O)P(O)[NHR^2]_2$, both linear and 2-D hydrogen-bonded arrangements were observed; for example, $C_6H_5OP(O)[NHC_6H_{11}]_2.CH_3OH$ (HIVLOO [67]), $4\text{-}CH_3\text{-}C_6H_4OP(O)[NHC_6H_4\text{-}4\text{-}CH_3]_2$ (MUBPIJ [36]), $4\text{-}CH_3\text{-}C_6H_4OP(O)[NHC_6H_4\text{-}2\text{-}CH_3]_2$ (YUPVEL [32]) and $4\text{-}CH_3\text{-}C_6H_4OP(O)X$ (X = $NHCH_2C(CH_3)_2CH_2NH$, NIBNOC [83]-a) exist as a linear H-bonded arrangement, whereas a 2-D array is found for instance in each of $4\text{-}CH_3-C_6H_4OP(O)X$ (X =

NHCH$_2$CH$_2$CH$_2$NH, KIVXIX [121]), $C_6H_5OP(O)X$ (X = NHCH$_2$C(CH$_3$)$_2$CH$_2$NH, KIVXOD [83]-a), $C_6H_5OP(O)(NH_2)_2$ (PPOSAM [122]) and $C_6H_5OP(O)X$ (X = NHNHP(O)(OC$_6$H$_5$) NHNH (FIMVUS [123])).

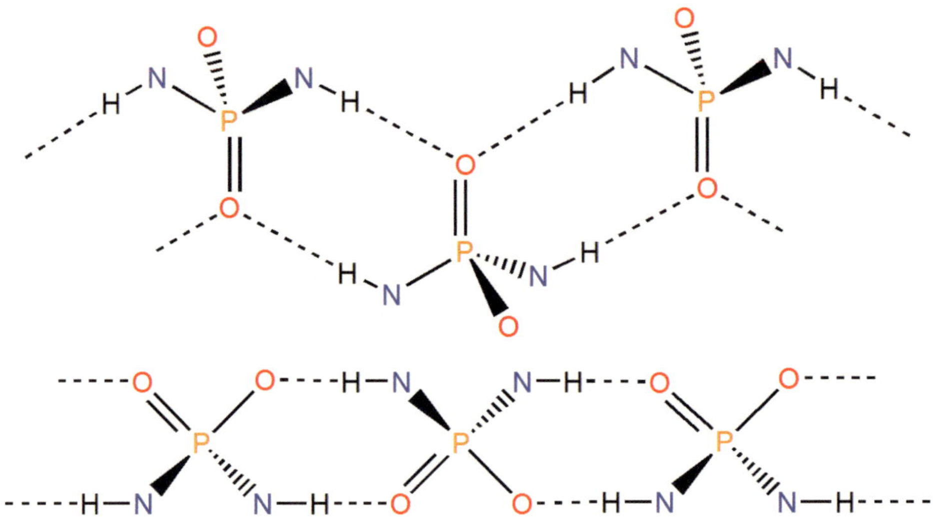

Scheme 31. A view of contribution of phosphoryl oxygen atom as a double-H acceptor (top) and a view of contribution of both oxygen atoms (bottom) in hydrogen bond pattern of compounds having a P(O)(O)(NH)(NH) skeleton

In summary, the cif files of all published compounds with the (O)P(=O)(NH)$_2$, (O)$_2$P(=O)(NH) and (O)P(=O)(NH)(N) skeletons were investigated and the following "empirical rules" were obtained:

1. In none of the reported structures, the nitrogen atom doesn't cooperate in HB interaction as an acceptor.
2. Almost in all of the compounds having the (O)$_2$P(=O)(NH) and (O)P(=O)(NH)(N) skeletons, the oxygen atom of the phenoxy (or alkoxy) group doesn't cooperate in the HB interaction, as it can't compete with the phosphoryl oxygen atom to H-accepting from the unique H-donor site in the molecule. There is only one example of hydrogen bond of the type N−H...O(R) in this family of compounds in one compound containing some H-acceptor centers in addition to one phosphoryl group.
3. The oxygen atom of OR moiety in some examples of compounds with a higher number of H-donor sites relative to the H-acceptor centers, such as compounds containing an (O)P(=O)(NH)$_2$ skeleton, however, it has a lower H-acceptability than the phosphoryl oxygen atom, is enforced to involve in the HB interaction.
4. In compounds having an (O)P(=O)(NH)$_2$ skeleton, the better H-acceptability of the phosphoryl O atom than that of the RO moiety, in some cases, leads to act it as a double-H acceptor.

Amidophosphoric acid and amido phosphate compounds

Compounds with refcodes PHOXBP (Scheme 32, left) [124] and TMPMET (Scheme 32, right) [125] are respectively the examples of an acid and a solvated acidic-salt belonging to the phosphorus-nitrogen compounds' family. The crystal packing of the latter compound contains some various HBs such as two very strong homo-conjugated [O−H…O]⁻ hydrogen bonds (O…O = 2.43 & 2.50 Å) and relatively strong hetero-conjugated [N−H…O]$^{\delta+}$ hydrogen bond (N…O = 2.93 Å). In the hydrated zwitterionic compound shown in Scheme 33 (refcode: GAHFUS) [126], the N^+ which is lack of the lone electron pair doesn't involve in H-bonding interaction; whereas, the $[P(O)(O)]^-$ unit cooperates in some N−H…O and O−H…O hydrogen bonds (Fig. 6). Compounds with refcodes IGASUF (Scheme 34) [127] and WIYFAL (Scheme 35) [128] are respectively a simple phosphate salt and an HCl-water absorbed phosphorus-nitrogen compound in which the Cl^- ion is hydrogen-bonded to the N−H units of two-neighboring phosphoramidates.

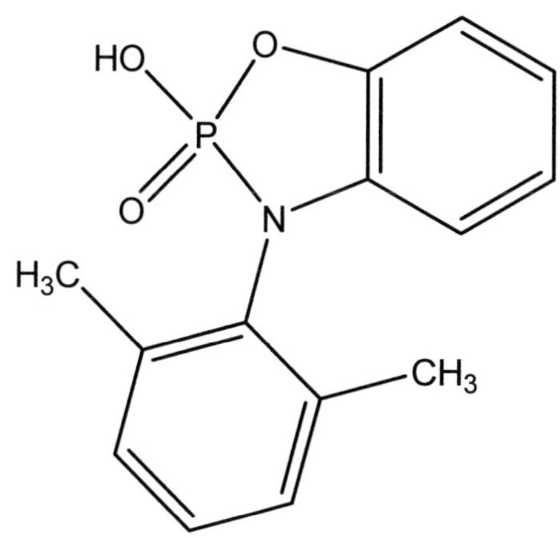

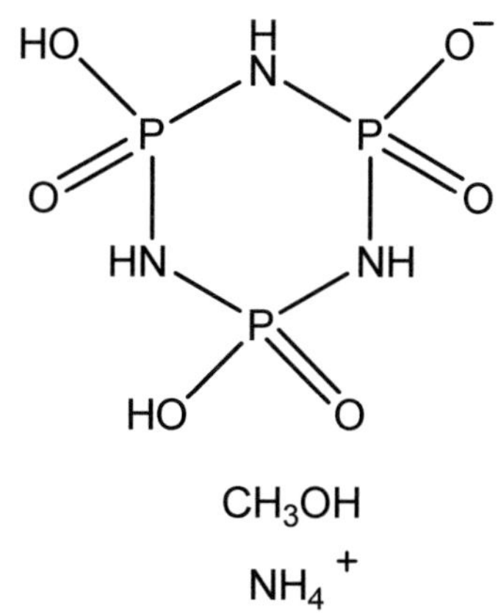

Scheme 32. Refcodes PHOXBP (left) [124] and TMPMET (right) [125]

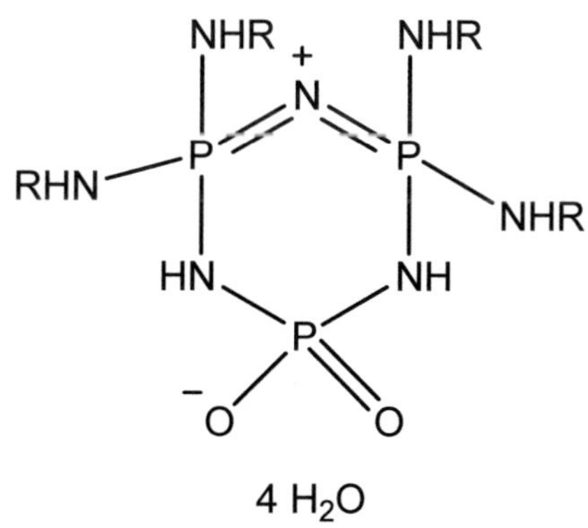

Scheme 33. Refcode GAHFUS, R = cyclo-hexyl [126]

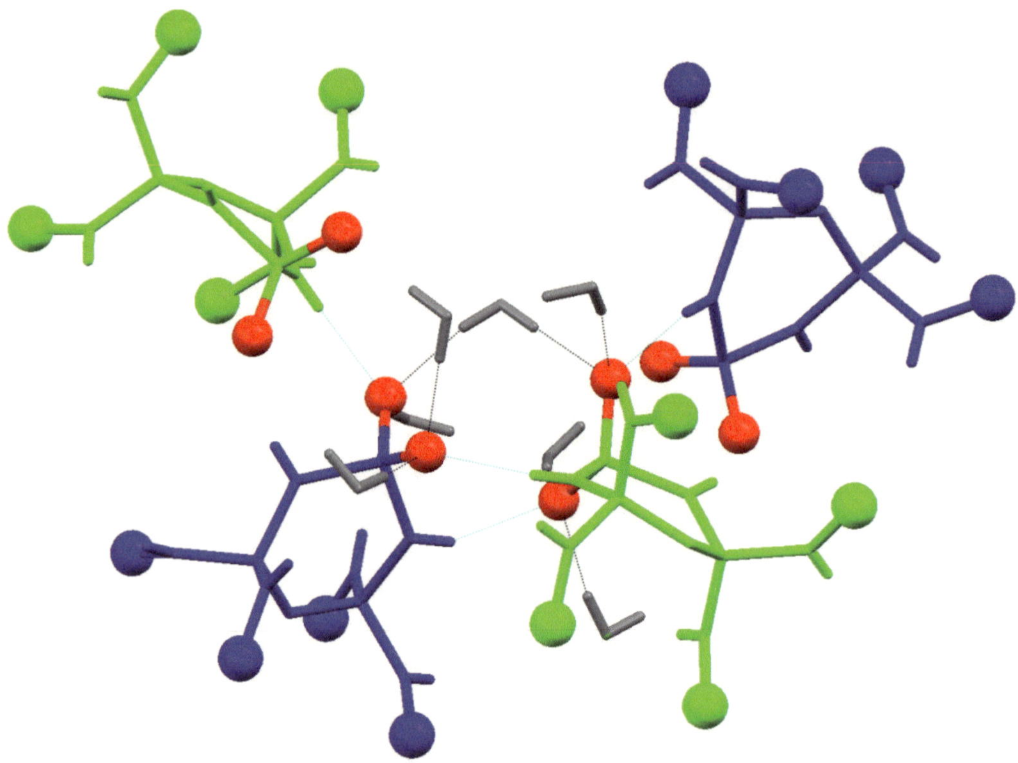

Fig. 6. Fragment of the crystal packing of the hydrated zwitterionic compound with refcode GAHFUS [126] showing the involvement of $[P(O)(O)]^-$ units in N–H...O (blue dotted lines) and O–H...O (black dotted lines) hydrogen bonds, two symmetrically independent zwitterionic compounds in the structure are shown as blue and green apart the oxygen atoms of $[P(O)(O)]^-$ units which are shown as red balls, the water molecules are represented with grey color. The cyclohexyl groups are shown as balls (green and blue).

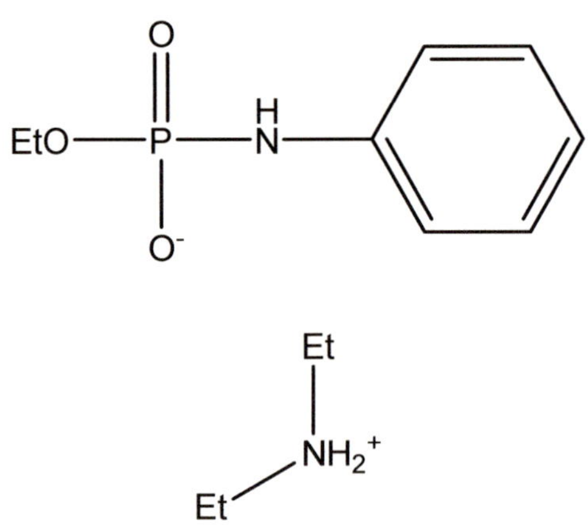

Scheme 34. Refcode IGASUF [127]

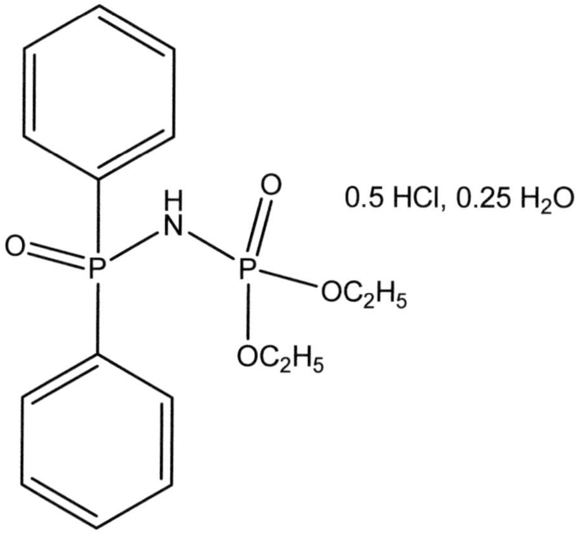

Scheme 35. Refcode WIYFAL [128]

5. Crystallographically independent molecules and ions

Different orientations resulting from non-rigid units in some molecules and ions and the presence of different H-bonds or the other short contacts may result in two or more conformers (or symmetrically independent molecules (or ions)) in solid state. Compound $C_6H_5C(O)NHP(O)[NH($*tert*-$C_4H_9)]_2$, exists as two conformers in crystalline lattice (which are detectable in solution, too by NMR experiment) [27]. They are due to different spatial orientations of *tert*-butyl amido groups. One of the two conformers has two NH units (of *tert*-butyl amido moieties) which are *syn*, but not in the other. Another example is the presence of disorder in the cyclic amido moiety. For example, $C_6H_5C(O)NHP(O)[NC_4H_8]_2$ appears as two crystallographically independent molecules [55]. This is based on the conformational forms of the pyrrolidinyl groups and the orientation of the phenyl ring. The dimmeric aggregate in this case, between two independent molecules, is not centrosymmetric. The structure of $[NH_2(C_6H_{11})($*tert*-$C_4H_9)][PO_2Cl_2]$ consists of two symmetrically independent dichlorophosphate anions as well as cyclohexyl-*tert*-butylammonium cations [75]. In the crystal structure of [*tert*-$C_4H_9NH_3][CF_3C(O)NHP(O)(O)($*tert*-$C_4H_9NH)].0.333CH_3CN.0.333H_2O$ [79], there are three symmetrically independent trifluoroacetyl-N-(*tert*-butylamino) phosphate anions and three independent cations of *tert*-butyl- ammonium; one of the anion indicates disorder in the *tert*-C_4H_9 moiety. There are some other examples of disordered components for the groups such as *tert*-C_4H_9, cyclopentyl and cyclohexyl etc. in the deposited cifs.

6. Hydrogen bond strengths in phosphoramidates

Histogram of the N...O distances in the N—H...O hydrogen bonds in compounds having a $P(O)(N)_n(O)_{3-n}$ (n = 1, 2, 3) skeleton is given in Fig. 7. In this figure, the distribution of H-bond strength in different families of phosphoramidates are shown with different colored columns: compounds having a $P(O)(NH)_n(N)_m(O)_{3-(n+m)}$ skeleton (n = 1, 2; n+m < 3) as black columns, C(O)NHP(O)(NH)$_2$ and C(O)NHP(O)(N)$_2$ as blue and P(O)(NH)$_x$(N)$_{3-x}$ (x = 1, 2, 3) as red columns.

In phosphoramidates having a P(O)(NH)$_n$(N)$_m$(O)$_{3-(n+m)}$ skeleton (balck), the strongest and weakest N−H...O hydrogen bonds are found for hydrogen bonds in the range of 2.65 to 2.75 Å and 3.20 to 3.30 Å.

In compounds containing a P(O)(NH)$_x$(N)$_{3-x}$ skeleton, the strongest N−H...O hydrogen bonds are seen for the HBs in the range of 2.70 to 2.80 Å. The phosphoryl group' involvement in a multi-centered P(O)···[H−N]$_n$ (n = 2 & 3) grouping may lead to some weak H-bonds; for example in P(O)[NHC(CH$_3$)$_3$]$_3$ (KABVEP [93]), N...O distances & N−H...O angles are 3.255(4) Å & 111.1(2)°, 3.294(4) Å & 93.4(2)° and 3.159(4) Å & 123.0(2)°, and in P(O)[NH(C$_6$H$_5$)]$_3$ (KEQLUO [95]) these parameters are 3.06 Å & 110° and 3.06 Å & 108°; this weakening of H-bond strength is attributed to the *anti*-cooperativity effect [89].

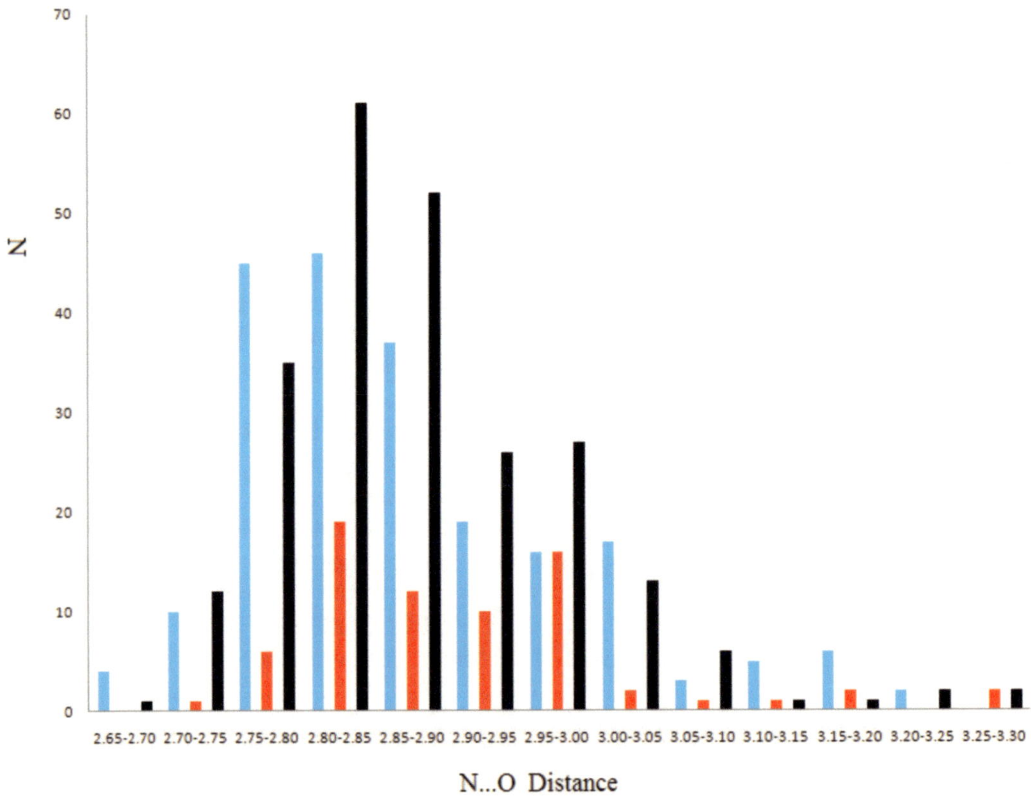

Fig. 7. Histogram of the N...O distances in the N−H...O hydrogen bonds in compounds having P(O)(NH)$_n$(N)$_m$(O)$_{3-(n+m)}$ (n = 1, 2; n+m < 3) (black), C(O)NHP(O)(NH)$_2$ and C(O)NHP(O)(N)$_2$ (blue), and P(O)(NH)$_x$(N)$_{3-x}$ (x = 1, 2, 3) (red) skeletons (the co-crystals and solvated compounds and the compounds having a disorder in the sites involving HB interaction were not enumerated).

In this family, a long donor···acceptor distance (3.477(2) Å) is observed with a relatively linear N−H...O angle (171 (2)°) for the N−H···O(CH_3) hydrogen bond in the packing of P(O)[NHC_6H_4(4-OCH_3)$]_3$ (WAWNIS [97]).

In compounds having a C(O)NHP(O) fragment, the strongest N−H...O hydrogen bonds are found for the P=O...H−$N_{C(O)NHP(O)}$ hydrogen bonds, especially in the $R_2^2(8)$ rings of some molecules [in the case of a *syn* orientation of P=O versus N−H which allows the building of the cyclic motif through a pair of P=O...H−$N_{C(O)NHP(O)}$ hydrogen bonds]. In this sub-category of compounds, the strongest and weakest hydrogen bonds are observed for the N...O distances in the ranges 2.70-2.80 Å [for $N_{C(O)NHP(O)}$−H...O hydrogen bonds] and 3.00-3.25 Å [for N_{amide}−H...O hydrogen bonds], while in the range 2.80-3.00 Å for donor-acceptor distances both types of hydrogen bonds are found. In a recently published structure [129] with a NHC(O)NHP(O)$[NH]_2$ skeleton, the intramolecular N−H...O hydrogen bonds are found in the range 2.65-2.70 Å (Fig. 7). The asymmetric unit contains four independent molecules and the hydrogen bond pattern is different from all phosphoric triamides having a C(O)NHP(O)$[NH]_2$ fragment.

7. Some future aims and proposals

Phosphorus has a very deep and widespread chemistry and understanding its nature in the compounds is noticeable interest. This may be achieved through the study on collective behaviors of phosphorus compounds in point of view of their different aspects.

The structural investigations and the study on the hydrogen-bond patterns may help to predict the molecular packing from the molecular structure. Moreover, as the biological activity of phosphorus compounds is very important, finding a relationship between the structure and a biological property is beneficial. For example, as well-known, a biological property may be related to the three important factors: lipophilicity, electronic and steric parameters. Probably, part of these factors could be well-understand by considering the crystal structure' study of such compounds; the electronic parameters may be related to the nature of chemical bonds or to the electron density or valence bond in different parts of the molecule. It is believed that, in the first step of interaction with acetylcholinesterase, the phosphoryl group is involving with the enzyme active site through a non-covalent bond; so, considering the non-covalent interaction of phosphoryl group with different atoms such as hydrogen helps to understand that the molecule how much could close to the enzyme active site. The steric parameters may be elucidated from the volume of molecule or considering the V(volume of unit cell)/Z(number of molecules in the unit cell). This may be in fact the practical volume which the molecule has, as the molecule usually cannot be closer to the neighboring molecules from this frontier boundary.

The solubility of molecules in different solvents must be checked in the crystal growth process; so, elucidation of lipophilicity is easy; however, the best method for measuring of this parameter is using of the spectrophotometer. And finally, the steric parameters can simply accessible through a structural study.

Preparation of phosphorus acids of the formula $RP(O)(OH)_2$ may develop the synthesis of the functionalized nano-phosphate materials and also polyoxometalate-based organic/inorganic hybrid compounds in which the phosphorus atom is trapping between the R group from one side and a cluster containing metal-oxygen framework in the other side. These acids may develop the extraction process of cations, too.

Application of phosphoryl donor ligands in preparation of oxo-centered clusters, in which their terminal ligands are replaced by phosphorus compounds, may be interesting for consideration. Preparation of single- enantiomer phosphoramidates by using a chiral primary or secondary amine is easy; it may extend the strategies for the synthesis of chiral phosphoramidates, phosphoric acids, nano-phosphates and so on.

Synthesis of phosphoramidate-based hybrid compounds by using polyoxoanions may be valuable for spending the time on its consideration and experiment. Some of the well-known hybrids contain the molecule-cation components of the type $[B-H...B]^+$, where the B may be a base such as amide. Designing of such molecule-cation pairs with phosphoramidates, $[PO-H...OP]^+$, may extend the experimental data about the ^{31}P-^{31}P coupling constant through the hydrogen-bond.

The NMR experiments on phosphoramidate-based compounds may develop the study on coupling constants of phosphorus and the other atoms, such as $^2J(^{31}P$-$^{127}Tl)$ or $^2J(^{31}P$-$^{39}K)$. These values may apply to evaluate the strengths of P=O−Tl or P=O−K bonds in their complexes.

We wish to develop the spectroscopic features and chemical calculations on phosphorus compounds, preparation of N-deuterated compounds in order to a good assignment of IR and Raman absorption bands, collecting the NMR data such as chemical shifts and short and long-range coupling constants, and finally chemical calculations on hydrogen-bonded molecules in the crystals.

8. Conclusion

In this chapter, the common methods for the synthesis and crystallization of phosphoramidates, their molecular structural features and the hydrogen-bond patterns and strengths were reviewed; the important structural aspects may be classified as follows:

1. The four different groups linked to the P atom result in a distorted tetrahedral configuration.
2. In the C(O)NHP(O) unit, the P=O is a better H-acceptor than the C=O counterpart, moreover, the *anti* orientation of P=O versus C=O is more common than the *gauche* orientation; in acyclic compounds with formula $RC(O)NHP(O)[NHR']_2$, a *gauche* situation has not been reported, so far. In the $C(O)NHP(O)[N]_2$ fragment, the P−N bond of the C(O)NHP(O) moiety is longer than the two other P−N bonds; whereas, in the $[C(O)NP(O)[N]_2]^-$ fragment, similar P−N bond are shorter than the two others.
3. The nitrogen atom of the P(O)N unit has a sp^2 character and virtually doesn't involve in hydrogen-bond pattern as an H-acceptor.
4. In the C−O−P(=O) fragment, the oxygen of phosphoryl is a better H-acceptor than the other oxygen atom; the C−O−P angle is about 120°.
5. In the diazaphosphorinane ring, the P=O bond is placed in an equatorial position.
6. The hydrogen bond in a neutral phosphoramidate is of the type polarization-assisted hydrogen bond; whereas, in the proton-transfer and phosphate compounds two factors help to strength of hydrogen bond: polarization-assisted and charge-assisted.

7. In the multi-centered hydrogen-bond of the type $P{=}O[...H{-}N]_n$ (n = 2 & 3) the hydrogen-bond is weak due to the anticooperativity effect.

We wish to collect more structural data about this class of compounds and study the collective behavior of this family in the other domains such as spectroscopy and chemical calculations.

9. Acknowledgements

M. P. wishes to thank Ferdowsi University of Mashhad for the Research University Grant (No. 15144/2). The authors also thank Dr. Karla Fejfarová for the CSD searches and Monireh Negari (MSc student) for some computer works.

10. References

[1] Allen, F. H. (2002). *Acta Cryst.* B58, 380.
[2] Müller, P. & Siegfried, B. (1972). *Helvetica Chimica Acta*, 55, 2965.
[3] (a) Le Carpentier, J.-M., Schlupp, R. & Weiss, R. (1972). *Acta Cryst.* B28, 1278. (b) Crawford, M.-J., Mayer, P., Nöth, H. & Suter M. (2004). *Inorg. Chem.* 43, 6860.
[4] (a) Aas, P. (2004). *Prehospital and Disaster Medicine*, 18, 208. (b) Bhattacharjee, A. K., Kuča, K., Musilek, K. & Gordon, R. K. (2010). *Chem. Res. Toxicol.* 23, 26.
[5] (a) Rodriguez, J. A. & Fernandez-Garcia, M. (2007). *Synthesis, properties and applications of oxide nanomaterials*, Wiley-interscience, A John Wiley & Sons, INC., Publication. (b) Hirakawa, T., Sato, K., Komano, A., Kishi, Sh., Nishimoto, Ch. K., Mera, N., Kugishima, M., Sano, T., Ichinose, H., Negishi, N., Seto, Y. & Takeuchi, K. (2010). *J. Phys. Chem.* C114, 2305.
[6] (a) Nguyen, C. & Kim, J. (2008). *Polym. Degrad. Stabil.* 93, 1037. (b) Levchik, S. V., Levchik, G. F. & Murashko, E. A. (2001). *Phosphorus-Containing Fire Retardants in Aliphatic Nylons, Fire and Polymers*, Chapter 17, pp 214.
[7] Roush, R. F., Nolan, E. M., Löhr, F. & Walsh, C. T. (2008). *J. Am. Chem. Soc.* 130, 3603.
[8] Gholivand, K., Ghadimi, S., Naderimanesh, H. & Forouzanfar, A. (2001). *Magn. Reson. Chem.* 39, 684.
[9] Gholivand, K., Mostaanzadeh, H., Koval, T., Dusek, M., Erben, M. F., Stoeckli-Evans, H. & Della Védova, C. O. (2010). *Acta Cryst.* B66, 441.
[10] Pourayoubi, M., Tarahhomi, A., Saneei, A., Rheingold, A. L. & Golen, J. A. (2011). *Acta Cryst.* C67, o265.
[11] Ghadimi, S., Mousavi, S. L. & Javani, Z. (2008). *J. Enz. Inhibit. Med. Chem.* 23, 213.
[12] Ekstrom, F., Akfur, C., Tunemalm, A. & Lundberg, S. (2006). *Biochemistry*, 45, 74.
[13] Pourayoubi, M., Rostami Chaijan, M., Torre-Fernández, L. & García-Granda, S. (2011). *Acta Cryst.* E67, o1360.
[14] Pourayoubi, M., Rostami Chaijan, M., Torre-Fernández, L. & García-Granda, S. (2011). *Acta Cryst.* E67, o1031.
[15] Pourayoubi, M. & Saneei, A. (2011). *Acta Cryst.* E67, o665.
[16] Pourayoubi, M., Tarahhomi, A., Rheingold, A. L. & Golen, J. A. (2010). *Acta Cryst.* E66, o2524.

[17] Pourayoubi, M., Tarahhomi, A., Rheingold, A. L. & Golen, J. A. (2010). *Acta Cryst.* E66, o3159.
[18] Pourayoubi, M., Tarahhomi, A., Rheingold, A. L. & Golen, J. A. (2011). *Acta Cryst.* E67, o934.
[19] Pourayoubi, M., Tarahhomi, A., Rheingold, A. L. & Golen, J. A. (2011). *Acta Cryst.* E67, o3027.
[20] Pourayoubi, M., Fadaei, H., Tarahhomi, A. & Parvez, M. (2011). *Acta Cryst.* E66, o2795.
[21] Pourayoubi, M., Tarahhomi, A., Rheingold, A. L. & Golen, J. A. (2011). *Acta Cryst.* E67, o2643.
[22] Pourayoubi, M., Tarahhomi, A., Rheingold, A. L. & Golen, J. A. (2011). *Acta Cryst.* E67, o2444.
[23] Pourayoubi, M., Sadeghi Seraji, S., Bruno, G. & Amiri Rudbari, H. (2011). *Acta Cryst.* E67, o1285.
[24] Pourayoubi, M., Toghraee, M. & Divjakovic, V. (2011). *Acta Cryst.* E67, o333.
[25] Tarahhomi, A., Pourayoubi, M., Rheingold, A. L. & Golen, J. A. (2011). *Struct. Chem.* 22, 201.
[26] Toghraee, M., Pourayoubi, M. & Divjakovic, V. (2011). *Polyhedron*, 30, 1680.
[27] Gholivand, K. & Pourayoubi, M. (2004). *Z. Anorg. Allg. Chem.* 630, 1330.
[28] Pourayoubi, M., Golen, J. A., Rostami Chaijan, M., Divjakovic, V., Negari, M. & Rheingold, A. L. (2011). *Acta Cryst.* C67, m160.
[29] Tarahhomi, A., Pourayoubi, M., Rheingold, A. L. & Golen, J. A. (2011). *Acta Cryst.* E67, o2643.
[30] Pourayoubi, M., Eshtiagh-Hosseini, H., Negari, M. & Nečas, M. (2011). *Acta Cryst.* E67, o2202.
[31] Sabbaghi, F., Pourayoubi, M., Toghraee, M. & Divjakovic, V. (2010). *Acta Cryst.* E66, o344.
[32] Sabbaghi, F., Mancilla Percino, T., Pourayoubi, M. & Leyva, M. A. (2010). *Acta Cryst.* E66, o1755.
[33] Pourayoubi, M., Ghadimi, S. & Ebrahimi Valmoozi, A. A. (2010). *Acta Cryst.* E66, o450.
[34] Pourayoubi, M., Ghadimi, S. & Ebrahimi Valmoozi, A. A. (2007). *Acta Cryst.* E63, o4631.
[35] Ghadimi, S., Ebrahimi Valmoozi, A. A. & Pourayoubi, M. (2007). *Acta Cryst.* E63, o3260.
[36] Pourayoubi, M., Ghadimi, S., Ebrahimi Valmoozi, A. A. & Banan, A. R. (2009). *Acta Cryst.* E65, o1973.
[37] Sabbaghi, F., Rostami Chaijan, M. & Pourayoubi, M. (2010). *Acta Cryst.* E66, o1754.
[38] Pourayoubi, M., Eshtiagh-Hosseini, H., Zargaran, P. & Divjakovic, V. (2010). *Acta Cryst.* E66, o204.
[39] Pourayoubi, M., Ghadimi, S. & Ebrahimi Valmoozi, A. A. (2007). *Acta Cryst.* E63, o4093.
[40] Pourayoubi, M. & Sabbaghi, F. (2007). *Acta Cryst.* E63, o4366.
[41] Sabbaghi, F., Pourayoubi, M., Negari, M. & Nečas, M. (2011). *Acta Cryst.* E67, o2512.

[42] Pourayoubi, M., Eshtiagh-Hosseini, H., Negari, M. & Nečas, M. (2011). *Acta Cryst.* E67, o1870.

[43] Pourayoubi, M., Karimi Ahmadabad, F. & Nečas, M. (2011). *Acta Cryst.* E67, o2523.

[44] Pourayoubi, M., Zargaran, P., Rheingold, A. L. & Golen, J. A. (2011). *Acta Cryst.* E67, o5.

[45] Raissi Shabari, A., Pourayoubi, M. & Saneei, A. (2011). *Acta Cryst.* E67, o663.

[46] Pourayoubi, M., Keikha, M. & Nečas, M. (2011). *Acta Cryst.* E67, o2439.

[47] Sabbaghi, F., Pourayoubi, M., Karimi Ahmadabad, F., Azarkamanzad, Z. & Ebrahimi Valmoozi, A. A. (2011). *Acta Cryst.* E67, o502.

[48] Pourayoubi, M., Padělková, Z., Rostami Chaijan, M. & Růžička, A. (2011). *Acta Cryst.* E67, o450.

[49] Pourayoubi, M., Fadaei, H. & Parvez, M. (2011). *Acta Cryst.* E67, o2046.

[50] Pourayoubi, M., Keikha, M. & Parvez, M. (2011). *Acta Cryst.* E67, o2792.

[51] Pourayoubi, M. & Sabbaghi, F. (2009). *J. Chem. Crystallogr.* 39, 874.

[52] Pourayoubi, M., Yousefi, M., Eslami, F., Rheingold, A. L. & Chen, C. (2011). *Acta Cryst.* E67, o3220.

[53] Gholivand, K., Pourayoubi, M., Shariatinia, Z. & Mostaanzadeh, H. (2005) *Polyhedron*, 24, 655.

[54] Gholivand, K., Shariatinia, Z. & Pourayoubi, M. (2005). *Z. Anorg. Allg. Chem.* 631, 961.

[55] Gholivand, K., Pourayoubi, M. & Mostaanzadeh, H. (2004). *Anal. Sci.* 20, x51.

[56] Gholivand, K., Hosseini, Z., Pourayoubi, M. & Shariatinia, Z. (2005). *Z. Anorg. Allg. Chem.* 631, 3074.

[57] Gholivand, K., Shariatinia, Z., Pourayoubi, M. & Farshadian, S. (2005). *Z. Naturforsch. Teil B*, 60, 1021.

[58] Gholivand, K., Shariatinia, Z. & Pourayoubi, M. (2006). *Z. Anorg. Allg. Chem.* 632, 160.

[59] Gholivand, K., Shariatinia, Z. & Pourayoubi, M. (2006). *Polyhedron*, 25, 711.

[60] Gholivand, K., Mahzouni, H. R., Pourayoubi, M. & Amiri, S. (2010). *Inorg. Chim. Acta*, 363, 2318.

[61] Pourayoubi, M. & Zargaran, P. (2010). *Acta Cryst.* E66, o3273.

[62] Raissi Shabari, A., Pourayoubi, M., Ghoreishi, F. & Vahdani, B. (2011). *Acta Cryst.* E67, o3401.

[63] Pourayoubi, M., Shoghpour, S., Torre-Fernández, L. & García-Granda, S. (2011). *Acta Cryst.* E67, o3425.

[64] Pourayoubi, M., Elahi, B. & Parvez, M. (2011). *Acta Cryst.* E67, o2848.

[65] Cameron, T. S., Cordes, R. E. & Jackman F. A. (1978). *Z. Naturforsch. Teil B*, 33, 728.

[66] Setzer, W. N., Black, B. G., Hovanes, B. A. & Hubbard, J. L. (1989). *J. Org. Chem.* 54, 1709.

[67] Gholivand, K., Della Védova, C. O., Erben, M. F., Mahzouni, H. R., Shariatinia, Z. & Amiri, S. (2008). *J. Molec. Struct.* 874, 178.

[68] Gholivand, K., Pourayoubi, M., Farshadian, S., Molani, S. & Shariatinia, Z. (2005). *Anal. Sci.* 21, x55.

[69] Sabbaghi, F., Pourayoubi, M., Karimi Ahmadabad, F. & Parvez, M. (2011). *Acta Cryst.* E67, o1502.
[70] Pourayoubi, M., Tarahhomi, A., Keikha, M. & Karimi Ahmadabad, F. unpublished results.
[71] Pourayoubi, M. & Negari, M. (2010). *Acta Cryst.* E66, o708.
[72] Pourayoubi, M., Eshtiagh-Hosseini, H. & Negari, M. (2010). *Acta Cryst.* E66, o1180.
[73] Yazdanbakhsh, M. & Sabbaghi, F. (2007). *Acta Cryst.* E63, o4318.
[74] Pourayoubi, M., Negari, M. & Nečas, M. (2011). *Acta Cryst.* E67, o332.
[75] Gholivand, K. & Pourayoubi, M. (2004). *Z. Kristallogr. New Cryst. Struct.* 219, 314.
[76] Pourayoubi, M., Tarahhomi, A., Karimi Ahmadabad, F., Fejfarová, K., van der Lee, A. & Dušek, M. (2012). Acta Cryst. C68, In press.
[77] Pourayoubi, M., Fadaei, H., Tarahhomi, A. & Parvez, M. (2012). *Z. Kristallogr. New Cryst. Struct.* 227, In press.
[78] Pourayoubi, M. & Shoghpour, S. unpublished results.
[79] Gholivand, K., Pourayoubi, M., Shariatinia, Z. & Molani, S. (2005). *Z. Kristallogr. New Cryst. Struct.* 220, 387.
[80] Yazdanbakhsh, M., Eshtiagh-Hosseini, H. & Sabbaghi, F. (2009). *Acta Cryst.* E65, o78.
[81] Pourayoubi, M. & Fadaei, H. unpublished results.
[82] Gholivand, K., Shariatinia, Z. & Tadjarodi, A. (2005). *Main Group Chem.* 4, 111.
[83] (a) Gholivand, K., Pourayoubi, M. & Shariatinia, Z. (2007). *Polyhedron,* 26, 837. (b) Gholivand, K., Mojahed, F., Mohamadi, L. & Bijanzadeh, H. R. (2007). *Phosphorus, Sulfur, Silicon, Relat. Elem.* 182, 631.
[84] Kadyrov, A. A., Neda, I., Kaukorat, T., Fischer, A., Jones, P. G. & Schmutzler, R. (1995). *J. Fluorine Chemistry,* 72, 29.
[85] Wan, H. & Modro, T. A. (1996). *Synthesis,* 1227.
[86] Mbianda, X. Y., Modro, T. A. & Van Rooyen, P. H. (1998). *Chem. Commun.* 741.
[87] Pourayoubi, M., Nečas, M. & Negari, M. (2012). *Acta Cryst.* C68, o51.
[88] Corbridge, D. E. C. (1995). *Phosphorus, an Outline of its Chemistry, Biochemistry and Technology,* 5th ed., p. 1179. New York: Elsevier Science.
[89] Steiner, T. (2002). *Angew. Chem. Int. Ed.* 41, 48.
[90] Amirkhanov, V. M., Ovchynnikov, V. A., Glowiak, T. & Kozlowski, H. (1997). *Z. Naturforsch. Teil B,* 52, 1331.
[91] Narula, P. M., Day, C. S., Powers, B. A., Odian, M. A., Lachgar, A., Pennington, W. T. & Noftle, R. E. (1999). *Polyhedron,* 18, 1751.
[92] Gilli, P., Bertolasi, V., Ferretti, V. & Gilli, G. (1994). *J. Am. Chem. Soc.* 116, 909.
[93] Chivers, T., Krahn, M., Schatte, G. & Parvez, M. (2003). *Inorg. Chem.* 42, 3994.
[94] Hartmann, F., Dahlems, T. & Mootz, D. (1998). *Z. Kristallogr. New Cryst. Struct.* 213, 639.
[95] Anjum, S., Atta-ur-Rahman & Fun, H.-K. (2006). *Acta Cryst.* E62, o4569.
[96] Gholivand, K., Mostaanzadeh, H., Shariatinia, Z. & Oroujzadeh, N. (2006). *Main Group Chem.* 5, 95.
[97] Li, C., Dyer, D. J., Rath, N. P. & Robinson, P. D. (2005). *Acta Cryst.* C61, o654.
[98] Romming, C. & Songstad, J. (1982). *Acta Chem. Scand.* A36, 665.

[99] Du, D.-M., Fang, T., Xu, J. & Zhang, S.-W. (2006). *Org. Lett.* 8, 1327.
[100] Gholivand, K., Alavi, M. D. & Pourayoubi, M. (2004). *Z. Kristallogr. New Cryst. Struct.* 219, 124.
[101] Benincori, T., Marchesi, A., Pilati, T., Ponti, A., Rizzo, S. & Sannicolô, F. (2009). *Chem. Eur. J.* 15, 94.
[102] Vyšvařil, M., Dastych, D., Taraba, J. & Nečas, M. (2009). *Inorg. Chim. Acta*, 362, 4899.
[103] Cameron, T. S., Magee, M. G. & McLean, S. (1976). *Z. Naturforsch. Teil B*, 31, 1295.
[104] Cameron, T. S. (1977). *Z. Naturforsch. Teil B*, 32, 1001.
[105] Li, N., Jiang, F., Chen, L., Li, X., Chen, Q. & Hong, M. (2011). *Chem. Commun.* 47, 2327.
[106] Liu, X., Ilankumaran, P., Guzei, I. A. & Verkade, J. G. (2000). *J. Org. Chem.* 65, 701.
[107] Bourne, S. A., Mbianda, X. Y., Modro, T. A., Nassimbeni, L. R. & Wan, H. (1998). *J. Chem. Soc., Perkin Trans. 2*, 83.
[108] Hsu, I.-N. & Craven, B. M. (1974). *Acta Cryst.* B30, 1299.
[109] Schmuck, C. & Wienand, W. (2003). *J. Am. Chem. Soc.* 125, 452.
[110] Fei, Z., Neda, I., Thönnessen, H., Jones, P. G. & Schmutzler, R. (1997). *Phosphorus, Sulfur, Silicon, Relat. Elem.* 131, 1.
[111] Gholivand, K., Shariatinia, Z., Yaghmaian, F. & Faramarzpour, H. (2006). *Bull. Chem. Soc. Jpn.* 79, 1604.
[112] Hutton, A. T., Modro, T. A., Niven, M. L. & Scaillet, S. (1986). *J. Chem. Soc., Perkin Trans. 2*, 17.
[113] Vasil'chenko, V. N., Mitkevich, V. V., Moiseenko, A. A., Khomenko, V. G. & Chernov, V. A. (1982). *Zh. Strukt. Khim.* 23, 107.
[114] Gholivand, K., Tadjarodi, A. & Ng, S. W. (2002). *Acta Cryst.* E58, o200.
[115] Kattuboina, A. & Li, G. (2008). *Tetrahedron Lett.* 49, 1573.
[116] Raissi Shabari, A., Pourayoubi, M., Taghizadeh, A., Ghoreishi, F. & Vahdani, B. (2011). *Acta Cryst.* E67, o2167.
[117] Rudd, M. D., Lindeman, S. V. & Husebye, S. (1996). *Acta Chem. Scand.* 50, 759.
[118] Hempel, A., Camerman, N., Mastropaolo, D. & Camerman, A. (1999). *Acta Cryst.* C55, 1173.
[119] Ghadimi, S., Pourayoubi, M. & Ebrahimi Valmoozi, A. A. (2009). *Z. Naturforsch. Teil B*, 64, 565.
[120] Sa, M. M., Silveira, G. P., Bortoluzzi, A. J. & Padwa, A. (2003). *Tetrahedron*, 59, 5441.
[121] Gholivand, K., Shariatinia, Z., Mahzouni, H. R. & Amiri, S. (2007). *Struct. Chem.* 18, 653.
[122] Bullen, G. J. & Dann, P. E. (1973). *Acta Cryst.* B29, 331.
[123] Engelhardt, U. & Franzmann, A. (1987). *Acta Cryst.* C43, 1313.
[124] Cadogan, J. I. G., Gould, R. O., Gould, S. E. B., Sadler, P. A., Swire, S. J. & Tait, B. S. (1975). *J. Chem. Soc., Perkin Trans. 1*, 2392.
[125] Attig, R. & Mootz, D. (1976). *Z. Anorg. Allg. Chem.* 419, 139.
[126] Ledger, J., Boomishankar, R. & Steiner, A. (2010). *Inorg. Chem.* 49, 3896.
[127] Fu, Z. & Liu, X. (2008). *Acta Cryst.* E64, o2171.
[128] Balazs, G., Drake, J. E., Silvestru, C. & Haiduc, I. (1999). *Inorg. Chim. Acta*, 287, 61.

[129] Gholivand, K., Dorosti, N., Ghaziany, F., Mirshahi, M. & Sarikhani, S. (2012). *Heteroatom Chem.* 23, 74

Synthesis and X-Ray Crystal Structure of α-Keggin-Type Aluminum-Substituted Polyoxotungstate

Chika Nozaki Kato[1], Yuki Makino[1], Mikio Yamasaki[2], Yusuke Kataoka[3], Yasutaka Kitagawa[3] and Mitsutaka Okumura[3]
[1]*Shizuoka University*
[2]*Rigaku Corporation*
[3]*Osaka University*
Japan

1. Introduction

Aluminum and its derivatives such as alloys, oxides, organometallics, and inorganic compounds have attracted considerable attention because of their extreme versatility and unique range of properties, including acidity, hardness, and electroconductivity (Cotton & Wilkinson, 1988). Since the properties and activities of an aluminum species are strongly dependent on the structures of the aluminum sites, the syntheses of aluminum compounds with structurally well-defined aluminum sites are considerably significant for the development of novel and efficient aluminum-based materials. However, the use of these well-defined aluminum sites is slightly limited by the conditions resulting from the hydrolysis of the aluminum species by water (Djurdjevic et al., 2000; Baes & Mesmer, 1976; Orvig, 1993; Akitt, 1989).

Polyoxometalates have been of particular interest in the fields of catalytic chemistry, surface science, and materials science because their chemical properties such as redox potentials, acidities, and solubilities in various media can be finely tuned by choosing appropriate constituent elements and countercations (Pope, 1983; Pope & Müller, 1991, 1994). In particular, the coordination of metal ions to the vacant site(s) of lacunary polyoxometalates is one of the most effective techniques used for constructing efficient and well-defined active metal centers. Among various lacunary polyoxometalates, a series of Keggin-type phosphotungstates is one of the most useful types of lacunary polyoxometalates. Fig. 1 shows some examples of lacunary Keggin-type phosphotungstates, i.e., *mono*-lacunary α-Keggin $[\alpha\text{-}PW_{11}O_{39}]^{7-}$ (Contant, 1987), *di*-lacunary γ-Keggin $[\gamma\text{-}PW_{10}O_{36}]^{7-}$ (Domaille, 1990; Knoth, 1981), and *tri*-lacunary α-Keggin $[A\text{-}\alpha\text{-}PW_9O_{34}]^{9-}$ (Domaille, 1990) phosphotungstates. Knoth and co-workers first synthesized the Keggin derivative $(Bu_4N)_4(H)ClAlW_{11}PO_{39}$ by the reaction of *mono*-lacunary α-Keggin phosphotungstate with $AlCl_3$ in dichloroethane (Knoth et al., 1983). However, only a few aluminum-coordinated polyoxometalates (determined by X-ray crystallographic analysis) have been reported, e.g., a monomeric, *di*-aluminum-substituted γ-Keggin polyoxometalate $TBA_3H[\gamma\text{-}SiW_{10}O_{36}\{Al(OH_2)\}_2(\mu\text{-}$

$OH)_2]\cdot 4H_2O$ (TBA = tetra-*n*-butylammonium) (Kikukawa et al., 2008), a monomeric, *mono*-aluminum-substituted α-Keggin polyoxometalate $K_6H_3[ZnW_{11}O_{40}Al]\cdot 9.5H_2O$ (Yang et al., 1997), and a dimeric aluminum complex having *mono*- and *di*-aluminum sites sandwiched by *tri*-lacunary α-Keggin polyoxometalate $K_6Na[(A\text{-}PW_9O_{34})_2\{W(OH)(OH_2)\}\{Al(OH)(OH_2)\}\{Al(\mu\text{-}OH)(OH_2)_2\}_2]\cdot 19H_2O$ (Kato et al., 2010); these structures are shown in Fig. 2.

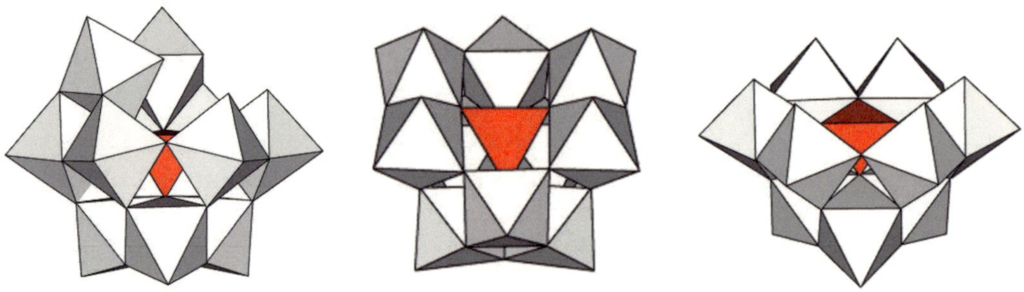

Fig. 1. Some examples of lacunary phosphotungstates. The polyhedral representations of *mono*-lacunary α-Keggin $[\alpha\text{-}PW_{11}O_{39}]^{7-}$ (left), *di*-lacunary γ-Keggin $[\gamma\text{-}PW_{10}O_{36}]^{7-}$ (center), and *tri*-lacunary α-Keggin $[A\text{-}\alpha\text{-}PW_9O_{34}]^{9-}$ (right) phosphotungstates. The WO_6 and internal PO_4 groups are represented by the white octahedra and red tetrahedron, respectively.

In this study, we successfully obtained a monomeric, α-Keggin *mono*-aluminum-substituted polyoxotungstate in the form of crystals (suitable for X-ray structure analysis) of $[(n\text{-}C_4H_9)_4N]_4[\alpha\text{-}PW_{11}\{Al(OH_2)\}O_{39}]$ that were fully characterized by X-ray crystallography; elemental analysis; thermogravimetric/differential thermal analysis; Fourier transform infrared spectroscopy; and solution ^{31}P, ^{27}Al, and ^{183}W nuclear magnetic resonance spectroscopies. Although the X-ray crystallography of $[\alpha\text{-}PW_{11}\{Al(OH_2)\}O_{39}]^{4-}$ showed that the *mono*-aluminum-substituted site was not identified because of the high symmetry in the compound, the bonding mode (bond lengths and bond angles) were significantly influenced by the insertion of aluminum ions into the *mono*-vacant sites. In addition, density-functional-theory (DFT) calculations showed a unique coordination sphere around the *mono*-aluminum-substituted site in $[\alpha\text{-}PW_{11}\{Al(OH_2)\}O_{39}]^{4-}$; this was consistent with the X-ray crystal structure and spectroscopic results. In this paper, we report the complete details of the synthesis, molecular structure, and characterization of $[(n\text{-}C_4H_9)_4N]_4[\alpha\text{-}PW_{11}\{Al(OH_2)\}O_{39}]$.

2. Experimental section

2.1 Materials

$K_7[\alpha\text{-}PW_{11}O_{39}]\cdot 11H_2O$ (Contant, 1987) and $Cs_7[\gamma\text{-}PW_{10}O_{36}]\cdot 19H_2O$ (Domaille, 1990; Knoth, 1981) were prepared as described in the literature. The number of solvated water molecules was determined by thermogravimetric/differential thermal analyses. Acetonitrile-soluble, tetra-*n*-butylammonium salts of $[\alpha\text{-}PW_{12}O_{40}]^{3-}$ and $[\alpha\text{-}PW_{11}O_{39}]^{7-}$ were prepared by the addition of excess tetra-*n*-butylammonium bromide to the aqueous solutions of $Na_3[\alpha\text{-}PW_{12}O_{40}]\cdot 16H_2O$ (Rosenheim & Jaenicke, 1917) and $K_7[\alpha\text{-}PW_{11}O_{39}]\cdot 11H_2O$. All the reagents and solvents were obtained and used as received from commercial sources. $Al(NO_3)_3\cdot 9H_2O$ (Aldrich, 99.997% purity) was used in the synthesis. The X-ray crystal structure of

$[(CH_3)_2NH_2]_4[\alpha\text{-}PW_{11}Re^{V}O_{40}]$ (Kato et al., 2010) was resolved by SHELXS-97 (direct methods) and re-refined by SHELXL-97 (Sheldrick, 2008). The crystal data are as follows: $C_8H_{32}N_3O_4PReW_{11}$: M = 3063.87, *trigonal*, space group *R-3m*, a = 16.53(2) Å, c = 25.21(4) Å, V = 5963(12) Å^3, Z = 6, D_c = 5.119 g/cm^3, R_1 = 0.0559 ($I > 2\sigma(I)$) and wR_2 = 0.1513 (for all data). The four dimethylammonium ions could not be identified due to the disorder (Nomiya et al., 2001, 2002; Weakley & Finke, 1990; Lin et al., 1993). CCDC number 851154.

2.2 Instrumentation/analytical procedures

The elemental analysis was carried out by using Mikroanalytisches Labor Pascher (Remagen, Germany). The sample was dried overnight at room temperature under pressures of 10^{-3} - 10^{-4} Torr before analysis. Infrared spectra were recorded on a Parkin Elmer Spectrum100 FT-IR spectrometer in KBr disks at room temperature. Thermogravimetric (TG) and differential thermal analyses (DTA) data were obtained using a Rigaku Thermo Plus 2 series TG/DTA TG 8120. TG/DTA measurements were performed in air by constantly increasing the temperature from 20 to 500 °C at a rate of 4 °C per min. The ^{31}P nuclear magnetic resonance (NMR) (242.95 MHz) spectra in acetonitrile-d_3 solution were recorded in tubes (outer diameter: 5 mm) on a JEOL ECA-600 NMR spectrometer. The ^{31}P NMR spectra were referenced to an external standard of 85% H_3PO_4 in a sealed capillary. Negative chemical shifts were reported on the δ scale for resonance upfields of H_3PO_4 (δ 0). The ^{27}Al NMR (156.36 MHz) spectrum in acetonitrile-d_3 was recorded in tubes (outer diameter: 5 mm) on a JEOL ECA-600 NMR spectrometer. The ^{27}Al NMR spectrum was referenced to an external standard of saturated $AlCl_3$-D_2O solution (substitution method). Chemical shifts were reported as positive on the δ scale for resonance downfields of $AlCl_3$ (δ 0). The ^{183}W NMR (25.00 MHz) spectra were recorded in tubes (outer diameter: 10 mm) on a JEOL ECA-600 NMR spectrometer. The ^{183}W NMR spectra measured in acetonitrile-d_3 were referenced to an external standard of saturated Na_2WO_4-D_2O solution (substitution method).

Fig. 2. The polyhedral representation of $K_6H_3[ZnW_{11}O_{40}Al]\cdot 9.5H_2O$ (left), $TBA_3H[\gamma\text{-}SiW_{10}O_{36}\{Al(OH_2)\}_2(\mu\text{-}OH)_2]\cdot 4H_2O$ (TBA = tetra-*n*-butylammonium) (center), and $K_6Na[(A\text{-}PW_9O_{34})_2\{W(OH)(OH_2)\}\{Al(OH)(OH_2)\}\{Al(\mu\text{-}OH)(OH_2)_2\}_2]\cdot 19H_2O$ (right). The aluminum groups are represented by the blue octahedra. The WO_6 groups are represented by white octahedra. The internal ZnO_4, SiO_4, and PO_4 groups are represented by green, yellow, and red tetrahedra, respectively.

Chemical shifts were reported as negative for resonance upfields of Na_2WO_4 (δ 0). Potentiometric titration was carried out with 0.4 mol/L tetra-*n*-butylammonium hydroxide as a titrant under argon atmosphere (Weiner et al., 1996). The compound $[(n\text{-}C_4H_9)_4N]_4[\alpha\text{-}PW_{11}\{Al(OH_2)\}O_{39}]$ (0.018 mmol) was dissolved in acetonitrile (30 mL) at 25 °C and the solution was stirred for approximately 5 min. The titration data were obtained with a pH meter (Mettler Toledo). Data points were obtained in milivolt. A solution of tetra-*n*-butylammonium hydroxide (9.0 mmol/L) was syringed into the suspension in 0.25-equivalent intervals.

2.3 Synthesis of $[(n\text{-}C_4H_9)_4N]_4[\alpha\text{-}PW_{11}\{Al(OH_2)\}O_{39}]$

$Cs_7[\gamma\text{-}PW_{10}O_{36}]\cdot 19H_2O$ (2.00 g; 0.538 mmol) was dissolved in water (600 mL) at 40 °C, and solid $Al(NO_3)_3\cdot 9H_2O$ (0.250 g, 0.666 mmol) was added to the colorless clear solution. After stirring for 1 h at 40 °C, a solid $[(n\text{-}C_4H_9)_4N]_4Br$ (12.14 g; 37.7 mmol) was added to the solution, followed by stirring at 25 °C for 3 days. The white precipitate was collected on a glass frit (G4) and washed with water (ca. 1 L). At this stage, a crude product was obtained in a 1.662 g yield. The crude product (1.662 g) was dissolved in acetonitrile (10 mL), followed by filtering through a folded filter paper (Whatman #5). After the product was left standing for a week at 25 °C, colorless platelet crystals were formed. The obtained crystals weighted 0.752 g (the yield calculated considering that [mol of $[(n\text{-}C_4H_9)_4N]_4[\alpha\text{-}PW_{11}\{Al(OH_2)\}O_{39}]$/[mol of $Cs_7[\gamma\text{-}PW_{10}O_{36}]\cdot 19H_2O$] × 100 was 36.9%). The elemental analysis results were as follows: C, 20.73; H, 4.00; N, 1.58; P, 0.84; Al, 0.77; W, 54.6; Cs, <0.1%. The calculated values for $[(n\text{-}C_4H_9)_4N]_4[\alpha\text{-}PW_{11}\{Al(OH_2)\}O_{39}] = C_{64}H_{146}AlN_4O_{40}PW_{11}$: C, 20.82; H, 3.99; N, 1.52; P, 0.84; Al, 0.73; W, 54.77; Cs, 0%. A weight loss of 2.16% was observed in the product during overnight drying at room temperature under 10^{-3}–10^{-4} Torr before the analysis, thereby suggesting the presence of two weakly solvated or adsorbed acetonitrile molecules (2.18%). TG/DTA under atmospheric conditions showed a weight loss of 31.0% with an exothermic peak at 337 °C was observed in the temperature range from 25 to 500 °C; our calculations indicated the presence of four $[(C_4H_9)_4N]^+$ ions, two acetonitrile molecules, and a water molecule (calcd. 28.4%). The results were as follows: IR soectroscopy results (KBr disk): 1078s, 964s, 887s, 818s, 749m, 702w, 518w cm^{-1}; ^{31}P NMR (25°C, acetonitrile-d_3): δ -12.5; ^{27}Al NMR (25 °C, acetonitrile-d_3): δ 16.1; ^{183}W NMR (25 °C, acetonitrile-d_3): δ -56.2 (2W), -93.1 (2W), -108.6 (2W), -115.8 (2W), -118.5 (1W), -153.9 (2W).

2.4 X-Ray crystallography

A colorless platelet crystal of $[(n\text{-}C_4H_9)_4N]_4[\alpha\text{-}PW_{11}\{Al(OH_2)\}O_{39}]$ (0.16 × 0.16 × 0.01 mm^3) was mounted on a MicroMount. All measurements were made on a Rigaku VariMax with a Saturn diffractometer using multi-layer mirror monochromated Mo Kα radiation (λ= 0.71075 Å) at 93 K. Data were collected and processed using CrystalClear for Windows, and structural analysis was performed using the CrystalStructure for Windows. The structure was solved by SHELXS-97 (direct methods) and refined by SHELXL-97 (Sheldrick, 2008). Since one aluminum atom was disordering over twelve tungsten sites in $[\alpha\text{-}PW_{11}\{Al(OH_2)\}O_{39}]^{4-}$, the occupancies for the aluminum and tungsten sites were fixed at 1/12 and 11/12 throughtout the refinement. Four tetra-*n*-butylammonium ions could not be modelled with disordered atoms. Accordingly, the residual electron density was removed using the SQUEEZE routine in PLATON (Spek, 2009).

2.5 Crystal data for $[(n\text{-}C_4H_9)_4N]_4[\alpha\text{-}PW_{11}\{Al(OH_2)\}O_{39}]$

$C_{64}H_{146}AlN_4O_{40}PW_{11}$; M = 3692.17, *cubic*, space group *Im-3m (#229)*, a = 17.665(2) Å, V = 5512.2(8) Å^3, Z = 2, D_c = 2.224 g/cm^3, μ(Mo-Kα) = 115.313 cm^{-1}. R_1 = 0.0220 ($I > 2\sigma(I)$) and wR_2 = 0.0554 (for all data). GOF = 1.093 (22662 total reflections, 652 unique reflections where $I > 2\sigma(I)$). CCDC number 851155.

2.6 Computational details

The optimal geometry of $[\alpha\text{-}PW_{11}\{Al(OH_2)\}O_{39}]^{4-}$ was computed by means of a DFT method. First, we optimized the crystal geometries and followed this up with single-point calculations with larger basis sets. All calculations were performed by a spin-restricted B3LYP on Gaussian09 program package (Frisch et al., 2009). The basis sets used for the geometry optimization were LANL2DZ for W atoms, 6-31+G* for P atoms and 6-31G* for H, O, and Al atoms. LANL2DZ and 6-31+G* were used for W and other atoms, respectively, for the single-point calculations. The geometry optimizations were started using the X-ray structure of $[\alpha\text{-}PW_{12}O_{40}]^{3-}$ as an initial geometry, and they were performed under the gas phase condition. The optimized geometries were confirmed to be true minima by frequency analyses. All atomic charges used in this text were obtained from Mulliken population analysis.

3. Results and discussion

3.1 Synthesis and molecular formula of $[(n\text{-}C_4H_9)_4N]_4[\alpha\text{-}PW_{11}\{Al(OH_2)\}O_{39}]$

The tetra-*n*-butylammonium salt of $[\alpha\text{-}PW_{11}\{Al(OH_2)\}O_{39}]^{4-}$ was formed by the direct reaction of aluminum nitrate with $[\gamma\text{-}PW_{10}O_{36}]^{7-}$ (the molar ratio of Al^{3+}:$[\gamma\text{-}PW_{10}O_{36}]^{7-}$ was ca. 1.0) in an aqueous solution at 40 °C under air, followed by the addition of excess tetra-*n*-butylammonium bromide. The crystallization was performed by slow-evaporation from acetonitrile at 25 °C. During the formation of $[\alpha\text{-}PW_{11}\{Al(OH_2)\}O_{39}]^{4-}$, the decomposition of a *di*-lacunary γ-Keggin polyoxotungstate, and isomerization of γ-isomer to α-isomer occurred in order to construct the *mono*-aluminum-substituted site in an α-Keggin structure. It was noted that the polyoxoanion $[\alpha\text{-}PW_{11}\{Al(OH_2)\}O_{39}]^{4-}$ was easily obtained by the stoichiometric reaction of aluminum nitrate with a *mono*-lacunary α-Keggin polyoxotungstate, $[\alpha\text{-}PW_{11}O_{39}]^{7-}$, in an aqueous solution; however, a single species of $[\alpha\text{-}PW_{11}\{Al(OH_2)\}O_{39}]^{4-}$ could not be obtained as a tetra-*n*-butylammonium salt by using $[\alpha\text{-}PW_{11}O_{39}]^{7-}$ as a starting polyoxoanion.[1] Thus, single crystals that were suitable for X-ray crystallography could be obtained for the crystallization of the tetra-*n*-butylammonium salt of $[\alpha\text{-}PW_{11}\{Al(OH_2)\}O_{39}]^{4-}$ synthesized by using a *di*-lacunary γ-Keggin polyoxotungstate.

---

[1] The ^{31}P NMR spectrum in acetonitrile-d_3 of the tetra-*n*-butylammonium salt of $[\alpha\text{-}PW_{11}\{Al(OH_2)\}O_{39}]^{4-}$ prepared by the stoichiometic reaction of $[\alpha\text{-}PW_{11}O_{39}]^{7-}$ with $Al(NO_3)_3\cdot 9H_2O$ in an aqueous solution showed two signals at -12.35 ppm and -12.48 ppm. The signal at -12.48 ppm was assigned to the internal phosphorus atom in $[\alpha\text{-}PW_{11}\{Al(OH_2)\}O_{39}]^{4-}$, whereas the signal at -12.35 ppm could not be identified; however, the signal was not due to the proton isomer, as reported for $[(CH_3)_2NH_2]_{10}[Hf(PW_{11}O_{39})_2]\cdot 8H_2O$ (Hou et al., 2007).

The sample for the elemental analysis was dried overnight at room temperature under a vacuum of 10^{-3} – 10^{-4} Torr. The elemental results for C, H, N, P, Al, and W were in good agreement with the calculated values for the formula without any absorbed or solvated molecules for $[(n\text{-}C_4H_9)_4N]_4[\alpha\text{-}PW_{11}\{Al(OH_2)\}O_{39}]$.

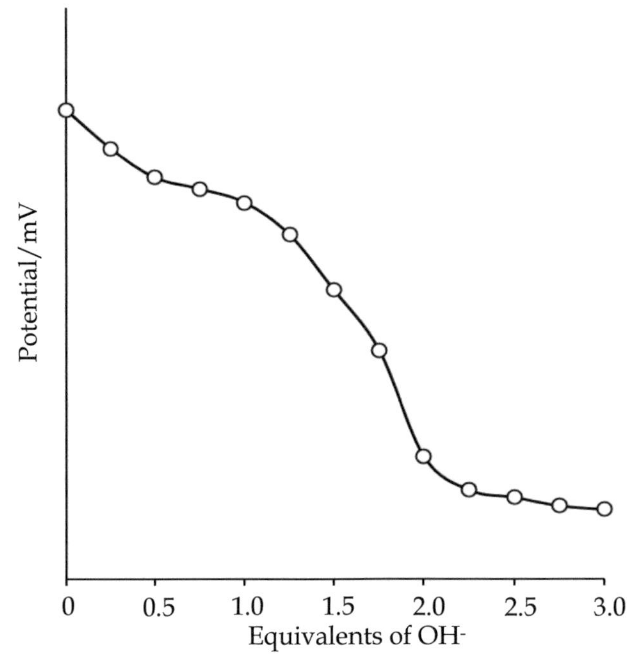

Fig. 3. Profile for the potentiometric titration of $[(n\text{-}C_4H_9)_4N]_4[\alpha\text{-}PW_{11}\{Al(OH_2)\}O_{39}]$ with tetra-*n*-butylammonium hydroxide as a titrant.

The Cs analysis revealed no contamination of cesium ions from $Cs_7[\gamma\text{-}PW_{10}O_{36}]\cdot 19H_2O$. The weight loss observed during the course of drying before the analysis was 2.16% for $[(n\text{-}C_4H_9)_4N]_4[\alpha\text{-}PW_{11}\{Al(OH_2)\}O_{39}]$; this corresponded to two weakly solvated or adsorbed acetonitrile molecules. On the other hand, in the TG/DTA measurement performed under atmospheric conditions, a weight loss of 31.0% observed in the temperature range from 25 to 500 °C corresponded to four tetra-*n*-butylammonium ions, two acetonitrile molecules, and a water molecule.

From the potentiometric titration, a break point at 2.0 equivalents of added base was observed, as shown in Fig. 3. The titration profile revealed that $[(n\text{-}C_4H_9)_4N]_4[\alpha\text{-}PW_{11}\{Al(OH_2)\}O_{39}]$ had two titratable protons dissociated from the Al-OH_2 group. This result was consistent with the elemental analysis result.

3.2 The molecular structure of $[(n\text{-}C_4H_9)_4N]_4[\alpha\text{-}PW_{11}\{Al(OH_2)\}O_{39}]$

The molecular structure of $[\alpha\text{-}PW_{11}\{Al(OH_2)\}O_{39}]^{4-}$ as determined by X-ray crystallography is shown in Figs. 4 and 5. The bond lengths and bond angles are summarized in appendix. The molecular structure of $[\alpha\text{-}PW_{11}\{Al(OH_2)\}O_{39}]^{4-}$ was identical to that of a monomeric, α-Keggin polyoxotungstate $[\alpha\text{-}PW_{12}O_{40}]^{3-}$ (Neiwert et al., 2002; Busbongthong & Ozeki, 2009). Due to the high symmetry space group, the eleven tungsten(VI) atoms were disordered and the *mono*-aluminum-substituted site was not identified, as observed for $[W_9ReO_{32}]^{5-}$ (Ortéga et al., 1997), $[\alpha\text{-}PW_{11}Re^{V}O_{40}]^{5-}$ (Kato et al., 2010), $[\{SiW_{11}O_{39}Cu(H_2O)\}\{Cu_2(ac)\text{-}(phen)_2(H_2O)\}]^{14-}$ (phen = phenanthroline, ac = acetate) (Reinoso et al., 2006), $(ANIH)_5[PCu(H_2O)W_{11}O_{39}](ANI)\cdot 8H_2O$ (ANI = aniline, $ANIH^+$ = anilinium ion) (Fukaya et al., 2011), $Cs_5[PMn(H_2O)W_{11}O_{39}]\cdot 4H_2O$ (Patel et al., 2011), and $Cs_5[PNi(H_2O)W_{11}O_{39}]\cdot 2H_2O$ (T. J. R. Weakley, 1987). However, the bond lengths of $[(n\text{-}C_4H_9)_4N]_4[\alpha\text{-}PW_{11}\{Al(OH_2)\}O_{39}]$ were clearly influenced by the insertion of aluminum ion into the vacant site as compared with those of $[CH_3NH_3]_3[PW_{12}O_{40}]\cdot 2H_2O$, $[(CH_3)_2NH_2]_3[PW_{12}O_{40}]$, and $[(CH_3)_3NH]_3\text{-}[PW_{12}O_{40}]$ (Busbongthong & Ozeki, 2009) (Table 1). Thus, the lengths of the oxygen atoms belonging to the central PO_4 tetrahedron (O_a) are longer than those of the three alkylammonium salts of $[PW_{12}O_{40}]^{3-}$; whereas, the lengths of the bridging oxygen atoms between corner-sharing MO_6 (M = W and Al) octahedra (O_c) and bridging oxygen atoms between edge-sharing MO_6 octahedra (O_e) are shorter than those of $[PW_{12}O_{40}]^{3-}$. For comparisons, the bond lengths of *mono*-metal-substituted α-Keggin phosphotungstates, e.g., $[(CH_3)_2NH_2]_4[\alpha\text{-}PW_{11}Re^{V}O_{40}]$, $(ANIH)_5[PCu(H_2O)W_{11}O_{39}](ANI)\cdot 8H_2O$ (ANI = aniline,

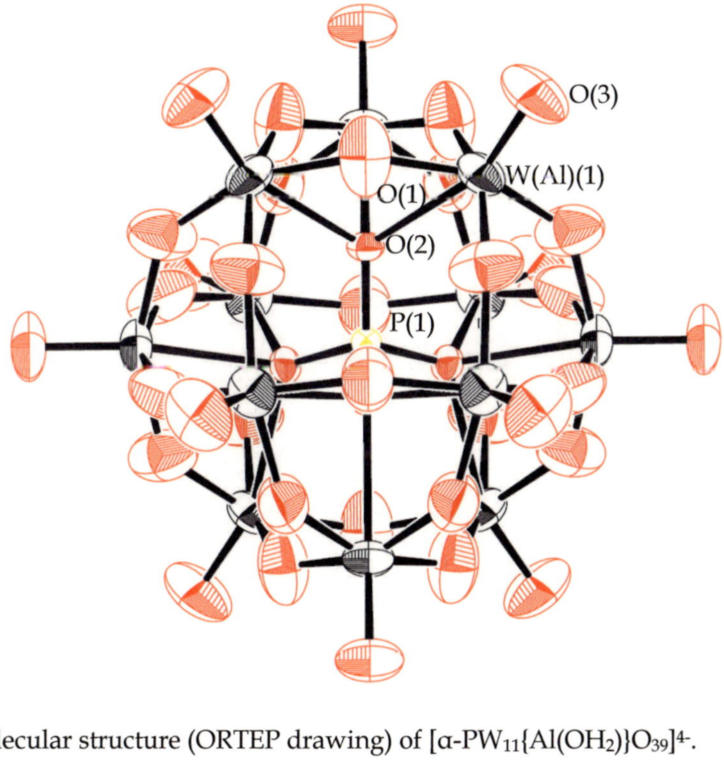

Fig. 4. The molecular structure (ORTEP drawing) of $[\alpha\text{-}PW_{11}\{Al(OH_2)\}O_{39}]^{4-}$.

	$[(n\text{-}C_4H_9)_4N]_4[\alpha\text{-}PW_{11}\{Al(OH_2)\}O_{39}]$
W(Al)-O_a	2.466 (2.466)
W(Al)-O_c	1.883 (1.883)
W(Al)-O_e	1.883 (1.883)
W(Al)-O_t	1.667 (1.667)
P-O	1.5206 (1.5206)
	$[CH_3NH_3]_3[\alpha\text{-}PW_{12}O_{40}]\cdot 2H_2O$
W-O_a	2.4077 - 2.4606 (2.4398)
W-O_c	1.8766 - 1.9407 (1.9076)
W-O_e	1.8808 - 1.9448 (1.9166)
W-O_t	1.6818 - 1.7068 (1.6951)
P-O	1.5286 - 1.5377 (1.5324)
	$[(CH_3)_2NH_2]_3[\alpha\text{-}PW_{12}O_{40}]$
W-O_a	2.4273 - 2.4568 (2.4430)
W-O_c	1.9044 - 1.9164 (1.9103)
W-O_e	1.9029 - 1.9234 (1.9158)
W-O_t	1.7000 - 1.7038 (1.7026)
P-O	1.5220 - 1.5348 (1.5313)
	$[(CH_3)_3NH]_3[\alpha\text{-}PW_{12}O_{40}]$
W-O_a	2.4313 – 2.4497 (2.4313)
W-O_c	1.8840 – 1.9286 (1.9127)
W-O_e	1.8996 – 1.9437 (1.9186)
W-O_t	1.6890 - 1.6970 (1.6933)
P-O	1.5296 – 1.5355 (1.5340)

Table 1. Ranges and mean bond distances (Å) for $[(n\text{-}C_4H_9)_4N]_4[\alpha\text{-}PW_{11}\{Al(OH_2)\}O_{39}]$, and the three alkylammonium salts of $[PW_{12}O_{40}]^{3-}$. The terms O_a, O_c, O_e, and O_t are explained in Fig. 5. The mean values are provided in parentheses.

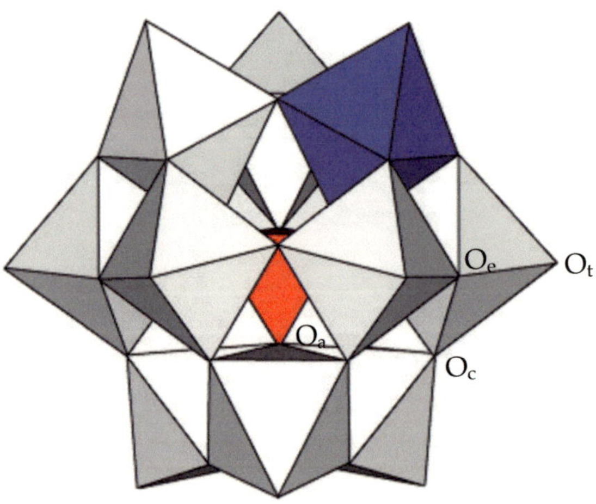

Fig. 5. The polyhedral representation of $[\alpha\text{-}PW_{11}\{Al(OH_2)\}O_{39}]^{4-}$. In the polyhedral representation, the AlO_6 and WO_6 groups are represented by blue and white octahedra, respectively. The internal PO_4 group is represented by the red tetrahedron. Further, O_a,

oxygen atoms belonging to the central PO_4 tetrahedron; O_c, bridging oxygen atoms between corner-sharing MO_6 (M = Al and W) octahedra; O_e, bridging oxygen atoms between edge-sharing MO_6 octahedra (M = Al and W); O_t, terminal oxygen atoms.

$ANIH^+$ = anilinium ion), $Cs_5[PMn(H_2O)W_{11}O_{39}]\cdot 4H_2O$, and $Cs_5[PNi(H_2O)W_{11}O_{39}]\cdot 2H_2O$ as determined by X-ray crystallography are summarized in Table 2. Although a simple comparison was difficult to draw, the following trends were observed: The W-O_a bond lengths of $[PCu(H_2O)W_{11}O_{39}]^{5-}$, $[PMn(H_2O)W_{11}O_{39}]^{5-}$, and $[PNi(H_2O)W_{11}O_{39}]^{5-}$ were significantly longer than those of $[\alpha\text{-}PW_{12}O_{40}]^{3-}$ and $[\alpha\text{-}PW_{11}Re^{V}O_{40}]^{4-}$, as observed for $[\alpha\text{-}PW_{11}\{Al(OH_2)\}O_{39}]^{4-}$ due to the presence of a water molecule coordinated to the *mono*-metal-substituted sites. The W(M)-O_c and W(M)-O_e (M = Re, Cu, Mn, and Ni) bond lengths of the four polyoxoanions mentioned in Table 2 were similar to those of $[\alpha\text{-}PW_{12}O_{40}]^{3-}$, whereas, the bond lengths of $[\alpha\text{-}PW_{11}\{Al(OH_2)\}O_{39}]^{4-}$ were clearly shorter than those of $[\alpha\text{-}PW_{12}O_{40}]^{3-}$.

	$[(CH_3)_2NH_2]_4[\alpha\text{-}PW_{11}Re^{V}O_{40}]$
W(Re)-O_a	2.418 – 2.441 (2.432)
W(Re)-O_c	1.896 – 1.914 (1.906)
W(Re)-O_e	1.895 – 1.922 (1.907)
W(Re)-O_t	1.647 – 1.694 (1.680)
P-O	1.538 – 1.540 (1.539)
	$(ANIH)_5[PCu(H_2O)W_{11}O_{39}](ANI)\cdot 8H_2O$
W(Cu)-O_a	2.4784 - 2.5044 (2.4916)
W(Cu)-O_c	1.8946 - 1.9277 (1.9077)
W(Cu)-O_e	1.8946 - 1.9277 (1.9077)
W(Cu)-O_t	1.7163 - 1.7220 (1.7178)
P-O	1.4925 - 1.5078 (1.4965)
	$Cs_5[PMn(H_2O)W_{11}O_{39}]\cdot 4H_2O$
W(Mn)-O_a	2.4220 - 2.5520 (2.4874)
W(Mn)-O_c	1.9223 - 1.8698(1.9051)
W(Mn)-O_e	1.8689 - 1.9620 (1.9079)
W(Mn)-O_t	1.6678 - 1.752(1.6889)
P-O	1.4902 - 1.602 (1.5265)
	$Cs_5[PNi(H_2O)W_{11}O_{39}]\cdot 2H_2O$
W(Ni)-O_a	2.4013 - 2.5152 (2.4792)
W(Ni)-O_c	1.8628 - 1.9430 (1.8974)
W(Ni)-O_e	1.8633 - 1.9421 (1.8964)
W(Ni)-O_t	1.6714 - 1.7354 (1.7010)
P-O	1.5150 - 1.5256 (1.5209)

Table 2. Ranges and mean bond distances (Å) for four *mono*-metal-substituted α-Keggin phosphotungstates. The terms O_a and O_t are explained in Fig. 5. The terms O_c and O_e indicate bridging oxygen atoms between corner- and edge-sharing MO_6 (M = W, Re, Cu, Mn, Ni) octahedra. The mean values are provided in parentheses.

To investigate the coordination sphere around the *mono*-aluminum-substituted site in $[\alpha\text{-}PW_{11}\{Al(OH_2)\}O_{39}]^{4-}$, the optimized geometry was computed by means of a DFT method, as shown in Figs. 6 and 7. The ranges and mean bond distances, and the Millken charges for the DFT-optimized $[\alpha\text{-}PW_{11}\{Al(OH_2)\}O_{39}]^{4-}$ are summarized in Tables 3 and 4. It was noted that the *mono*-aluminum-substituted site was uniquely concave downward, which caused the extension of the P-O bond linkaged to the aluminum atom (1.5654 Å), whereas the Al-O bond linkaged to the internal phosphorus atom was shortened due to the insertion of the Al^{3+} ion that has a smaller ionic radius (0.675 Å) than that of W^{6+} (0.74 Å) into the *mono*-vacant site (Shannon, 1976). The lengths of Al-O bonds at the corner- and edge-sharing Al-O-W bondings were shorter than those of W-O bonds at the corner- and edge-sharing W-O-W bondings, which caused shortening of the average W(Al)-O bond lengths, as observed by X-ray crystallography.

The Mulliken charges of all oxygen atoms linkaged to aluminum atoms in $[\alpha\text{-}PW_{11}\{Al(OH_2)\}O_{39}]^{4-}$ were more positive than those linkaged to tungsten atoms in $[\alpha\text{-}PW_{12}O_{40}]^{3-}$; whereas the charges of oxygen atoms linkaged to tungsten atoms in $[\alpha\text{-}PW_{11}\{Al(OH_2)\}O_{39}]^{4-}$ were similar to those in $[\alpha\text{-}PW_{12}O_{40}]^{3-}$. In addition, the atomic charge of the phosphorus atom in $[\alpha\text{-}PW_{11}\{Al(OH_2)\}O_{39}]^{4-}$ was more negative than that in $[\alpha\text{-}PW_{12}O_{40}]^{3-}$. In the case of *mono*-vanadium(V)-substituted Keggin silicotungstate $[SiW_{11}VO_{40}]^{5-}$, the net charge associated with the inner tetrahedron was very similar to that supported by SiO_4 in $[SiW_{12}O_{40}]^{4-}$ (Maestre et al., 2001). Thus, the difference in the charge on the internal phosphorus atom for $[\alpha\text{-}PW_{11}\{Al(OH_2)\}O_{39}]^{4-}$ and $[\alpha\text{-}PW_{12}O_{40}]^{3-}$ might be due to the gravitation of aluminum atoms towards the internal PO_4 group.

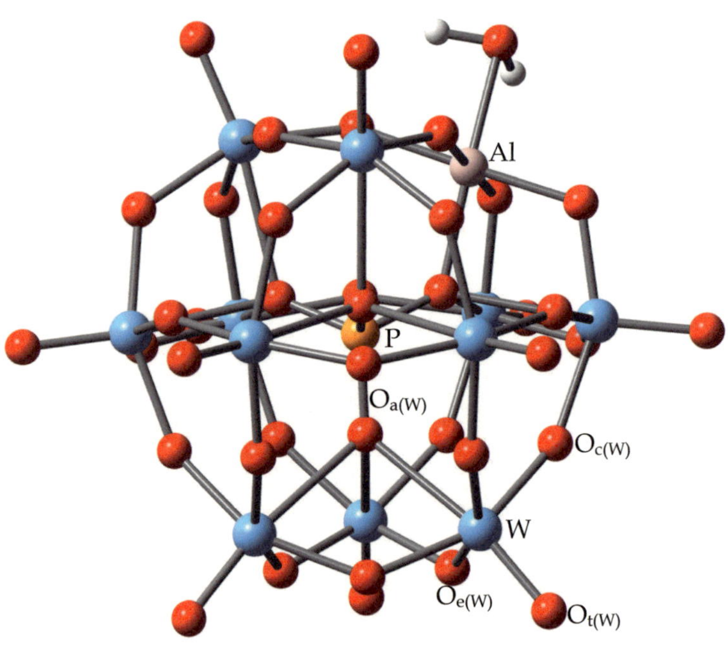

Fig. 6. The DFT-optimized geometry of $[\alpha\text{-}PW_{11}\{Al(OH_2)\}O_{39}]^{4-}$. The phosphorus, oxygen, aluminum, tungsten, and hydrogen atoms are represented by orange, red, pink, blue, and white balls, respectively.

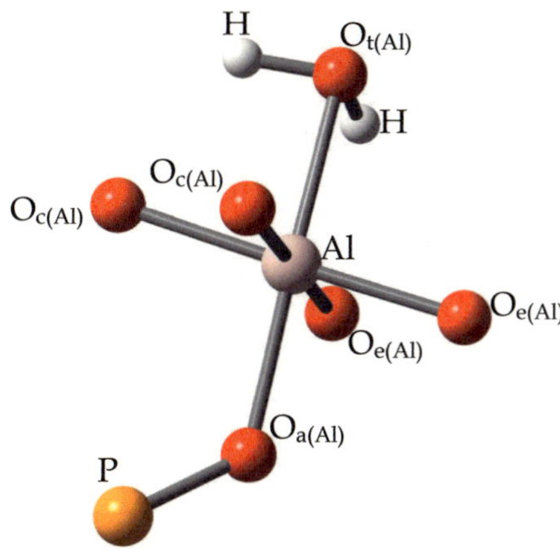

Fig. 7. The coordination sphere around the *mono*-aluminum-substituted site in DFT-optimized $[\alpha\text{-}PW_{11}\{Al(OH_2)\}O_{39}]^{4-}$.

	$[\alpha\text{-}PW_{11}\{Al(OH_2)\}O_{39}]^{4-}$	$[\alpha\text{-}PW_{12}O_{40}]^{3-}$
W-O_a	2.4422 – 2.5140 (2.4702)	2.4568 – 2.4579 (2.4574)
W-O_c	1.8311 – 1.9828 (1.9206)	1.9202 – 1.9216 (1.9209)
W-O_e	1.8373 – 1.9918 (1.9267)	1.9262 – 1.9276 (1.9267)
W-O_t	1.7196 – 1.7246 (1.7210)	1.7103 – 1.7106 (1.7105)
P-O	1.5450 – 1.5654 (1.5517)	1.5530 – 1.5535 (1.5533)
Al-O_a	1.9487 (1.9487)	-
Al-O_c	1.8519, 1.8955 (1.8737)	-
Al-O_e	1.8723, 1.9215 (1.8969)	-
Al-OH_2	2.0983 (2.0983)	-

Table 3. Ranges and mean bond distances (Å) for $[\alpha\text{-}PW_{11}\{Al(OH_2)\}O_{39}]^{4-}$ and $[\alpha\text{-}PW_{12}O_{40}]^{3-}$ optimized by DFT calculations. The terms O_a, O_c, O_e, and O_t are explained in Fig. 5. The average values are provided in parentheses.

	$[\alpha\text{-}PW_{11}\{Al(OH_2)\}O_{39}]^{4-}$	$[\alpha\text{-}PW_{12}O_{40}]^{3-}$
$O_{a\,(W)}$	-0.7356 – -0.8445 (-0.7734)	-0.8951 – -0.8990 (-0.8968)
$O_{c\,(W)}$	-1.226 – -1.345 (-1.317)	-1.353 – -1.355 (-1.353)
$O_{e\,(W)}$	-1.030 – -1.160 (-1.074)	-1.085 – -1.087 (-1.086)
$O_{t\,(W)}$	-0.6757 – -0.6991 (-0.6882)	-0.6273 – -0.6277 (-0.6275)
P	7.255 (7.255)	9.256 (9.256)
W	2.101 – 2.343 (2.257)	2.343 – 2.346 (2.345)
$O_{a(Al)}$	-0.1495 (-0.1495)	-
$O_{c(Al)}$	-0.3332, -0.5920 (-0.4626)	-
$O_{e(Al)}$	-0.4910, -0.7848 (-0.6379)	-
$O_{t(Al)}$	-0.5553 (-0.5553)	-
Al	-0.5307 (-0.5307)	-
H	0.5754, 0.5796 (0.5775)	-

Table 4. Mulliken charges computed for $[\alpha\text{-}PW_{11}\{Al(OH_2)\}O_{39}]^{4-}$ and $[\alpha\text{-}PW_{12}O_{40}]^{3-}$. The terms $O_{a(M)}$, $O_{c(M)}$, $O_{e(M)}$, and $O_{t(M)}$ (M = Al and W) are explained in Figs. 6 and 7. The average values are provided in parentheses.

3.2 Spectroscopic data for $[(n\text{-}C_4H_9)_4N]_4[\alpha\text{-}PW_{11}\{Al(OH_2)\}O_{39}]$

The FTIR spectra measured as a KBr disk of $[(n\text{-}C_4H_9)_4N]_4[\alpha\text{-}PW_{11}\{Al(OH_2)\}O_{39}]$, $K_7[\alpha\text{-}PW_{11}O_{39}]\cdot 11H_2O$, $Cs_7[\gamma\text{-}PW_{10}O_{36}]\cdot 19H_2O$, and $Na_3[\alpha\text{-}PW_{12}O_{40}]\cdot 16H_2O$ are shown in Fig. 8. For

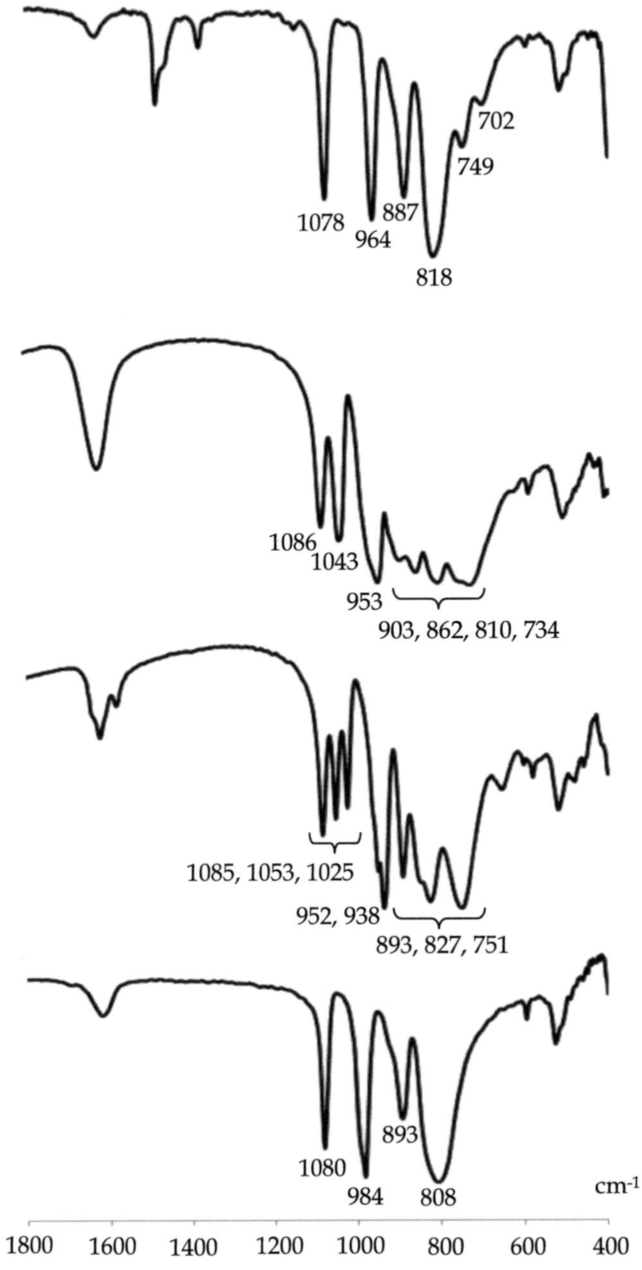

Fig. 8. FTIR spectra (as KBr disks) in the range of 1800 – 400 cm^{-1} for $[(n\text{-}C_4H_9)_4N]_4[\alpha\text{-}PW_{11}\{Al(OH_2)\}O_{39}]$ (top), $K_7[\alpha\text{-}PW_{11}O_{39}]\cdot 11H_2O$ (the second top), $Cs_7[\gamma\text{-}PW_{10}O_{36}]\cdot 19H_2O$ (the third top), and $Na_3[\alpha\text{-}PW_{12}O_{40}]\cdot 16H_2O$ (bottom)

$[(n\text{-}C_4H_9)_4N]_4[\alpha\text{-}PW_{11}\{Al(OH_2)\}O_{39}]$, the P-O band was observed at 1078 cm^{-1}, and the W-O bands were observed at 964, 887, 818, 749, and 702 cm^{-1}, these were different from those of $K_7[\alpha\text{-}PW_{11}O_{39}]\cdot 11H_2O$ (1086, 1043, 953, 903, 862, 810, and 734 cm^{-1}) and $Cs_7[\gamma\text{-}PW_{10}O_{36}]\cdot 19H_2O$ (1085, 1053, 1025, 952, 938, 893, 827, and 751 cm^{-1}) (Rocchiccioli-Deltcheff et al., 1983; Thouvenot et al., 1984). This result suggested that the aluminum atom was coordinated into the vacant site in the polyoxometalate. It should be noted that the bands observed for $[(n\text{-}C_4H_9)_4N]_4[\alpha\text{-}PW_{11}\{Al(OH_2)\}O_{39}]$ were significantly different from those of $Na_3[\alpha\text{-}PW_{12}O_{40}]\cdot 16H_2O$ (1080, 984, 893, and 808 cm^{-1}). This was consistent with the results observed by X-ray crystallography and DFT calculations, as mentioned above.

The ^{31}P NMR spectrum of $[(n\text{-}C_4H_9)_4N]_4[\alpha\text{-}PW_{11}\{Al(OH_2)\}O_{39}]$ in acetonitrile-d_3 at ~25 °C was a clear single line spectrum at -12.5 ppm due to the internal phosphorus atom, thereby confirming the compound's purity and homogeneity, as shown in Fig. 9. The signal exhibited a shift from the signals of tetra-*n*-butylammonium salts of $[\alpha\text{-}PW_{12}O_{40}]^{3-}$ (δ -14.6) and $[\alpha\text{-}PW_{11}O_{39}]^{7-}$ (δ -12.0), suggesting the insertion of aluminum ion into the vacant site.

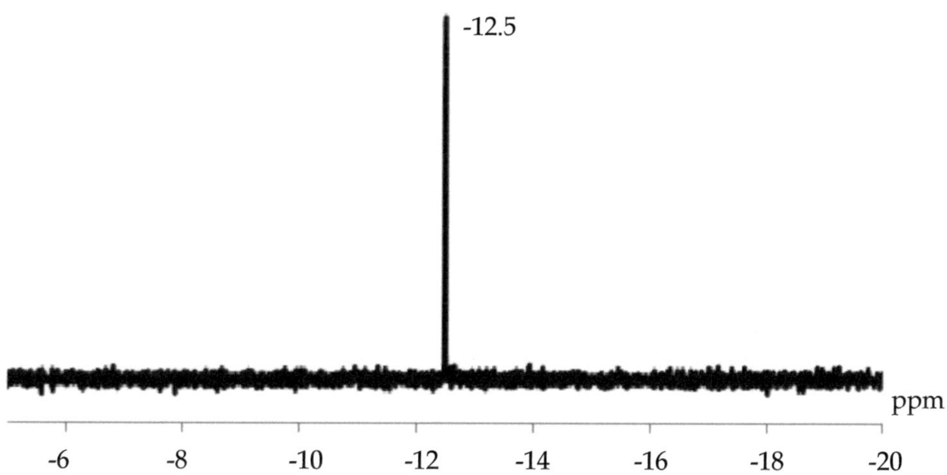

Fig. 9. ^{31}P NMR spectrum in acetonitrile-d_3 of $[(n\text{-}C_4H_9)_4N]_4[\alpha\text{-}PW_{11}\{Al(OH_2)\}O_{39}]$.

The ^{27}Al NMR spectrum (Fig. 10) of $[(n\text{-}C_4H_9)_4N]_4[\alpha\text{-}PW_{11}\{Al(OH_2)\}O_{39}]$ in acetonitrile-d_3 at ~25 °C showed a broad signal at 16.1 ppm due to the *mono*-aluminum-substituted site in $[\alpha\text{-}PW_{11}\{Al(OH_2)\}O_{39}]^{4-}$.

The ^{183}W NMR spectrum (Fig. 11) of $[(n\text{-}C_4H_9)_4N]_4[\alpha\text{-}PW_{11}\{Al(OH_2)\}O_{39}]$ in acetonitrile-d_3 at ~25 °C was a six-line spectrum of (δ -56.2, -93.1, -108.6, -115.8, -118.5, -153.9) with 2:2:2:2:1:2 intensities, which were in accordance with the presence of eleven tungsten atoms with *Cs* symmetry. These spectral data were completely consistent with the X-ray structure and the optimized structure, suggesting that the solid structure was maintained in the solution.

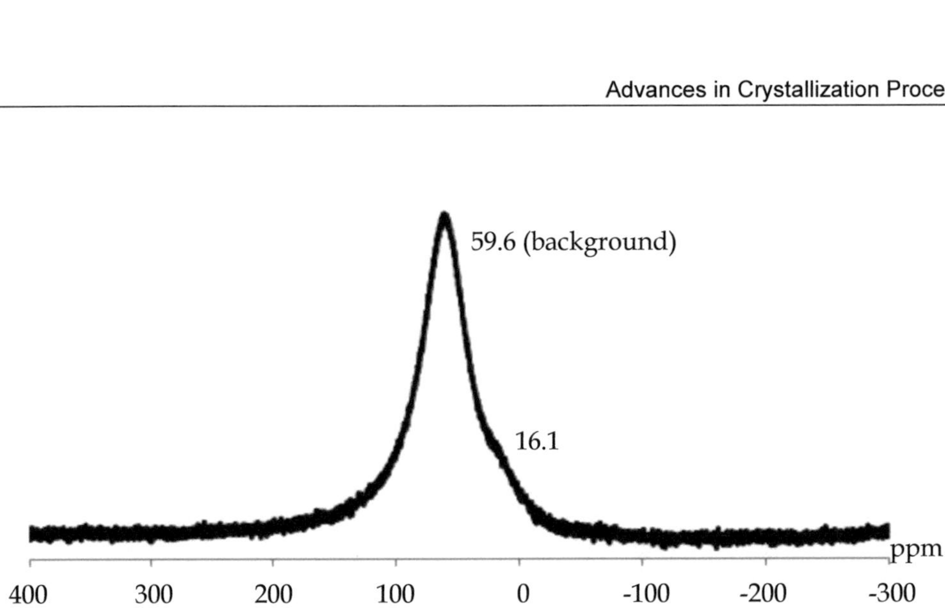

Fig. 10. ^{27}Al NMR spectrum in acetonitrile-d_3 of $[(n\text{-}C_4H_9)_4N]_4[\alpha\text{-}PW_{11}\{Al(OH_2)\}O_{39}]$.

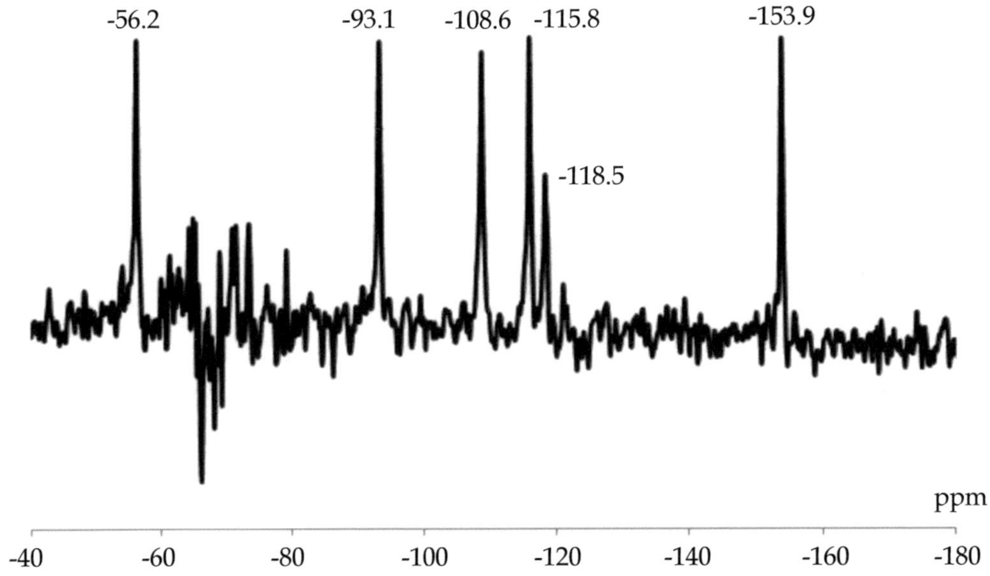

Fig. 11. ^{183}W NMR spectrum in acetonitrile-d_3 of $[(n\text{-}C_4H_9)_4N]_4[\alpha\text{-}PW_{11}\{Al(OH_2)\}O_{39}]$.

4. Conclusion

The synthesis of a monomeric, *mono*-aluminum-substituted α-Keggin polyoxometalate is described in this study. We successfully obtained single crystals of acetonitrile-soluble tetra-*n*-butylammonium salt $[(n\text{-}C_4H_9)_4N]_4[\alpha\text{-}PW_{11}\{Al(OH_2)\}O_{39}]$ by reacting aluminum nitrate with a *di*-lacunary γ-Keggin phosphotungstate. The characterization of $[(n\text{-}C_4H_9)_4N]_4[\alpha\text{-}PW_{11}\{Al(OH_2)\}O_{39}]$ was accomplished by X-ray crystallography, elemental analysis, thermogravimetric/differential thermal analysis, Fourier transform infrared spectra, and solution ^{31}P, ^{27}Al, and ^{183}W nuclear magnetic resonance spectroscopy. The single-crystal X-ray structure analysis, revealed as $[(n-C_4H_9)_4N]_4[\alpha-PW_{11}\{Al(OH_2)\}O_{39}]$, was a monomeric, α-Keggin structure, and the *mono*-aluminum-substituted site could not be identified due to the high symmetry in the product. In contrast, the DFT-optimized geometry of $[\alpha-PW_{11}\{Al(OH_2)\}O_{39}]^{4-}$ showed that the *mono*-aluminum-substituted site was uniquely concave downward, which caused the extension of the P-O bond linkaged to the aluminum atom, whereas the Al-O bond linkaged to the phosphorus atom was shortened. This structural difference strongly influenced the bonding mode (bond lengths and bond angles) as determined by X-ray crystallography. In addition, the Mulliken charges clearly exhibited the effect caused by the insertion of aluminum atoms into the *mono*-vacant sites.

5. Acknowledgment

This work was supported by a Grant-in-Aid for Scientific Research on Innovative Areas (No. 21200055) of the Ministry of Education, Culture, Sports, Science and Technology, Japan. Y. Kataoka acknowledges the JSPS Research Fellowship for Young Scientist. Y. Kitagawa also has been supported by Grant-in-Aid for Scientific Research on Innovative Areas ("Coordination Programming" area 2170, No. 22108515) from the Ministry of Education, Culture, Sports, Science and Technology (MEXT). This research was partially carried out using equipment at the Center for Instrumental Analysis, Shizuoka University.

6. Appendix

Bond lengths (Å) of $[(n-C_4H_9)_4N]_4[\alpha-PW_{11}\{Al(OH_2)\}O_{39}]$: W(1)-O(1) 1.883(4); W(1)-O(1)^1 1.883(4); W(1)-O(1)^2 1.883(4); W(1)-O(1)^3 1.883(4); W(1)-O(2) 2.465(5); W(1)-O(2)^4 2.465(5); W(1)-O(3) 1.667(4); P(1)-O(2) 1.522(5); P(1)-O(2)^5 1.522(5); P(1)-O(2)^6 1.522(5); P(1)-O(2)^7 1.522(5); P(1)-O(2)^4 1.522(5); P(1)-O(2)^8 1.522(5); P(1)-O(2)^9 1.522(5); P(1)-O(2)^{10} 1.522(5); Al(1)-O(1) 1.883(4); Al(1)-O(1)^1 1.883(4); Al(1)-O(1)^2 1.883(4); Al(1)-O(1)^3 1.883(4); Al(1)-O(3) 1.667(4). Symmetry operators: (1) X,Z,Y (2) Z,Y,-X+1 (3) Z,-X+1,Y (4) Y,Z,-X+1 (5) Y,Z,X (6) Z,X,Y (7) X,Y,-Z+1 (8) Z,X,-Y+1 (9) -Z+1,X,-Y+1 (10) -Y+1,-Z+1,-X+1.

Bond angles (°) of $[(n-C_4H_9)_4N]_4[\alpha-PW_{11}\{Al(OH_2)\}O_{39}]$: O(1)-W(1)-O(1)^1 87.5(2); O(1)-W(1)-O(1)^2 87.08(18); O(1)-W(1)-O(1)^3 154.8(2); O(1)-W(1)-O(2) 63.32(19); O(1)-W(1)-O(2)^4 92.40(18); O(1)-W(1)-O(3) 102.58(17); O(1)^1-W(1)-O(1)^2 154.8(2) O(1)^1-W(1)-O(1)^3 87.08(18); O(1)^1-W(1)-O(2) 63.32(19); O(1)^1-W(1)-O(2)^4 92.40(18); O(1)^1-W(1)-O(3) 102.58(17); O(1)^2-W(1)-O(1)^3 87.5(2); O(1)^2-W(1)-O(2) 92.40(18); O(1)^2-W(1)-O(2)^4 63.32(19); O(1)^2-W(1)-O(3) 102.58(17); O(1)^3-W(1)-O(2) 92.40(18); O(1)^3-W(1)-O(2)^4 63.32(19); O(1)^3-W(1)-O(3) 102.58(17); O(2)-W(1)-O(2)^4 41.76(15); O(2)-W(1)-O(3) 159.12(11); O(2)^4-W(1)-O(3) 159.12(11); O(2)-P(1)-O(2)^5 109.5(3); O(2)-P(1)-O(2)^6 109.5(3); O(2)-P(1)-O(2)^7 70.5(3); O(2)-P(1)-O(2)^4 70.5(3); O(2)-P(1)-O(2)^8 180.0(4); O(2)-P(1)-O(2)^9 109.5(3); O(2)-P(1)-O(2)^{10} 70.5(3); O(2)^5-P(1)-O(2)^6 109.5(3); O(2)^5-P(1)-O(2)^7 70.5(3); O(2)^5-P(1)- O(2)^4 70.5(3); O(2)^5-P(1)-O(2)^8 70.5(3); O(2)^5-P(1)-O(2)^9 109.5(3); O(2)^5-P(1)-O(2)^{10} 180.0(4); O(2)^6-P(1)-O(2)^7 180.0(4); O(2)^6-P(1)-O(2)^4 70.5(3); O(2)^6-P(1)-O(2)^8 70.5(3); O(2)^6-P(1)-O(2)^9 109.5(3); O(2)^6-P(1)-O(2)^{10} 70.5(3); O(2)^7-P(1)-O(2)^4 109.5(3); O(2)^7-P(1)-O(2)^8 109.5(3); O(2)^7-P(1)-O(2)^9 70.5(3); O(2)^7-P(1)-O(2)^{10} 109.5(3); O(2)^4-P(1)-O(2)^8 109.5(3); O(2)^4-P(1)-O(2)^9 180.0(4); O(2)^4-P(1)-O(2)^{10} 109.5(3); O(2)^8-P(1)-O(2)^9 70.5(3); O(2)^8-P(1)-O(2)^{10} 109.5(3); O(2)^9-P(1)-O(2)^{10} 70.5(3); O(1)-Al(1)-O(1)^1 87.5(2); O(1)-

Al(1)-O(1)^{2} 87.08(18); O(1)-Al(1)-O(1)^{3} 154.8(2); O(1)-Al(1)-O(3) 102.58(17); O(1)^{1}-Al(1)-O(1)^{2} 154.8(2); O(1)^{1}-Al(1)-O(1)^{3} 87.08(18); O(1)^{1}-Al(1)-O(3) 102.58(17); O(1)^{2}-Al(1)-O(1)^{3} 87.5(2); O(1)^{2}-Al(1)-O(3) 102.58(17); O(1)^{3}-Al(1)-O(3) 102.58(17); W(1)-O(1)-W(1)^{11} 140.7(3); W(1)-O(1)-Al(1)^{11} 140.7(3); W(1)^{11}-O(1)-Al(1) 140.7(3); Al(1)-O(1)-Al(1)^{11} 140.7(3); W(1)-O(2)-W(1)^{11} 91.97(16); W(1)-O(2)-W(1)^{12} 91.97(16); W(1)-O(2)-P(1) 123.9(3); W(1)-O(2)-O(2)^{7} 131.4(3); W(1)-O(2)-O(2)^{4} 69.1(3); W(1)-O(2)-O(2)^{10} 131.4(3); W(1)^{11}-O(2)-W(1)^{12} 91.97(16); W(1)^{11}-O(2)-P(1) 123.9(3); W(1)^{11}-O(2)-O(2)^{7} 69.1(3) W(1)^{11}-O(2)-O(2)^{4} 131.4(3); W(1)^{11}-O(2)- O(2)^{10} 131.4(3); W(1)^{12}-O(2)-P(1) 123.9(3); W(1)^{12}-O(2)-O(2)^{7} 131.4(3); W(1)^{12}-O(2)-O(2)^{4} 131.4(3); W(1)^{12}-O(2)-O(2)^{10} 69.1(3); P(1)-O(2)-O(2)^{7} 54.7(3); P(1)-O(2)-O(2)^{4} 54.7(3); P(1)-O(2)-O(2)^{10} 54.7(3); O(2)^{7}-O(2)-O(2)^{4} 90.0(3); O(2)^{7}-O(2)-O(2)^{10} 90.0(3); O(2)^{4}-O(2)-O(2)^{10} 90.0(3). Symmetry operators: (1) X,Z,Y (2) Z,Y,-X+1 (3) Z,-X+1,Y (4) Y,Z,-X+1 (5) Y,Z,X (6) Z,X,Y (7) X,Y,-Z+1 (8) Z,X,-Y+1 (9) -Z+1,X,-Y+1 (10) -Y+1,-Z+1,-X+1 (11) -Y+1,Z,-X+1 (12) -Z+1,-X+1,Y.

7. References

Akitt, J. W. (1988). Multinuclear Studies of Aluminum Compounds. *Prog. Nucl. Magn. Res. Spectr.*,Vol.21, No.1-2, pp. 1–149

Baes, C. F. Jr. & Mesmer, R. E. (1976). *The Hydrolysis of Cations*, pp. 112–123, John Wiley, New York, 1976

Busbongthong, S. & Ozeki, T. (2009). Structural Relationships among Methyl-, Dimethyl-, and Trimethylammonium Phosphdodecatungstates. *Bull. Chem. Soc. Jpn.*, Vol.82, No.11, pp. 1393–1397

Contant, R. (1987). Relation between Tungstophosphates Related to the Phosphorus Tungsten Oxide Anion ($PW_{12}O_{40}^{3-}$). Synthesis and Properties of a New Lacunary Potassium Polytungstophosphate ($K_{10}P_2W_{20}O_{70}\cdot 24H_2O$). *Can. J. Chem.*, Vol.65, No.3, pp. 568–573

Cotton, F. A. & Wilkinson, G. (1988). *Advanced Inorganic Chemistry, Fifth Edition*, John Wiley & Sons, New York

Djurdjevic P.; Jelic, R. & Dzajevic, D. (2000). The Effect of Surface Active Substances on Hydrolysis of Aluminum(III) Ion. *Main Metal Chemistry*, Vol.23, No.8, pp. 409–421

Domaille, P. J. (1990). Vanadium(V) Substituted Dodecatungstophosphates. *Inorg. Synth.*, Vol.27, pp. 96–104

Frisch, M. J.; Trucks, G. W.; Schlegel, H. B.; Scuseria, G. E.; Robb, M. A.; Cheeseman, J. R.; Scalmani, G.; Barone, V.; Mennucci, B.; Petersson, G. A.; Nakatsuji, H.; Caricato, M.; Li, X.; Hratchian, H. P.; Izmaylov, A. F.; Bloino, J.; Zheng, G.; Sonnenberg, J. L.; Hada, M.; Ehara, M.; Toyota, K.; Fukuda, R.; Hasegawa, J.; Ishida, M.; Nakajima, T.; Honda, Y.; Kitao, O.; Nakai, H.; Vreven, T.; Montgomery, Jr., J. A.; Peralta, J. E.; Ogliaro, F.; Bearpark, M.; Heyd, J. J.; Brothers, E.; Kudin, K. N.; Staroverov, V. N.; Kobayashi, R.; Normand, J.; Raghavachari, K.; Rendell, A.; Burant, J. C.; Iyengar, S. S.; Tomasi, J.; Cossi, M.; Rega, N.; Millam, N. J.; Klene, M.; Knox, J. E.; Cross, J. B.; Bakken, V.; Adamo, C.; Jaramillo, J.; Gomperts, R.; Stratmann, R. E.; Yazyev, O.; Austin, A. J.; Cammi, R.; Pomelli, C.; Ochterski, J. W.; Martin, R. L.; Morokuma, K.; Zakrzewski, V. G.; Voth, G. A.; Salvador, P.; Dannenberg, J. J.; Dapprich, S.; Daniels, A. D.; Farkas, Ö.; Foresman, J. B.; Ortiz, J. V.; Cioslowski, J. & Fox, D. J. (2009). *Gaussian 09, Revision B.1*, Gaussian, Inc., Wallingford CT

Fukaya, K.; Srifa, A.; Ishikawa, E. & Naruke, H. (2010). Synthesis and Structural Characterization of Polyoxometalates Incorporating with Anilinium Cations and Facile Preparation of Hybrid Film. *J. Mol. Struc.* Vol.979, pp. 221–226

Hou, Y.; Fang, X. & Hill, C. L. (2007). Breaking Symmetry: Spontaneous Resolution of a Polyoxometalate. *Chem. Eur. J.* Vol.13, pp. 9442–9447

Kato, C. N.; Hara, K.; Kato, M.; Amano, H.; Sato, K.; Kataoka, Y. & Mori, W. (2010). EDTA-Reduction of Water to Molecular Hydrogen Catalyzed by Visible-Light-Response TiO_2-Based Materials Sensitized by Dawson- and Keggin-Type Rhenium(V)-Containing Polyoxotungstates. *Materials*, Vol.3, pp. 897–917

Kato, C. N.; Katayama, Y.; Nagami, M.; Kato, M. & Yamasaki, M. (2010). A Sandwich-type Aluminium Complex Composed of Tri-lacunary Keggin-type Polyoxotungstate: Synthesis and X-Ray Crystal Structure of $[(A\text{-}PW_9O_{34})_2\{W(OH)(OH_2)\}\text{-}\{Al(OH)(OH_2)\}\{Al(\mu\text{-}OH)(OH_2)_2\}_2]^{7-}$. *Dalton Trans.*, Vol.39, pp. 11469–11474

Kikukawa, Y.; Yamaguchi, S.; Nakagawa, Y.; Uehara, K.; Uchida, S.; Yamaguchi, K. & Mizuno, N. (2008). Synthesis of a Dialuminum-Substituted Silicotungstate and the Diasteroselective Cyclization of Citronellal Derivatives. *J. Am. Chem. Soc.*, Vol.130, No.47, 15872–15878

Knoth, W. H.; Domaille, P. J. & Roe, D. C. (1983). Halometal Derivatives of $W_{12}PO_{40}^{3-}$ and Related ^{183}W NMR Studies. *Inorg. Chem.* Vol.22, 198–201

Knoth, W. H. & Harlow, R. L. (1981). New Tungstophosphates: $Cs_6W_5P_2O_{23}$, $Cs_7W_{10}PO_{36}$, and $Cs_7Na_2W_{10}PO_{37}$. *J. Am. Chem. Soc.*, Vol.103. No.7, pp. 1865–1867

Lin, Y.; Weakley, T. J. R.; Rapko, B. & Finke, R. G. (1993). Polyoxoanions Derived from Tungstosilicatie ($A\text{-}\beta\text{-}SiW_9O_{34}^{10-}$): Synthesis, Single-crystal Structural Determination, and Solution Structural Characterization by Tungsten-183 NMR and IR of Titanotungstosilicate ($A\text{-}\beta\text{-}Si_2W_{18}Ti_6O_{77}^{14-}$). *Inorg. Chem.*, Vol.32, No.23, pp. 5095–5101

Maestre, J. M.; Lopez, X.; Bo, C.; Poblet, J.-M. & Casan-Pastor N. (2001). Electronic and Magnetic Properties of α-Keggin Anions: A DFT Study of $[XM_{12}O_{40}]^{n-}$, (M = W, Mo; X = Al^{III}, Si^{VI}, P^{V}, Fe^{III}, Co^{II}, Co^{III}) and $[SiM_{11}VO_{40}]^{m-}$ (M = Mo and W). *J. Am. Chem. Soc.*, Vol.123, pp. 3749–3758

Neiwert, W. A.; Cowan, J. J.; Hardcastle, K. I.; Hill, C. L. & Weinstock, I. A. (2002). Stability and Structure in α- and β-Keggin Heteropolytungstates, $[X^{n+}W_{12}O_{40}]^{(8-n)-}$, X = *p*-Block Cation. *Inorg. Chem.*, Vol.41, 6950–6952

Nomiya, K.; Takahashi, M.; Ohsawa, K. & Widegren, J. A. (2001). Synthesis and Characterization of Tri-titanium(IV)-1,2,3-substituted α-Keggin Polyoxotungstates with Heteroatoms P and Si. Crystal Structure of the Dimeric, Ti-O-Ti Bridged Anhydride Form $K_{10}H_2[\alpha,\alpha\text{-}P_2W_{18}Ti_6O_{77}]\cdot 17H_2O$ and Confirmation of Dimeric Forms in Aqueous Solution by Ultracentrifugation Molecular Weight Measurements. *J. Chem. Soc., Dalton Trans.* No.19, pp. 2872-2878

Nomiya, K.; Takahashi, M. Widegren, J. A.; Aizawa, T.; Sakai, Y. & Kasuga, N. C. (2002). Synthesis and pH-Variable Ultracentrifugation Molecular Weight Measurements of the Dimeric, Ti-O-Ti Bridged Anhydride Form of a Novel Di-Ti^{IV}-substituted α-Keggin Polyoxotungstate. Molecular Structure of the $[(\alpha\text{-}1,2\text{-}PW_{10}Ti_2O_{39})_2]^{10-}$ Polyoxoanion. *J. Chem. Soc., Dalton Trans.* No.19, pp. 3679–3685

Ortéga, F.; Pope, M. T. & Evans, H.T., Jr. (1997). Tungstorhenate Heteropolyanions. 2. Synthesis and Characterization of Enneatungstorhebates(V), -(VI) and -(VII). *Inorg. Chem.*, Vol.36, No.10, pp. 2166–2169

Orvig, C. (1993). The Aqueous Coordination Chemistry of Aluminum In: *Coordination Chemistry of Aluminum*, G.H. Robinson, (Ed.), 85–121, VCH, Weinheim

Patel, K.; Shringarpure, P. & Patel, A. (2011). One-step Synthesis of a Keggin-type Manganese(II)-substitited Phosphotungstate: Structrural and Spectroscopic

Characterization and Non-solvent Liquid Phase Oxidation of Styrene. *Transition Met. Chem.*, Vol.36, pp. 171–177

Pope, M. T. (1983). *Heteropoly and Isopoly Oxometalates*, Springer-Verlag, Berlin

Pope, M. T. & Müller, A. (1991). Chemistry of Polyoxometallates. Actual Variation on an Old Theme with Interdisciplinary References. *Angew. Chem. Int. Ed. Engl.*, Vol.30, No.1, pp. 34–48

Pope, M. T. & Müller, A. (Eds.), (1994). *Polyoxometalates: From Platonic Solids to Anti-Retroviral Activity*, Kluwer Academic Publishers, Dordrecht, The Netherlands

Reinoso, S.; Vitoria, P.; Felices, L. S.; Lezama, L. & Gutiérrez-Zorrilla, J. M. (2006). Analysis of Weak Interactions in the Crystal Packing of Inorganic Metalorganic Hybrids Based on Keggin Polyoxometalates and Dinuclear Copper(II)-Acetate Complexes. *Inorg. Chem.*, Vol.45, pp. 108–118

Rocchiccioli-Deltcheff, C.; Fournier, M.; Franck, R. & Thouvenot, R. (1983). Vibrational Investigations of Polyoxometalates. 2. Evidence for Anion-Anion Interactions in Molybdenum(VI) and Tungsten(VI) Compounds Related to the Keggin Structure. *Inorg. Chem.*, Vol.22, pp. 207–216

Rosenheim, A. & Jaenicke, J. Z. (1917). Iso- and Heteropoly Acids. XV. Heteropoly tungstates and Some Heteropoly Molybdates. *Anorg. Allg. Chem.*, Vol.101, pp. 235–275

Shannon, R. D. (1976). Revised Effective Ionic Radii and Systematic Studies of Interatomic Distances in Halides and Chalcogenides. *Acta Crystallogr., Sect. A*, Vol.A32, pp. 751–767

Sheldrick, G. M. (2008). A Short History of SHELX. *Acta Crystallogr., Sect. A*, Vol.A46, No.1, pp. 112–122

Spek, A. L. (2009). Structure Validation in Chemical Crystallography. *Acta Crystallogr., Sect. D*, Vol.D65, No.2, pp. 148–155

Thouvenot, R.; Fournier, M.; Franck, R. & Rocchiccioli-Deltcheff, C. (1984). Vibrational Investigations of Polyoxometalates. 3. Isomerism in Molybdenum(VI) and Tungsten(VI) Compounds Related to the Keggin Structure. *Inorg. Chem.*, Vol.23, pp. 598–605

Weakley, T. J. R. (1987). Crystal Structure of Cesium Aquanickelo(II)undecatungstophsphate Dihydrate. J. Cryst. Spectro. Res., Vol.17, No.3, pp. 383–391

Weakley, T. J. R. & Finke, R. G. (1990). Single-crystal X-Ray Structures of the Polyoxotungstate Salts $K_{8.3}Na_{1.7}[Cu_4(H_2O)_2(PW_9O_{34})_2]\cdot 24H_2O$ and $Na_{14}Cu[Cu_4(H_2O)_2(P_2W_{15}O_{56})_2]\cdot 53H_2O$. *Inorg. Chem.*, Vol.29, No.6, pp. 1235–1241

Weiner, H.; Aiken III, J. D. & Finke, R. G. (1996). Polyoxometalate Catalyst Precursors. Improved Synthesis, H^+-Titration Procedure, and Evidence for ^{31}P NMR as a Highly Sensitive Support-Site Indicator for the Prototype Polyoxoanion-Organometallic-support System $[(n\text{-}C_4H_9)_4N]_9P_2W_{15}Nb_3O_{62}$. *Inorg. Chem.*, Vol.35, pp. 7905–7913

Yang, Q. H.; Zhou, D. F.; Dai, H. C.; Liu, J. F.; Xing, Y.; Lin, Y. H. & Jia, H. Q. (1997). Synthesis, Structure and Properties of Undecatungstozincate Containing 3A Elements. *Polyhedron*, Vol.16, No.23, 3985–3989

24

The Diffusion Model of Grown-In Microdefects Formation During Crystallization of Dislocation-Free Silicon Single Crystals

V. I. Talanin and I. E. Talanin
Classic Private University
Ukraine

1. Introduction

Dislocation-free silicon single crystals are the basic material of microelectronics and nanoelectronics. Physical properties of semiconductor silicon are determined by the structural perfection of the crystals grown by the Czochralski and float-zone processes (Huff, 2002). In such crystals during their growth are formed grown-in microdefects.

Grown-in microdefects degrade the electronic properties of microdevices fabricated on silicon wafers. Optimizing the number and size of grown-in microdefects is crucial to improving processing yield of microelectronic devices. Many of the advances in integrated-circuit manufacturing achieved in recent years would not have been possible without parallel advances in silicon-crystal quality and defect engineering (Yang et al., 2009). The problem of defect formation in dislocation-free silicon single crystals during their growth is a fundamental problem of physics and chemistry of silicon. In particular it is the key to solving the problem engineering applications of silicon crystals. This is connected with the transformation grown-in microdefects during the technological treatment of silicon monocrystals.

Formation of grown-in microdefects occurs as a result of the interaction of point defects during crystal cooling. The distribution of grown-in microdefects in a growing crystal is influenced by its temperature field and the boundary conditions defined by its surfaces. Until recently it was assumed that the formation of grown-in microdefects is due to condensation of intrinsic point defects (Voronkov et. al., 2011). Recombination-diffusion model assumes fast recombination of intrinsic point defects at the initial moment of cooling the grown crystal. Fast recombination determines the type of dominant intrinsic point defects in the crystal. In this model was first used mathematical tool which allows you to associate the defect structure of crystal with distribution in the crystal thermal fields during the growth (Prostomolotov et al., 2011). It has been suggested that the fast recombination of intrinsic point defects near the crystallization front as a function of the growth parameter V_g/G (where V_g is the rate of crystal growth; G is the axial temperature gradient) leads to the formation of microvoids or interstitial dislocation loops (Voronkov, 2008). It is assumed that in the case $V_g/G < \xi_{crit}$ formed only interstitial A-microdefects as a result of aggregation of intrinsic interstitial silicon atoms. It is assumed that in the case $V_g/G > \xi_{crit}$ formed only microvoids as a result of aggregation of vacancies (Goethem et al., 2008; Kulkarni, 2008a). In this physical model, the interaction between the impurities and intrinsic point defects is not considered (Kulkarni et. al., 2004).

Recent versions of this model have suggested that part of the vacancies (v) in the temperature range 1683 ... 1373 K, due to the interaction with oxygen (O) and nitrogen (N) impurities, are bound into complexes of the vO, vO_2, and vN types (Kulkarni 2007; 2008b). After the formation of microvoids, the aforementioned complexes grow and take up vacancies. This model has ignored the growth of the complexes by means of the injection of intrinsic interstitial silicon atoms and the interaction of an impurity with intrinsic interstitial silicon atoms (Kulkarni 2007; 2008b).

In the general case recombination-diffusion model assumes that the process of defect formation in dislocation-free silicon single crystals occurs in four stages: (i) fast recombination of intrinsic point defects near the crystallization front; (ii) the formation in the narrow temperature range 1423...1223 K depending on the value of V_g/G microvoids or interstitial dislocation loops; (iii) the formation of oxygen clusters in the temperature range 1223...1023 K; (iv) growth of precipitates as a result of subsequent heat treatments.

Recombination-diffusion model is the physical basis for models of the dynamics of point defects. The mathematical model of point defect dynamics in silicon quantitatively explains the homogeneous mechanism of formation of microvoids and dislocation loops. It should be noted that, in the general case, the model of point defect dynamics includes three approximations: rigorous, simplified, and discrete–continuum approaches (Sinno, 1999; Dornberger et. al., 2001; Wang & Brown, 2001; Kulkarni et. al., 2004; Kulkarni, 2005; Prostomolotov & Verezub, 2009). The rigorous model requires the solution to integro-differential equations for point defect concentration fields, and the distribution of grown-in microdefects in this model is a function of the coordinates, the time, and the time of evolution of the size distribution of microdefects. A high consumption of time and cost for the performance of calculations required the development of a simplified model in which the average defect radius is approximated by the square root of the average defect area. This approximation is taken into account in the additional variable, which is proportional to the total area of the defect surface. The simplified model is effective for calculating the two-dimensional distribution of grown-in microdefects. Both models use the classical nucleation theory and suggest the calculation of the formation of stable nuclei and the kinetics of diffusion-limited growth of defects. The discrete–continuum approximation suggests a complex approach: the solution to discrete equations for the smallest defects and the solution to the Fokker–Planck equation for large-sized defects.

Recently, we proposed a new model for the formation grown-in microdefects. The physical model of the formation of grown-in microdefects assumes that the defect formation in dislocation-free Si single crystals upon cooling occurs in three stages: (i) the formation of impurity aggregates near the crystallization front, (ii) the formation and growth of impurity precipitates upon cooling from the crystallization temperature, and (iii) the formation of microvoids or dislocation loops (depending on the growth parameter V_g/G) – in a narrow temperature range of 1423…1223 K (V.I. Talanin & I.E. Talanin, 2006a; V.I. Talanin & I.E. Talanin, 2010b). This model on the experimentally and theoretically established fact the absence of recombination of intrinsic point defects near the crystallization front of the crystal is based (V.I. Talanin & I.E. Talanin, 2006a; V.I. Talanin & I.E. Talanin, 2007a).With the help of the diffusion model of formation grown-in microdefects was calculated process of high-temperature precipitation (V.I. Talanin & I.E. Talanin, 2010a). The processes of formation and growth of precipitates during cooling of the crystal is a controlling stage in the formation of the grown-in defect structure of dislocation-free silicon single crystals. At this stage, the formation and growth of oxygen and carbon precipitates occur in the temperature range from 1682 to 1423 K (V.I. Talanin & I.E. Talanin, 2010a).

The mathematical model of point defect dynamics can be adequately used on the basis of the physical model in which the impurity precipitation process occurs before the formation of microvoids or dislocation loops (V.I. Talanin & I.E. Talanin, 2010b). The model of point defect dynamics can be considered as component of the diffusion model for formation grown-in microdefects.

The aim of this paper is to present a diffusion model of formation grown-in microdefects in general and to discuss the possibility of its use as a tool for building the defect structure of dislocation-free silicon single crystal and device structures based on them.

2. Classification of grown-in microdefects

Currently, there are three classifications of grown-in microdefects: experimental classification, technological classification and physical classification.

Experimental classification of grown-in microdefects is based on the use of methods of selective etching, X-ray topography and transmission electron microscopy (Kock, 1970; Petroff & Kock, 1975; Foll & Kolbesen, 1975; Veselovskaya et al., 1977; Sitnikova et al., 1984; Sitnikova at al., 1985). A.J.R. de Kock entered the name of A-microdefects and B-microdefects, whereas E.G. Sheikhet entered the name C-microdefects, D-microdefects. We are entered the name (I+V)-microdefects (V.I.Talanin et al., 2002a, 2002b). These research allowed to establish the physical nature of A-microdefects, B-microdefects, C-microdefects, D-microdefects and (I+V)-microdefects. Experimental results indicated the identity of the processes of defect formation in crystals of FZ-Si and CZ-Si (Kock et al., 1979; V.I. Talanin & I.E. Talanin, 2003). This means that the classifications of grown-in microdefects in both types of crystals should also be identical (V.I. Talanin & I.E. Talanin, 2004).

Technological classification is used for large-scale crystals. The larger the diameter of the growing crystal, the lower growth rate, at which the same type of grown-in microdefects is formed. This occurs by reducing the axial temperature gradient in the crystal (Ammon et al., 1999). This leads to the appearance of a new type of grown-in microdefects (microvoids) and dislocation-free crystal growth in a narrow range of growth rates (Voronkov & Falster, 1998). In large crystals of interstitial dislocation loops and microvoids are considered as major grown-in microdefects in dislocation-free silicon crystals (Kulkarni et al., 2004).

Analysis of the experimental results of investigations of grown-in microdefects indicates that there are only three types of grown-in microdefects: precipitates of impurities ((I+V)-microdefects, D(C)-microdefects, B-microdefects), dislocation loops (A-microdefects) and microvoids (V.I. Talanin et. al., 2011b). We established that the basic elements of defect formation are primary oxygen-vacancy and carbon-interstitial agglomerates, which are formed at impurity centers near the crystallization front (V.I. Talanin & I.E. Talanin, 2006a).

An excess concentration of intrinsic point defects (vacancies or silicon self-interstitials) arises when the crystal is cooled under certain thermal conditions (Cho et al., 2006). This process leads to the formation of secondary grown-in microdefects (A-microdefects or microvoids) (V.I. Talanin & I.E. Talanin, 2004). We have proposed the physical classification of grown-in microdefects. It is based on the differences in the physical nature of the formation of primary and secondary grown-in microdefects (V.I. Talanin & I.E. Talanin, 2006a).

3. The diffusion model for formation of grown-in microdefects in dislocation-free silicon single crystals

We propose a new diffusion model of the formation and transformation of grown-in microdefects. It is based on the experimental studies of undoped dislocation-free Si single crystals grown by the floating zone and Czochralski methods. The diffusion model combines the physical model (the heterogeneous mechanism for the formation of grown-in microdefects), the physical classification of grown-in microdefects, and mathematical models of the formation of primary and secondary grown-in microdefects (Fig. 1).

Physical model based on the assumption about the absence of recombination intrinsic point defects at high temperatures. This assumption was confirmed in several experimental works (Talanin et al., 2002a; Talanin et al., 2002b; Talanin et al., 2003). In paper (V.I. Talanin & I.E. Talanin, 2007a) we first theoretically proved the absence of recombination of intrinsic point defects at high temperatures and fast recombination at low temperatures. The experimental data and the results obtained from thermodynamic calculations have demonstrated that the process of aggregation of point defects dominates over the process of recombination of intrinsic point defects. At high temperatures, the process of recombination makes an insignificant contribution to the process of aggregation. Consequently, vacancies and intrinsic interstitial atoms coexist in thermal equilibrium. As a result, intrinsic point defects of both types are simultaneously involved in the process of aggregation. The decomposition of a supersaturated solid solution of point defects occurs upon cooling through two mechanisms, namely, the vacancy and interstitial mechanisms, with the formation of oxygen-vacancy and carbon- interstitial agglomerates.

Absence of recombination intrinsic point defects at high temperatures allows us to propose the physical model of the formation grown-in microdefects. The basic concepts of the physics model for the formation of grown-in microdefects imply the following (V.I. Talanin & I.E. Talanin, 2006a): (i) the recombination of intrinsic point defects at high temperatures can be neglected; (ii) background carbon and oxygen impurities are involved in the defect formation as nucleation centers; (iii) the decay of the supersaturated solid solution of point defects when the crystal is cooled from the crystallization temperature occurs in two independent ways (branches): vacancy and interstitial; (iv) the defect formation is based on primary agglomerates formed as the crystal is cooled from the crystallization temperature due to the interaction between the impurities and intrinsic point defects; (v) when the crystal is cooled at temperatures below 1423 K, depending on the thermal growth conditions, secondary grown-in microdefects are formed due to the interaction between intrinsic point defects; (vi) the secondary grown-in microdefects are formed due to the coagulation (microvoids and A-microdefects) and deformation (A-microdefects) effects; (vii) the vacancy and interstitial branches of the heterogeneous mechanism have a symmetry, which implies simultaneous processes of defect formation during the decay of supersaturated solid solution of point defects; and (viii) the consequence of this symmetry is the formation of vacancy and interstitial grown-in microdefects of the same type and, correspondingly, the growth of dislocation-free Si single crystals in the same vacancy-interstitial mode (V.I. Talanin & I.E. Talanin, 2006b). It was revealed that the growth parameter $V_g/G = \xi_{crit}$ describes the conditions under which the (111) face appears on the crystallization front (V.I. Talanin & I.E. Talanin, 2006a). On the basis of physical model and concepts of primary and secondary grown-in microdefects we developed of physical classification of the grown-in microdefects (V.I. Talanin & I.E. Talanin, 2006a).

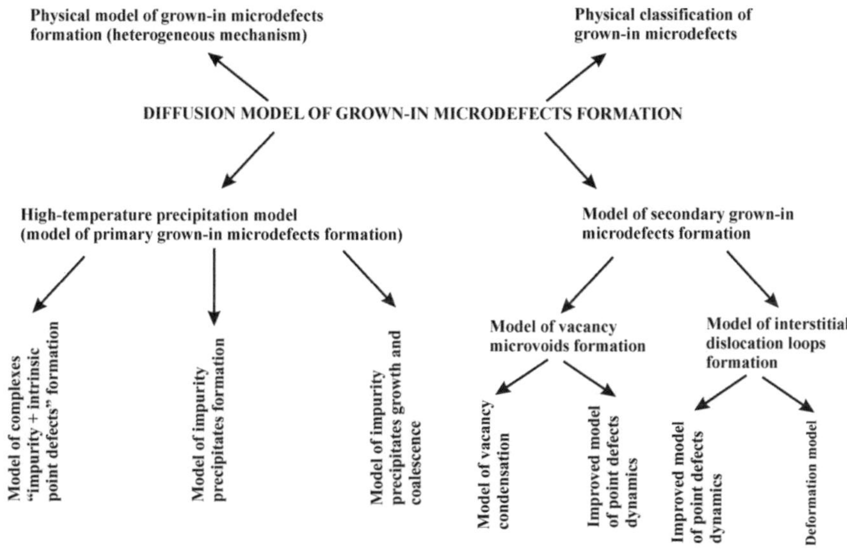

Fig. 1. Diffusion model of grown-in microdefects formation

A detailed description of the heterogeneous mechanism formation of grown-in microdefects and its correspondence to the results of experimental researches are presented in the articles (V.I. Talanin & I.E. Talanin, 2004; V.I. Talanin & I.E. Talanin, 2006a).

4. Diffusion kinetic of high-temperature precipitation

The calculation of the precipitation is carried out within the framework of the classical theory of nucleation, growth and coalescence of precipitates. For the calculation of formation and growth of precipitates are used analytic and approximate calculations. In the case of analytical calculations applied solution of differential equations of the dissociative diffusion (Talanin et al., 2007b, 2008). In the case of approximate calculations, the solution is sought in the form of systems of interconnected discrete differential equations of quasi-chemical reactions to describe the initial stages of nucleation of new phases and a similar system of continuous differential equations of the Fokker-Planck (V.I. Talanin & I.E. Talanin, 2010a).

4.1 Model of formation complex "impurity + intrinsic point defect"

The solution is sought within the model of dissociative diffusion–migration of impurities (Bulyarskii & Fistul', 1997). In this case, the difference from the decomposition phenomenon is that during diffusion (as a technological process), a diffusant is supplied to the sample from an external source, whereas in the case of decomposition it is produced by an internal source (lattice sites).

Vas'kin & Uskov are considered the problem of successive diffusion of a component *A* into a sample singly doped with a component *B*, taking into account the complex formation at the initial and boundary conditions (Vas'kin & Uskov, 1968). We are conducted similar consideration for our conditions (Talanin et al., 2007b; 2008). Under physical-model conditions (heterogeneous mechanism of grown-in microdefect formation), we assume that the component *A* is the background impurity (oxygen *O* or carbon *C*) and the component *B* is intrinsic point defects (vacancies *V* or interstitials *I*). For the vacancy and interstitial mechanisms, we consider, respectively, the oxygen+vacancy (*O*+*V*) and carbon+interstitial (*C* +*I*) interactions. The calculations performed in the framework of approximation of strong complex formation have demonstrated that the edge of the reaction front of the formation of a complex (i.e., the "oxygen+vacancy" and "carbon + self-interstitials" complex) is located at a distance of ~ $3 \cdot 10^{-4}$ mm from the crystallization front (Talanin et al., 2007b). We have shown that complex formation occurs near the crystallization front. Detailed calculations are presented in the articles (Talanin et al., 2007b, 2008).

4.2 Model for the formation of precipitates

Let us consider a system of a growing undoped dislocation-free silicon single crystal. The concentrations of all point defects at the crystallization front are assumed to be equilibrium, and both the vacancies and the intrinsic interstitial silicon atoms are present in comparable concentrations. During cooling of the crystal after passing through the diffusion zone, an excessive (nonequilibrium) concentration of intrinsic point defects appears. Excess intrinsic point defects disappear on sinks whose role in this process is played by uncontrollable (background) impurities of oxygen and carbon (V.I. Talanin & I.E. Talanin, 2006a). In real silicon crystals, the concentrations of carbon and oxygen impurities are higher than the concentrations of the intrinsic point defects. The formation of complexes between the intrinsic point defects and impurities is governed, on the one hand, by the fact that both the intrinsic point defects and the impurities are sources of internal stresses in the lattice (elastic interaction) and, on the other hand, by the Coulomb interaction between them (provided the defects and the impurities are present in the charged state). The mathematical model under consideration allows for the elastic interaction and the absence of the recombination of intrinsic point defects in the high-temperature range (V.I. Talanin & I.E. Talanin, 2007a). The concentrations of intrinsic point defects $C_{i,v}(r,t)$ in the growing crystal satisfy the diffusion equation $\partial C_{i,v}/\partial t = D_{i,v}\Delta(C_{i,v} - C_{ie,ve})$ where r is the coordinate and t is the time. In the vicinity of the sinks (oxygen and carbon atoms), the concentration of intrinsic point defects $C_{ie,ve}$ is kept equilibrium, whereas the diffusion coefficients $D_{i,v}$ and the concentrations $C_{ie,ve}$ of intrinsic point defects decrease exponentially with decreasing temperature. Under these conditions, the formation of microvoids and interstitial dislocation loops is possible only at significant supersaturations of intrinsic point defects, which take place at a temperature $T = T_m - 300K$ (where T_m is the crystallization temperature). For the formation of precipitates in the high-temperature range T ~ 1683...1403 K has been calculated using the model of dissociative diffusion (Talanin et al., 2007b; 2008). This approximation is valid at the initial stages of the formation of nuclei, when their sizes are small and the use of Fokker–Planck continuity differential equations is impossible. The calculations performed in the framework of this approximation have demonstrated that the edge of the reaction front of the formation of a complex is located at a distance of ~$3 \cdot 10^{-4}$ mm from the crystallization front. This spacing represents a diffusion layer in which an excessive concentration of intrinsic point defects appears.

We have considered the modern approach based on solving systems of coupled discrete differential equations of quasi-chemical reactions for the description of the initial stages of the formation of nuclei of new phases and a similar system of Fokker–Planck continuity differential equations.

In order to describe the kinetics of the simultaneous nucleation and growth (dissolution) of a new phase particles of several types in a supersaturated solid solution of an impurity in silicon was considered a system consisting of oxygen and carbon atoms, vacancies, and intrinsic interstitial silicon atoms. The interaction in this system during cooling of the crystal from 1683 K results in the formation of oxygen and carbon precipitates. In order to perform the computational experiments and to interpret their results was conducted a dimensional analysis of the kinetic equations and the conservation laws with the use of characteristic time constants and critical sizes of defects. This is made it possible to perform a comparative analysis of the joint evolution of oxygen and carbon precipitates and to optimize the computational algorithm for the numerical solution of the equations.

For example, for the case of a thin plane-parallel crystal plate of a large diameter, when the conditions in the plane parallel to the surface of the crystal can be considered to be uniform and the diffusion can be treated only along the normal to the surface (the z coordinate axis), the mass balance of point defects in the crystal is described by the system of diffusion equations for intrinsic interstitial silicon atoms, oxygen atoms, carbon atoms, and vacancies:

$$\begin{aligned}
\frac{\partial C_o}{\partial t} &= D_o \frac{\partial^2 C_o}{\partial z^2} - \frac{\partial C_o^{SiO_2}}{\partial t} \\
\frac{\partial C_c}{\partial t} &= D_c \frac{\partial^2 C_c}{\partial z^2} - \frac{\partial C_c^{SiC}}{\partial t} \\
\frac{\partial C_i}{\partial t} &= D_i \frac{\partial^2 C_i}{\partial z^2} + \frac{\partial C_i^{SiO_2}}{\partial t} - \frac{\partial C_i^{SiC}}{\partial t} \\
\frac{\partial C_v}{\partial t} &= D_v \frac{\partial^2 C_v}{\partial z^2} - \frac{\partial C_v^{SiO_2}}{\partial t} + \frac{\partial C_v^{SiC}}{\partial t}
\end{aligned} \tag{1}$$

where C_o, C_c, C_i, C_v are the concentrations of oxygen, carbon, self-interstitials and vacancies respectively; D_o, D_c, D_i, D_v are the diffusion coefficients of oxygen, carbon, self-interstitials and vacancies respectively.

In the system of equations (1), we took into account that the oxygen precipitates serves as sinks for oxygen atoms and vacancies and as sources of interstitial silicon atoms. At the same time, the carbon precipitates, in turn, also serve as sinks for carbon atoms and interstitial silicon atoms and as sources for vacancies. Kinetic model of decomposition of solid solutions of oxygen and carbon impurities not only allows one to simulate the processes of precipitation during cooling of the as-grown silicon crystal to a temperature of 300 K but also adequately describes the available experimental data on the oxygen and carbon precipitation (V.I. Talanin & I.E. Talanin, 2011a).

The algorithm used for solving the problem of simulation of the simultaneous growth and dissolution of the oxygen and carbon precipitates due to the interaction of point defects during cooling of the crystal from the crystallization temperature is based on the monotonic explicit difference scheme of the first-order accuracy as applied to the Fokker–Planck equations.

Detailed calculations are presented in the articles (V.I. Talanin & I.E. Talanin, 2010a). These calculations demonstrate that intrinsic point defects (vacancies and intrinsic interstitial silicon atoms) exert a significant influence on the dynamics of mass exchange and mass transfer of point defects between the oxygen and carbon precipitates. The absorption of vacancies by the growing oxygen precipitates leads to the emission of silicon atoms into interstitial positions. The intrinsic interstitial silicon atoms, in turn, interact with the growing carbon precipitates, which, in the process of growth, supply vacancies for growing oxygen precipitates. This interaction leads to such a situation that, first, the growth of the precipitates is suppressed more weakly because of the slower increase in the supersaturation of the intrinsic point defects in the bulk of the growing crystal and, second, the critical radius of the formation of carbon precipitates increases more slowly, which favors a more rapid growth of the carbon precipitates. The higher rate of the evolution of the size distribution function for carbon precipitates can be associated with the higher mobility of interstitial silicon atoms as compared to vacancies in the high-temperature range. It can be assumed that the mutual formation and growth of oxygen and carbon precipitates result in a lower rate of the evolution of the size distribution function of the oxygen precipitates, regardless of their smaller critical size at the initial instant of time, owing to the effect of the carbon impurity.

4.3 Model of growth and coalescence of precipitates

In the classical theory of nucleation and growth of new-phase particles, the process of precipitation in a crystal is treated as a first-order phase transition and the kinetics of this process is divided into three stages: the formation of new-phase nuclei, the growth of clusters, and the coalescence stage. At the second stage of the precipitation process, clusters grow without a change in their number. At the third stage of the precipitation process, when the particles of the new phase are sufficiently large, the supersaturation is relatively low, new particles are not formed and the decisive role is played by the coalescence, which is accompanied by the dissolution of small-sized particles and the growth of large-sized particles. The condition providing for changeover to the coalescence stage is the ratio $u(t) = R(t) / R_{cr}(t) \approx 1$, where $R_{cr}(t)$ is the critical radius of the precipitate.

Detailed calculations stages of the growth and the coalescence are presented in the article (V.I. Talanin & I.E. Talanin, 2011a). The analysis was carried out under the assumption that precipitates grow at a fixed number of nucleation centers according to the diffusion mechanism of growth. The model corresponds to the precipitation uniform in the volume. An analysis of the results obtained and the data taken from (Talanin et al., 2007b; V.I.

Talanin & I.E. Talanin, 2010a) has demonstrated that the phase transition occurs according to the mechanism of nucleation and growth of a new phase so that these two processes are not separated in time and proceed in parallel.

The condition providing changeover to the stage of the coalescence is written in the form $R(t) \approx R_{cr}(t)$, which is satisfied for large-sized crystals at the temperature T ≈ 1423 K. Taking into account the computational errors, this temperature for large-sized crystals corresponds to the initial point of the range of the formation of microvoids (at V_g = 0.6 mm/min). In this range, all impurities are bound and there arises a supersaturation with respect to vacancies, which is removed as a result of the formation of microvoids. With a change in the thermal conditions of the growth (for example, at V_g = 0.3 mm/min), there arises a supersaturation with respect to interstitial silicon atoms, which leads to the formation of dislocation loops. In this case, the condition $R(t) \approx R_{cr}(t)$ is satisfied at $T \approx 1418$ K. Consequently, the stage of the coalescence in large-sized silicon single crystals begins at temperatures close to the temperatures of the formation of clusters of intrinsic point defects (depending on the thermal growth conditions, these are microvoids or dislocation loops).

The absorption of vacancies by growing oxygen precipitates results in the emission of silicon atoms in interstitial sites. In turn, the intrinsic interstitial silicon atoms interact with growing carbon precipitates, which, in the course of their growth, supply vacancies for growing oxygen precipitates. This interplay between the processes leads to an accelerated changeover of the subsystems of oxygen and carbon precipitates to the stage of the coalescence as compared to the independent evolution of these two subsystems. The change in the thermal conditions for the growth of small-sized FZ-Si single crystals (high growth rates and axial temperature gradients) leads to the fact that the stage of the coalescence begins far in advance (at $T \approx T_m - 20K$). The results of theoretical calculations have demonstrated that a decrease in the concentrations of oxygen and carbon in small-sized single crystals leads to a further decrease in the time of occurrence of the growth stage of precipitates. The change in thermal conditions of crystal growth (in particular, an increase in the growth rate and in the axial temperature gradient in the crystal) substantially affects the stage of the growth of precipitates. In turn, the decrease in the time of occurrence of the growth stage of precipitates is associated, to a lesser extent, with the decrease in the concentration of impurities in crystals. Eventually, these factors are responsible for the decrease in the average size of the precipitates.

The kinetic model of growth and coalescence of oxygen and carbon precipitates in combination with the kinetic models describing their formation represents a unified model of the process of precipitation in dislocation-free silicon single crystals.

5. Diffusion kinetic of formation of the microvoids and dislocation loops

As mentioned earlier the defect formation processes in a semiconductor crystal, in general, and in silicon, in particular, have been described using the model of point defect dynamics; in this case, the crystal has been considered a dynamic system and real boundary conditions have been specified. However, the model of point defect dynamics has not been used for calculating the formation of interstitial dislocation loops and microvoids under the assumption that the recombination of intrinsic point defects is absent in the vicinity of the crystallization front. This fact is evidenced by experimental and theoretical investigations (V.I. Talanin & I.E. Talanin, 2006a, 2007a).

5.1 Kinetics of formation of microvoids

The experimentally determined temperature range of the formation of microvoids in crystals with a large diameter is 1403...1343 K (Kato et al., 1996; Itsumi, 2002). In this respect, the approximate calculations for the solution in terms of the model of point defect dynamics were performed at temperatures in the range 1403...1073 K. The computational model uses the classical theory of nucleation and formation of stable clusters and, in strict sense, represents the size distribution of clusters (microvoids) reasoning from the time process of their formation and previous history.

The calculations were carried out in the framework of the model of point defect dynamics, i.e., for the same crystals with the same parameters as in already the classical work on the simulation of microvoids and interstitial dislocation loops (A-microdefects) (Kulkarni et al., 2004). According to the analysis of the modern temperature fields used when growing crystals by the Czochralski method, the temperature gradient was taken to be $G = 2.5$ K/mm (Kulkarni et al., 2004). The simulation was performed for crystals 150 mm in diameter, which were grown at the rates $V_g = 0.6$ and 0.7 mm/min. These growth conditions correspond to the growth parameter $V_g/G > \xi_{crit}$.

Detailed calculations are presented in the articles (V.I. Talanin & I.E. Talanin, 2010b).Our results somewhat differ from those obtained in (Kulkarni et al., 2004). These differences are as follows: (i) the nucleation rate of microvoids at the initial stage of their formation is low and weakly increases with a decrease in the temperature and (ii) a sharp increase in the nucleation rate, which determines the nucleation temperature, occurs at a temperature T ~ 1333 K. These differences result from the fact that the recombination factor in our calculations was taken to be $k_{IV} = 0$. For $k_{IV} \neq 0$, consideration of the interaction between impurities and intrinsic point defects in the high-temperature range becomes impossible, which is accepted by the authors of the model of point defect dynamics (Kulkarni et al., 2004). In this case, in terms of the model, there arises a contradiction between the calculations using the mathematical model and the real physical system, which manifests itself in the ignoring of the precipitation process (Kulkarni et al., 2004).

5.2 Kinetics of formation of dislocation loops (A-microdefects)

The computational experiment was performed similarly to the calculations of the formation of microvoids. The simulation was performed for crystals 150 mm in diameter, which were grown at the rates $V_g = 0.10$ and 0.25 mm/min for the temperature gradient $G = 2.5$ K/mm. These growth conditions correspond to the growth parameter $V_g/G < \xi_{crit}$.

Detailed calculations are presented in the articles (V.I. Talanin & I.E. Talanin, 2010b). The temperature of the formation of A-microdefects corresponds to ~1153 K. An increase in the crystal growth rate weakly decreases the critical radius of A-microdefects and slightly affects the nucleation temperature. An increase in the crystal growth rate leads to an almost twofold decrease in the concentration of introduced defects.

The data of the computational experiment on the determination of the microvoid concentration correlate well with the experimentally observed results (10^4 ... 10^5 cm^{-3}) (Itsumi, 2002). For the A-microdefects, for which the concentration according to the experimental data is ~10^6...10^7 cm^{-3} (Petroff & Kock, 1975; Foll & Kolbesen, 1975), the discrepancy is as large as three orders of magnitude. This can be explained by the fact that, unlike microvoids, which are formed only through the coagulation mechanism, the formation of A-microdefects occurs according to both the coagulation mechanism and the mechanism of prismatic extrusion (deformation mechanism) (V.I. Talanin & I.E. Talanin, 2006a). The results of the calculations suggest that the main contribution to the formation of A-microdefects is made by the mechanism of prismatic extrusion when the formation of interstitial dislocation loops is associated with the relieving of stresses around the growing precipitate. Consequently, the impurity precipitation processes that proceed during cooling of the crystal from the crystallization temperature are fundamental (primary) in character and determine the overall defect formation process in the growth of dislocation-free silicon single crystals.

The calculations of the formation of microvoids and dislocation loops (A-microdefects) demonstrated that the above assumptions do not lead to substantial differences from the results of the previous calculations in terms of the model of point defect dynamics. This circumstance indicates that the mathematical model of point defect dynamics can be adequately used on the basis of the physical model in which the impurity precipitation process occurs before the formation of microvoids or interstitial dislocation loops. Moreover, the significant result of the calculations is the confirmation of the coagulation mechanism of the formation of microvoids and the deformation mechanism of the formation of interstitial dislocation loops. Therefore the model of the dynamics of point defects can be considered as component part of the diffusion model for formation grown-in microdefects.

5.3 Model of the vacancy coalescence

Model vacancy coalescence is a simplified model for the analysis of individual parameters of process of the formation microvoids. Detailed calculations are presented in the articles (V.I. Talanin & I.E. Talanin, 2010c). The fundamental interaction between impurities and intrinsic point defects upon crystal cooling under certain thermal conditions (T < 1423 K) leads to impurity depletion and the formation of a supersaturated solid solution of intrinsic point defects. The decay of this supersaturated solid solution causes the coagulation of intrinsic point defects in the form of microvoids.

An analysis of the experimental and calculated data within model of the vacancion coalescence in accordance with the heterogeneous diffusion model of the formation of grown-in microdefects revealed the following reasons for the occurrence of microvoids in dislocation-free silicon single crystals: (i) a sharp decrease in the concentration of background impurity that was not associated into impurity agglomerates (formed in the cooling range of 1683...1423 K); (ii) a large (over 80 mm) crystal diameter (in this case vacancies fail to drain from the central part of the crystal to the lateral surface); (iii) crystals of large diameter generally contain a ring of D-microdefects which forms due to the emergence of the (111) face on the crystallization front and which depletes the region inside with impurity atoms.

The growth parameter V_g / G describes the fundamental reasons related to the systematic nonuniform impurity distribution during crystal growth from a melt. Based on an analysis of the experimental results, one can suggest that the parameter V_g / G controls the growth because it describes the condition for the emergence of the (111) face at the crystallization front. Therefore, the impurity depletion inside the ring of D-microdefects upon crystal cooling at $T < 1423$ K is caused by two things: the impurity bonding during the formation of primary grown-in microdefects ((I+V)-microdefects) and the impurity drift to the (111) face, which is equivalent to the annular distribution of primary D-type grown-in microdefects. In this case, excess vacancies arise within the ring of D-microdefects to form a supersaturated solid solution with its subsequent decay and the formation of vacancy microvoids. In contrast, excess silicon self-interstitials arise beyond the D-ring to form a supersaturated solid solution with its subsequent decay and the formation of interstitial dislocation loops (A-microdefects) (V.I. Talanin & I.E. Talanin, 2010c).

The experimental classification of grown-in microdefects employs the terms such as A-microdefects, B-microdefects, D(C)-microdefects, (I+V)-microdefects and microvoids (V.I. Talanin & I.E. Talanin, 2006a). It was found that A-microdefects constitute interstitial-type dislocation loops, and B-microdefects, D(C)-microdefects, (I+V)-microdefects constitute precipitates of background oxygen and carbon impurities at different stages of their evolution (V.I. Talanin & I.E. Talanin, 2006a; V.I. Talanin & I.E. Talanin, 2011a)). At present, it is difficult to apply the experimental classification, since it is necessary to interpret the terms of every type of the grown-in microdefects for each publication. At the same time, from the physical point of view there are only three types of grown-in microdefects, i.e. impurity precipitates, dislocation loops and microvoids. Besides, when considering the formation of defects in silicon after processing (post-growth microdefects) the terms such as precipitates, dislocation loops and microvoids are also employed. Therefore, in order to harmonize a defect structure, we propose to switch to the physical classification of grown-in microdefects (V.I. Talanin & I.E. Talanin, 2004).

5.4 Kinetic model for the formation and growth of dislocation loops

Kinetics of high-temperature precipitation involves three stages: (i) the nucleation of a new phase, (ii) the growth stage and (iii) the stage coalescence. Precipitates originate from elastic interaction between point defects. They are, initially, present in coherent, elastic and deformable state, when lattice distortions close to the precipitate-matrix boundary are not large, and one atom of the precipitate corresponds to one atom of the matrix (Goldstein et al., 2011). Elastic deformations and any mechanical stress connected with them cause a transfer of excessive (deficient) substance from the precipitate or vice versa. Storage of elastic strain energy during the precipitate growth results in a loss of coherence by matrix. In this case it is impossible to establish one-to-one correspondence between atoms at different sides of the boundary. It results in structural relaxation of precipitates which occurs due to formation and movement of dislocation loops.

To simulate a stress state of the precipitate and the matrix surrounding it, it is sufficient to observe the precipitate which is simple spherical in shape. There can be found analytical solutions in respect of spherical precipitates (Kolesnikova & Romanov, 2004). Let us take the theoretical and experimental researches of stress relaxation at volume quantum dots as initial model (Chaldyshev et al., 2002; Kolesnikova & Romanov, 2004; Chaldyshev et al., 2005; Kolesnikova et al., 2007). According to these representations, as far as the precipitate grows, its elastic field induces the formation of a circular interstitial dislocation loop of mismatch. This process contributes to the decrease in total strain energy of the system. A growing precipitate displaces the matrix material in the crystal volume. Interstitial atoms form an interstitial dislocation loop near to the precipitate. At the same time, a mismatch dislocation loop is formed on the very precipitate (Kolesnikova et al., 2007). At the same time, the critical sizes of precipitates, at which formation of dislocations is energy favorable, have the same order as the critical size of dislocation loops (Kolesnikova et al., 2007).

In the volume of silicon the precipitate produces a stress field caused by mismatch between the lattice parameters of precipitate (a_1) and the surrounding matrix (a_2) (Kolesnikova et al., 2007). Then, the intrinsic deformation of the precipitate is defined as described bellow

$$\varepsilon = \frac{a_1 - a_2}{a_1} \tag{2}$$

In general, the precipitate intrinsic deformation in the matrix volume can be expressed as follows

$$\varepsilon^* = \begin{pmatrix} \varepsilon_{xx} & \varepsilon_{xy} & \varepsilon_{xz} \\ \varepsilon_{xy} & \varepsilon_{yy} & \varepsilon_{yz} \\ \varepsilon_{zx} & \varepsilon_{zy} & \varepsilon_{zz} \end{pmatrix} \delta\left(\Omega_{pr}\right) \tag{3}$$

where the diagonal terms constitute a dilatation mismatch between the precipitate and matrix lattices; the other terms are shear components; $\delta\left(\Omega_{pr}\right)$ is the Kronecker symbol. Elastic fields of precipitate (stresses σ_{ij} and deformation ε_{ij}) and field of full displacements are calculated taking into account their own deformation (3) and region of localization of the precipitate $\delta\left(\Omega_{pr}\right)$. The calculation of elastic fields of the precipitate is carried out by well-known scheme by using the elastic modules, Green's function of an elastic medium or its Fourier transform (Kolesnikova & Romanov, 2004).

Consider the simplest model of a spherical precipitate with equiaxed own deformation $\varepsilon_{ii}^* = \varepsilon, \varepsilon_{ij}^* = 0\left(i \neq j; i, j, = x, y, z\right)$. The elastic strain energy of spheroidal defect with increasing radius of precipitate $\left(R_{pr}\right)$ increases as a cubic law (Kolesnikova et al., 2007):

$$E_{pr} = \frac{32 \cdot \pi}{45 \cdot (1 - \upsilon)} \cdot J \cdot \varepsilon^2 \cdot R_{pr}^3 \tag{4}$$

where J is the shear modulus; υ is the Poisson's ratio. From a certain critical radius R_{crit} takes effect mechanism for resetting the elastic energy of the precipitate. This mechanism leads to the formation of circular interstitial dislocation loop. Energy criterion of this mechanism is the condition $E^{initial} \geq E^{final}$, here $E^{initial}, E^{final}$ is the elastic energy of the system with the precipitate before and after relaxation, respectively (Kolesnikova et al., 2007).

In respect of a spherical precipitate with equiaxial intrinsic deformation, the calculation of elastic fields of the precipitate is substantially simplified. Let us assume that the intrinsic elastic strain energy of the precipitate before and after the formation of a dislocation loop of mismatch remains constant $E_{pr}^{initial} = E_{pr}^{final}$. Then the criterion of nucleation loop of misfit dislocation can be represented by the condition $0 \geq E_D + E_{prD}$, where E_D is the energy of loop of misfit dislocation; E_{prD} is the energy of interaction of precipitate with the dislocation loop (Kolesnikova et al., 2007).

To estimate believe that loop of misfit dislocation is the equatorial location on the spheroidal precipitate $R_D = R_{pr}$ the self-energy prismatic loop (Kolesnikova et al., 2007)

$$E_{loop} = \frac{J \cdot b^2 \cdot R_D}{2 \cdot (1-\upsilon)} \cdot \left(\ln \frac{2 \cdot R_D}{f} - 2 \right) \tag{5}$$

where f is the radius of the core loop; b is the magnitude of the Burgers vector. The critical radius of precipitate for the formation of dislocation loop is determined from the expression (Kolesnikova et al., 2007)

$$R_{crit} = \frac{3b}{8\pi(1+\nu)\varepsilon} \left(\ln \frac{1.08\alpha R_{crit}}{b} \right) \tag{6}$$

where α is a constant contribution of the dislocation core. Expression (6) is approximate and can only be used to determine the value critical radius R_{crit}.

This paper (Bonafos et al., 1998) theoretically considers the increase kinetics for dislocation loops at the stages of loop growth and coalescence. It is assumed that, in general, the growth is either controlled by energy barrier when atom is captured by the loop, or by activation energy of interstitial atom diffusion. In conditions of cooling the crystal after being grown, we presume that the diffusion processes play a core role. The model (Burton & Speight, 1985) is further used in the calculations for evolution in size-dependant distribution of loops and for evolution in loop density.

The dislocation loops with a radius of $R > R_{crit}$ become bigger in size at the coalescence stage, while small dislocation loops with a radius of $R < R_{crit}$ will dissolve (Bonafos et al., 1998; Burton & Speight, 1985). The growth of dislocation loops during cooling after the growth of single crystal silicon occurs as due to dissolution of small loops with sizes less than critical, and as a result supersaturation for intrinsic interstitial silicon atoms. In this case, the crystal growth ratio is $V_g/G < \xi_{crit}$. When oversaturation of vacancies ($V_g/G > \xi_{crit}$) occurs, the interstitial dislocation loops start to dissolve. Increase in the radius of interstitial dislocation loop can be defined by the formula depending on the crystal cooling time (Burton & Speight, 1985):

$$R(t) = \sqrt{R_{crit}^2 + j \cdot D(t) \cdot t} \tag{7}$$

where $D(t)$ is the diffusion coefficient of intrinsic interstitial silicon atoms; t is the time cooling the crystal; j is the proportionality factor. The value of the cooling time of the crystal is determined from the dependence: $T(t) = \frac{T_m^2}{T_m + U \cdot t}$, where T_m is the crystallization temperature (melting) of silicon; $U = V_g \cdot G$ is the cooling rate of the crystal. The loop concentration depends on the crystal cooling time (Burton & Speight, 1985):

$$N(t) = \frac{M(t)}{1 + D(t) \cdot t / 2 \cdot R_{crit}^2} \quad (8)$$

where $M(t)$ is the concentration of precipitates.

Initially, the precipitates act as stoppers for the dislocation loops restraining their distribution and generation. Then, the precipitates facilitate the formation of dislocation loops due to the action of Bardeen-Herring or Frank-Read sources (Gyseva et al., 1986). These processes lead to the formation and growth of complex dislocation loops. Formation and development of dislocation loops caused by the high-temperature precipitation of background impurities (oxygen and carbon). Growth and coalescence of dislocation loops are generally maintained due to the generation of growing precipitates instead of silicon self-interstitials, and as well to the dissolution of small dislocation loops.

If the parameter of crystal growth $V_g / G < \xi_{crit}$, for stress relaxation precipitate generates interstitial silicon atoms. If the parameter of crystal growth $V_g / G > \xi_{crit}$, for stress relaxation precipitate adsorbs vacancies. In this case is suppressed the formation of dislocation loops.

6. Construction of the defect structure of dislocation-free silicon single crystal and device structures on their base

Experimental studies require large material and time costs, while theoretical studies are carried out for single crystals with selected fixed parameters of their growth. It is necessary to develop a new method for studying the defect structure of silicon without these drawbacks. In the diffusion model of formation grown-in microdefects all the parameters of precipitates, microvoids and dislocation loops are determined through the thermal conditions of growth. Therefore, definition the type of defect structure and calculation of the formation of microdefects is conducted depending on the values of crystal growth rate, temperature gradients and cooling rate of the crystal. On this basis, we have developed a new method for studying the defect structure of silicon. This method allows to simulate a real experiment by the software (V.I. Talanin et. al., 2011b).

Electronic equivalent of an object for direct test on the computer are programs that converted the mathematical models and algorithms to the available computer language (C++). The program is written high-level language programming in C++ compiler Borland C++ Builder. Program complex consists of two consecutive parts: (i) the unit determination the type of defect structure and (ii) the unit of calculation and graphs.

At the stage of determining the type of the structure defect of software system works as follows. Initially is the choice of method of growing dislocation-free silicon single crystals (Czochralski method or the floating-zone method) and then is the choice of certain diameter of the crystal. The ratio of $V_{gcrit} / G = \xi$ theoretically and experimentally determined in a certain range of values (0.06 mm²/K·min ≤ ξ ≤ 0.3 mm²/K·min). Therefore, we choose the certain value ξ for the calculation. For a given diameter in a certain range of values are selected: the value of the axial temperature gradient in the center of the crystal (G_a) and the value of the minimum ($V_{\min}$) and maximum ($V_{\max}$) crystal growth rate. These values are determined from the analysis of experimental and theoretical data for different diameters of the crystal. Then produced choice of the axial temperature gradient at the edge of the crystal (G_e) in the range G_e / G_a = 1.0 ... 2.5. The reliability and accuracy of the computational experiments can be experimentally verified by means of selective etching crystal of the plane which passes through the center of the crystal and parallel to the direction of growth. In case of deviations can by using of the selection of parameters G_a, G_e, ξ achieve full compliance with theoretical and experimental data. In addition, this technique avoids the difficulties of experimental determination of the G_a and G_e, especially for large diameter crystals (V.I. Talanin et. al., 2011b). At the last stage of determining the type of defect structure on the resulting dependence of the critical growth rate $V_{gcrit}(r) = \xi \cdot G(r)$ is imposed value of real crystal growth rate $V = const$ (Fig. 2). This procedure allows determining the type of defect structure of a real crystal. Depending on the position of the line $V = const$ relative to the curve $V_{gcrit}(r) = \xi \cdot G(r)$ may be three areas of the defect structure. Calculation of these areas produced the block of calculation and graphs (Fig. 3).

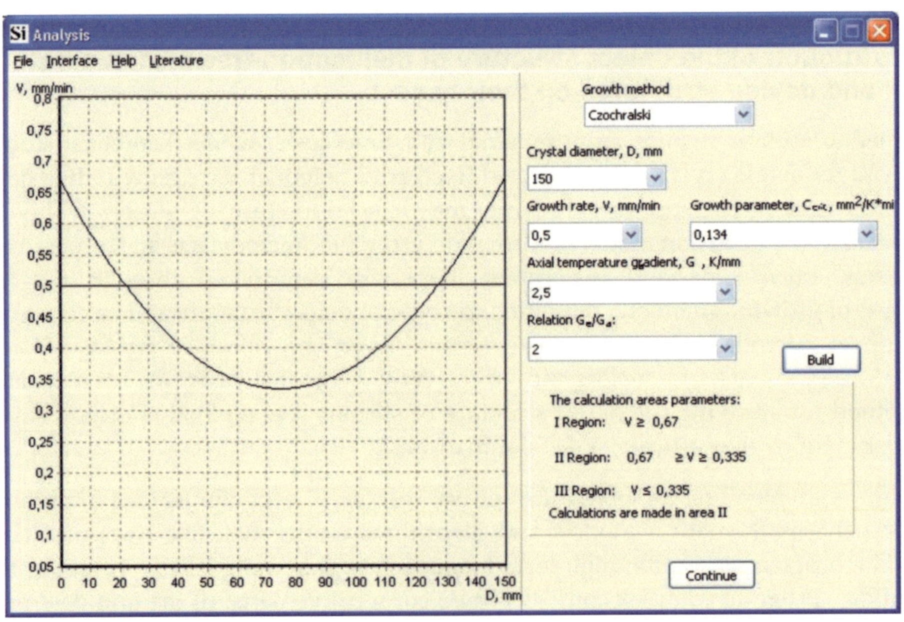

Fig. 2. Shape analysis of the defect structure

The first area of the calculation is characterized by high rates of crystal growth, when the above V-shaped distribution of precipitates formed only microvoids and precipitates. The second area of calculation is characterized by the average growth rate of the crystal, when a ring of precipitate in the plane perpendicular to the direction of growth crystal is formed. In this case inside the ring are formed precipitates and microvoids, outside the ring are formed precipitates and interstitial dislocation loops. The third area of calculation is characterized by low rates of crystal growth, when the below V-shaped distribution of precipitates are formed precipitates and interstitial dislocation loops. Calculation of the precipitates is carried out within the classical theory of nucleation, growth and coalescence of precipitates by means of the analytical and approximate calculations. Critical radius of precipitates, the distribution of precipitates in size, change in the average size of precipitates during the cooling of the crystal, and other parameters of the precipitation of carbon and oxygen are determined (Fig. 3). Mathematical models and calculation parameters are given in (V.I. Talanin et al., 2008; V.I. Talanin & I.E. Talanin, 2010a; 2011a). When calculating the vacancy microvoids initially are tested of conditions their formation, since microvoids are not formed at a cooling rate of the crystal $V_{cool} \geq 40K/\min$ (Nakamura et al., 2002) and in crystals with a diameter less than 70 mm (V.I. Talanin et al., 2010c). Calculation of the microvoids and interstitial dislocation loops to determines for each of these types of defects such parameters as the critical radii and the concentrations (V.I. Talanin & I.E. Talanin, 2010b).

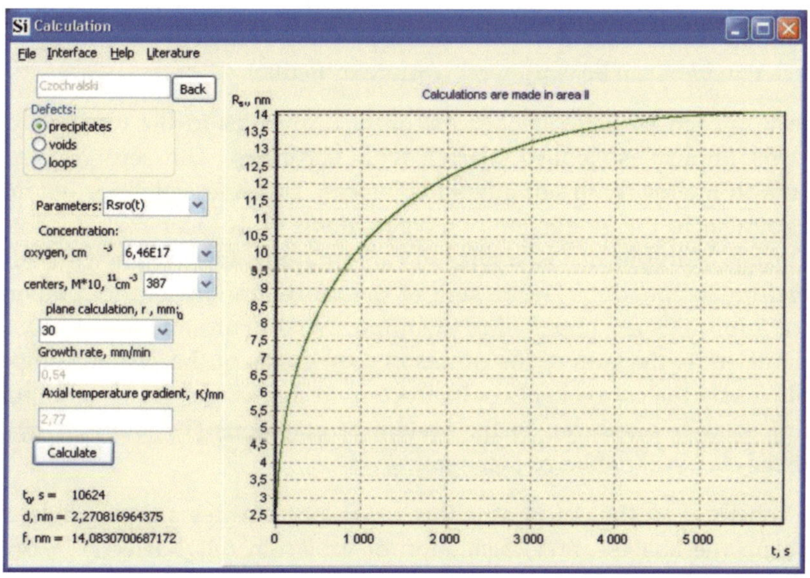

Fig. 3. Form of calculation of the defect structure

The software complex performs imitation of a real experiment and with the maximum precision reproduces the thermal characteristics of silicon single crystal growing. The software complex allows to determine the thermal conditions of crystal growth, to predict and control the defect structure of the crystal. Calculations formations of grown-in microdefects are in good agreement with the experimental results of research (V.I. Talanin et. al., 2011b).

The program complex is the first experience of a virtual experimental device for research the real structure of dislocation-free silicon single crystals. Currently, it can be used for the analysis and calculation of the defect structure of undoped single-crystal silicon. Depending on the thermal conditions of crystal growth can determine crucial parameters such as size and concentration of grown-in microdefects. Approach to the analysis and calculation of formation grown-in microdefects has an important advantage in simplicity, accessibility and sufficient adequacy of mathematical modeling in comparison with other methods. For its implementation does not require supercomputers, and can effectively use the experimental data, experience and intuition of physicists, materials scientists and technologists for the analytical calculation and design of the defect structure. The program complex is easy to implement on a personal computer in technology and research practices.

Disadvantages software system is determined deficiencies of the diffusion model of formation grown-in microdefects. These include: (i) one-dimensional model; (ii) failure to account for the width of the V-shaped distribution of precipitates; (iii) uncertainty in determining of thermal conditions of growth; (iv) the error of approximate numerical methods. Elimination of these deficiencies will increase the accuracy of the calculations.

The influence of other impurities (e.g., dopants, nitrogen, hydrogen, iron, and others) on the formation of the defect structure of silicon can be taken into account by using two approaches: a rigorous approach and a simplified approach. A rigorous approach requires accurate accounting of all the components in equations (1) mass balance of point defects in the crystal. In this case, the solution of the corresponding system of coupled equations of the Fokker-Planck equation can be considerably more difficult.

A simplified approach assumes the separation of impurities in the two groups. The first group contains impurities, which interact with vacancies. The second group contains impurities, which interact with self-interstitial atoms. This interaction for the first group of impurities is similar to the interaction of oxygen atoms with vacancies. This interaction for the second group of impurities is similar to carbon atoms interaction with self-interstitial atoms.. Therefore, in equations (1) the sum of the concentrations of the components of the first group of impurities is equivalent to the impurity concentration of oxygen, and oxygen diffusion coefficient is the sum of the diffusion coefficients of the system components. Are defined analogously the components of interaction in the second group of impurities. Then the system of equations reduced to the system of equations (1), followed by the task of determining the chemical nature of precipitates.

The task of construction the defect structure of dislocation-free silicon single crystal is an inverse problem the analysis and calculation of grown-in microdefects. In this case, you must first specify the type, size and concentration of grown-in microdefects. Parameters of grown-in microdefects are selected based on the requirements for defect structure of devices and integrated circuits. In the next stage are calculated parameters of crystal growth (the growth rate for a certain diameter of the crystal and temperature gradients), which provide presence of given defects of structure. We get that the defect structure determines thermal conditions of crystal growth. To automate the process of crystal growth need to carry out the development of software products based on the diffusion model in combination with known software products for modelling crystal growth.

Production technologies of devices based on silicon technology are connected with various impacts on the material. Heat treatments, ionizing radiations and mechanical effects have a critical impact on the initial defect structure of dislocation-free silicon single crystals. Technological impacts are lead to the transformation of grown-in microdefects. In the general case technological impacts are leads to: (i) the growth of initial grown-in microdefects; (ii) formation on grown-in microdefects of new defects (e.g., stacking faults); (iii) the formation of new defects. In contrast to grown-in microdefects these defects of crystal structure can be defined as postgrowth microdefects.

At the present time for the description of the formation of postgrowth microdefects are used a variety of models. Overview of the main models has been presented in (Sadamitsu et. al., 1993). The purpose of all models is to consider the formation and growth of defects as a result of technological impacts (e.g., by heating the crystal). A common deficiency of these models is the lack of consideration of the influence of grown-in microdefects, which are in the initial wafer.

Diffusion model of formation grown-in microdefects makes it possible to calculate of the defect structure of the initial silicon wafers. So from our point of view of theoretical analysis of the formation of postgrowth microdefects must be based on the diffusion model. It should be noted that taking into account all components of the general solution may be difficult. However, in some cases (e.g., heat treatment of silicon with a certain type of grown-in microdefects) can be solved in the near future. Building a general model for the formation of postgrowth microdefects will help optimize technological processes for production of devices. In this case, we can construct the defect structure of devices during their manufacture.

7. Conclusion

The diffusion model of the formation grown-in microdefects provides the unity and adequacy of physical and mathematical modeling. This model simulates of the defect structure of dislocation-free silicon single crystals of any diameters. The model of point defects dynamics can be considered as component of the diffusion model for formation grown-in microdefects. The diffusion model allowed to create software for personal computer. With the help of software can be conducted analytical researches which replace the expensive experimental researches.

Further development and modification of the software will lead to the development of information system of formation grown-in microdefects in dislocation-free silicon single crystals. The combinations of an information system with software for control the crystal growth will allow construct the defect structure of crystals during their growth. In turn, application a diffusion model of formation grown-in microdefects in the calculation of the formation of postgrowth microdefects allow to calculate the defect structure of silicon-based devices. In this case, it is possible to adequately construction the defect structure of silicon devices. We believe that the proposed in an article algorithm for the engineering of the defect structure of silicon can be used for other materials.

8. References

Ammon von, W.; Dornberger, E. & Hansson P.O. (1999). Bulk properties of very large diameter silicon single crystal. *Journal Crystal Growth*, Vol. 198-199, No. 1-4, pp. 390-398, ISSN 0022-0248.

Bonafos, C.; Mathiot, D. & Claverie A. (1998). Ostwald ripening of end-of-range defects in silicon. *Journal of Applied Physics*, Vol. 83, No. 6, pp. 3008-3018, ISSN 0021-8979.

Bulyarskii, S.V. & Fistul, V.I. (1997). *Thermodynamics and kinetics of interacting defects in semiconductors*, Nauka, ISBN 5-02-015164-5, Moscow, Russia.

Burton, B. & Speight M.V. (1985). The coarsening and annihilation kinetics of dislocation loops. *Philosophical Magazine A*, Vol. 53, No. 3, pp. 385-402, ISSN 0141-8610.

Chaldyshev, V.V.; Bert, N.A.; Romanov, A.E.; Suvorova, A.A.; Kolesnikova, A.L.; Preobrazhenskii, V.V.; Putyato, M.A.; Semyagin, B.R.; Werner, P.; Zakharov, N.D. & Claveria, A. (2002). Local stresses induced by nanoscale As-Sb clusters in GaAs matrix. *Applied Physics Letters*, Vol. 80, No. 3, pp. 377-381, ISSN 0003-6951.

Chaldyshev, V.V.; Kolesnikova, A.L.; Bert, N.A. & Romanov, A.E. (2005). Investigation of dislocation loops associated with As-Sb nanoclusters in GaAs. *Journal of Applied Physics*, Vol. 97, No. 2, pp. 024309-024319, ISSN 0021-8979.

Cho, H.-J.; Sim, B.-C. & Lee, J.Y. (2006). Asymmetric distributions of grown-in microdefects in Czochralski silicon. *Journal Crystal Growth*, Vol. 289, No. 2, pp. 458-463, ISSN 0022-0248.

Dornberger, E.; Ammon, von W.; Virbulis, J.; Hanna, B. & Sinno T. (2001). Modeling of transient point defect dynamics in Czochralski silicon crystal. *Journal Crystal Growth*, Vol. 230, No. 1-2, pp. 291-299, ISSN 0022-0248.

Föll, H. & Kolbesen B.O. (1975). Formation and nature of swirl defects in silicon. *Journal of Applied Physics*, Vol. 8, No. 3, pp. 319-331, ISSN 0021-8979.

Goethem, van N.; Potter, de A.; Bogaert, van den N. & Dupret, F. (2008). Dynamic prediction of point defects in Czochralski silicon growth. An attempt to reconcile experimental defect diffusion coefficients with the criterion V/G. *Journal of Physics and Chemistry of Solids*, Vol. 69, No. 2-3, pp. 320-324, ISSN 0022-3697.

Goldstein, R.V.; Mezhennyi, M.V.; Mil'vidskii, M.G.; Reznik, V.Ya.; Ustinov, K.B. & Shushpannikov, P.S. (2011). Experimental and theoretical investigation of formation of the oxygen-containing precipitate-dislocation loop system in silicon. *Physics of the Solid State*, Vol. 53, No. 3, pp. 527-538, ISSN 1063-7834.

Gyseva, N.B.; Sheikhet, E.G.; Shpeizman, V.V. & Shulpina I.L. (1986). Dislokazionnay aktivnost mikrodefektov v monokristalax kremnia. *Fizika tverdogo tela*, Vol. 28, No. 10, pp. 3192-3194, ISSN 0367-3294.

Huff, H.R. (2002). An electronics division retrospective (1952-2002) and future opportunities in the twenty-first century. *Journal of the Electrochemical Society*, Vol. 149, No. 5, pp. S35-S58, ISSN 0013-4651.

Itsumi, M. (2002). Octahedral void defects in Czochralski silicon. *Journal Crystal Growth*, Vol. 237-239, No. 3, pp. 1773-1778, ISSN 0022-0248.

Kato, M.; Yoshida, T.; Ikeda, Y. & Kitagawara, Y. (1996). Transmission electron microscope observation of "IR scattering defects" in as-grown Czochralski Si crystals. *Japanese Journal Applied Physics*, Vol. 35, No. 11, pp. 5597-5601, ISSN 0021-4922.

Kock de, A.J.R. (1970). Vacancy clusters in dislocation-free silicon. *Applied Physics Letters*, Vol. 16, No. 3, pp. 100-102, ISSN 0003-6951.

Kock de, A.J.R.; Stacy, W.T. & Wijgert van de, W.M. (1979). The effect of doping on microdefect formation in as-grown dislocation-free Czochralski silicon crystals. *Applied Physics Letters*, Vol. 34, No. 9, pp. 611-616, ISSN 0003-6951.

Kolesnikova, A.L. & Romanov, A.E. (2004). Misfit dislocation loops and critical parameters of quantum dots and wires. *Philosophical Magazine Letters*, Vol. 84, No. 3, pp. 501-506, ISSN 0950-0839.

Kolesnikova, A.L.; Romanov, A.E. & Chaldyshev V.V. (2007). Elastic-energy relaxation in heterostructures with strained nanoinclusions. *Physics of the Solid State*, Vol. 49, No. 4, pp. 667-674, ISSN 1063-7834.

Kulkarni, M.S.; Voronkov, V.V. & Falster, R. (2004). Quantification of defect dynamics in unsteady-state and steady-state Czochralski growth of monocrystalline silicon. *Journal Electrochemical Society*, Vol. 151. - No. 5, pp. G663-G669, ISSN 0013-4651.

Kulkarni, M.S. (2005). A selective review of the quantification of defect dynamics in growing Czochralski silicon crystals. *Ind. Eng. Chem. Res.*, Vol. 44, No. 16, pp. 6246-6263, ISSN 0888-5885.

Kulkarni, M.S. (2007). Defect dynamics in the presence of oxygen in growing Czochralski silicon crystals. *Journal Crystal Growth*, Vol. 303, No. 2, pp. 438-448, ISSN 0022-0248.

Kulkarni, M.S. (2008a). Lateral incorporation of vacancies in Czochralski silicon crystals. *Journal Crystal Growth*, Vol. 310, No. 13, pp. 3183-3191, ISSN 0022-0248.

Kulkarni, M.S. (2008b). Defect dynamics in the presence of nitrogen in growing Czochralski silicon crystals. *Journal Crystal Growth*, Vol. 310, No. 2, pp. 324-335, ISSN 0022-0248.

Nakamura, K.; Saishoji, T. & Tomioka, J. (2002). Grown-in defects in silicon crystals. *Journal Crystal Growth*, Vol. 237-239, No. 1-4, pp. 1678-1684, ISSN 0022-0248.

Petroff, P.M. & Kock de, A.J.R. (1975). Characterization of swirl defects in floating-zone silicon crystals. *Journal Crystal Growth*, Vol. 30, No. 1, pp. 117-124, ISSN 0022-0248.

Prostomolotov, A.I. & Verezub, N.A. (2009). Simplistic approach for 2D grown-in microdefects modeling. *Physica status solidi (c)*, Vol. 6, No. 8, pp. 1878-1881, ISSN 1610-1642.

Prostomolotov, A.I.; Verezub, N.A.; Mezhennii, M.V. & Reznik, V.Ua. (2011). Thermal optimization of CZ bulk growth and wafer annealing for crystalline dislocation-free silicon. *Journal of Crystal Growth*, Vol. 318, No. 1, pp. 187-192, ISSN 0022-0248.

Sadamitsu, S.; Umeno, S.; Koike, Y.; Hourai, M.; Sumita, S. & Shigematsu, T. (1993). Dislocations, precipitates and other defects in silicon crystals. *Japanese Journal Applied Physics*, Vol. 32, No. 9, pp. 3675-3679, ISSN 0021-4922.

Sinno, T. (1999). Modeling microdefect formation in Czochralski silicon. *Journal Electrochemical Society*, Vol. 146, No. 6, pp. 2300-2312, ISSN 0013-4651.

Sitnikova, A.A.; Sorokin, L.M.; Talanin, I.E.; Sheikhet, E.G. & Falkevich E.S. (1984). Electron-microscopic study of microdefects in silicon single crystals grown at high speed. *Physica Status Solidi (a)*, Vol. 81, No. 2, pp. 433-439, ISSN 1862-6300.

Sitnikova, A.A.; Sorokin, L.M.; Talanin, I.E.; Sheikhet, E.G. & Falkevich, E.S. (1985). Vacancy type microdefects in dislocation-free silicon single crystals. *Physica Status Solidi (a)*, Vol. 90, No. 1, pp. K31-K35, ISSN 1862-6300.

Talanin, V.I.; Talanin, I.E. & Levinson, D.I. (2002a). Physical model of paths of microdefects nucleation in dislocation-free single crystals float-zone silicon. *Cryst. Res. & Technol.*, Vol. 37, No. 9, pp. 983-1011, ISSN 0232-1300.

Talanin, V.I.; Talanin, I.E. & Levinson, D.I. (2002b). Physics of the formation of microdefects in dislocation-free monocrystals of float-zone silicon. *Semicond. Sci. Cryst. Res. & Technol.*, Vol. 17, No. 2, pp. 104-113, ISSN 0268-1242.

Talanin, V.I. & Talanin, I.E. (2003). Physical nature of grown-in microdefects in Czochralski-grown silicon and their transformation during various technological effects . *Physica Status Solidi (a)*, Vol. 200, No. 2, pp. 297-306, ISSN 1862-6300.

Talanin, V.I. & Talanin, I.E. (2004). Mechanism of formation and physical classification of the grown-in microdefects in semiconductor silicon. *Defect & Diffusion Forum*, Vol. 230-232, No. 1, pp. 177-198, ISSN 1012-0386.

Talanin, V.I. & Talanin, I.E. (2006a). Formation of grown-in microdefects in dislocation-free silicon monocrystals, In: *New research on semiconductors*, T.B. Elliot,(Ed.), 31-67, Nova Science Publishers, Inc., ISBN 1-59454-920-6, New York, USA.

Talanin, V.I. & Talanin, I.E. (2006b). On the formation of vacancy microdefects in dislocation-free silicon single crystals. *Ukranian Journal of Physics*, Vol. 51, No. 11-12, pp. 108-112, ISSN 0503-1265.

Talanin, V.I. & Talanin, I.E. (2007a). On the recombination of intrinsic point defects in dislocation-free silicon single crystals. *Physics of the Solid State*, Vol. 49, No. 3, pp. 467-470, ISSN 1063-7834.

Talanin, V.I.; Talanin, I.E. & Voronin, A.A. (2007b). About formation of grown-in microdefects in dislocation-free silicon single crystals. *Canadian Journal of Physics*, Vol. 85, No. 12, pp. 1459-1471, ISSN 1208-6045.

Talanin, V.I.; Talanin, I.E. & Voronin A.A. (2008). Modeling of the defect structure in dislocation-free silicon single crystals. *Crystallography Reports*, Vol. 53, № 7, pp. 1124-1132, ISSN 1063-7745.

Talanin, V.I. & Talanin, I.E. (2010a). Kinetic of high-temperature precipitation in dislocation-free silicon single crystals. *Physics of the Solid State*, Vol. 52, No. 10, pp. 2063-2069, ISSN 1063-7834.

Talanin, V.I. & Talanin, I.E. (2010b). Kinetics of formation of vacancy microvoids and interstitial dislocation loops in dislocation-free silicon single crystals. *Physics of Solid State*, Vol. 52, No. 9, pp. 1880-1886, ISSN 1063-7834.

Talanin, V.I. & Talanin, I.E. (2010c). Modeling of defect formation processes in dislocation-free silicon single crystals. *Crystallography Reports*, Vol. 55, No. 4, pp. 675-681, ISSN 1063-7745.

Talanin, V.I. & Talanin, I.E. (2011a). Kinetic model of growth and coalescence of oxygen and carbon precipitates during cooling of as-grown silicon crystals. *Physics of the Solid State*, Vol. 53, No. 1, pp. 119-126, ISSN 1063-7834.

Talanin, V.I.; Talanin, I.E. & Ustimenko N.Ph. (2011b). A new method for research of grown-in microdefects in dislocation-free silicon single crystals. *Journal of Crystallization Process & Technologys*, Vol.1, № 2, pp. 13-17, ISSN 2161-7678.

Yang, D.; Chen, J.; Ma, X. & Que, D. (2009). Impurity engineering of Czochralski silicon used for ultra large-scaled-integrated circuits. *Journal Crystal Growth*, Vol. 311, No. 3, pp. 837-841, ISSN 0022-0248.

Vas'kin, V.V. & Uskov V.A. (1968). Vlianie kompleksoobrazovania na diffyziu primesei v polyprovodnikax. *Fizika Tverdogo Tela*, Vol. 10, No. 6, pp. 1239-1241, ISSN 0367-3294.

Veselovskaya, N.V.; Sheikhet, E.G.; Neimark, K.N. & Falkevich, E.S. (1977). Defecty tipa klasterov v kremnii. *Proceedings of IV simposiuma Rost i legirovanie polyprovodnikovyx kristalov i plenok*, Vol. 2, pp. 284-288, Novosibirsk, USSR, June 1975.

Voronkov, V.V. & Falster, R. (1998). Vacancy-type microdefect formation in Czochralski silicon. *Journal Crystal Growth*, Vol. 194, No.1, pp. 76-88, ISSN 0022-0248.

Voronkov, V.V. (2008). Grown-in defects in silicon produced by agglomeration of vacancies and self-interstitial. *Journal Crystal Growth*, Vol. 310, No. 7-9, pp. 1307-1314, ISSN 0022-0248.

Voronkov, V.V.; Dai B. & Kulkarni M.S. (2011). Fundamentals and engineering of the Czochralski growth of semiconductor silicon crystals. *Comprehensive Semiconductor science and Technology*, Vol. 3, pp. 81-169, ISBN 978-0-444-53153-7.

Wang, Z. & Brown, R.A. (2001). Simulation of almost defect-free silicon crystal growth. *Journal Crystal Growth*, Vol. 231, No. 2, pp. 442-452, ISSN 0022-0248.

25

Preparation of Carvedilol Spherical Crystals Having Solid Dispersion Structure by the Emulsion Solvent Diffusion Method and Evaluation of Its *in vitro* Characteristics

Amit R. Tapas, Pravin S. Kawtikwar and Dinesh M. Sakarkar
Sudhakarrao Naik Institute of Pharmacy, Pusad, Dist Yavatmal, Maharashtra India

1. Introduction

Solid dispersion is one of the most efficient techniques to improve the dissolution rate of poorly water-soluble drugs, leading to an improvement in the relative bioavailability of their formulations. At present, the solvent method and the melting method are widely used in the preparation of solid dispersions. In general, subsequent grinding, sieving, mixing and granulation are necessary to produce the different desired formulations.

The spherical agglomeration technique has been used as an efficient particle preparation technique developed by Kawashima in the 1980s (Kawashima et al., 1994). Initially, spherical agglomeration technique was used to improve powder flowability, packability, and compressibility (Usha et al. 2008; Yadav and Yadav, 2008; Bodmeier and Paeratakul et al., 1989). Then polymers were introduced in this system to modify their release (Di Martino et al., 1999). Currently, this technique is used more frequently for the solid dispersion preparation of water-insoluble drugs in order to improve their solubility, dissolution rate and simplify the manufacturing process (Cui et al., 2003, Tapas et al. 2009, 2010). Spherical crystallization has been developed by Yoshiaki Kawashima and co-workers as a novel particulate design technique to improve processibility such as mixing, filling, tableting characteristics and dissolution rate of pharmaceuticals (Kawashima et al., 1974, 1976, 1981, 1982, 1983, 1984, 1985, 1989, 1991, 1994, 1995, 2002, 2003). The resultant crystals can be designated as spherical agglomerates (Kulkarni and Nagavi, 2002). Spherical crystallization is an effective alternative to improve dissolution rate of drugs (Sano et al., 1992). Now days functional drug devices such as microspheres, microcapsules, microballoons and biodegradable nanospheres were developed using the emulsion solvent diffusion techniques involving the introduction of a functional polymer into the system (Di Martino et al., 1999; Marshall and York, 1991; Garekani and Ford, 1999). This can be achieved by various methods such as

1. Spehrical Agglomeration (SA)
2. Quasi Emulsion Solvent Diffusion (QESD)
3. Ammonia Diffusion System (ADS)
4. Neutralization (NT)

Out of which first two are the most common methods in practice.

In the spherical crystallization process, crystal formation, growth and agglomeration occur simultaneously within the same system. In this method, a third solvent called the bridging liquid is added in a smaller amount to purposely induce and promote the formation of agglomerates. Crystals are agglomerated during the crystallization process and large spherical agglomerates are produced. A near saturated solution of the drug in a good solvent is poured into a poor solvent. The poor and good solvents are freely miscible and the "affinity" between the solvents is stronger than the affinity between drug and good solvent, leading to precipitation of crystals immediately. Under agitation, the bridging liquid (the wetting agent) is added, which is immiscible with the poor solvent and preferentially wet the precipitated crystals. As a result of interfacial tension effects and capillary forces, the bridging liquid acts to adhere the crystals to one another and facilitates them to agglomerate (Fig. 1).

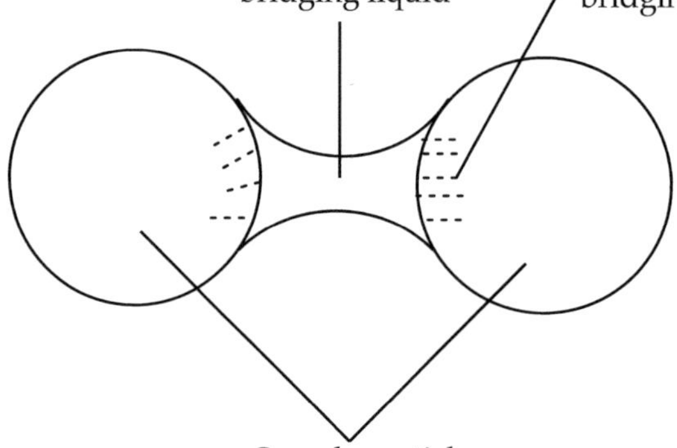

Fig. 1. Two sample particles joined together by a liquid bridge

In spherical agglomeration method, when a drug solution (in good solvent) was poured into a poor solvent under agitation, the drug crystals were formed immediately and agglomerated with a bridging liquid dispersed in the poor solvent, because the bridging liquid has a preference for wetting the drug crystals.

Quasi emulsion solvent diffusion method is also known as transient emulsion method. Firstly the drug was dissolved in a mixed solvent of good solvent and bridging liquid. Because of the increased interfacial tension between the two solvents, the solution is dispersed into the poor solvent producing emulsion (quasi) droplets, even though the pure solvents are miscible. The good solvent diffuses gradually out of the emulsion droplets into the surrounding poor solvent phase, and the poor solvent diffuses into the droplets by which the drug crystallizes inside the droplets. The method is considered to be simpler than the SA method, but it can be difficult to find a suitable additive to keep the system emulsified and to improve the diffusion of the poor solute into the dispersed phase. Especially hydrophilic/hydrophobic additives are used to improve the diffusion remarkably. In this method the shape and the structure of the agglomerate depend strongly on the good solvent to poor solvent ratio and the temperature difference between the two solvents when the drug solution was introduced into the poor solvent under certain temperature and stirring, the drug solution was dispersed immediately to form quasi o/w emulsion droplets, the emulsion droplets were gradually solidified, forming spherical agglomerates along with the diffusion of the good solvent from the droplets into the poor solvent.

Spherical agglomeration has got more importance than other methods because it is easy to operate and the selection of the solvents is easier than in the other methods. Quasi emulsion solvent diffusion method has the second importance.

An ammonia diffusion system is applicable to amphoteric drug substances. In this method, the mixture of three partially immiscible solvent i.e. acetone, ammonia water, dichloromethane was used as crystallization system. In this system ammonia water acted as bridging liquid as well as good solvent, acetone as the water miscible but poor solvent, thus drug precipitated by solvent change without forming ammonium salt. Water immiscible solvent such as hydrocarbon or halogenated hydrocarbons e.g. dichloromethane induced liberation of ammonia water.

In neutralization method sodium hydroxide acts as a good solvent and hydrochloric acid as a poor solvent or vice-versa. These solutions were added to each other in order to get neutralization. The bridging liquid was added drop wise under agitation to form spherical agglomerates.

In this study special attention was given to improving the solubility and dissolution rate of poorly water soluble drug carvedilol using quasi emulsion solvent diffusion method. Carvedilol (CAR) (fig. 2), (±)-1-(carbazol-4yloxy)-3-[[2-(o-methoxyphenoxy)ethyl]amino]-2-propanol is an α_1, β_1 and β_2 adrenergic receptor antagonist (Sweetman, 2002). It is used to treat mild to moderate essential hypertension, mild to severe heart failure, and patients with systolic dysfunction after myocardial infraction. Carvedilol is practically insoluble in water and exhibits pH dependent solubility. Its solubility is <1μg/ml above pH 9.0, 23 μg/ml at pH 7, and about 100 μg/ml at pH 5 at room temperature. It's extremely low solubility at alkaline pH levels may prevent the drug from being available for absorption in the small intestine and colon, thus making it poor candidate for an extended release dosage form. In the present study, to overcome the problems related to solubility, dissolution rate, flowability, and compressibility, the microspheres having solid dispersions structure of Carvedilol were prepared by emulsion solvent diffusion method by using a poloxamer (polxamer F68 and poloxamer F127) as a hydrophilic polymer.

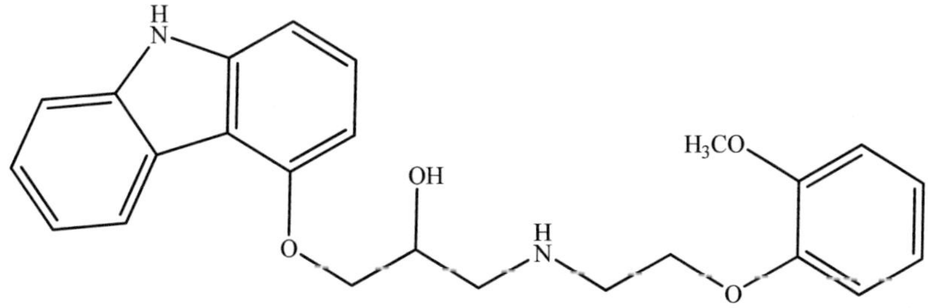

1-(9*H*-carbazol-4-yloxy)-3-(2-(2-methoxyphenoxy)ethylamino)propan-2-ol

Fig. 2. Chemical structure of Carvedilol (CAR)

2. Materials and methods

2.1 Materials

Carvedilol was supplied by Dr. Reddy's Laboratory, Hyderabad, India as a gift sample. Poloxamer F68 and F127 were supplied by Lupin Research Park, Pune, India. All other chemicals used were of analytical grade.

2.2 Methods

Carvedilol (1.0 g) with poloxamer was dissolved in good solvent methanol (12.0 mL). The bridging liquid dichloromethane (2.0 mL) was added to it. The resulting solution was then poured dropwise in to the poor solvent distilled water (100 mL) containing Aerosil 200 Pharma (0.1 g). The mixture was stirred continuously for a period of 0.5 h using a controlled speed mechanical stirrer (Remi motors, India) at 1000 rpm. As the good solvent diffused into the poor solvent, droplets gradually solidified. Finally the coprecipitated microspheres of the drug-polymer were filtered through Whatman filter paper (No.1) and dried in desiccator at room temperature. The amount of poloxamers was altered to get desired microspheres. The composition is given in Table 1.

Composition/Parameters	CP681	CP682	CP1271	CP1272
CAR (g)	1.0	1.0	1.0	1.0
Methanol (ml)	12.0	12.0	12.0	12.0
DCM (ml)	2.0	2.0	2.0	2.0
Water (ml)	100	100	100	100
Poloxamer F68 (g)	1.5	3.0	--	--
Poloxamer F127 (g)	--	--	1.5	3.0
Aerosil 200 pharma (g)	0.1	0.1	0.1	0.1
Stirring speed (rpm)	1000	1000	1000	1000

Table 1. Composition of spherical crystals

2.2.1 Drug content study

The drug content study of agglomerates was determined by dissolving 100 mg of crystals in 3 ml methanol and diluting further with distilled water (100 ml) followed by measuring the absorbance of appropriately diluted solution spectrophotometrically (PharmaSpec UV-1700, UV-Vis spectrophotometer, Shimadzu) at 286 nm.

2.2.2 Fourier Transform Infrared Spectroscopy (FTIR)

The FTIR spectra of powder CAR, and their agglomerates were recorded on an FTIR spectrophotometer (JASCO, FTIR V-430 Plus).

2.2.3 Differential Scanning Calorimetry (DSC)

DSC analysis was performed using a DSC 823 calorimeter (Mettler Toledo model) operated by STARe software. Samples of CAR and its agglomerates were sealed in an aluminium crucible and heated at the rate of 10 °C/min up to 300 °C under a nitrogen atmosphere (40 ml/min).

2.2.4 Powder X-ray diffraction studies

Powder X-ray diffraction patterns (XRD) of the CAR and its spherical agglomerates were monitored with an x-ray diffractometer (Philips Analytical XRD) using copper as x-ray target, a voltage of 40 KV, a current of 25 mA and with 2.28970 Å wavelength. The samples were analyzed over 2θ range of 10.01-99.99^0 with scanning step size of 0.02^0 (2θ) and scan step time of 0.8 second.

2.2.5 Scanning electron microscopy

The surface morphology of the agglomerates was accessed by SEM. The crystals were splutter coated with gold before scanning.

2.2.6 Micromeritic properties

The size of agglomerates was determined by microscopic method using stage and eyepiece micrometers. The shape of the agglomerates was observed under an optical microscope (×60 magnification) attached to a computer. Flowability of untreated carvedilol and agglomerates was assessed by determination of angle of repose, Carr's index (CI) and Hausner's ratio (HR) (Wells, 2002). Angle of repose was determined by fixed funnel method (Martin et al., 2002). The mean of three determinations was reported. The CI and HR were calculated from the loose and tapped densities. Tapped density was determined by tapping the samples into a 10 ml measuring cylinder. The CI and HR were calculated according to the following equation 1 and 2.

$$\text{C.I.} = \frac{\text{Tapped density} - \text{Bulk density}}{\text{Tapped density}} \times 100 \tag{1}$$

$$\text{H.R.} = \frac{\text{Tapped density}}{\text{Bulk density}} \tag{2}$$

2.2.7 Solubility studies

A quantity of crystals (about 100 mg) was shaken with 10 mL distilled water in stoppered conical flask at incubator shaker for 24 h at room temperature. The solution was then passed through a whatmann filter paper (No. 42) and amount of drug dissolved was analyzed spectrophotometrically.

2.2.8 Dissolution rate studies

The dissolution rate studies of carvedilol alone and its spherical agglomerates were performed in triplicate in a dissolution apparatus (Electrolab, India) using the paddle method (USP Type II). Dissolution studies were carried out using 900 ml of 0.1N HCl (pH 1.2) at 37 ± 0.5 ^{0}C at 50 rpm as per US FDA guidelines (U.S. Food and drug administration [USFDA], 2010 and Bhutani et al., 2007.). 12.5 mg of carvedilol or its equivalent amount of spherical agglomerates were added to 900 ml of 0.1N HCl (pH 1.2). Samples (5 ml) were withdrawn at time intervals of 10, 20, 30, and 60 min. The volume of dissolution medium was adjusted to 900 ml by replacing each 5 ml aliquot withdrawn with 5 ml of fresh 0.1N HCl (pH 1.2). The solution was immediately filtered, suitably diluted and the concentrations of carvedilol in samples were determined spectrophotometrically at 286 nm. The results obtained from the dissolution studies were statistically validated using ANOVA.

2.2.9 Dissolution efficiency studies

The dissolution efficiency (DE) of the batches was calculated by the method mentioned by Khan (Khan, 1975). It is defined as the area under the dissolution curve between time points t_1 and t_2 expressed as a percentage of the curve at maximum dissolution, y100, over the same time period or the area under the dissolution curve up to a certain time, t, (measured using trapezoidal rule) expressed as a percentage of the area of the rectangle described by 100% dissolution in the same time. DE_{60} values were calculated from dissolution data and used to evaluate the dissolution rate (Anderson et al., 1998).

$$\text{Dissolution Efficiency} = \frac{\int_0^t y\,dt}{y100(t_2 - t_1)} \times 100 \qquad (3)$$

3. Result and discussion

3.1 Preparation of spherical crystals

A typical spherical crystallization system involved a good solvent, a poor solvent for a drug and a bridging liquid. The selection of these solvent depends on miscibility of the solvents and solubility of drug in individual solvents. Accordingly acetone, dichloromethane, water were selected as a good solvent, bridging liquid, and poor solvent, respectively. CAR is soluble in methanol, but poorly soluble in water. Also it is soluble in dichloromethane which is immiscible in water. Hence, this solvent system was used in the present study. When drug polymer solution was poured into the poor solvent under agitation at selected temperature, the drug polymer solution became immediately semitransparent due to the presence of small sized emulsion droplets. Gradually emulsion droplets solidified along with diffusion of the good solvents, as bridging liquid dichloromethane was commixed with good solvent, when the good solvent in the droplets diffused into the poor solvent, the residual dichloromethane in the droplets bridged the Aerosil, coprecipitated drug, and polymer to form spherical crystals. The Aerosil acts as a dispering agent and mass compactor, because coacervation droplets formed from the drug-polymer droplets during the solidifying period were sticky and readily coalesced, while the introduction of Aerosil efficiently prevented coalescence and produced compact spherical crystals.

3.2 Optimization of process variables for preparation of spherical crystals

To optimize the Carvedilol spherical crystallization by methanol, water, dichloromethane solvent system following parameters considered amount and mode of addition of bridging liquid, stirring speed and temperature. (Table 2).

S.No.	Parameter	Variables	Observation
1	Conc. of Bridging Liquid (Dichloromethane)	1 ml	No agglomeration
		2 ml	Agglomeration
		3 ml	No agglomeration
2	Agitation speed	800 rpm	Spherical but large
		1000 rpm	Spherical
		1200 rpm	Irregular shape and small
3	Agitation time	15 min	Incomplete agglomerates
		30 min	Spherical agglomerates
4	Mode of addition of bridging liquid	Whole at a time	Crystals of irregular geometry
		Drop wise	Spherical agglomerates

Table 2. Parameters affecting spherical agglomeration

3.3 Drug content study

Percent drug content was found to be in the range of 92.12±1.60 to 94.4±2.37 (Table 3).

3.4 Micromeritic properties

Pure CAR could not pass through the funnel during the angle of repose experiment which could be due to the irregular shape and high fineness of the powder, which posed hurdles in the uniform flow from the funnel. It exhibited poor flowability and packability as indicated by Hausner ratio (1.52) and Carr's Index (34.37%). All agglomerates showed excellent flowability and packability (Angle of repose: 23-28^0; Carr's Index: 15-18%; Hausner ratio: 1.13-1.17) when compared to pure CAR. The improved flowability of agglomerates may be due to good sphericity and larger size of agglomerates. During the tapping process, smaller agglomerates might have infiltrated into the voids between larger particles, which could result improved packability (lower CI). The results of micromeritic properties are shown in Table 3.

S.No.	Samples	Drug Content	Carr's Index	Hausner Ratio	Angle of Repose	Particle Size (μm)	Aqueous Solubility (μgmL^{-1})
1.	CAR	100.0±0.0	34.37±1.79	1.52±1.26	---	71.55±5.37	21.0 ± 0.51
2.	CP681	94.4±2.37	17.53±1.49^b	1.15±2.67^b	27.12±2.43	112.7±31.26^b	32.20 ± 1.32
3.	CP682	92.31±2.02	15.64±1.43^b	1.13±2.21^b	25.36±0.67	136.43±18.57^b	39.04 ± 1.65^b
4.	CP1271	94.19±1.15	18.63±2.90^b	1.17±1.82^b	28.42±0.23	88.92±25.38^b	58.08 ± 1.28^b
5.	CP1272	92.12±1.60	17.24±1.87^b	1.14±1.53^b	23.05±0.89	102.03±20.11^b	64.25 ± 1.31^b

^{a}Mean ± SD, n = 3; ^{b}Significantly different compared to pure CAR ($p<0.05$).

Table 3. Micromeritics, Drug Content, Particle size, aqueous solubility data of carvedilol and its spherical crystals

3.5 FTIR, DSC and powder X-ray studies

The FTIR spectra of CAR as well as its spherical crytals are presented in Figure 3. FTIR of CAR showed a characteristic peaks at 3343.96 (N-H str. Aromatic Amines), 3062.41 (C-H str.

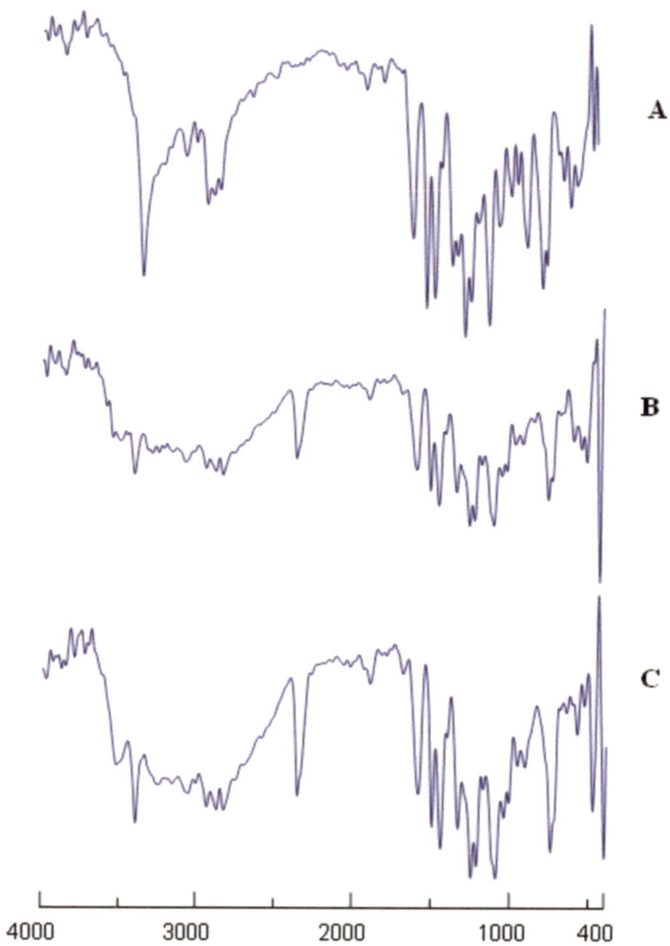

Fig. 3. FTIR spectra of (A) - Carvedilol, (B) - CP682, (C) - CP1272

Aromatic Hydrocarbon), 2923.56 (C-H str. in $-CH_3$/ $-CH_2$), 1592.91 (C=C str. Aromatic), 1253.5, 1214.93, 1099.23 (C-O str. in Ar C=C-O-C) cm^{-1}. There was no considerable change in the IR peaks of the spherical agglomerates when compaired with pure CAR, which revealed that no chemical interaction had occurred between drug and polymer during agglomeration process.

Figure 4 shows the DSC thermogram of pure CAR and its spherical crystals. DSC thermogram of CAR showed endothermic peak at 120.47^{0}C, which represented melting of carvedilol. There was negligible change in the melting point endotherms of prepared spherical crystals compared to pure drug (CP682 = 115.47^{0}C, CP1272 = 118.67^{0}C). The endotherms at 57.05^{0}C and 59.07^{0}C ascribed to the melting of Poloxamer F68 and Poloxamer F127 respectively. This observation further supports the IR spectroscopy results, which indicated the absence of any interactions between the drug and additives used in the preparation. However, there was a decrease, although very small, in the melting point of the drug in the spherical crystals compared to that of pure carvedilol. This indicates the little amorphization of carvedilol when prepared in the form of spherical crystals.

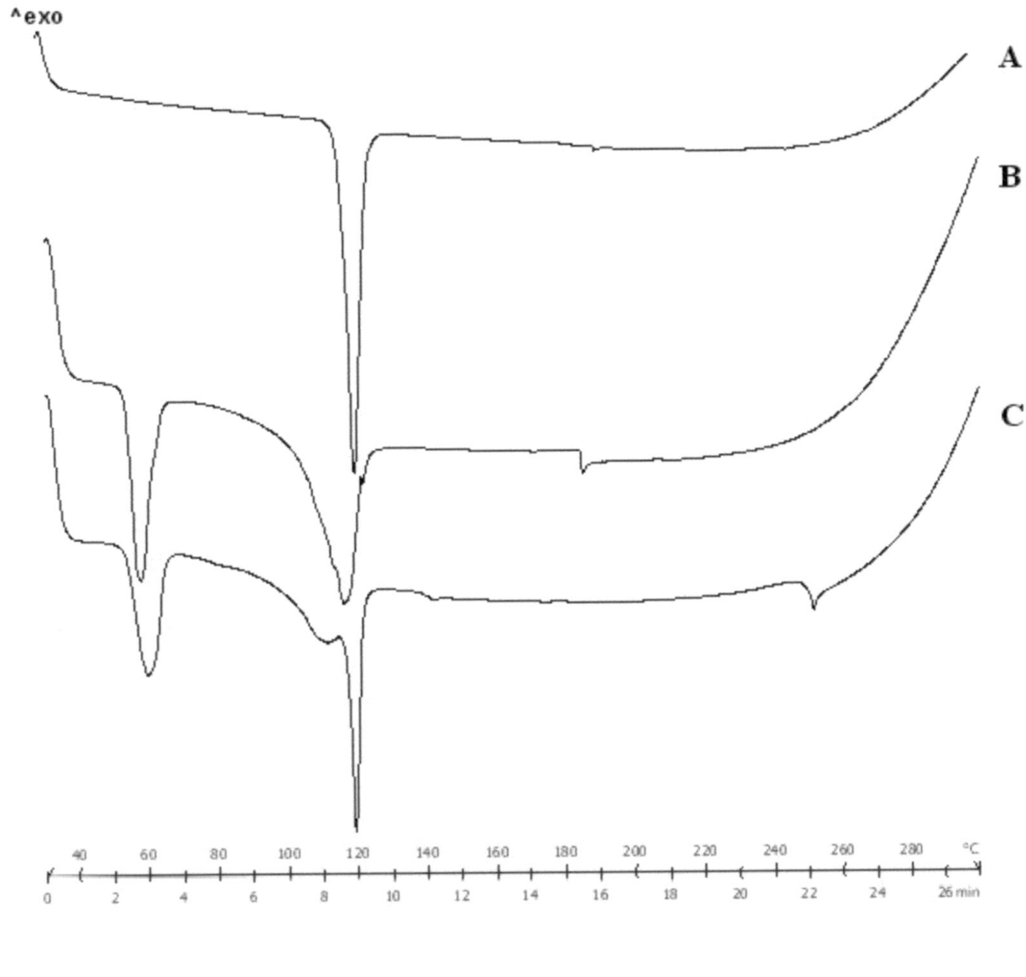

Fig. 4. DSC thermogram of (A) - Carvedilol, (B) - CP682, (C) - CP1272.

The XRD patterns of CAR shown in figure 5. The intense peaks at 2θ of 26.16^0, 27.48^0, 36.47^0 and 39.34^0 with peak intensities (counts) 310, 256, 228 and 135 respectively obtained from CAR confirmed the crystalline form of CAR. The PXRD patterns of CAR spherical crystals could be distinguished from the pure CAR. Peaks at around 8.4, 17, 22^0 2θ confirms the change in crystal arrangements of CAR in its spherical crystal form.

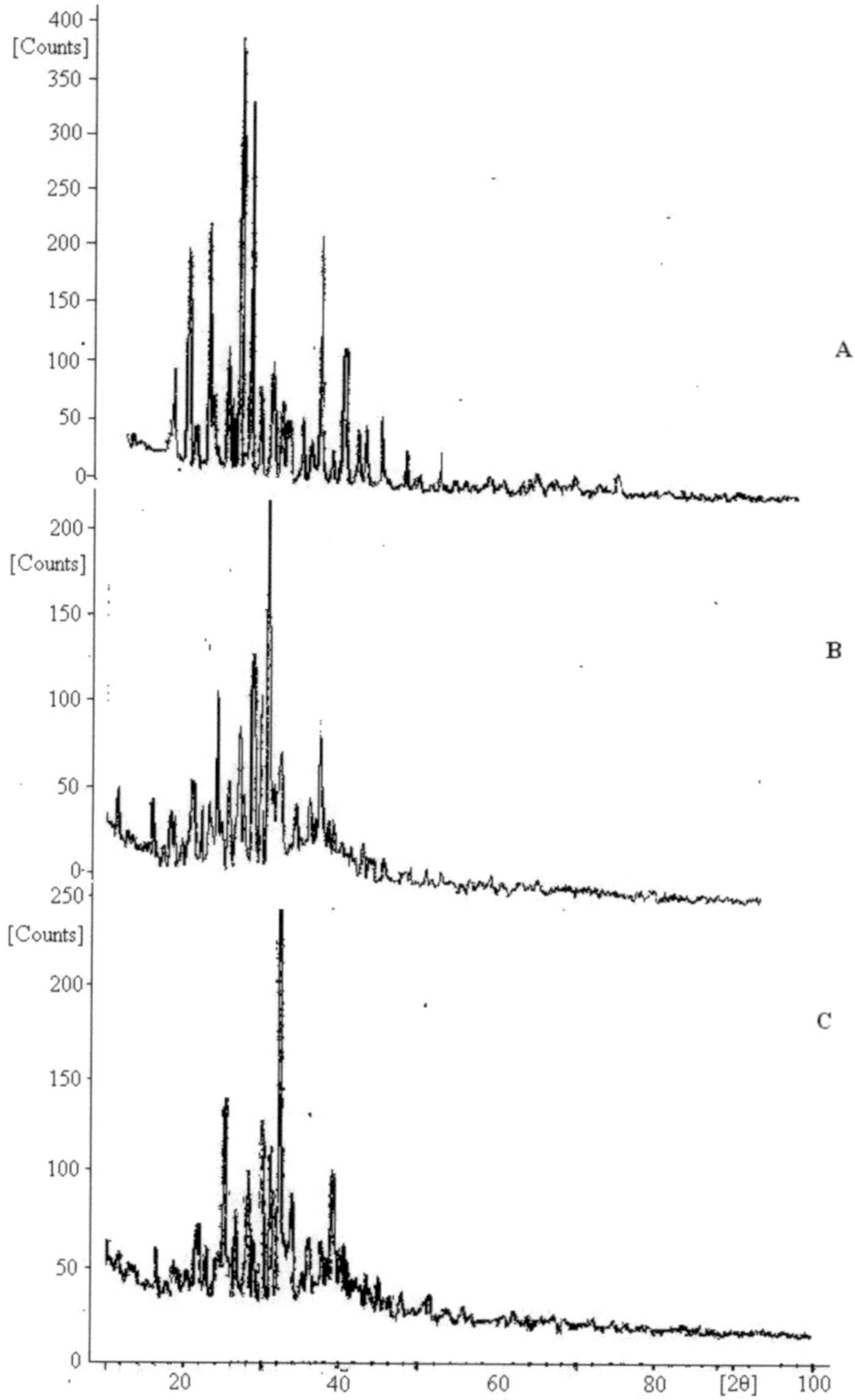

Fig. 5. XRD patterns of (A) – Carvedilol, (B) – CP682, (C) – CP1272.

3.6 Scanning electron microscopy

The results of surface morphology studies are shown in Figure 6. The SEM results revealed the spherical structure of agglomerates. The surface morphology studies also revealed that the agglomerates were formed by very small crystals, which were closely compacted into spherical form. These photo-micrographs show that the prepared agglomerates were spherical in shape which enabled them to flow very easily.

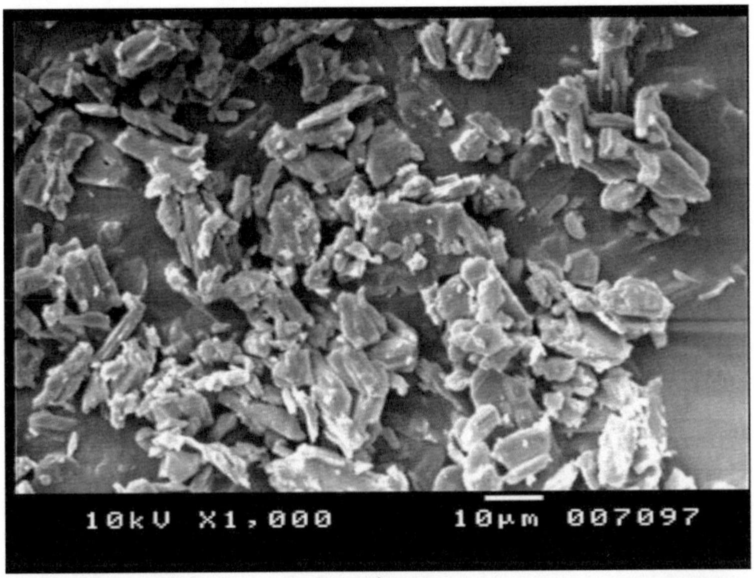

A

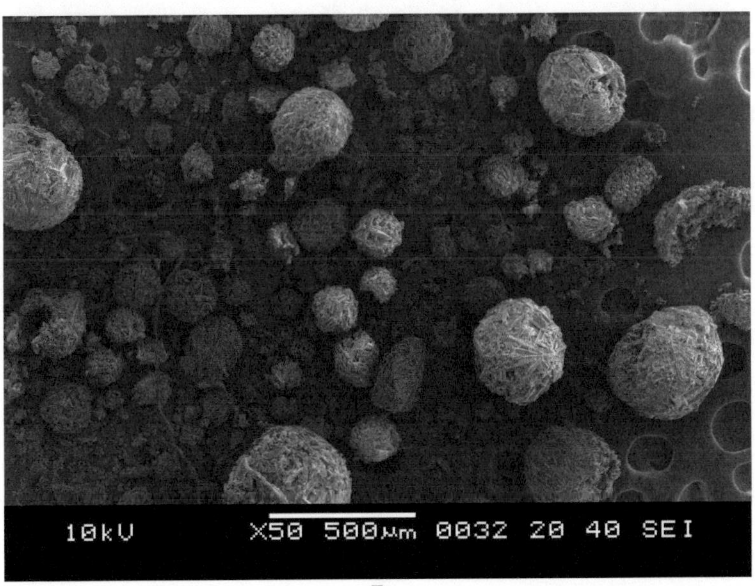

B

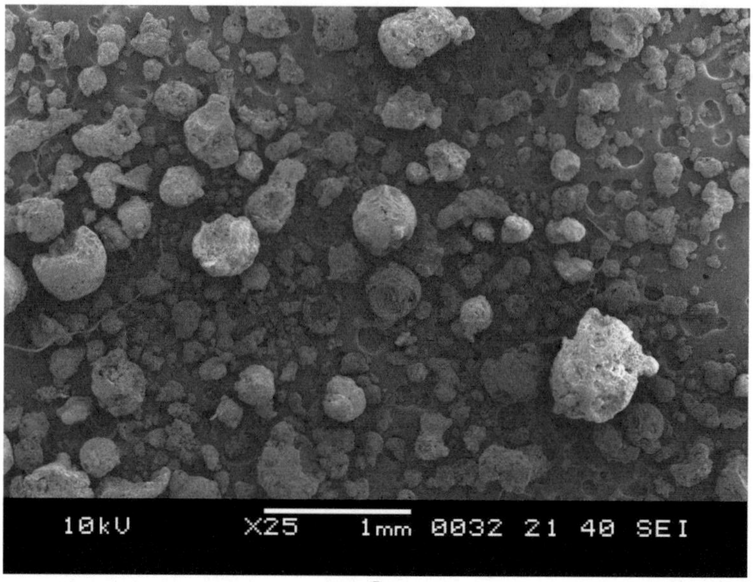

C

Fig. 6. SEM of (A) - Carvedilol, (B) - CP682, (C) - CP1272

3.7 Dissolution rate studies

The dissolution curves of pure carvedilol and its spherical crystals in 0.1 N HCl (pH 1.2) are shown in fig. 7. The release rate profiles were expressed as the percentage drug released vs. time. Table 4 shows % drug dissolved in 1h (DP_{60}) and dissolution efficiency values at 30 min (DE_{30}) for carvedilol and its spherical crystals. These values are tested statistically through one way ANOVA and are found significantly different ($p<0.05$) from pure carvedilol. As indicated carvedilol was dissolved more than 80% from spherical crystals CP681, CP1271 and CP1272 after 1h and more than 90% from spherical crystal CP682 while the pure CAR powder was just dissolved 34.37% at comparable time. The results revealed that the spherical crystals caused significant increase ($P<0.05$) in drug release compared to the pure drug. Enhancement in dissolution rate of spherical agglomerates as compared to pure drug may be due the presence of hydrophilic polymer, Poloxamer. The mechanism

Sample	Carvedilol Release after 1h	Dissolution Efficiency at 30 min
CAR	34.37±0.19	20.56±0.09
CP681	83.42±0.25[b]	66.12±0.31[b]
CP682	95.52±0.25[b]	70.63±1.38[b]
CP1271	82.26±0.96[b]	65.90±0.14[b]
CP1272	84.42±0.07[b]	67.39±0.39[b]

[a]Mean ± SD, n = 3
[b]Significantly different compared to carvedilol ($p < 0.05$).

Table 4. Drug Release and Dissolution Efficiency[a]

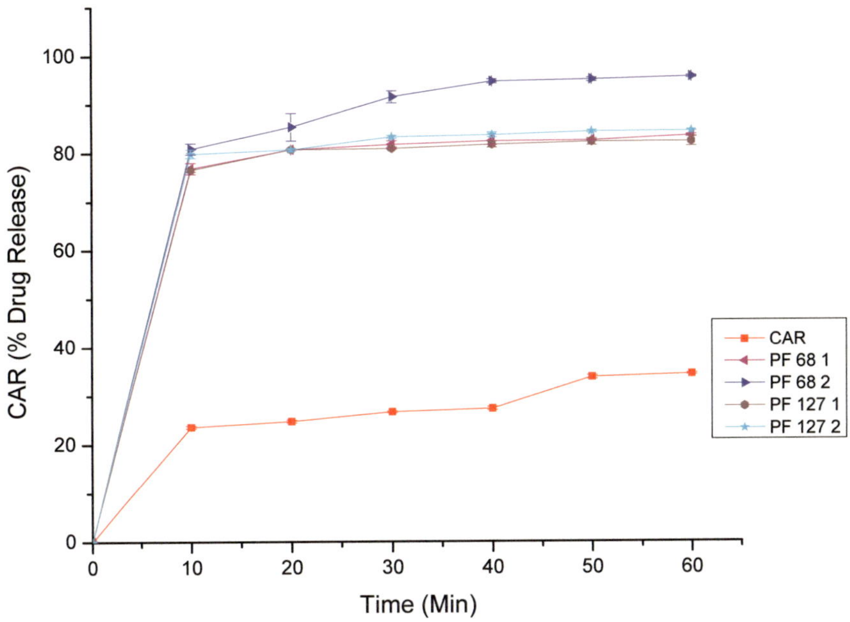

Fig. 7. Dissolution profile of CAR and its agglomerates. (Mean ±SD, n = 3.)

behind the greater solubility and dissolution of CAR from its agglomerated form may resemble the solid dispersion mechanism despite the larger particle size of agglomerates. This effect may be due to improved wettability of the surface of agglomerates by the adsorption of poloxamer onto the surfaces of crystals. These results confirm that the dissolution rate of carvedilol was increased in form of spherical crystals when compared to its pure form.

4. Conclusion

CAR-poloxamer spherical crystals were prepared successfully by ESD method. The resultant crystals have the desired micromeritic properties, such as flowability and packability. In the present investigation Poloxamer F68 and Poloxamer F127 has significantly improved dissolution rate of carvedilol. However *in vivo* bioavailability studies are required to ensure whether, the results obtain in this investigation can be extrapolated to the *in vivo* conditions.

5. Acknowledgments

The authors gratefully acknowledge Dr. Reddy's Laboratory, Hyderabad, India for the gift sample of Carvedilol. The authors are thankful to AISSMS college of Pharmacy, Pune, India for providing FTIR and DSC facilities. Also the authors would like to thank Shivaji University, Kolhapur, India for providing PXRD facility.

6. References

Anderson, N.H., Bauer, M., Boussac, N., Khan-Malek, R., Munden, P., Sardaro, M. (1998). An evaluation of fit factors and dissolution efficiency for the comparison of in vitro dissolution profiles. *Journal of Pharmaceutical and Biomedical Analysis*, 17, 811–822.

Bhutani, S., Hiremath, S.N., Swamy, P.V., Raju, S.A. (2001). Preparation and evaluation of inclusion complexes of carvedilol. *Journal of Scientific and Industrial Research*, 66, 830-834.

Bodmeier R., Paeratakul R. (1989). Spherical agglomerates of water-insoluble drugs. *Journal of Pharmaceutical Sciences*, 78, 964-967.

Cui F., Yang M., Jiang Y., Cun D., Lin W., Fan Y., Kawashima Y (2003). Design of sustained-release nitrendipine microspheres having solid dispersion structure by quasi-emulsion solvent diffusion method. *Journal of ControlledRelease*, 91, 375-384.

Di Martino, P., Barthelemy, C., Piva, F., Joiris, E., Palmieri, G.F., Martelli, S. (1999). Improved dissolution behavior of fenbufen by spherical crystallization. *Drug Delivery and Industrial Pharmacy*, 25, 1073-1081.

Garekani, H.A., Ford, J.L., Rubinstein, M.H., Rajabi-Siahboomi, A.R. (1999). Formation and Compression Properties of Prismatic Polyhedral and Thin Plate-like Crystals of Paracetamol. *International Journal of Pharmaceutics*, 187, 77-89.

Kawashima Y, Aoki S, Takenaka H. Spherical agglomeration of aminophylline crystals during reaction in liquid by the spherical crystallization technique. *Chem. Pharm. Bull.* 1982; 30, 1900-2.

Kawashima Y, Capes CE. Experimental study of the kinetics of spherical agglomeration in as stirred vessel. *Powder Technol.* 1974; 10, 85-92.

Kawashima Y, Capes CE. Further studies of the kinetics of spherical agglomeration in a stirred vessel. *Powder Technol.* 1976; 13, 279-288.

Kawashima Y, Cui F, Takeuchi H, Niwa T, Hino T, Kiuchi K. Improvements in flowability and compressibility of pharmaceutical crystals for direct tabletting by spherical crystallization with a 2 solvent system. *Powder Technol.* 1994; 78, 151-157.

Kawashima Y, Cui F, Takeuchi H, Niwa T, Hino T, Kiuchi K. Improved static compression behaviors and tablettabilities of spherically agglomerated crystals produced by the spherical crystallization technique with a two-solvent system. *Pharma. Res.* 1995; 12, 1040-4.

Kawashima Y, Cui F, Takeuchi H, Niwa T, Hino T, Kiuchi K. Parameters determining the agglomeration behavior and the micromeritic properties of spherically agglomerated crystals prepared by the spherical crystallization technique with miscible solvent systems. *Int. J. Pharm.* 1995; 119, 139-47.

Kawashima Y, Furukawa K, Takenaka H. The physicochemical parameters determining the size of agglomerate prepared by the wet spherical agglomeration technique. *Powder Technol.* 1981; 30, 211-16.

Kawashima Y, Imai M, Takeuchi H, Yamamoto H, Kamiya K, Hino T. Improved flowability and compactibility of spherically agglomerated crystals of ascorbic acid for direct tabletting designed by spherical crystallization process. *Powder Technol.* 2003; 130, 283-289.

Kawashima Y, Imai M, Takeuchi H, Yamamoto H, and Kamiya, K, Development of Agglomerated Crystals of Ascorbic acid by the Spherical Crystallization Technique for Direct Tabletting, and Evaluation of their Compactibilities, *Kona.* 2002; 20, 251-261.

Kawashima Y, Lin SY, Ogawa M, Handa T, Takenaka H. Preparations of agglomerated crystals of polymorphic mixtures and a new complex of indomethacin-epirizole by the spherical crystallization technique. *J. Pharm. Sci.* 1985; 74, 1152-6.

Kawashima Y, Morishima K, Takeuchi H, Niwa T, Hino T, Kawashima Y. Crystal design for direct tabletting and coating by the spherical crystallization technique. *AIChE Symposium Series*. 1991; 284, 26-32.

Kawashima Y, Naito M, Lin SY, Takenaka H. An experimental study of the kinetics of the spherical crystallization of sodium theophylline monohydrate. *Powder Technol.* 1983; 34, 255-60.

Kawashima Y, Okumura M, Takenaka H. Spherical crystallization: direct spherical agglomeration of salicylic acid crystals during crystallization. *Science*. 1982; 4550(216), 1127-8.

Kawashima Y, Okumura M, Takenaka H. The effects of temperature on the spherical crystallization of salicylic acid. *Powder Technol.* 1984; 39, 41-47.

Kawashima Y, Takeuchi H, Niwa T, Hino T, Yamakoshi M, Kihara K. Preparation of spherically agglomerated crystals of an antibacterial drug for direct tabletting by a novel spherical crystallization technique. *Congr. Int. Technol. Pharm.* 1989; 5, 228-34.

Kulkarni, P.K. and Nagavi B.G. (2002). Spherical crystallization. *Indian Journal of Pharmaceutical. Education*, 36, 66-71.

Khan, K.A. (1975). The concept of dissolution efficiency. *Journal Pharmacy and Pharmacology*, 27, 48-49.

Marshall, P.V., York, P. (1991). Compaction Properties of Nitrofurantoin Samples Crystallised from Different Solvents. *International Journal of Pharmaceutics*, 67, 59-65.

Martin, A., Bustamante, P., Chun, A. (2002). Micromeritics, In: *Physical Pharmacy- physical chemical principles in the pharmaceutical sciences*, 4th ed., pp. 423-452, Lippincott Williams amd Wilkins, Baltimore.

Sano, A., Kuriki T., Kawashima Y., Takeuchi H., Hino T., and Niwa T. (1992). Particle design of tolbutamide by spherical crystallization technique. V. Improvement of dissolution and bioavailability of direct compressed tablets prepared using tolbutamide agglomerated crystals. *Chemical and Pharmaceutical. Bulletin*, 40, 3030-3035.

Sweetman, S.C. (Ed(s).). (2002). *Martindale The Complete Drug Reference*. 33rd ed., Pharmaceutical Press, London, 2002, pp. 855-856.

Tapas, A.R., Kawtikwar, P.S., Sakarkar, D.M. (2009). Enhanced dissolution rate of felodipine using spherical agglomeration with Inutec SP1 by quasi emulsion solvent diffusion method. *Research in Pharmaceutical Sciences*, 4, 77-84.

Tapas, A.R., Kawtikwar, P.S., Sakarkar, D.M. (2010). Spherically agglomerated solid dispersions of valsartan to improve solubility, dissolution rate and micromeritic properties. *International Journal of Drug Delivery*, 2, 304-313.

Usha, A.N., Mutalik, S., Reddy, M.S., Rajith, A.K., Kushtagi, P., Udupa, N. (2008). Preparation and in vitro, preclinical and clinical studies of aceclofenac spherical agglomerates. *European Journal of Pharmaceutics and Biopharmaceutics*, 70, 674-683.

U.S. Food and drug administration [Internet]. Dissolution methods for drug products; 2010. Available from: http://www.accessdata.fda.gov/scripts/cder/dissolution/dsp_SearchResults_Dissolutions.cfm?PrintAll=1.

Wells, J. (2002). Pharmaceutical preformulation, the physicochemical properties of drug substances, In: *Pharmaceutics- the science of dosage form design*, M.E. Aulton (Ed), pp. 113-138, Churchill Livingstone, London.